Photonics and Laser Engineering

Cover photo: End mirror of a multipass-cavity femtosecond Ti:sapphire laser (Koç University Laser Research Laboratory; photo by Aref Mostafazadeh).

Photonics and Laser Engineering

Principles, Devices, and Applications

Alphan Sennaroğlu
Departments of Physics and Electrical-Electronics Engineering
Koç University
İstanbul, Turkey

McGraw Hill

New York Chicago San Francisco
Lisbon London Madrid Mexico City
Milan New Delhi San Juan
Seoul Singapore Sydney Toronto

The McGraw-Hill Companies

Cataloging-in-Publication Data is on file with the Library of Congress

1 2 3 4 5 6 7 8 9 0 DOC/DOC 1 6 5 4 3 2 1 0

ISBN 978-0-07-160608-0
MHID 0-07-160608-4

Sponsoring Editor
Michael Penn

Editorial Supervisor
Stephen M. Smith

Production Supervisor
Richard C. Ruzycka

Acquisitions Coordinator
Michael Mulcahy

Project Manager
Arushi Chawla, Glyph International

Copy Editor
Bev Weiler

Proofreader
Surendra Nath Shivam, Glyph International

Indexer
Arc Films, Inc.

Art Director, Cover
Jeff Weeks

Composition
Glyph International

Printed and bound by RR Donnelley.

This book is printed on acid-free paper.

To the memory of Özdemir M. Sennaroğlu and Hasan Behçet Kıvılcım
and
to Dr. Aydın Sennaroğlu

About the Author

Alphan Sennaroğlu is a professor in the Departments of Physics and Electrical-Electronics Engineering at Koç University in İstanbul, Turkey, and founded and leads the Koç University Laser Research Laboratory. He received B.S., M.S., and Ph.D. degrees in electrical engineering from Cornell University. Dr. Sennaroğlu was a visiting researcher at the Massachusetts Institute of Technology during the 2002–2003 and 2009–2010 academic years and in the summers of 2005 to 2008. In 2010 he was the program chair of the Advanced Solid-State Photonics (ASSP) Topical Meeting organized by the Optical Society of America (OSA), and in 2004 and 2006 he was the program chair of the Solid-State Lasers and Amplifiers Conference during the Photonics Europe meeting in Strasbourg, France. Dr. Sennaroğlu has also served as a member of the technical program committees for CLEO-America: Ultrafast Optics (2005–2007), CLEO-Europe: Ultrafast Optics (2005, 2009), Photonics Europe—Solid-State Lasers (2004, 2006, 2008), and ASSP (2008–2009). He has been elected an associate member of the Turkish Academy of Sciences, and is a senior member of IEEE and a member of OSA and SPIE. His research interests include solid-state lasers, femtosecond lasers, nonlinear optics, and spectroscopy.

Contents

Preface

The field of optics underwent an unprecedented revival following the invention of the laser in 1960. With the subsequent development of semiconductor lasers, which enabled the miniaturization of coherent light sources, and, later, the development by the 1970s of ultralow-loss optical fibers, new disciplines with widespread technological impact soon emerged. Understanding and exploiting the quantum aspects of radiation—in other words, the photon nature of light—clearly played a central role in this technological development. Today, development still continues at a fascinating rate, and we can see the outcome of this "optics revolution" in both science and technology. For example, use of ultrafast lasers and sophisticated optical detection schemes now make it possible to measure extremely fast atomic events in the subpicosecond (1 picosecond = 10^{-12} s or one-trillionth of a second) and even subfemtosecond (10^{-15} s or one-thousandth of a picosecond) time scales. Furthermore, it has become possible to probe, manipulate, and control systems with micrometer (1 micrometer = 10^{-6} m) and nanometer (1 nanometer = 10^{-9} m) resolution using light. Technological products based on photon sources and devices are also becoming commonplace in our lives. Semiconductor lasers, detectors, and optical fibers are routinely used in fiber-optic communication systems for fast, low-noise communication. Compact disc players and display systems are further examples of this technology. Today, we call this multifaceted technological area "photonics" and define it as the discipline that provides solutions to scientific and technological problems by using devices which generate, transmit, or detect photons.

This text is intended to provide an introduction to the principles and some of the main areas of photonics and laser engineering. It has originated from several courses I taught at Koç University, Istanbul, Turkey, on lasers, photonics, and photonic devices for more than a decade. Most of the attendees were seniors, as well as graduate students in physics and electrical engineering. In addition, I often had students from other technical fields who became interested in this material through their involvement in photonics-related research.

In the latter case, to overcome the barrier due to their lack of prior exposure to classical electromagnetism and elementary quantum theory, I would include basic review of these topics at various points throughout the courses. I adopted the same philosophy in the preparation of this text and included two extensive background chapters, one on classical electromagnetism and one on elementary quantum mechanics. This allows the book to be self-contained and enables readers from a diverse range of technical fields to easily grasp the concepts in photonics and lasers by simply using the background material.

In the preparation of this manuscript, I have greatly benefited from many valuable references including textbooks, scientific articles, and reviews. These are listed at the end of each chapter. All of the graphical data in the text and computation problems at the ends of the chapters were prepared by using Mathematica™.

Chapter 1 begins with a brief description of photonics and gives a historical review of some of the milestones. The rest of this chapter is devoted to an in-depth review of classical electromagnetism and electromagnetic wave phenomena, serving two important purposes: First, for those students who have studied these subjects before, it gives a complete review and introduces the notation that will be used in the remainder of the book. Second, and more important, it provides a thorough elementary introduction to electromagnetism for those students who had training in technical fields other than electrical engineering or physics. The only assumed prerequisites are university-level general physics, differential equations, and vector calculus. Maxwell's equations and Poynting's theorem are derived. Several wave phenomena, including reflection/refraction, interference, total internal reflection, and dipole radiation, are discussed.

Chapter 2 takes a step back and discusses geometrical optics where propagation of light in optical systems is explored by neglecting the wave aspects of light and treating it as rays. For the most part, geometrical optics in the paraxial limit is investigated. ABCD matrix formalism is introduced and applied to imaging problems. Ray tracing in continuous media and stability analysis of optical resonators are further treated.

The ray picture developed in Chap. 2 does not account for several important aspects of light beams such as transverse confinement and diffractive spreading. These topics are addressed in Chap. 3 by introducing the ideal Gaussian beam. The transformation rule for a Gaussian beam in a general optical system is derived, and spatial as well as spectral characteristics of resonator modes are further discussed.

Chapter 4 is devoted to light-matter interactions and consists of two parts. In the first part, we give a classical description of light-matter interactions based on the Lorentz model. The classical model is used to discuss material dispersion (Sellmeier equation) and absorption resonance phenomena. Broadening mechanisms, scattering, and dispersive pulse broadening are also covered. In the second part of the chapter, elementary quantum ideas are introduced to discuss blackbody radiation, basic radiative transitions, stimulated and spontaneous emission, and rate equations.

Chapter 5 provides a concise review of quantum mechanics. While in Chap. 4 radiative decay rates are calculated based on a phenomenological model first introduced by Einstein, Chap. 5 provides a more rigorous treatment of light-matter interactions. In this chapter, we derive the Einstein *A* and *B* coefficients by using perturbation theory. Postulates of quantum mechanics, operator formalism, the quantum mechanical harmonic oscillator, raising/lowering operators, semiclassical treatment of light-matter interactions, and quantization of electromagnetic radiation are further discussed.

Chapter 6 uses the material developed in the earlier chapters to discuss the concept of feedback and explain how light oscillation can be obtained based on stimulated emission. General characteristics of lasers such as monochromaticity and coherence are first reviewed. Continuous-wave and pulsed operation of lasers are then discussed. Most of the analysis presented in this chapter is general and can be applied to a wide variety of laser systems. As a specific example, we examine solid-state lasers and describe their continuous-wave, pulsed, and mode-locked operation.

In Chap. 7, we give a concise review of semiconductor fundamentals and describe the operation of semiconductor lasers. In the discussion of semiconductors, Fermi-Dirac statistics, density of states, and light-matter interactions in semiconductors are reviewed. The condition for optical gain is derived. We then discuss different types of semiconductor lasers and review some of the emerging technologies.

Up to this point in the text it is assumed that the medium in which light propagates is isotropic. In Chap. 8, we relax this assumption and examine light propagation in anisotropic crystals where the index of refraction is a function of direction. The Fresnel equation is derived, and the index ellipsoid method is discussed. Jones calculus is introduced and the operation of polarization control devices such as waveplates is described. Finally, electro-optic modulation and acousto-optic modulation are briefly explained.

The concept of noise plays an important role in light detection. We start Chap. 9 by introducing the reader to general properties of noise. This is then followed by a discussion of types of noise. We also examine different types of detectors and study their noise performance.

Chapter 10 is devoted to the analysis of translationally invariant wave guiding structures. We start out with the dielectric slab waveguide to discuss the general characteristics of wave guiding in a relatively simple geometry. We then provide a rigorous treatment of the step-index fiber and discuss the classification of the modes. The condition for single-mode operation is also derived. Practical aspects of fibers, such as coupling, the ray picture of multimode fibers, and numerical aperture, are also examined. We end the chapter by giving an overview of emerging fiber technologies such as fiber lasers and photonic crystal fibers.

In the last chapter of the text, we provide a condensed overview of some of the important nonlinear interactions in photonics. Three-wave mixing, second-harmonic generation, and optical parametric oscillation are discussed, followed by a description of phase matching in uniaxial crystals. We then briefly examine third-order nonlinearities such as self-focusing, pulse propagation in nonlinear dispersive media, self-phase modulation, optical solitons, and the Raman effect.

Alphan Sennaroğlu

Acknowledgments

First and foremost, I would like to thank my wife Figen Ecer Sennaroğlu, my daughter Canan Sennaroğlu, and my son Özalp Demir Sennaroğlu for their never-ending support and encouragement during the writing of this book. I would like to further thank Prof. Clifford R. Pollock from Cornell University for reading the proposal and for his valuable comments. Special thanks also go to Profs. James G. Fujimoto, Erich P. Ippen, and Franz X. Kaertner of the Massachusetts Institute of Technology for their hospitality and for providing the resources and facilities of the Optics and Quantum Electronics Group, where I spent two sabbatical terms during the 2002–2003 and 2009–2010 academic years. I have dedicated the book to three individuals who had a profound impact on my upbringing as well as my education: my father Özdemir M. Sennaroğlu (1931–1985), my grandfather Hasan Behçet Kıvılcım (1909–1991), and my mother Dr. Aydın Sennaroğlu.

I am especially indebted to my student assistant Hilmi Enes Eğilmez, who put such a diligent effort into the preparation of the figures and typing of the draft. His superb organizational skills helped in getting things completed on time. I would also like to thank my students Erman Timurdoğan and Mehmet Özerk Memiş for their assistance in typing the text and preparing the figures. Special thanks also go to my long-time friend and collaborator Adnan Kurt, as well as my graduate students Ümit Demirbaş and Hüseyin Çankaya, for collecting the experimental data that appear in some of the figures of Chap. 6. The experimental data were collected in the Laser Research Laboratory of Koç University (KULRL), İstanbul, Turkey. Finally, I thank my graduate student Aref Mostafazadeh, who took the cover picture for the book, showing our home-built multipass-cavity femtosecond Ti^{3+}:sapphire laser at KULRL. My research at KULRL has been supported by the Vehbi Koç Foundation, the Turkish Academy of Sciences (TUBA), the Scientific and Technological Council of Turkey (TUBITAK), the National Science Foundation (NSF), and the European Union 6th Framework Program.

The completion of this project would not have been possible without the support and assistance of the McGraw-Hill publishing team. In particular, I would like to thank my former editor Taisuke Soda for initiating the project and his friendly support all along. Further thanks go to Steve Chapman, Michael Penn, Michael Mulcahy, Richard Ruzycka, and Stephen Smith, and Arushi Chawla and the production team at Glyph International, for all of their valuable help and support during the preparation of the book.

Notation

A list of the symbols used in the text is provided here. Unfortunately, the same letter may have different meanings in different circumstances. For example, the symbol R can represent the radius of curvature, reflectivity, responsivity, electrical resistance, and so on. With practice, it is quite easy to recognize the correct meaning from the context. To minimize confusion, next to each symbol is indicated the physical quantity(ies) it represents.

Symbols and functions beginning with lower-case Latin letters:

a: area

$a(k)$: amplitude distribution function in terms of the wave number

$a(z,t)$: pulse envelope function

c: speed of light in vacuum

d: distance/length, d-coefficient in second-order nonlinear interactions

dB: decibel ratio

d_g: length of a gain medium

df_{MB}: Maxwell-Boltzmann velocity distribution

d_0: sample length

e: electron charge

$\hat{e}$: polarization vector, unit vector

$\hat{e}_r$: radial unit vector

$\hat{e}_{SP}$: unit vector from source coordinates to observation coordinates

f: frequency, oscillator strength, focal length

f_{eff}: effective focal length

f_m: modulation frequency

f_{rep}: repetition frequency of a pulse train

$f(E)$: Fermi-Dirac distribution function

g: resonator g parameters, differential parametric gain coefficient

g_k: degeneracy factor of state $|k\rangle$

g_R: differential Raman gain coefficient
g_T: maximum extractable small-signal gain
$g(\nu)$: lineshape function
$g(\omega)$: spectral amplitude distribution
$g_D(\nu)$: Doppler lineshape function
$g_L(\nu)$: Lorentzian lineshape function
h: Planck's constant
$\hbar$: $h/2\pi$
i: $\sqrt{-1}$/current
i_e: electron current
i_h: hole current
i_s: signal current
$\overline{i_n^2}$: mean squared noise current
k: wave number
k_0: vacuum wave number
k_B: Boltzmann constant
k_z: z-component of the wave vector
$\vec{k}$: wave vector
$\vec{k}_s$: wave vector for sound waves
l: length, integer
l_c: coherence length
$\vec{l}$: displacement vector
ℓ: single-pass fractional power loss
ℓ_A: single-pass fractional amplitude loss
ℓ_i: internal material loss
ℓ_m: output coupling loss
m: mass, index of longitudinal modes, integer
m_a: atomic mass
m_e: electron mass
m_e^*: effective electron mass
m_h^*: effective hole mass
$\vec{m}$: magnetic moment vector
n: refractive index, real part of the complex refractive index $\overline{n}$, density of electrons, integer
n_0: background refractive index, equilibrium electron concentration
n_i: intrinsic carrier concentration
$\overline{n}$: complex refractive index, $\overline{n} = n - in''$
n'': imaginary part of the complex refractive index $\overline{n}$
$\hat{n}$: unit normal vector
p: momentum, hole concentration, probability
p_0: equilibrium hole concentration
$\vec{p}$: dipole moment vector

q: point charge
q_a: induced charge on the anode
q_c: induced charge on the cathode
q_f: total free charge
$q(z)$: complex beam parameter of a Gaussian beam
r: radius, roots of an equation, normalized pumping parameter
r, r': amplitude reflection coefficients
r_{ij}: electro-optic tensor
r_{TE}, r_{TM}: amplitude reflection coefficients for TE and TM waves
$\hat{r}$: radial unit vector
$\vec{r}$: position vector/ray vector
$\hat{s}$: unit vector parallel to the propagation direction of an electromagnetic wave in a crystal
t: time
t, t': amplitude transmission coefficients
u: speed, average drift velocity of the charge carriers, electromagnetic energy density
u_e: electric energy density, electron velocity
u_{em}: total electromagnetic energy density
u_g: group velocity, wave packet velocity
u_h: hole velocity
u_m: magnetic energy density
u_p: phase velocity
$\vec{u}$: velocity, drift velocity
$u(r,z)$: transverse amplitude distribution
$u(x)$: spatial part of the quantum mechanical wave function
$u_E(x)$: energy eigenfunction of the hamiltonian operator
w_0: beam waist
$w(z)$: spotsize function
x: distance
z_0: Rayleigh range
z_D: characteristic dispersion length
z_f: beam focus location

Symbols and functions beginning with capital Latin letters:

A, B, C, D: elements of the ray transformation matrix
A_{21}: Einstein's A coefficient for spontaneous emission
A_{eff}: effective beam area
$\vec{A}$: vector magnetic potential
$A(\omega)$: spectral amplitude distribution function
B: spectral brightness
B_{12}: Einstein's B coefficient for absorption
B_{21}: Einstein's B coefficient for stimulated emission
$\vec{B}$: magnetic induction vector, magnetic flux density

C: closed curve, closed path, center of curvature, capacitance
C_N^M: N combinations from a set of M members
D: diffusion coefficient, aperture size, beam diameter, group delay dispersion parameter
D^*: detectivity
$\vec{D}$: electric displacement vector
D_{GV}: group velocity dispersion parameter
D_n: electron diffusion coefficient
D_p: hole diffusion coefficient
E: energy
$\bar{E}$: average energy per mode
$\vec{E}$: electric field vector
E_A: energy level introduced by acceptor atoms
E_c: energy at bottom of the conduction band
E_D: energy level introduced by donor atoms
E_f: Fermi level
E_{fn}: Quasi Fermi level for electrons
E_{fp}: Quasi Fermi level for holes
E_g: bandgap energy of a semiconductor
E_i: Fermi level of an intrinsic semiconductor
$\vec{E}_o$: local electric field
$\vec{E}_s$: signal electric field
E_v: energy at top of the valance band
E_z: electric field component in a p-n junction
$\bar{E}(\lambda)$: average energy per mode
F: finesse of the Fabry-Perot interferometer
$\vec{F}$: electrostatic force
$\vec{F}_{21}$: electrostatic force between charges q_2 and q_1
$\vec{F}_d$: damping force
$\vec{F}_{dj}$: damping force on the jth electron
$\vec{F}_L$: Lorentz force
F_N: noise figure
$\vec{F}_r$: restoring force acting on the electron
$\vec{F}_{rj}$: linear restoring force on the jth electron
G: gain, fractional single-pass power gain, rate of generation of carriers, total amplification
G_0: small-signal fractional single-pass power gain
G_f: final fractional power gain
G_i: initial fractional power gain

G_T: thermal generation rate
$\hat{H}$: hamiltonian operator
$\vec{H}$: magnetic field intensity
$H_m(y)$: Hermite polynomial of order m
I: total current, light intensity
I_0: reverse leakage current, initial intensity
I_B: background induced current
I_c: average cathode photocurrent generated by incoming light
I_d: dark current
I_L: laser intensity
I_p: pump intensity
I_s: saturation intensity, photo induced signal current
I_S: Stokes signal intensity
$\bar{I}_s$: average signal current
$I_\ell(x)$: modified Bessel function of the first kind of order ℓ
$I_s(\omega)$: Fourier transform of $i_s(t)$
$I_{sT}(\omega)$: Fourier transform of $i_s(t)$ measured over the interval $\left[-T/2, T/2\right]$
J: total angular momentum number
$\vec{J}$: total volume current density, Jones vector
$\vec{J}_b$: bound volume current density
$\vec{J}_d$: displacement current density
$\vec{J}_f$: free volume current density
$\vec{J}_n$: electron current density
$\vec{J}_p$: hole current density
$J_\ell(x)$: Bessel function of the first kind of order ℓ
K: kinetic energy
$\vec{K}_b$: bound surface current density
$\vec{K}_f$: free surface current density
$K_\ell(x)$: modified Bessel function of the second kind of order ℓ
L: total loss of the cavity, self inductance, number of layers of thin films, lens, thickness, total orbital angular momentum, total round trip material loss, length
L_0: length of a laser resonator
L_N: characteristic diffusion length for electrons

L_p: characteristic diffusion length for holes
M: mirror
M^2: M-square beam parameter
$\vec{M}$: magnetization vector
$\mathbf{M}$: the ray transformation matrix
$\mathbf{M}_j(\alpha,\Gamma)$: overall Jones transformation matrix
$\mathbf{M}_{jo}$: Jones matrix for a birefringent plate
$\mathbf{M}_T$: the ray transformation matrix
N: population density, volume density of the atoms, concentration of atoms, total atomic density
N_1, N_2: population densities of a two-level atomic system
N_B: steady-state population density of radiation induced photo-carriers
N_g: ground state population
N_S: steady-state concentration of induced charge carriers
N_t: total population density of atoms
$N(k)dk$: number of modes per unit volume in the wavenumber interval k to $k+dk$
$N(\lambda)d\lambda$: number of modes per unit volume in the wavelength interval λ to $\lambda+d\lambda$
$N_m^{TE}(\theta), N_m^{TM}(\theta)$: effective direction dependent index (TE, TM wave) of the mth layer
$N.A.$: numerical aperture
NEP: noise equivalent power, $(P_s)_{\min}$
$\hat{O}$: quantum mechanical operator of the physical observable O
P: total power, average power, observation point
$\vec{P}$: polarization vector
P_1: noise power delivered to a resistor by one mode
P_{av}: average power over the integration time T
P_b: background radiation
P_B: Boltzmann probability distribution
P_C: probability of collisions
P_{cr}: critical focusing power
P_g: gas pressure
P_i: probability of not seeing an electron in the interval $\left[t_i, t_i+\Delta t\right]$
$\hat{P}_i$: momentum operator for the ith ion
P_j: Johnson noise power
P_L: laser beam power inside the resonator
P_{NC}: probability of no collisions

P_{out}: output power of a laser
P_p: pump power of the laser
P_{rad}: radiated power
P_s: incident signal power of the light beam
P_{shot}: shot noise power
Q: quality factor, fractional power loss due to a saturable absorber, dimensionless diffraction parameter
R: radius of curvature, intensity reflection coefficient, responsivity, resistance
R_a: rate of absorption
R_e: rate of emission
$\mathbf{R}(\alpha)$: rotation matrix
$R(\lambda)$: spectral radiant intensity function
$R_T(\lambda)$: integrated radiant intensity
S: magnitude of the Poynting vector, intensity, closed surface, source, total spin
$\vec{S}$: Poynting vector
T: absolute temperature in degrees Kelvin, intensity transmission, power transmission coefficient, sample transmission, integration time, time measured in the reference frame that moves at the group velocity of a pulse, output coupler transmission
T_2: mean free time between collisions
T_f: transmission factor for a lens
T_p: pulse transit time
T_R: cavity round-trip time, temporal separation between adjacent pulses in a pulse train
T_r: resonant transmission of a sample, intensity transmission
$T(t)$: amplitude transmission function
V: volume, electrostatic potential, fringe visibility, V-number of an optical fiber
V_{emf}: electromotive force
V_L: inter-ionic potential
V_π: π-voltage
W: work
W_{nr}: spontaneous nonradiative decay rate
W_p: pumping rate per atom
$Y_\ell(x)$: Bessel function of the second kind of order ℓ
Z: atomic number
Z_0: characteristic impedance

Symbols and functions beginning with lower-case Greek letters:

α: differential absorption coefficient, attenuation coefficient, rotation angle for the Jones vector
α_L: differential loss coefficient
α_{p0}: small-signal pump absorption coefficient
α_R: differential loss coefficient due to Rayleigh scattering
α_r: resonant differential absorption coefficient
α_T: total differential loss coefficient
β: quadratic index coefficient, recombination parameter of a semiconductor, propagation constant of a guided wave
γ: differential gain coefficient, damping parameter, Kerr nonlinearity constant
γ_0: small-signal differential gain coefficient
$\gamma(\nu)$: frequency-dependent differential gain coefficient
$\gamma(\tau)$: normalized self-coherence function
$\gamma_{12}(\tau)$: normalized mutual coherence function
δ: penetration depth, relative phase difference between TE and TM waves, round-trip phase shift
δ_{ij}: Kroenecker delta symbol
$\delta(x)$: Dirac delta function
ε: medium permittivity, complex dielectric constant, $\varepsilon = \varepsilon' + i\varepsilon''$
ε_0: permittivity of free space
η: wave impedance, slope efficiency, quantum efficiency of a photodetector
η_0: vacuum impedance
η_a: fractional pump absorption
η_s: quantum efficiency at signal wavelength
η_{SHG}: second harmonic generation conversion efficiency
θ: angle
θ_B: Brewster's angle
θ_c: critical angle
θ_i: angle of incidence
θ_r: angle of reflection
θ_t: angle of transmission
κ: heat conductivity
κ_0: spring constant
κ_1: nonlinear spring constant

λ: vacuum wavelength, de Broglie wavelength
λ_L: laser wavelength
λ_p: pump wavelength
λ_s: effective sound wavelength
μ: medium permeability, mobility
μ_0: permeability of free space
μ_n: electron mobility
μ_p: hole mobility
ν: frequency
ν_B: average frequency of background radiation
ν_{FSR}: free spectral range
ν_L: laser frequency, classical frequency of ionic vibration modes
ν_m: frequency of the mth mode
ν_p: pump frequency
ν_R: relaxation oscillation frequency
ν_s: sound frequency, signal frequency
ρ: volume charge density, electromagnetic energy density
ρ_b: bound volume charge density
ρ_f: free volume charge density
ρ_L: line charge density
ρ_ν: electromagnetic energy density of a monochromatic wave at frequency ν
ρ_V: volume charge density
$\vec{\rho}$: ray vector
$\rho(k), \rho(E)$: density-of-states functions
$\rho(k)dk$: number of allowed electron states per unit volume with wave numbers in the range k to $k+dk$
$\rho(\lambda)$, $\rho(\nu)$, $\rho(\omega)$: spectral energy density distribution, noise spectral power density
$\rho_c(k), \rho_c(E), \rho_v(k), \rho_v(E)$: density-of-state functions for the conduction band and valance band
$\rho_j(\nu)$: joint optical density of states
$\bar{\rho}_c(E), \bar{\rho}_v(E)$: weighted density of states for electrons and holes
σ: surface charge density, electrical conductivity
σ_a: absorption cross section
$(\sigma_a)_T$: integrated absorption cross section
σ_b: bound surface charge density
σ_e: stimulated emission cross section
σ_f: free surface charge density
σ_{SB}: Stefan-Boltzmann constant

τ: mean temporal separation between randomly generated electrons
τ_0: electron-hole recombination time
τ_A: decay time of absorber atoms
τ_c: coherence time
τ_f: fluorescence lifetime
τ_n: electron recombination time
τ_p: cavity photon lifetime, hole recombination time
τ_{pulse}: pulsewidth
τ_r: radiative lifetime, radiative recombination time
τ_{stim}: stimulated lifetime
τ_T: transit time, average drift time
τ_{TE}, τ_{TM}: amplitude transmission coefficient for TE and TM waves
ϕ: phase, angular advance of the spot pattern after one round trip
χ_b: background susceptibility
χ_e: electric susceptibility
χ_{ij}: susceptibility tensor
χ_m: magnetic susceptibility
χ_r: resonant susceptibility
ψ: wave function
ω: angular frequency (real part of $\bar{\omega}$)
ω_{AS}: anti-Stokes frequency
ω_F: difference in the angular frequency of two adjacent modes
ω_i: imaginary part of $\bar{\omega}$
ω_m: modulation frequency
ω_o: local oscillator frequency, resonance frequency
ω_S: Stokes frequency
ω_s: angular frequency of the sound wave, signal frequency
$\bar{\omega}$: complex angular frequency, $\bar{\omega} = \omega + i\omega_i$

Symbols and functions beginning with capital Greek letters:

Γ: relative phase difference, retardation
$\Gamma_{\ell \to m}$: transition rate
Γ_{total}: total transition rate
$\Gamma_{12}(\tau)$: mutual coherence function
Δn: deviation in the electron concentration from equilibrium
Δp: deviation in the hole concentration from equilibrium
Δt: temporal width

ΔN: population inversion
ΔN_0: small signal population inversion
$\Delta \nu$: detector bandwidth
$\Delta \nu_D$: spectral width (FWHM) of the Doppler lineshape function
$\Delta \nu_L$: spectral width (FWHM) of the Lorentzian lineshape function
Φ: phase of the reflected wave, work function
Φ_{NL}: nonlinear phase shift
Ψ_m: magnetic flux
Ω : solid-angle, angular frequency, bandwidth
Ω_R: Rabi frequency
$\langle \cdot \rangle$: average, expectation value

Photonics and Laser Engineering

CHAPTER 1

Electromagnetic Wave Theory of Light with Applications

This chapter provides an in-depth review of electromagnetism and electromagnetic wave phenomena, serving two important purposes. First, for those who have studied these subjects before, it gives a complete, self-contained review and introduces the notation that will be used in the remainder of the book. Second and more importantly, it provides a thorough introduction to electromagnetism for those who had training in technical fields other than electrical engineering or physics. The only assumed prerequisites are university-level general physics and vector calculus. Those readers with a prior background in electromagnetism can skip the introductory sections and start from Sec. 1.6.

We start out with a concise overview of photonics and discuss its scope. Three examples of photonic systems are mentioned to introduce some of the issues that will be discussed in the remainder of the text. We then mention some of the milestones in photonics. The remainder of the chapter is devoted to a detailed review of the classical electromagnetic wave theory of light from electrostatics to Maxwell equations and conservation of energy principle for electromagnetic fields. As applications, we investigate electromagnetic waves in uniform media, reflection and transmission of waves at a dielectric discontinuity, design of optical coatings, and dipole radiation. References 1 to 6 given at the end of the chapter provide additional discussion on electromagnetic theory.

1.1 What Is Photonics?

I believe it is not at all an exaggeration to call photonics "the Renaissance age of optics." Optics, which studies the generation and propagation of light, is an old and matured discipline, historically extending over centuries. Prior to 1960, optical technologies had a limited scope and mainly involved the design and development of traditional optical systems including eyeglasses, telescopes, microscopes, and so on. However, following the invention of the laser in 1960,[5] development of the semiconductor light sources in the early 1960s,[6] and the realization of the ultralow-loss optical fibers in the 1960s and 1970s,[7,8] optics underwent an unprecedented revival, leading to the emergence of new disciplines with widespread technological impact. Understanding and exploiting the quantum aspects of radiation, in other words, the photon nature of light, played a critical role in this development.[8] Nowadays, we see the outcome in both science and technology. In science, for example, use of lasers and sophisticated detection systems has made it possible to measure extremely fast atomic events in the picosecond (1 picosecond = 10^{-12} second), femtosecond (1 femtosecond = 10^{-15} s), and more recently, attosecond (1 attosecond = 10^{-18} s) time scales.[9,10] Furthermore, it has become possible to use lasers in biomedical imaging to obtain images in living tissue with cellular-level resolution.[9] Technological products based on photon sources and devices are also becoming commonplace in our lives. For example, semiconductor lasers, detectors, and optical fibers are routinely used in fiber-optic communication systems for fast, low-noise communication. Compact disc players using semiconductor lasers for optical information processing are another example. Today, we call this multifaceted technological area "photonics" and define it as the discipline that aims at providing solutions to scientific and technological problems by using devices which generate, transmit, or detect photons.

As a specific example, let us consider a fiber-optic communication link sketched in Fig. 1.1. What I like most about this example is that nearly all of the topics discussed in the text will be needed to understand its various components. Here, a light source, typically a semiconductor laser, is sent through a modulator that varies the light wave to transmit a message. The modulated optical signal is then coupled into the fiber and travels over a long distance. Due to attenuation, the signal gets weaker as the propagation distance increases, so one or more amplifier units are needed to boost the signal. At the end of the fiber, the signal is detected by a detector and demodulated to recover the original message. In the design of this particular photonic system, several important questions of interest arise:

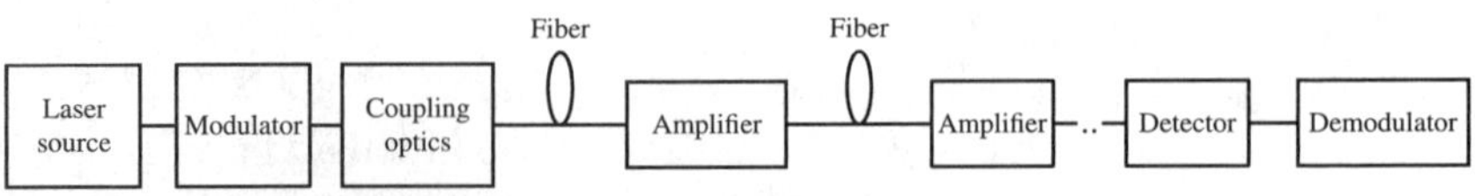

FIGURE 1.1 Sketch of a fiber-optic communication link.

1. How does the modulator operate? What materials or techniques are available to vary the amplitude, phase, or the frequency of a light wave?
2. What type of optics is needed to couple the light into an optical fiber?
3. How does light travel down the fiber?
4. Is the wavelength of light traveling in the fiber critical? Can we find a wavelength that is optimum for fiber transmission?
5. What are the mechanisms that lead to attenuation as light travels inside the fiber?
6. How does an optical amplifier operate?
7. At the receiving end, how do we detect the optical signal? What factors influence the performance of the detector? How low a signal level can we detect?

As a second example, let us mention an important emerging application from biomedical optics based on imaging using two-photon microscopy.[10, 11] Here, tissues doped with fluorescent organic dyes are excited with high-intensity femtosecond laser pulses and the fluorescence obtained as a result of two-photon excitation is measured to obtain high resolution images. To master this technique, it is important to gain an understanding of the basic principles of absorption/emission spectroscopy, nonlinear optics, lasers, mode locking of lasers to generate ultrashort pulses, and optical detection systems.

Both of the examples mentioned above use lasers to generate light. As a third example, we look at the basic building blocks of a generic laser system as shown in Fig. 1.2. In a nutshell, this is a device based on light oscillation by stimulated emission of radiation. It is important to understand how the following four components of the laser are designed:

1. Active medium: This is the medium where optical amplification takes place. How can we model light-matter interactions?

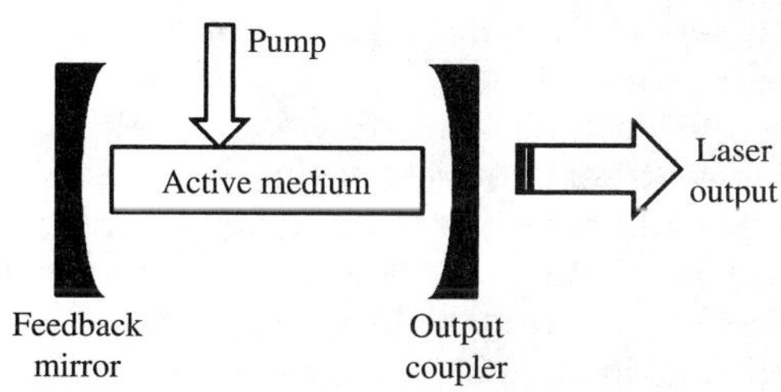

FIGURE 1.2 Basic components of a laser.

Which physical characteristics of the medium are important in determining the level of amplification in the medium?

2. Pump source: We need an external pumping mechanism to excite the atoms to higher levels and to achieve optical amplification. What is the most effective pumping scheme for a particular gain medium?
3. Feedback: The mirrors form a resonator which provides feedback to enable oscillation. How should we design the resonator so that light remains confined after many round trips? Can we determine the spotsize of the light beam trapped between the mirrors?
4. Output coupling: One of the mirrors has partial reflection and lets some of the circulating beam out. This is known as the output coupler of the laser. How should we choose the reflectivity of this output coupler? Is there an optimum value where the output power is maximized? How do we coat these mirrors to make them partially reflecting or highly reflecting?

These are some of the questions which we will address in different parts of the text. Here is a final ironic remark before ending the discussion on lasers. As you probably know already, the acronym "laser" stands for *l*ight *a*mplification by *s*timulated *e*mission of *r*adiation. Strictly speaking, however, the device is a light oscillator. Since the acronym for the latter becomes "loser," researchers instead found it more appropriate, from the early days on, to use the name "laser."

1.2 Some Milestones in Photonics

It is certainly very difficult, if not impossible, to provide a complete and accurate account of the historical developments that took place in the field of Photonics. Here, we present a sketchy summary of some of the key achievements since the seventeenth century.

Following the experimental discovery of the law of refraction by Willebrord Snell (1591–1626) in 1621, two competing theories were put forward to explain the law of refraction and to understand the fundamental nature of light: the particle theory and the wave theory of light. The particle theory championed by René Descartes (1596–1650) and later by Sir Isaac Newton (1642–1727) was based on the assumption that light consists of tiny particles which obey the classical laws of motion. One of the most important triumphs of the particle theory was that it could explain the mathematical form of Snell's law. However, it predicted that the speed of light should be larger in water than in air, which turned out to be exactly the opposite, as was later verified experimentally through the terrestrial speed of light measurements conducted by Léon Foucault in 1850.

One of the earliest advocates of the wave theory was Christiaan Huygens (1629–1695), who in the seventeenth century used the wave theory to explain the laws of reflection and refraction of light. However, due to Newton's authority stemming from his success in developing a general theory of terrestrial and celestial motions, Huygens' wave theory did not find large support in the scientific community. This began to change with the work of Thomas Young (1773–1829), who in 1803 experimentally demonstrated the wave aspects of light by measuring the wavelength of light waves in his famous double-slit experiment. Further support for the wave theory came from the work of Augustin-Jean Fresnel (1788–1827), who developed a rigorous theory of diffraction based on the wave picture of light. One of the most successful predictions of Fresnel's wave theory is the famous "Poisson spot," the bright point which is formed at the center of the otherwise dark shadow of an opaque circular disk, an effect which could be precisely accounted for by using Fresnel's wave theory.

No doubt, the most important figure in the development of the wave theory of light was James Clerk Maxwell (1831–1879), who combined the work of Gauss, Ampere, and Faraday, among others, to develop a unified theory of electricity and magnetism.[12] Maxwell's theory correctly explained the coupling between electric and magnetic phenomena and showed that electromagnetic waves consisting of coupled electric and magnetic fields could propagate in free space and matter. By using electric permittivity and magnetic permeability measurements performed before him, Maxwell further predicted that the speed of the electromagnetic waves in vacuum is equal to the speed of light, a fact which resulted in the exciting discovery that light itself is an electromagnetic wave. Maxwell's prediction of electromagnetic waves was later experimentally demonstrated by Heinrich Hertz (1857–1894) in 1888.

Around the turn of the twentieth century, we see the development of the quantum theory of light and a return to the particle picture. In 1900, Max Planck (1858–1947) correctly explained the spectral characteristics of a black body by assuming that the emitted electromagnetic energy is quantized; in other words, it comes in integral multiples of a basic lump of energy, $h\nu$, where h is Planck's constant and ν is the light frequency. This remarkable discovery marked the birth of quantum theory. Later, in 1905, Albert Einstein (1879–1955) used the quantum picture to explain the photoelectric effect, the emission of light from illuminated metallic surfaces. He later received the 1921 Nobel Prize in Physics for his discovery of the theory of the photoelectric effect. In 1917, Einstein further provided the quantum theory of light-matter interactions and identified the processes of absorption, spontaneous emission, and stimulated emission.[13] These three processes all play an important role in the operation of a laser, as we will discuss in later chapters. The modern theory of light is based on the principle of wave-particle duality where both particle and wave aspects of light complement each other. However, these particles, called photons, are nonclassical in the sense that they do not follow definite, continuous trajectories contrary to what was originally envisioned by Newton. There is still ongoing research aimed at understanding the true nature of light (see, for example, Ref. 14 for a recent account).

It took more than four decades, following Einstein's treatment of light-matter interactions, to develop the first laser. In the meantime, research continued in the design of resonator geometries that provide positive feedback, and also in the search for suitable materials for obtaining light amplification. In the late 1950s, C. H. Townes (b. 1915), N. G. Basov (1922–2001), and A. M. Prokorov (1916–2002) formulated the theory of light amplification by stimulated emission of radiation and were awarded the 1964 Nobel Prize in physics.[15-17] The idea was first applied in the microwave region, leading to the invention of the ammonia maser (*m*icrowave *a*mplification by *s*timulated *e*mission of *r*adiation),[18] and later extended by A. Shawlow and C. H. Townes, to the optical domain.[15] Lasing at optical frequencies was first demonstrated in 1960 by T. H. Maiman[19] (1927–2007), who used a ruby crystal as the optical gain medium. As was discussed in the previous section, development of semiconductor lasers in the 1960s, and low-loss optical fibers by the early 1970s, further revolutionized optical communications systems and led to the emergence of many new disciplines in Photonics. Today, extensive research effort still continues in the development of high-performance lasers, photonic materials, micro- and nanophotonic systems, novel photonic devices, ultrafast photonics, and much more. Some of the references cited at the end of the chapter provide further discussion on the historical developments in Photonics.[13, 20]

1.3 Electrostatics

1.3.1 The Electric Field

Now, we take a step back and concentrate on the fundamental concepts needed to develop an understanding of the classical electromagnetic theory of light. We start with electrostatics. Electrostatics studies the interaction of electric charges at rest. Let us consider two point charges q_1 and q_2 with the respective position vectors $\vec{r}_1$ and $\vec{r}_2$, as shown in Fig. 1.3. When the charges are stationary, Coulomb's law states that the force between them is proportional to the product of the charges and inversely proportional to the square of the separation distance. Quantitatively, the force $\vec{F}_{21}$ on charge q_2 due to q_1 is given by

$$\vec{F}_{21} = \frac{1}{4\pi\varepsilon_0}\frac{q_1 q_2}{r^2}\hat{e}_r \tag{1.1}$$

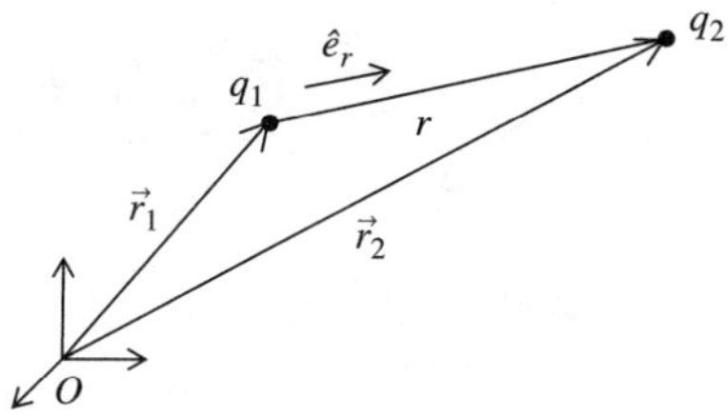

FIGURE 1.3 Two point charges q_1 and q_2, separated by a distance r.

where r is the separation distance and $\hat{e}_r$ is the unit vector along the line from q_1 to q_2. Note that Eq. (1.1) is written in MKS units with q_1 and q_2 in coulombs, r in meters, and $\vec{F}_{21}$ in newtons (MKS system of units uses the fundamental units of *m*eter, *k*ilogram, and *s*econd to represent physical measurements). The constant ε_0 is called the permittivity of free space and has the value $\varepsilon_0 = 8.854 \times 10^{-12}\ \mathrm{C^2/N \cdot m^2}$. Also, we have, from Newton's third law, $\vec{F}_{21} = -\vec{F}_{12}$.

Here, we see an example of "action at a distance." In other words, the two charges exert a force on each other even when they are not in contact. This effect can be conveniently described by introducing the concept of the electric field. As such, the charge q_1 produces an electric field which occupies the whole space around it. When the second charge q_2 is in the vicinity of q_1, it interacts with the electric field produced by q_1 and hence experiences the force $\vec{F}_{21}$. To define the electric field, let us assume that a small test charge of size q is brought to a region where an electric field exists. If the force acting on q is $\vec{F}$ (when q is stationary), then the electric field at this location is defined as

$$\vec{E} \equiv \frac{\vec{F}}{q} \tag{1.2}$$

As an example, the electric field due to the charge q_1 will be given by

$$\vec{E} = \frac{1}{4\pi\varepsilon_0}\frac{q_1}{r^2}\hat{e}_r \tag{1.3}$$

What will the electric field be if, instead of a point charge, we have a continuous distribution of charge in a region of space as sketched in Fig. 1.4? First, let's focus on a small section of the charged object located at the position vector $\vec{r}'$ with volume d^3r'. The primed position vector $\vec{r}'$ represents the source coordinates. The infinitesimal contribution $d\vec{E}$ to the electric field at the observation point P with position vector $\vec{r}$ is given by

$$d\vec{E} = \frac{\rho(\vec{r}')d^3r'}{4\pi\varepsilon_0}\frac{\hat{e}_{SP}}{|\vec{r} - \vec{r}'|^2} \tag{1.4}$$

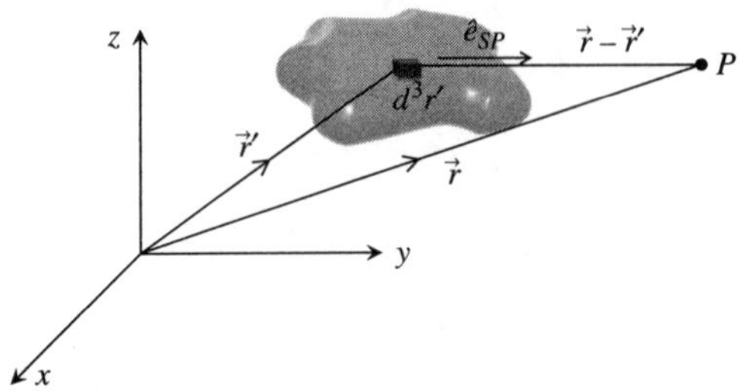

FIGURE 1.4 Continuous distribution of volume charge in space. The element of volume at the source location $\vec{r}'$ is d^3r'.

where $\rho(\vec{r}')$ is the volume charge density, and the unit vector $\hat{e}_{SP}$, from the source to the observation point is parallel to $\vec{r}-\vec{r}'$. The volume charge density $\rho(\vec{r}')$ can be due to free charges or bound charges as will be further elucidated in the discussion of dielectric materials. Using the principle of superposition, the total electric field at $\vec{r}$ will be the vector sum of the electric field contributions due to charges over the whole source. Since the charge is continuously distributed, the sum is given by the volume integral

$$\vec{E}(\vec{r}) = \int_V d^3r' \frac{\rho(\vec{r}')}{4\pi\varepsilon_0} \frac{\hat{e}_{SP}}{|\vec{r}-\vec{r}'|^2} \tag{1.5}$$

From the integral expression given by Eq. (1.5), we can derive general expressions for the divergence and curl of the static electric field. First, by taking the divergence with respect to the unprimed observation coordinates, we get

$$\nabla \cdot \vec{E}(\vec{r}) = \frac{1}{4\pi\varepsilon_0} \int_V d^3r' \rho(\vec{r}') \nabla \cdot \frac{\hat{e}_{SP}}{|\vec{r}-\vec{r}'|^2} \tag{1.6}$$

Note that the charge density is a function of $\vec{r}'$ only and can be treated as a constant in the divergence operation. As is well known (see Prob. 1.1),

$$\nabla \cdot \left(\frac{\hat{e}_{SP}}{|\vec{r}-\vec{r}'|^2} \right) = \nabla \cdot \left(\frac{\vec{r}-\vec{r}'}{|\vec{r}-\vec{r}'|^3} \right) = 4\pi\delta^3(\vec{r}-\vec{r}') \tag{1.7}$$

where $\delta^3(\vec{r}-\vec{r}')$ is the three-dimensional Dirac delta function which satisfies

$$\int_V d^3r' f(\vec{r}')\delta^3(\vec{r}-\vec{r}') = f(\vec{r}) \tag{1.8}$$

for an arbitrary, well-behaved function $f(\vec{r})$. Hence the divergence of the electrostatic field becomes

$$\nabla \cdot \vec{E} = \frac{\rho(\vec{r})}{\varepsilon_0} \tag{1.9}$$

Equation (1.9) is the well-known Gauss' law in differential form. We will shortly generalize it to include material effects as well. It is also straightforward to derive the integral form of Gauss' law by starting with Eq. (1.9) and by applying the divergence theorem (see Prob. 1.2):

$$\oint_S d\vec{a} \cdot \vec{E} = \frac{q_{\text{encl}}}{\varepsilon_0} \tag{1.10}$$

Above, S is a closed surface that encloses the volume V and q_{encl} is the total charge enclosed within S. The integral form of Gauss' law is especially useful when the symmetry of the problem allows the calculation of the electric field from the surface integral, as the following example illustrates.

Example Let's calculate the electric field due to an infinite, uniform, line charge with linear charge density of ρ_L (C/m) as shown in Fig. 1.5. Due to the cylindrical symmetry of the problem, it is convenient to work in cylindrical coordinates (r, ϕ, z). The electric field will have a radial component only along the $\hat{r}$ direction and will be constant as a function of ϕ and z at a given value of r. To use the integral form of Gauss' law, we choose a cylinder of radius r and length l, concentric with the infinite line charge. Since $\vec{E} = E_r\hat{r}$, we get no contribution to the surface integral from the side surfaces. Over the cylinder, the element of area is given by $d\vec{a} = \hat{r}rd\phi dz$. Hence, we get

$$\oint_S d\vec{a} \cdot \vec{E} = \int_0^{2\pi} d\phi \int_0^{l} dz E_r r = 2\pi l r E_r = \frac{q_{\text{encl}}}{\varepsilon_0} = \frac{\rho_L l}{\varepsilon_0}$$

Simplifying the above relation, the electric field becomes $\vec{E} = \hat{r}\dfrac{(\rho_L)}{(2\pi\varepsilon_0 r)}$.

1.3.2 The Electrostatic Potential

We now go back to Eq. (1.5) which gives the general expression for the electric field due to a volume charge density ρ. It can be shown

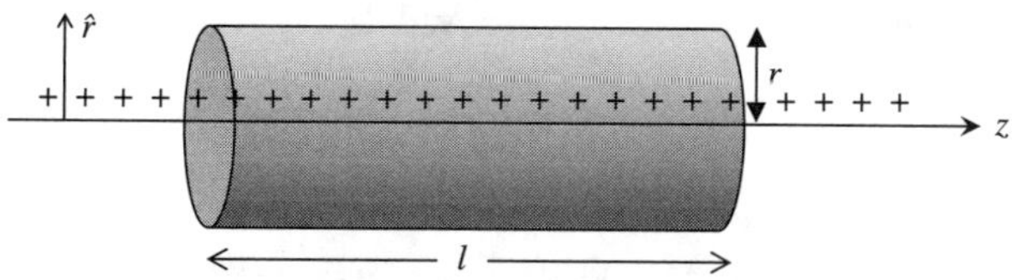

Figure 1.5 Infinite line charge.

(Prob. 1.4) that

$$\nabla\left(\frac{1}{|\vec{r}-\vec{r}'|}\right) = -\nabla'\left(\frac{1}{|\vec{r}-\vec{r}'|}\right) = -\frac{\vec{r}-\vec{r}'}{|\vec{r}-\vec{r}'|^3} = -\frac{\hat{e}_{SP}}{|\vec{r}-\vec{r}'|^2} \tag{1.11}$$

where ∇' designates the gradient operator with respect to the primed (source) coordinates. Hence, $\vec{E}$ can be written as

$$\vec{E}(\vec{r}) = -\int_V d^3r' \frac{\rho(\vec{r}')}{4\pi\varepsilon_0} \nabla\left(\frac{1}{|\vec{r}-\vec{r}'|}\right) = -\nabla \frac{1}{4\pi\varepsilon_0} \int_V d^3r' \left(\frac{\rho(\vec{r}')}{|\vec{r}-\vec{r}'|}\right) = -\nabla V \tag{1.12}$$

where V is the scalar electrostatic potential given by

$$V = \frac{1}{4\pi\varepsilon_0} \int_V d^3r' \left(\frac{\rho(\vec{r}')}{|\vec{r}-\vec{r}'|}\right) \tag{1.13}$$

Note that in Eq. (1.12), we again made use of the fact that $\rho(\vec{r}')$ is constant as far as the del (∇) operation is concerned. For an arbitrary scalar function f, we know that $\nabla \times (\nabla f) = 0$. Since $\vec{E}(\vec{r}) = -\nabla V$ it follows that $\nabla \times \vec{E} = 0$. In other words, the electric field due to a static charge distribution is conservative, i.e., its line integral is path-independent and only depends on the end points. Since the electrostatic force $\vec{F}$ is further related to $\vec{E}$ through $\vec{F} = q\vec{E}$ [Eq. (1.2)], it follows that the work W done by the electrostatic force ($W = \int d\vec{l} \cdot \vec{F}$, where $d\vec{l}$ is the infinitesimal element of displacement) is also path independent. Equivalently, the work done by the electrostatic force field over a closed path is zero.

Example Equation (1.13) can be used to obtain the expression for the electrostatic potential due to a point charge q located at the origin. In this case, the corresponding volume charge density will be zero everywhere except at the origin where it is infinitely large. Hence, $\rho(\vec{r}')$ becomes $\rho(\vec{r}') = q\delta^3(\vec{r}')$, giving

$$V(\vec{r}) = \frac{q}{4\pi\varepsilon_0 r} \tag{1.14}$$

What does the electrostatic potential physically correspond to? To gain physical insight about this, consider moving a point charge q from an initial position $\vec{r}_0$ to the final position $\vec{r}_1$ in the presence of the electrostatic field. This is schematically shown in Fig. 1.6. The work done against the electrostatic force $\vec{F}$ will be

$$W = \int_{\vec{r}_0}^{\vec{r}_1} -d\vec{l} \cdot \vec{F} = q\int_{\vec{r}_0}^{\vec{r}_1} d\vec{l} \cdot \nabla V = q[V(\vec{r}_1) - V(\vec{r}_0)] \tag{1.15}$$

Stated in words, the potential difference between two points is equal to the work done per unit charge in moving the charge from the initial to the final point.

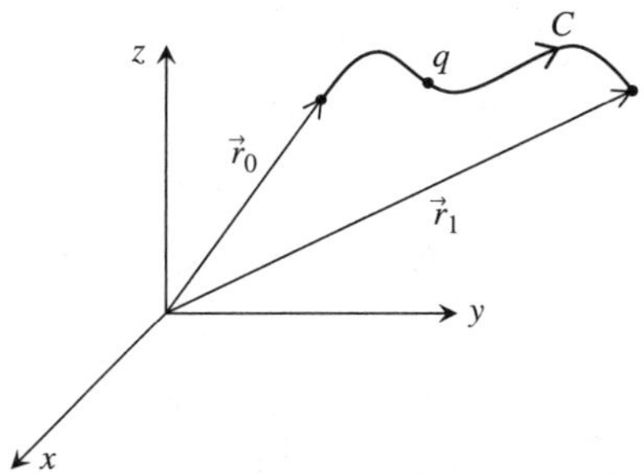

Figure 1.6 A point charge q is moved in the presence of the electrostatic force $\vec{F}$ along the curve C. The end points have the position vectors $\vec{r}_0$ and $\vec{r}_1$.

Example In this example, we derive the electrostatic potential due to an electric dipole sketched in Fig. 1.7. The result obtained here will be used in the next section to discuss the electrical properties of dielectric materials. An electric dipole simply consists of two point charges of size $+q$ and $-q$, separated by a small distance. Figure 1.7 shows such a dipole oriented along the z-axis. The separation between the charges is d. From the superposition principle, the potential at the observation point P will be given by

$$V(\vec{r}) = \frac{q}{4\pi\varepsilon_0}\left(\frac{1}{r_1} - \frac{1}{r_2}\right) \tag{1.16}$$

By using the law of cosines, r_1 and r_2 can be expressed in terms of r, d, and θ as

$$\begin{aligned} r_1 &= r\sqrt{1 + \frac{d^2}{4r^2} - \frac{d}{r}\cos\theta} \\ r_2 &= r\sqrt{1 + \frac{d^2}{4r^2} + \frac{d}{r}\cos\theta} \end{aligned} \tag{1.17}$$

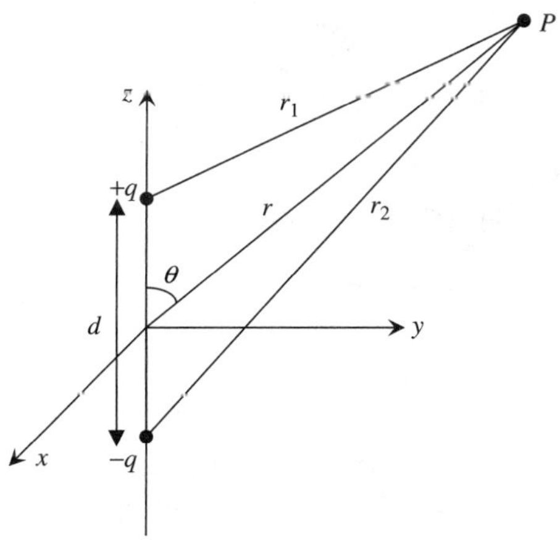

Figure 1.7 An electric dipole consisting of two equal and opposite charges $+q$ and $-q$, separated by a small distance d.

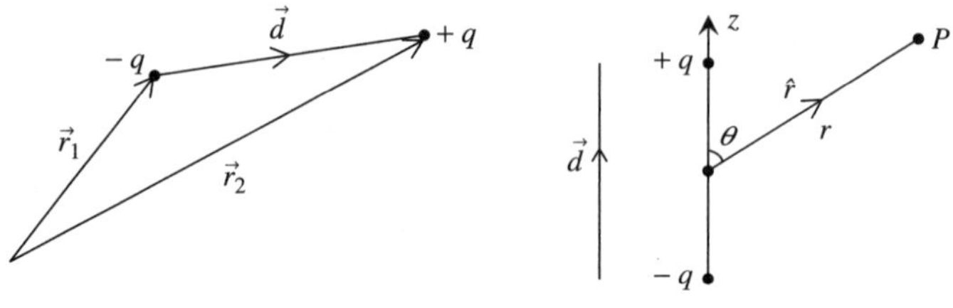

(a) (b)

FIGURE 1.8 (*a*) The dipole moment of two point charges $-q$ and $+q$ at the position vectors $\vec{r}_1$ and $\vec{r}_2$ is given by $q\vec{d}$, where $\vec{d} = \vec{r}_2 - \vec{r}_1$. (*b*) A dipole oriented along the *z*-axis.

By assuming that r is much larger than d, we do a Taylor series expansion of the square root terms in Eq. (1.17) to obtain

$$\frac{1}{r_1} - \frac{1}{r_2} = \frac{2}{r}\left(\frac{d}{2r}\cos\theta\right) + \mathrm{O}(d^2) \tag{1.18}$$

In Eq. (1.18), $\mathrm{O}(d^2)$ indicates that the neglected terms are at least quadratic or higher-order in d. Hence, at large distances away from the dipole, the potential becomes

$$V(\vec{r}) = \frac{qd}{4\pi\varepsilon_0 r^2}\cos\theta \tag{1.19}$$

We can also express the potential in Eq. (1.19) in terms of the electric dipole moment. For a collection of point charges, the electric dipole moment $\vec{p}$ is defined by

$$\vec{p} = \sum_{\mathrm{i}} q_{\mathrm{i}}\vec{r}_{\mathrm{i}} \tag{1.20}$$

where $\vec{r}_{\mathrm{i}}$ is the position vector of the ith point charge q_i. In the case of two point charges as sketched in Fig. 1.8*a*, the dipole moment will be

$$\vec{p} = -q\vec{r}_1 + q\vec{r}_2 = q(\vec{r}_2 - \vec{r}_1) = q\vec{d} \tag{1.21}$$

The sketch of the dipole in Fig. 1.8*b* shows its orientation with respect to the observation coordinates. The unit vector $\hat{r}$ points in the radial direction from the origin to the observation point. In particular, note that $\vec{d}\cdot\hat{r} = d\cos\theta$, hence giving

$$V(\vec{r}) = \frac{1}{4\pi\varepsilon_0}\frac{\vec{p}\cdot\hat{r}}{r^2} \tag{1.22}$$

1.3.3 Electrostatic Fields in Dielectric Materials

We now consider a neutral dielectric material in which there are no free charge carriers. If the constituents are atoms or nonpolar molecules, we can crudely model them as consisting of a positively charged

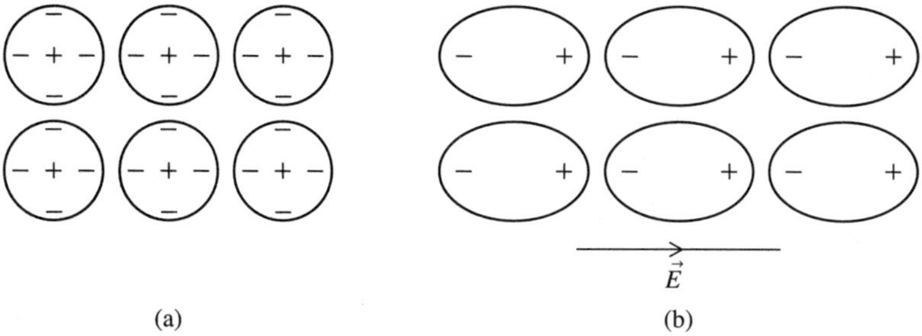

Figure 1.9 (*a*) Schematic representation of nonpolar atoms in the absence of an external electric field with spherical electron clouds around the positive nuclei. (*b*) Induced atomic dipoles in the presence of an applied external electric field.

nucleus surrounded by a negative electron cloud. Overall, the material will be charge neutral. Furthermore, we assume that in the absence of an external electric field, the electron charge is symmetric around the ion, giving rise to no dipole moment. This is schematically shown in Fig. 1.9*a*. In the presence of an external field, although the electrons will not aid in charge transport, the charge distributions will be stretched in space. In particular, the external field will tend to push the nucleus in the direction of the electric field while the electron cloud will be pushed in the opposite direction until the built-in electric field between the displaced nucleus and the electron cloud is exactly balanced by the applied local field. This is schematically shown in Fig. 1.9*b*. Hence, the net effect of the applied external field will be the formation of electric dipoles inside the material. Alternatively, the dielectric material may consist of polar molecules which already have built-in dipole moments. Typically, these will be randomly oriented in the absence of an external field, giving rise to no net macroscopic dipole moment. Application of the external field on a polar material will tend to align the otherwise randomly oriented dipoles along the direction of the electric field and again produce a net dipole moment. In short, we can say that application of an external electric field on a dielectric material results in the formation of an electric dipole distribution inside the material. Understanding the general properties of the induced dipole distribution is critical in developing the models for optical materials as well as light–matter interaction.

Next we consider a dielectric material which has inside it a distribution of electric dipole moments. Such a material is said to be polarized. We will now obtain the expression for the electrostatic potential produced by a polarized dielectric. Figure 1.10 shows a sketch of a polarized material in which an infinitesimal element of volume has a dipole moment of $d\vec{p}$. The distribution of electric dipole moments

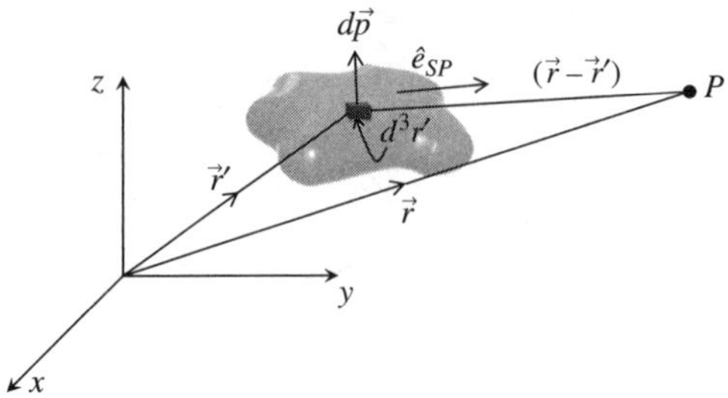

FIGURE 1.10 Sketch of a polarized object, showing an infinitesimal region of volume d^3r' and dipole moment $d\vec{p}$ at the source coordinate $\vec{r}'$.

within the dielectric can be characterized by introducing the polarization vector $\vec{P}$, which gives the electric dipole moment per unit volume, or simply the dipole density. Hence, the dipole moment of an infinitesimal element at the source coordinate $\vec{r}'$ can be written as $d\vec{p} = \vec{P}d^3r'$. The infinitesimal contribution dV of this dipole element to the potential at the observation point $\vec{r}$ will be

$$dV = \frac{d\vec{p}\cdot\hat{e}_{SP}}{4\pi\varepsilon_0|\vec{r}-\vec{r}'|^2} = \frac{d^3r'\vec{P}\cdot(\vec{r}-\vec{r}')}{4\pi\varepsilon_0|\vec{r}-\vec{r}'|^3} \tag{1.23}$$

To obtain the resultant potential, we need to integrate over the whole object. By using Eq. (1.11), V becomes

$$V = \frac{1}{4\pi\varepsilon_0}\int_V d^3r'\vec{P}\cdot\nabla'\left(\frac{1}{|\vec{r}-\vec{r}'|}\right) \tag{1.24}$$

where the del operator inside the integrand acts on the primed source coordinates only. By using the vector identity

$$\nabla\cdot(f\vec{G}) = \nabla f\cdot\vec{G} + f\nabla\cdot\vec{G} \tag{1.25}$$

with $f = 1/|\vec{r}-\vec{r}'|$ and $\vec{G} = \vec{P}$, the integrand in Eq. (1.24) can be written as

$$\vec{P}\cdot\nabla'\left(\frac{1}{|\vec{r}-\vec{r}'|}\right) = \nabla'\cdot\left(\frac{\vec{P}}{|\vec{r}-\vec{r}'|}\right) - \frac{\nabla'\cdot\vec{P}}{|\vec{r}-\vec{r}'|} \tag{1.26}$$

giving

$$V = \frac{1}{4\pi\varepsilon_0}\int_V d^3r'\left[\nabla'\cdot\left(\frac{\vec{P}}{|\vec{r}-\vec{r}'|}\right) - \frac{\nabla'\cdot\vec{P}}{|\vec{r}-\vec{r}'|}\right] \tag{1.27}$$

Applying the divergence theorem on the first term and noting that the element of area $d\vec{a}' = \hat{n}da'$ with $\hat{n}$ = unit normal vector on the surface S, we get

$$V = \frac{1}{4\pi\varepsilon_0}\left[\oint_S da' \frac{\vec{P}\cdot\hat{n}}{|\vec{r}-\vec{r}'|} + \int_V d^3r' \frac{(-\nabla\cdot\vec{P})}{|\vec{r}-\vec{r}'|}\right] \tag{1.28}$$

Equation (1.28) is a very important result. To interpret the result, let's recall the integral expression [Eq. (1.13)] we derived earlier for the potential due to a volume charge. Similarly, we see that if the source charge is distributed on the surface, represented by the surface charge density $\sigma(\vec{r}')$, the potential will be given by

$$V = \frac{1}{4\pi\varepsilon_0}\int_S da' \frac{\sigma(\vec{r}')}{|\vec{r}-\vec{r}'|} \tag{1.29}$$

So, going back to Eq. (1.28), we note that the potential due to a polarized dielectric is identical to that produced by two types of effective charge distributions:

1. Surface charge density $\sigma_b = \vec{P}\cdot\hat{n}$. This is commonly referred to as the bound surface charge density or the polarization-induced surface charge density.
2. Volume charge density $\rho_b = -\nabla\cdot\vec{P}$. This is commonly referred to as the bound volume charge density or the polarization-induced volume charge density. The subscript b indicates that these are the bound charge densities. The term "bound" refers to the fact that these charges are spatially constrained due to the binding between the electron cloud and the nucleus of the atoms and do not aid in charge conduction. Another general characteristic of bound charges is that the total bound charge due to volume and surface contributions is zero for a polarized object (see Prob. 1.5).

1.3.4 Electric Displacement Vector

How do we use Gauss' law inside a dielectric? Following the analysis of the previous section, we should now distinguish between two types of charges: bound and free charges. We can describe the difference between these two types of charges as follows. Free charges would be those which, we could distribute in any way we wish within the material. However, we cannot control the spatial distribution of bound charges since they arise solely due to the polarization present inside the material. The total volume charge density $\rho(\vec{r})$ appearing in Eq. (1.9) contains both types of charges. We thus rewrite Eq. (1.9) as

$$\nabla\cdot\vec{E} = \frac{\rho(\vec{r})}{\varepsilon_0} = \frac{\rho_b + \rho_f}{\varepsilon_0} = \frac{\rho_f - \nabla\cdot\vec{P}}{\varepsilon_0} \tag{1.30}$$

Above, ρ_f represents the free volume charge density. By examining Eq. (1.30), we see that we can introduce a vector field whose divergence depends only on the free charge density. This is known as the electric displacement vector $\vec{D}$ and is defined according to

$$\vec{D} = \varepsilon_0 \vec{E} + \vec{P} \tag{1.31}$$

From Eq. (1.30), $\vec{D}$ satisfies

$$\nabla \cdot \vec{D} = \rho_f \tag{1.32}$$

Equation (1.32) is the generalized form of Gauss' law which also includes material effects through the dependence of the displacement vector on polarization. Furthermore, this is the first of the four Maxwell's equations. In integral form, we have

$$\oint_S d\vec{a} \cdot \vec{D} = q_f \tag{1.33}$$

where q_f is the total free charge within the volume enclosed by the surface S.

In linear, isotropic dielectrics, the induced polarization is directly proportional to the applied electric field. This proportionality is written in equation form as

$$\vec{P} = \varepsilon_0 \chi_e \vec{E} \tag{1.34}$$

Above, the dimensionless parameter χ_e is known as the electric susceptibility and quantifies the tendency of a material to become polarized in the presence of an applied electric field. Combining Eqs. (1.31) and (1.34), we can express the electric displacement vector as

$$\vec{D} = \varepsilon \vec{E} \tag{1.35}$$

in terms of the electric field. ε is known as the material permittivity and is given by

$$\varepsilon = \varepsilon_0 (1 + \chi_e) \tag{1.36}$$

1.4 Magnetostatics

1.4.1 Magnetic Induction Vector

In Sec. 1.3, we saw that static electric charges produce electric fields. Observations show that moving charges which constitute electric currents give rise to a different type of vector field known as the magnetic induction vector $\vec{B}$. $\vec{B}$ is also referred to as the magnetic flux density. Different than the electric field, $\vec{B}$ produces a velocity-dependent force on electric charges. The resulting force on a charge q moving with velocity $\vec{u}$ in the presence of an electric and a magnetic field is given by the Lorentz force law:

$$\vec{F}_L = q(\vec{E} + \vec{u} \times \vec{B}) \tag{1.37}$$

Above, $\vec{F}_L$ is the Lorentz force acting on the charged particle. Note that if the charge is stationary, Eq. (1.37) reduces to Eq. (1.2).

The law of Biot and Savart is an empirical relation from which the magnetic induction can be calculated for a given electric current distribution. Let us suppose that the electric current flow in a region is described by the total volume current density $\vec{J}$. The direction of $\vec{J}$ corresponds to the direction of current flow and its magnitude gives the current flow per unit area perpendicular to the direction of flow. The units of $\vec{J}$ are A/m^2. Hence, the total current I flowing through a surface S is given by the surface integral of $\vec{J}$:

$$I = \int_S d\vec{a} \cdot \vec{J} \tag{1.38}$$

Let us consider an infinitesimal element of volume at the source coordinate $\vec{r}'$, containing electric current flow represented by the volume current density $\vec{J}$. This is schematically shown in Fig. 1.11. The infinitesimal contribution $d\vec{B}$ to the magnetic induction at the observation point with position vector $\vec{r}$ will be given by

$$d\vec{B} = \frac{\mu_0}{4\pi} \frac{d^3r'\vec{J}(\vec{r}') \times \hat{e}_{SP}}{|\vec{r} - \vec{r}'|^2} \tag{1.39}$$

Equation (1.39) is the MKS version of the law of Biot and Savart, where, the proportionality constant μ_0 is known as the permeability

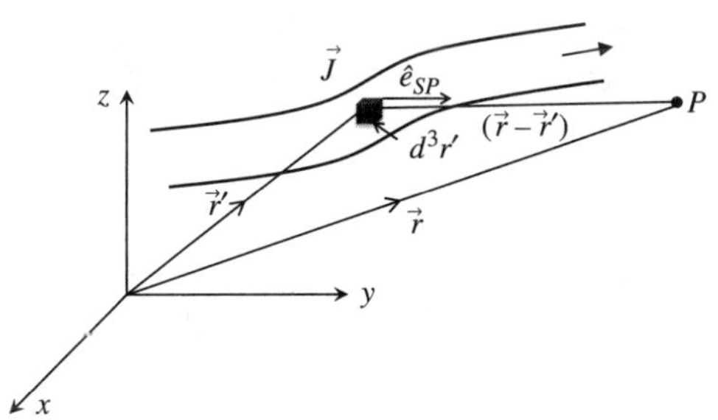

FIGURE 1.11 Infinitesimal element of volume d^3r' at the source coordinate $\vec{r}'$, containing electric current flow represented by the volume current density $\vec{J}$.

of free space and has the value $4\pi \times 10^{-7}$ N/A^2. As before, $\hat{e}_{SP}$ is the unit vector from the source coordinate to the observation point. We leave it as an exercise for the reader to show that when Eq. (1.39) is integrated over the volume of the source region, the expression for the resultant magnetic field at $\vec{r}$ can be expressed as (Prob. 1.6)

$$\vec{B} = \nabla \times \vec{A} \tag{1.40}$$

where $\vec{A}$ given by

$$\vec{A} = \frac{\mu_0}{4\pi} \int_V d^3\vec{r}' \frac{\vec{J}(\vec{r}')}{|\vec{r} - \vec{r}'|} \tag{1.41}$$

is known as the magnetic vector potential. Since $\nabla \cdot (\nabla \times \vec{V}) = 0$ for any vector field $\vec{V}$, we conclude from Eq. (1.40) that

$$\nabla \cdot \vec{B} = 0 \tag{1.42}$$

Equation (1.42) is sometimes referred to as the law of no magnetic monopoles.

As can be seen, our approach in deriving the key relations of magnetostatics parallels that used in Sec. 1.3 for electrostatics. Having found the general result about the divergence of the magnetic induction vector, we next move on to calculating its curl. The curl of $\vec{B}$ is given by

$$\nabla \times \vec{B} = \nabla \times (\nabla \times \vec{A}) = \nabla(\nabla \cdot \vec{A}) - \nabla^2 \vec{A} \tag{1.43}$$

In Eq. (1.43), $\nabla^2 \vec{A}$ is the vector Laplacian of $\vec{A}$. It can be shown after somewhat tedious algebra (which we skip) that for a steady-state current distribution, $\nabla(\nabla \cdot \vec{A}) = 0$. See, for example, Refs. 1 and 2. To calculate $\nabla^2 \vec{A}$, we note that

$$\nabla^2 \vec{A} = \nabla^2 \left(\frac{\mu_0}{4\pi} \int_V d^3r' \frac{\vec{J}(\vec{r}')}{|\vec{r} - \vec{r}'|} \right) = \frac{\mu_0}{4\pi} \int_V d^3r' \vec{J}(\vec{r}') \nabla^2 \frac{1}{|\vec{r} - \vec{r}'|} \tag{1.44}$$

where

$$\nabla^2 \left(\frac{1}{|\vec{r} - \vec{r}'|} \right) = \nabla \cdot \left(\nabla \left(\frac{1}{|\vec{r} - \vec{r}'|} \right) \right) = -4\pi \delta^3(\vec{r} - \vec{r}') \tag{1.45}$$

Hence, we get

$$\nabla \times \vec{B} = \mu_0 \vec{J}(\vec{r}) \tag{1.46}$$

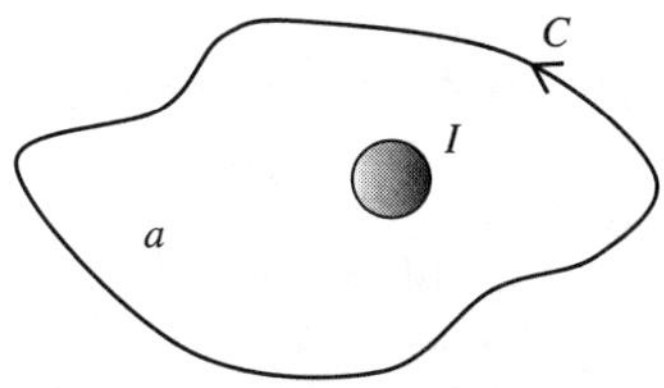

FIGURE 1.12 Schematic of the geometry used in the derivation of the integral form of Ampere's law. The area *a* is enclosed within the closed curve *C*. The total current (flowing out of the paper) enclosed inside *C* is *I*.

Eq. (1.46) is the differential form of Ampere's law for magnetostatics. It turns out that it is not valid for time varying magnetic fields, and we will revisit this relation in the next section when we discuss Maxwell's Equations.

The integral form of Ampere's law can be readily derived by taking the surface integral of both sides of Eq. (1.46). A schematic of the geometry is shown in Fig. 1.12. The closed curve C encloses the area *a*. Straightforward application of Stokes' theorem gives

$$\oint_C d\vec{l} \cdot \vec{B} = \mu_0 I \tag{1.47}$$

Here, *I* is the total current enclosed within C due to all sources of current.

1.4.2 Vector Magnetic Potential of a Magnetic Dipole

As an example of the above formulas, let us calculate the vector magnetic potential of a magnetic dipole in the form of a circular loop carrying a current *I*. We will calculate the magnetic vector potential at large distances. The result will be used in the discussion of the magnetic properties of materials. Figure 1.13 shows a sketch of the current

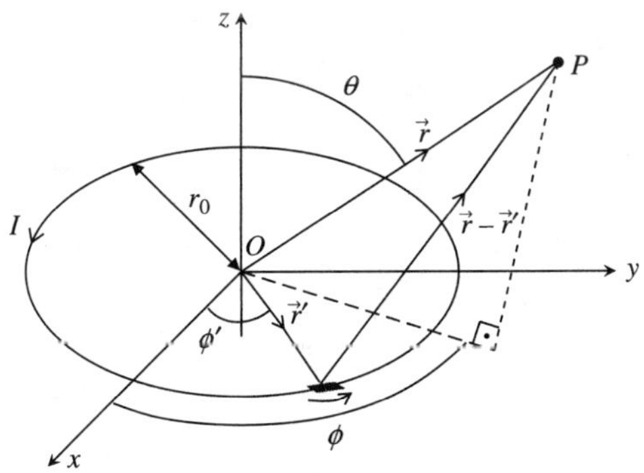

FIGURE 1.13 Sketch of the current loop with a radius of r_0 situated in the *x-y* plane and centered at the origin. The current carried by the loop is *I*.

loop with a radius of r_0 situated in the x-y plane and centered at the origin. We will use cylindrical coordinates (ρ',ϕ',z') to represent the source coordinates and spherical coordinates (r,θ,ϕ) for the observation point. In order to use Eq. (1.41) to calculate $\vec{A}$, we need to find the expressions for the volume current density $\vec{J}(\vec{r}')$ and the separation distance $|\vec{r}-\vec{r}'|$. Since the current is confined to flow in a circular loop, there is a singularity in the value of the volume current density at $z'=0$ and $\rho'=r_0$. Furthermore, the current flows along the $\hat{\phi}'$ direction. Hence we take the volume current density to be of the form

$$\vec{J}(\vec{r}')=J_0\delta(r'-r_0)\delta(z')\hat{\phi}' \tag{1.48}$$

where the constant J_0 is chosen such that the total current carried across the cross section of the loop is I. Taking the element of area in the $\hat{\phi}'$direction as $d\vec{a}=\hat{\phi}'dr'dz'$, we get

$$\int d\vec{a}\cdot\vec{J}=\int_0^{\infty}dr'\int_{-\infty}^{+\infty}dz'J_0\delta(r'-r_0)\delta(z')\hat{\phi}'\cdot\hat{\phi}'=J_0=I \tag{1.49}$$

Figure 1.14 further shows the overhead view of the geometry and the projection of the radial vector $\vec{r}$ in the x-y plane. Decomposing $\hat{\varphi}'$ into its components along the observation unit vectors $\hat{\varphi}$ and $\hat{\rho}$, we get the following expression for the volume current density:

$$\vec{J}(\vec{r}')=I\delta(r'-r_0)\delta(z')[\sin(\varphi-\varphi')\hat{\rho}+\cos(\varphi-\varphi')\hat{\varphi}] \tag{1.50}$$

The separation distance $|\vec{r}-\vec{r}'|$ can be calculated by expressing the observation and source coordinates in Cartesian components. We

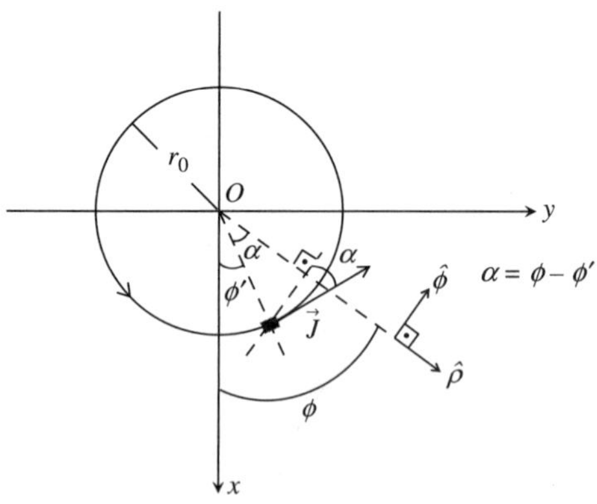

FIGURE 1.14 Overhead view of the current loop, showing the orientation of the current element.

leave it as an exercise (Prob. 1.7) for the reader to show that $|\vec{r}-\vec{r}'|$ becomes

$$|\vec{r}-\vec{r}'| = r\left(1+\left(\frac{r_0}{r}\right)^2 - \frac{2r_0}{r}\sin\theta\,\cos(\varphi-\varphi')\right)^{1/2} \tag{1.51}$$

Since we want the magnetic vector potential at large distances, we expand $1/|\vec{r}-\vec{r}'|$ up to first order in r_0 and obtain

$$\frac{1}{|\vec{r}-\vec{r}'|} = \frac{1}{r} + \frac{r_0}{r^2}\,\sin\theta\,\cos\,(\varphi-\varphi') + O(r_0^2) \tag{1.52}$$

By substituting Eqs. (1.50) and (1.52) into Eq. (1.41), we get after some algebra,

$$\vec{A} = \frac{\mu_0}{4\pi}\frac{\pi r_0^{\,2} I}{r^2}\sin\theta\hat{\varphi} \tag{1.53}$$

The magnitude of the magnetic vector potential varies linearly with the product of the loop area and the current. It also depends on the angular orientation of the observation point through the $\sin\theta$ term. We can express the above result in vector form by introducing the magnetic moment $\vec{m}$ which is defined according to

$$\vec{m} \equiv \pi r_0^{\,2} I\hat{z} = I\vec{a} \tag{1.54}$$

where the direction of the area vector $\vec{a}$ can be determined by using the right-hand rule: when the fingers of the right hand are curled in the direction of current flow, the thumb points in the direction of the area vector. This vector, in turn, is perpendicular to the plane containing the current loop. Note that choosing the direction of $\vec{m}$ along the z-direction in Eq. (1.54) is in agreement with this right-hand rule (see also Fig. 1.13). Finally, since $\hat{z}\times\hat{r} = \sin\theta\hat{\varphi}$, we get

$$\vec{A} = \frac{\mu_0}{4\pi}\frac{\vec{m}\times\hat{r}}{r^2} \tag{1.55}$$

1.4.3 Magnetic Effects in Matter

You may be wondering why we went through the heavy dose of mathematics in the previous section to obtain the magnetic vector potential of a dipole. Well, this turns out to serve as a crude but at the same time, useful model to describe the general properties of magnetic materials. Classically, we can view the electrons orbiting atoms as constituting small current loops. Hence, each electron will have a magnetic moment. In the absence of an external magnetic field, these magnetic moments are randomly oriented except in permanently magnetic materials such as ferromagnets. After the application of an external magnetic field, the magnetic moments experience a torque and become aligned parallel (in the case of paramagnetic materials) or anti-parallel (in the case of diamagnetic materials) to $\vec{B}$. In this case, the material is said to become magnetized and develops a magnetic field around it. In other words, a magnetic moment distribution is established inside the material. As we did in the earlier sections for extended charged or polarized objects, we characterize the distribution of the magnetic moments by introducing the magnetization vector $\vec{M}$. $\vec{M}$ gives the magnetic moment per unit volume. So, for an extended magnetized object described by the magnetic moment $\vec{M}$, the vector magnetic potential at the observation position $\vec{r}$ will be given by

$$\vec{A}(\vec{r}) = \frac{\mu_0}{4\pi}\int d^3r' \frac{\vec{M}(\vec{r}\,')\times(\vec{r}-\vec{r}\,')}{\left|\vec{r}-\vec{r}\,'\right|^3} = \frac{\mu_0}{4\pi}\int d^3r'\ \vec{M}(\vec{r}\,')\times\nabla'\left(\frac{1}{\left|\vec{r}-\vec{r}\,'\right|}\right) \tag{1.56}$$

Here, $\vec{r}\,'$ is the usual source coordinate. By using the identity

$$\nabla\times(f\vec{V}) = f\nabla\times\vec{V} - \vec{V}\times(\nabla f) \tag{1.57}$$

with $\vec{V} = \vec{M}(\vec{r}\,')$ and $f = 1/\left|\vec{r}-\vec{r}\,'\right|$, we get

$$\vec{A}(\vec{r}) = \frac{\mu_0}{4\pi}\int d^3r' \left[\frac{1}{\left|\vec{r}-\vec{r}\,'\right|}\nabla'\times\vec{M}(\vec{r}\,') - \nabla'\times\left(\frac{\vec{M}(\vec{r}\,')}{\left|\vec{r}-\vec{r}\,'\right|}\right)\right] \tag{1.58}$$

At this point, we use a less common vector identity that can be derived from the divergence theorem:

$$\int_V d^3r'(\nabla\times\vec{V}) = -\oint_S (\vec{V}\times d\vec{a}\,') \tag{1.59}$$

Above, S is the closed surface that encloses the volume V and $d\vec{a}\,' = \hat{n}da'$, where $\hat{n}$ is the unit outward normal vector on S. By using the relation (1.59) in Eq. (1.58), we get

$$\vec{A}(\vec{r}) = \frac{\mu_0}{4\pi}\int_V d^3r' \left[\frac{\nabla'\times\vec{M}}{\left|\vec{r}-\vec{r}\,'\right|}\right] + \frac{\mu_0}{4\pi}\oint_S da' \frac{\vec{M}(\vec{r}\,')\times\hat{n}}{\left|\vec{r}-\vec{r}\,'\right|} \tag{1.60}$$

Remembering the general expression for the vector magnetic potential [Eq. (1.41)], we see that the potential of a magnetized object appears to arise from two types of current densities:

1. a bound volume current density $\vec{J}_b = \nabla\times\vec{M}$ and
2. a bound surface current density $\vec{K}_b = \vec{M}\times\hat{n}$

Hence, inside a magnetic material we encounter two types of volume current densities and write the total volume current density $\vec{J}$ as

$$\vec{J} = \vec{J}_f + \vec{J}_b = \vec{J}_f + \nabla \times \vec{M} \tag{1.61}$$

Above, $\vec{J}_f$ is the free current density which consists of charge flow that can be externally controlled, independent of the magnetization. Application of Ampere's law gives

$$\nabla \times \vec{B} = \mu_0(\vec{J}_f + \nabla \times \vec{M}) \tag{1.62}$$

Eq. (1.62) can be rearranged to give

$$\nabla \times \vec{H} = \vec{J}_f \tag{1.63}$$

where the magnetic field intensity vector $\vec{H}$ is defined according to

$$\vec{H} \equiv \frac{\vec{B}}{\mu_0} - \vec{M} \tag{1.64}$$

The unit of the magnetic field intensity is A/m. For most materials in the linear response range, the induced magnetization varies linearly with the applied $\vec{H}$ field according to

$$\vec{M} = \chi_m \vec{H} \tag{1.65}$$

where χ_m is known as the magnetic susceptibility and gives a quantitative measure of the induced magnetization per unit applied magnetic field intensity. Combining Eqs. (1.64) and (1.65), we have

$$\vec{B} = \mu_0(\vec{M} + \vec{H}) = \mu\vec{H} \tag{1.66}$$

where μ is the permeability of the medium and is given by

$$\mu = \mu_0\,(1 + \chi_m) \tag{1.67}$$

1.5 Faraday's Law of Electromagnetic Induction

In a series of experiments, the famous English scientist Michael Faraday discovered a very important principle that describes the coupling between electric and magnetic fields. The results can be summarized as follows:

1. When a coil is moved through a magnetic field, current flow is induced in the coil.
2. When the coil is stationary and the magnet is moved instead, again, a current flow is obtained in the coil.
3. With the coil and the magnet at rest, a time varying magnetic field can also produce a current flow in the coil.

This phenomenon is known as electromagnetic induction. As a result of these experiments, Faraday came to the conclusion that whenever the magnetic flux linking a coil varied in time, it would induce an electromotive force V_{emf} which, just like a battery, can push electric charges around the circuit and create an electric current. The magnetic flux Ψ_m is defined according to

$$\Psi_m = \int_S d\vec{a} \cdot \vec{B} \tag{1.68}$$

The SI unit of Ψ_m is the weber (Wb). According to the results of Faraday's experiments, the induced electromotive force is equal to the time rate of change of the magnetic flux. In equation form, we have

$$V_{\text{emf}} = \oint_C d\vec{l} \cdot \vec{E} = -\frac{d\Psi_m}{dt} \tag{1.69}$$

So, note that the induced electric field is no longer conservative. Applying Stokes' theorem to Eq. (1.69), we get

$$\oint_C d\vec{l} \cdot \vec{E} = \int_S d\vec{a} \cdot (\nabla \times \vec{E}) = -\frac{d}{dt}\int_S d\vec{a} \cdot \vec{B} = -\int_S d\vec{a} \cdot \frac{\partial \vec{B}}{\partial t} \tag{1.70}$$

Since the above relation is valid for an arbitrary surface S bounded by the closed path C, we conclude that the integrands must be identical, giving

$$\nabla \times \vec{E} = -\frac{\partial \vec{B}}{\partial t} \tag{1.71}$$

Eq. (1.71) is one of the four Maxwell's equations which we discuss next. If there are no time varying magnetic fields, we get $\nabla \times \vec{E} = 0$ as was already obtained in electrostatics.

1.6 Maxwell's Equations

So far, we have derived four fundamental equations [Eqs. (1.32), (1.42), (1.63), (1.71)] that summarize the important characteristics of electric and magnetic fields. Before we list them, we focus on Ampere's law once again:

$$\nabla \times \vec{H} = \vec{J}_f \tag{1.72}$$

This is not yet in its final, correct form. The problem with this equation can be seen by taking the divergence of both sides. For any vector field $\vec{V}$, we know that $\nabla \cdot (\nabla \times \vec{V}) = 0$. So, after taking the divergence of both sides, the left-hand side becomes zero, whereas $\nabla \cdot \vec{J}_f$ is not

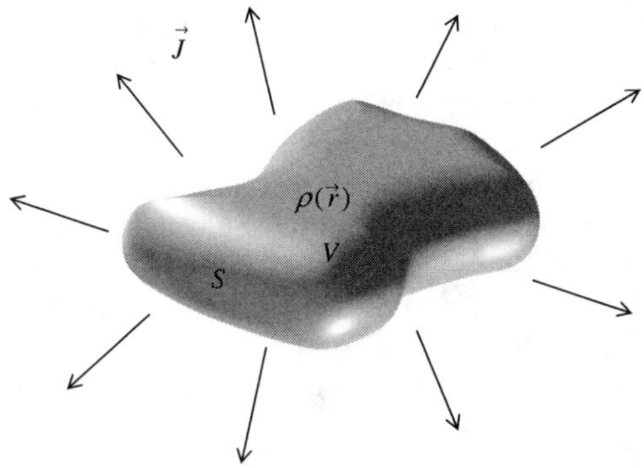

FIGURE 1.15 A closed surface *S* enclosing a volume *V*. This geometry is used in the derivation of the charge continuity equation.

necessarily zero. In fact, we can find the value of this divergence by using the principle of conservation of charge.

To use the principle of charge conservation, we consider a closed surface *S* which encloses a region of volume *V*, as shown in Fig. 1.15. Let us also assume that the region contains some electric charge characterized by the volume charge density $\rho(\vec{r})$. The principle of charge conservation states that the net electric charge cannot be created or destroyed. In other words, if the charge stored within the volume *V* decreases, this should be accompanied by an outward current flow. Since the current *I* is equal to the rate of change of electric charge, we have

$$-\frac{d}{dt}\int_V d^3r\rho(\vec{r}) = -\int_V d^3r\frac{\partial\rho}{\partial t} = I = \oint_S d\vec{a}\cdot\vec{J} = \int_V d^3r\nabla\cdot\vec{J} \tag{1.73}$$

Because the above relations hold for an arbitrary enclosed region, the integrands inside the volume integrals should be equal, leading to

$$\nabla\cdot\vec{J} + \frac{\partial\rho}{\partial t} = 0 \tag{1.74}$$

Eq. (1.74) summarizes the principle of conservation of electric charge and is known as the continuity equation. To correct Eq. (1.72), we add an additional current density $\vec{J}_d$ on the right-hand side to satisfy the continuity equation. This gives

$$\begin{aligned}\nabla\cdot(\nabla\times\vec{H}) = 0 = \nabla\cdot\vec{J}_f + \nabla\cdot\vec{J}_d &= -\frac{\partial\rho_f}{\partial t} + \nabla\cdot\vec{J}_d \\ &= -\frac{\partial\nabla\cdot\vec{D}}{\partial t} + \nabla\cdot\vec{J}_d = \nabla\cdot\left(\vec{J}_d - \frac{\partial\vec{D}}{\partial t}\right)\end{aligned} \tag{1.75}$$

Above, we used Gauss' law in differential form [Eq. (1.32)]. Hence, charge conservation is satisfied by adding $\vec{J}_d = \partial\vec{D}/\partial t$ to the free current density in Ampere's law. $\vec{J}_d$ is referred to as the displacement current. Historically, this was first introduced by James Clerk Maxwell. Since the inclusion of the displacement current played such a major role in the later development of electromagnetic interactions and the theoretical demonstration that light is a form of electromagnetic wave, the four key equations which summarize the important properties of the electric and magnetic fields became known as Maxwell's equations. The complete set of Maxwell's equations are:

Gauss' law $$\nabla \cdot \vec{D} = \rho_f \tag{1.76}$$

Law of no magnetic monopoles $$\nabla \cdot \vec{B} = 0 \tag{1.77}$$

Faraday's law $$\nabla \times \vec{E} = -\frac{\partial \vec{B}}{\partial t} \tag{1.78}$$

Ampere-Maxwell law $$\nabla \times \vec{H} = \vec{J}_f + \frac{\partial \vec{D}}{\partial t} \tag{1.79}$$

Equations (1.76) to (1.79) form the theoretical framework that is used in the analysis of all known electromagnetic phenomena. We note that Maxwell's equations alone are not sufficient in uniquely determining the electric and magnetic fields. In addition, constitutive equations which relate $\vec{D}$ to $\vec{E}$ and $\vec{B}$ to $\vec{H}$ are needed. Additional physics is needed to obtain these relations as we will discuss in later chapters on light-matter interactions. Finally, note that all of the fields appearing in Maxwell's equations have position as well as time dependence. We have not explicitly shown this in the above equations for the sake of simplicity.

1.7 Electromagnetic Boundary Conditions

In many problems, we encounter interfaces which separate two media with different electric or magnetic properties. For example, consider the interface between air and a glass slab. It is of interest to know how the electric and magnetic fields on either side of the interface are related. For example, will there be a discontinuity in the field components when the medium characteristics abruptly change? It turns out that Maxwell's equation alone tell us precisely how the field components on either side of an interface are related. These relations are referred to electromagnetic boundary conditions and can be derived by using the integral form of Maxwell's equations.

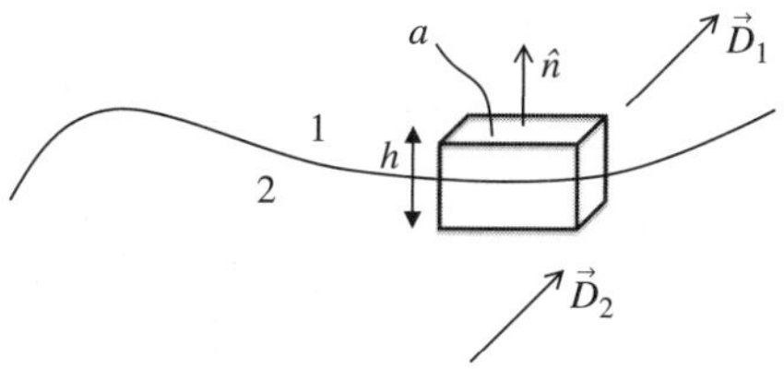

FIGURE 1.16 Two media separated by an interface. The Gaussian pillbox of width h and surface area a is used to derive the boundary condition for the normal component of the displacement vector.

First, let us use Gauss' law (Eq. 1.76) to derive the continuity relation for the electric displacement vector $\vec{D}$. Integrating both sides of Eq. (1.76) and using the divergence theorem, we have

$$\oint_S d\vec{a} \cdot \vec{D} = q_f \tag{1.80}$$

where q_f is the total free charge within the volume enclosed by the surface S. Suppose that there are two media separated by an interface as shown in Fig. 1.16. The unit normal vector on the surface is $\hat{n}$. Let us also assume that the interface has some free surface charge represented by σ_f. We consider a Gaussian pillbox which is partly above and partly below the interface. The height and the cross-sectional area of the pillbox are taken as h and a, respectively. For a finite value of h, any volume charge ρ inside the box will also contribute to q_f. However, in the limit as h tends to zero, the only enclosed charge within the volume will be due to surface charges. In addition, when the height h approaches zero, the only contribution to the surface integral will come from the normal components of the $\vec{D}$ vector. Let us also assume that the cross sectional area a is small enough so that the displacement vector does not vary significantly over the area. Hence, for vanishing h, we will have

$$\vec{D}_1 \cdot \hat{n}a - \vec{D}_2 \cdot \hat{n}a = a(D_{1n} - D_{2n}) = \sigma_f a \tag{1.81}$$

giving

$$D_{1n} - D_{2n} = \sigma_f \tag{1.82}$$

We see from this analysis that the normal component of the displacement vector is continuous across an interface, provided that no free surface charge exists. If surface charges are present, the discontinuity in the normal component of the electric displacement vector is equal to the surface charge density σ_f. Similarly, it can be shown in a straightforward way (Prob. 1.8) by using Eq. (1.77) that the normal

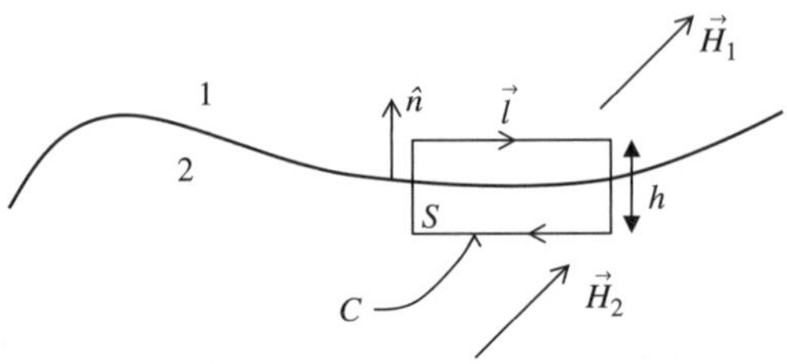

FIGURE 1.17 Sketch of the interface with a closed amperian loop C, used to derive the boundary condition for the magnetic field intensity $\vec{H}$. The width of the loop is h and the unit vector normal to the interface is $\hat{n}$. S is the area enclosed by C.

component of the magnetic induction vector $\vec{B}$ will be continuous across the interface:

$$B_{1n} = B_{2n} \tag{1.83}$$

To obtain the boundary conditions for the vector fields $\vec{H}$ and $\vec{E}$, we integrate the curl equations [Eqs. (1.78) and (1.79)] by using Stokes' theorem. Let us first investigate the continuity of $\vec{H}$. Figure 1.17 shows the interface with the unit normal vector $\hat{n}$ and a closed amperian loop C which is partly above and partly below the interface. The width of the loop is h and $\vec{l}$ is the displacement vector corresponding to the upper section of the loop. We take the surface integral of Eq. (1.79) over the surface S which is enclosed by C. Using Stokes theorem, we get

$$\oint_C d\vec{l} \cdot \vec{H} = I_f + \frac{d}{dt}\int_S d\vec{a} \cdot \vec{D} \tag{1.84}$$

Above, I_f is the total current within the closed loop C. As the width h of the loop shrinks to zero, the contribution from the surface integral of $\vec{D}$ goes to zero. In this limit, the only contribution to I_f may come from currents that are confined to the surface. Let the corresponding surface current density (units A/m) be denoted by $\vec{K}_f$. As such, if $\vec{K}_f$ is perpendicular to the plane of the loop, then $K_f l$ will be the current enclosed inside the loop. If the line integral is computed in the clockwise sense as indicated by the arrows around the loop, it will be positive for a surface current directed into the plane containing the loop. In the general case, where the direction of the surface current density vector is arbitrary, we need to use its component that is perpendicular to the plane of the loop. Note that $\hat{n} \times \hat{e}_l$ ($\hat{e}_l$ is the unit vector along $\vec{l}$) is the unit vector in this direction. So, the total enclosed current can be found by finding the component of $\vec{K}_f$ along $\hat{n} \times \hat{e}_l$ and then multiplying by the length l of the interface. Hence, in the limit as h tends to zero and when l is small enough so that $\vec{H}$ is nearly constant, we have from Eq. (1.84)

$$\oint_C d\vec{l}\cdot\vec{H} = \vec{H}_1\cdot\vec{l} - \vec{H}_2\cdot\vec{l} = l(H_{1t} - H_{2t}) = \vec{K}_f\cdot(\hat{n}\times\hat{e}_l)l \qquad (1.85)$$

resulting in

$$H_{1t} - H_{2t} = \vec{K}_f\cdot(\hat{n}\times\hat{e}_l) \qquad (1.86)$$

The subscript t denotes the tangential component of the magnetic field vector at the interface along $\hat{e}_l$. Hence, we see that the presence of free surface currents at an interface can cause a discontinuity in the tangential component of the magnetic field vector. In most cases of interest in photonics, we will be concerned with interfaces between two dielectric media with no free surface currents. In this special but very important case, the tangential component of the magnetic field vector remains continuous across the interface.

Faraday's law in integral form [Eq. (1.70)] can be used to obtain the last of the continuity equations for the electric field. The derivation is very similar to that for the magnetic field vector. Leaving this as an exercise for the reader (Prob. 1.9), we get

$$E_{1t} = E_{2t} \qquad (1.87)$$

In words, the tangential component of the electric field is always continuous across an interface between different media. We will use the continuity equations in later sections to derive the reflection and transmission coefficients for light waves incident at an interface between different dielectric media.

1.8 Conservation of Energy Principle for Electromagnetic Fields

In this section, we first derive the general expressions for the electrostatic and magnetostatic energy densities. We then use Maxwell's equations to derive a general conservation of energy principle for electromagnetic fields. In particular, we identify a very important vector, referred to as the Poynting vector, which represents the power density carried by an electromagnetic wave.

1.8.1 Electrostatic Energy Density u_e

To obtain the expression for the electrostatic energy density, we first consider N static point charges located in space and calculate the work necessary in forming this configuration of charges. Recall that if the electrostatic potential at the position $\vec{r}$ is $V(\vec{r})$, then $qV(\vec{r})$ equals

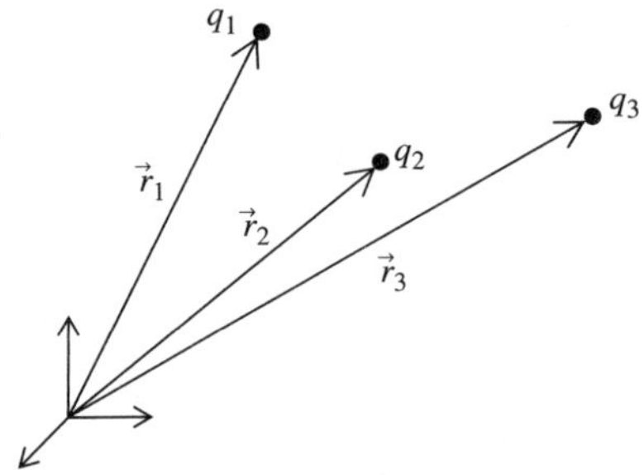

FIGURE 1.18 Three point charges located at the positions $\vec{r}_1$, $\vec{r}_2$, and $\vec{r}_3$.

the work done in moving the point charge q from infinity to $\vec{r}$. To see the general result for the case of N charges, we consider a simpler case with only three point charges as shown in Fig. 1.18. To find the energy of the distribution, we compute the work done in moving the charges one by one to their respective positions. Let's first concentrate on the case where we first move the charge q_1 to location $\vec{r}_1$, then q_2 to location $\vec{r}_2$, and finally q_3 to $\vec{r}_3$. As q_1 is brought to $\vec{r}_1$, no work is done because there is no electrostatic field to begin with in the region. Once q_1 is located at $\vec{r}_1$, the work W_2 needed in bringing q_2 to $\vec{r}_2$ will be

$$W_2 = \frac{1}{4\pi\varepsilon_0}\frac{q_2 q_1}{r_{12}} \tag{1.88}$$

Finally, bringing q_3 to $\vec{r}_3$ requires an additional work of W_3 given by

$$W_3 = \frac{1}{4\pi\varepsilon_0}\left(\frac{q_3 q_1}{r_{13}} + \frac{q_3 q_2}{r_{23}}\right) \tag{1.89}$$

The total work W needed to form the charge distribution will be

$$W = \frac{1}{4\pi\varepsilon_0}\left(\frac{q_2 q_1}{r_{12}} + \frac{q_3 q_1}{r_{13}} + \frac{q_3 q_2}{r_{23}}\right) \tag{1.90}$$

It is difficult to see the general formula from this expression for the case of N point charges. To obtain a result that can be easily generalized, we note that moving the charges in the reverse order, starting with q_3 first, should lead to the same expression for the work as in Eq. (1.90). Following a similar procedure, we get

$$W = \frac{1}{4\pi\varepsilon_0}\left(\frac{q_2 q_3}{r_{23}} + \frac{q_1 q_3}{r_{13}} + \frac{q_1 q_2}{r_{12}}\right) \tag{1.91}$$

Since the sum of the right-hand sides of Eqs. (1.90) and (1.91) equals $2W$, we get, after rearranging the terms,

$$W = \frac{1}{2}(q_1 V_1 + q_2 V_2 + q_3 V_3) \tag{1.92}$$

where V_i denotes the potential at the location of the charge q_i due to the other charges. For N discrete point charges, W will be given by

$$W = \frac{1}{2}\sum_{i=1}^{N} q_i V_i \tag{1.93}$$

Eq. (1.93) is the starting point for the generalization of the energy expression for a continuous charge distribution. Replacing the point charge q by $dq = \rho_f(\vec{r})d^3r$ and doing a continuous sum, we obtain the expression for the energy of a continuous charge distribution:

$$W = \frac{1}{2}\int_V d^3r \rho_f(\vec{r})V(\vec{r}) \tag{1.94}$$

We can come up with an expression for W involving the electrostatic fields only. The idea behind this approach is that the work we do in assembling a particular charge distribution is equal to the energy stored in the fields produced by this charge distribution. By using Gauss' law, W can be written as

$$W = \frac{1}{2}\int_V d^3r(\nabla\cdot\vec{D})V(\vec{r}) \tag{1.95}$$

By using the vector identity

$$\nabla\cdot(V\vec{D}) = \nabla V\cdot\vec{D} + V\nabla\cdot\vec{D} \tag{1.96}$$

and the relation $\vec{E} = -\nabla V$ between the electric field and the electrostatic potential, we obtain

$$W - \frac{1}{2}\left(\oint_S d\vec{a}\cdot(V\vec{D}) + \int_V d^3r\vec{E}\cdot\vec{D}\right) \tag{1.97}$$

In Eq. (1.97) we employed the divergence theorem to convert the volume integral of $\nabla\cdot(V\vec{D})$ to a surface integral of $V\vec{D}$ over the surface S that encloses the region. Here, the choice of the closed surface S is arbitrary. In fact, S can be extended to cover all space, including regions where the charge density is zero. If the continuous charge density is localized, at distances far from the charged region, V will vary as $1/r$, while $\vec{D}$ and $d\vec{a}$ vary as $(1/r^2)$ and r^2, respectively. So, as the surface S is extended to cover all space, the contribution coming from the surface integral tends to zero and we get

$$W = \frac{1}{2}\int_{\text{allspace}} d^3r\vec{E}\cdot\vec{D} \tag{1.98}$$

From Eq. (1.98), we see that the electrostatic energy density u_e which corresponds to the energy stored in the electrostatic fields per unit volume is given by

$$u_e = \frac{1}{2}\vec{E}\cdot\vec{D} \tag{1.99}$$

1.8.2 Magnetic Energy Density u_m

To derive the expression for the magnetic energy density, let us consider a current carrying loop. When the current $i(t)$ flows through the loop, it establishes a magnetic flux $\Psi_m = L\,i(t)$, where L is the proportionality constant related to the geometry of the loop and referred to as the self inductance. The SI unit for L is (weber/A) or henry. Going back to Faraday's law of induction, we see that an electromotive force (emf) V_{emf} will be generated across the loop as the current changes with time according to

$$V_{\text{emf}} = \oint_C d\vec{l}\cdot\vec{E} = -\frac{d\Psi_m}{dt} = -L\frac{di}{dt} \tag{1.100}$$

Now, work needs to be done against this induced electromotive force to increase the current to its steady state value. So, to push an additional charge dq against this emf, the work dW that needs to be done is given by

$$dW = -V_{\text{emf}}dq = L\frac{di}{dt}i(t)dt = Li(t)di \tag{1.101}$$

We assume that the current is increased from 0 to I. Integrating both sides of Eq. (1.101), we obtain

$$W = \frac{1}{2}LI^2 \tag{1.102}$$

Eq. (1.102) represents the energy stored in the inductive coil. Equivalently, we can view this energy as being stored in the magnetic fields set up by the current carrying loop. Similar to the electric case analyzed in the last section, we now express W in terms of the magnetic fields. Noting that

$$\Psi_m = LI = \int_S d\vec{a}\cdot\vec{B} = \int_S d\vec{a}\cdot(\nabla\times\vec{A}) = \oint_S d\vec{l}\cdot\vec{A} \tag{1.103}$$

the expression for W becomes

$$W = \frac{1}{2}ILI = \frac{1}{2}I\oint_C d\vec{l}\cdot\vec{A} = \frac{1}{2}\oint_C I\,d\vec{l}\cdot\vec{A} \tag{1.104}$$

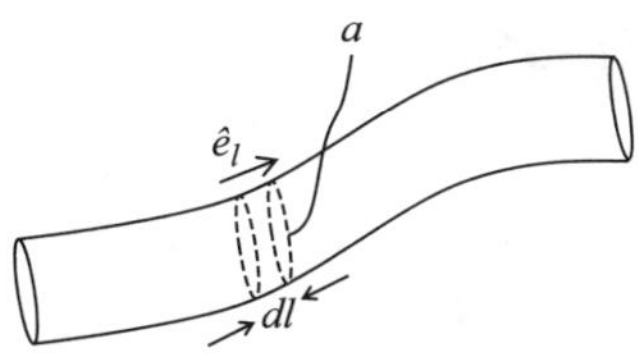

FIGURE 1.19 Current flow in a filament. $\hat{e}_l$ is the unit vector along the current flow and the cross sectional area of a thin length of dl is a.

The line integral involving the filamentary current I can be converted to a volume integral involving the volume current density $\vec{J}$. To see this, consider the three-dimensional flow of charge as sketched in Fig. 1.19. Over an infinitesimal displacement $d\vec{l} = dl\hat{e}_l$ ($\hat{e}_l$ = unit vector parallel to the current flow), we have

$$I d\vec{l} = J_f a dl \hat{e}_l = (J_f \hat{e}_l) a dl = \vec{J}_f d^3 r \tag{1.105}$$

Hence,

$$dW = \frac{1}{2}\vec{A}\cdot\vec{J}_f d^3 r \tag{1.106}$$

To calculate the energy W in steady state after the transients have died out, we set the time rate of change of the displacement current to zero in the Ampere-Maxwell equation and obtain

$$W = \frac{1}{2}\int_V d^3 r[\vec{A}\cdot(\nabla\times\vec{H})] \tag{1.107}$$

By using the vector identity

$$\nabla\cdot(\vec{A}\times\vec{H}) = \vec{H}\cdot(\nabla\times\vec{A}) - \vec{A}\cdot(\nabla\times\vec{H}) \tag{1.108}$$

W can be expressed as

$$W = \frac{1}{2}\int_V d^3 r[\vec{H}\cdot\vec{B} - \nabla\cdot(\vec{A}\times\vec{H})] = \frac{1}{2}\int_V d^3 r\vec{H}\cdot\vec{B} - \frac{1}{2}\oint_S d\vec{a}\cdot(\vec{A}\times\vec{H}) \tag{1.109}$$

As in the electric case, the surface integral contribution goes to zero as S is extended to cover all space, giving the magnetic energy density u_m:

$$u_m = \frac{1}{2}\vec{H}\cdot\vec{B} \tag{1.110}$$

1.8.3 Poynting's Theorem

Poynting's theorem expresses the conservation of energy principle for electromagnetic fields. To derive the general result, we consider the work done by the Lorentz force on a free point charge q_f over a time period of dt. If the instantaneous velocity is $\vec{u}$, the infinitesimal displacement over a time dt will be $d\vec{r} = \vec{u}\ dt$. The work done by the Lorentz force $\vec{F}_L = q_f(\vec{E} + \vec{u} \times \vec{B})$ will be given by

$$dW = \vec{F}_L \cdot \vec{u}\ dt = q_f \vec{u} \cdot \vec{E}\ dt \tag{1.111}$$

Extending this result to the case of a continuous charge distribution represented by the volume charge density ρ_f, the rate at which work is done by the Lorentz force becomes

$$\frac{dW}{dt} = \int_V d^3r \rho_f \vec{u} \cdot \vec{E} \tag{1.112}$$

The term $\rho_f \vec{u}$ is equal to the free volume current density $\vec{J}_f$. To see this relation, we consider the flow of free charges in a region with cross sectional area a and infinitesimal width dl as shown in Fig. 1.20. Amount of charge dq displaced over time dt will be given by

$$dq = \rho_f a |\vec{u}| dt \tag{1.113}$$

Hence, the amount of charge displaced per unit time per unit cross-sectional area in the direction of flow will be equal to the magnitude of the volume current density and will be given by

$$|\vec{J}_f| = \frac{\rho_f a |\vec{u}| dt}{a dt} \tag{1.114}$$

In vector form, $\vec{J}_f = \rho_f |\vec{u}| \hat{e}_l = \rho_f \vec{u}$. Hence, we end up with the following expression for the work done per unit time by the Lorentz force

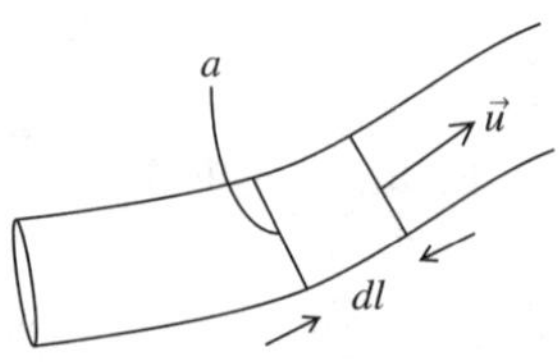

FIGURE 1.20 Flow of free charges through a thin slice of cross sectional area *a* and infinitesimal width *dl*.

(in other words, the power exerted by the force):

$$\frac{dW}{dt} = \int_V d^3r\vec{E}\cdot\vec{J}_f \tag{1.115}$$

We now go back to Maxwell's equations to express this work in terms of the electric and magnetic energies stored in the region. By using the Ampere-Maxwell equation and the vector identity in Eq. (1.108), $\vec{E}\cdot\vec{J}_f$ can be written as

$$\begin{aligned}\vec{E}\cdot\vec{J}_f &= \vec{E}\cdot\left(\nabla\times\vec{H} - \frac{\partial\vec{D}}{\partial t}\right)\\ &= \vec{H}\cdot(\nabla\times\vec{E}) - \nabla\cdot(\vec{E}\times\vec{H}) - \vec{E}\cdot\frac{\partial\vec{D}}{\partial t}\\ &= -\vec{H}\cdot\frac{\partial\vec{B}}{\partial t} - \vec{E}\cdot\frac{\partial\vec{D}}{\partial t} - \nabla\cdot(\vec{E}\times\vec{H})\end{aligned} \tag{1.116}$$

Since $\vec{D} = \varepsilon\vec{E}$, we have

$$\vec{E}\cdot\frac{\partial\vec{D}}{\partial t} = \frac{\varepsilon}{2}\frac{\partial}{\partial t}(\vec{E}\cdot\vec{E}) = \frac{\partial}{\partial t}\left(\frac{1}{2}\vec{E}\cdot\vec{D}\right) = \frac{\partial u_e}{\partial t} \tag{1.117}$$

where u_e is the electric energy density derived in Sec. 1.8.1. Similarly

$$\vec{H}\cdot\frac{\partial\vec{B}}{\partial t} = \frac{1}{2}\frac{\partial}{\partial t}(\vec{H}\cdot\vec{B}) = \frac{\partial u_m}{\partial t} \tag{1.118}$$

At this point, we introduce the Poynting vector $\vec{S}$ and the total electromagnetic energy density u_{em} which are defined respectively according to

$$\vec{S} = \vec{E}\times\vec{H} \tag{1.119}$$

and

$$u_{em} = u_e + u_m \tag{1.120}$$

To obtain a conservation law, we rewrite Eq. (1.116) as

$$-\frac{\partial u_{em}}{\partial t} = \vec{E}\cdot\vec{J}_f + \nabla\cdot\vec{S} \tag{1.121}$$

and integrate over the volume of the region to obtain

$$-\frac{d}{dt}\int_V d^3r u_{em} = \int_V d^3r\vec{E}\cdot\vec{J}_f + \oint_S d\vec{a}\cdot\vec{S} \tag{1.122}$$

Equation (1.122) is Poynting's theorem. It states that the total electromagnetic energy stored in a region of space can decrease in two possible ways: either through the work done by the electromagnetic fields on the free charges or by an outward flux of energy represented by $\oint_s d\vec{a}\cdot\vec{S}$. Hence, $\vec{S}$ given by Eq. (1.119), represents the energy transported by the electromagnetic fields per unit time per unit area, or in other words, the power density, in units of W/m^2 or W/cm^2. The magnitude $|\vec{S}|$ of the Poynting vector, commonly denoted by I, corresponds to the intensity of the electromagnetic field. $\vec{S}$ is a very important physical quantity which we will often encounter in the remainder of the text. In particular, we will use Eq. (1.119) in the following sections to obtain the expression for the intensity of electromagnetic waves.

1.9 Transverse Electromagnetic Waves in a Uniform Medium

1.9.1 Vector Wave Equation

We can use Maxwell's equations to obtain the wave solutions in a uniform medium. Let us suppose that we have a linear, isotropic medium with no free charges ($\rho_f = 0$) or free currents ($\vec{J}_f = 0$). If the medium is linear, the induced polarization inside the medium is proportional to the electric field. In Chap. 11, we will relax this assumption when we investigate the optics of nonlinear media. Second, if the medium is isotropic, the permittivity ε and permeability μ are scalars, in other words the induced polarization or magnetization are linearly proportional to the applied fields. If the material is anisotropic, meaning that the induced polarization, as an example, is not parallel to the applied electric field, we need to use the permittivity tensor to describe the response of the medium. Certain classes of crystals will have this property. We will study this topic later in Chap. 8 when we look at the optics of anisotropic media.

Under the assumption of no free charges or currents, Maxwell's equations become

$$\nabla\cdot\vec{D} = 0 \tag{1.123}$$

$$\nabla\cdot\vec{B} = 0 \tag{1.124}$$

$$\nabla\times\vec{E} = -\frac{\partial\vec{B}}{\partial t} \tag{1.125}$$

$$\nabla\times\vec{H} = \frac{\partial\vec{D}}{\partial t} \tag{1.126}$$

with $\vec{B} = \mu\vec{H}$ and $\vec{D} = \varepsilon\vec{E}$. Taking the curl of (1.125) and using (1.126), we obtain

$$\nabla \times \nabla \times \vec{E} = -\nabla \times \frac{\partial \vec{B}}{\partial t} = -\mu \frac{\partial}{\partial t}(\nabla \times \vec{H}) = -\mu\varepsilon \frac{\partial^2 \vec{E}}{\partial t^2}$$

$$= \nabla(\nabla \cdot \vec{E}) - \nabla^2 \vec{E} \tag{1.127}$$

Let us first investigate $\nabla \cdot \vec{E}$. Assuming that the permittivity ε may have position dependence, we have

$$\nabla \cdot \vec{E} = \nabla \cdot \left(\frac{\vec{D}}{\varepsilon} \right) = \frac{1}{\varepsilon} \nabla \cdot \vec{D} - \frac{1}{\varepsilon^2} \nabla \varepsilon \cdot \vec{D} = -\frac{\nabla \varepsilon}{\varepsilon} \cdot \vec{E} \tag{1.128}$$

In what follows, we will assume that even if the permittivity function has position dependence, it is a slowly varying function of position so that its fractional change is negligible. Under this assumption, we can take $\nabla \cdot \vec{E} = 0$. Hence, the electric field satisfies

$$\nabla^2 \vec{E} - \mu\varepsilon \frac{\partial^2 \vec{E}}{\partial t^2} = 0 \tag{1.129}$$

This is the well-known vector wave equation for the electric field. Note that the first term involves the vector Laplacian of the electric field which becomes

$$\nabla^2 \vec{E} = \nabla^2 E_x \hat{e}_x + \nabla^2 E_y \hat{e}_y + \nabla^2 E_z \hat{e}_z \tag{1.130}$$

in Cartesian coordinates. We will use the same vector wave equation [Eq. (1.129)] later in Chap. 3 to obtain beam solutions with transverse confinement. Instead of eliminating the magnetic field from Maxwell's equations, we could have just as well eliminated the electric field. By doing so, it can be shown (Prob. 1.10) that the magnetic field intensity $\vec{H}$ satisfies a similar wave equation given by

$$\nabla^2 \vec{H} - \mu\varepsilon \frac{\partial^2 \vec{H}}{\partial t^2} = 0 \tag{1.131}$$

Equations (1.129) and (1.131) are examples of partial differential equations which require the specification of boundary conditions before they can be solved. In this book, we will investigate three important classes of solutions of Eq. (1.129):

1. Plane-wave solutions in a homogeneous unbounded medium in Cartesian coordinates
2. Gaussian beam solutions with spherical phase fronts in cylindrical coordinates

3. Propagating guided modes of a cylindrically symmetric optical fiber

In this chapter, we will first look at the plane-wave solutions and defer the discussion of Gaussian beams and optical fibers to Chaps. 3 and 10, respectively.

1.9.2 Plane-Wave Solutions of Maxwell's Equations in a Uniform Medium

First, we investigate the solution of the wave equation in a uniform medium where the permittivity ε and the permeability μ are both constants. In other words, they have no position or frequency dependence. At this point, we have not yet introduced the concept of frequency, but we will come back to this issue in the next section when we discuss periodic waves. In Cartesian coordinates, (Eq. 1.129) can be decomposed into three scalar wave equations, each of the form

$$\nabla^2 g - \mu\varepsilon \frac{\partial^2 g}{\partial t^2} = 0 \tag{1.132}$$

where the scalar function g represents one of the Cartesian components of the electric or the magnetic field. We will first concentrate on the solution for the electric field and then derive the expression for the associated magnetic field. We also assume that the electric field has only one Cartesian component, say the x-component. In this case, the field is said to be linearly polarized along the x-direction. Different states of polarization will be discussed in Sec. 1.9.6. Let us also assume that the field has no variation in the x and y directions, hence the term plane-wave. We can therefore express the electric field as

$$\vec{E}(\vec{r},t) = \hat{e}_x E_x(z,t) \tag{1.133}$$

By substituting Eq. (1.133) into Eq. (1.132), we obtain the one-dimensional wave equation

$$\frac{\partial^2 E_x}{\partial z^2} - \mu\varepsilon \frac{\partial^2 E_x}{\partial t^2} = 0 \tag{1.134}$$

It can be shown by direct substitution that the general solution of Eq. (1.134) is of the form

$$E_x(z,t) = e_x^+(z - ut) + e_x^-(z + ut) \tag{1.135}$$

Above, e_x^+ and e_x^- are arbitrary functions. The above solution is known as the d'Alembert solution, named after the French mathematician Jean Le Rond d'Alembert (1717–1783).[21] This is an important property of the linear wave equation. An electric field of any arbitrary shape satisfies the wave equation, provided that the argument of the function is of the form $z \pm ut$. Furthermore, it can be shown by direct substitution that the parameter u, appearing in Eq. (1.135), can be expressed in terms of ε and μ as (Prob. 1.11)

$$u = \frac{1}{\sqrt{\mu\varepsilon}} \tag{1.136}$$

What does the parameter u physically represent? To see this, let's take a look at $e_x^+(z-ut)$. As an example, let's suppose that $e_x^+(x)$ is a triangular function as sketched in Fig. 1.21*a*. Figure 1.21*b* and 1.21*c* show $e_x^+(z-ut)$ plotted at the times $t = 0$ and $t = d/u$, respectively. Note that, as time progresses, the shape is preserved but the waveform is displaced along the positive z direction by a distance d over a time interval of $t = d/u$. In other words, we see that u represents the rate of displacement or the velocity of the waveform. Hence $e_x^+(z-ut)$ represents an arbitrary waveform propagating with a velocity u in the positive z direction. So, the superscript + was used to indicate that it is the forward propagating solution of the wave equation. Similarly, it can be shown that $e_x^-(z+ut)$ represents an arbitrary waveform moving in the negative z direction with a velocity of u (subscript – for backward propagating).

What is the corresponding magnetic field? To find the relation between the electric and the magnetic fields, we go back to Faraday's law. By assuming that the electric field is polarized along the x-axis and depends only on z and t, we have

$$\nabla \times \vec{E} = \begin{vmatrix} \hat{e}_x & \hat{e}_y & \hat{e}_z \\ \dfrac{\partial}{\partial x} & \dfrac{\partial}{\partial y} & \dfrac{\partial}{\partial z} \\ E_x(z,t) & 0 & 0 \end{vmatrix} = \hat{e}_y \frac{\partial E_x}{\partial z} = -\mu\left(\hat{e}_x \frac{\partial H_x}{\partial t} + \hat{e}_y \frac{\partial H_y}{\partial t} + \hat{e}_z \frac{\partial H_z}{\partial t}\right) \tag{1.137}$$

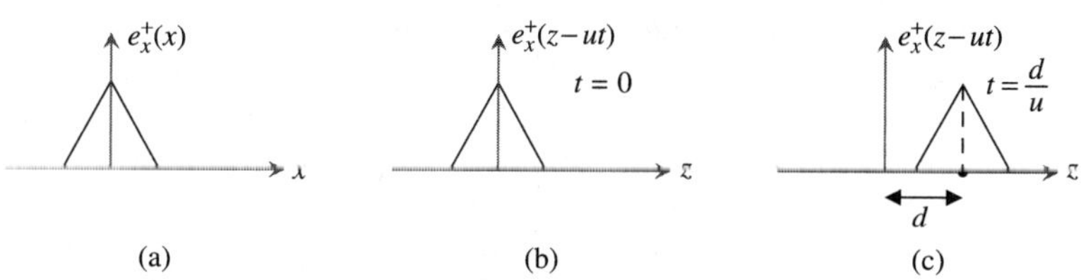

FIGURE 1.21 (*a*) $e_x^+(x)$ in the form of a triangular function. (*b*) The function $e_x^+(z-ut)$ at $t = 0$. (*c*) The function $e_x^+(z-ut)$ at $t = d/u$.

As can be seen from above, the x- and z-components of the magnetic field vector $\vec{H}$ are at most constant. We will take these to be zero and concentrate on the y-component which will have both space and time dependence as E_x. Since $\vec{H}$ satisfies the same wave equation as the electric field, H_y will be of the form

$$H_y(z,t) = h_y^+(z - ut) + h_y^-(z + ut) \tag{1.138}$$

where, again, h_y^+ and h_y^- are arbitrary functions. Let us assume for a moment that we only have the forward propagating field components e_x^+ and h_y^+. By making a change of variables to $w = z - ut$, we get from Eq. (1.137),

$$\frac{\partial e_x^+}{\partial w} = \mu u \frac{\partial h_y^+}{\partial w} \tag{1.139}$$

Integrating Eq. (1.139), neglecting the constant of integration, and using Eq. (1.136), we find that the electric and magnetic field components are related through

$$\frac{e_x^+}{h_y^+} = \eta = \sqrt{\frac{\mu}{\varepsilon}} \tag{1.140}$$

In Eq. (1.140), η is known as the wave impedance and has the units of ohms (Ω). We will comment further on η below. Similarly, for propagation along the negative z-direction, we find that $e_x^- / h_y^- = -\eta$. At this point, we can summarize the important features of the solution of Maxwell's equations in a uniform medium:

1. A time varying electric field is always accompanied by a magnetic field. For propagation along the positive z-axis, the electric field polarized along the x-axis has an associated magnetic field along the y-axis. In other words, the electric and magnetic fields are perpendicular to each other and also perpendicular to the direction of propagation. Equivalently, we see that there is no component of the electric or magnetic field in the direction of propagation. This is referred to as a transverse electromagnetic wave or TEM wave. Also note that the Poynting vector $\vec{S} = \vec{E} \times \vec{H}$ is along the direction of wave propagation (z axis). The right-hand rule can be used to find the relative orientation of the field components and the direction of wave motion. Point your right-hand fingers along the electric field vector and curl them toward the magnetic field. The thumb then points along the direction of wave motion. The relative orientation of the $\vec{E}$, $\vec{H}$, and $\vec{S}$ vectors are indicated in Fig. 1.22.

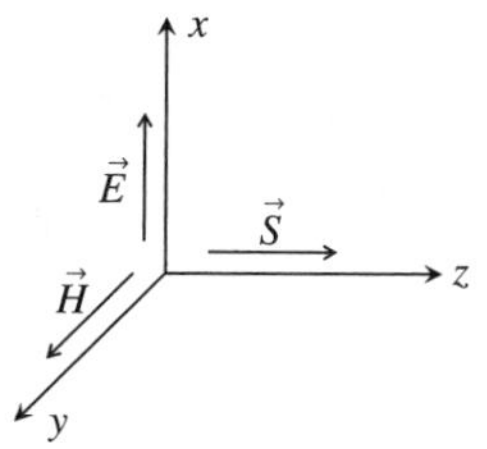

FIGURE 1.22 Relative orientation of the $\vec{E}$, $\vec{H}$, and $\vec{S}$ vectors for a transverse electromagnetic (TEM) wave in a uniform medium.

2. The electromagnetic wave propagates at the velocity u which only depends on the electrical and magnetic properties of the medium. In vacuum, where $\varepsilon = \varepsilon_0$ and $\mu = \mu_0$, u is given by

$$u = \frac{1}{\sqrt{\varepsilon_0 \mu_0}} = 2.99793 \times 10^8 m/s = c \tag{1.141}$$

   In other words, an electromagnetic waveform propagates at the speed of light in vacuum. It was exactly this discovery by Maxwell that led him to conclude that light is a form of electromagnetic wave.

3. The electric field of an electromagnetic wave is not independent of the magnetic field. In fact, the ratio of the electric field component to the magnetic field is given by the wave impedance η which has the units of ohms. In free space, η_0 is given by $\eta_0 = \sqrt{\mu_0/\varepsilon_0} = 377\ \Omega$.

1.9.3 Monochromatic TEM Waves

In many cases, we will be working with periodic, monochromatic (single-frequency) electromagnetic waves which can be expressed by choosing the following specific functional form for the electric field:

$$e_x^+(z - ut) = E_0 \cos[k(ut - z)] \tag{1.142}$$

In the above equation, k is known as the wave number (or wave vector) and has the units m^{-1} or cm^{-1}. Furthermore, the angular frequency ω is defined as

$$\omega = ku \tag{1.143}$$

so that the electric field becomes

$$e_x^+(z - ut) = E_0 \cos(\omega t - kz) \tag{1.144}$$

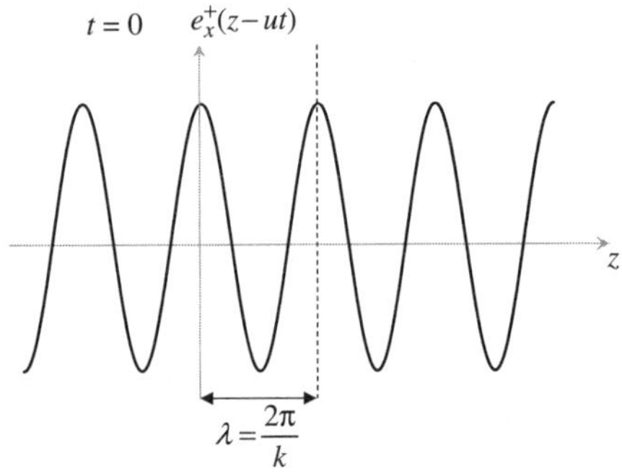

FIGURE 1.23 Spatial variation of a monochromatic sinusoidal wave frozen in time at $t = 0$.

The mathematical relation between the wave number k and the angular frequency ω is often called the dispersion relation. To understand the physical meaning of the parameters k and ω, we consider two cases. In the first case, we capture the wave at a particular instant in time and look at the variation with position. Let us say that we do this at $t = 0$. This gives a waveform that varies as $\cos(kz)$, as sketched in Fig. 1.23. Notice that the waveform is periodic in space. What is more, there is a characteristic minimum distance, known as the wavelength and denoted as λ, after which the shape of the waveform repeats. As an example, the length corresponding to one wavelength is shown between two consecutive peaks of the wave. Since the cosine function repeats when its argument is increased by 2π, we have $k\lambda = 2\pi$ or

$$\lambda = \frac{2\pi}{k} \tag{1.145}$$

In the second case, we observe the wave at a particular point in space and look at the temporal variation. This also gives a waveform periodically varying in time as shown in Fig. 1.24. Here, the pattern repeats after a minimum time T known as the period. Similar to the above case, $\omega T = 2\pi$, or

$$T = \frac{2\pi}{\omega} \tag{1.146}$$

The frequency f of the wave is further defined as the number of full cycles of the waveform per unit time, which is given by $1/T$. The unit of frequency is hertz, which is abbreviated as Hz (1/sec). Hence, the angular frequency ω and the frequency f are related through

$$\omega = 2\pi f \tag{1.147}$$

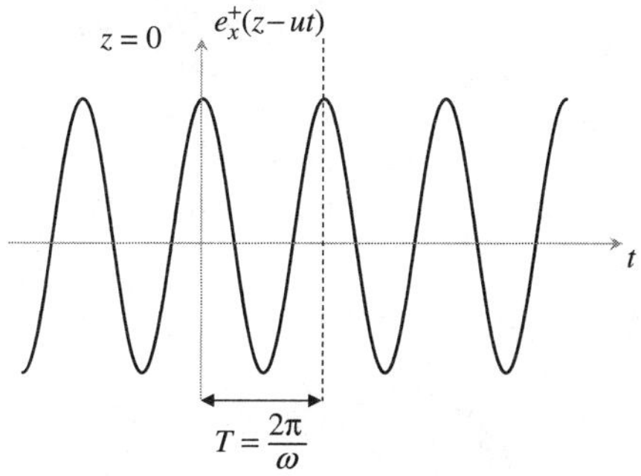

Figure 1.24 Temporal variation of a monochromatic sinusoidal wave at the particular location $z = 0$.

The unit of angular frequency is radian/sec. In vacuum, $kc = \omega$, giving the well-known relation between the vacuum wavelength and the frequency:

$$c = \lambda f \tag{1.148}$$

The variation of the electric and magnetic fields of a monochromatic plane wave is sketched in Fig. 1.25. Note that the directions of oscillation of the electric and magnetic fields are orthogonal to each other. Depending on the vacuum wavelength (or equivalently, the frequency) of the wave, the electromagnetic spectrum is divided into different regions. For example, the visible range spans from approximately 400 nm (blue) to 700 nm (red). The corresponding frequency range is 7.5×10^{14} Hz–4.3×10^{14} Hz.

The monochromatic electromagnetic wave derived above consists of electric and magnetic fields which are real functions of space

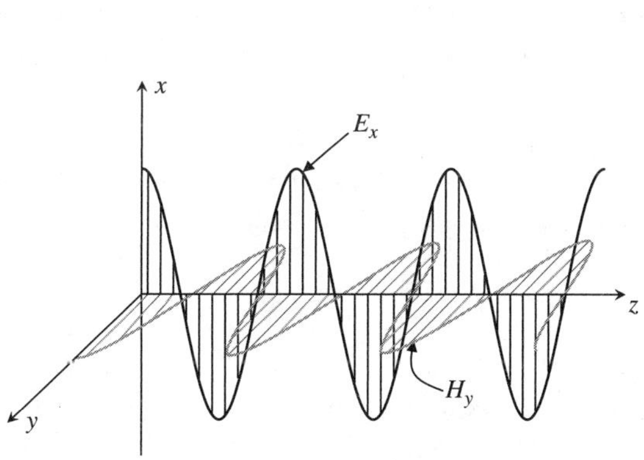

Figure 1.25 Electric and magnetic fields of a monochromatic plane wave moving in the $+z$ direction.

and time. However, in order to facilitate the manipulation of vector operations or differential equations involving the periodic functions, we choose to represent the monochromatic TEM waves as complex exponentials, keeping in mind that when calculations are done, it is the real part of the complex exponential that gives the fields. So, we will use the notation

$$e_x^+(z, t) = E_0 \exp[i(\omega t - kz)] \tag{1.149}$$

to represent the monochromatic plane wave.

1.9.4 General Plane-Wave Solutions in a Uniform Medium

For the general case, the electric field of the propagating wave will be in an arbitrary direction with x-, y-, and z-components. Since the vector wave equation (Eq. 1.129) can be separated into three scalar wave equations, we only consider the equation for the x-component E_x to see the general form of the solution. E_x satisfies

$$\nabla^2 E_x - \mu\varepsilon \frac{\partial^2 E_x}{\partial t^2} = 0 \tag{1.150}$$

To solve Eq. (1.150), we employ the method of separation of variables and assume that the solution is of the form

$$E_x(x,y,z,t) = X(x)Y(y)Z(z)T(t) \tag{1.151}$$

Substituting into the wave equation, dividing by $XYZT$ and rearranging, we obtain

$$\frac{X''}{X} = -\frac{Y''}{Y} - \frac{Z''}{Z} + \mu\varepsilon \frac{T''}{T} \tag{1.152}$$

Above, the notation X'' represents the second (ordinary) derivative of X with respect to the x-coordinate. Note that each term in Eq. (1.152) is a function of only one variable and hence, in order for the equality to hold, must be a constant. To obtain periodic wave solutions, we set

$$\frac{X''}{X} = -k_x^2 \tag{1.153}$$

where k_x is a separation constant, giving

$$X(x) = X_0 \exp(\pm\, ik_x x) \tag{1.154}$$

Rearranging Eq. (1.152) with the help of Eq. (1.153), we obtain the following differential equation for $Y(y)$:

$$\frac{Y''}{Y} = -\frac{Z''}{Z} + \mu\varepsilon \frac{T''}{T} + k_x^2 = -k_y^2 \tag{1.155}$$

Similar to $X(x)$, $Y(y)$ has the solution

$$Y(y) = Y_0 \exp(\pm i k_y y) \tag{1.156}$$

Finally, by introducing the separation constant k_z for the solution of $Z(z)$, we obtain the following equation for $T(t)$:

$$\frac{T''}{T} = \frac{-1}{\mu\varepsilon}\left(k_x^2 + k_y^2 + k_z^2\right) = -\omega^2 \tag{1.157}$$

Putting these solutions together, the general solution for the x-component of the electric field becomes

$$E_x(x,y,z,t) = E_0 \exp[i(\omega t - k_x x - k_y y - k_z z)] \tag{1.158}$$

where the constants k_x, k_y, and k_z, can be positive, negative, or zero. Hence, for the general case, where the direction of the electric field is arbitrary, the plane-wave solution can be expressed as

$$\vec{E}(\vec{r},t) = \hat{e}_1 E_0 \exp[i(\omega t - k_x x - k_y y - k_z z)] = \hat{e}_1 E_0 \exp[i(\omega t - \vec{k}\cdot\vec{r})] \tag{1.159}$$

where E_0 is the amplitude of the electric field, $\hat{e}_1$ is the unit vector in the direction of the electric field, and the position dependence was written as a dot product by introducing the wave vector $\vec{k}$ given by

$$\vec{k} = k_x \hat{e}_x + k_y \hat{e}_y + k_z \hat{e}_z \tag{1.160}$$

In general, E_0 can be complex as we will see in several specific examples. Also, note that for the general plane wave, the wave number k defined earlier becomes the magnitude of the wave vector $\vec{k}$.The corresponding magnetic field has the form

$$\vec{H}(\vec{r},t) = \hat{e}_2 H_0 \exp[i(\omega t - \vec{k}\cdot\vec{r})] \tag{1.161}$$

where, $\hat{e}_2$ is the unit vector in the direction of the magnetic field and H_0 is the wave amplitude. In fact, the vectors $\hat{e}_1$, $\hat{e}_2$, and $\vec{k}$ are not independent of each other. Similar orthogonality relations between the electric and magnetic fields and the direction of propagation can be derived. We leave it as an exercise for the reader (Prob. 1.12) to show that

$$\vec{k}\cdot\hat{e}_1 = 0 \tag{1.162}$$

$$\vec{k}\cdot\hat{e}_2 = 0 \tag{1.163}$$

and

$$\hat{e}_1 \times \hat{e}_2 = \hat{e}_k \tag{1.164}$$

Hence, in a homogeneous medium, the electric field, magnetic field, and the direction of propagation of a TEM wave form a right-handed coordinate system.

1.9.5 Phase and Group Velocities

While discussing the solution of the wave equation in a uniform medium, we assumed that the permittivity (also known as the dielectric constant) and the permeability are both constants. As a result, whether we look at a monochromatic wave or an arbitrary waveform, in other words a pulse, we encounter the same speed of propagation. This is no longer the case if the dielectric constant is a function of frequency, which is the case in most common materials of interest in photonics. In this case, the medium is said to be dispersive and we have to distinguish between two types of velocities for a propagating wave.

First, we start with the phase velocity denoted as u_p. Suppose that a monochromatic wave of the form $e_x^+(z, t) = E_0 \exp[i(\omega t - kz)]$ propagates along the positive z-axis. We define the phase $\Phi(z, t)$ as

$$\Phi(z, t) = \omega t - kz \tag{1.165}$$

A specific value of Φ corresponds to a point of definite phase on the periodic waveform. For example, the point A on the waveform depicted in Fig. 1.26 is a point of fixed phase. Other special points of fixed phase are the crest and trough of a periodic wave, also shown in Fig. 1.26. The phase velocity u_p can be defined as the velocity with which a point of fixed phase moves. Since we want the phase to remain constant with respect to position, we get, by differentiating Eq. (1.165),

$$\frac{\partial \Phi}{\partial z} = 0 = \omega \frac{\partial t}{\partial z} - k \tag{1.166}$$

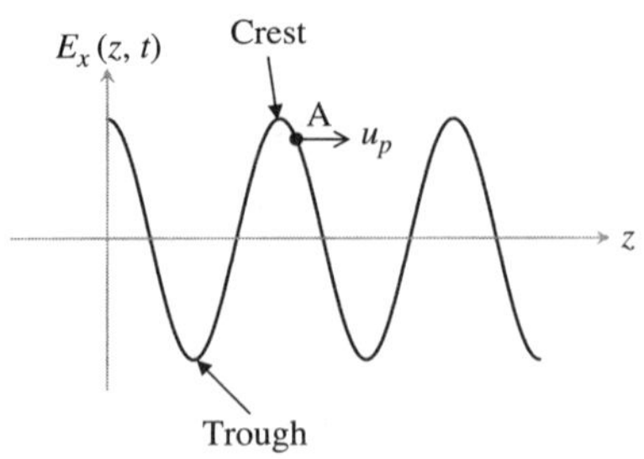

FIGURE 1.26 The phase velocity is the velocity with which a point of fixed phase on a periodic wave (such as point A) moves.

or

$$u_p = \frac{\partial z}{\partial t} = \frac{\omega}{k} \qquad (1.167)$$

Let us consider a nonmagnetic, dielectric medium. If there is negligible wave attenuation in the medium, the dielectric constant is approximately real. The case where absorption leads to a complex-valued dielectric constant will be analyzed later in Chap. 4. When losses in the medium are neglected, we can define the index of refraction (also called the refractive index) n as

$$n = \sqrt{\frac{\varepsilon}{\varepsilon_0}} \qquad (1.168)$$

For a lossy medium, we will define n as the real part of the square root of the complex relative dielectric constant as will be discussed in Chap. 4. In general, n varies with the frequency (or the vacuum wavelength) of the wave. By using Eq. (1.136), k can be expressed in terms of ω as

$$k = \omega\sqrt{\mu\varepsilon} = \frac{\omega}{c}\sqrt{\frac{\varepsilon}{\varepsilon_0}} = \frac{\omega}{c}n(\omega) \qquad (1.169)$$

The phase velocity of an electromagnetic wave in a uniform, lossless medium is therefore given by

$$u_p = \frac{c}{n} \qquad (1.170)$$

As we noted earlier, Eq. (1.169) is referred to as the dispersion relation of a medium. The refractive index of the medium tells us by what factor the phase velocity of a light wave is reduced with respect to its vacuum value c. As an example, n is around 1.5 for glass. This means that the phase velocity of light waves in glass is about two-thirds of the speed of light in vacuum. The frequency of the wave determines the particular color of the light. We know from everyday experience that, when light passes through a linear transparent material such as glass, its color remains the same. This means that the frequency of the light does not change as it goes from one medium to another as long as the two media are linear. Physically, a change in the frequency would further lead to energy buildup either at the air-glass or glass-air interface, which we do not observe. Thus, we see from Eq. (1.169) that the effective wavelength λ_{eff} inside the dielectric becomes

$$\lambda_{\text{eff}} = \frac{\lambda}{n} \qquad (1.171)$$

If the index of refraction has frequency dependence, the medium is said to be dispersive. In this case, we need to distinguish between the velocity of an ideal monochromatic plane wave, which extends over all time and space, and that of a localized waveform. The latter will be referred to as a pulse. Typically, a pulse can be thought of as a superposition of many plane waves with different frequencies. This superposition of the plane waves produces an envelope that moves at a different velocity than the individual plane-wave components. The rate of displacement of the envelope is called the group velocity u_g. To gain a better insight about what happens when plane waves are superimposed, we consider two equal-amplitude plane waves with angular frequencies of $\omega \pm \Delta\omega$ and wave numbers of $k \pm \Delta k$. The total electric field will be given by

$$\begin{aligned} E_x(z,t) &= E_0\{\exp[i(\omega+\Delta\omega)t - i(k+\Delta k)z] + \exp[i(\omega-\Delta\omega)t - i(k-\Delta k)z]\} \\ &= 2E_0 \exp[i(\omega t - kz)]\cos(\Delta\omega t - \Delta kz) \end{aligned} \tag{1.172}$$

As a specific example, Fig. 1.27 shows the resulting waveform at $t = 0$ for $\Delta k = 0.2k$. Note that the plane wave part $\exp[i(\omega t - kz)]$ is modulated by the envelope function $\cos(\Delta\omega t - \Delta kz)$. As the plane wave component travels with the phase velocity $u_p = \omega/k$, we see that the velocity of a fixed point on the envelope will be given by

$$u_g = \frac{\Delta\omega}{\Delta k} \tag{1.173}$$

This result can be extended to the case where we have a continuous superposition of infinitely many plane wave components. In fact, any realistic wave will have a finite spread of frequencies. In this case, let us assume that the amplitude of each plane wave component is itself frequency-dependent and given by $g(\omega) \cdot g(\omega)$ is called the spectral

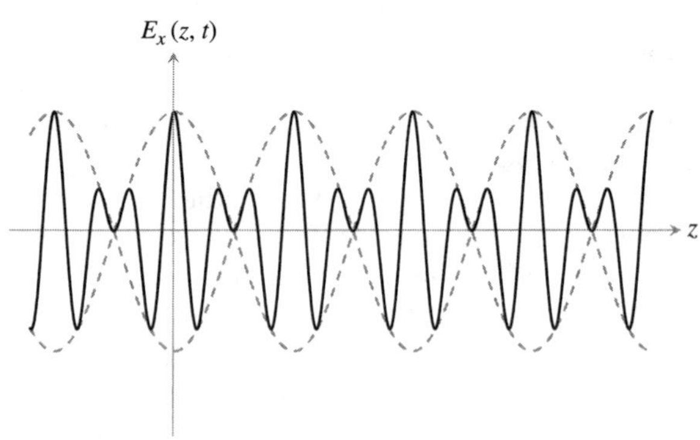

FIGURE 1.27 Superposition of two plane waves with closely spaced frequencies with $\Delta k = 0.2k$.

amplitude distribution. Taking the electric field waveform along the x direction, the field $E_x(z, t)$ at the location z will be a continuous superposition of the plane waves with the amplitude distribution $g(\omega)$:

$$E_x(z,t) = \int_{-\infty}^{+\infty} d\omega g(\omega) \exp[i(\omega t - k(\omega)z)] \tag{1.174}$$

Let us assume that $g(\omega)$ is a locally peaked, continuous distribution about the center frequency ω_0 as sketched in Fig. 1.28. Expanding the wave number $k(\omega)$ as a Taylor series around ω_0, we get

$$k(\omega) = k_0 + (\omega - \omega_0)\left.\frac{dk}{d\omega}\right|_0 + \frac{1}{2}(\omega - \omega_0)^2 \left.\frac{d^2k}{d\omega^2}\right|_0 + h.o.t \tag{1.175}$$

In Eq. (1.175), the subscript 0 indicates that the quantity is evaluated at $\omega = \omega_0$. The abbreviation *h.o.t* denotes higher-order terms involving cubic and higher powers of $(\omega - \omega_0)$. Let us assume that the distribution $g(\omega)$ has a narrow spread around the center frequency so that only the zeroth and first-order terms in Eq. (1.175) are sufficient in the expansion for k. Neglecting the quadratic and higher-order contributions, the electric field $E_x(z, t)$ can be written as

$$\begin{aligned} E_x(z,t) &= \int_{-\infty}^{+\infty} d\omega g(\omega) \exp\left[i\left(\omega t - k_0 z - (\omega - \omega_0)\left.\frac{dk}{d\omega}\right|_0 z\right)\right] \\ &= \exp\left[i\left(k_0 \quad \left.\frac{dk}{d\omega}\right|_0 \omega_0\right)z\right]\int_{-\infty}^{+\infty} d\omega g(\omega)\exp\left[i\omega\left(t - \left.\frac{dk}{d\omega}\right|_0 z\right)\right] \\ &= \exp\left[-i\left(k_0 - \left.\frac{dk}{d\omega}\right|_0 \omega_0\right)z\right]E_x\left(z = 0, t - \left.\frac{dk}{d\omega}\right|_0 z\right) \end{aligned} \tag{1.176}$$

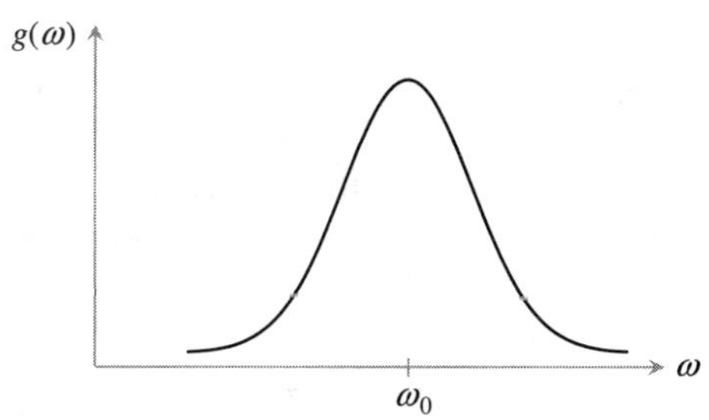

FIGURE 1.28 Example of a locally peaked, continuous spectral amplitude distribution $g(\omega)$ about the center frequency ω_0.

The last line of Eq. (1.176) shows a very important result. When the second and higher-order contributions to k are negligible, the pulse acquires a phase shift but the initial shape of the pulse at $z = 0$ is preserved as it propagates in the dispersive medium. As time progresses, the envelope moves with the group velocity u_g given by

$$u_g = \frac{d\omega}{dk} \tag{1.177}$$

When higher order terms in Eq. (1.175) are not negligible, the shape of the pulse gets distorted during propagation. We will come back to this issue later in Chaps. 4 and 11 to discuss the propagation of a pulse subject to second-order dispersion. Since k is related to ω through $k = (\omega/c)n(\omega)$, it can be shown that (Prob. 1.13) the group velocity can also be expressed as

$$u_g = \frac{c}{n(\omega) + \omega \dfrac{dn}{d\omega}} \tag{1.178}$$

1.9.6 Polarization of TEM Light Waves

By the polarization of an electromagnetic wave, we mean the orientation of the electric (or the magnetic) field as a function of time at a fixed location. To explain this in more detail we will look at three specific cases.

Linear Polarization

Again, we will consider a wave propagating in the positive z-direction. The electric field vector which will be perpendicular to the z-axis will in general have x- and y-components. If the wave is linearly polarized, the x- and y-components are in phase and the electric field can be expressed as follows:

$$\bar{E}(\bar{r},t) = (\hat{x}E_{x0} + \hat{y}E_{y0})\exp[i(\omega t - kz)] \tag{1.179}$$

Above, E_{x0} and E_{y0} are real. To see how the electric field orientation varies over time, we take $z = 0$. At $t = 0$, $\bar{E}(\bar{r},0) = (\hat{x}E_{x0} + \hat{y}E_{y0})$ as shown in Fig. 1.29. As time increases, the magnitude of the electric vector decreases until $t = (\pi/2)/\omega$, where the electric field vanishes. As time further progresses, the electric field arrow again grows in size but this time in the opposite direction. When $t = \pi/\omega$, the electric field becomes $\bar{E}(\bar{r}, t = \pi/\omega) = -(\hat{x}E_{x0} + \hat{y}E_{y0})$. The arrow comes back to its original configuration when $t = 2\pi/\omega$. In this case, we see that the tip of the electric field vector traces the straight line which makes an angle of $\theta = \tan^{-1}(E_{y0}/E_{x0})$ with the x-axis. Hence, we refer to such a wave as a linearly polarized electromagnetic wave.

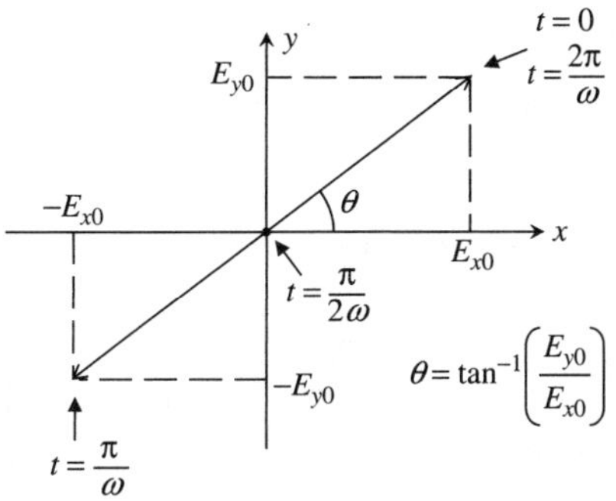

FIGURE 1.29 Locus of the tip of a linearly polarized electric field moving along the *z* axis.

Circular Polarization

Let us again consider a wave moving along the positive *z*-direction. In this case, the tip of the electric field vector moves along a circle. This occurs when the *x*- and *y*-components of the electric field have equal amplitudes but are out of phase by $\pi/2$. The tip can trace the circle in two opposite senses, giving rise to counterclockwise (ccw) and clockwise (cw) circularly polarized light.

We adopt the convention where the counterclockwise circularly polarized electric field $\vec{E}_{\text{ccw}}$ propagating along the positive *z* direction is given by

$$\vec{E}_{\text{ccw}}(\vec{r},t) = E_0(\hat{x} - i\hat{y})\exp[i(\omega t - kz)] \tag{1.180}$$

The actual real electric field will have the components

$$\vec{E}_{\text{ccw}}(\vec{r},t) = \hat{x}E_0\cos(\omega t - kz) + \hat{y}E_0\sin(\omega t - kz) \tag{1.181}$$

If we plot the electric field given by Eq. (1.181) at $z = 0$, we see that the tip rotates in the counterclockwise sense with an angular frequency of ω. This is sketched in Fig. 1.30*a*. Note that if we direct the thumb of our right hand along the direction of propagation (+*z*-direction), the fingers curl along the direction of rotation of the electric field vector. Equivalently, an observer facing the arriving wave sees the electric field vector rotate in the counterclockwise sense. We will refer to this particular wave as the right circular or counterclockwise circular wave. This is easy to remember. At a fixed location in space, the right circular wave rotates in the same sense as a right-handed cork screw.

The clockwise or left circularly polarized wave $\vec{E}_{\text{cw}}$ will be given by

$$\vec{E}_{\text{ccw}}(\vec{r},t) = E_0(\hat{x} + i\hat{y})\exp[i(\omega t - kz)] \tag{1.182}$$

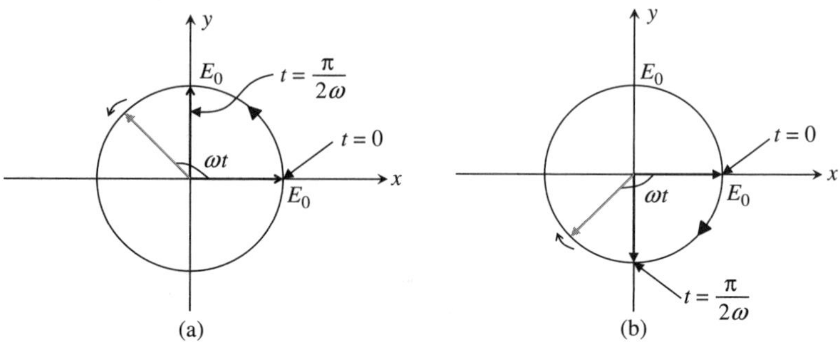

FIGURE 1.30 Locus of the tip of a (*a*) right circular (counterclockwise) and (*b*) left circular (clockwise) electric field moving along the +*z*-direction.

Different from the above case, the tip of the left circular wave rotates with an angular frequency of ω in the clockwise direction as sketched Fig. 1.30*b*.

Elliptic Polarization

The third type of polarization is elliptical where the tip of the electric field vector moves around an ellipse. The electric field of an elliptically polarized wave can be written as

$$\vec{E}(\vec{r},t) = (\hat{x}E_{x0} + \hat{y}E_{y0}\exp(i\phi))\exp[i(\omega t - kz)] \tag{1.183}$$

where ϕ is the arbitrary phase difference between the x- and y-components of the electric field, having the respective amplitudes E_{x0} and E_{y0}. It can be shown (Prob. 1.14) that for this case, the electric field moves around an ellipse which is tilted with respect to the x-axis with an angle θ that satisfies

$$\tan 2\theta = \frac{2E_{x0}E_{y0}}{E_{x0}^2 - E_{y0}^2}\cos\phi \tag{1.184}$$

This is schematically shown in Fig. 1.31*a*. In the special case where $\phi = \pi/2$, the major and minor axes of the ellipse become parallel to the x and y axes as shown in Fig. 1.31*b*.

1.9.7 Poynting Vector for Plane Waves

As we discussed earlier, the Poynting vector $\vec{S} = \vec{E} \times \vec{H}$ gives the power density carried by the electromagnetic waves. Since light waves oscillate at high frequencies, it is of interest to find the cycle-averaged Poynting vector whose magnitude corresponds to the average intensity of the waves. Furthermore, because complex exponential notation is used to represent the electromagnetic waves, care is

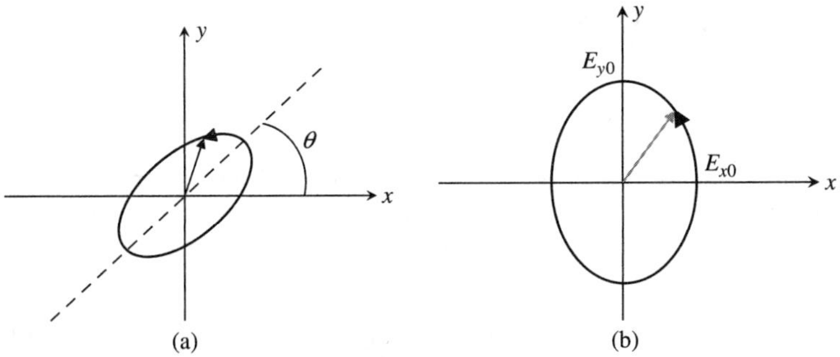

FIGURE 1.31 Locus of the tip of an elliptically polarized electric field with (*a*) an arbitrary angle of inclination θ and (*b*) with an angle of inclination of $\pi/2$ with respect to the *x*-axis.

needed in obtaining the correct expression for the average intensity. In this section, we will derive the expression for the cycle-averaged Poynting vector for plane electromagnetic waves represented by using the complex exponential notation.

Let us assume that the electric and magnetic fields are given by

$$\vec{E}(\vec{r},t) = \vec{E}_0 \exp[i(\omega t - \vec{k}\cdot\vec{r} + \phi_1)] \tag{1.185}$$

and

$$\vec{H}(\vec{r},t) = \vec{H}_0 \exp[i(\omega t - \vec{k}\cdot\vec{r} + \phi_2)] \tag{1.186}$$

where ϕ_1 and ϕ_2 are arbitrary phase factors and the amplitudes $\vec{E}_0$ and $\vec{H}_0$ are assumed to be real. The actual real electric fields will then be given by

$$\vec{E}(\vec{r},t) = \vec{E}_0 \cos(\omega t - \vec{k}\cdot\vec{r} + \phi_1) \tag{1.187}$$

and

$$\vec{H}(\vec{r},t) = \vec{H}_0 \cos(\omega t - \vec{k}\cdot\vec{r} + \phi_2) \tag{1.188}$$

The instantaneous Poynting vector will be given by

$$\vec{S}(\vec{r},t) = \vec{E}_0 \times \vec{H}_0 \cos(\omega t - \vec{k}\cdot\vec{r} + \phi_1)\cos(\omega t - \vec{k}\cdot\vec{r} + \phi_2) \tag{1.189}$$

By using the trigonometric identity

$$\cos A \cos B = \frac{1}{2}[\cos(A+B) + \cos(A-B)] \tag{1.190}$$

the Poynting vector becomes

$$\vec{S} = \frac{1}{2}\vec{E}_0 \times \vec{H}_0[\cos[2(\omega t - \vec{k}\cdot\vec{r}) + \phi_1 + \phi_2] + \cos(\phi_1 - \phi_2)] \tag{1.191}$$

The first term in Eq. (1.191) oscillates at twice the wave frequency. At optical frequencies, this is too rapid to be detected by an ordinary detector. Hence, the cycle-averaged Poynting vector becomes

$$\langle \vec{S} \rangle = \frac{1}{2}\vec{E}_0 \times \vec{H}_0 \cos(\phi_1 - \phi_2) \tag{1.192}$$

By observing that

$$\vec{E}(\vec{r},t) \times \vec{H}(\vec{r},t)^* = \vec{E}_0 \times \vec{H}_0 \exp[i(\phi_1 - \phi_2)] \tag{1.193}$$

the average Poynting vector becomes

$$\langle \vec{S} \rangle = \frac{1}{2}\mathrm{Re}\{\vec{E}(\vec{r},t) \times \vec{H}(\vec{r},t)^*\} \tag{1.194}$$

Above, Re and $*$ denote the real part and the complex conjugate, respectively.

Example As an example, let us consider a plane electromagnetic wave whose electric and magnetic fields have the respective amplitudes of E_x and H_y. By using Eq. (1.194), the intensity I (i.e., the magnitude of the average Poynting vector) becomes

$$I = \frac{1}{2}\mathrm{Re}\left[E_x H_y^*\right] = \frac{|E_x|^2}{2\eta} \tag{1.195}$$

Suppose that we have a plane wave in vacuum with an electric field amplitude of 500 V/m. The corresponding intensity will be 332 W/m^2.

1.10 Reflection and Transmission of Light Waves

1.10.1 Derivation of the Reflection Coefficients

A very important photonic application of the electromagnetic theory of light is in reflection/transmission calculations of light waves from an interface separating two homogeneous dielectric regions. To pose the problem, let us look at Fig. 1.32. Here, two homogeneous dielectric regions with refractive indices of n_1 and n_2 are separated by an interface ($z = 0$ plane in Fig. 1.32). As we have already seen before, the phase velocity as well as the effective wave vector will be different in each region. For example, in region 1, the phase velocity is c/n_1,

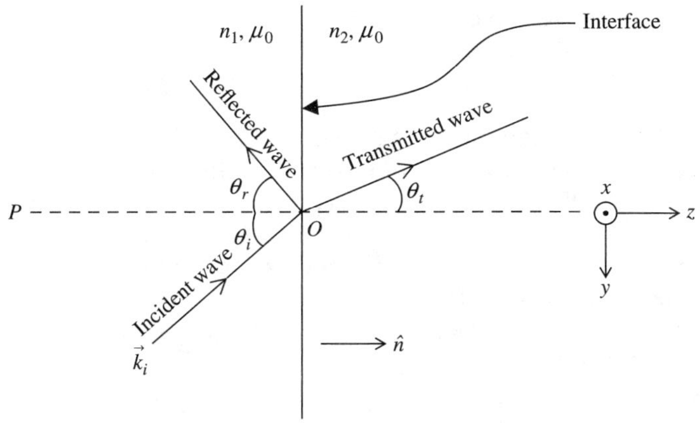

FIGURE 1.32 Two homogeneous media separated by an interface at $z = 0$. The incident, reflected, and transmitted waves make angles of θ_i, θ_r, and θ_t with the normal *OP*.

whereas it changes to c/n_2 in region 2. This abrupt change in the dielectric properties of the medium causes partial reflection of the electromagnetic waves. We will calculate the reflection coefficient of electromagnetic waves for two orthogonal states of polarization of the incident wave. To describe these polarization states, we first introduce the concept of the plane of incidence. In Fig. 1.32, note that the wave vector $\vec{k}_i$ of the incident wave lies in the y-z plane. In addition, the interface coincides with the $z = 0$ plane and the unit normal vector $\hat{n}$ to the interface is along the z-direction (see Fig. 1.32). We define the plane of incidence as the plane which contains both $\vec{k}_i$ and n. Note that the wave vectors of the reflected and transmitted waves also lie in this plane. At the point where the incident ray meets the interface, the line drawn perpendicular to the interface (shown as *OP* in Fig. 1.32) is called the normal. In the case of oblique incidence, the inclination of the incident wave is specified by giving the angle made with the normal. The two linearly independent polarizations that we will consider are as follows:

1. TE waves: For this case, the electric field of the wave is perpendicular to the plane of incidence. This is also referred to as the s-polarization.
2. TM waves: Here, the magnetic field of the wave is perpendicular to the plane of incidence. This case is also known as the p-polarization.

An arbitrary TEM wave can be expressed as a superposition of the TE and TM waves. In the derivation of the reflection coefficients, we will consider these two polarization states separately and use the fact that the tangential components (in this particular case the x- and y-components) of the electric and magnetic field vectors remain continuous across the interface.

Reflection Coefficient for the TE Wave (s-polarization)

Figure 1.33 shows the details of the incident, reflected, and transmitted TE waves. Note that for each wave, the electric field is assumed to be perpendicular to the plane of incidence. Since the plane of incidence is the y-z plane, the electric field is hence along the x-direction. Since the factor $\exp(i\omega t)$ is common to all fields, we will drop this and concentrate on the spatial part. For the incident wave, the wave vector $\vec{k}_i$ is given by

$$\vec{k}_i = k_1(-\hat{y}\sin\theta_i + \hat{z}\cos\theta_i) \tag{1.196}$$

where θ_i is the angle of incidence and the magnitude k_1 of the wave vector is $k_1 = k_0 n_1$, k_0 being the vacuum wave vector. Since $\vec{r} = \hat{x}x + \hat{y}y + \hat{z}z$, the electric field of the incident TE wave can be written as

$$\vec{E}_i = \hat{x}E_i \exp(-i\vec{k}_i \cdot \vec{r}) = \hat{x}E_i \exp[-ik_0 n_1(-y\sin\theta_i + z\cos\theta_i)] \tag{1.197}$$

Above, E_i is the amplitude of the incident electric field. The direction of the associated magnetic field is indicated in Fig. 1.33. Recalling that the magnetic field amplitude is given by E_i/η_1, where the medium

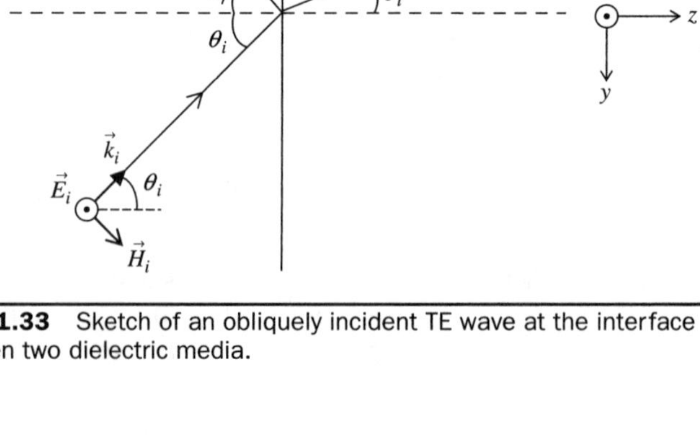

FIGURE 1.33 Sketch of an obliquely incident TE wave at the interface between two dielectric media.

wave impedance $\eta_1 = \eta_0/n_1$, we can express the incident magnetic field as

$$\vec{H}_i = \frac{E_i}{\eta_1}(\hat{y}\cos\theta_i + \hat{z}\sin\theta_i)\exp[-ik_0 n_1(-y\sin\theta_i + z\cos\theta_i)] \quad (1.198)$$

How about the reflected wave? In this case, the main difference is that the z-component of the wave vector is reversed. In addition, we assume that the angle of reflection is θ_r and the reflected wave has a different amplitude E_r. Hence, the electric and magnetic fields of the reflected TE wave become

$$\vec{E}_r = \hat{x}E_r \exp[-ik_0 n_1(-y\sin\theta_r - z\cos\theta_r)] \quad (1.199)$$

and

$$\vec{H}_r = \frac{E_r}{\eta_1}(-\hat{y}\cos\theta_r + \hat{z}\sin\theta_r)\exp[-ik_0 n_1(-y\sin\theta_r - z\cos\theta_r)] \quad (1.200)$$

The transmitted TE wave looks similar in functional form to the incident wave except the angle made with the normal is different (θ_t) and the medium refractive index is n_2. The corresponding fields become

$$\vec{E}_t = \hat{x}E_t \exp[-ik_0 n_2(-y\sin\theta_t + z\cos\theta_t)] \quad (1.201)$$

and

$$\vec{H}_t = \frac{E_t}{\eta_2}(\hat{y}\cos\theta_t + \hat{z}\sin\theta_t)\exp[-ik_0 n_2(-y\sin\theta_t + z\cos\theta_t)] \quad (1.202)$$

Now, we can apply the boundary conditions to come up with relations between the amplitudes of the electric and magnetic fields. At the interface ($z = 0$), the tangential component of the electric field is continuous. Therefore,

$$E_i \exp(ik_0 n_1 y\sin\theta_i) + E_r \exp(ik_0 n_1 y\sin\theta_r) = E_t \exp(ik_0 n_2 y\sin\theta_t) \quad (1.203)$$

Note that Eq. (1.203) is valid anywhere along the interface, in other words, for any value of y. Setting y to zero, we obtain

$$E_i + E_r = E_t \quad (1.204)$$

Substituting for E_t from Eq. (1.204) into Eq. (1.203) and rearranging, we obtain

$$E_i[\exp(ik_0 n_1 y\sin\theta_i) - \exp(ik_0 n_2 y\sin\theta_t)] = E_r[\exp(ik_0 n_2 y\sin\theta_t) - \exp(ik_0 n_1 y\sin\theta_r)] \quad (1.205)$$

For Eq. (1.205) to hold for any arbitrary, non-zero values of E_i and E_r, the bracketed terms must each be zero or equivalently, the exponents must be equal. This gives

$$k_0 n_1 y \sin\theta_i = k_0 n_2 y \sin\theta_t \tag{1.206}$$

and

$$k_0 n_1 y \sin\theta_r = k_0 n_2 y \sin\theta_t \tag{1.207}$$

Simplifying Eqs. (1.206) and (1.207), we obtain

$$\theta_i = \theta_r \tag{1.208}$$

and

$$n_1 \sin\theta_i = n_2 \sin\theta_t \tag{1.209}$$

Both of the results summarized in Eqs. (1.208) and (1.209) are very important. First of all, note that we did not need to make a prior assumption about the equality of the angles of incidence and reflection. This has been rigorously proven by using Maxwell's equations and the associated boundary conditions. Second, we have also derived Snell's law [Eq. (1.209)] which can be used to determine the direction of propagation of the transmitted wave, in other words, the refracted wave. Snell's law plays an important role in ray tracing and we will see various examples in the next chapter on Geometrical Optics.

Requiring the continuity of the tangential component of the magnetic field across the interface provides another relation between the field components. In our case, the tangential magnetic field is in the y-direction, giving

$$\frac{E_i}{\eta_1}\cos\theta_i - \frac{E_r}{\eta_1}\cos\theta_r = \frac{E_t}{\eta_2}\cos\theta_t \tag{1.210}$$

at $z = 0$ and $y = 0$. Typically, we know the incident field amplitude or intensity and we are interested in determining how much is reflected and how much is transmitted. So, we need to determine the ratios E_r/E_i and E_t/E_i. By expressing the wave impedances in terms of the indices of refraction and by using Eqs. (1.210) and (1.204), we obtain the well-known Fresnel amplitude coefficients r_{TE} and τ_{TE} for reflection and transmission:

$$r_{TE} = \frac{E_r}{E_i} = \frac{n_1\cos\theta_i - n_2\cos\theta_t}{n_1\cos\theta_i + n_2\cos\theta_t} \tag{1.211}$$

$$\tau_{TE} = \frac{E_t}{E_i} = \frac{2n_1\cos\theta_i}{n_1\cos\theta_i + n_2\cos\theta_t} \tag{1.212}$$

Reflection Coefficient for the TM Wave (p-Polarization)

The orientation of the fields in the case of an incident TM wave is shown in Fig. 1.34. The boundary conditions can be applied as was done above to find the relations between the incident, reflected, and transmitted field components. We leave it as an exercise for the reader (Prob. 1.15) to show that in the TM case, the Fresnel amplitude reflection and transmission coefficients become

$$r_{TM} = \frac{E_r}{E_i} = \frac{n_2 \cos\theta_i - n_1 \cos\theta_t}{n_2 \cos\theta_i + n_1 \cos\theta_t} \tag{1.213}$$

and

$$\tau_{TM} = \frac{E_t}{E_i} = \frac{2n_1 \cos\theta_i}{n_2 \cos\theta_i + n_1 \cos\theta_t} \tag{1.214}$$

The next important issue is about the calculation of the power reflection coefficients. Again, we leave it as an exercise for the reader to show that (Prob. 1.16) for TE or TM waves, the cycle-averaged incident power P_i on the dielectric interface of area a is given by

$$P_i = \frac{1}{2}\frac{|E_i|^2}{\eta_1} a \cos\theta_i \tag{1.215}$$

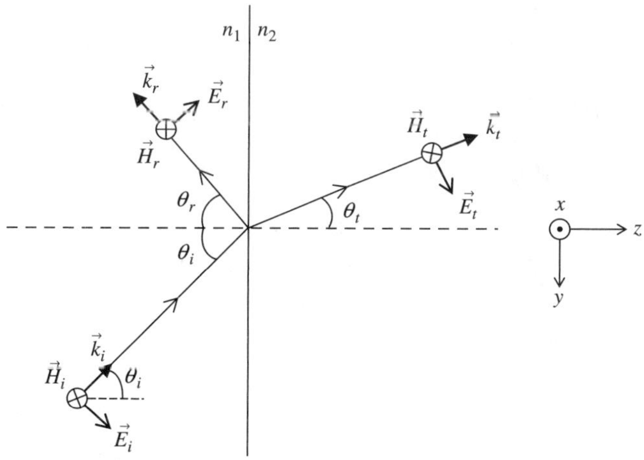

FIGURE 1.34 Sketch of an obliquely incident TM wave at the interface between two dielectric media.

Similarly, the reflected and transmitted average powers P_r and P_t are given by

$$P_r = \frac{1}{2}\frac{|E_r|^2}{\eta_1} a \cos\theta_r \tag{1.216}$$

and

$$P_t = \frac{1}{2}\frac{|E_t|^2}{\eta_2} a \cos\theta_t \tag{1.217}$$

The power reflection coefficient R is defined as the fraction of the incident power that is reflected at the interface. In general, it is a function of the refractive indices and the angle of incidence. By using Eqs. (1.215) and (1.216), and the fact that $\theta_r = \theta_i$, R becomes

$$R = \frac{P_r}{P_i} = \frac{|E_r|^2}{|E_i|^2} = |r|^2 \tag{1.218}$$

where r is the amplitude reflection coefficient for TE or TM waves. Similarly, the power transmission coefficient T is given by

$$T = \frac{P_t}{P_i} = |\tau|^2 \frac{n_2}{n_1}\frac{\cos\theta_t}{\cos\theta_i} \tag{1.219}$$

Note that T is not just $|\tau|^2$, since the refractive index and the direction of wave propagation are both different in the second medium. It can be easily shown that

$$T = 1 - R \tag{1.220}$$

Based on the above results, we can draw several important conclusions about the general characteristics of reflection for TE and TM waves. Figure 1.35*a* and *b* show the variation of the power reflection coefficients $R_{TE}(\theta_i)$ and $R_{TM}(\theta_i)$ as a function of the angle of incidence θ_i for the TE and TM waves for the case $n_1 = 1$ and $n_2 = 1.5$. First, note that, in general, $R_{TE}(\theta_i) \neq R_{TM}(\theta_i)$. However, at normal incidence ($\theta_i = 0$), the power reflection coefficient is the same for both TE and TM waves and is given by

$$R_{TE}(\theta_i = 0) = R_{TM}(\theta_i = 0) = \left(\frac{n_1 - n_2}{n_1 + n_2}\right)^2 \tag{1.221}$$

From Fig. 1.35*a*, we further see that there is no angle of incidence where the TE reflectivity becomes zero. This is not the case for the TM wave shown in Fig. 1.35*b*, where the power reflection coefficient vanishes

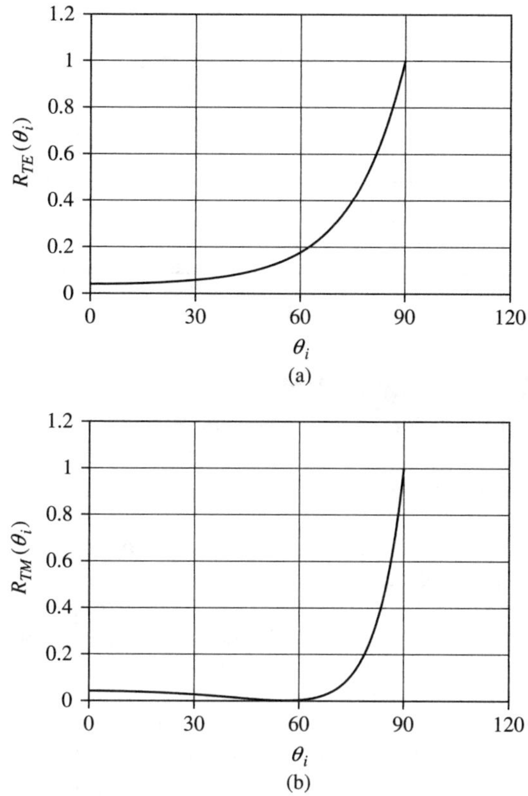

FIGURE 1.35 Variation of the power reflection coefficients (*a*) $R_{TE}(\theta_i)$ and (*b*) $R_{TM}(\theta_i)$ as a function of the angle of incidence θ_i for TE and TM waves ($n_1 = 1$ and $n_2 = 1.5$).

at a special angle known as the Brewster's angle which we denote as θ_B. It can be shown that (Prob. 1.17)

$$\theta_B = \tan^{-1}\left(\frac{n_2}{n_1}\right) \tag{1.222}$$

Example Let us consider the passage of light from air ($n_1 = 1$) into BK-7 glass with $n_2 = 1.51$. At normal incidence, we find from Eq. (1.221) that 4.1% of the incident power is reflected from each interface. Furthermore, the Brewster angle θ_B at the air-glass interface equals 56.5°.

Furthermore, the phase of the reflected wave is in general different from that of the incident wave. The phase can be easily determined from the amplitude reflection coefficients. For the case where $n_1 < n_2$, the variation of the phase of the reflected wave as a function of the incidence angle is shown in Fig. 1.36*a* and *b* for the TE and TM waves, respectively.

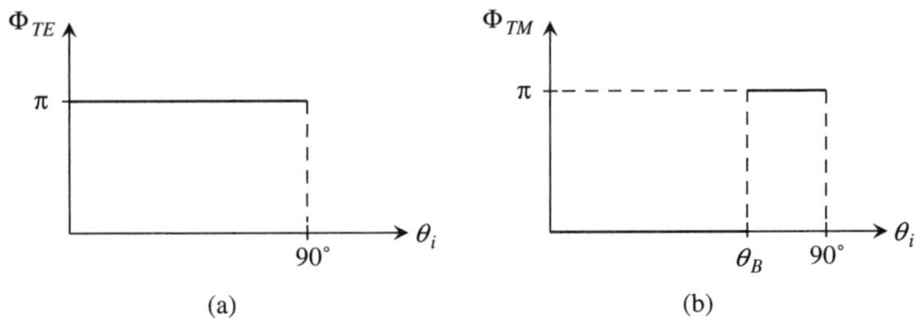

FIGURE 1.36 Variation of the phase of the reflected wave as a function of the incidence angle for (*a*) TE and (*b*) TM waves ($n_1 < n_2$).

1.10.2 Total Internal Reflection

Total internal reflection occurs when the refractive index of the first medium is larger than that of the second medium, and when the angle of incidence in the first medium exceeds a critical value. In this case, the angle of refraction is larger than the angle of incidence as sketched in Fig. 1.37*a*. When the angle of incidence equals the critical angle denoted as θ_c, the corresponding angle of refraction becomes $\pi/2$, in which case, the transmitted wave moves parallel to the interface (Fig. 1.37*b*). If the angle of incidence exceeds the critical angle, all

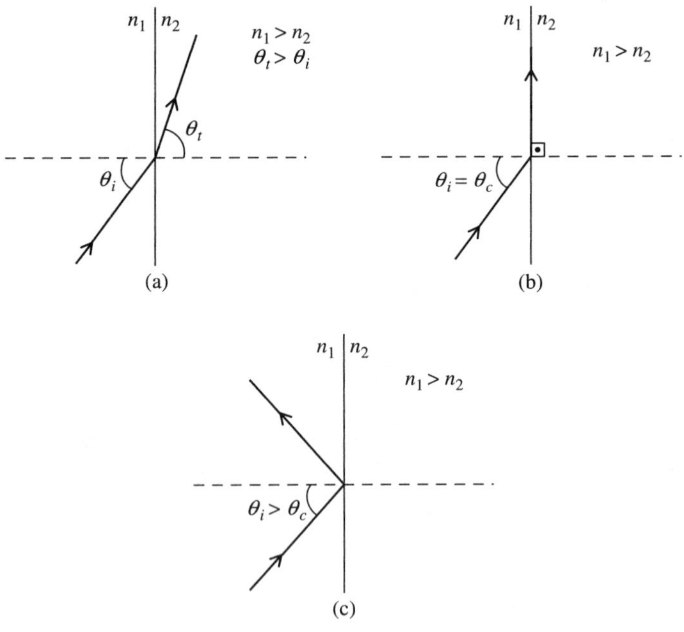

FIGURE 1.37 Passage of light from a medium of higher refractive index into another with a lower refractive index for the cases where the angle of incidence is (*a*) smaller than, (*b*) equal to, and (*c*) greater than the critical angle.

of the incident wave is reflected back into the first medium (Fig. 1.37*c*). From Snell's law, the critical angle satisfies $n_1 \sin\theta_c = n_2 \sin(\pi/2)$ and is given by

$$\theta_c = \sin^{-1}\left(\frac{n_2}{n_1}\right) \tag{1.223}$$

We can provide a simple description of total internal reflection by using the electromagnetic wave model for light. We consider the TE case. Similar results can be obtained for the TM case as well. For angles greater than the critical angle, $\cos\theta_t$ appearing in the expressions for the transmitted electric and magnetic fields [see Eqs. (1.201) and (1.202)], becomes imaginary:

$$\cos\theta_t = \sqrt{1-\sin^2\theta_t} = \sqrt{1-\left(\frac{n_1}{n_2}\right)^2 \sin^2\theta_i} = \frac{-i\sqrt{\sin^2\theta_i - \sin^2\theta_c}}{\sin\theta_c} \tag{1.224}$$

Hence, the transmitted fields in the second medium decay exponentially with distance. At $y = 0$, the fields in the second medium ($z > 0$) can be written as

$$\vec{E}_t = \hat{x}\tau\, E_i \exp[-z/\delta] \tag{1.225}$$

and

$$\vec{H}_t = \frac{\tau E_i}{\eta_2}\left(-\hat{y}\frac{i\sqrt{\sin^2\theta_i - \sin^2\theta_c}}{\sin\theta_c} + \hat{z}\sin\theta_t\right)\exp[-z/\delta] \tag{1.226}$$

where the penetration depth δ for total internal reflection is given by

$$\delta = \frac{\sin\theta_c}{k_0 n_2\sqrt{\sin^2\theta_i - \sin^2\theta_c}} \tag{1.227}$$

If we calculate the *z*-component of the cycle-averaged Poynting vector, we obtain

$$\langle S_z\rangle = \frac{1}{2}\mathrm{Re}\left\{\vec{E}_t \times \vec{H}_t^*\right\}_z = \frac{1}{2}\mathrm{Re}\left\{i\frac{|\tau E_i|^2}{\eta_2}\frac{\sqrt{\sin^2\theta_i - \sin^2\theta_c}}{\sin\theta_c}\exp(-2z/\delta)\right\} = 0 \tag{1.228}$$

Note that both the electric and magnetic fields decay exponentially in the second medium. The electric field component in the *x*-direction and the associated magnetic field component in the y-direction are out of phase by $\pi/2$, resulting in no average power flow in the second medium at incidence angles greater than the critical angle. The electromagnetic field we encounter in the second medium ($z > 0$) is referred to as an evanescent field. In such a case, we do have energy stored in the vicinity of the interface for $z > 0$, but no wave propagation is possible. Hence, we prefer to name this as an evanescent field rather than an evanescent wave.

We can also use the reflection coefficient derived earlier to show that all of the incident wave will be reflected at the interface for incidence angles larger than the critical angle. Since $\cos\theta_t$ is imaginary in this case, the amplitude of the reflection coefficient becomes unity. For example, for the TE wave, the reflection coefficient r_{TE} can be written as

$$r_{TE} = \frac{A - iB}{A + iB} \tag{1.229}$$

where

$$A = n_1 \cos\theta_i \tag{1.230}$$

and

$$B = n_2 \sqrt{\frac{\sin^2\theta_i - \sin^2\theta_c}{\sin^2\theta_c}} \tag{1.231}$$

We note from Eq. (1.229) that $|r_{TE}|^2 = 1$. Hence, the incident wave is totally reflected back into the first medium. As can be further seen from Eq. (1.229), upon reflection, the reflected wave acquires a phase shift with respect to the incident wave. In the case of the TE wave, the phase shift Φ_{TE} becomes

$$\Phi_{TE} = 2\tan^{-1}\left(\frac{B}{A}\right) = 2\tan^{-1}\left(\frac{\sqrt{\sin^2\theta_i - \sin^2\theta_c}}{\cos\theta_i}\right) \tag{1.232}$$

We leave it as an exercise for the reader to show that, for the TM case,

$$\Phi_{TM} = 2\tan^{-1}\left(\frac{B}{A}\right) = 2\tan^{-1}\left(\frac{\sqrt{\sin^2\theta_i - \sin^2\theta_c}}{\cos\theta_i \sin^2\theta}\right) \tag{1.233}$$

Figure 1.38 shows that variation of the phase for the TE and TM waves as a function of the incidence angle θ_i when $n_1 = 1.51$ and $n_2 = 1$. Here $\theta_c = 41.5°$.

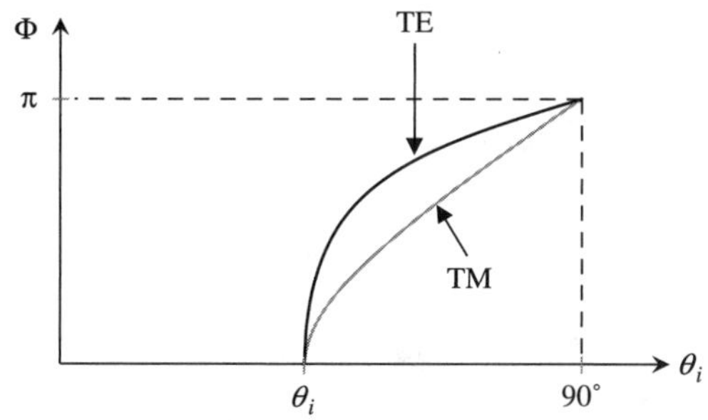

FIGURE 1.38 Variation of the phase Φ for the TE and TM waves during total internal reflection as a function of the incidence angle θ_i when $n_1 = 1.51$ and $n_2 = 1$.

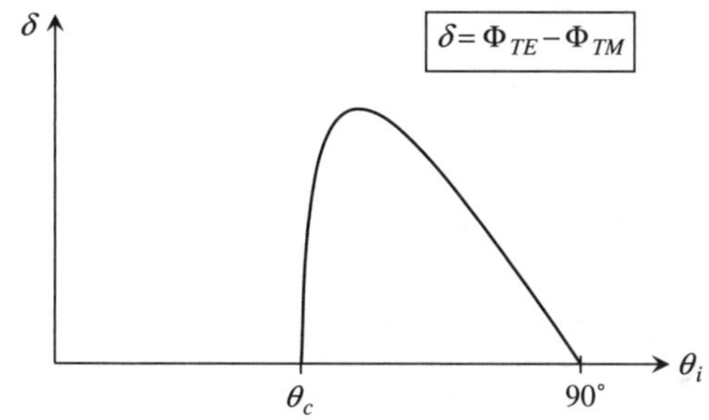

FIGURE 1.39 Variation of the relative phase difference δ between the TE and TM waves after total internal reflection as a function of the incidence angle θ_i when $n_1 = 1.51$ and $n_2 = 1$.

Note that for incidence angles between θ_c and 90°, $\Phi_{TE} > \Phi_{TM}$. Figure 1.39 further shows the variation of the relative phase difference δ between the TE and TM waves, defined according to

$$\delta \equiv \Phi_{TE} - \Phi_{TM} \tag{1.234}$$

In this calculation, $n_1 = 1.51$ and $n_2 = 1$. When the angle of incidence is approximately 54.6°, the relative phase difference between the TE and TM waves upon reflection is 45° (Prob. 1.18). This result can be used in the ingenious device known as the Fresnel rhomb to convert linearly polarized light into circularly polarized light. The Fresnel rhomb is a parallelogram-shaped prism, as shown in Fig. 1.40, in which the angle θ_0 is set to the angle of incidence where $\delta = 45°$. The exact value of θ_0 depends on the refractive index of the glass slab at the design wavelength. For the case where the slab has an index of

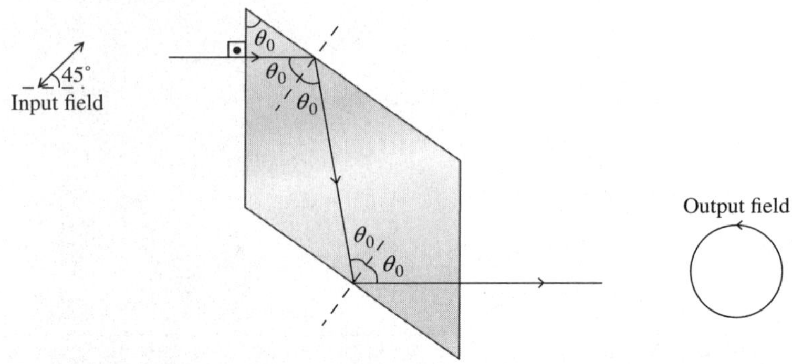

FIGURE 1.40 Sketch of a Fresnel rhomb. The sketches of the input and output fields show the locus of the tip of the electric field in a plane that is perpendicular to the direction of wave motion.

1.51, θ_0 will be about 54.6°. If the input electric field is inclined at an angle of 45°, (in a plane perpendicular to the direction of wave motion), it will have equal TE and TM components. After each internal reflection, the TE component will acquire a relative phase shift of 45° with respect to the TM component, and the emerging wave after two internal reflections will be circularly polarized.

1.11 Multilayer Dielectric Coatings

Multilayer dielectric coatings are used in many photonic applications. The most common example is that of eyeglasses, which typically have antireflection coating to prevent back reflections. The high-quality mirrors used inside laser resonators are also made by coating dielectric layers on top of a suitable transparent substrate. The basic principles used in the design of dielectric coatings can be understood by analyzing the reflection and transmission of electromagnetic waves from layered media. In this section, we analyze the reflection and transmission of electromagnetic waves from multiple layers by introducing a matrix method. We will first derive the characteristic matrix which relates the total tangential fields at two adjacent interfaces separated by a thin layer. The matrix approach can be readily applied to the case of many dielectric layers, and the coefficients of the matrix can be used in reflection/transmission calculations. In what follows, we assume that the thin-film layers are nonmagnetic and nonabsorbing. We will treat the TE case in detail and provide the results for the TM case. As applications, we will look at the design of antireflection and high reflection coatings.

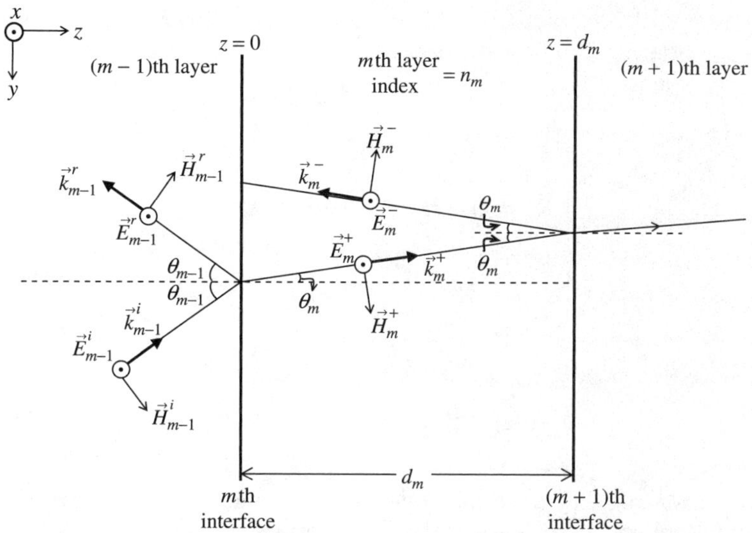

FIGURE 1.41 Sketch of the incident and reflected waves inside the *m*th dielectric layer for the TE case. All of the multiple reflections are not shown for the sake of clarity. See the text for a description of the parameters.

1.11.1 Electromagnetic Analysis of Multilayer Coatings: The TE Case

Let us consider the *m*th dielectric layer of thickness d_m and refractive index n_m. The *m*th layer is sandwiched between the (*m*)th and ($m + 1$)th interfaces as shown in Fig. 1.41. First, we treat the TE case where the incident electric field is perpendicular to the plane of incidence. The angle of incidence of the electric field at the *m*th interface is θ_{m-1}, whereas the angle made by the *k*-vector in the *m*th layer is θ_m. With respect to the coordinate system shown in Fig. 1.41, the wave vector and the fields of the forward moving TE wave in the *m*th layer are given by the following expressions (omitting the common time dependence):

$$\vec{k}_m^+ = k_0 n_m (\cos\theta_m \hat{z} - \sin\theta_m \hat{y}) \tag{1.235}$$

$$\vec{E}_m^+(y,z) = \hat{x} E_m^+ \exp(-i\Phi_m^+(y,z)) \tag{1.236}$$

$$\vec{H}_m^+(y,z) = \frac{n_m E_m^+}{\eta_0}\left[\sin\theta_m \hat{z} + \cos\theta_m \hat{y}\right]\exp(-i\Phi_m^+(y,z)) \tag{1.237}$$

In Eqs. (1.235) to (1.237), the phase $\Phi_m^+(y,z)$ is given by

$$\Phi_m^+(y,z) = k_0 n_m(-y\sin\theta_m + z\cos\theta_m) \tag{1.238}$$

where k_0 is the vacuum wave number ($k_0 = 2\pi/\lambda$, λ = vacuum wavelength). Similarly, the backward moving TE wave (see Fig. 1.41) is represented by the following vectors:

$$\vec{k}_m^- = k_0 n_m(-\cos\theta_m \hat{z} - \sin\theta_m \hat{y}) \tag{1.239}$$

$$\vec{E}_m^-(y,z) = \hat{x} E_m^- \exp(-i\Phi_m^-(y,z)) \tag{1.240}$$

$$\vec{H}_m^-(y,z) = \frac{n_m E_m^-}{\eta_0}\left[\sin\theta_m \hat{z} - \cos\theta_m \hat{y}\right]\exp(-i\Phi_m^-(y,z)) \tag{1.241}$$

where

$$\Phi_m^-(y,z) = k_0 n_m(-y\sin\theta_m - z\cos\theta_m) \tag{1.242}$$

Next, we calculate the total tangential components of the electric (E_m^T) and magnetic (H_m^T) fields at the mth interface ($z = 0$):

$$E_m^T = E_m^+ + E_m^- \tag{1.243}$$

$$\eta_0 H_m^T = N_m^{TE}(\theta)\left(E_m^+ - E_m^-\right) \tag{1.244}$$

where the effective direction dependent index $N_m^{TE}(\theta)$ of the mth layer is given by

$$N_m^{TE}(\theta) = n_m \cos\theta_m \tag{1.245}$$

We then calculate the total tangential components of the electric (E_{m+1}^T) and magnetic (H_{m+1}^T) fields at the $(m + 1)$th interface ($z = d_m$) at $y = 0$. We obtain

$$E_{m+1}^T = E_m^+ \exp(-i\Phi_m) + E_m^- \exp(i\Phi_m) \tag{1.246}$$

where the phase Φ_m acquired along the z-direction is given by

$$\Phi_m = k_0 n_m d_m \cos\theta_m \tag{1.247}$$

In the case of the magnetic field H_{m+1}^T, we get

$$\eta_0 H_{m+1}^T = N_m^{TE}(\theta)\left[E_m^+ \exp(-i\Phi_m) - E_m^- \exp(i\Phi_m)\right] \tag{1.248}$$

Equations (1.246) and (1.248) can be written in matrix form as

$$\begin{bmatrix} E_{m+1}^T \\ \eta_0 H_{m+1}^T \end{bmatrix} = \begin{bmatrix} \exp(-i\Phi_m) & \exp(i\Phi_m) \\ N_m^{TE}(\theta)\exp(-i\Phi_m) & -N_m^{TE}(\theta)\exp(i\Phi_m) \end{bmatrix}\begin{bmatrix} E_m^+ \\ E_m^- \end{bmatrix} \tag{1.249}$$

We can solve for E_m^+ and E_m^- in terms of the total tangential fields at the $(m+1)$th interface by employing Cramer's rule in Eq. (1.249). This gives, after simplifying the algebra,

$$E_m^+ = \frac{\left[N_m^{TE}(\theta)E_{m+1}^T + \left(\eta_0 H_{m+1}^T\right)\right]\exp(i\Phi_m)}{2N_m^{TE}(\theta)} \tag{1.250}$$

and

$$E_m^- = \frac{\left[N_m^{TE}(\theta)E_{m+1}^T - \left(\eta_0 H_{m+1}^T\right)\right]\exp(-i\Phi_m)}{2N_m^{TE}(\theta)} \tag{1.251}$$

By substituting Eqs. (1.250) and (1.251) for the forward and backward wave amplitudes E_m^+ and E_m^- into Eqs. (1.243) and (1.244), we can finally express the total tangential fields at the mth interface in terms of the total tangential fields at the $(m+1)$th interface as

$$\begin{bmatrix} E_m^T \\ \eta_0 H_m^T \end{bmatrix} = \begin{bmatrix} \cos\Phi_m & \dfrac{i\sin\Phi_m}{N_m^{TE}(\theta)} \\ iN_m^{TE}(\theta)\sin\Phi_m & \cos\Phi_m \end{bmatrix} \begin{bmatrix} E_{m+1}^T \\ \eta_0 H_{m+1}^T \end{bmatrix} \tag{1.252}$$

1.11.2 Electromagnetic Analysis of Multilayer Coatings: The TM Case

In the TM case, the incident magnetic field is perpendicular to the plane of incidence. Figure 1.42 shows a sketch of the forward and backward moving waves inside the mth layer. We can carry out a similar analysis as was done for the TE case in the previous section. It can be shown (Prob. 1.19) that the total tangential electric and magnetic fields at the adjacent interfaces (mth and $(m+1)$th) can be related through a similar matrix equation given by

$$\begin{bmatrix} E_m^T \\ \eta_0 H_m^T \end{bmatrix} = \begin{bmatrix} \cos\Phi_m & \dfrac{-i\sin\Phi_m}{N_m^{TM}(\theta)} \\ -iN_m^{TM}(\theta)\sin\Phi_m & \cos\Phi_m \end{bmatrix} \begin{bmatrix} E_{m+1}^T \\ \eta_0 H_{m+1}^T \end{bmatrix} \tag{1.253}$$

where the phase factor Φ_m is the same as in Eq. (1.247) and the effective index $N_m^{TM}(\theta)$ for the TM case is given by

$$N_m^{TM}(\theta) = \frac{n_m}{\cos\theta_m} \tag{1.254}$$

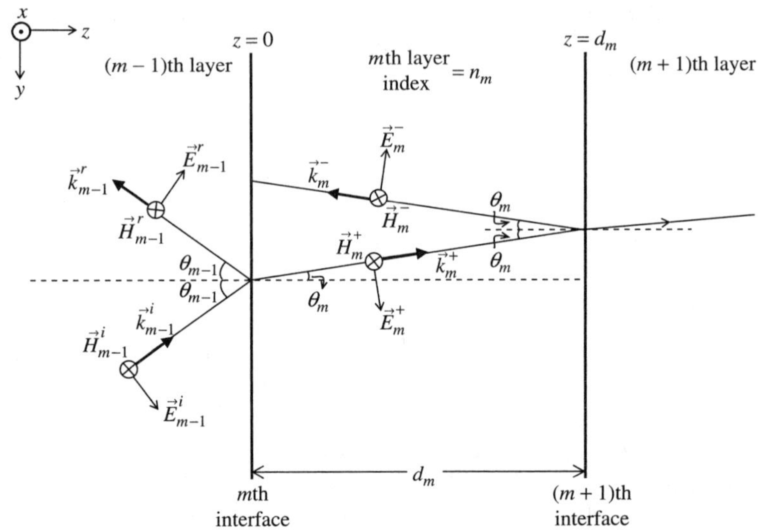

FIGURE 1.42 Sketch of the incident and reflected waves inside the *m*th dielectric layer for the TM case.

1.11.3 Reflection Coefficient of a Multilayer Dielectric Coating

Let us suppose that a substrate with a refractive index of n_s is coated with L (L is an integer) layers of thin films as illustrated in Fig. 1.43. The wave is incident from a region with an index of refraction of n_0. This region is typically air. Notice that according to our convention, the mth film is sandwiched between the (m)th and $(m + 1)$th interfaces. The mth thin film can be represented by a characteristic matrix $\mathbf{M}_n$ of the form given in Eqs. (1.252) or (1.253). The overall characteristic matrix $\mathbf{M}$ of a multilayer coating with L films will therefore be given by

$$\mathbf{M} = \mathbf{M}_1\mathbf{M}_2 \ldots \mathbf{M}_l = \prod_{i=1}^{l} \mathbf{M}_i = \begin{bmatrix} \mathrm{M}_{11} & \mathrm{M}_{12} \\ \mathrm{M}_{21} & \mathrm{M}_{22} \end{bmatrix} \tag{1.255}$$

The four elements of $\mathbf{M}$ are designated as M_{11}, M_{12}, M_{21}, and M_{22}. We assume that the substrate is infinite in extent, and hence there is no back-reflected wave in the last region. We consider the TE case. Based on the definition of the characteristic matrix, we can express the total tangential fields at the first and $(L + 1)$th interfaces as

$$\begin{bmatrix} E_1^T \\ \eta_0 H_1^T \end{bmatrix} = \begin{bmatrix} E_i + E_r \\ (E_i - E_r)N_0 \end{bmatrix} = \begin{bmatrix} \mathrm{M}_{11} & \mathrm{M}_{12} \\ \mathrm{M}_{21} & \mathrm{M}_{22} \end{bmatrix} \begin{bmatrix} E_{L+1}^T \\ \eta_0 H_{L+1}^T \end{bmatrix} = \begin{bmatrix} \mathrm{M}_{11} & \mathrm{M}_{12} \\ \mathrm{M}_{21} & \mathrm{M}_{22} \end{bmatrix} \begin{bmatrix} E_t \\ N_s E_t \end{bmatrix} \tag{1.256}$$

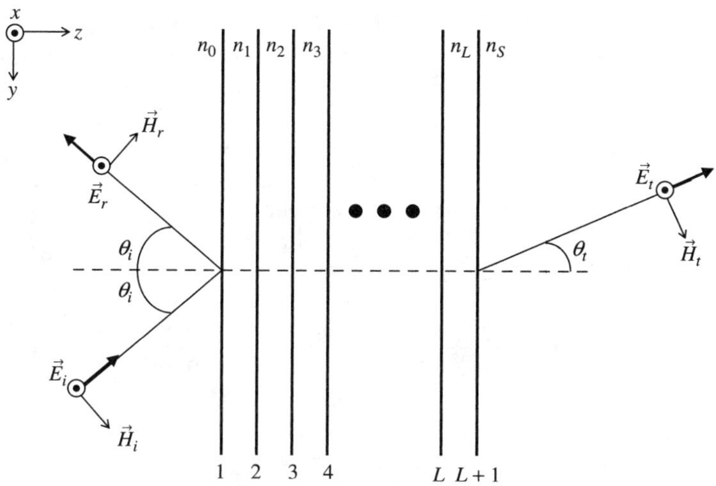

FIGURE 1.43 A substrate with a refractive index of n_s, coated with L layers of thin films.

where $N_0 = n_0 \cos\theta_i$ and $N_s = n_s \cos\theta_t$ for the TE case. By defining the amplitude reflection and transmission coefficients as

$$r = \frac{E_r}{E_i} \tag{1.257}$$

and

$$\tau = \frac{E_t}{E_i} \tag{1.258}$$

We find, using Eq. (1.256), that the equations satisfied by r and τ become

$$(1+r) = (\mathrm{M}_{11} + \mathrm{M}_{12}N_s)\tau \tag{1.259}$$

and

$$N_0(1-r) = (\mathrm{M}_{21} + \mathrm{M}_{22}N_s)\tau \tag{1.260}$$

From the solution of Eqs. (1.259) and (1.260), we find that the power reflection coefficient R becomes

$$R = |r|^2 = \left|\frac{\mathrm{M}_{11}N_0 + \mathrm{M}_{12}N_0N_s - \mathrm{M}_{21} - \mathrm{M}_{22}N_s}{\mathrm{M}_{11}N_0 + \mathrm{M}_{12}N_0N_s + \mathrm{M}_{21} + \mathrm{M}_{22}N_s}\right|^2 \tag{1.261}$$

Below, we investigate special multilayer designs to obtain antireflection and high-reflection coatings.

1.11.4 Single- and Multilayer Antireflection Coatings

In this section, we will discuss possible single- and multilayer coating designs that can be used to reduce the amount of reflection from a transparent substrate. In all of the examples, we will concentrate on normal incidence where $N_m^{TM} = N_m^{TE} = n_m$. Let us start with the single-layer case. We assume that the refractive index and the thickness of the single layer are n_1 and d_1, respectively. By using Eq. (1.252) or (1.253), it can be shown (Prob. 1.20) that at normal incidence, the power reflection coefficient R can be expressed as

$$R = \frac{n_1^2(n_0 - n_s)^2 \cos^2 \Phi_1 + (n_0 n_s - n_1^2)^2 \sin^2 \Phi_1}{n_1^2(n_0 + n_s)^2 \cos^2 \Phi_1 + (n_0 n_s + n_1^2)^2 \sin^2 \Phi_1} \tag{1.262}$$

where

$$\Phi_1 = k_0 n_1 d_1 = \frac{2\pi}{\lambda_0} n_1 d_1 \tag{1.263}$$

at normal incidence [see Eq. (1.247)]. Our task now is to determine n_1 and d_1 such that R becomes zero or very small at the design wavelength of λ_0. It turns out that this can be achieved for the single-layer case by investigating special values of Φ_1. As an example, let us choose $\Phi_1 = 0$. In this case, R becomes

$$R = \frac{(n_0 - n_s)^2}{(n_0 + n_s)^2} \tag{1.264}$$

in which case, R can be zero only when $n_s = n_0$. In other words, this says that the substrate index should be equal to the index of air. This condition will not be met in most practical cases, and even if it were the case, we would not need to put any antireflection coating on the substrate! Next, we investigate the case where the thickness of the thin film is adjusted to make $\cos\Phi_1 = 0$. This relation is satisfied by infinitely many possible values of Φ_1, given by

$$\Phi_1 = (2m+1)\frac{\pi}{2} \tag{1.265}$$

where m is an integer. The corresponding film thicknesses are given by

$$d_1 = (2m+1)\frac{\lambda_0}{4n_1} \tag{1.266}$$

At a particular design wavelength of λ_0, the smallest value of d_1 which satisfies this condition is $\lambda_0/4n_1$. This particular choice of thickness leads to what is known as a "quarter-wave layer" and is widely used in coating designs. If we choose the layer thickness to be a quarter layer, we see from Eq. (1.262) that R now becomes

$$R = \frac{(n_0 n_s - n_1^2)^2}{(n_0 n_s + n_1^2)^2} \tag{1.267}$$

Clearly, R will vanish if $n_1^2 = n_0 n_s$. So, to summarize, two conditions need to be simultaneously satisfied for the reflectivity to vanish: n_1 should be chosen as $\sqrt{n_0 n_s}$, and d_1 should be adjusted to $\lambda_0/4n_1$.

Example As an example, let us assume that we have a BK-7 glass substrate with a refractive index of approximately $n_s = 1.52$ at 500 nm. To make the reflectivity zero, n_1 must be around 1.23. In practice, it is next to impossible to find a coating material that will have an index exactly equal to $\sqrt{n_0 n_s}$. A commonly used low-index antireflection coating material is MgF_2 whose refractive index is around 1.38. If a quarter layer of MgF_2 is used, the residual reflectivity at the design wavelength will be [by using Eq. (1.267)] 1.25%. If the design wavelength of the antireflection coating is 500 nm, the corresponding layer thickness should be 90.6 nm.

It is also of interest to see the wavelength range over which the antireflection coating works. By using Eq. (1.261), we can now calculate how reflectivity R varies with the wavelength of incident radiation. Figure 1.44 shows the variation of the reflectivity at normal incidence as a function of wavelength between 300 and 800 nm. Here, the layer thickness is 90.6 nm, $n_s = 1.52$, and $n_1 = 1.38$. Note that the minimum reflectivity of 1.25% is attained at 500 nm. Further note that the reflectivity is less than 2% over the wavelength range 377–743 nm. If no antireflection coating were used, the reflectivity from a single interface would be about 4.3%, meaning that a single quarter-wave layer MgF_2 will reduce the reflectivity by more than 50% over the whole visible spectrum.

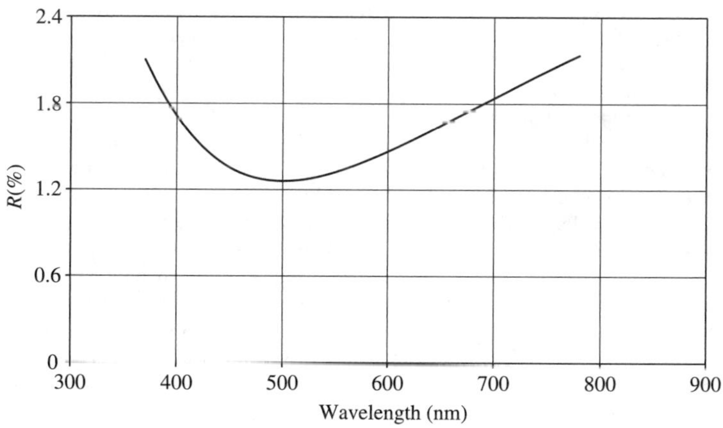

FIGURE 1.44 Variation of the reflectivity at normal incidence as a function of wavelength for a single, quarter-wave layer of MgF_2 on BK-7 glass ($n_0 = 1$, $n_1 = 1.38$, $n_s = 1.52$, $d_1 = 90.6$ nm, $\lambda_0 = 500$ nm).

We can reduce the reflectivity even further by using two quarter-wave layers of different materials. Let's assume that we pick two different coating materials with indices of n_1 and n_2 with quarter-wave thicknesses d_1 and d_2 at the design wavelength of λ_0 ($d_1 = \lambda_0/4n_1$ and $d_2 = \lambda_0/4n_2$). Again, we will take λ_0 as 500 nm. At the design wavelength, the overall characteristic matrix of the two layers is given by

$$\mathbf{M} = \begin{bmatrix} -\dfrac{n_2}{n_1} & 0 \\ 0 & -\dfrac{n_1}{n_2} \end{bmatrix} \tag{1.268}$$

resulting in the following expression for the reflectivity R:

$$R = \left[\frac{n_2^2 n_0 - n_s n_1^2}{n_2^2 n_0 + n_s n_1^2}\right]^2 \tag{1.269}$$

By using Eq. (1.269), we see that if n_1 and n_2 are chosen so that

$$\left(\frac{n_2}{n_1}\right)^2 = \frac{n_s}{n_0} \tag{1.270}$$

R becomes zero at the design wavelength. A similar practical problem occurs in this case, since finding two different coating materials with indices that exactly satisfy Eq. (1.269) is unlikely. Taking the substrate as BK-7 with $n_s = 1.52$, we see that n_1 should be less than n_2 and (n_2/n_1) must be around 1.23. Among commonly used coating materials, we choose the first layer as Al_2O_3 (n_1 = 1.62) and the second layer as ZrO_2 ($n_2 = 2.1$). Both are chosen to be quarter-wave in thickness at 500 nm. Using Eq. (1.269), this gives a residual reflectivity of 0.25% at 500 nm, significantly lower than that for a single-layer MgF_2. Figure 1.45*a* further shows the variation of the reflectivity as a function of wavelength for the two-layer coating with the parameters, $n_1 = 1.62$, $d_1 = 77$ nm, $n_2 = 2.1$, $d_2 = 60$ nm, $n_0 = 1$, and $n_s = 1.52$. Note that although the two quarter-wave coating performs much better than the single quarter-wave layer at the design wavelength, the wavelength range over which the reflectivity is less than 2% is much narrower. As can be seen from Fig. 1.45*b*, R is less than 2% only in the 453–557 nm range. This particular coating based on two quarter-wave layers has a sharp minimum and is commonly referred to as a V coating.

It is possible to add a third layer to lower the reflectivity as well as to increase the spectral bandwidth over which low reflectivity can be attained. Beyond two layers, it is easier to optimize the designs by using numerical simulations. We show the wavelength-dependent reflectivity of one possible design in Fig. 1.46, where a quarter-wave layer of Al_2O_3 grown on the BK-7 substrate is followed by a half-wave layer of ZrO_2 ($n_2 = 2.1$) and then a quarter-wave layer of MgF_2. Half- and quarter-wave thicknesses are set at the design wavelength of 500 nm. Note that with this configuration, the reflectivity is lower than 0.5% over the wavelength range 402–660 nm. This type of coating is commonly referred to as a W coating. At the wavelengths of 432 and 592 nm, R is zero, and actually between these wavelengths, R is less than 0.24%.

In the above calculations, we made an approximation and assumed that the refractive indices of the coating layers and the substrate are constant as a function of wavelength. This is not the case in reality, and the wavelength dependence of the indices should be taken into account to obtain more realistic results. However, the above examples serve as very useful guidelines in the design of dielectric coatings. I highly recommend the reader to develop a computer code to do such reflectivity calculations. It is always very exciting to try different practical material combinations and different numbers of layers to design coatings with a targeted reflectivity at a particular wavelength. Finally, the interested reader should consult Refs. 22 to 31 at the end of the chapter for further discussion on reflection/transmission calculations and the design of dielectric coatings.

1.11.5 High Reflection Coatings

Especially in the design of laser resonators, we need mirrors that have high reflectivity at a particular wavelength or a range of wavelengths. In comparison with metallic mirrors, dielectric-coated mirrors often

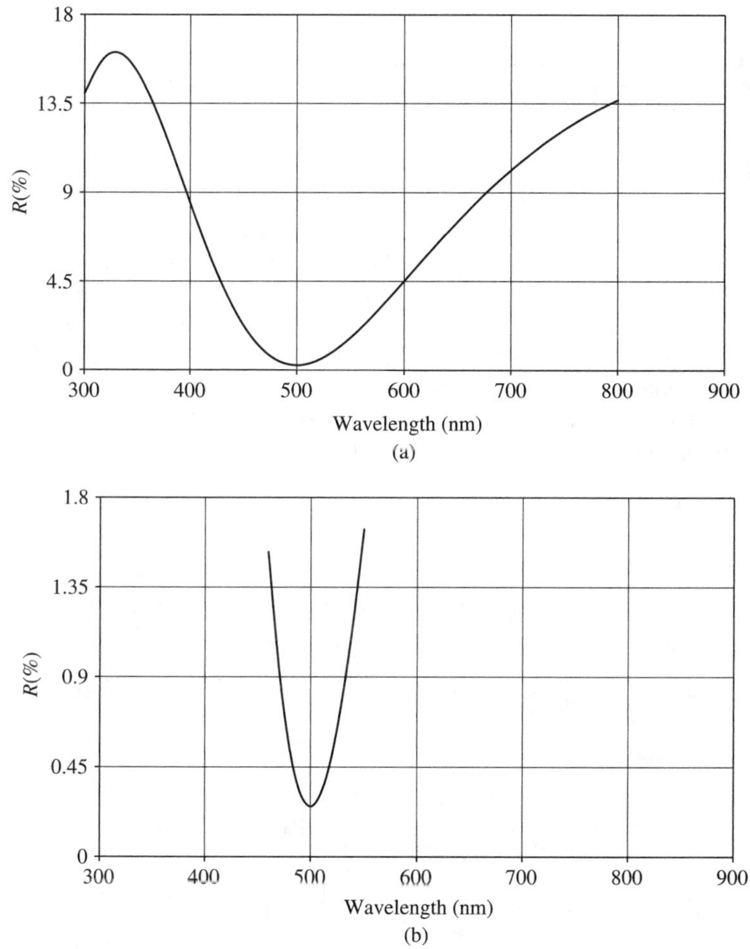

FIGURE 1.45 (*a*) Variation of the reflectivity as a function of wavelength for the two-layer coating with the parameters, $n_1 = 1.62$, $d_1 = 77$ nm, $n_2 = 2.1$, $d_2 = 60$ nm, $n_0 = 1$, and $n_s = 1.52$. (*b*) Reflectivity variation for the range less than 1.8%.

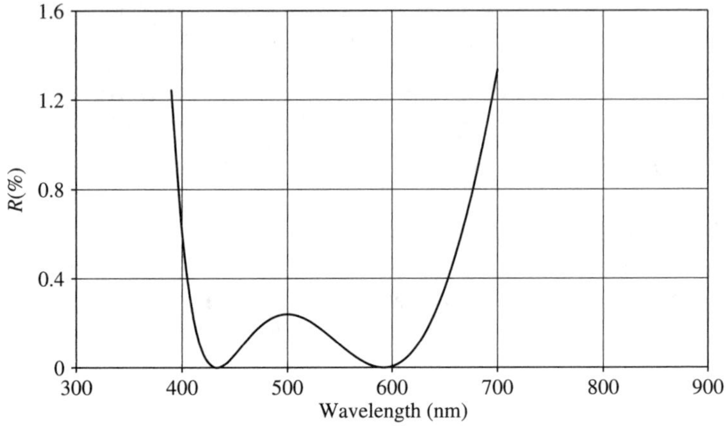

FIGURE 1.46 Wavelength dependence of the reflectivity for a three-layer coating with $n_0 = 1$, $n_1 = 1.38$ (MgF_2), $n_2 = 2.1$(ZrO_2), $n_3 = 1.62$ (Al_2O_3), $n_s = 1.52$, $d_1 = \lambda_0/4n_1$, $d_2 = \lambda_0/2n_2$, $d_3 = \lambda_0/4n_3$, and $\lambda_0 = 500$ nm.

perform much better because they can provide higher reflectivity and lower loss. We can follow a similar strategy and use quarter-wave layers to see how the reflectivity of a coated substrate can be increased. It turns out that to obtain a highly reflecting coating we can use pairs of quarter-wave layers with a high-index and a low-index film. We will call each pair a stack. If we assume that L (L is again an integer) quarter-wave stacks are deposited on top of a substrate, the characteristics matrix at the design wavelength becomes

$$\mathbf{M} = \begin{bmatrix} \left(-\dfrac{n_2}{n_1}\right)^L & 0 \\ 0 & \left(-\dfrac{n_1}{n_2}\right)^L \end{bmatrix} \tag{1.271}$$

The corresponding power reflectivity at the design wavelength of λ_0 becomes

$$R = \left[\frac{\left(-\dfrac{n_2}{n_1}\right)^L - \left(-\dfrac{n_1}{n_2}\right)^L \dfrac{n_s}{n_0}}{\left(-\dfrac{n_2}{n_1}\right)^L + \left(-\dfrac{n_1}{n_2}\right)^L \dfrac{n_s}{n_0}}\right]^2 \tag{1.272}$$

To increase the reflectivity of the coating, we choose $n_1 > n_2$. In this case, note that as L is increased, $(n_2/n_1)^L$ tends to zero and R approaches 100%. This is illustrated in Fig. 1.47*a* to *d* which shows the wavelength

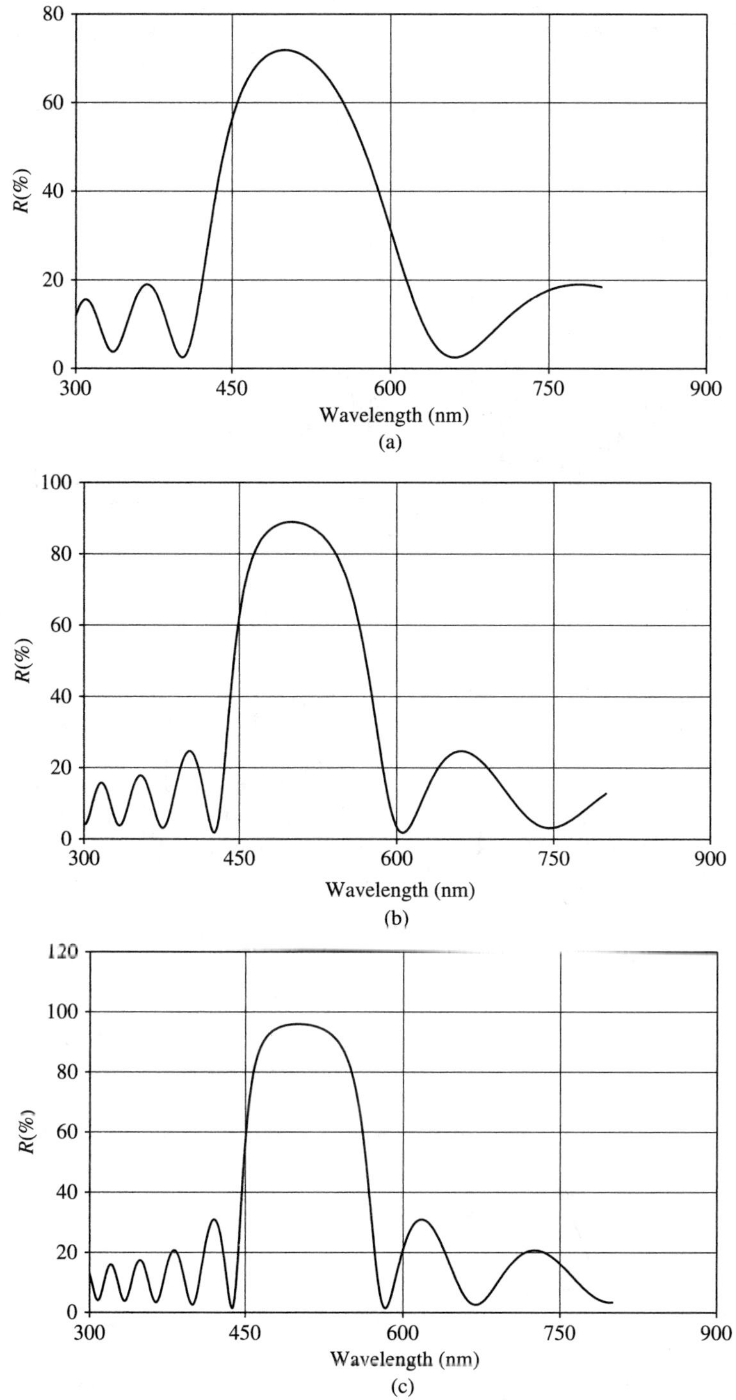

FIGURE 1.47 Wavelength dependence of the reflectivity for a high reflection coating containing (*a*) 4, (*b*) 6, (*c*) 8, and (*d*) 16 stacks. Each stack consists of quarter-wave ZrO_2 ($n_1 = 2.1$) and Al_2O_3 ($n_2 = 1.62$) layers around the design wavelength of 500 nm.

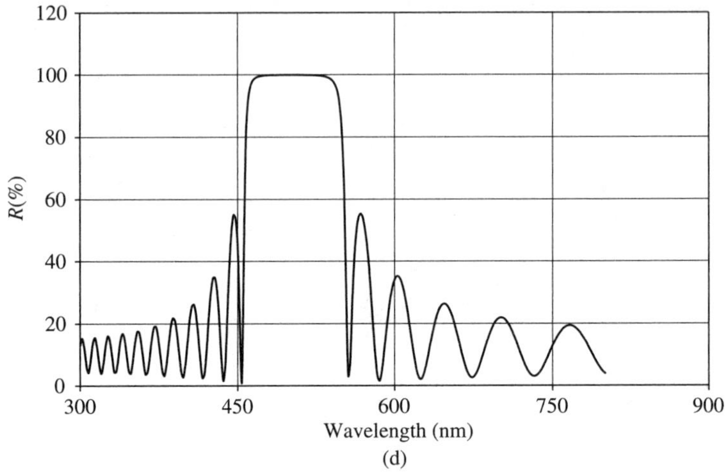

FIGURE 1.47 (*Continued*)

dependence of the reflectivity for different numbers of stacks L. Each stack consists of quarter-wave ZrO_2 ($n_1 = 2.1$) and Al_2O_3 ($n_2 = 1.62$) layers and the design wavelength λ_0 is 500 nm. Note that as L increases, the reflectivity bandwidth increases and reflectivity at 500 nm approaches 100%. This is further delineated in Fig. 1.48 which shows the dependence of the reflectivity at 500 nm on the number of stacks L.

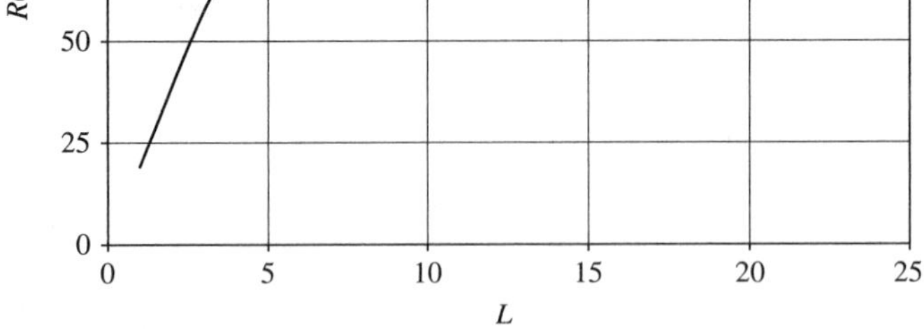

FIGURE 1.48 Variation of the reflectivity as a function of the number of stacks L at the design wavelength of 500 nm. Each stack consists of quarter-wave ZrO_2 ($n_1 = 2.1$) and Al_2O_3 ($n_2 = 1.62$) layers.

1.12 Oscillating Dipoles: Sources of Electromagnetic Waves

In this section, we will discuss how electromagnetic waves are generated by time-varying sources. Starting with Maxwell's equations, we will introduce the scalar and vector potentials and derive the expression for the retarded potentials. We will then look at the specific case of a sinusoidally varying electric dipole and derive the expressions for the generated electromagnetic fields as well as the radiated power.

Let us start with Faraday's law given by Eq. (1.78). Since $\vec{B}$ can be written in terms of the vector magnetic potential $\vec{A}$ as $\vec{B} = \nabla \times \vec{A}$, we have

$$\nabla \times \vec{E} + \frac{\partial(\nabla \times \vec{A})}{\partial t} = \nabla \times \left(\vec{E} + \frac{\partial \vec{A}}{\partial t} \right) = 0 \tag{1.273}$$

Since $\nabla \times (\nabla f) = 0$ for any well-behaved scalar function, it follows from the above equation that

$$\vec{E} + \frac{\partial \vec{A}}{\partial t} = -\nabla V \tag{1.274}$$

where V is defined as the time-dependent scalar potential. Since $\nabla \cdot \vec{E} = \rho/\varepsilon_0$ from Gauss's law, we can write, by using Eq. (1.274),

$$-\nabla^2 V - \frac{\partial}{\partial t}(\nabla \cdot \vec{A}) = \frac{\rho}{\varepsilon_0} \tag{1.275}$$

In free space, we recall that the magnetic induction vector $\vec{B}$ satisfies

$$\nabla \times \vec{B} = \mu_0 \vec{J} + \mu_0 \varepsilon_0 \frac{\partial \vec{E}}{\partial t} \tag{1.276}$$

Now, expressing $\vec{E}$ and $\vec{B}$ in terms of the scalar and vector potentials, we obtain

$$\left(\nabla^2 \vec{A} - \mu_0 \varepsilon_0 \frac{\partial^2 \vec{A}}{\partial t^2} \right) - \nabla \left(\nabla \cdot \vec{A} + \mu_0 \varepsilon_0 \frac{\partial V}{\partial t} \right) = -\mu_0 \vec{J} \tag{1.277}$$

In the derivation of the above result, we made use of the vector identity

$$\nabla \times \nabla \times \vec{A} = \nabla(\nabla \cdot \vec{A}) - \nabla^2 \vec{A} \tag{1.278}$$

In the solution of time-dependent electromagnetic problems, it can be shown that we have the freedom in setting the divergence of the vector magnetic potential to any convenient value without affecting the result for the actual fields $\vec{E}$ and $\vec{B}$. This is usually referred to as choosing the gauge. We have two commonly used gauge choices. In the Coulomb gauge, $\nabla \cdot \vec{A}$ is set to zero. In time-dependent problems, the Lorentz gauge is employed according to which we set

$$\nabla \cdot \vec{A} + \mu_0 \varepsilon_0 \frac{\partial V}{\partial t} = 0 \tag{1.279}$$

This convenient choice allows us to come up with the following set of decoupled partial differential equations for the potentials V and $\vec{A}$:

$$\nabla^2 V - \mu_0 \varepsilon_0 \frac{\partial^2 V}{\partial t^2} = -\frac{\rho}{\varepsilon_0} \tag{1.280}$$

$$\nabla^2 \vec{A} - \mu_0 \varepsilon_0 \frac{\partial^2 \vec{A}}{\partial t^2} = -\mu_0 \vec{J} \tag{1.281}$$

We note that in Eq. (1.281), we have the vector Laplacian. The above equations can be solved by employing rigorous mathematical techniques based on time-dependent Green's functions.[1] Instead of pursuing that approach, we will solve for the potential by introducing the retarded time concept. Let us suppose that the charge or the current sources are localized. As we discussed earlier, in vacuum, electromagnetic effects propagate at the speed of light c. Hence, the effect of a charge or current distribution at time t will be felt at the observation point after a delay of d/c, where c is the vacuum speed of light and d is the separation between the source and the observation point. In this case the potentials are said to be retarded. Earlier, we obtained the expression

$$\vec{A} = \frac{\mu_0}{4\pi} \int_V d^3 r' \frac{\vec{J}(\vec{r}')}{|\vec{r} - \vec{r}'|} \tag{1.282}$$

for the vector magnetic potential $\vec{A}$ due to a steady volume current density. Now, for a time-dependent current density $\vec{J}(\vec{r},t)$, the retarded magnetic vector potential will be given by

$$\vec{A}(\vec{r},t) = \frac{\mu_0}{4\pi} \int_V d^3 r' \frac{\vec{J}(\vec{r}',t')}{|\vec{r} - \vec{r}'|} \tag{1.283}$$

where t' is given by

$$t' = t - \frac{|\vec{r} - \vec{r}'|}{c} \tag{1.284}$$

This is schematically shown in Fig. 1.49.

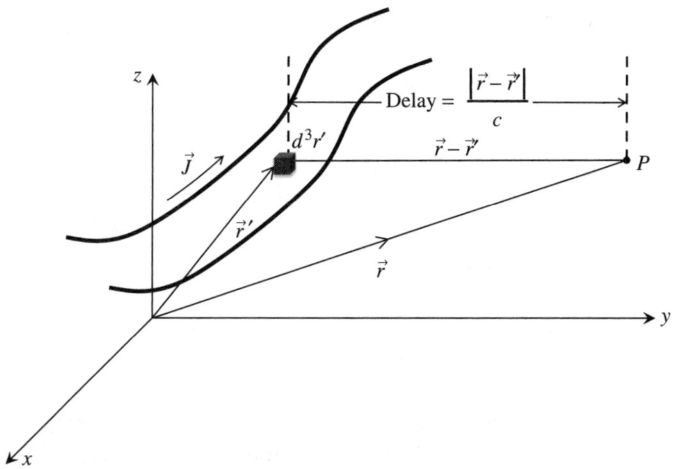

FIGURE 1.49 Time-dependent current source in three dimensions. The delay between the source and the observation points is $|\vec{r}-\vec{r}'|/c$.

As an example, let us consider an oscillating dipole as sketched in Fig. 1.50 to calculate the radiated fields. We assume that the charges q and $-q$ are located symmetrically about the origin along the z axis, and the distance of separation varies sinusoidally as $d\cos(\omega t)$, where ω is the angular frequency of oscillation. Equivalently, we can view the charges as fixed at the locations $z = \pm d/2$ and assume that the amount of charge at each location varies as $q\cos(\omega t) = \text{Re}\{q\exp(i\omega t)\}$. In either case, the electric dipole of the point charges will be given by

$$\vec{p}(\vec{r},t) = qd\exp(i\omega t)\hat{z} \tag{1.285}$$

If we assume that the amount of each charge varies sinusoidally in time, the corresponding filamentary current $I(\vec{r}',t)$ can be written as

$$I(\vec{r}',t) = i\omega q\exp(i\omega t) \tag{1.286}$$

FIGURE 1.50 Time-dependent dipole oriented along the z axis.

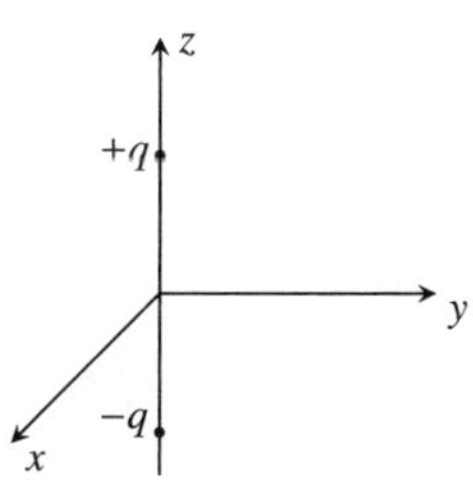

We will be calculating the vector magnetic potential in the far field where $r >> r'$ so that $|\vec{r}-\vec{r}'| \cong r$. Also, replacing $\vec{J}d^3r'$ by $Id\vec{l}'$ in the case of a filamentary current, $\vec{A}$ becomes

$$\vec{A}(\vec{r},t) = \frac{\mu_0}{4\pi r}\int_{-d/2}^{+d/2} dz' I(\vec{r}',t')\hat{z}$$
$$= \frac{i\omega\mu_0}{4\pi r} p \exp\left[i\omega\left(t-\frac{r}{c}\right)\right]\hat{z}$$
$$= \frac{i\omega\mu_0}{4\pi r} p \exp[i(\omega t - kr)](\cos\theta\hat{r} - \sin\theta\hat{\theta}) \tag{1.287}$$

In the above equation, the amplitude p of the electric dipole moment is equal to qd, and in the last line, the unit vector along the z-axis is expressed in terms of the radial and angular unit vectors (see Fig. 1.51). To calculate the associated magnetic induction vector, we need to calculate the curl of the vector magnetic potential. Here, we need to retain only the terms that vary as $1/r$. These are the so-called radiation terms that give rise to a nonvanishing Poynting vector in the far field. We leave it as an exercise for the reader (Prob. 1.23) to show that in this case, the only relevant term to be calculated in the curl expression is $(1/r)(\partial/\partial r)(rA_\theta)\hat{\phi}$, giving

$$\vec{B} = -\hat{\phi}\frac{\mu_0\omega^2 p}{4\pi c}\frac{\sin\theta}{r}\exp[i(\omega t - kr)] \tag{1.288}$$

In the far field, where we assume that there are no current sources, the magnetic induction and electric field are further related through

$$\nabla \times \vec{B} = i\omega\mu_0\varepsilon_0\vec{E} \tag{1.289}$$

if we assume that both fields have harmonic time dependence $\exp(i\omega t)$. In the calculation of $\nabla\times\vec{B}$, the dominant radiation term with $(1/r)$ dependence turns out to be $-(1/r)(\partial/\partial r)(rB_\phi)\hat{\theta}$ (see Prob. 1.23),

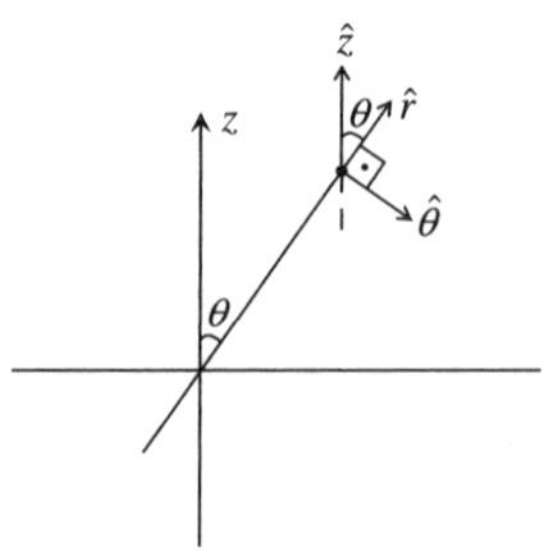

FIGURE 1.51 The unit vectors $\hat{z}$, $\hat{r}$, and $\hat{\theta}$.

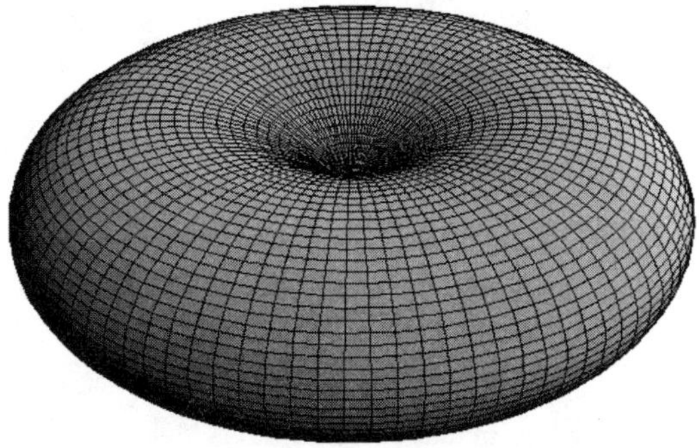

FIGURE 1.52 Radiation pattern due to an oscillating dipole. This plot shows the dependence of the radiated power density on direction. Note that there is no radiated power along the dipole axis. The radiated power density is maximum in the equatorial plane.

resulting in

$$\vec{E} = -\hat{\theta}\frac{\omega^2 \mu_0 p}{4\pi}\frac{\sin\theta}{r}\exp[i(\omega t - kr)] \tag{1.290}$$

The resulting cycle-averaged Poynting vector is in the radial direction and can be expressed as

$$\langle \vec{S} \rangle = \frac{\mu_0 p^2 \omega^4}{32\pi^2 c}\frac{\sin^2\theta}{r^2}\hat{r} \tag{1.291}$$

As can be seen from the sketch of the radiation pattern due to an ideal oscillating dipole in Fig. 1.52, the radiated power density is maximum in the equatorial plane if the dipole is oriented along the z-axis. Furthermore, there is no radiation along the dipole axis. Finally, it can be shown that the total power radiated by an oscillating dipole is given by (Prob. 1.24)

$$P = \frac{1}{4\pi\varepsilon_0}\frac{p^2\omega^4}{3c^3} \tag{1.292}$$

We will later use these results in the analysis of Rayleigh scattering.

Problems

1.1 Derive the result given by Eq. (1.7).

1.2 Use the differential form of Gauss' law to obtain the integral form given by Eq. (1.10).

1.3 A long cylinder of radius r_0 has a volume charge density given by

$$\rho(r) = \rho_0\left(\frac{r}{r_0}\right)$$

where ρ_0 is a constant. Calculate how the electric field E_r in the radial direction varies for $r < r_0$ and $r > r_0$.

1.4 Derive Eq. (1.11).

1.5 Show that the total bound charge due to volume and surface contributions is zero for a polarized object.

1.6 Show, starting from the law of Biot and Savart, that the magnetic induction vector can be written as $\vec{B} = \nabla \times \vec{A}$ where $\vec{A}$ is given by Eq. (1.41).

1.7 Show that the separation distance $|\vec{r} - \vec{r}'|$ between the source and the observation points for the circular current loop is given by Eq. (1.51).

1.8 Show, by using Eq. (1.77), that the normal component of the magnetic induction vector $\vec{B}$ is continuous across the interface between two different media.

1.9 Show that the tangential component of the electric field is always continuous across an interface between different media.

1.10 Show, by using Maxwell's equations, that the magnetic field intensity satisfies the vector wave equation [Eq. (1.131)].

1.11 Show, by direct substitution of the general solution [Eq. (1.135)] into the wave equation (1.134), that u is given by Eq. (1.136).

1.12 Derive the orthogonality relations given in Eqs. (1.162), (1.163), and (1.164).

1.13 Show that the group velocity is given by Eq. (1.178).

1.14 Show that for an elliptically polarized wave given by Eq. (1.183), the angle of inclination of the ellipse with respect to the x-axis satisfies Eq. (1.184).

1.15 Derive the amplitude reflection and transmission coefficients [Eqs. (1.213) and (1.214)] for an oblique TM wave at the interface between two dielectric media.

1.16 Show that for TE or TM waves, the cycle-averaged incident power P_i on the dielectric interface of area a is given by Eq. (1.215).

1.17 Prove that the Brewster angle is given by Eq. (1.222).

1.18 Write a computer program to calculate the variation of the relative phase difference between the TE and TM waves after total internal reflection by using Eq. (1.234). Calculate the angle of incidence for which the relative phase difference is 45°. Use $n_1 = 1.51$ and $n_2 = 1$.

1.19 Derive the characteristic matrix equation [Eq. (1.253)] for a thin dielectric layer for the TM case.

1.20 Show that at normal incidence, the power reflection coefficient for a single dielectric layer is given by Eq. (1.262).

1.21 Write a computer program to reproduce the reflectivity calculation results displayed in Figs. 1.44 to 1.48.

1.22 Suppose that you want to design a dielectric mirror with better than 80% reflectivity at a wavelength of 500 nm. The indices of the coating layers that are available are 2.1 and 1.62. How many stacks would you require? What will be the thickness of each layer?

1.23 Consider an electric dipole oriented along the *z*-axis and oscillating in time. Show that the dominant contribution to the generated electric and magnetic fields with $(1/r)$ dependence in the far field are given by Eqs. (1.288) and (1.290).

1.24 Show that the total power radiated by the oscillating electric dipole is given by Eq. (1.292).

1.25 We can apply the boundary conditions discussed in this chapter to analyze the propagation of surface waves at the interface between two media. These are referred to as surface plasmon polaritons and further discussion on their properties can be found in Refs. 32 and 33. Consider the interface between two nonmagnetic, dielectric media as shown in Fig. 1.53. ε_i is the relative dielectric constant of the *i*th medium. First, we want to investigate whether or not it is possible to satisfy Maxwell's equations and the boundary conditions without the need for a reflected wave!

(*a*) Start with the TE case. In the *i*th region $(i = 1,2)$, the magnetic field can be expressed as

$$\vec{H}_i = (H_{iy}\hat{y} + H_{iz}\hat{z})\exp[i(\omega t - k_{iy}y - k_{iz}z)]$$

Do not bother to express the components in terms of the angles. By using Maxwell's equations and the boundary conditions, derive four equations for the four unknown components of the magnetic field. Assume that there are no free volume charges or surface charges. Also, drop the time dependence of the waves. Then, solve for the characteristic matrix equation to find the condition satisfied by the wave vector components for non-trivial field solutions. Show from these that it is not possible to satisfy the boundary conditions without having a reflected wave in the TE case.

(*b*) Repeat the above exercise for the TM case. Now, work with the electric field components. Prove that it is possible to satisfy the boundary

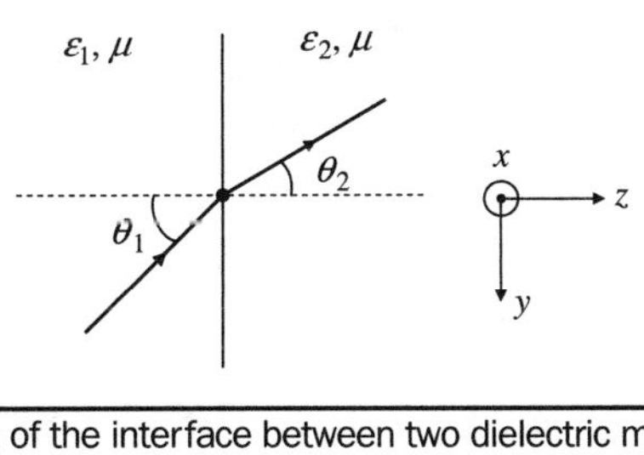

FIGURE 1.53 Sketch of the interface between two dielectric media for Prob. 1.25.

conditions without a reflected wave for the TM case. In particular, prove that k_{1y} is given by

$$k_{1y}^2 = \frac{\varepsilon_1 \varepsilon_2}{\varepsilon_1 + \varepsilon_2}\left(\frac{\omega}{c}\right)^2$$

Also derive the expressions for k_{1z} and k_{2z} in terms of ω and dielectric constants.

(*c*) Can you find under what conditions it is possible to have a propagating *surface* wave? This is your surface plasmon polariton.

References

1. J. D. Jackson, *Classical Electrodynamics,* 3d ed. (John Wiley & Sons, Inc., New York, 1998).
2. D. J. Griffiths, *Introduction to Electrodynamics,* 3d ed. (Prentice Hall, Upper Saddle River, N.J. 1999).
3. A. Yariv and P. Yeh, *Optical Waves in Crystals* (John Wiley & Sons, Inc., New York, 1984).
4. A. Yariv, *Optical Electronics in Modern Communications,* 5th ed. (Oxford University Press, New York, 1997), p. 744.
5. J. A. Stratton, *Electromagnetic Theory* (McGraw-Hill, New York, 1941).
6. J. A. Kong, *Electromagnetic Wave Theory* (EMW, Cambridge, Mass. 2000).
7. F. P. Kapron, D. B. Keck, and R. D. Maurer, "Radiation losses in glass optical waveguides," *Applied Physics Letters,* **17**, 423 (1970).
8. K. C. Kao and G. A. Hockham, "Dielectric-Fibre surface waveguides for optical frequencies," *Proceedings of the Institute of Electrical Engineers—London,* **113**(7), 1151 (1966).
9. D. Huang, E. A. Swanson, C. P. Lin, J. S. Schuman, W. G. Stinson, W. Chang, M. R. Hee, T. Flotte, K. Gregory, C. A. Puliafito, and J. G. Fujimoto, "Optical coherence Tomography," *Science,* **254**, 1178–1181 (1991).
10. W. Denk, J. H. Strickler, and W. W. Webb, "2-Photon laser scanning fluorescence microscopy," *Science,* **248**(4951), 73–76 (1990).
11. S. Sakadzic, U. Demirbas, T. R. Mempel, A. Moore, S. Ruvinskaya, D. A. Boas, A. Sennaroglu, F. X. Kartner, and J. G. Fujimoto, "Multi-photon microscopy with a low-cost and highly efficient Cr:LiCAF laser," *Optics Express,* **16**(25), 20848–20863 (2008).
12. J. C. Maxwell, "A dynamical theory of the electromagnetic field," *Proceedings of the Royal Society of London,* **13**, 531–536 (1863–1864).
13. A. Pais, *Subtle Is the Lord: The Science and the Life of Albert Einstein* (Oxford University Press, Oxford, 1982).
14. C. Roychoudhuri and E. R. Roy, "The nature of light: What is a photon?" *Optics and Photonics News,* **14** (51) (2003).
15. A. L. Schawlow and C. H. Townes, "Infrared and optical masers," *Physical Review,* **112**(6), 1940–1949 (1958).
16. A. M. Prokhorov, "Molecular amplifier and generator for submillimeter waves," *Soviet Physics JETP-USSR,* **7**(6), 1140–1141 (1958).
17. N. G. Basov, O. N. Krokhin, and I. M. Popov, "Generation, amplification, and indication of infrared and optical radiation by means of quantum systems," *Uspekhi Fizicheskikh Nauk,* **72**(2), 161–209 (1960).
18. J. P. Gordon, H. J. Zeiger, and C. H. Townes, "Maser—new type of microwave amplifier, frequency standard, and spectrometer," *Physical Review,* **99**(4), 1264–1274 (1955).
19. T. H. Maiman, "Stimulated optical radiation in ruby," *Nature,* **187**(4736), 493–494 (1960).
20. R. Baierlein, *Newton to Einstein* (Cambridge University Press, Cambridge, 2002).

21. C. R. Wylie and L. C. Barrett, *Advanced Engineering Mathematics* (McGraw Hill, New York, 1982).
22. M. Born and E. Wolf, *Principles of Optics*, 6th ed. (Cambridge University Press, Cambridge, 1997).
23. P. Yeh, *Optical Waves in Layered Media* (Wiley, Hoboken, N.J., 2005).
24. H. A. Macleod, *Thin-Film Optical Filters* (Institute of Physics Publishing, Bristol, 2001).
25. E. Hecht, *Optics*, 4th ed. (Addison-Wesley, Reading, Mass. 2002).
26. G. R. Fowles, *Introduction to Modern Optics* (Dover, New York, 1975).
27. C. R. Pollock, *Fundamentals of Optoelectronics* (Irwin, Chicago, 1995).
28. P. Yeh, A. Yariv, and C.-S. Hong, "Electromagnetic propagation in periodic stratified media. I. General theory," *Journal of the Optical Society of America*, **67**, 423–438 (1977).
29. B. E. A. Saleh and M. C. Teich, *Fundamentals of Photonics*, 2d ed. (Wiley, Hoboken, N.J., 2007).
30. A. Yariv and P. Yeh, *Photonics: Optical Electronics in Modern Communications*, 6th ed. (Oxford University Press, New York, 2007).
31. P. Markos and C. M. Soukoulis, *Wave Propagation: From Electrons to Photonic Crystals and Left-Handed Materials* (Princeton University Press, Princeton, N.J., 2008).
32. L. Novotny and B. Hecht, *Principles of Nano-Optics* (Cambridge University Press, Cambridge, 2006).
33. E. Ozbay, "Plasmonics: Merging photonics and electronics at nanoscale dimensions," *Science*, **311**(5758), 189–193 (2006).

CHAPTER 2

Geometrical Optics

2.1 Introduction

In Chap. 1, we discussed the solution of the wave equation and developed techniques to analyze different phenomena and devices based on the wave aspects of light, such as reflection and refraction of light and optical interference coatings. We will use these techniques in later chapters to investigate other important wave effects in photonics, including diffractive spreading of confined laser beams, light propagation in optical fibers, pulse broadening in dispersive media, nonlinear coupling of waves, and so on. In fact, we can also employ wave methods in the analysis of imaging systems. For example, suppose that we place a candle in front of a converging lens and want to find the location and magnification of the image. After all, light emanating from each point of the candle can be described by electromagnetic waves and Maxwell's equations give us the full recipe to determine how light will propagate through the optical system. This approach falls within the scope of diffraction theory which we will not discuss here. In such a case, keeping track of the electromagnetic field distribution between the object and its image can turn out to be quite a difficult task and often requires numerical computation. However, if the wavelength of light is much smaller than the characteristic system dimensions, such as the length of the candle or the aperture of the lens, as in this particular example, the location and the magnification of the image can be determined by using a much simpler approach based on geometrical optics. This is what we will be exploring in this chapter.

In geometrical optics, wave aspects of light such as wavelength, electric field distribution, and polarization are not taken into account. Light is simply modeled as consisting of rays. As such, it is assumed to travel in a straight line without spreading as long as propagation takes place in a homogeneous medium. The direction of the ray is the same as the direction of the Poynting vector for the corresponding light wave. The ray direction will change only when a discontinuity such as a dielectric interface or a lens is encountered. In this case, Snell's law [see Eq. (1.209)] can be used to find the new direction of the ray. Alternatively, ray trajectories can also be computed by application of the Fermat principle which states that light will follow the specific trajectory along which the transit time is minimized. This approach will not be discussed here. Let us emphasize once again, how simplified the geometrical optic approach is: we will make no reference to the wavelength, electric field strength, or the polarization of light waves, and simply keep track of the direction of the Poynting vector. The method works especially well in the calculation of image position and magnification so long as the wavelength is much smaller than the system dimensions.

In this chapter, we will develop the methods of geometrical optics and apply them to the analysis of common optical systems. Typically, the components of an optical system are centered on an optical axis. If the light ray coincides with the optical axis, it goes through the system without an offset or deflection. As an arbitrary light ray moves through the optical system, its height r and angle of inclination θ will change with respect to the optical axis. In this chapter, we will concentrate on the case where θ is very small so that the approximation $\sin\theta \approx \tan\theta \approx \theta$ is always valid. This regime is commonly referred to as paraxial geometrical optics. In this case, nonlinear terms involving the trigonometric functions of θ are negligible and the effect of the optical system can be described by linear transformations on r and θ. Specifically, each optical component can be modeled by a 2×2 matrix also referred to as the ABCD matrix. In Sec. 2.2, we will introduce the ABCD matrices and describe how the matrix elements can be determined. We will then discuss how the image location and magnification can be determined in the paraxial limit. By using the concepts of the principal planes, we will also show in Sec. 2.5 that a general paraxial optical system can be represented in terms of displacements and a thin lens. The general representation of a paraxial system developed in this chapter will be used later in Sec. 3.3 to derive the transformation rule for a Gaussian light beam. Ray tracing in continuous media, the Eikonal equation, graded-index rods, and stability criteria for optical resonators will also be discussed. Further discussion of geometrical optics can be found in Refs. 1 to 8 listed at the end of the chapter.

2.2 Matrix Formulation of Paraxial Optics

Let us consider a general optical system sketched in Fig. 2.1. The z axis corresponds to the optical axis. The input and output planes of the optical system are located at $z = z_i$ and $z = z_o$, respectively. So, as an example, we may have a telescope between $z = z_i$ and $z = z_o$, as shown in Fig. 2.1. In this particular case, the telescope consists of two lenses of focal lengths f_1 and f_2 separated by a distance d. An incident light ray enters the system at $z = z_i$ and exists at $z = z_o$. We are interested in developing a simple method to determine how the characteristics of the light ray will be modified by the optical system.

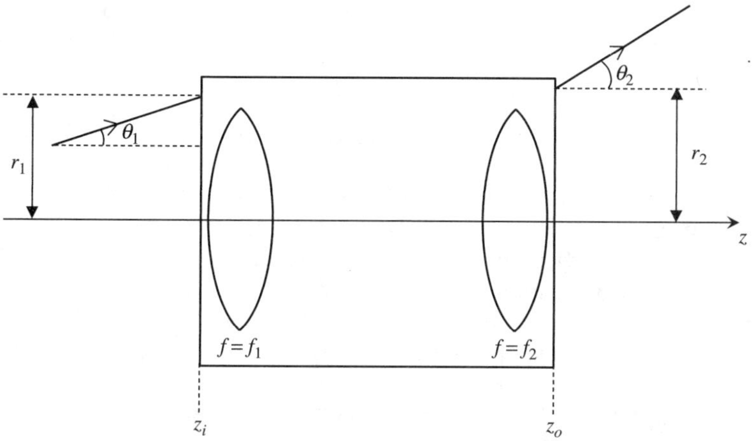

FIGURE 2.1 Sketch of a general optical system. The incident ray has the height r_1 and the slope angle θ_1 at the input plane $z = z_i$. The ray emerging at the output plane $z = z_o$ has the height r_2 and the slope angle θ_2.

In general, two parameters are required to completely describe a ray of light: its height (shown as r_1 on the input side of Fig. 2.1) and slope (θ_1 on the input side of Fig. 2.1) with respect to the optical axis. It is therefore convenient to describe this ray as a 2×1 column vector:

$$\vec{r} = \begin{bmatrix} r \\ \theta \end{bmatrix} \tag{2.1}$$

An optical system will in general modify these ray parameters so that the emerging ray at the output plane will have a different height r_2 and a different slope θ_2. Our goal is to determine how the optical system maps the input ray to the output ray. Hence, we will represent the optical system with a set of transformation rules which, given an initial ray height and slope, produces the final height and slope. Provided that we stay in the paraxial optics regime, where the rays maintain a small angle of inclination with the optical axis, this transformation can be represented by a 2×2 matrix $\mathbf{M}_\mathrm{T}$ with elements that are independent of the ray height and slope. The parameters of the output ray are hence related to those of the input ray according to

$$\vec{r}_f = \mathbf{M}_\mathrm{T}\vec{r}_i \tag{2.2}$$

where $\vec{r}_f$ and $\vec{r}_i$ are the respective final and initial ray vectors given by

$$\vec{r}_f = \begin{bmatrix} r_f \\ \theta_f \end{bmatrix} \quad \text{and} \quad \vec{r}_i = \begin{bmatrix} r_i \\ \theta_i \end{bmatrix} \tag{2.3}$$

In Eq. (2.2), $\mathbf{M}_T$ is a 2×2 matrix conventionally expressed as

$$\mathbf{M}_T = \begin{bmatrix} A & B \\ C & D \end{bmatrix} \tag{2.4}$$

We will refer to $\mathbf{M}_T$ as the ray transformation matrix or simply the ABCD matrix.

How do we determine the ray transfer matrix of a given optical system? Since $\mathbf{M}_T$ is a 2×2 matrix, it turns out that use of two special input rays is sufficient to fully determine the four elements of $\mathbf{M}_T$. These rays, designated as $\vec{\rho}_1$ and $\vec{\rho}_2$, are shown in Fig. 2.2. Ray $\vec{\rho}_1$ (Fig. 2.2*a*) is parallel to the optical axis at a finite offset of r_0 from the optical axis, whereas the second ray $\vec{\rho}_2$ (Fig. 2.2*b*) goes through the optical axis at the input plane $z = z_i$ with zero initial height and finite slope angle θ_0. In general, the parameters of the emerging ray are different for these two cases.

Expanding the ray transformation given in Eq. (2.2), we get the following explicit expressions for r_f and θ_f:

$$r_f = \mathrm{A}r_i + \mathrm{B}\theta_i \tag{2.5}$$

$$\theta_f = \mathrm{C}r_i + \mathrm{D}\theta_i \tag{2.6}$$

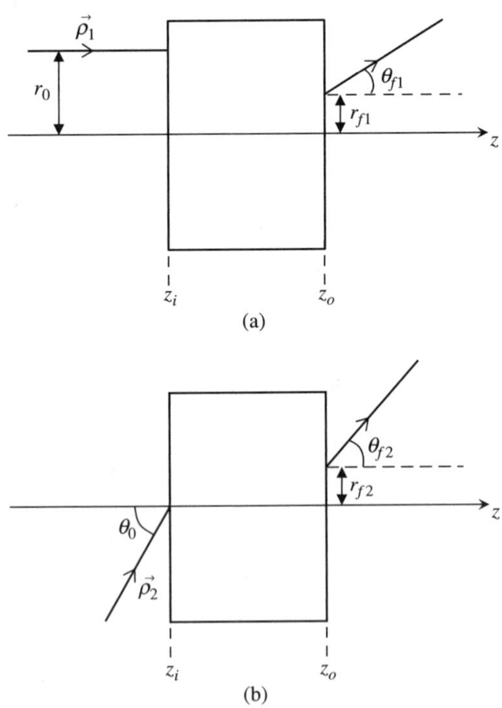

FIGURE 2.2 Sketch of the special incident rays (*a*) $\vec{\rho}_1$ and (*b*) $\vec{\rho}_2$ which are used to determine the elements of the ray transfer matrix $\mathbf{M}_T$.

Let us first consider the ray $\vec{\rho}_1$ which has a finite height and zero slope ($\theta_i = 0$). In this case, the output ray has a height of r_{f1} and a slope angle of θ_{f1}. Since $\theta_i = 0$, Eqs. (2.5) and (2.6) reduce to

$$A = \frac{r_{f1}}{r_0} \quad \text{and} \quad C = \frac{\theta_{f1}}{r_0} \tag{2.7}$$

immediately, giving the elements A and C of $\mathbf{M_T}$. Similarly, the initial height is zero for the second ray $\vec{\rho}_2$, in which case, Eqs. (2.5) and (2.6) can be used to determine B and D from

$$B = \frac{r_{f2}}{\theta_0} \quad \text{and} \quad D = \frac{\theta_{f2}}{\theta_0} \tag{2.8}$$

By using the special rays $\vec{\rho}_1$ and $\vec{\rho}_2$, the four elements of the ray transformation matrix $\mathbf{M_T}$ are determined. Once $\mathbf{M_T}$ is known, the effect of the optical system on any arbitrary ray can be calculated by using Eq. (2.2). Note that the choice of the reference plane locations is critical. If z_i and z_o are changed, the elements of $\mathbf{M_T}$ will change. Below, we provide several examples to demonstrate the application of this technique.

Example The ray transformation matrix for a displacement of length d

We start with the simplest possible example and use the above method to determine the ray transformation matrix corresponding to a displacement of length d shown in Fig. 2.3. Here, the optical system which is located between the input ($z = z_i$) and output ($z = z_o$) planes is simply a segment of space of length d.

First, consider the ray vector $\vec{\rho}_1$ which corresponds to a parallel ray incident at a height of r_0 (see Fig. 2.4*a*). Since there is only free space, the ray will emerge parallel to the optical axis at the same height, so that $r_{f1} = r_0$ and $\theta_{f1} = 0$. This gives A = 1 and C = 0 using Eq. (2.7). In the case of $\vec{\rho}_2$, where the incident ray has an arbitrary slope of θ_0 and an initial height of $r_1 = 0$ (Fig. 2.4*b*), the emerging ray will have the same slope ($\theta_{f2} = \theta_0$) and the ray height will be $r_{f2} = d\tan\theta_0 \approx d\theta_0$. Note that here, we have used the paraxial approximation by assuming that $\tan\theta_0 \approx \theta_0$.

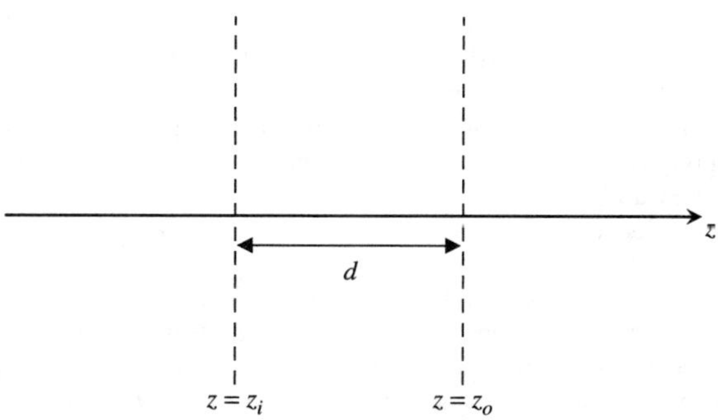

FIGURE 2.3 Empty space of length *d*.

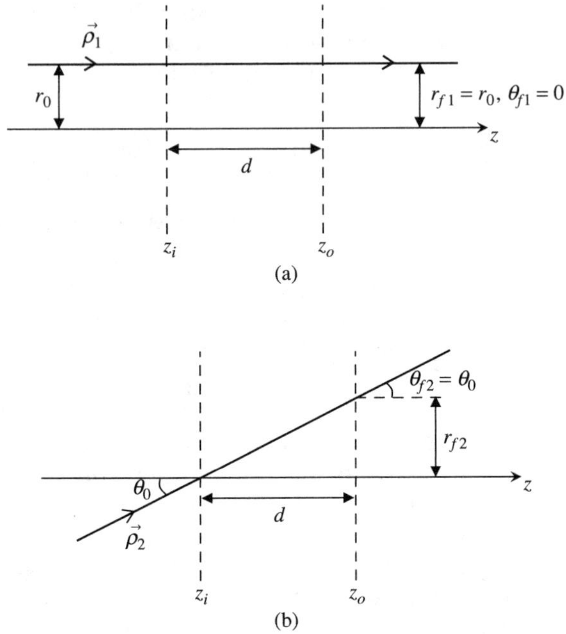

FIGURE 2.4 Use of the special rays (*a*) $\vec{\rho}_1$ and (*b*) $\vec{\rho}_2$ in the determination of the ray transformation matrix for a displacement of *d*.

By using Eq. (2.8), this gives $B = d$ and $D = 1$. Therefore, the ray transfer matrix of a displacement of length *d* is given by

$$\mathbf{M}_T = \begin{bmatrix} 1 & d \\ 0 & 1 \end{bmatrix} \tag{2.9}$$

Note that the determinant of this matrix is unity. We will find that if the index of refraction of the medium on both sides of the optical system is the same, the determinant will always be unity. This is a useful fact to check whether or not you have calculated the ray transfer matrix properly.

Example The ray transformation matrix for a thin lens of focal length *f*
As a second example, let us derive the ray transformation matrix for a thin lens of focal length *f*. Under the thin-lens aproximation, we simply neglect the finite thickness of the lens. The input and output reference planes are chosen to coincide with the front and back surfaces, respectively. Figures 2.5*a* and *b* show the action of the thin lens on the rays $\vec{\rho}_1$ and $\vec{\rho}_2$.

First, consider the incident ray vector $\vec{\rho}_1$. It is well known from elementary optics that a parallel ray of light passing through a thin lens of focal length *f* is bent to cross the optical axis at the focal point (shown as *F* in Fig. 2.5*a*), which is at a distance of *f* from the lens. Since the thin lens has a negligible thickness, the ray height remains essentially the same, so $r_{f1} = r_0$. On the other hand, the ray is bent downward and the new slope becomes $\theta_{f1} \approx \tan\theta_{f1} = -r_0/f$. This gives $A = 1$ and $C = -1/f$ using Eq. (2.7). Similarly, the effect of the thin lens on $\vec{\rho}_2$ can be analyzed by using the fact that a light ray going through the center of a thin

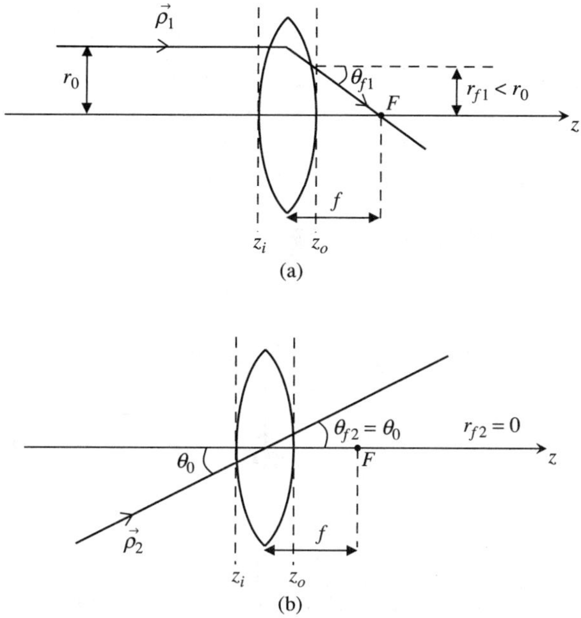

FIGURE 2.5 Effect of a thin lens of focal length f on the special rays (*a*) $\vec{\rho}_1$ and (*b*) $\vec{\rho}_2$.

lens emerges without any deflection (see Fig. 2.5*b*). In this case, we have $r_{f2} = 0$ and $\theta_{f2} = \theta_0$. Hence, B = 0 and D = 1 from Eq. (2.8). Hence, the overall ray transfer matrix for the thin lens becomes

$$\mathbf{M}_\mathrm{T} = \begin{bmatrix} 1 & 0 \\ -\dfrac{1}{f} & 1 \end{bmatrix} \tag{2.10}$$

Another important optical system is the spherical refracting surface shown in Figs. 2.6*a* and *b*. Here, the two sides of the curved interface have different indices of refraction and this leads to bending of the light rays. The procedure outlined above can also be used in this case to derive the ray transformation matrix. We leave it as an exercise for the reader (Prob. 2.1) to show that $\mathbf{M}_\mathrm{T}$ has the form

$$\mathbf{M}_\mathrm{T} = \begin{bmatrix} 1 & 0 \\ \dfrac{n_1 - n_2}{n_2 R} & \dfrac{n_1}{n_2} \end{bmatrix} \tag{2.11}$$

where R is the radius of curvature of the interface, and n_1 and n_2 are the indices of refraction on the input and output sides. Here, it is also necessary to establish a sign convention for the radius of the surface. For a ray propagating in the positive z-direction, we take the radius of curvature R to be positive if the center of curvature is to the right of the curved surface as in Fig. 2.6*a*, and negative otherwise (Fig. 2.6*b*). When R is negative as in Fig. 2.6*b*, an incident parallel ray will diverge for $n_2 > n_1$. This can be easily verified by the application of Snell's law. Also note that when the refractive index before and after the optical system is not the same, the determinant of the corresponding ray transformation matrix

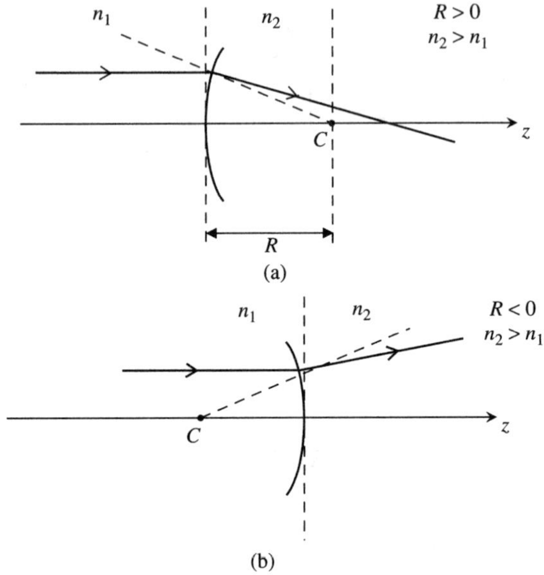

FIGURE 2.6 Effect of a curved refracting surface with a (*a*) positive and (*b*) negative radius on a parallel ray of light. For both cases, $n_2 > n_1$.

is no longer unity. In general, the determinant AD-BC becomes (n_1/n_2), where n_1 and n_2 are the indices of refraction on the input and output sides of the optical system, respectively. In the next section, we will use the ray transformation matrix of the curved refracting surface to obtain an expression for the focal length of a thin lens in terms of the radii of curved surfaces and the refractive index of the lens material.

2.3 Cascaded Optical Systems

A general optical system such as a telescope may consist of several curved surfaces, displacements, and so on. This is represented as a block diagram in Fig. 2.7. In such a case, the overall ray transfer

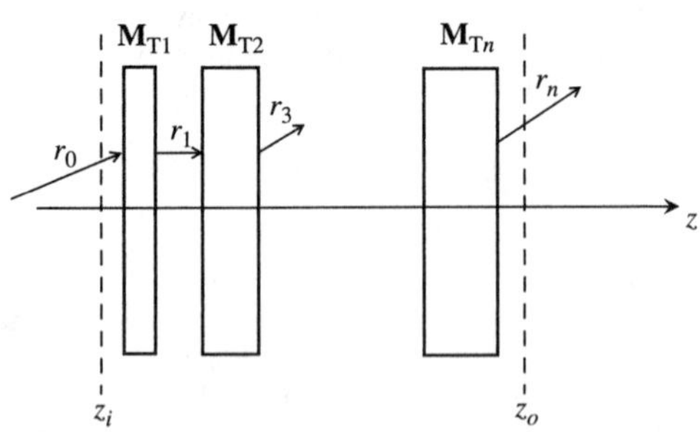

FIGURE 2.7 Block diagram of a cascaded optical system. The ray transformation matrix for the composite system is the product of the matrices for the individual elements. The correct ordering of matrix multiplication is shown in Eq. (2.13).

matrix $\mathbf{M}_T$ of the composite system is the product of the transformation matrices representing the individual optical elements. The order in which the matrices are multiplied is particularly important. To see this better, let us refer to Fig. 2.7 and assume that the light ray is propagating in the positive z-direction. The incident ray represented by $\vec{r}_0$ will first encounter the optical element with the ray transformation matrix $\mathbf{M}_{T1}$. The emerging ray $\vec{r}_1$ will be given by $\vec{r}_1 = \mathbf{M}_{T1}\vec{r}_0$. Now, $\vec{r}_1$ is incident on the second optical element, and the output ray $\vec{r}_2$ will become $\vec{r}_2 = \mathbf{M}_{T2}\vec{r}_1 = \mathbf{M}_{T2}\mathbf{M}_{T1}\vec{r}_0$. It is clear that the ray $\vec{r}_n$ emerging from the nth element will be related to the incident ray through

$$\vec{r}_n = \mathbf{M}_T\vec{r}_0 \tag{2.12}$$

where

$$\mathbf{M}_T = \mathbf{M}_{Tn}\mathbf{M}_{Tn-1}\ldots\mathbf{M}_{T1} = \prod_{i=n}^{1}\mathbf{M}_{Ti} \tag{2.13}$$

Hence, this analysis shows that we can describe the entire paraxial optical system with one 2 × 2 matrix. When several matrices are involved, it is always a good strategy to check the determinant of the overall matrix. This should be unity if the refractive index before and after the optical system is the same. The next two examples illustrate the use of Eq. (2.13) in the analysis of cascaded optical systems.

Example Consider the optical system consisting of two thin lenses L_1 and L_2 and a displacement of d in between, as shown in Fig. 2.8. $f_1 = 10$ cm, $f_2 = 20$ cm, and $d = 8$ cm. The ray incident on L_1 has a height of 3 mm and a slope angle of 1°. Calculate the height and the slope angle of the output ray immediately after L_2.

Solution The input and output planes of the cascaded optical system are located at L_1 and L_2. First, note that we need to express the incidence angle in radians so that the approximation $\sin\theta \approx \tan\theta \approx \theta$ applies. A slope angle of 1° corresponds to 0.017 radian. Expressing lengths in cm and angles in radians, the incident ray vector becomes

$$\vec{r}_0 = \begin{bmatrix} 0.3 \\ 0.017 \end{bmatrix}$$

FIGURE 2.8 Example involving a cascaded optical system.

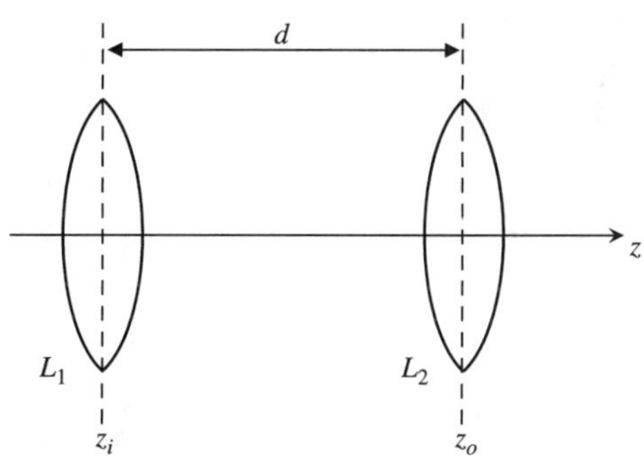

The cascaded optical system consists of the lens L_1, followed by the displacement d which is followed by the lens L_2. Therefore, the ray transformation matrix $\mathbf{M}_\mathrm{T}$ which describes the optical system between z_i and z_o becomes

$$\mathbf{M}_\mathrm{T} = \begin{bmatrix} 1 & 0 \\ -1/f_2 & 1 \end{bmatrix}\begin{bmatrix} 1 & d \\ 0 & 1 \end{bmatrix}\begin{bmatrix} 1 & 0 \\ -1/f_1 & 1 \end{bmatrix}$$

$$= \begin{bmatrix} 1 & 0 \\ -0.05 & 1 \end{bmatrix}\begin{bmatrix} 1 & 8 \\ 0 & 1 \end{bmatrix}\begin{bmatrix} 1 & 0 \\ -0.1 & 1 \end{bmatrix}$$

$$= \begin{bmatrix} 0.2 & 8 \\ -0.11 & 0.6 \end{bmatrix}$$

Note that the determinant of $\mathbf{M}_\mathrm{T}$ is unity as expected. The output ray $\vec{r}_f$ becomes

$$\vec{r}_f = \begin{bmatrix} 0.2 & 8 \\ -0.11 & 0.6 \end{bmatrix}\begin{bmatrix} 0.3 \\ 0.0175 \end{bmatrix} = \begin{bmatrix} 0.2 \\ -0.0225 \end{bmatrix}$$

So, the output ray has a height of 2 mm and a slope angle of –1.29°.

Example Lens maker's formula
Equation (2.13) can be used to obtain a useful expression for the focal length of a thin lens in terms of the radii of curvature of the surfaces and the refractive index of the lens material. Consider the thin lens sketched in Fig. 2.9. It consists of two curved surfaces with radii R_1 and R_2 and the refractive index of the material is n. We can treat this lens as a cascaded optical system composed of three optical elements: a curved refracting surface with a radius of R_1, followed by a displacement of d, followed by a second refracting surface of radius R_2. We will make the thin lens approximation and neglect the thickness of the lens. Using Eq. (2.13), we can then write the ray transformation matrix $\mathbf{M}_\mathrm{T}$ as

$$\mathbf{M}_\mathrm{T} = \begin{bmatrix} 1 & 0 \\ \dfrac{(n-1)}{R_2} & n \end{bmatrix}\begin{bmatrix} 1 & 0 \\ 0 & 1 \end{bmatrix}\begin{bmatrix} 1 & 0 \\ \dfrac{(1-n)}{nR_1} & \dfrac{1}{n} \end{bmatrix} = \begin{bmatrix} 1 & 0 \\ -(n-1)\left(\dfrac{1}{R_1} - \dfrac{1}{R_2}\right) & 1 \end{bmatrix} \tag{2.14}$$

Above, we have used the ray transfer matrix of a curved refracting surface given in Eq. (2.11). Because the thickness of the lens is neglected, the second matrix

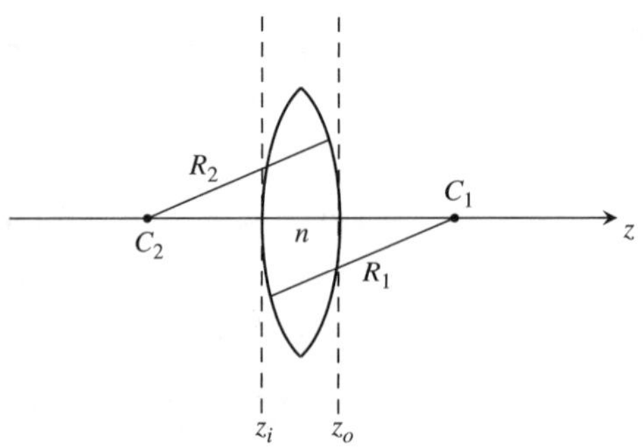

FIGURE 2.9 Sketch of a thin lens consisting of spherical refracting surfaces.

in Eq. (2.14) becomes the identity matrix. Note that the result of Eq. (2.14) looks identical to the ray transformation matrix of a thin lens whose focal length f is given by

$$\frac{1}{f}=(n-1)\left(\frac{1}{R_1}-\frac{1}{R_2}\right) \tag{2.15}$$

Equation (2.15) is the well-known "lens maker's formula" that is widely used in geometrical optics. This formula simply tells us that a dielectric slab bounded by spherical surfaces will act as a lens and its focusing characteristics are related to the refractive index of the medium, and the radii of the curved surfaces. Furthermore, by carefully choosing the signs of these radii, it is possible to make lenses with positive (convex) as well as negative (concave) effective focal lengths. The inverse of the focal length gives the focusing power of the lens, typically measured in m^{-1} or diopters. It is fulfilling to see that the matrix formulation of paraxial optics can be used to obtain this important result in an elegant way.

2.4 Imaging by a Paraxial Optical System

Using the matrix formalism of paraxial optics, we can determine the location of an image and the magnification produced by an optical system. In this section, we want to establish a criterion for finding the image location. Consider Fig. 2.10 where an optical system creates the image of an object as shown. Let us assume that the ray transformation matrix $\mathbf{M}_\mathrm{T}$ connecting the object to the image is given by

$$\mathbf{M}_\mathrm{T}=\begin{bmatrix} \mathrm{A} & \mathrm{B} \\ \mathrm{C} & \mathrm{D} \end{bmatrix} \tag{2.16}$$

We are interested in finding a general relation satisfied by the elements of $\mathbf{M}_\mathrm{T}$ when the object and the image are located at the input and output reference planes, respectively, as shown in Fig. 2.10. To obtain this relation, consider rays originating from a particular point P on the object. P is at a distance of r_1 from the optical axis. Every ray originating from P converges to a point P' on the image, irrespective

FIGURE 2.10 Imaging in a general paraxial optical system.

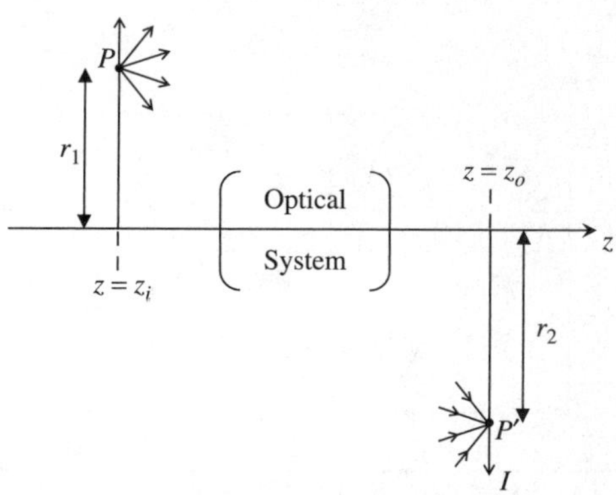

of the initial slope angle θ_1 at P. If P' is at a distance of r_2 from the optical axis, then, by definition, r_2 and r_1 are related by

$$r_2 = \mathrm{A}r_1 + \mathrm{B}\theta_1 \tag{2.17}$$

Since r_2 is independent of θ_1, B must be zero. This is an important result in imaging calculations. One can determine the location of the image by requiring that the element B of the ray transformation matrix between the object and the image planes vanish. In addition, notice that when the imaging condition is satisfied, the matrix element A gives the magnification of the optical system, in other words, the ratio of the image height to the object height.

Example To illustrate the versatility of the above method, let us revisit the cascaded optical system example shown in Fig. 2.8. An object with a height of 4 mm is placed at a distance of 30 cm from L_1 as shown in Fig. 2.11. Calculate the location and the height of the image.

Solution The ray transformation matrix of the cascaded system shown in Fig. 2.8 and consisting of the lenses L_1 and L_2, separated by d, was calculated earlier. Expressing all distances in centimeters, the ray transformation matrix $\mathbf{M}_\mathrm{T}$ connecting the image and the object is given by

$$\mathbf{M}_\mathrm{T} = \begin{bmatrix} 1 & d_I \\ 0 & 1 \end{bmatrix} \begin{bmatrix} 0.2 & 8 \\ -0.11 & 0.6 \end{bmatrix} \begin{bmatrix} 1 & 30 \\ 0 & 1 \end{bmatrix}$$

$$= \begin{bmatrix} 0.2 - 0.11 d_I & 14 - 2.7 d_I \\ -0.11 & -2.7 \end{bmatrix}$$

At the location of the image, B = 0 and hence $d_\mathrm{I} = 14/2.7 = 5.2$ cm. To find the height of the image, we calculate A at $d_\mathrm{I} = 5.2$ cm. This gives A = –0.37. The negative sign of A indicates that the image is inverted. Since the initial height of the object was 4 mm, the image height becomes 1.48 mm (inverted).

Example Imaging by a thin lens
Let us use the imaging result to find the general relation between the image and object positions in the case of a thin lens. Figure 2.12 shows an object which is

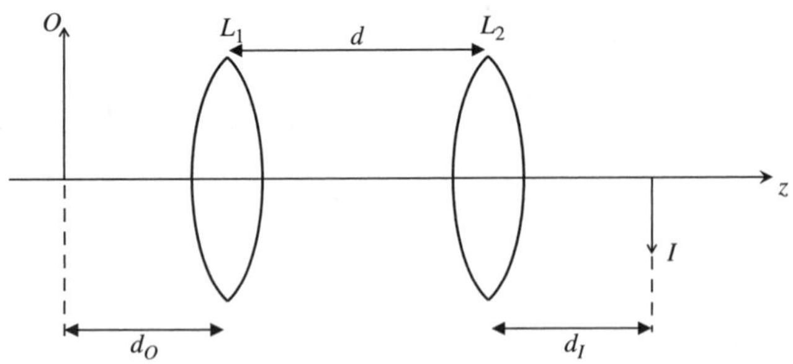

FIGURE 2.11 Example of imaging by a cascaded optical system.

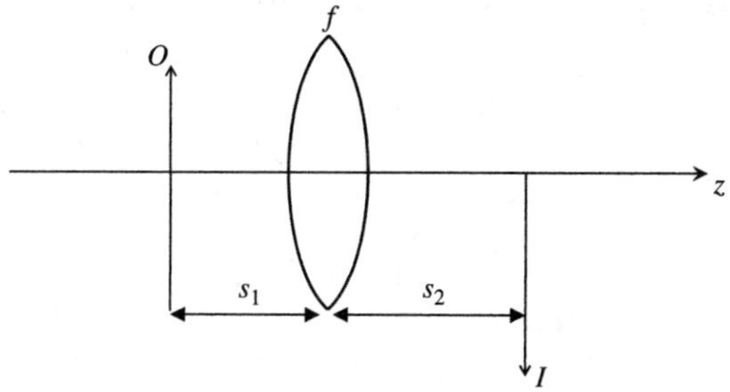

FIGURE 2.12 Imaging by a thin lens.

placed at a distance of s_1 from a thin lens of focal length f. We are interested in finding the distance s_2 of the image from the thin lens and the resulting magnification. We leave it as an exercise (Prob. 2.5) for the reader to show that the ray transformation matrix from the object to the image is given by

$$\mathbf{M}_T = \begin{bmatrix} 1-\frac{s_2}{f} & s_1\left(1-\frac{s_2}{f}\right)+s_2 \\ -\frac{1}{f} & 1-\frac{s_1}{f} \end{bmatrix} \tag{2.18}$$

The condition B = 0 readily gives the relation between s_1 and s_2, a result again well known in elementary optics:

$$\frac{1}{s_2} = \frac{1}{f} - \frac{1}{s_1} \tag{2.19}$$

Furthermore, the magnification becomes

$$A = 1 - \frac{s_2}{f} = -\frac{s_2}{s_1} \tag{2.20}$$

The above example also shows that one does not need to remember many of the specific formulas of imaging introduced in courses on elementary optics. The condition B = 0 provides a general approach for imaging calculations in any paraxial optical system. If the output side of the lens is immersed in a medium with refractive index n, the distance s_2' of the image becomes ns_2, where s_2 is given by Eq. (2.19). The effect of having a different refractive index on the output side of the imaging system is further explored in Prob. 2.6.

2.5 General Decomposition of a Paraxial Optical System

In this section, we show that an arbitrary paraxial system can always be decomposed into a cascaded optical system consisting of a thin lens sandwiched between two displacements. We will derive this important result by introducing the concept of principal planes and focal points. The parameters of the equivalent cascaded system will be expressed in terms of the elements of the ray transfer matrix. The results obtained here will be used later in the next chapter to derive the transformation rule for Gaussian beams.

2.5.1 Second Principal Plane and Second Focal Point

Let us consider a general paraxial optical system shown in Fig. 2.13. In general, an incident ray parallel to the optical axis leaves the system at an arbitrary slope angle and height which are different from the initial values. Note that we could, in principle, produce the same output ray height and slope angle by using a thin lens. How can we determine the focal length and the location of this lens? To answer this question, we assume that the incident light ray moves parallel to the optical axis up to an imaginary plane H_2. At H_2, it is bent and crosses the optical axis at the point F_2. H_2 and F_2 are called the second principal plane and the second focal point, respectively, of the optical system. H_2 simply gives the location of this effective thin lens that will produce the same output ray height and the output slope angle as the original optical system. We can easily obtain expressions for the positions of H_2 and F_2 by defining the following displacements:

δ_{F2} = displacement from the output plane $z = z_o$ to the focal point F_2
f_2 = displacement from the second principal plane H_2 to the second focal point F_2
δ_2 = displacement from the output plane to the second principal plane H_2

Because the incident light ray is parallel to the optical axis, $\theta_1 = 0$, giving $r_2 = Ar_1$ and $\theta_2 = Cr_1$. From Fig. 2.13, one therefore sees that

$$\delta_{F2} = -\frac{r_2}{\theta_2} = -\frac{A}{C} \tag{2.21}$$

$$f_2 = -\frac{r_1}{\theta_2} = -\frac{1}{C} \tag{2.22}$$

and

$$\delta_2 = \delta_{F2} - f_2 = \frac{(1-A)}{C} \tag{2.23}$$

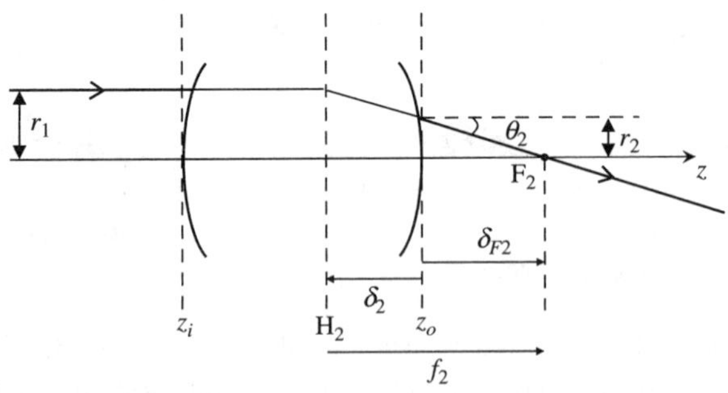

FIGURE 2.13 Second principal plane and second focal length of a paraxial optical system.

To summarize, a thin lens with an effective focal length of f_2, placed at H_2 has the same action on an incident parallel light ray as the original system. Note that the displacements given in Eqs. (2.21) to (2.23) can be positive or negative, depending on the specific configuration of the optical system. Hence, it is possible to have H_2 on either side of the output plane. Similarly, the effective focal length f_2 of the optical system can be positive or negative.

2.5.2 First Principal Plane and First Focal Point

We next investigate an oblique incident ray as shown in Fig. 2.14. If the ray passes through a special point, shown as F_1, it exits the system parallel to the optical axis. F_1 is defined as the first focal point. As before, we want to determine the focal length and location of a thin lens which would produce the same effect. We assume that the light ray moves in a straight line up to an imaginary plane H_1, known as the first principal plane. At H_1, the ray is refracted to move parallel to the optical axis. To find the location of H_1 and F_1, we define the following displacements:

δ_{F1} = displacement from the input plane z_i to the first principal point F_1
f_1 = displacement from the first principal focal point F_1 to the first principal plane H_1
δ_1 = displacement from the input plane z_i to the first principal plane H_1

We leave it as an exercise (Prob. 2.7) for the reader to show that δ_{F1}, f_1, and δ_1 are given by

$$\delta_{F1} = \frac{D}{C} \tag{2.24}$$

$$f_1 = -\frac{(AD - BC)}{C} \tag{2.25}$$

$$\delta_1 = \frac{D - (AD - BC)}{C} \tag{2.26}$$

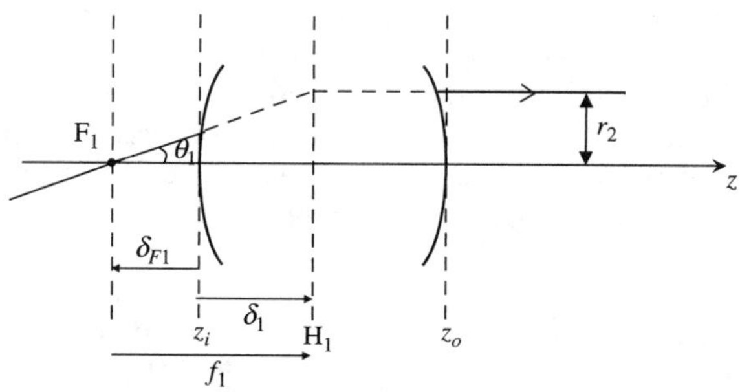

FIGURE 2.14 First principal plane and first focal point of a paraxial optical system.

Note that if the refractive index before and after the optical system is the same, then AD – BC = 1 and $f_1 = f_2 = -1/C$.

2.5.3 Ray Transformation Matrix Between H_1 and H_2

By using Figs. 2.13 and 2.14, we see that the ray transformation matrix $\mathbf{M}_T(H_1 \rightarrow H_2)$ between H_1 and H_2 consists of three components:

1. Displacement of $-\delta_1$ from H_1 to the input plane z_i
2. The optical system between z_i and z_o
3. Displacement of δ_2 from z_o to H_2

Written in full, $\mathbf{M}_T(H_1 \rightarrow H_2)$ becomes

$$\mathbf{M}_T(H_1 \rightarrow H_2) = \begin{pmatrix} 1 & \delta_2 \\ 0 & 1 \end{pmatrix} \begin{pmatrix} A & B \\ C & D \end{pmatrix} \begin{pmatrix} 1 & -\delta_1 \\ 0 & 1 \end{pmatrix}$$

$$= \begin{pmatrix} 1 & \dfrac{1-A}{C} \\ 0 & 1 \end{pmatrix} \begin{pmatrix} A & B \\ C & D \end{pmatrix} \begin{pmatrix} 1 & -\dfrac{D-(AD-BC)}{C} \\ 0 & 1 \end{pmatrix} \tag{2.27}$$

$$= \begin{pmatrix} 1 & 0 \\ C & AD-BC \end{pmatrix}$$

Recall from before that $AD - BC = n_1/n_2$, where n_1 and n_2 are the respective refractive indices at the input and output planes of the optical system. Hence, $\mathbf{M}_T(H_1 \rightarrow H_2)$ becomes

$$\mathbf{M}_T(H_1 \rightarrow H_2) = \begin{pmatrix} 1 & 0 \\ C & \dfrac{n_1}{n_2} \end{pmatrix} = \begin{pmatrix} 1 & 0 \\ C & 1 \end{pmatrix} \begin{pmatrix} 1 & 0 \\ 0 & \dfrac{n_1}{n_2} \end{pmatrix} \tag{2.28}$$

The above equation summarizes a very important result about paraxial optical systems. In the most general case, where the input and output planes of an optical system have different refractive indices, the ray transformation matrix between the principal planes is equivalent to that of a cascaded optical system consisting of a dielectric interface followed by a thin lens of focal length $f = -1/C$. If the input and exit planes both have the same refractive index, then the ray transfer matrix reduces to that of a thin lens of focal length $f = -1/C$. In fact, one can use this as an alternative method in imaging calculations as the following example illustrates.

Example Let us solve the imaging problem shown in Fig. 2.11 by using the method of principal planes. Recall that the ray transformation matrix $\mathbf{M}_T$ between the input and output planes was calculated to be

$$\mathbf{M}_T = \begin{bmatrix} 0.2 & 8 \\ -0.11 & 0.6 \end{bmatrix}$$

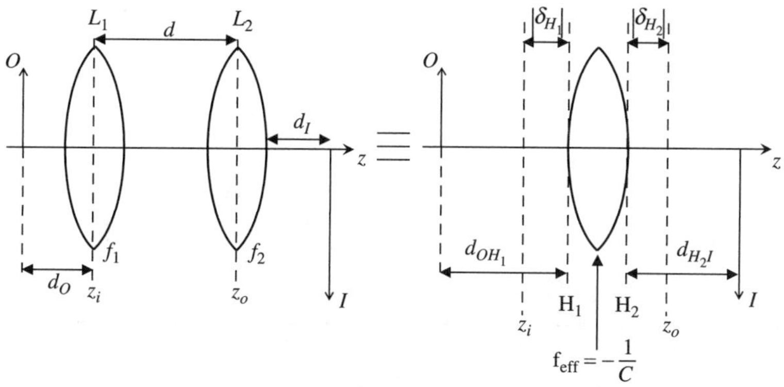

Figure 2.15 Application of principal planes to imaging problems.

By using Eqs. (2.26) and (2.23), we calculate δ_{H_1} and δ_{H_2} to determine the locations of the principal planes H_1 and H_2:

$$\delta_{H_1} = \frac{D-(AD-BC)}{C} = \frac{0.6-1}{-0.11} = 3.64 \text{ cm}$$

$$\delta_{H_2} = \frac{1-A}{C} = \frac{1-0.2}{-0.11} = -7.27 \text{ cm}$$

So, we can represent the optical system with a thin lens of effective focal length $f_{eff} = -1/C = 9.09$ cm, situated between H_1 and H_2 as shown in Fig. 2.15. The distance d_{OH_1} between the object and the thin lens becomes $d_{OH_1} = 3.64 + 30 = 33.64$ cm. The image distance d_{H_2I} can be calculated by using the imaging relation for a thin lens:

$$\frac{1}{d_{OH_1}} + \frac{1}{d_{H_2I}} = \frac{1}{f_{eff}}$$

This gives $d_{H_2I} = 12.46$ cm. So, the distance of the image from the second output plane becomes $12.46 - 7.27 = 5.2$ cm, as calculated before. Finally, the magnification M is given by

$$M = -\frac{d_{H_2I}}{d_{OH_1}} = -0.37$$

2.5.4 Representation of a Paraxial Optical System in Terms of Displacements and a Lens

Let us consider a general paraxial optical system described by the ray transfer matrix

$$\mathbf{M_T} = \begin{pmatrix} A & B \\ C & D \end{pmatrix} \tag{2.29}$$

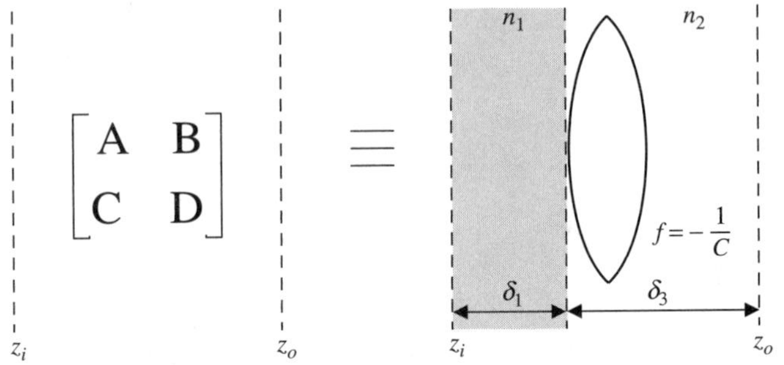

FIGURE 2.16 General representation of a paraxial optical system.

By using Eqs. (2.27) and (2.28), $\mathbf{M}_\mathrm{T}$ can be expressed as

$$\mathbf{M}_\mathrm{T} = \begin{pmatrix} \mathrm{A} & \mathrm{B} \\ \mathrm{C} & \mathrm{D} \end{pmatrix} = \begin{pmatrix} 1 & \delta_3 \\ 0 & 1 \end{pmatrix} \begin{pmatrix} 1 & 0 \\ \mathrm{C} & 1 \end{pmatrix} \begin{pmatrix} 1 & 0 \\ 0 & \dfrac{n_1}{n_2} \end{pmatrix} \begin{pmatrix} 1 & \delta_1 \\ 0 & 1 \end{pmatrix} \tag{2.30}$$

where

$$\delta_3 = -\delta_2 = \frac{\mathrm{A}-1}{\mathrm{C}} \tag{2.31}$$

and

$$\delta_1 = \frac{\mathrm{D}-(\mathrm{AD}-\mathrm{BC})}{\mathrm{C}} \tag{2.32}$$

In other words, a general paraxial optical system can always be represented as a cascaded system consisting of a displacement (δ_1), a dielectric interface, a thin lens of focal length $f = -1/\mathrm{C}$, and a second displacement (δ_3). This is schematically shown in Fig. 2.16. In the next chapter, we will use this result to derive the general transformation rule for the q-parameter of Gaussian beams.

2.6 Geometrical Optics in Continuous Media

In the previous sections, we analyzed the effect of discrete optical systems on the propagation of light rays. Here, we will consider optical systems where the refractive index changes continuously in space. We will assume during the analysis that, as before, the z-axis defines the optical axis and the change in the refractive index occurs in the radial direction as shown in Fig. 2.17. As can be seen from Fig. 2.17, the ray makes an angle of θ with the optical axis at the point A which is at a distance of r from the optical axis. The refractive index at A is $n(r)$. After propagating an infinitesimal distance dz along the optical axis, the ray moves to point B which is $r + dr$ away from the axis and the angle made with the optical axis becomes $\theta + d\theta$. Because n changes continuously with r, the ray will in general follow a curved

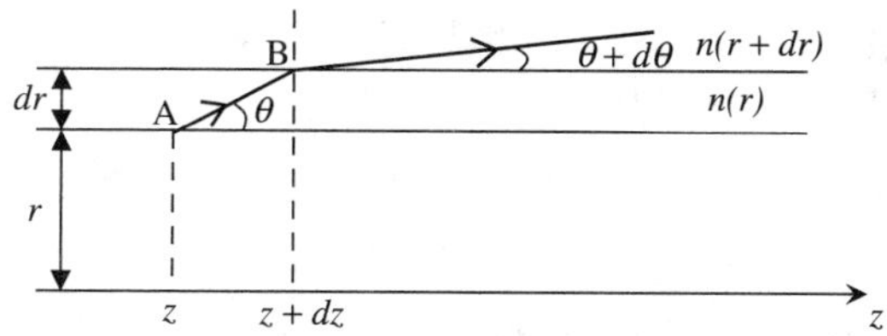

FIGURE 2.17 Propagation of a light ray in a medium with continuously changing refractive index in the radial direction.

trajectory. In what follows, we will assume that in the radial direction, the region is divided into narrow segments of thickness dr and in each segment, the refractive index is approximately constant so that the ray follows a straight path. At the point B, the refractive index changes to $n(r + dr)$, causing the ray to refract. Applying Snell's law at B, we have,

$$n(r)\cos\theta = n(r + dr)\cos(\theta + d\theta) \tag{2.33}$$

Note that $\cos\theta$ is used in the above equation since the angle made with the normal to the interface is $(\pi/2 - \theta)$. We expand Eq. (2.33) in a Taylor series, retaining terms which are first-order in dr and $d\theta$ to obtain

$$n(r)\cos\theta = n(r)\cos\theta - n(r)\sin\theta d\theta + \frac{dn}{dr}\cos\theta dr + O(drd\theta) \tag{2.34}$$

Note that above, we made the approximations $\cos(d\theta) \sim 1$ and $\sin(d\theta) \sim d\theta$. The term $O(drd\theta)$ includes all of the remaining terms of order $drd\theta$ or higher. Neglecting higher-order terms in Eq. (2.34) leads to

$$\frac{dn}{dr} = n(r)\tan\theta\frac{d\theta}{dr} \tag{2.35}$$

The ray trajectory is defined by the function $r = r(z)$. This tells the ray height at the location z along the optical axis. The ray slope is further given by

$$\tan\theta = \frac{dr}{dz} \tag{2.36}$$

By using Eqs. (2.35) and (2.36), we get

$$\frac{dn}{dr} = n(r)\frac{d\theta}{dz} \tag{2.37}$$

If we now differentiate both sides of Eq. (2.36) with respect to z and use the paraxial approximation ($\cos\theta \sim 1$), we come up with the following useful approximation:

$$\sec^2\theta\frac{d\theta}{dz} \approx \frac{d\theta}{dz} = \frac{d^2r}{dz^2} \tag{2.38}$$

Substituting the above result into Eq. (2.37), we obtain a second-order differential equation satisfied by the ray height r:

$$\frac{d^2r}{dz^2} = \frac{1}{n(r)}\frac{dn(r)}{dr} \tag{2.39}$$

Equation (2.39) is known as the Eikonal equation and describes how the ray height changes in an inhomogeneous optical medium under the paraxial approximation.

Example Ray propagation in a GRIN lens
Graded-index (GRIN) lenses are commonly used in coupling light into optical fibers. In its simplest form, a GRIN lens is a cylindrically symmetric glass rod with a radially varying refractive index, as sketched in Fig. 2.18*a* and *b*. If $n(r)$ decreases with r as depicted in Fig. 2.18*b*, the rod acts like a continuous lens and confines light rays near the optical axis. In this example, we will derive the ray transformation matrix of a GRIN lens by using the Eikonal equation. Suppose that the radial index variation is given by

$$n(r) = n_0\left(1 - \frac{\beta^2 r^2}{2}\right) \tag{2.40}$$

In our analysis, we will assume that β is very small so that even at the edge of the GRIN lens, the departure of $n(r)$ from n_0 is very small. Under this approximation, we see that

$$\frac{1}{n(r)}\frac{dn}{dr} \approx \frac{1}{n_0} n_0\left(-\frac{2\beta^2 r}{2}\right) = -\beta^2 r \tag{2.41}$$

giving

$$\frac{d^2r}{dz^2} + \beta^2 r = 0 \tag{2.42}$$

Equation (2.42) can be easily solved to give r as a function of z:

$$r(z) = \alpha_1 \cos(\beta z) + \alpha_2 \sin(\beta z) \tag{2.43}$$

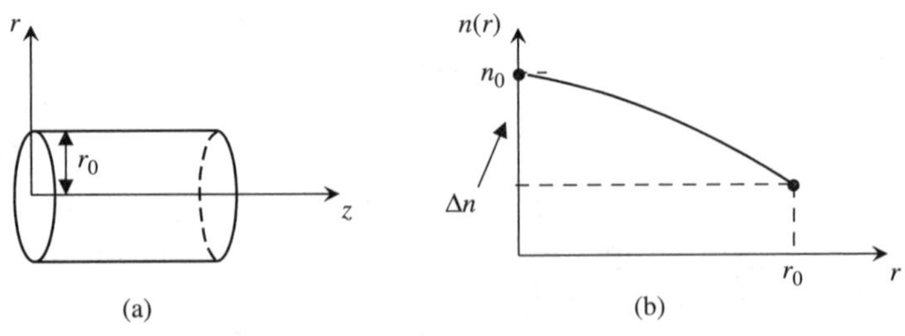

Figure 2.18 (*a*) Sketch of a GRIN lens and (*b*) typical variation of the refractive index with the radial coordinate *r*.

The constants α_1 and α_2 in Eq. (2.43) can be determined from initial conditions. Let us assume that r_i and θ_i are the initial ray height and slope, respectively. In the paraxial approximation,

$$\theta_i \approx \tan\theta_i = \left.\frac{dr}{dz}\right|_{z=0} \tag{2.44}$$

From Eq. (2.43), $r(0) = \alpha_1 = r_i$. Also,

$$\left.\frac{dr}{dz}\right|_{z=0} = \left[-\beta\alpha_1 \sin(\beta z) + \beta\alpha_2 \cos(\beta z)\right]\Big|_{z=0} = \beta\alpha_2 = \theta_i \tag{2.45}$$

In other words, we can express the integration constants α_1 and α_2 in terms of the initial ray parameters r_i and θ_i. Note further that by differentiating Eq. (2.43), we can obtain the equation for the ray slope $\theta(z) = dr/dz$ as a function of the ray position z. The expressions for $r(z)$ and $\theta(z)$ hence become

$$r(z) = r_i \cos(\beta z) + \frac{1}{\beta}\sin(\beta z)\theta_i \tag{2.46}$$

and

$$\theta(z) = \frac{dr}{dz} = -\beta \sin(\beta z) r_i + \cos(\beta z)\theta_i \tag{2.47}$$

Note that Eqs. (2.46) and (2.47) already give us the rule for transforming the initial ray vector $\vec{r}_i$ with components r_i and θ_i to the output ray vector $\vec{r}_f$ with $r_f = r(z)$ and $\theta_f = \theta(z)$ [see Eq. (2.3)]. Hence, the ray transformation matrix for the GRIN lens of length d becomes

$$\mathbf{M}_\mathrm{T} = \begin{bmatrix} \cos(\beta d) & \frac{1}{\beta}\sin(\beta d) \\ -\beta\sin(\beta d) & \cos(\beta d) \end{bmatrix} \tag{2.48}$$

Let us illustrate the lensing action of the GRIN lens with a simple example. Suppose that a parallel ray ($\theta_i = 0$) is incident on the GRIN lens. If the parameters β and d are adjusted to give $\beta d = \pi/2$, then $r_f = 0$, indicating that the GRIN lens focuses all the incident rays to the output plane, irrespective of their initial height from the optical axis.

2.7 Ray Analysis of Optical Resonators

One of the most important components of a laser is the optical resonator. In its simplest form, an optical resonator consists of two reflecting mirrors aligned along a common axis. Throughout the book, we will use the terms "optical resonator" and "optical cavity" interchangeably to mean the same thing. Let us look at the simplified sketch of a laser shown in Fig. 2.19 to better appreciate the function of the optical resonator. Here, an amplifying medium is placed inside an optical resonator consisting of mirrors M_1 and M_2. Without these mirrors, the light beam can only make a single pass through the amplifier. On the other hand, M_1 and M_2 allow multiple passes of the beam,

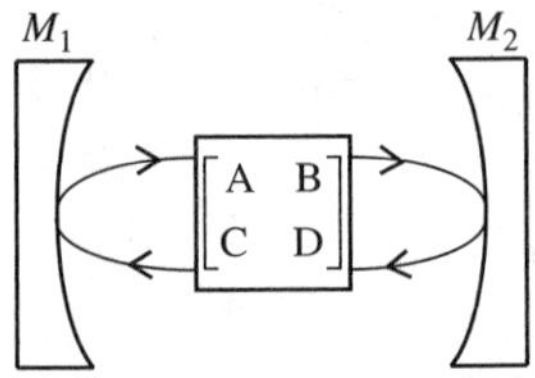

FIGURE 2.19 Sketch of a two-mirror optical resonator.

leading to more effective amplification. If the net gain over one round trip exceeds unity, light oscillation starts and the beam intensity inside the resonator can build up to very large values in steady state. Here, we say that the resonator consisting of M_1 and M_2 produces positive feedback. In Chap. 6, we will provide a detailed discussion of how optical amplification can be produced in an atomic system and how the positive feedback mechanism is designed to achieve laser action. In this section, we will concentrate on understanding some of the important characteristics of optical resonators by using the tools of geometrical optics.

With regard to optical resonators, we need to answer two important questions:

1. Under what conditions is the optical resonator stable? In other words, what cavity configurations guarantee that the circulating light beam remains confined inside the cavity?
2. Assuming that the resonator supports a confined light beam, can we determine the size of the beam at any location inside the cavity?

Both of these questions are very important in the design of optical resonators. It turns out, however, that the second question is not possible to answer using geometrical optics alone. We will revisit this issue in the next chapter after discussing the electromagnetic aspects of light beams. In this section, we will apply the matrix formalism that was developed in the previous sections to tackle the first question about resonator stability. To gain a qualitative understanding, let us first examine the cavity shown in Fig. 2.20. Here, we have two

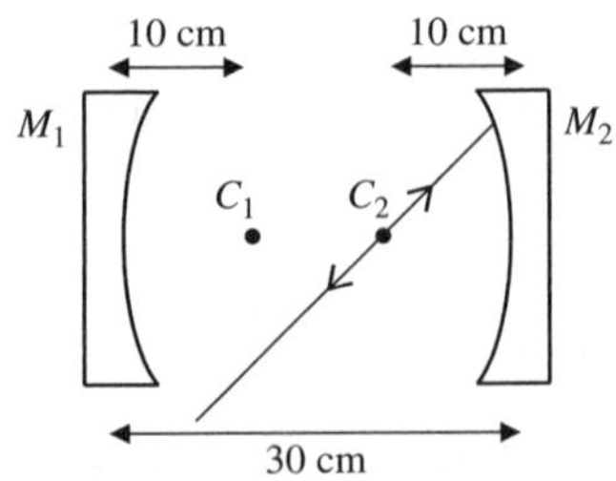

FIGURE 2.20 Example of an unstable resonator.

concave reflecting mirrors, each with a radius of 10 cm, and they are separated by 30 cm. If the resonator configuration is stable, then a paraxial ray of arbitrary initial parameters remains confined in the cavity after an indefinite number of round trips. To test the stability, let us consider, as an example, a ray that originates from the center of curvature C_2 of mirror M_2. Upon reflection from M_2, it will be reftro-reflected back in the same direction and, assuming that the mirrors do not have very large apertures, will miss mirror M_1 (see Fig. 2.20). Hence, the ray will not remain in the cavity due to the large mirror separation. This tells us right away that the mirror separation and radii are critical parameters that affect the resonator stability. In fact, it turns out in this particular case that if the separation between the mirrors is reduced to less than twice the radius of curvature, then the cavity becomes stable.

As a second example, suppose that we look at two flat reflecting mirrors separated by a certain distance as shown in Fig. 21*a*. In this case, if the mirrors are perfectly parallel to begin with, and if the initial light ray exactly coincides with the optical axis, then it will remain within the cavity. On the other hand, even a slight misalignment of one of the mirrors will cause the ray to escape from the cavity (see Fig. 21*b*). The resonator of this example is said to be conditionally stable since perfect alignment of the mirrors is essential to confine the circulating ray.

We will analyze the stability of a general paraxial resonator by using two different methods: the difference equation method and eigenvalue method. These are discussed next in Secs. 2.7.1 and 2.7.2.

2.7.1 Difference Equation Approach for Stability Analysis

To obtain the stability condition, we will first derive a difference equation that relates the ray heights after successive round trips. Consider the generalized resonator sketched in Fig. 2.22. Starting at the location P, the ray transformation matrix that describes one round

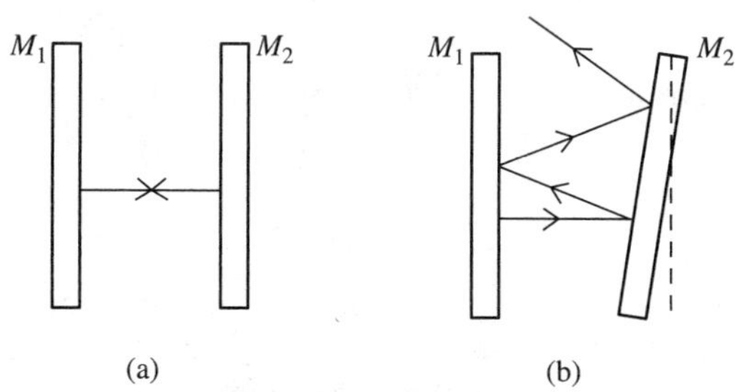

Figure 2.21 (*a*) A conditionally stable optical resonator consisting of two perfectly aligned flat mirrors. (*b*) A slight misalignment of the flat mirrors causes the circulating ray to escape.

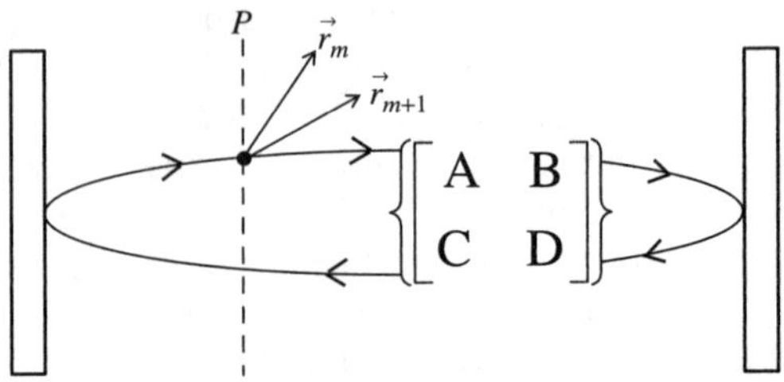

FIGURE 2.22 Sketch of a generalized resonator. Starting from P, the ray transformation matrix that describes one round trip is $\mathbf{M_T}$. The ray vector after m round trips is r_m.

trip is given by $\mathbf{M_T}$. This means that we can relate the ray vector $\vec{r}_{m+1}$ after $(m+1)$ round trips to the ray vector $\vec{r}_m$ after m round trips through the relation

$$\vec{r}_{m+1} = \mathbf{M}_\mathrm{T}\vec{r}_m \tag{2.49}$$

Let r_m and θ_m be defined according to

r_m = the ray height at P after the mth round trip
θ_m = angle of inclination of the ray at P after the mth round trip

The ray parameters after the $(m + 1)$th round trip are related to those after the mth round trip according to

$$r_{m+1} = \mathrm{A}r_m + \mathrm{B}\theta_m \tag{2.50}$$

and

$$\theta_{m+1} = \mathrm{C}r_m + \mathrm{D}\theta_m \tag{2.51}$$

We can eliminate the angle of inclination θ and obtain a second-order difference equation involving the ray heights in successive round trips. Using Eq. (2.50), θ_m can be expressed in terms of r_m and r_{m+1} as

$$\theta_m = \frac{(r_{m+1} - \mathrm{A}r_m)}{\mathrm{B}} \tag{2.52}$$

Changing the index from m to $m + 1$, we can obtain a similar relation for θ_{m+1}:

$$\theta_{m+1} = \frac{(r_{m+2} - \mathrm{A}r_{m+1})}{\mathrm{B}} \tag{2.53}$$

Note that we have two equivalent expressions for θ_{m+1}, given in Eqs. (2.51) and (2.53). If we equate them and substitute for θ_m by using Eq. (2.52), we obtain a second-order difference equation satisfied by the ray height:

$$r_{m+2} - (\mathrm{A}+\mathrm{D})r_{m+1} + (\mathrm{AD}-\mathrm{BC})r_m = 0 \tag{2.54}$$

After one round trip, the ray necessarily comes back to the same region in the cavity and the starting and the ending refractive indices are the same. Therefore, AD – BC = 1. If the cavity is stable, the ray height must remain bounded. Therefore, we guess a complex exponential solution for Eq. (2.54) of the form

$$r_m = r_{m0} \exp(im\phi) \tag{2.55}$$

Above, m is the number of round trips that the ray has completed, r_{m0} is the ray amplitude, and the angle ϕ is assumed to be real. Substituting this solution into Eq. (2.54) yields

$$r_{m0} \exp(im\phi)[\exp(i\phi)^2 - (\mathrm{A}+\mathrm{D})\exp(i\phi) + 1] = 0 \tag{2.56}$$

For a nontrivial solution, we require that $r_{m0} \exp(im\phi)$ is non-zero so that the expression appearing inside the square bracket of Eq. (2.56) should vanish. This leads to

$$\exp(i\phi) = \frac{\mathrm{A}+\mathrm{D}}{2} \pm i\sqrt{1 - \left(\frac{\mathrm{A}+\mathrm{D}}{2}\right)^2} \tag{2.57}$$

If ϕ is real, $\exp(i\phi)$ further satisfies the well-known Euler relation

$$\exp(i\phi) = \cos\phi + i\sin\phi \tag{2.58}$$

with $|\cos\phi| \leq 1$. Hence, the form given in Eq. (2.57) for $\exp(i\phi)$ will be valid if the inequality

$$\left|\frac{\mathrm{A}+\mathrm{D}}{2}\right| \leq 1 \tag{2.59}$$

is satisfied. In other words, the condition given in Eq. (2.59) guarantees that the ray height is bounded and the optical resonator is stable. Equation (2.59) is a very important result which applies to any general optical resonator in the paraxial limit. When the equality is satisfied, the resonator is conditionally stable, meaning that a slight misalignment of one of the mirrors will cause the ray to leave the resonator.

Before discussing the second method, let us comment on the form of the solution that we have assumed for r_m in Eq. (2.55). The reader may be disturbed by the fact that a complex-valued function was assumed as a solution for the ray height, which is known to be always real. This was simply done for the sake of convenience since the complex exponential form is a solution of the second-order difference equation [Eq. (2.54)] with constant coefficients. Because the ray height is a real function, the constants appearing in Eq. (2.55) must satisfy certain requirements. Note that A and D are also real, so the two solutions of $\exp(i\phi)$ are complex conjugates of each other. Therefore, we can express the ray height after the mth round trip as a combination of the two linearly independent solutions as

$$r_m = r_{m1}\exp(im\phi) + r_{m2}\exp(-im\phi) \tag{2.60}$$

Since r_m and ϕ are real, we require r_{m1} to be the complex conjugate of r_{m2}:

$$r_{m1} = r_{m2}^* \tag{2.61}$$

This gives

$$r_m = 2\,\mathrm{Re}\{r_{m1}\exp(im\phi)\} \tag{2.62}$$

Letting

$$r_{m1} = |r_{m1}|\exp(i\phi_0) = \frac{r_{\max}}{2}\exp(i\phi_0) \tag{2.63}$$

we get

$$r_m = r_{\max}\cos(m\phi + \phi_0) \tag{2.64}$$

Above, $r_{\max}$, and ϕ_0 are constants. So, we see that in a stable resonator, the ray height remains bounded and assumes the same value at a particular location after each round trip.

2.7.2 Eigenvalue Method for Stability Analysis

Suppose that the initial ray vector at the starting location P (see Fig. 2.22) is $\vec{r}_0$. By using Eq. (2.49), we see that the ray vector $\vec{r}_m$ after m round trips becomes

$$\vec{r}_m = \mathbf{M}_\mathrm{T}^m \vec{r}_0 \tag{2.65}$$

The mth power of the matrix $\mathbf{M}_\mathrm{T}$ can be computed from

$$\mathbf{M}_\mathrm{T}^m = \mathbf{\Pi}^{-1}\begin{bmatrix} \lambda_1^m & 0 \\ 0 & \lambda_2^m \end{bmatrix}\mathbf{\Pi} \tag{2.66}$$

where λ_1 and λ_2 are the eigenvalues of $\mathbf{M}_\mathrm{T}$, and the diagonalizing matrix $\mathbf{\Pi}$ has the corresponding eigenvectors as its columns. The eigenvalues of $\mathbf{M}_\mathrm{T}$ are determined by solving the characteristic equation

$$\det[\mathbf{M}_\mathrm{T} - \lambda\mathbf{I}] = 0 \tag{2.67}$$

In Eq. (2.67), $\mathbf{I}$ is the identity matrix. While solving the characteristic equation, we again assume that $\mathrm{AD} - \mathrm{BC} = 1$. Depending on the value of $(\mathrm{A} + \mathrm{D})/2$, we have two different cases:

Case 1: $\left|\frac{\mathrm{A}+\mathrm{D}}{2}\right|<1$

In this case, the two eigenvalues are $\lambda_{\pm}=\exp(\pm i\phi)$, where $\cos\phi=$ (A + D)/2. The magnitude of any power of these eigenvalues always remains below 1. In addition, the corresponding eigenvectors $\vec{V}_{\pm}$ are given by

$$\vec{V}_{+}=\frac{1}{\mathrm{B}(\mathrm{B}-\mathrm{C})}\begin{bmatrix}\mathrm{B}\\-(\mathrm{A}-\exp(i\phi))\end{bmatrix}$$
$$\vec{V}_{-}=\frac{1}{\mathrm{B}(\mathrm{B}-\mathrm{C})}\begin{bmatrix}\mathrm{B}\\-(\mathrm{A}-\exp(-i\phi))\end{bmatrix} \tag{2.68}$$

In this case, the matrix Π consists of bounded elements. Hence, the components of the ray vector $\vec{r}_m$ remain bounded after any number of round trips. This corresponds to the stable resonator configuration. After lengthy algebraic manipulations, $\mathbf{M}_{\mathrm{T}}^m$ can be written in this case as

$$\mathbf{M}_{\mathrm{T}}^m=\begin{bmatrix}\frac{\mathrm{A}-\mathrm{D}}{2}\frac{\sin m\phi}{\sin\phi}+\cos m\phi & \mathrm{B}\frac{\sin m\phi}{\sin\phi}\\ \mathrm{C}\frac{\sin m\phi}{\sin\phi} & \frac{\mathrm{D}-\mathrm{A}}{2}\frac{\sin m\phi}{\sin\phi}+\cos m\phi\end{bmatrix} \tag{2.69}$$

The ray height r_m after m round trips becomes

$$r_m=r_0\cos m\phi+\left(\frac{r_o(\mathrm{A}-\mathrm{D})+2B\theta_0}{2\sin\phi}\right)\sin m\phi \tag{2.70}$$

Again, Eq. (2.68) shows that the ray height remains bounded after an arbitrary number of round trips.

Case 2: $\left|\frac{\mathrm{A}+\mathrm{D}}{2}\right|>1$

Now, the eigenvalues become $\lambda_{1,2}=\exp(\pm\mu)$, where $\cosh\mu=$ (A+D)/2. Note that, the magnitude of $\lambda^m=\exp(m\mu)$ grows without a bound as the number of round trips m increases. Therefore, the components of the ray vector $\vec{r}_m$ become progressively larger as m increases and the ray escapes the cavity. This corresponds to the unstable resonator configuration. When (A + D)/2 = 1, we again have the conditionally stable resonator.

2.8 Two-Mirror Optical Resonators

We will use the results of Sec. 2.7 to analyze the stability of two-mirror resonators. For this, we need to know the ray transformation matrix of a curved reflecting mirror. It is left as an exercise for the reader to show that in the paraxial limit, an incident ray parallel to the optical axis of a concave mirror with a radius of R crosses the optical axis at a distance of $f=R/2$ (see Prob. 2.10). So, the curved mirror with a radius R is equivalent to a thin lens whose focal length f equals $R/2$. This is schematically shown in Fig. 2.23.

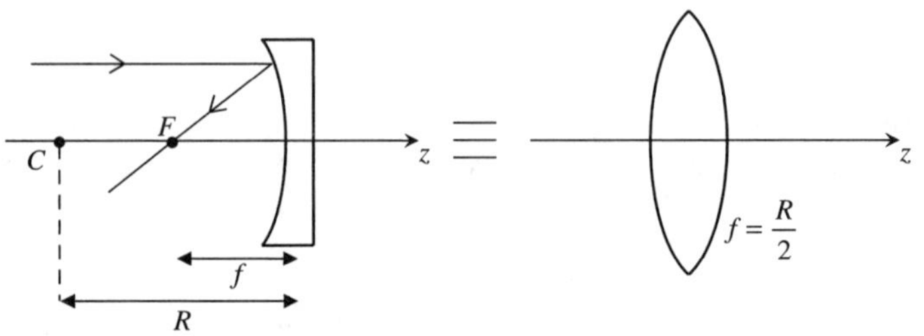

Figure 2.23 A concave mirror with a radius of R is equivalent to a thin converging lens of focal length $f = R/2$.

Next, let us consider the two-mirror optical cavity shown in Fig. 2.24. The radii of the mirrors M_1 and M_2 are R_1 and R_2, respectively. The separation between M_1 and M_2 is d. The equivalent optical system that represents one round trip is also shown in Fig. 2.24. Starting at the right side of the mirror M_1, one round trip consists of a displacement of d, followed by a thin lens of focal length $f_1 = R_1/2$, followed by a displacement of d, and finally a second thin lens of focal length $f_2 = R_2/2$.

By using the techniques discussed earlier, the round-trip ray transformation matrix $\mathbf{M}_\mathrm{T}$ becomes (see Prob. 2.11)

$$\mathbf{M}_\mathrm{T} = \begin{pmatrix} 1 - \dfrac{d}{f_2} & d + d\left(1 - \dfrac{d}{f_2}\right) \\ -\dfrac{1}{f_1} - \dfrac{1}{f_2}\left(1 - \dfrac{d}{f_1}\right) & \left(1 - \dfrac{d}{f_1}\right)\left(1 - \dfrac{d}{f_2}\right) - \dfrac{d}{f_1} \end{pmatrix} \tag{2.71}$$

where $f_1 = R_1/2$ and $f_2 = R_2/2$. The stability condition derived earlier can be written in the equivalent form

$$-1 \le \frac{\mathrm{A} + \mathrm{D}}{2} \le 1 \tag{2.72}$$

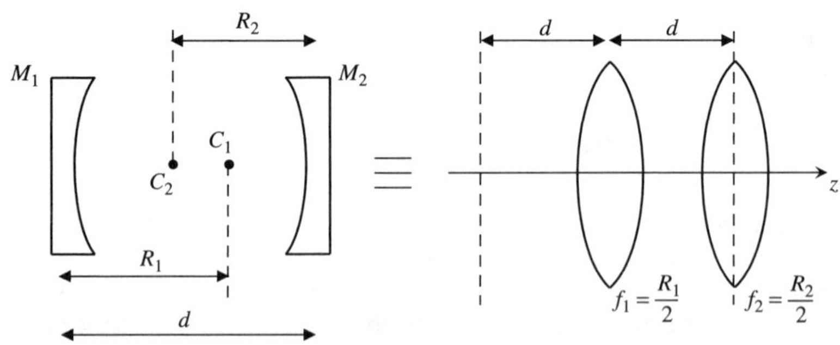

Figure 2.24 A two-mirror optical resonator and its equivalent lens system that describes one round trip.

Adding 1 to both sides and diving by 2, we obtain

$$0 \le \frac{A+D+2}{4} \le 1 \tag{2.73}$$

Use of the ray transformation matrix given in Eq. (2.71) gives

$$\frac{A+D+2}{4} = \left(1-\frac{d}{2f_1}\right)\left(1-\frac{d}{2f_2}\right) \tag{2.74}$$

Conventionally, the stability condition of the two-mirror resonator is expressed in terms of the resonator g parameters which are defined according to

$$g_m \equiv 1-\frac{d}{R_m} \tag{2.75}$$

where g_m and R_m are the g parameter and the radius of curvature of the mth mirror, respectively, and d is the mirror separation. In this case, m can be 1 or 2. In terms of the g parameters, the stability condition can be written as

$$0 \le g_1 g_2 \le 1 \tag{2.76}$$

Eq. (2.76) is very useful in the design of two-mirror laser resonators. The stability condition is schematically shown in Fig. 2.25 as a function of the g parameters. Any point inside the shaded region bounded by the axes and the curves $g_1g_2 = 1$ represents a stable resonator configuration. For example, let us pick the point with coordinates of (1/2, 1/2). At this point, $g_1g_2 = 1/4$, suggesting that the resonator is stable. This particular stable resonator configuration is symmetric with $R_1 = R_2 = 2d$.

Finally, we note that in calculating the round-trip ray transformation matrix in Eq. (2.69), we chose a particular starting point. It turns out that no matter which starting point is chosen inside the cavity, the value of A + D remains invariant and the same stability condition is obtained (see Prob. 2.13).

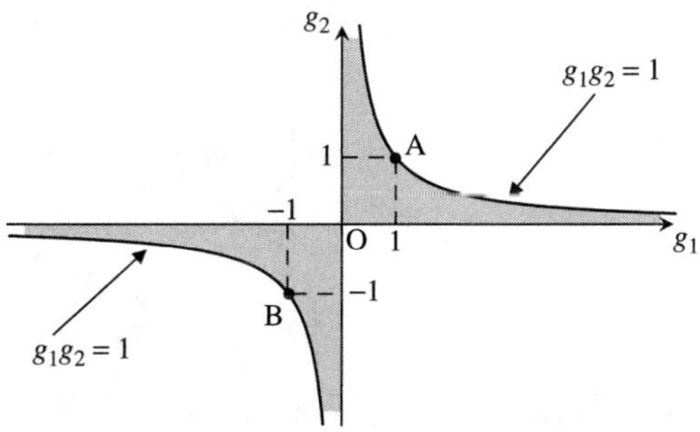

FIGURE 2.25 Stability diagram of a two-mirror resonator. The shaded region corresponds to stable resonator configurations.

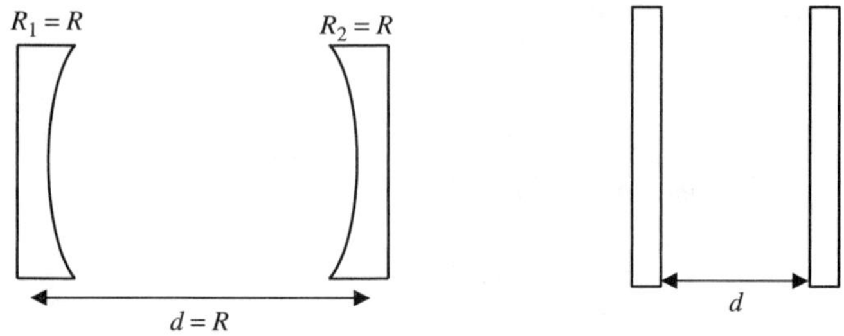

FIGURE 2.26 Examples of conditionally stable resonators: (*a*) a confocal cavity and (*b*) plane-mirror cavity.

Points on the boundaries of the stability region, i.e. where $g_1 g_2 = 0$ or 1, represent cavity configurations which are conditionally stable. Two examples are worth mentioning. Consider the origin of the graph with $g_1 = g_2 = 0$ (point O in Fig. 2.25). This point corresponds to the confocal cavity with $d = R_1 = R_2 = R$ sketched in Fig. 2.26*a*. The confocal cavity, so named because the foci of the mirrors are exactly coincident, is widely used as an optical spectrum analyzer as we will discuss later in Chap 3. Point A, on the other hand, with $g_1 = g_2 = 1$ corresponds to a cavity with two plane mirrors, since $R_1 = R_2 = \infty$ in this case. This is sketched in Fig. 2.26*b*. In both cases, the mirrors have to be exactly parallel in order to confine the ray inside the resonator.

Problems

2.1 (*a*) Derive Eq. (2.11) for the the ray transformation matrix of a curved dielectric refracting surface with a radius of curvature of R.

(*b*) Obtain the ray transformation matrix for the flat dielectric interface shown in Fig. 2.27.

2.2 What is the ray transformation matrix for a dielectric slab of refractive index n and length d?

2.3 Derive the ray transformation matrix for the optical system consisting of a thin lens of focal length f followed by a displacement of d.

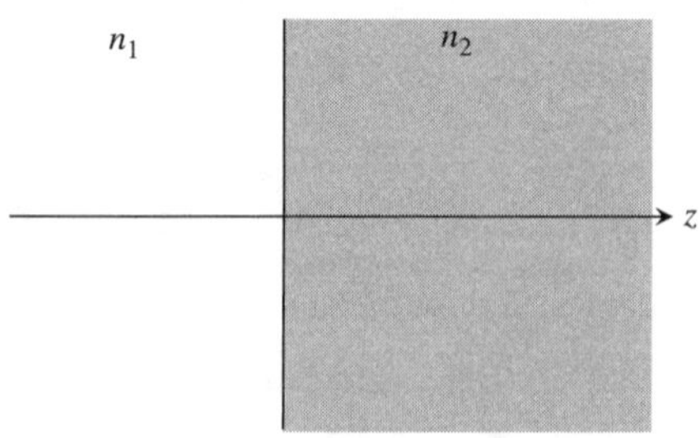

FIGURE 2.27 Flat dielectric interface for Prob. 2.1(*b*).

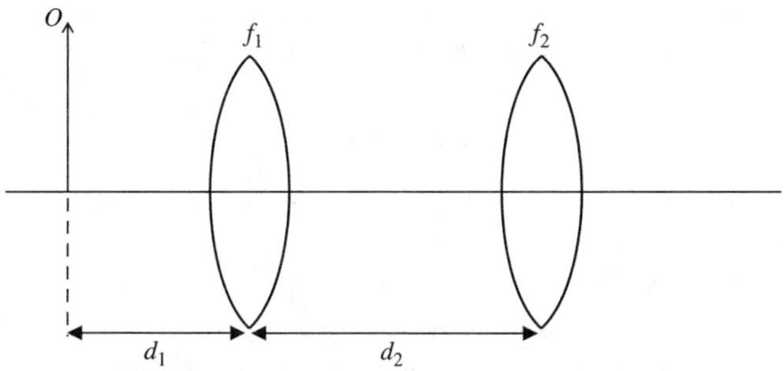

FIGURE 2.28 Optical system for Prob. 2.4.

2.4 Use matrix methods to find the image location and magnification in the optical system shown in Fig. 2.28 ($f_1 = 5$ cm, $f_2 = 8$ cm, $d_1 = 4$ cm, $d_2 = 7$ cm).

2.5 Derive the ray transformation matrix from the object to the image given in Eq. (2.18).

2.6 An object is placed at a distance of s_1 from the input plane of a paraxial optical system. Show that if the input and output sides of the imaging system have the same refractive index, then the distance s_2 of the image from the output plane is given in general by

$$s_2 = -\frac{\mathrm{A}s_1 + \mathrm{B}}{\mathrm{C}s_1 + \mathrm{D}}$$

Also show that if the input and the output sides of the imaging system have different refractive indices of n_1 and n_2, then the image distance s_2' becomes

$$s_2' - \frac{n_2}{n_1} s_2$$

2.7 Derive the expressions for the displacements δ_{F1}, f_1, and δ_1 given in Eqs. (2.24) to (2.26).

2.8 (*a*) Find the effective focal length of a ball lens of radius R and refractive index n. Here, the focal length corresponds to the distance between one of the principal planes and the corresponding focal point.

(*b*) What is the working distance of the ball lens, i.e., how far is the focal point from the exit end of the ball lens?

2.9 Newton's imaging equation

(*a*) What is the ray transfer matrix between the focal points F_1 and F_2?

(*b*) Suppose that an object is placed such that the displacement from the first focal point to the object is d_1. In addition, d_2 is the displacement from the second focal point to the image. Show that

$$d_1 d_2 = -f_1 f_2.$$

(*c*) Show that in the special case where $n_1 = n_2$ one recovers the Newton's imaging equation

$$d_1 d_2 = -f^2.$$

(*d*) Express the magnification in terms of d_1 and f_1.

2.10 Show that in the paraxial limit, a concave mirror with a radius of R is equivalent to a thin lens of focal length f given by $f = R/2$.

2.11 Derive the ray transformation matrix for the two-mirror resonator given in Eq. (2.69).

2.12 Suppose that a convex reflecting mirror with a radius of curvature of R is placed in front of a flat mirror at a distance of d. Investigate the stability of the resonator that is formed.

2.13 Show that the trace (A + D) of the ray transformation matrix which represents one round-trip of the two-mirror cavity is independent of the starting point.

2.14 Thermal lensing in a cylindrical dielectric rod: Solid-state crystals and glasses which are used in lasers are often susceptible to thermal lensing. Here, heat load due to unused pump power produces a temperature distribution inside the medium leading to a position dependent refractive index. The thermally loaded rod then behaves like a lens. In this problem, we will use geometrical optics to investigate how the strength of thermal lensing varies with power. For further discussion of thermal lensing, see Refs. 9 to 11.

(*a*) Consider a cylindrical dielectric rod of length L_0 and radius r_0. Assume that heat generation rate per unit volume is h (units = W/cm^3). Show that in a cylindrically symmetric geometry, the steady-state temperature distribution $T(r)$ satisfies the differential equation

$$\frac{1}{r}\frac{\partial}{\partial r}\left(r\frac{\partial T}{\partial r}\right)+\frac{\partial^2 T}{\partial r^2}=-\frac{h}{\kappa}$$

where κ is the heat conductivity of the medium.

(*b*) Assume that h is a constant. Also, assume that heat conduction in the longitudinal direction is negligible. Find the temperature distribution $T(r)$ as a function of r. Take $T(r_0) = T_b$, where T_b is the boundary temperature.

(*c*) What is the maximum temperature rise inside the crystal? Note that the maximum temperature occurs at $r = 0$.

(*d*) Typically, the refractive index n of a dielectric varies as a function of temperature T according to

$$n(T) = n_0 + n_T(T - T_r)$$

where T_r is some reference temperature at which $n = n_0$ and $n_T = \dfrac{dn}{dT}$ is the thermal index coefficient. In the presence of the heat source h, find the effective quadratic index coefficient β. Take $T_r = T_b$ and assume that $n(r)$ departs by a small amount from n_0.

(*e*) Show that at low pump powers, the effective thermal focal length varies inversely with the absorbed pump power.

2.15 Find the eigenvalues and the eigenvectors [Eq. (2.68)] of the round-trip ray transformation matrix $\mathbf{M_T}$ for the case where AD – BC = 1 and the stability condition is met. By using standard diagonalization techniques of matrix algebra, show that the mth power of $\mathbf{M_T}$ is given by Eq. (2.69).

References

1. A. Gerrard and J. M. Burch, *Introduction to Matrix Methods in Optics* (Dover, New York, 1975).
2. A. E. Siegman, *Lasers* (University Science Books, Mill Valley, Calif., 1986).
3. E. Hecht, *Optics*, 4th ed. (Addison-Wesley, Reading, Mass., 2002).
4. M. Born and E. Wolf, *Principles of Optics*, 6th ed. (Cambridge University Press, Cambridge, 1997).
5. J. T. Verdeyen, *Laser Electronics*, 3d ed. (Prentice Hall, Upper Saddle River, NJ, 1995).
6. A. Yariv and P. Yeh, *Photonics: Optical Electronics in Modern Communications*, 6th ed. (Oxford University Press, New York, 2007).
7. B. E. A. Saleh and M. C. Teich, *Fundamentals of Photonics*, 2d ed. (Wiley, Hoboken, N. J., 2007).
8. C. C. Davis, *Lasers and Electro-Optics* (Cambridge University Press, Cambridge, 1996).
9. M. E. Innocenzi, H. T. Yura, C. L. Fincher, and R. A. Fields, "Thermal modeling of continuous-wave end-pumped solid state lasers," *Applied Physics Letters*, **56**, 1831–1833 (1990).
10. A. K. Cousins, "Temperature and thermal stress scaling in finite-length end-pumped laser rods," *IEEE Journal of Quantum Electronics*, **28**, 1057–1069 (1992).
11. A. Sennaroglu, "Continuous wave thermal loading in saturable absorbers: Theory and experiment," *Applied Optics*, **36**, 9528–9535 (1997).

CHAPTER 3

Laser Beams and Resonators

The ray picture developed in the previous chapter gives a simple description of optical systems and can be used to determine image position or magnification. However, this simplistic approach provides an incomplete picture because it does not account for several important characteristics of realistic light beams. First and foremost, light beams are confined in the transverse direction and hence experience diffractive spreading during propagation. The beam typically has a color or a distribution of colors described by the wavelength. Furthermore, there is an electric field distribution across the transverse extent of the beam. In this chapter, we use Maxwell's equations to address these issues and develop the electromagnetic theory of confined light beams. We place a particular emphasis on the cylindrically symmetric Gaussian beam solution, also referred to as the TEM_{00} mode. By investigating the effect of a thin lens and a displacement on the propagation characteristics of a Gaussian beam, we develop a general description of how a paraxial optical system modifies the parameters of a Gaussian beam, the well-known ABCD transformation law for Gaussian beams. Higher-order modes are further discussed.

We then turn our attention to resonators with spherical mirrors which were first introduced in Sec. 2.7. It is well known that the lowest-order mode produced by such resonators is cylindrically symmetric and confined in the transverse direction. Hence, it is important to investigate the solutions of the wave equation which:

1. Satisfy boundary conditions on spherical surfaces
2. Are confined in the transverse direction
3. Have cylindrical symmetry

It turns out that the Gaussian beam discussed in the first part of this chapter naturally satisfies the boundary conditions on spherical surfaces because of the fact that it has spherical phase fronts in the paraxial limit. We will use the ABCD transformation law and determine the self consistent Gaussian beam solution for a stable resonator. Illustrative examples of mode calculations for two-mirror and four-mirror cavities will be provided. Finally, we will investigate the spectral characteristics of the two-mirror Fabry-Perot resonator and determine its resonance frequencies. Further discussion on laser beams and resonators may be found in Refs. 1 to 9.

3.1 Fundamental Gaussian Beam

3.1.1 Differential Equations for the Beam Parameters *P*(*z*) and *q*(*z*)

We start with Maxwell's equations and employ the slowly varying envelope approximation to show how the cylindrically symmetric Gaussian beam solution, also referred to as the TEM_{00} solution, is obtained in a medium with quadratic index variation. We then discuss Gaussian beam solutions in free space, the q-parameter, Rayleigh range, radius of curvature, and spot-size functions.

Let us assume that we have a homogeneous, isotropic, nonmagnetic dielectric with no free charges or free currents. We further assume that there is a weak index variation in the plane perpendicular to the direction of propagation as shown in Fig. 3.1. The actual index variation can be a complicated function of radial position. If it has a local maximum on the axis of propagation (r = 0) and monotonically decreases with increasing radial coordinate r, for small values of r, we can approximate it as a quadratic function given by

$$n(r) = n_0\left(1 - \frac{\beta^2 r^2}{2}\right) \tag{3.1}$$

where n_0 is the axial value of the refractive index and β is the quadratic index coefficient introduced earlier in Sec. 2.6.

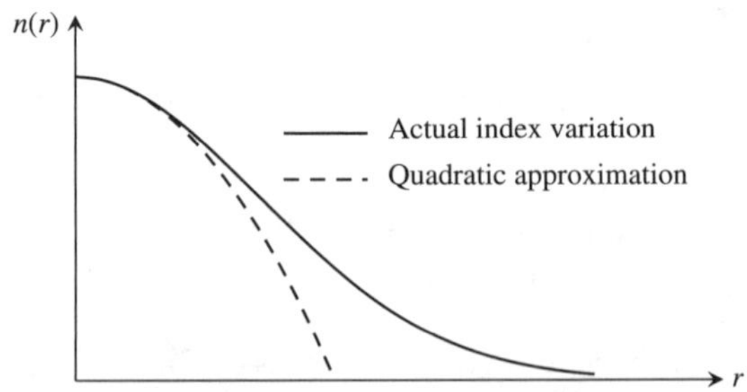

FIGURE 3.1 The variation of the actual refractive index and its quadratic approximation near the axis of propagation.

Let us also assume that the electric field vector is linearly polarized along the x direction. The x-component E_x then satisfies the wave equation

$$\nabla^2 E_x - \mu_0 \varepsilon \frac{\partial^2 E_x}{\partial t^2} = 0 \tag{3.2}$$

where the dielectric function ε is given by

$$\varepsilon = n(r)^2 \varepsilon_0 \tag{3.3}$$

If we now assume that E_x has harmonic time dependence given by

$$E_x(\vec{r},t) = E_x(\vec{r})\exp(i\omega t) \tag{3.4}$$

and substitute Eq. (3.4) into Eq. (3.2), we obtain

$$\nabla^2 E_x + \omega^2 \mu_0 \varepsilon_0 n(r)^2 E_x = 0 \tag{3.5}$$

By approximating $n(r)^2$ as

$$n(r)^2 = n_0^2\left(1 - \frac{\beta^2 r^2}{2}\right)^2 \cong n_0^2(1 - \beta^2 r^2) \tag{3.6}$$

we can express the wave equation in the form

$$\nabla^2 E_x + k^2(1 - \beta^2 r^2)E_x = 0 \tag{3.7}$$

where

$$k^2 = \omega^2 \mu_0 \varepsilon_0 n_0^2 = \left(\frac{\omega}{c}\right)^2 n_0^2 \tag{3.8}$$

To solve Eq. (3.7), we express the Laplacian operator in cylindrical coordinates. By assuming cylindrical symmetry with no ϕ dependence ($\partial/\partial\phi = 0$), the Laplacian operator ∇^2 becomes

$$\nabla^2 = \frac{1}{r}\frac{\partial}{\partial r}\left(r\frac{\partial}{\partial r}\right) + \frac{1}{r^2}\frac{\partial^2}{\partial \phi^2} + \frac{\partial^2}{\partial z^2} = \frac{1}{r}\frac{\partial}{\partial r}\left(r\frac{\partial}{\partial r}\right) + \frac{\partial^2}{\partial z^2} = \frac{\partial^2}{\partial r^2} + \frac{1}{r}\frac{\partial}{\partial r} + \frac{\partial^2}{\partial z^2} \tag{3.9}$$

Later in Sec. 3.2, we will relax the cylindrical symmetry requirement to discuss the higher-order mode solutions of Eq. (3.7). To obtain the cylindrically symmetric solution of Eq. (3.7), we assume that the electric field E_x is nearly a plane wave propagating along the optic axis (z axis in this geometry) with some transverse structure given by

$$E_x(\vec{r},t) = u(r,z)\exp[i(\omega t - kz)] \tag{3.10}$$

Hence, omitting the time dependence, Eq. (3.7) becomes

$$\left(\frac{\partial^2}{\partial r^2}+\frac{1}{r}\frac{\partial}{\partial r}+\frac{\partial^2}{\partial z^2}\right)[u(r,z)\exp(-ikz)]+k^2(1-\beta^2 r^2)u(r,z)\exp(-ikz)=0 \tag{3.11}$$

If we employ the slowly varying envelope approximation where we neglect the second derivative $\partial^2 u(r,z)/\partial z^2$ of $u(r,z)$ in relation to $k\partial u(r,z)/\partial z$, it can be shown that (see Prob. 3.1) the differential equation satisfied by $u(r,z)$ becomes

$$\left(\frac{\partial^2}{\partial r^2}+\frac{1}{r}\frac{\partial}{\partial r}\right)u(r,z)-2ik\frac{\partial u(r,z)}{\partial z}-k^2\beta^2 r^2 u(r,z)=0 \tag{3.12}$$

To solve Eq. (3.12), we then assume that $u(r,z)$ consists of an axial and quadratic radial phase factor of the form

$$u(r,z)=\exp\left[-i\left(P(z)+\frac{k}{2q(z)}r^2\right)\right] \tag{3.13}$$

Hence, our problem is reduced to the determination of the functions $P(z)$ and $q(z)$. The procedure for obtaining the differential equations satisfied by $P(z)$ and $q(z)$ is straightforward but requires a bit of algebra. We leave it as an exercise for the reader see (Prob. 3.2) to show that by substituting the ansatz of Eq. (3.13) into Eq. (3.12) and by comparing the coefficients of terms containing r^2 and r^0 (i.e., terms independent of r), the functions $P(z)$ and $q(z)$ satisfy two coupled ordinary differential equations of the form

$$\left(\frac{1}{q(z)}\right)^2+\frac{d}{dz}\left(\frac{1}{q(z)}\right)+\beta^2=0 \tag{3.14}$$

and

$$\frac{dP(z)}{dz}=-\frac{i}{q(z)} \tag{3.15}$$

We can obtain an equivalent, alternative form of Eq. (3.14) by making the change of variables

$$\frac{1}{q(z)}=\frac{1}{u(z)}\frac{du(z)}{dz} \tag{3.16}$$

Since

$$\frac{d}{dz}\left(\frac{1}{q}\right)=\frac{d}{dz}\left(\frac{1}{u}\frac{du}{dz}\right)=-\frac{1}{u^2}\left(\frac{du}{dz}\right)^2+\frac{1}{u}\frac{d^2u}{dz^2} \tag{3.17}$$

Equation (3.14) can be re-expressed as

$$\frac{d^2u(z)}{dz^2} + \beta^2 u(z) = 0 \tag{3.18}$$

3.1.2 Gaussian Beam Solutions in a Homogeneous Medium

As an application, we first consider the solution of Eqs. (3.18) and (3.15) in a homogeneous medium where the refractive index $n(r)$ has no spatial variation. This particular solution that we will obtain is, in fact, one of the most important results that we use in modeling laser beams. Since there is no quadratic variation in $n(r)$, we take $\beta = 0$, which simplifies Eq. (3.18) to

$$\frac{d^2u}{dz^2} = 0 \tag{3.19}$$

Equation (3.19) can be readily solved to yield

$$u(z) = c_1 z + c_2 \tag{3.20}$$

where c_1 and c_2 are constants. The corresponding q parameter $q(z)$ becomes

$$q(z) = z + \frac{c_2}{c_1} \tag{3.21}$$

As the initial condition, we take q equal to some constant q_0 at the location $z = z_f$, which gives

$$q_0 = z_f + \frac{c_2}{c_1} \tag{3.22}$$

resulting in

$$q(z) = (z - z_f) + q_0 \tag{3.23}$$

Going back to Eq. (3.15) for $P(z)$, we have

$$\frac{dP}{dz} = -\frac{i}{q} = \frac{-i}{(z - z_f) + q_0} \tag{3.24}$$

which can be integrated to give

$$P(z) = -i\ln((z - z_f) + q_0) + c_3 \tag{3.25}$$

Choosing the initial condition $P(z = z_f) = 0$, we have

$$c_3 = i\ln(q_0) \tag{3.26}$$

and $P(z)$ becomes

$$P(z) = -i\ln[(z - z_f) + q_0] + i\ln(q_0) = -i\ln\left[1 + \frac{z - z_f}{q_0}\right] \tag{3.27}$$

Putting the solution for $P(z)$ and $q(z)$ together, the transverse amplitude distribution $u(r,z)$, which was introduced in Eqs. (3.10) and (3.13), can be written in the form

$$u(r,z) = \exp\left[-i\left(-i\ln\left(1 + \frac{z - z_f}{q_0}\right) + \frac{k}{2(q_0 + z - z_f)}r^2\right)\right] \tag{3.28}$$

We need to choose the constant q_0 in such a way as to guarantee the transverse confinement of the beam. Otherwise, we do not have a physically meaningful solution. The easiest way to see how we should choose q_0 is by going to the location $z = z_f$ where we set our initial conditions. By using Eq. (3.28), we see that $u(r, z = z_f)$ becomes

$$u(r,z) = \exp\left(-\frac{ik}{2q_0}r^2\right) \tag{3.29}$$

To guarantee that the beam has transverse confinement, we choose $ik/2q_0 = 1/w_0^2$, where w_0 is commonly referred to as the "beam waist." Note that q_0 can now be written as

$$q_0 = q(z = z_f) = iz_0 \tag{3.30}$$

where z_0 is known as the Rayleigh range and can be expressed in terms of the vacuum wavelength λ, medium refractive index n_0, and the beam waist w_0 as

$$z_0 = \frac{kw_0^2}{2} = \frac{n_0 \pi w_0^2}{\lambda} \tag{3.31}$$

Having determined the functions $P(z)$ and $q(z)$, we are now in a position to write down the full expression for the electric field of a confined beam in a homogeneous medium. This part involves quite a bit of algebra which we leave as an exercise for the reader. By using the identity

$$\ln(a + ib) = \ln\left(\sqrt{a^2 + b^2}\, e^{i\tan^{-1}\left(\frac{b}{a}\right)}\right) = \ln\left(\sqrt{a^2 + b^2}\right) + i\tan^{-1}\left(\frac{b}{a}\right) \tag{3.32}$$

it can be shown that (see Prob. 3.3) the electric field can be expressed in the form

$$E_x(\vec{r},t) = E_0 \frac{\omega_0}{w(z)} \exp\left[i(\Phi(r,z) + \omega t) - \left(\frac{r}{w(z)}\right)^2\right] \tag{3.33}$$

where the phase factor $\Phi(r,z)$ and the spot-size function $w(z)$ are given respectively by

$$\Phi(r,z) = \tan^{-1}\left(\frac{z-z_f}{z_0}\right) - kz - \frac{kr^2}{2R(z)} \tag{3.34}$$

and

$$w(z) = w_0\sqrt{1+\left(\frac{z-z_f}{z_0}\right)^2} \tag{3.35}$$

The function $R(z)$ appearing in Eq. (3.34) is known as the radius of curvature function and has the functional form

$$R(z) = (z-z_f)\left[1+\left(\frac{z_0}{z-z_f}\right)^2\right] \tag{3.36}$$

3.1.3 Physical Interpretation of *w*(*z*): The Spot-Size Function

As can be seen from Eq. (3.33), the spot-size function $w(z)$ provides a measure of the confinement in the transverse plane of the Gaussian beam. In particular, it gives the radial position where the field decays to $(1/e)$ or 36.8% of its maximum value on the axis. The variation of the electric field as a function of the radial coordinate is shown in Fig. 3.2.

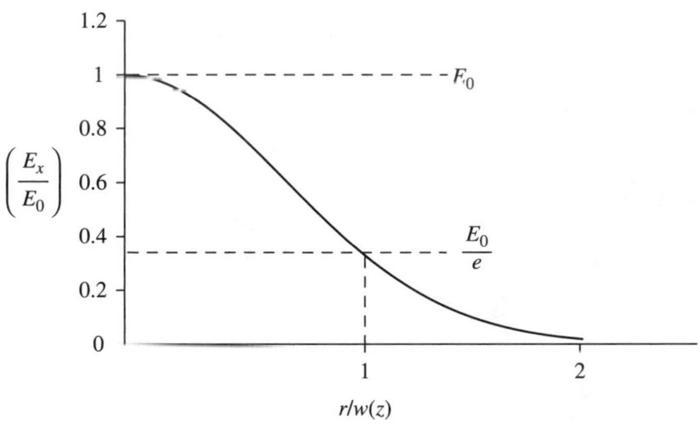

FIGURE 3.2 The variation of the electric field of a Gaussian beam as a function of the radial coordinate at a fixed position along the axis of propagation.

From Eq. (3.35), we further see that the spot-size function $w(z)$ is not constant and varies as a function of position along the axis of propagation. Figure 3.3 further gives a sketch of $w(z)$ as a function of z. Note that with the chosen boundary conditions, $w(z)$ assumes its minimum value of w_0 (hence the term beam waist) at the location $z = z_f$. The subscript f was intentionally chosen earlier to refer to this position as the "focus" location. As the distance from the focus location increases, the spotsize monotonically increases on either side due to diffraction. The Rayleigh range defined earlier in Eq. (3.31) gives a quantitative measure of the amount of diffractive beam spreading. In particular, we note, by using Eq. (3.35) that the spotsize increases to $\sqrt{2}\omega_0$ after one Rayleigh range. The width $2z_0$ gives the distance over which the beam remains nearly focused with intensity larger than half the maximum intensity, and is commonly referred to as the depth of focus or the confocal parameter of the beam. Looking at the expression for the Rayleigh range [Eq. (3.31)], we see that the smaller the beam waist is to begin with, the larger the amount of diffractive spreading, since z_0 is linearly proportional to w_0^2. In fact, when we look at the asymptotic form of Eq. (3.35), we see that the spot-size function grows linearly with distance according to

$$w(z) \approx w_0 \frac{z - z_f}{z_0} \tag{3.37}$$

and the divergence angle θ (see Fig. 3.3) in the paraxial regime is given by

$$\theta \approx 2\tan\frac{\theta}{2} = 2\frac{w(z)}{z - z_f} = 2\frac{w_0}{z_0} = 2\frac{w_0}{\frac{n_0 \pi w_0^2}{\lambda}} = \frac{2\lambda}{n_0 \pi w_0} \tag{3.38}$$

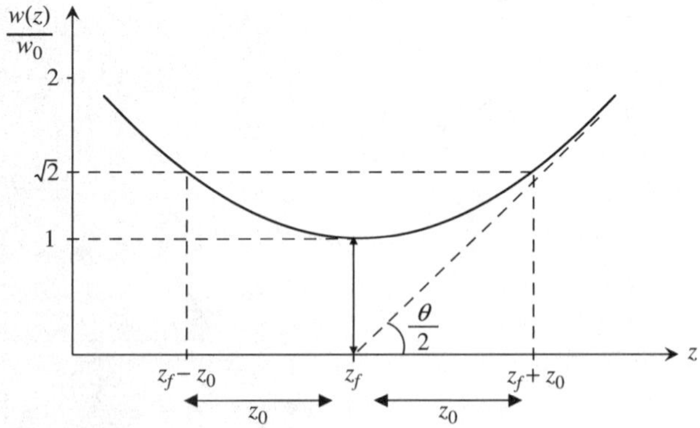

FIGURE 3.3 The variation of the spotsize of a Gaussian beam as a function of the propagation distance z around the focus location $z = z_f$. The Rayleigh range gives the distance from the focus where the spotsize increases to $\sqrt{2}w_0$.

The result given in Eq. (3.38) is in agreement with standard diffraction theory where the divergence angle of a wave confined to an aperture of size D is of the order of (λ/D).

Example Calculate the Rayleigh range and divergence angle of a He-Ne laser with a beam waist of 1 mm. Take the medium to be air and $\lambda = 6328\text{Å}$.

Solution By using Eqs. (3.31) and (3.38), we calculate z_0 and θ as

$$\text{Rayleigh range:}\quad z_0 = \frac{n_0 \pi w_0^2}{\lambda} = \pi\left(\frac{10^3}{0.6328}\right)(0.1)\ \text{cm} = 496\ \text{cm} \cong 5\ \text{m}$$

$$\text{Divergence angle:}\ \theta = \frac{2\lambda}{n_0 \pi w_0} = \frac{2(0.6328)}{\pi(10^3)} = 4\times 10^{-4}\ \text{rad} = 0.4\ \text{mrad}$$

If the beam waist were 10 mm (1 cm), the corresponding Rayleigh range would become approximately 500 m.

3.1.4 Physical Interpretation of $R(z)$: The Radius of Curvature Function

From Eq. (3.34), we see that the electric field has the radial phase term $\exp(-ikr^2/2R(z))$. In other words, the plane z = constant is no longer an equiphase surface. Therefore, the Gaussian beam is not really a plane wave, but instead, as we will show below, has a spherical equiphase surface and $R(z)$ corresponds to the radius of curvature of this equiphase surface at the location z. We will refer to $R(z)$ as the radius of curvature. Also note that $R(z)$ varies with distance of propagation. To see the interpretation of $R(z)$, let us examine the electric field variation of a spherical wave at large distances, near the axis of propagation. A schematic of the spherical wave is shown in Fig. 3.4.

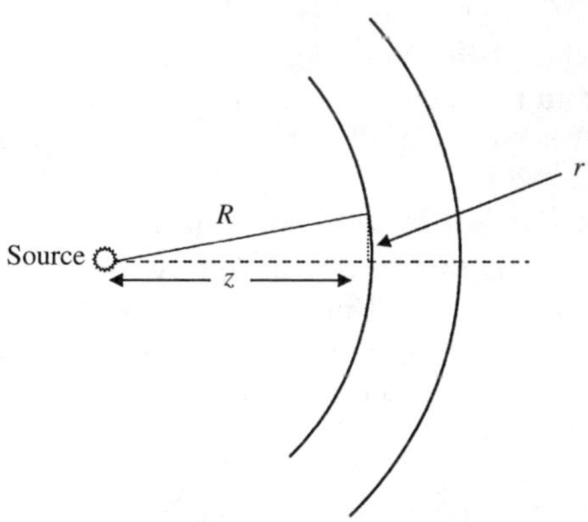

FIGURE 3.4 A schematic of a spherical wave near the axis of propagation.

In general, the electric field of an isotropic spherical wave varies as

$$E(r,z) = \frac{E_0 R_0}{R} \exp(-ikR) \tag{3.39}$$

where E_0 and R_0 are constants, and R is the radius of curvature of the wave given by

$$R = \sqrt{r^2 + z^2} \tag{3.40}$$

In the paraxial limit, where $r << z$ and $R \approx z$, we can approximate R as

$$R = z\sqrt{1+\frac{r^2}{z^2}} \approx z\left(1+\frac{r^2}{2z^2}\right) = z + \frac{1}{2}\frac{r^2}{z} \approx z + \frac{1}{2}\frac{r^2}{R} \tag{3.41}$$

This is a rather crude approximation but goes to show that, near the axis of propagation, the electric field of a spherical wave does indeed vary as

$$E(r,z) \approx \frac{E_0 R_0}{R} \exp(-ikz)\exp\left(-i\frac{kr^2}{2R}\right) \tag{3.42}$$

Equation (3.42) is similar in form to the electric field of a Gaussian beam and hence justifies the reason why we have identified $R(z)$ as the radius of curvature of the spherical equiphase surface. Note further that the radius of curvature for a Gaussian beam also varies with distance of propagation. At the location of the beam waist ($z = z_f$), $R(z = z_f) = \infty$, meaning that the beam has a planar wavefront. In such a case, note that the spot-size function also has a local minimum. We say, in general, that when the Gaussian beam has a planar wavefront, it is collimated. We can further assume that the beam will remain nearly collimated over a distance of $2z_0$ around the focus point. Figure 3.5 shows the variation of the radius of curvature $R(z)$ as a function of position around the beam waist location $z = z_f$. Note that at large distances away from the focus location, the radius again becomes very large in which case the phase fronts again become planar.

3.1.5 Power Carried by the Gaussian Beam

We can also calculate the power carried by the Gaussian beam in a medium where attenuation is negligible. Here, the phase factors will play no role. We leave it as an exercise for the reader (see Prob. 3.4) to show that the total power P crossing any plane is constant and can be expressed as

$$P = \frac{1}{2}\frac{E_0^2}{\eta_0} n_0 \frac{\pi w_0^2}{2} \tag{3.43}$$

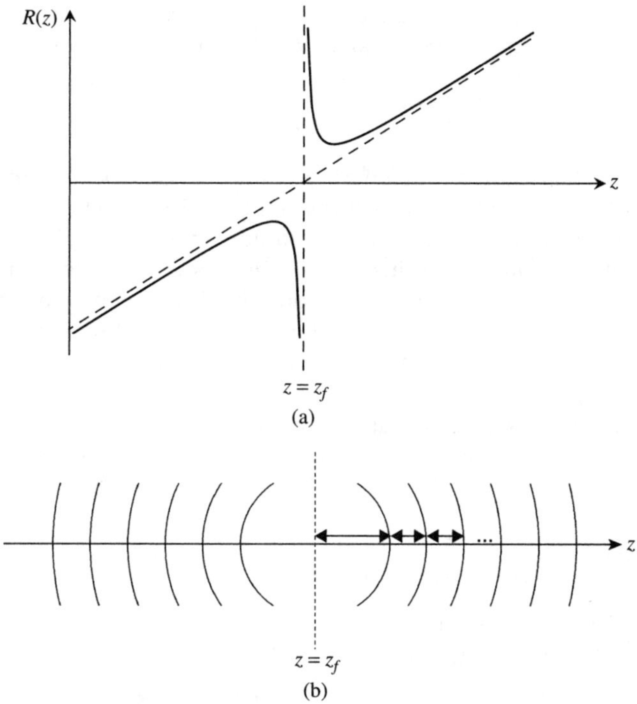

FIGURE 3.5 (*a*) Variation of the radius of curvature $R(z)$ as a function of position around the beam waist location $z = z_f$. (*b*) Constant-phase surfaces around the focus location.

Above, η_0 is the vacuum impedance. Furthermore, the Poynting vector $\vec{S}(r,z)$ for a Gaussian beam can be expressed as

$$\vec{S}(r,z) = \frac{1}{2\eta}\left|\vec{E}\right|^2 \hat{z} = \hat{z}\frac{1}{2}\frac{n_0}{\eta_0}E_0^2\frac{w_0^2}{w^2(z)}\exp\left(-\frac{2r^2}{w^2(z)}\right)$$
$$= \hat{z}\frac{2P}{\pi w^2(z)}\exp\left(-\frac{2r^2}{w^2(z)}\right) \qquad (3.44)$$

In a more realistic scenario, the field intensity decays exponentially due to various sources of attenuation such as absorption or scattering. These effects will be discussed in more detail in the next chapter.

3.1.6 Complex Beam Parameter $q(z)$

If we look, once again, at the expression for the electric field in Eqs. (3.33) and (3.34), we see that the exponential terms containing r^2 terms can be compactly written in the form $\exp(-ikr^2/2q(z))$, where the function $q(z)$, commonly referred to as the Gaussian q-parameter, is related to the radius of curvature and spot-size functions via

$1/2q(z) = 1/2R(z) + 1/ikw^2(z)$. Expressing k in terms of the vacuum wavelength, the q-parameter can be written in the more familiar form

$$\frac{1}{q(z)} = \frac{1}{R(z)} - i\frac{\lambda}{n_0 \pi w^2(z)} \tag{3.45}$$

In practical design problems involving Gaussian beams, Eq. (3.45) proves to be one of the most important equations. This is because knowledge of the complex beam parameter $q(z)$ contains the full information about the radius of curvature as well as the spotsize of the beam at a particular location along the direction of propagation. Also note that $q(z)$ has the dimensions of length. Let us illustrate the use of Eq. (3.45) with a simple numerical example.

Example Let us suppose that at a particular location in air, $q = (5 + i0.5)$ m. Let us assume that the wavelength is 1.5 μm. In order to determine R and w, we first need to express the complex beam parameter in the standard form given by Eq. (3.45). In this case, we have

$$\frac{1}{q} = \frac{1}{5+i0.5} = \frac{5-i0.5}{25+(0.5)^2} = 0.198 - i0.198\,(\text{m}^{-1})$$

Hence, $R = 0.198^{-1} = 5.05$ m.
Since $\lambda = 1.5\,\mu\text{m}$ and $n_0 = 1$, we further get

$$\frac{\lambda}{\pi w^2} = 0.0198$$

or

$$w = \sqrt{\frac{1.5\times 10^{-6}}{\pi(0.0198)}} = 0.00491\,\text{m} \cong 4.9\,\text{mm}$$

As we can see from the above example, full knowledge of $q(z)$ tells us about two of the most important parameters of the Gaussian beam. Now, the question is, can we determine the effect of an arbitrary optical system on the parameters of a Gaussian beam? In other words, suppose that we send a Gaussian beam through a series of lenses, displacements, or slabs of different refractive index, is it possible to calculate the effect on the q-parameter for the general case? It turns out that this can be done in an elegant way by using what is known as the ABCD law for Gaussian beams. In fact, we will need the results developed earlier in Chap. 2 for the ABCD matrices. Before we examine the general case, let us first of all see how the q-parameter is modified as a result of propagation in a quadratic medium. This will already give us some clues about the general form of the transformation rule.

3.1.7 Propagation of a Gaussian Beam in a Quadratic Medium

We start with Eq. (3.18) and assume that the quadratic index coefficient β is constant throughout the medium. We need to solve the second-order differential equation

$$\frac{d^2u(z)}{dz^2} + \beta^2 u(z) = 0 \tag{3.46}$$

The solution can be readily written in the form

$$u(z) = c_1 \cos \beta z + c_2 \sin \beta z \tag{3.47}$$

where c_1 and c_2 are constants. The derivative $u'(z)$ becomes

$$u'(z) = -\beta c_1 \sin \beta z + \beta c_2 \cos \beta z \tag{3.48}$$

Now, let's apply the initial conditions. Let us suppose that the initial values of $u(z)$ and $u'(z)$ are u_0 and u_0'. We have

$$\begin{aligned} u(0) &= c_1 \\ u'(0) &= \beta c_2 \end{aligned} \tag{3.49}$$

giving

$$\begin{aligned} u(z) &= u(0) \cos \beta z + \frac{u'(0)}{\beta} \sin \beta z \\ u'(z) &= -\beta u(0) \sin \beta z + u'(0) \cos \beta z \end{aligned} \tag{3.50}$$

By using Eq. (3.16), which relates $q(z)$ to $u(z)$, we obtain

$$\frac{u(z)}{u'(z)} - q(z) = \frac{u(0)\cos\beta z + \dfrac{u'(0)}{\beta}\sin\beta z}{-\beta u(0)\sin\beta z + u'(0)\cos\beta z} - \frac{q_0 \cos\beta z + \dfrac{1}{\beta}\sin\beta z}{-q_0\beta\sin\beta z + \cos\beta z} \tag{3.51}$$

Above, we used $q_0 = u(0)/u'(0)$ as the initial value of $q(z)$ at $z = 0$. Let us note that the result in Eq. (3.51) can be written in the form

$$q(z) = \frac{\mathrm{A} q_0 + \mathrm{B}}{\mathrm{C} q_0 + \mathrm{D}} \tag{3.52}$$

where A, B, C, and D are the standard elements of the ray transformation matrix for the quadratic lens given by

$$\mathbf{M}_\mathrm{T} = \begin{bmatrix} \cos\beta z & \dfrac{1}{\beta}\sin\beta z \\ -\beta\sin\beta z & \cos\beta z \end{bmatrix} \tag{3.53}$$

Our next task will be to justify that this transformation law indeed holds for any paraxial optical system that can be represented by a ray transformation matrix.

3.1.8 Transformation of a Gaussian Beam by a General Paraxial System: The ABCD Law

As we saw back in Sec. 2.5, a general paraxial optical system described by a ray transformation matrix $\mathbf{M}_T$ can always be represented by a thin lens, displacements, and an interface. So, we first examine how each one of these optical elements transforms the q-parameter of a Gaussian beam. First, let us investigate a thin lens, schematically shown in Fig. 3.6.

We assume that an optical beam represented by the electric field $\vec{E}_i(\vec{r})$ is incident on the lens. Let the transformed beam have the electric field $\vec{E}_T(\vec{r})$. We will have

$$\vec{E}_T(\vec{r}) = T_f \vec{E}_i(\vec{r}) \tag{3.54}$$

where T_f is the transmission factor for the lens. Assuming that the lens is lossless, we have $|T_f| = 1$. T_f will therefore consist of a radially varying phase factor. At the distance r from the optic axis, we can calculate T_f from

$$T_f = \exp\{-ik_0[d_1 + n_0(d_{10} - d_1) + d_2 + n_0(d_{20} - d_2)]\} \tag{3.55}$$

Above, n_0 is the refractive index of the lens material. For small values of r, the distance d_1 becomes

$$d_1 = R_1 - \sqrt{R_1^2 - r^2} = R_1 - R_1\left(1 - \frac{r^2}{R_1^2}\right)^{1/2} \cong \frac{r^2}{2R_1} \tag{3.56}$$

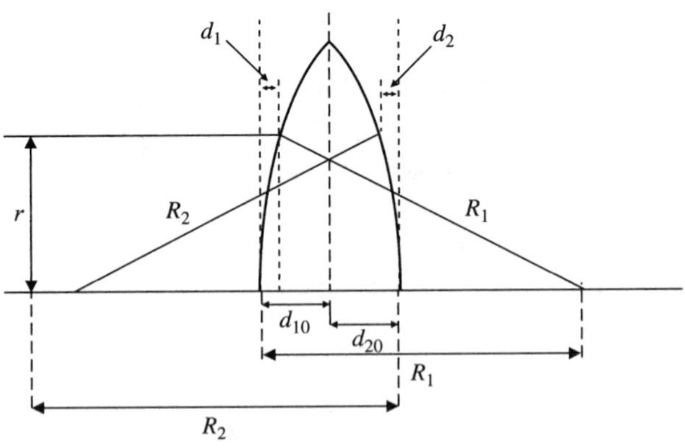

FIGURE 3.6 Schematic of a thin lens showing the radii of curvature and other dimensions needed to calculate the phase transmission factor for a Gaussian beam.

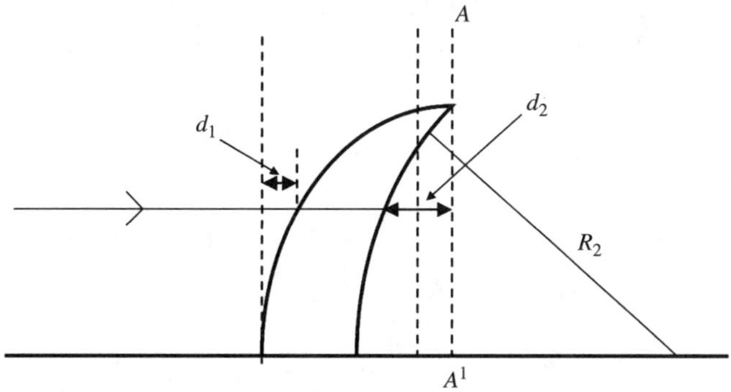

FIGURE 3.7 Section of the thin lens for the case where both radii of curvature are positive. Note that the width of the glass section is reduced in this case.

Similarly, we have

$$d_2 = -\frac{r^2}{2R_2} \tag{3.57}$$

Note that according to our lens convention from Chap. 2, the sketch shown in Fig. 3.6 has a negative radius of curvature for the second surface ($R_2 < 0$), giving additional glass material. When both R_1 and R_2 are positive, as shown in Fig. 3.7, the width of the glass section is reduced.

The expression for the transmission factor at small values of r becomes

$$\begin{aligned}
T_f &= \exp\left[-ik_0\left(\frac{r^2}{2R_1} + n_0 d_{10} - \frac{r^2}{2R_1}n_0 - \frac{r^2}{2R_2} + n_0 d_{20} + \frac{r^2}{2R_2}n_0\right)\right] \\
&= \exp[-ik_0 n_0(d_{10}+d_{20})]\exp\left[ik_0\left(\frac{1}{2}(n_0-1)\left(\frac{1}{R_1}-\frac{1}{R_2}\right)\right)r^2\right] \\
&= (\text{constant phase})\exp\left[i\frac{k_0 r^2}{2f}\right]
\end{aligned} \tag{3.58}$$

In Eq. (3.58), f is the effective focal length of the lens given by

$$\frac{1}{f} = (n_0 - 1)\left(\frac{1}{R_1} - \frac{1}{R_2}\right) \tag{3.59}$$

We have already obtained this relation back in Chap. 2 [see Sec. 2.3, Eq. (2.15)]. Hence, an incident Gaussian beam with a radial field distribution given by

$$\vec{E}_i(\vec{r}) = E_i(z)\exp\left[-\frac{ik_0 r^2}{2q(z)}\right] \tag{3.60}$$

will be transformed to the field (excluding the phase factors independent of r)

$$\vec{E}_T(\vec{r}) = \exp\left[-\frac{ik_0 r^2}{2}\left(\frac{1}{q(z)} - \frac{1}{f}\right)\right] \tag{3.61}$$

after passing through the thin lens. Therefore, we see that the q-parameter of the Gaussian beam is transformed by a thin lens according to

$$\frac{1}{q_2} = \frac{1}{q_1} - \frac{1}{f} \tag{3.62}$$

where q_1 is the incident q-parameter before the lens, q_2 is the transformed beam parameter after the lens, and f is the focal length. Equation (3.62) is one of the important results we will use in the general derivation of the beam transformation equation.

Now, we turn our attention to an interface between two dielectric media shown in Fig. 3.8. We consider a Gaussian beam, far from its focus point, incident on the interface. The refractive indices on two sides of the interface are n_1 and n_2. Let us approximately take the angle of incidence θ_1 in the ray picture as

$$\theta_1 = \frac{r}{R(z)} \tag{3.63}$$

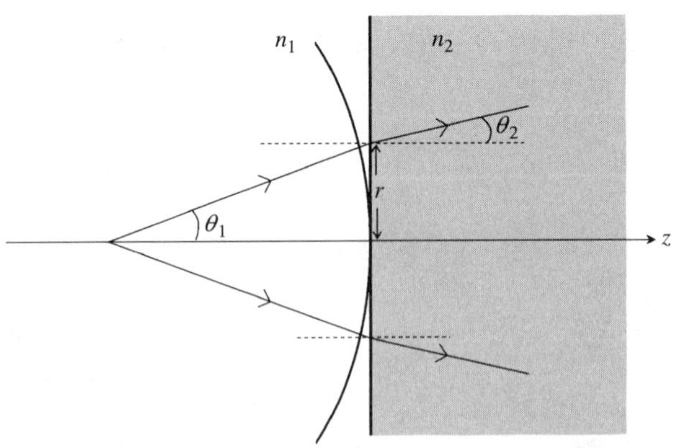

FIGURE 3.8 Sketch of a Gaussian beam incident at the interface between two different dielectric media.

where $R(z)$ is the radius of curvature. In the paraxial limit, the angles θ_1 and θ_2 will obey Snell's law given by

$$\theta_2 = \frac{n_1\theta_1}{n_2} \tag{3.64}$$

This gives

$$R_2 = R_1\frac{n_2}{n_1} \tag{3.65}$$

The q-parameters on the left and right sides of the interface are given respectively by

$$\frac{1}{q_1} = \frac{1}{R_1} - i\frac{\lambda}{n_1\pi w_1^2} \tag{3.66}$$

and

$$\frac{1}{q_2} = \frac{1}{R_2} - i\frac{\lambda}{n_2\pi w_2^2} \tag{3.67}$$

The spotsizes will not be changed, so we have $w_2 = w_1$. Hence, we get

$$\begin{aligned}\frac{1}{q_2} &= \frac{1}{R_2} - i\frac{\lambda}{n_2\pi w_2^2} = \frac{n_1}{n_2}\frac{1}{R_1} - i\frac{\lambda}{n_2\pi w_2^2}\\ &= \frac{n_1}{n_2}\left[\frac{1}{R_1} - i\frac{\lambda}{n_1\pi w_2^2}\right] = \frac{n_1}{n_2}\left[\frac{1}{R_1} - i\frac{\lambda}{n_1\pi w_1^2}\right]\end{aligned} \tag{3.68}$$

or

$$q_2 = q_1\frac{n_2}{n_1} \tag{3.69}$$

The third result we need is the effect of a simple displacement of d on the Gaussian beam (see Fig. 3.9). In fact, we have already analyzed

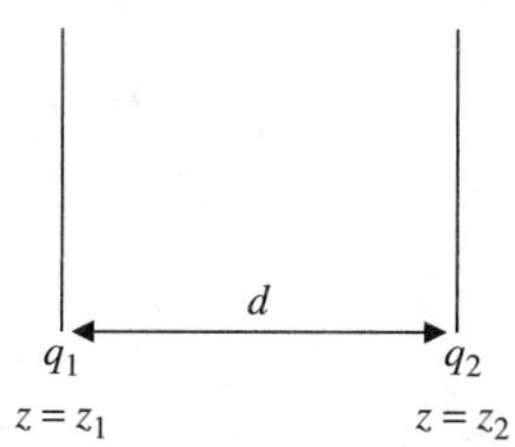

FIGURE 3.9 A Gaussian beam undergoing a displacement of d.

this case in detail in Sec. 3.1.2, Eq. (3.23). Let the optical system consist of a displacement of length d between the planes $z = z_1$ and $z = z_2 = z_1 + d$. At $z = z_1$ and $z = z_2$, we have

$$q_1(z = z_1) = q_0 + (z_1 - z_f) \tag{3.70}$$

and

$$q_2(z = z_2) = q_0 + (z_2 - z_f) \tag{3.71}$$

Subtracting Eq. (3.70) from Eq. (3.71), we get

$$q_2 = q_1 + (z_2 - z_1) = q_1 + d \tag{3.72}$$

Now, we can analyze a general paraxial system described by the ray transfer matrix $\mathbf{M}_T$ given by

$$\mathbf{M}_T = \begin{bmatrix} A & B \\ C & D \end{bmatrix} \tag{3.73}$$

As was discussed before, the equivalent optical system consists of two displacements, an interface and a thin lens, as shown in Fig. 3.10.

The equivalent optical system consists of the following optical elements: (displacement of δ_4)(dielectric interface from n_1 to n_2)(thin lens of focal length $f = -1/C$)(displacement of δ_5).

The displacements δ_4 and δ_5 are in turn given by

$$\delta_4 = \frac{D - \Delta}{C} \tag{3.74}$$

and

$$\delta_5 = \frac{A - 1}{C} \tag{3.75}$$

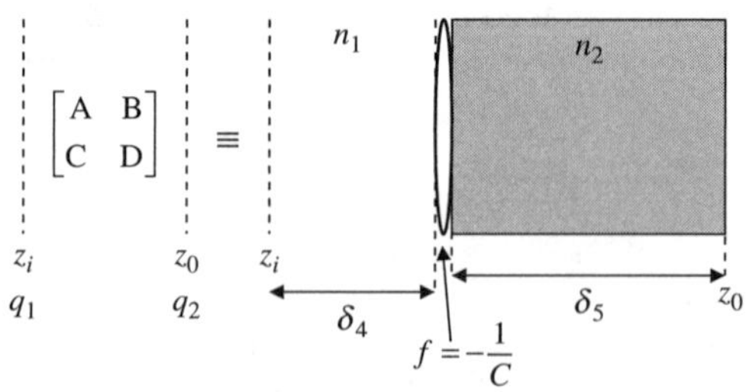

FIGURE 3.10 A general paraxial optical system and its equivalent representation in terms of two displacements, a thin lens and an interface.

Above, Δ is the determinant of the matrix given by $\Delta = AD - BC = n_1/n_2$. After the displacement of δ_4, the initial q-parameter is transformed to q_1', given by

$$q_1' = q_1 + \delta_4 = q_1 + \frac{D - \Delta}{C} \tag{3.76}$$

After the interface, the transformed q-parameter q_1'' becomes

$$q_1'' = \frac{n_2}{n_1} q_1' = \frac{1}{\Delta}\left(q_1 + \frac{D - \Delta}{C}\right) \tag{3.77}$$

Next, the thin lens of focal length $f = -1/C$ transforms q_1'' to q_1''' given by

$$\frac{1}{q_1'''} = \frac{1}{q_1''} - \frac{1}{f} = \frac{C\Delta}{Cq_1 + D - \Delta} + C = \frac{C\Delta + C^2 q_1 + CD - C\Delta}{Cq_1 + D - \Delta} = \frac{C^2 q_1 + CD}{Cq_1 + D - \Delta} \tag{3.78}$$

Finally, we go through one more displacement of δ_5 to get

$$\begin{aligned} q_2 = q_1''' + \delta_5 &= \frac{Cq_1 + D - \Delta}{C^2 q_1 + CD} + \frac{A - 1}{C} \\ &= \frac{C^2 q_1 + DC - \Delta C + AC^2 q_1 + ACD - C^2 q_1 - CD}{C(C^2 q_1 + CD)} \\ &= \frac{-\Delta + ACq_1 + AD}{C(Cq_1 + D)} = \frac{-AD + BC + ACq_1 + AD}{C(Cq_1 + D)} \\ &= \frac{Aq_1 + B}{Cq_1 + D} \end{aligned} \tag{3.79}$$

The result given by Eq. (3.79) is applicable to the beam analysis of all paraxial optical systems which can be represented by 2×2 ray transformation matrices. So, the ray transformation matrices we discussed back in Chap. 2 will be very useful in studying the transformation characteristics of Gaussian beams as well. Several illustrative examples will be given in the next section.

3.1.9 Examples on Gaussian Beam Transformation

The first example we will discuss is the effect of free-space displacement on the Gaussian beam parameter. This is a useful example that shows how we can actually use Eq. (3.79) to recover the expressions for the position dependence of the spot-size and radius of curvature functions given respectively by Eqs. (3.35) and (3.36).

Let us assume that the Gaussian beam is collimated at the location $z = z_f$. At this point, we take the radius of curvature to be $R = \infty$

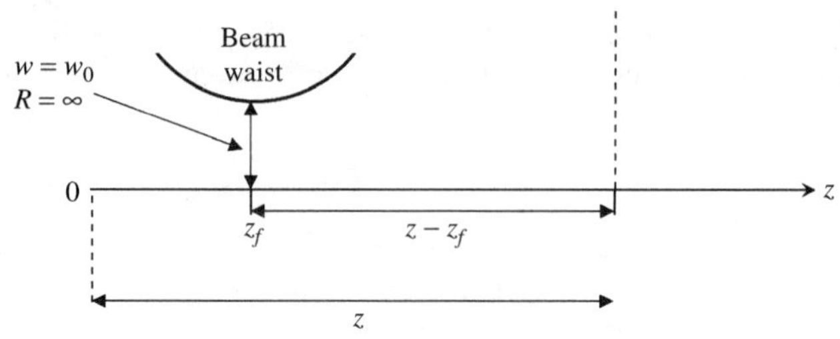

FIGURE 3.11 The transformation of a collimated Gaussian beam over a displacement of $(z - z_f)$.

and the spotsize as $w = w_0$. This is schematically shown in Fig. 3.11. The initial q-parameter q_1 satisfies

$$\frac{1}{q_1} = -i\frac{\lambda}{n_0 \pi w_0^2} \tag{3.80}$$

or, equivalently

$$q_1 = i\frac{n_0 \pi w_0^2}{\lambda} = iz_0 \tag{3.81}$$

To find $w(z)$ and $R(z)$, we need the ray transformation matrix $\mathbf{M}_\mathrm{T}$ for the displacement $(z - z_f)$, which is given by

$$\mathbf{M}_\mathrm{T} = \begin{bmatrix} 1 & (z - z_f) \\ 0 & 1 \end{bmatrix} \tag{3.82}$$

Hence, $q(z)$ will be given by

$$q(z) = q_2 = \frac{\mathrm{A}q_1 + \mathrm{B}}{\mathrm{C}q_1 + \mathrm{D}} = \frac{iz_0 + (z - z_f)}{0 + 1} = (z - z_f) + iz_0 \tag{3.83}$$

We now need to take the inverse of $q(z)$ and express it in the standard form given by Eq. (3.45). This yields

$$\frac{1}{q(z)} = \frac{1}{(z - z_f) + iz_0} = \frac{z - z_f - iz_0}{(z - z_f)^2 + z_0^2} = \frac{z - z_f}{(z - z_f)^2 + z_0^2}$$
$$-i\frac{z_0}{(z - z_f)^2 + z_0^2} = \frac{1}{R} - i\frac{\lambda}{n_0 \pi w^2(z)} \tag{3.84}$$

Comparing the real parts of $1/q(z)$, we see that $R(z)$ becomes

$$R = \frac{(z - z_f)^2 + z_0^2}{z - z_f} = (z - z_f)\left[1 + \left(\frac{z_0}{z - z_f}\right)^2\right] \tag{3.85}$$

In addition, from a comparison of the imaginary parts of $q(z)$, we obtain

$$w^2(z) = \frac{\lambda}{n_0 \pi z_0}[(z - z_f)^2 + z_0^2] = \frac{z_0 \lambda}{n_0 \pi}\left[1 + \left(\frac{z - z_f}{z_0}\right)^2\right] = w_0^2\left(1 + \left(\frac{z - z_f}{z_0}\right)^2\right) \tag{3.86}$$

The next example we consider provides a very useful result that can be used in determining the waist as well as the waist location of a Gaussian beam after a general paraxial optical system. Let us assume that a collimated Gaussian beam with an initial q-parameter $q_1 = iz_0$ is incident on an optical system represented by the ray transformation matrix given by Eq. (3.73). This is shown in Fig. 3.12. We assume that the medium immediately after the optical system has a refractive index of n.

We want to determine the distance d_f of the next beam waist from the output plane of the optical system, and the Rayleigh range z_2 at the focus position. This is a straightforward application of the transformation rule given by Eq. (3.79). Note that the overall transformation matrix from the input plane of the optical system to the location of the next beam waist can be written in the form

$$\mathbf{M}_{\mathrm{T}} = \begin{bmatrix} 1 & d_f \\ 0 & 1 \end{bmatrix} \begin{bmatrix} 1 & 0 \\ 0 & \dfrac{1}{n} \end{bmatrix} \begin{bmatrix} \mathrm{A} & \mathrm{B} \\ \mathrm{C} & \mathrm{D} \end{bmatrix}$$

$$= \begin{bmatrix} \mathrm{A} + \mathrm{C}\dfrac{d_f}{n} & \mathrm{B} + \mathrm{D}\dfrac{d_f}{n} \\ \dfrac{\mathrm{C}}{n} & \dfrac{\mathrm{D}}{n} \end{bmatrix} \tag{3.87}$$

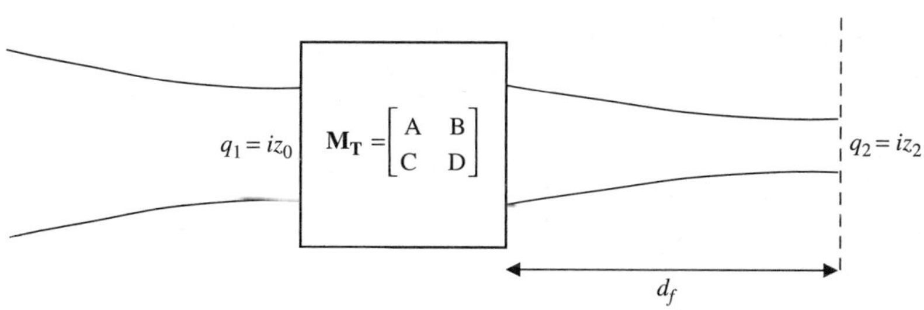

Figure 3.12 Transformation of a collimated Gaussian beam by a general paraxial optical system.

Here, $(1/q_2)$ will be given by

$$\frac{1}{q_2} = \frac{\frac{C}{n} iz_0 + \frac{D}{n}}{\left(B + D \frac{d_f}{n}\right) + \left(A + C \frac{d_f}{n}\right) iz_0} \tag{3.88}$$

To determine d_f, we require that

$$\mathrm{Re}\left[\frac{1}{q_2}\right] = 0 \tag{3.89}$$

since the beam will be collimated at this location with an infinite radius of curvature. We leave it as an exercise for the reader (see Prob. 3. 5) to show that d_f is given by the formula

$$d_f = -n \frac{DB + ACz_0^2}{D^2 + z_0^2 C^2} \tag{3.90}$$

Furthermore, the corresponding Rayleigh range z_2 becomes

$$z_2 = \frac{nz_0}{D^2 + z_0^2 C^2} \tag{3.91}$$

An important special case of Eqs. (3.90) and (3.91) is the focusing by a thin lens of focal length f, schematically shown in Fig. 3.13.

Here, the ray transformation matrix of the optical system is [see Eq. (2.10)]

$$\mathbf{M}_T = \begin{bmatrix} 1 & 0 \\ -1/f & 1 \end{bmatrix} \tag{3.92}$$

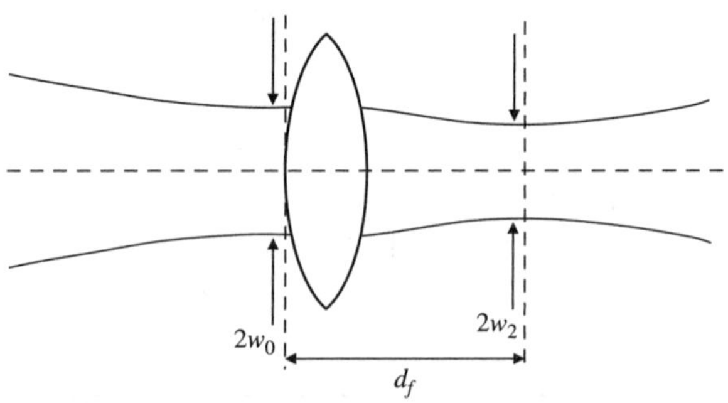

FIGURE 3.13 Focusing of a Gaussian beam by a thin lens of focal length f.

with $A = 1$, $B = 0$, $C = -1/f$, and $D = 1$. Hence, the distance of the focused beam from the thin lens will be given by

$$d_f = \frac{-n\left(0 + z_0^2(-1/f)\right)}{\left(1 + z_0^2(1/f)^2\right)} = nf \frac{\left(\dfrac{z_0}{f}\right)^2}{1 + \left(\dfrac{z_0}{f}\right)^2} \tag{3.93}$$

Note that d_f becomes approximately equal to f only when the initial Rayleigh range z_0 is much larger than the focal length f, for the case where $n = 1$. The ratio of the Rayleigh ranges becomes

$$\frac{z_2}{z_0} = n \frac{1}{1 + \left(\dfrac{z_0}{f}\right)^2} \tag{3.94}$$

The corresponding magnification which can be defined as the ratio (w_2/w_0) of the beam waists at the focus location and at the input of the lens, further becomes

$$\left(\frac{w_2}{w_0}\right) = \frac{1}{\sqrt{1 + \left(\dfrac{z_0}{f}\right)^2}} \tag{3.95}$$

Note that the magnification, in the limit of large input Rayleigh ranges becomes f/z_0.

3.1.10 Non-Ideal Gaussian Beams: The M^2 Factor

In real-life situations, a practical laser beam may deviate from being an ideal, diffraction-limited Gaussian beam. This is typically caused by the presence of higher-order modes, which we will discuss next. Under most circumstances, the transverse intensity distribution at each location still resembles that of a Gaussian beam, but the resulting Rayleigh range is typically smaller than that of an ideal beam. This deviation from the ideal case is quantified by introducing the M^2 parameter. For an ideal diffraction limited beam, the M^2 parameter is taken as unity. The larger the value of the M^2 parameter, the smaller the corresponding Rayleigh range, which gives rise to larger diffractive beam spreading. The Rayleigh range of a real beam is given in terms of the M^2 parameter by

$$z_0 = \frac{n_0 \pi w_0^2}{M^2 \lambda} \tag{3.96}$$

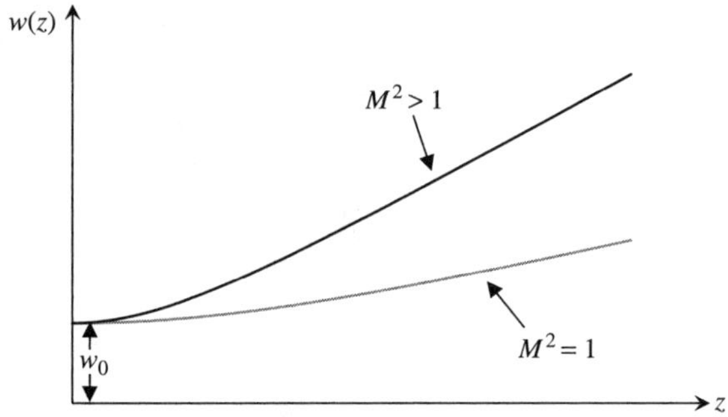

Figure 3.14 Variation of the spotsize for a diffraction-limited ($M^2 = 1$) and a non-ideal ($M^2 > 1$) beam as a function of position.

where $M^2 \geq 1$. Figure 3.14 provides a sketch of how the spotsize of a real beam with $M^2 \geq 1$ expands in relation to a diffraction limited Gaussian beam having $M^2 = 1$. For light from incoherent sources, M^2 may be significantly larger than 1.

3.2 Higher-Order Modes

So far, we have only concentrated on the cylindrically symmetric, TEM_{00} solution of the wave equation. In reality, higher-order modes which lack cylindrical symmetry are also encountered as, for example, in the case of resonators with a misaligned end mirror. To derive the higher-order mode solutions, we start with the scalar wave equation in a uniform medium:

$$\nabla^2 E_x - \mu_0 \varepsilon \frac{\partial^2 E_x}{\partial t^2} = 0 \tag{3.97}$$

Assuming $\exp(i\omega t)$ dependence gives

$$\nabla^2 E_x + \omega^2 \mu_0 \varepsilon E_x = 0 \tag{3.98}$$

Letting $k^2 = \omega^2 \mu_0 \varepsilon$, we get

$$\nabla^2 E_x + k^2 E_x = 0 \tag{3.99}$$

In what follows, we will assume that ε, and hence the wave vector magnitude k, are constants. Furthermore, we choose to work in Cartesian coordinates so that Eq. (3.99) becomes

$$\frac{\partial^2 E_x}{\partial x^2} + \frac{\partial^2 E_x}{\partial y^2} + \frac{\partial^2 E_x}{\partial z^2} + k^2 E_x = 0 \tag{3.100}$$

To solve for E_x, we assume a solution of the form

$$E_x = u(x,y,z)\exp(-ikz) \tag{3.101}$$

where the amplitude $u(x,y,z)$ describes the transverse field distribution. Note that the second derivative of E_x with respect to the propagation coordinate z becomes

$$\frac{\partial^2 E_x}{\partial z^2} = \left(\frac{\partial^2 u}{\partial z^2} - 2ik\frac{\partial u}{\partial z} - k^2 u\right)\exp(-ikz) \tag{3.102}$$

Under the slowly varying envelope approximation, we again neglect $\partial^2 u/\partial z^2$ in relation to $k(\partial u/\partial z)$ to obtain

$$\frac{\partial^2 u}{\partial x^2} + \frac{\partial^2 u}{\partial y^2} - 2ik\frac{\partial u}{\partial z} = 0 \tag{3.103}$$

To solve Eq. (3.103), we assume that $u(x,y,z)$ is given by the Ansatz:

$$u(x,y,z) = X\left(\frac{\sqrt{2}x}{w(z)}\right) Y\left(\frac{\sqrt{2}y}{w(z)}\right)\exp\left[-i\left(P(z) + \frac{k}{2q(z)}(x^2+y^2) + \Phi(z)\right)\right] \tag{3.104}$$

where, $w(z)$ is the spot-size function obtained earlier in Eq. (3.35), and $P(z)$ and $q(z)$ are the functions introduced in Eq. (3.13). To solve for the functions X, Y, and $\Phi(z)$, we will substitute $u(x,y,z)$ given by Eq. (3.104) into Eq. (3.103). Dividing each term by the common exponential factor, and then multiplying by $w(z)^2/2XY$, we obtain:

$$\frac{X''}{X} - i\sqrt{2}kx\left(\frac{w(z)}{q} - \frac{dw}{dz}\right)\frac{X'}{X} + \frac{Y''}{Y} - i\sqrt{2}ky\left(\frac{w(z)}{q} - \frac{dw}{dz}\right)\frac{Y'}{Y}$$
$$+\frac{k^2 w(z)^2}{2}(x^2+y^2)\left(-\frac{1}{q^2} + \frac{1}{q^2}\frac{dq}{dz}\right) - ikw(z)^2\left(\frac{1}{q} - i\frac{dP}{dz}\right) - kw(z)^2\frac{d\Phi}{dz} = 0 \tag{3.105}$$

Note that above, $X'(v)$ denotes dX/dv, in other words, the derivative of the function X with respect to the argument v. By using Eq. (3.35) derived earlier for $w(z)$, we find that

$$-i\sqrt{2}kx\left(\frac{w(z)}{q} - \frac{dw}{dz}\right) = -2\frac{\sqrt{2}x}{w(z)} \tag{3.106}$$

Furthermore, by using Eqs. (3.14) and (3.15) for $q(z)$ and $P(z)$, the differential equation involving X and Y simplifies to

$$\frac{X''}{X} - 2\left(\frac{\sqrt{2}x}{w(z)}\right)\frac{X'}{X} + \frac{Y''}{Y} - 2\left(\frac{\sqrt{2}y}{w(z)}\right)\frac{Y'}{Y} - kw(z)^2\Phi'(z) = 0 \tag{3.107}$$

We next introduce two separation constants ℓ and m according to

$$\frac{X''}{X} - 2\left(\frac{\sqrt{2}x}{w(z)}\right)\frac{X'}{X} = -2\ell \tag{3.108}$$

and

$$\frac{Y''}{Y} - 2\left(\frac{\sqrt{2}y}{w(z)}\right)\frac{Y'}{Y} = -2m \tag{3.109}$$

From Eq. (3.108), the differential equation satisfied by X can be recast into the form

$$X'' - 2\left(\frac{\sqrt{2}x}{w(z)}\right)X' + 2\ell X = 0 \tag{3.110}$$

which looks identical to the differential equation for a Hermite polynomial of order ℓ and for the argument $\sqrt{2}x/w(z)$. More on Hermite polynomials and their mathematical properties can be found in App. A. Therefore, X can be expressed as

$$X = H_\ell\left(\frac{\sqrt{2}x}{w(z)}\right) \tag{3.111}$$

A similar analysis can be used to show that Y becomes

$$Y = H_m\left(\frac{\sqrt{2}y}{w(z)}\right) \tag{3.112}$$

From Eq. (3.107), $\Phi(z)$ now satisfies

$$\Phi'(z) = \frac{-2(\ell + m)}{kw^2(z)} \tag{3.113}$$

By using Eq. (3.35) for $w(z)$ together with the boundary condition $\Phi(z = z_f) = 0$, we can readily solve for $\Phi(z)$ to get

$$\Phi(z) = -(\ell + m)\tan^{-1}\left(\frac{z - z_f}{z_0}\right) \tag{3.114}$$

Putting the solutions we obtained earlier for $P(z)$ and $q(z)$ into the solution for $u(x,y,z)$, we can write down the general solution for $E_x(x,y,z)$ that represents the $TEM_{\ell,m}$ mode in the form

$$E_x(x,y,z) = E_{\ell,m} H_\ell\left(\frac{\sqrt{2}x}{w(z)}\right) H_m\left(\frac{\sqrt{2}y}{w(z)}\right) \frac{w_0}{w(z)} \exp\left(-\frac{x^2+y^2}{w^2(z)}\right)$$
$$\times \exp\left[-i\left(kz - (1+\ell+m)\tan^{-1}\left(\frac{z-z_f}{z_0}\right)\right)\right] \exp\left[-\frac{ik(x^2+y^2)}{2R(z)}\right] \tag{3.115}$$

Above, $R(z)$ is the position-dependent radius of curvature function given by Eq. (3.36). At a practical level, the ℓth-order Hermite polynomial $H_\ell(v)$ can be calculated by using the well-known Rodrigue's formula

$$H_\ell(v) = (-1)^\ell \exp(v^2) \frac{d^\ell}{dv^\ell} \exp(-v^2) \tag{3.116}$$

The first three Hermite polynomials are given by $H_0(v) = 1$, $H_1(v) = v$, and $H_2(v) = 2v^2 - 1$, respectively. To gain insight into the visual characteristics of modes, we show, in Fig. 3.15, the three-dimensional intensity distributions proportional to $|E_x(x,y,z)|^2$ for the (*a*) $TEM_{0,0}$, (*b*) $TEM_{1,0}$, (*c*) $TEM_{0,1}$, (*d*) $TEM_{1,1}$, (*e*) $TEM_{0,2}$, and (*f*) $TEM_{1,3}$ modes. The corresponding cross-sectional intensity contour plots at a constant z

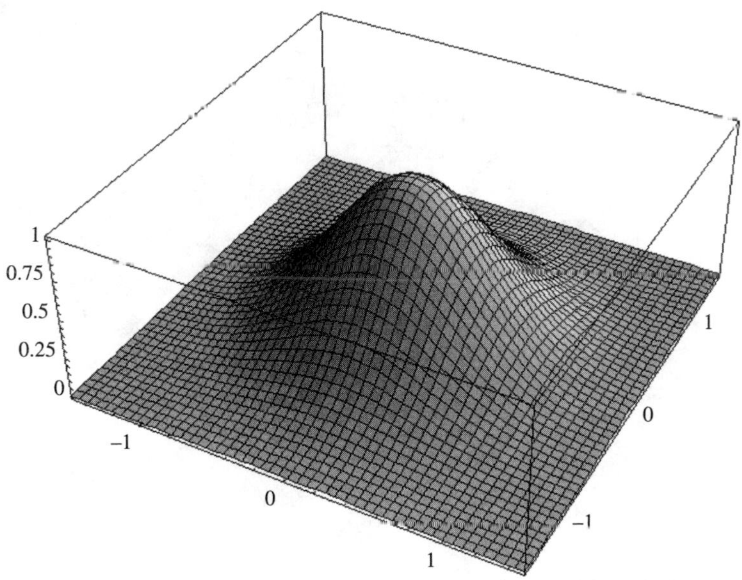

(a)

FIGURE 3.15 The three-dimensional intensity distributions for the (*a*) $TEM_{0,0}$, (*b*) $TEM_{1,0}$, (*c*) $TEM_{0,1}$, (*d*) $TEM_{1,1}$, (*e*) $TEM_{0,2}$, and (*f*) $TEM_{1,3}$ modes.

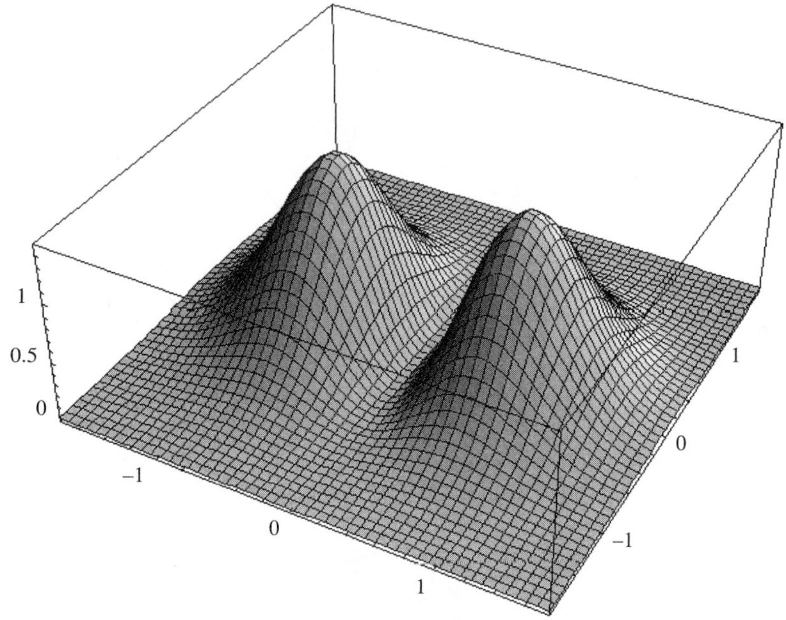

(b)

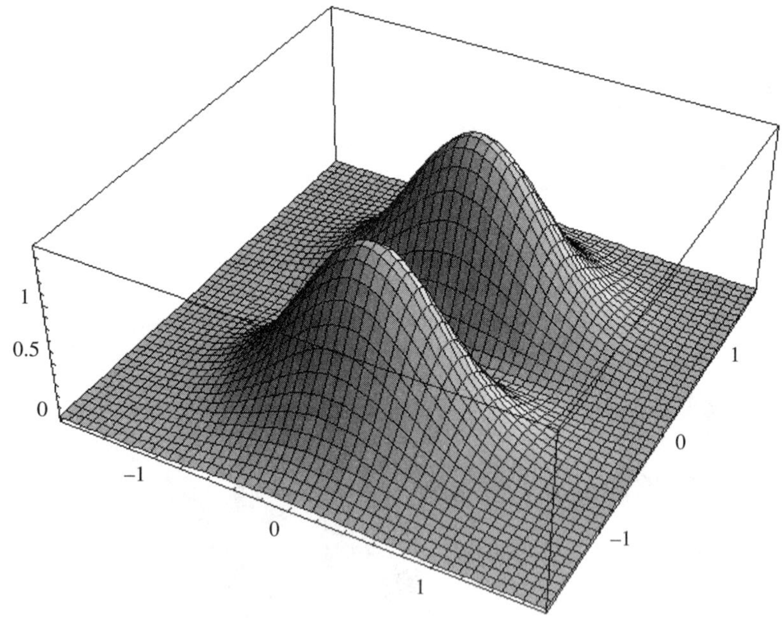

(c)

Figure 3.15 *(Continued)*

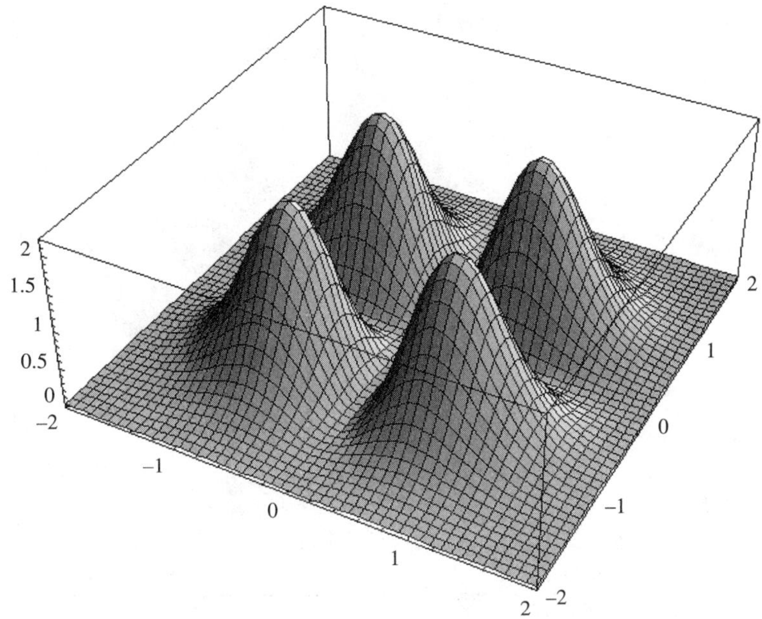

(d)

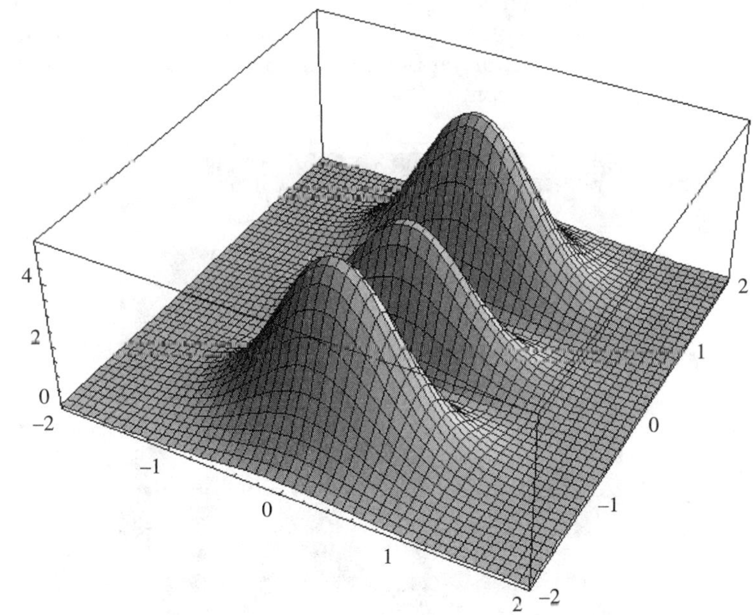

(e)

FIGURE 3.15 (*Continued*)

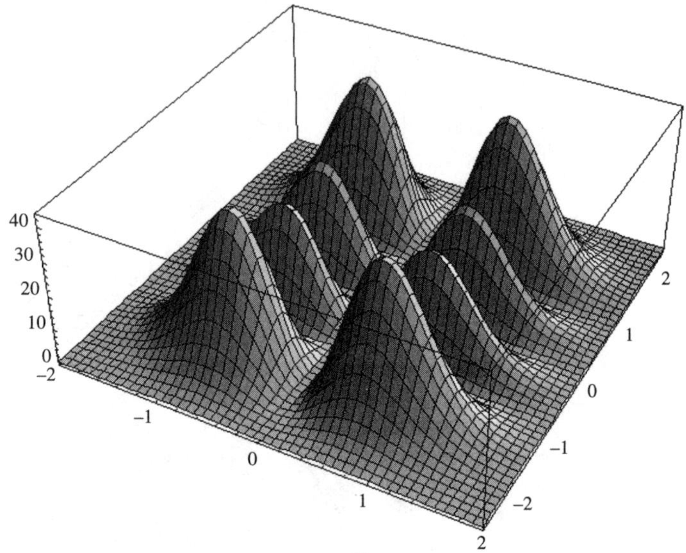

(f)

FIGURE 3.15 *(Continued)*

plane are further shown in Fig. 3.16. As a general rule, the reader can also confirm from these plots that the $\text{TEM}_{\ell,m}$ mode has ℓ intermediate nodes along the x direction (at finite values of x) and m intermediate nodes along the y direction.

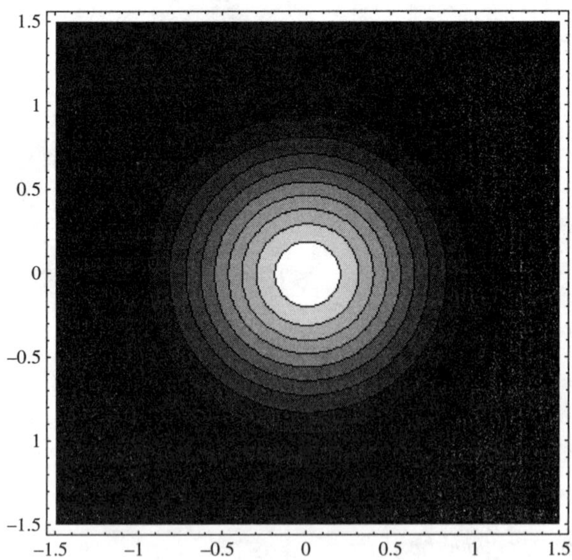

(a)

FIGURE 3.16 Intensity contour plots for the (*a*) $\text{TEM}_{0,0}$, (*b*) $\text{TEM}_{1,0}$, (*c*) $\text{TEM}_{0,1}$, (*d*) $\text{TEM}_{1,1}$, (*e*) $\text{TEM}_{0,2}$, and (*f*) $\text{TEM}_{1,3}$ modes.

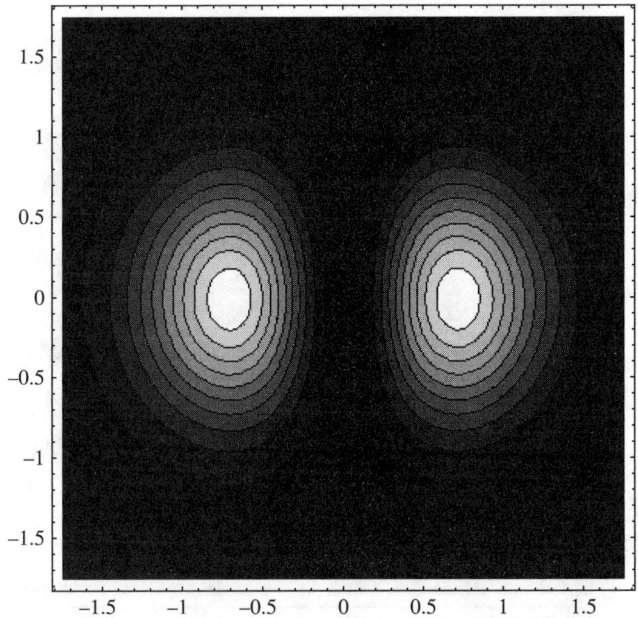

(b)

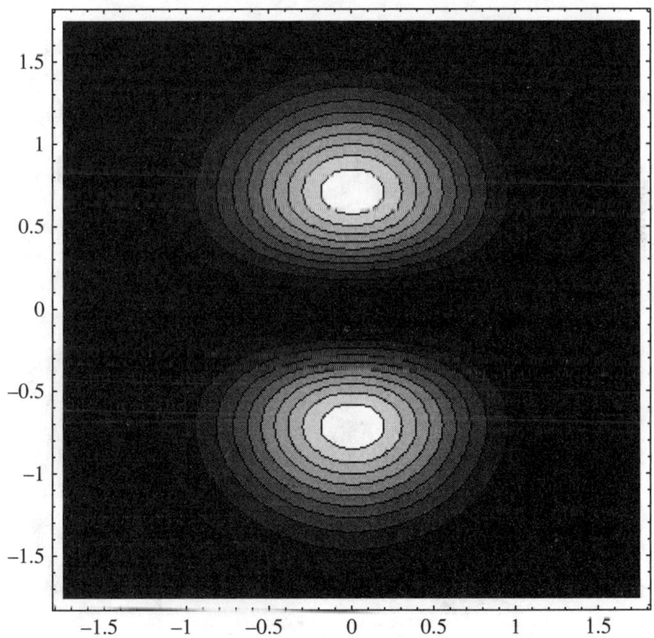

(c)

FIGURE 3.16 *(Continued)*

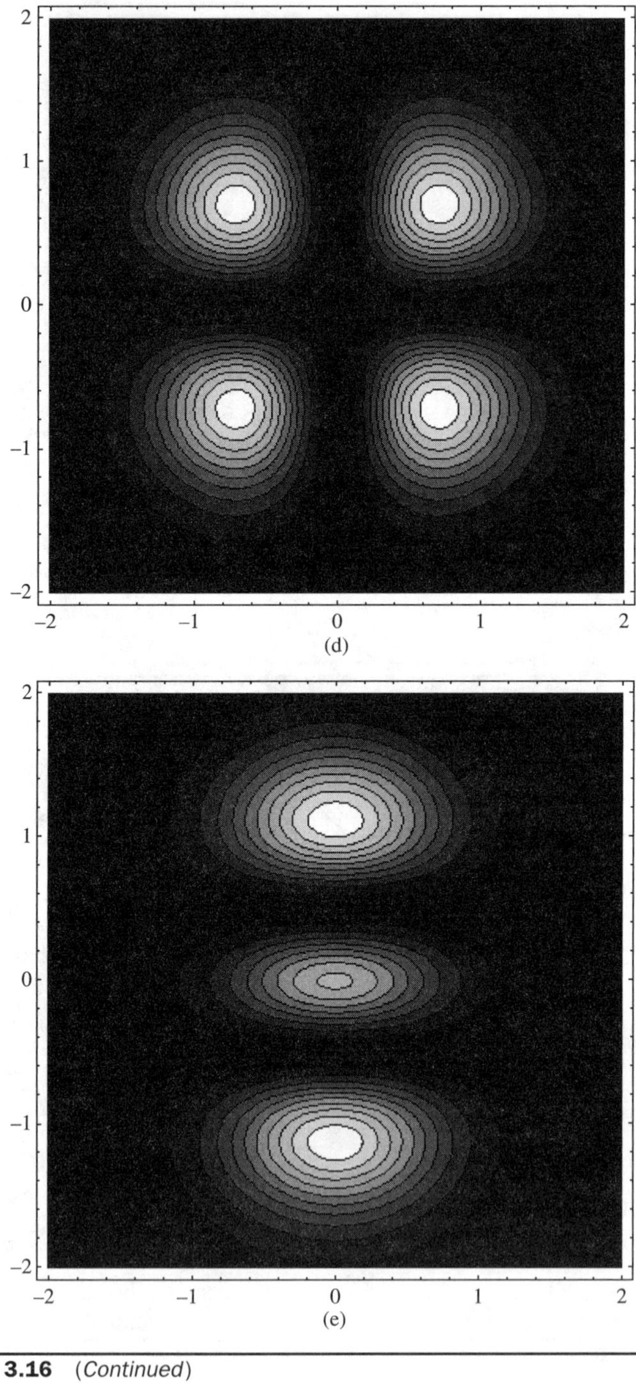

FIGURE 3.16 (*Continued*)

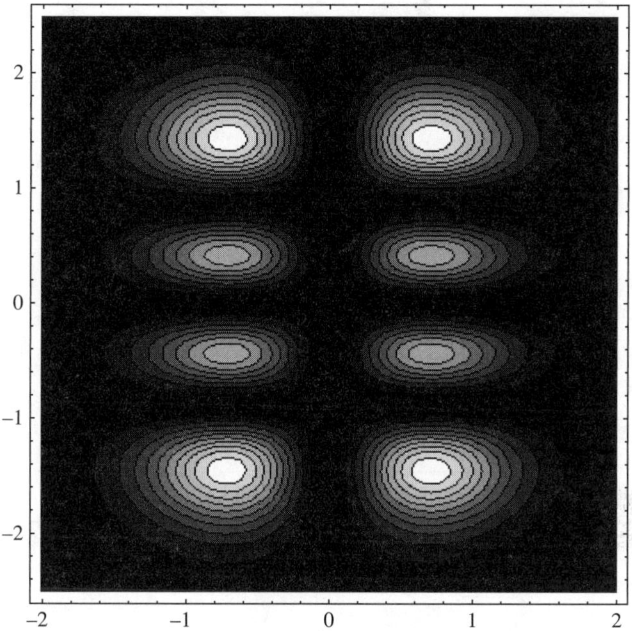

(f)

Figure 3.16 (*Continued*)

3.3 Optical Resonators

3.3.1 Two-Mirror Optical Resonators

In a laser, optical amplification is combined with positive feedback to obtain light oscillation. As we discussed back in Chap. 2, optical resonators consisting of highly reflecting end mirrors are used to provide positive feedback. The matrix formalism developed in Chap. 2 provides a simple method of determining whether or not an optical resonator is stable. In other words, the stability analysis tells us whether or not a ray, starting near the optical axis of a resonator, remains confined after an indefinite number of round trips. Now, let us suppose that the resonator is indeed stable. The next important task is the calculation of the parameters of the light beam which is supported by the resonator. To be specific, we will first focus on the simple two-mirror resonator depicted in Fig. 3.17. Here, the end mirrors M_1 and M_2 are separated by a distance d and have the respective radii of curvature of R_1 and R_2. It turns out that in the stable region of the resonator, d, R_1, and R_2 uniquely determine the parameters of the beam, in other words, its spotsize and radius of curvature at each location within the resonator. Because the Gaussian beam has a spherical

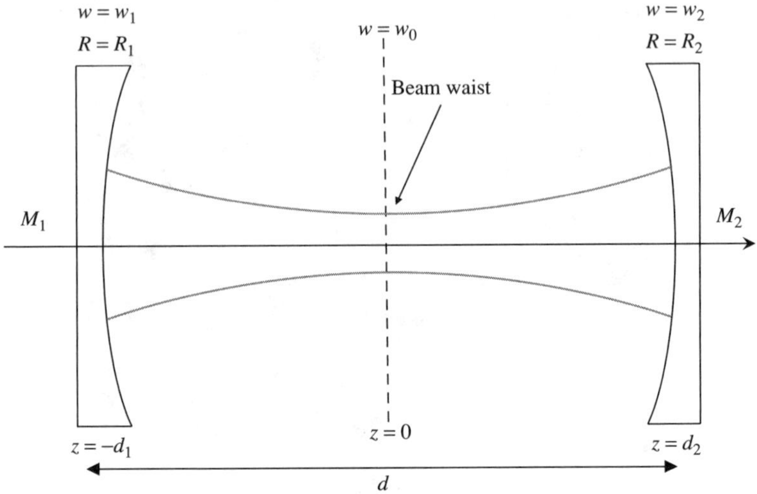

FIGURE 3.17 Two-mirror resonator. The beam waist is located at $z = 0$.

equiphase surface and can be matched to the boundary conditions on spherical surfaces, the electromagnetic mode supported by the resonator consisting of spherical mirrors will be a Gaussian beam. So, all the characteristics of the Gaussian beams we learned in the previous sections can be used in the beam analysis of the optical resonator. Here, we will be mainly concerned with stable optical resonators. For a thorough treatment of unstable optical resonators, the reader is referred to Refs. 1, 10, and 11.

In the standard analysis of the two-mirror resonator, we assume that the beam waist of the Gaussian beam is located at $z = 0$. This places the mirrors M_1 and M_2 at the locations $z = -d_1$ and $z = d_2$, respectively. The separation d between the mirrors will be given by

$$d = d_2 - (-d_1) = d_1 + d_2 \tag{3.117}$$

Let us assume that the beam waist at $z = 0$ is w_0. To meet the boundary conditions, we require that the radii of curvature of the beam should be $-R_1$ and $+R_2$ at $z = -d_1$ and $z = d_2$, respectively. The radius of curvature function for the case where $z_f = 0$ will be given [using Eq. (3.35)] by

$$R(z) = (z - z_f)\left[1 + \left(\frac{z_0}{z - z_f}\right)^2\right] = z\left(1 + \left(\frac{z_0}{z}\right)^2\right) \tag{3.118}$$

where z_0 is the Rayleigh range of the self-consistent solution of the resonator mode. For the mirror M_1, we will therefore have

$$-R_1 = -d_1 - \frac{z_0^2}{d_1} \tag{3.119}$$

Similarly, the boundary condition on mirror M_2 gives

$$R_2 = d_2 + \frac{z_0^2}{d_2} \tag{3.120}$$

We have three equations for the three unknown parameters d_1, d_2, and z_0. From Eqs. (3.119) and (3.120), we get two equivalent expressions for z_0:

$$z_0^2 = d_2(R_2 - d_2) = d_1(R_1 - d_1) = (d - d_2)(R_1 - d + d_2) \tag{3.121}$$

Solving for d_2, we get

$$d_2 = \frac{d(d - R_1)}{2(d - R_2)} = \frac{d(R_1 - d)}{R_1 + R_2 - 2d} \tag{3.122}$$

By following a similar approach, we can determine d_1 and z_0 as well. Often, the solution for d_1, d_2, and z_0 are expressed in terms of the resonator g-parameters which are defined according to

$$g_1 \equiv 1 - \frac{d}{R_1} \tag{3.123}$$

and

$$g_2 \equiv 1 - \frac{d}{R_2} \tag{3.124}$$

Hence, for example, R_1 can be expressed as $R_1 = d/(1 - g_1)$. We leave the rest of this analysis as an exercise (see Prob. 3.6) for the reader to show that the solution for d_1, d_2, and z_0, in terms of d, g_1, and g_2 becomes

$$d_1 = d\frac{g_2(1 - g_1)}{g_1 + g_2 - 2g_1g_2} \tag{3.125}$$

$$d_2 = d\frac{g_1(1 - g_2)}{g_1 + g_2 - 2g_1g_2} \tag{3.126}$$

and

$$z_0 = d\sqrt{\frac{g_1g_2(1 - g_1g_2)}{(g_1 + g_2 - 2g_1g_2)^2}} \tag{3.127}$$

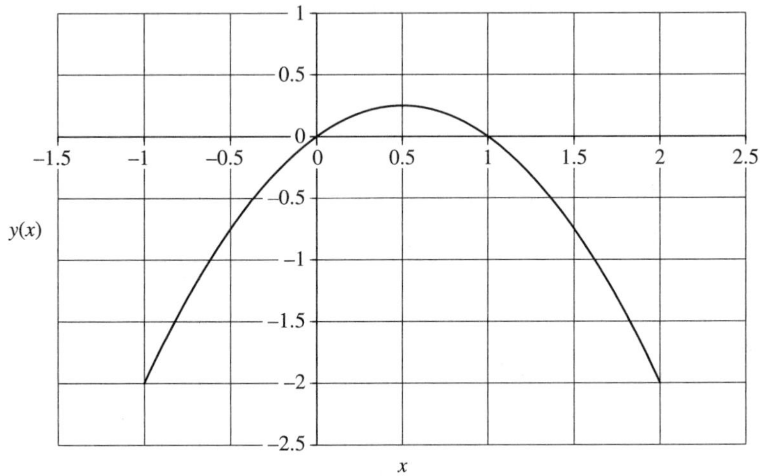

FIGURE 3.18 The plot of the function $y = x(1 - x)$.

Since $z_0 = \pi w_0^2/\lambda$, the beam waist of the resonator mode (at $z = 0$) becomes

$$w_0 = \text{beam waist} = \sqrt{\frac{\lambda d}{\pi}\sqrt{\frac{g_1 g_2(1-g_1 g_2)}{(g_1+g_2-2g_1 g_2)^2}}} \tag{3.128}$$

Note from Eq. (3.127) that the stability condition previously derived for a two-mirror resonator is already built in. In particular, we require z_0 to be a real number in order to have a confined beam. This means that the condition $g_1 g_2(1-g_1 g_2) \geq 0$ must be satisfied. Examining the graph of the function $y = x(1-x)$ (see Fig. 3.18), we see that for $y \geq 0$, x must satisfy $0 \leq x \leq 1$. In other words, for the two-mirror cavity, we must have $0 \leq g_1 g_2 \leq 1$ for a confined Gaussian beam solution.

As a specific example, we analyze the flat-curved resonator shown in Fig. 3.19. In this case, since $R_1 = \infty$, we have $g_1 = 1$. We further take $g_2 = g$. First of all, we note that the beam waist is now located on mirror M_1. The expression for w_0 [Eq. (3.128)] further simplifies to

$$w_0 = \sqrt{\frac{\lambda d}{\pi}\sqrt{\frac{g(1-g)}{(1+g-2g)^2}}} = \sqrt{\frac{\lambda d}{\pi}\sqrt{\frac{g}{1-g}}} \tag{3.129}$$

As an example, Fig. 3.20 shows the calculated variation of the beam waist w_0 as a function of the mirror separation d for the case where

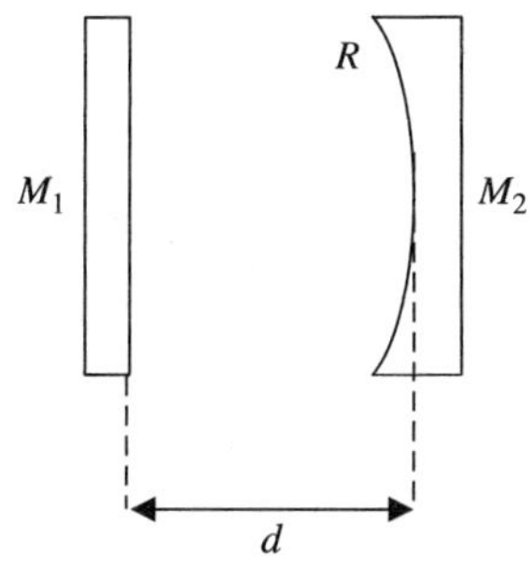

FIGURE 3.19 Flat-curved optical resonator.

$R = 0.5$ m. Two plots are shown for (a) $\lambda = 1.25$ µm and (b) $\lambda = 2$ µm. Here are a few important characteristics of the two-mirror flat-curved resonator. First, note that the beam waist tends to zero at both ends of the stability region ($d = 0$ and $d = R$). Furthermore, the maximum beam waist at the flat mirror occurs when $d = R/2$. The beam waist increases with wavelength but the dependence ($\sqrt{\lambda}$) is not very strong. For example, increasing the wavelength from 1.25 to 2 µm changes the maximum value of w_0 (at $d = R/2$) from 282 to 399 µm.

We can also investigate the dependence of the maximum beam waist $(w_0)_{max}$ (which occurs at $d = R/2$) as a function of the radius of curvature. This is shown in Fig. 3.21 for the range where R varies from 1 cm to 10 cm. The wavelength is taken as 1.25 µm. The corresponding maximum beam waist changes from 45 to 141 µm. We leave the generation of these plots as an exercise for the reader (Prob. 3.8).

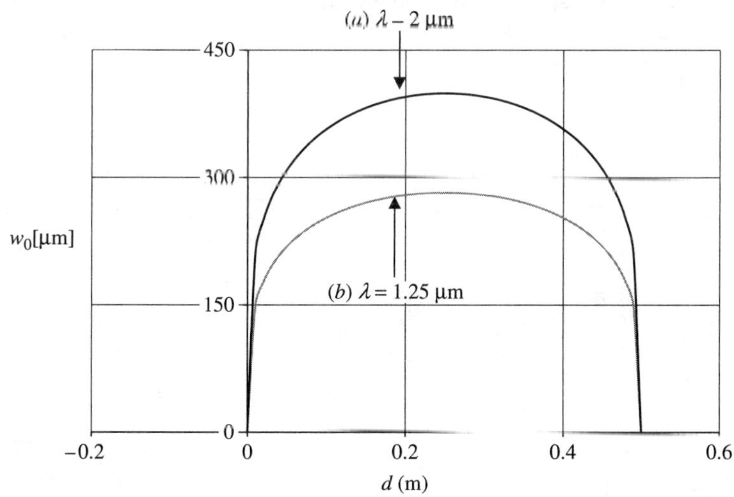

FIGURE 3.20 Calculated variation of the beam waist w_0 on the flat mirror as a function of the mirror separation d for the flat-curved optical resonator. $R = 0.5$ m. (a) $\lambda = 1.25$ µm and (b) $\lambda = 2$ µm.

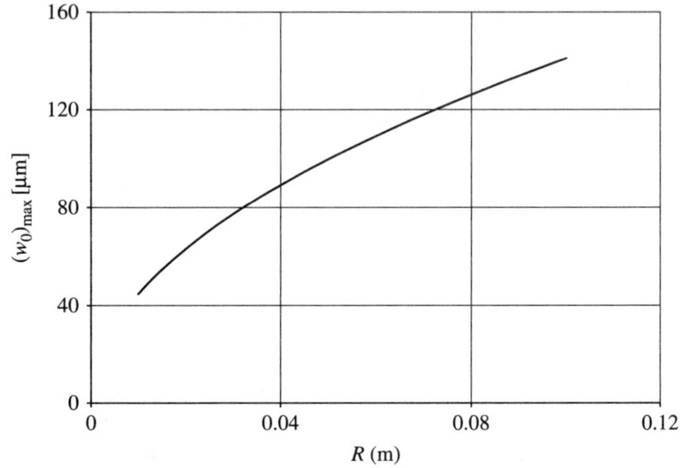

FIGURE 3.21 Calculated variation of the maximum beam waist $(w_0)_{max}$ (which occurs at $d = R/2$) as a function of the radius of curvature R for the wavelength of 1.25 μm.

3.3.2 Gaussian Beam Analysis of a General Paraxial Resonator

Having gained some insight about the beam analysis of two-mirror resonators, we now turn our attention to a general paraxial resonator sketched in Fig. 3.22. For the general case, in addition to the end mirrors M_1 and M_2, we may have other optical elements inside the cavity. Let us focus on a specific position within the resonator coinciding with the plane AA' (see Fig. 3.22). We assume that the ray transformation

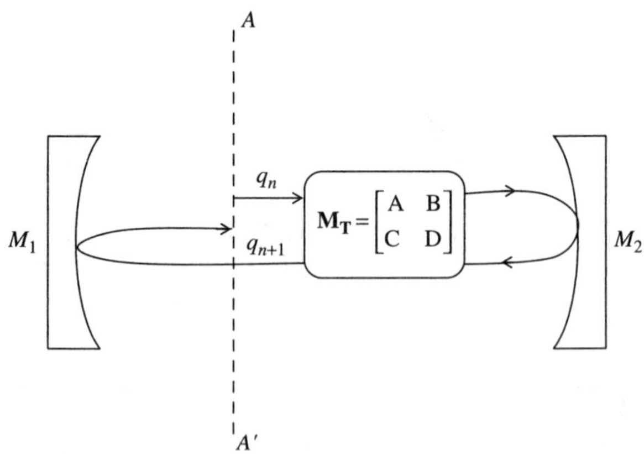

FIGURE 3.22 A general paraxial resonator, where q_n and q_{n+1} represent the q-parameters of the self-consistent Gaussian beam at the plane AA' after the nth and $(n + 1)$th round trips.

matrix starting at the plane AA' and returning to the same plane after one round trip is $\mathbf{M}_T$, given by

$$\mathbf{M}_T = \begin{bmatrix} A & B \\ C & D \end{bmatrix} \tag{3.130}$$

Clearly, the elements of $\mathbf{M}_T$ depend on where we start in the cavity. However, two quantities will always remain invariant. The first one is the determinant of $\mathbf{M}_T$. Because we start and end at the same location with the same refractive index, it is always the case that for a round-trip ray transformation matrix, the elements satisfy

$$AD - BC = 1 \tag{3.131}$$

Second, the trace (A + D) remains invariant because, after all, the value of (A + D)/2 tells us whether or not the resonator is stable, and this does not depend on where we start in the cavity. Let us suppose that q_n and q_{n+1} represent the q-parameters of the self-consistent Gaussian beam at the plane AA' after the nth and $(n + 1)$th round trips, respectively. For the self-consistent solution, we require that $q_n = q_{n+1}$. We know from the transformation characteristics of Gaussian beams that q_n and q_{n+1} are further related through

$$q_{n+1} = \frac{Aq_n + B}{Cq_n + D} \tag{3.132}$$

Hence, the q-parameter of the self-consistent beam solution at a particular location should satisfy the condition (irrespective of which round trip we are talking about)

$$q = \frac{Aq + B}{Cq + D} \tag{3.133}$$

The above condition gives the following quadratic relation for q:

$$Cq^2 + Dq - Aq - B = 0 \tag{3.134}$$

To put this in the standard form we obtained earlier, we divide out by q^2 to obtain

$$B\left(\frac{1}{q}\right)^2 + (A - D)\left(\frac{1}{q}\right) - C = 0 \tag{3.135}$$

Solving for $(1/q)$ and making use of Eq. (3.131), we obtain

$$\frac{1}{q} = \frac{D - A}{2B} \pm i\frac{\sqrt{1 - \left(\frac{A + D}{2}\right)^2}}{B} \tag{3.136}$$

Only the negative imaginary part gives a confined Gaussian beam as can be seen from the standard form for $(1/q)$:

$$\frac{1}{q} = \frac{1}{R} - i\frac{\lambda}{n_0 \pi w^2} \tag{3.137}$$

Comparing Eqs. (3.136) and (3.137), we therefore see that the expressions for the radius of curvature R and the spotsize w at the plane AA' are given by

$$\frac{1}{R} = \frac{\mathrm{D} - \mathrm{A}}{2\mathrm{B}} \tag{3.138}$$

and

$$\frac{\lambda}{n_0 \pi w^2} = \frac{\sqrt{1 - \left(\frac{\mathrm{A}+\mathrm{D}}{2}\right)^2}}{2|B|} \tag{3.139}$$

From Eq. (3.138), we see that whenever A = D, the beam will be collimated. This occurs, for example, on the flat end mirrors of a cavity.

3.3.3 ABCD Analysis of Four-Mirror Resonators

As an application, we will analyze a four-mirror resonator, depicted in Fig. 3.23, which is often used to generate laser output with solid-state gain media. Typically, a four-mirror laser resonator in z-configuration (as shown in Fig. 3.23) consists of two flat end mirrors (M_1 and M_4), two intracavity curved mirrors (M_2 and M_3), and a gain medium (typically a solid-state crystal) positioned between the curved mirrors. We will assume that the folding angles of the curved mirrors are small, so that astigmatic effects are negligible (see Prob. 3.18 for astigmatic compensation). Also, we assume that both of the curved mirrors have the same radius of curvature R.

In the stability region, the self-consistent Gaussian beam of the four-mirror resonator is collimated at three locations within the

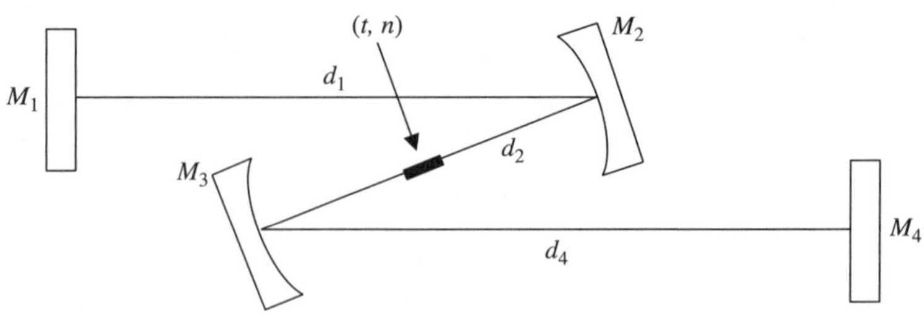

FIGURE 3.23 A four-mirror resonator in z-configuration.

resonator: on each of the flat end mirrors and in between the curved mirrors. Our task will be to determine the beam waist and its location between the curved mirrors when the resonator configuration is stable. We will outline the procedure for doing these calculations and leave the detailed analysis as an exercise for the reader (see Prob. 3.9).

The first step is the calculation of the round trip ray transfer matrix $\mathbf{M}_T$. As we discussed before, $\mathbf{M}_T$ depends on the location of the input plane AA' in the cavity (see Fig. 3.22). Let us suppose that we start on the flat end mirror M_1. We know that the Gaussian beam will be collimated at this location so all we need to determine is the Rayleigh range z_0 of the spot on M_1. $\mathbf{M}_T$ will consist of the product of the following ray transformation matrices:

$$\mathbf{M}_T = \begin{pmatrix} A & B \\ C & D \end{pmatrix} = \begin{pmatrix} 1 & d_1 \\ 0 & 1 \end{pmatrix} \begin{pmatrix} 1 & 0 \\ -\dfrac{2}{R} & 1 \end{pmatrix} \begin{pmatrix} 1 & d_2 \\ 0 & 1 \end{pmatrix} \begin{pmatrix} 1 & t/n \\ 0 & 1 \end{pmatrix} \begin{pmatrix} 1 & d_3 \\ 0 & 1 \end{pmatrix} \begin{pmatrix} 1 & 0 \\ -\dfrac{2}{R} & 1 \end{pmatrix} \begin{pmatrix} 1 & 2d_4 \\ 0 & 1 \end{pmatrix}$$
$$\times \begin{pmatrix} 1 & 0 \\ -\dfrac{2}{R} & 1 \end{pmatrix} \begin{pmatrix} 1 & d_3 \\ 0 & 1 \end{pmatrix} \begin{pmatrix} 1 & t/n \\ 0 & 1 \end{pmatrix} \begin{pmatrix} 1 & d_2 \\ 0 & 1 \end{pmatrix} \begin{pmatrix} 1 & 0 \\ -\dfrac{2}{R} & 1 \end{pmatrix} \begin{pmatrix} 1 & d_1 \\ 0 & 1 \end{pmatrix} \tag{3.140}$$

The slab between the curved mirrors has a length of t and a refractive index of n. The cavity will be stable when $|(A + D)/2| \le 1$. For a given set of resonator dimensions, this can be numerically checked. When the resonator has a stable configuration, the Rayleigh range on M_1 can be calculated [see Eq. (3.139)] from

$$z_0 = \frac{2|B|}{\sqrt{1 - \left(\dfrac{A + D}{2}\right)^2}} \tag{3.141}$$

Now, we can use Eqs. (3.90) and (3.91) to calculate the location and size of the beam waist within the crystal. The Rayleigh range calculated in Eq. (3.141) can be thought of as the Rayleigh range of a collimated Gaussian beam incident on an optical system shown in Fig. 3.24. The ray transformation matrix $\mathbf{M}_{T1}$ for this optical system is given by

$$\mathbf{M}_{T1} = \begin{pmatrix} A_1 & B_1 \\ C_1 & D_1 \end{pmatrix} = \begin{pmatrix} 1 & d_2 \\ 0 & 1 \end{pmatrix} \begin{pmatrix} 1 & 0 \\ -\dfrac{2}{R} & 1 \end{pmatrix} \begin{pmatrix} 1 & d_1 \\ 0 & 1 \end{pmatrix} \tag{3.142}$$

The displacement of d_2 is followed by a medium of refractive index n as we assumed in the derivation of Eqs. (3.90) and (3.91) earlier.

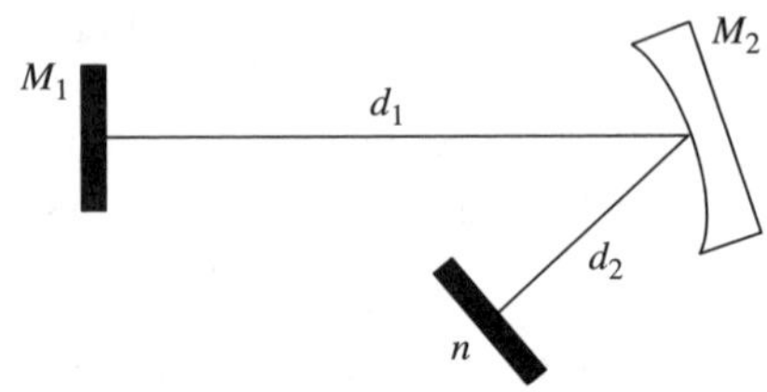

FIGURE 3.24 The optical system from the location of the collimated beam on the end mirror to the front edge of the crystal.

Hence, the location d_f and the Rayleigh range z_2 of the collimated beam in the crystal can be calculated by using

$$d_f = -n\frac{\left[D_1B_1 + A_1C_1z_0^2\right]}{\left(D_1^2 + z_0^2C_1^2\right)} \tag{3.143}$$

and

$$z_2 = \frac{nz_0}{D_1^2 + z_0^2C_1^2} \tag{3.144}$$

where A_1, B_1, C_1, and D_1 are the elements of the ray transfer matrix $\mathbf{M}_{T1}$ in Eq. (3.142). From a knowledge of z_2, the corresponding beam waist w_{02} can also be determined by using Eq. (3.31).

As an illustration, we first consider a symmetric cavity with $R = 10$ cm, $d_1 = d_4 = 50$ cm, $d_3 = 5.2$ cm, $t = 1$ cm, $n = 1.6$, and $\lambda = 1.25$ μm. Figure 3.25 shows the variation of the beam waist location d_f as a function of the separation d_2 between the curved mirror M_1 and the crystal edge. Whenever the resonator configuration is unstable, d_f is set to 0. The corresponding variation of the beam waist w_{02} is further shown in Fig. 3.26. Again, we took $w_{02} = 0$ when the cavity is unstable. Note that when the four-mirror resonator is symmetric, there occurs only one stability region as d_2 is varied. The beam waist reaches a maximum value (in this case 33 μm) at the center of the stability range and then tends to zero at the edges of the stability region.

Next, we numerically investigate how the beam waist w_{02} varies as a function of the arm length (d_1 or d_4). Keeping $d_1 = d_4$, we show this dependence in Fig. 3.27 for three cases where (a) $d_1 = 50$ cm, (b) $d_1 = 75$ cm, and (c) $d_1 = 100$ cm. For all three cases, $R = 10$ cm, $d_3 = 5.2$ cm, $t = 1$ cm, $n = 1.6$, and $\lambda = 1.25$ μm. As can be seen from Fig. 3.27, the maximum value $(w_{02})_{max}$ of the beam waist at the center of the stability range decreases with increasing arm length. In particular, as d_1 is increased from 50 to 100 cm, $(w_{02})_{max}$ decreases from 33 to 23 μm. Finally, introducing asymmetry in the resonator splits the stability

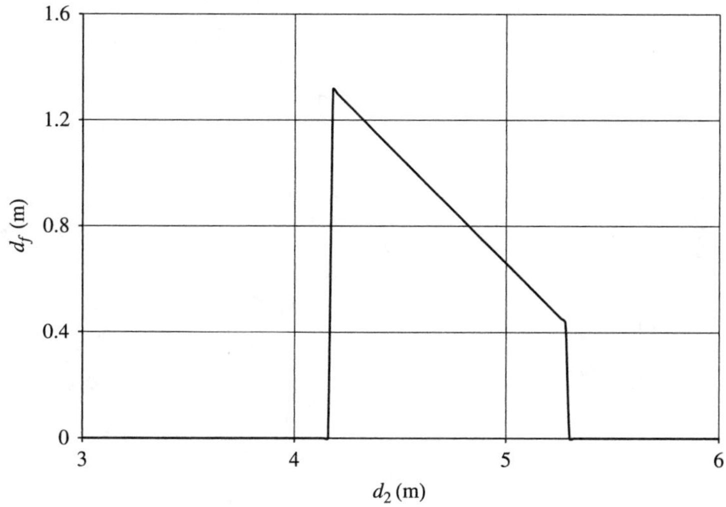

Figure 3.25 Variation of the beam waist location d_f as a function of the separation d_2 between the curved mirror M_2 and the crystal edge for a four-mirror resonator. The fixed resonator parameters are $R = 10$ cm, $d_1 = d_4 = 50$ cm, $d_3 = 5.2$ cm, $t = 1$ cm, $n = 1.6$, and $\lambda = 1.25$ μm.

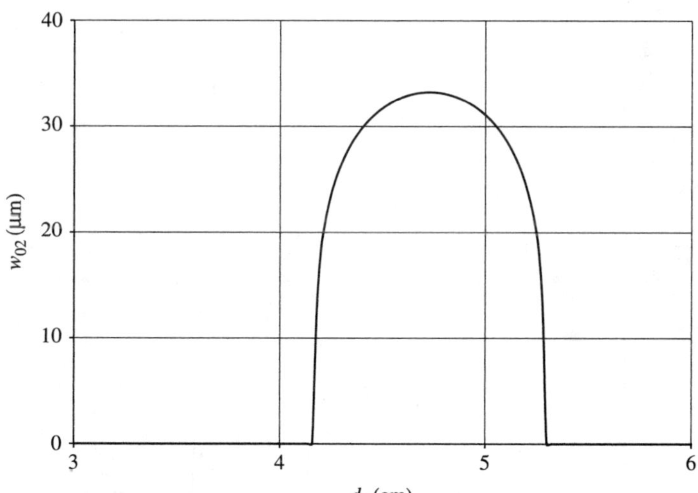

Figure 3.26 Variation of the beam waist w_{02} as a function of the separation d_2 between the curved mirror M_2 and the crystal edge for a four-mirror resonator. The fixed resonator parameters are $R = 10$ cm, $d_1 = d_4 = 50$ cm, $d_3 = 5.2$ cm, $t = 1$ cm, $n = 1.6$, and $\lambda = 1.25$ μm.

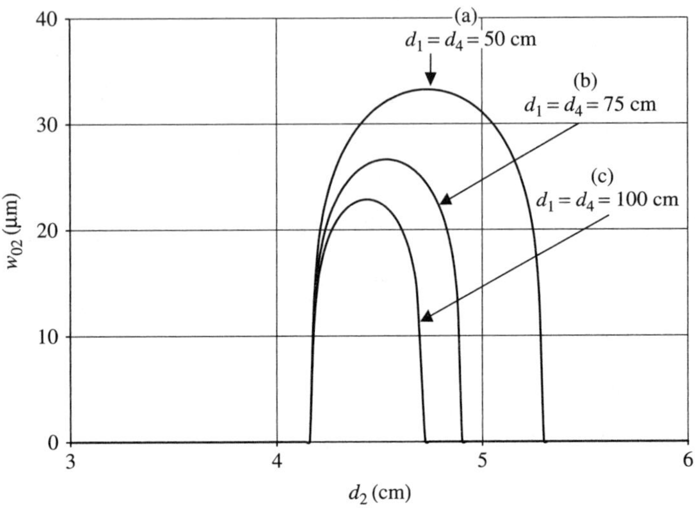

FIGURE 3.27 Variation of the beam waist w_{02} as a function of d_2 for a symmetric ($d_1 = d_4$) four-mirror cavity for (*a*) $d_1 = 50$ cm, (*b*) $d_1 = 75$ cm, and (*c*) $d_1 = 100$ cm. The fixed parameters are $R = 10$ cm, $d_3 = 5.2$ cm, $t = 1$ cm, $n = 1.6$, and $\lambda = 1.25$ μm.

region into two parts. To see this, we show the variation of w_{02} as a function of d_2 in Fig. 3.28 for the case $d_1 = 75$ cm, $d_2 = 50$ cm, $R = 10$ cm, $d_3 = 5.2$ cm, $t = 1$ cm, $n = 1.6$, and $\lambda = 1.25$ μm. In addition, also note that the maximum value of w_{02} at the center of the stability regions depends

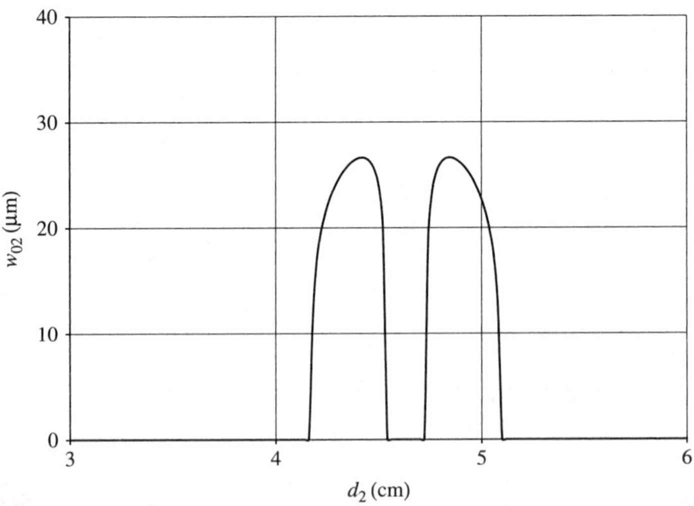

FIGURE 3.28 Variation of the beam waist w_{02} as a function of d_2 for an asymmetric 4-mirror cavity with $d_1 = 75$ cm and $d_4 = 50$ cm. The other fixed parameters are $R = 10$ cm, $d_3 = 5.2$ cm, $t = 1$ cm, $n = 1.6$, and $\lambda = 1.25$ μm.

on the length of the longer arm. Varying the length of the shorter arm simply shifts the location (in terms of d_2) of the stability regions.

3.3.4 Multi-Pass Cavities

Multi-pass cavities (MPC), first introduced by Herriott and coworkers in 1964,[12] can be used to obtain long effective optical paths with a compact arrangement of mirrors. They may come in various configurations. The simplest version will have two highly reflecting end mirrors and a scheme for beam injection and extraction. One possibility is to use notches cut on the end mirrors as shown in Fig. 3.29. When the separation and the radii of the mirrors are properly chosen, an incident beam with the correct offset and tilt with respect to the optical axis will bounce many times between the end mirrors before exiting through the second notch. If the separation between the end mirrors is d, and if the bouncing beam completes n full round trips within the MPC, the total effective optical path length will be $2nd$. In this section, we will analyze the optical characteristics of MPCs by using the ray and beam analysis methods developed for paraxial optical systems. We will be generally following the treatment given in Refs. 13 to 15.

Let us assume that a single round trip within the MPC is represented by the ray transformation matrix $\mathbf{M}_\mathrm{T}$ given by

$$\mathbf{M}_\mathrm{T} = \begin{bmatrix} \mathrm{A} & \mathrm{B} \\ \mathrm{C} & \mathrm{D} \end{bmatrix} \tag{3.145}$$

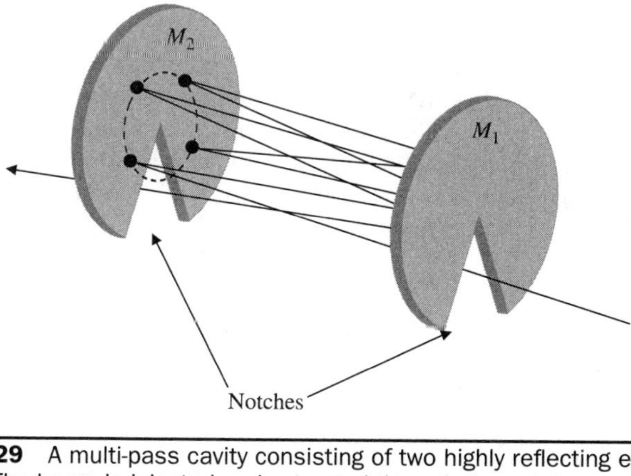

FIGURE 3.29 A multi-pass cavity consisting of two highly reflecting end mirrors. The beam is injected and extracted through the notches cut on the end mirrors.

If the elements A and D of $\mathbf{M}_\mathrm{T}$ satisfy the inequality

$$\left|\frac{\mathrm{A}+\mathrm{D}}{2}\right| \leq 1 \tag{3.146}$$

then the MPC will be stable in the sense that a ray originating near the optical axis will remain close to the optical axis after an indefinite number of round trips. Since the bouncing beam returns to the same location after each round trip, the determinant of $\mathbf{M}_\mathrm{T}$ is further unity. If the stability condition given by Eq. (3.146) is satisfied, the two eigenvalues $\lambda_\pm$ of $\mathbf{M}_\mathrm{T}$ have unit magnitude and can be expressed as $\lambda_\pm = \exp(\pm i\phi)$. Here, $\cos\phi = (\mathrm{A}+\mathrm{D})/2$. As was discussed in Chap. 2 [Eqs. (2.68) to (2.69)], the nth power $\mathbf{M}_\mathrm{T}^n$ of $\mathbf{M}_\mathrm{T}$, which gives the ray transformation matrix for n full round-trips inside the MPC becomes

$$\mathbf{M}_\mathrm{T}^n = \begin{bmatrix} \frac{\mathrm{A}-\mathrm{D}}{2}\frac{\sin n\phi}{\sin\phi} + \cos n\phi & \mathrm{B}\frac{\sin n\phi}{\sin\phi} \\ \mathrm{C}\frac{\sin n\phi}{\sin\phi} & \frac{\mathrm{D}-\mathrm{A}}{2}\frac{\sin n\phi}{\sin\phi} + \cos n\phi \end{bmatrix} \tag{3.147}$$

We can use Eq. (3.147) to investigate the effect of the MPC on light rays. Let us assume that the incident light ray $\vec{r}_i$ is represented by the 2×1 column vector

$$\vec{r}_i = \begin{bmatrix} r_0 \\ \theta_0 \end{bmatrix} \tag{3.148}$$

In Eq. (3.148), r_0 is the initial offset of the ray from the optical axis and θ_0 denotes the initial angle of inclination. After n full round trips, the ray vector $\vec{r}_n$ becomes

$$\vec{r}_n = \mathbf{M}_\mathrm{T}^n \vec{r}_i \tag{3.149}$$

To see what sort of a spot pattern is formed on the end mirrors by the bouncing beam, we assume that one of the mirrors coincides with the input plane, and the initial offset and angle of inclination of the ray along the x and y directions are (x_0, θ_{x0}) and (y_0, θ_{y0}), respectively. By using Eq. (3.147), the transverse offsets x_n and y_n of the ray after the nth round trip become

$$\begin{aligned} x_n &= x_0 \cos n\phi + \left(\frac{x_o(\mathrm{A}-\mathrm{D}) + 2\mathrm{B}\theta_{x0}}{2\sin\phi}\right)\sin n\phi \\ y_n &= y_0 \cos n\phi + \left(\frac{y_o(\mathrm{A}-\mathrm{D}) + 2\mathrm{B}\theta_{y0}}{2\sin\phi}\right)\sin n\phi \end{aligned} \tag{3.150}$$

For example, if we want to find the location of the spot on the mirror, say after the third round trip, we can evaluate x_n and y_n given by Eq. (3.150) by using $n = 3$. Since both sine and cosine terms appear mixed in Eq. (3.150), the spot pattern formed by the bouncing beam will in general be elliptical (see, for example, the picture on the front cover of this book). However, through a careful adjustment of the initial offset and angle of inclination along each of the transverse directions, one can obtain a circular spot pattern. In particular, consider the case, where the initial ray parameters are chosen according to

$$
\begin{aligned}
y_0 &= 0 \\
\theta_{y0} &= \frac{x_0 \sin\phi}{\mathrm{B}} \\
\theta_{x0} &= \frac{x_0}{2\mathrm{B}}(\mathrm{D} - \mathrm{A})
\end{aligned}
\tag{3.151}
$$

Then, the coordinates (x_n, y_n) of the spot pattern on the end mirrors after the nth round trip become

$$
\begin{aligned}
x_n &= x_0 \cos n\phi \\
y_n &= x_0 \sin n\phi
\end{aligned}
\tag{3.152}
$$

This is a very important special case which also provides a physical interpretation of what the angle ϕ represents. The resulting spot pattern is sketched in Fig. 3.30. The spots all fall on a circle of radius x_0, the initial offset of the beam in the x direction. Furthermore, the angle ϕ corresponds to the angular advance of the spot pattern after one round trip. The value of ϕ depends on the matrix elements A and D, in other words, on the specific choice of the radii of curvature of the mirrors and the mirror separation.

What can we say about the effect of the MPC on the Gaussian beam parameter? This is also a very important design issue, because

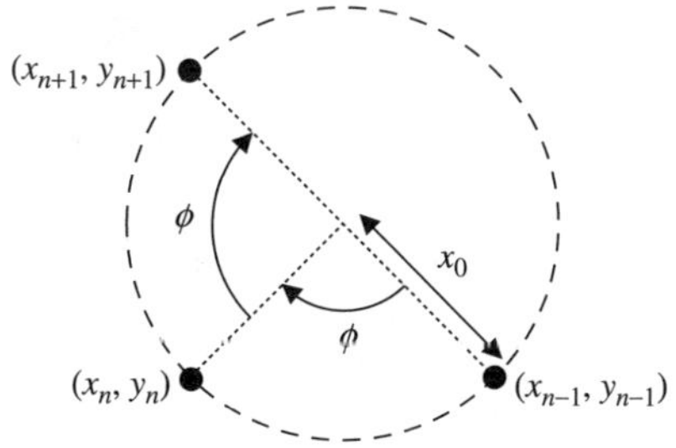

FIGURE 3.30 The resulting circular spot pattern on one of the end mirrors when the conditions in Eq. (3.151) are met. The angle ϕ gives the angular advance of the spot pattern after each round trip.

one expects, in general, that after a long effective optical length involving periodic focusing elements, the q-parameter of the exiting beam will be quite different than that of the incident beam. Often, it is desirable to use an MPC in which the exiting beam has the same q-parameter as the input beam. Such configurations are referred to as q-preserving MPCs. q-preserving MPCs have gained importance in the development of extended cavity mode-locked lasers.[15, 16] Here, the output pulse energy of the laser is directly proportional to the cavity length and MPCs can be used to achieve resonator extension with a compact arrangement of mirrors. Furthermore, when q-preserving MPCs are used, the original spot-size distribution of the short cavity remains the same even after the MPC is added.

To derive the condition for q-preserving MPCs, we examine the passage of a Gaussian beam through the MPC after n full round trips. If the elements of $\mathbf{M}_\mathrm{T}^n$ are $(\bar{\mathrm{A}},\bar{\mathrm{B}},\bar{\mathrm{C}},\bar{\mathrm{D}})$, then, the q-parameter q_2 of the emerging Gaussian beam will be related to the input q-parameter q_1 through

$$q_2 = \frac{\bar{\mathrm{A}}q_1 + \bar{\mathrm{B}}}{\bar{\mathrm{C}}q_1 + \bar{\mathrm{D}}} \tag{3.153}$$

If $\mathbf{M}_\mathrm{T}^n$ is $\pm\mathbf{I}$, where $\mathbf{I}$ is the 2×2 unity matrix, then we see from Eq. (3.153) that $q_2 = q_1$ and the exiting beam will have the same q-parameter as the input beam. Examining Eq. (3.147), we see that $\mathbf{M}_\mathrm{T}^n = (-1)^m\mathbf{I}$ when the angular advance ϕ satisfies the relation

$$\phi = \frac{m\pi}{n} \tag{3.154}$$

Above, both n and m are integers. Physically, as noted earlier, n gives the number of round trips that the bouncing beam completes. The integer m, on the other hand, represents the number of semicircular paths that the spot pattern produced by the bouncing beam traverses on one of the end mirrors before exiting the cavity. Figure 3.31 shows the case with $m = 2$ and $n = 6$. As can be seen, the bouncing beam undergoes a full angular sweep of 360° after six round trips before coming out of the MPC.

Example We now discuss how to design a specific two-mirror, q-preserving MPC which has the spot pattern depicted in Fig. 3.31. A sketch of the two-mirror MPC is shown in Fig. 3.32. We assume that the radii of curvature of the mirrors M_1 and M_2 are R_1 and R_2, respectively. The separation between the mirrors is d. Starting to the right of the mirror M_1 and proceeding to the right, the overall ray matrix $\mathbf{M}_\mathrm{T}$ for one round trip can be calculated from

$$\mathbf{M}_\mathrm{T} = \begin{pmatrix} 1 & 0 \\ -\dfrac{2}{R_1} & 1 \end{pmatrix}\begin{pmatrix} 1 & d \\ 0 & 1 \end{pmatrix}\begin{pmatrix} 1 & 0 \\ -\dfrac{2}{R_2} & 1 \end{pmatrix}\begin{pmatrix} 1 & d \\ 0 & 1 \end{pmatrix} \tag{3.155}$$

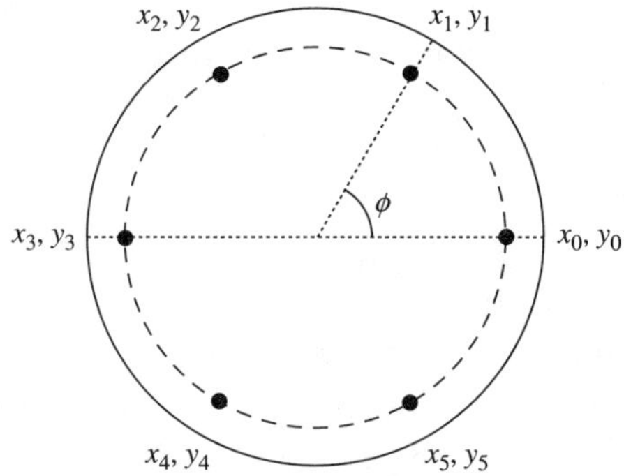

FIGURE 3.31 The spot pattern on one of the end mirrors of a q-preserving MPC for the case where $m = 2$ and $n = 6$.

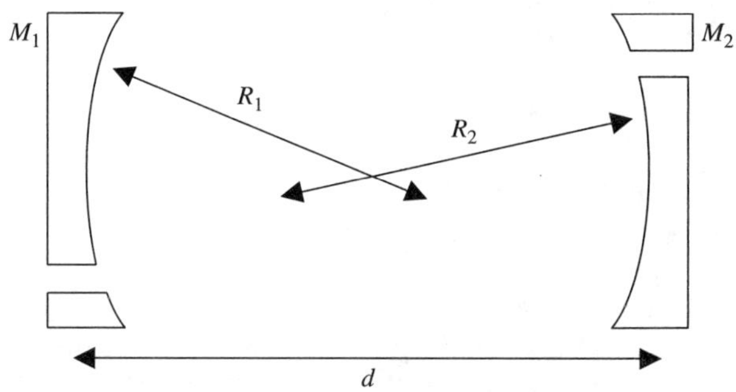

FIGURE 3.32 A two mirror multi-pass cavity with a mirror separation of d.

We leave it as an exercise for the reader to show that for the particular geometry, (A + D)/2 is given by

$$\frac{\mathrm{A+D}}{2} = 2\left(1-\frac{d}{R_1}\right)\left(1-\frac{d}{R_2}\right)-1 \tag{3.156}$$

We further assume that the mirror M_1 is flat with $R_1 = \infty$. The mirror separations d^* giving q-preserving MPC configurations become (see Prob. 3.10)

$$d^* = \frac{R_2}{2}\left(1-\cos\left[\frac{m\pi}{n}\right]\right) \tag{3.157}$$

Let us consider a flat-curved MPC with $R_2 = 4$ meters. If we choose $m = 2$ and $n = 6$, the mirror separation giving this q-preserving configuration should be $R_2/4$ or 1 m. The resulting spot pattern is the same as shown in Fig. 3.31. The effective optical path length of 12 m is hence obtained by using a MPC configuration with a mirror separation of only 1 m.

3.4 Resonant Frequencies of Optical Cavities

Consider a simple two-mirror resonator. There are two important issues regarding the modes supported by the optical resonator:

1. The mode has a particular spatial pattern which satisfies the boundary conditions on the curved mirrors. We have already discussed this in detail and seen how to determine the Rayleigh range and beam waist of the mode.
2. The cavity allows the growth of self-consistent modes only at selected frequencies. We will refer to these as the resonant frequencies of the cavity.

In this section, our objective is to determine the resonant frequencies of a two-mirror optical resonator. To gain further insight about the cavity resonances, we first investigate the transmission resonances of a simple but very important structure known as the Fabry-Perot interferometer. We will then look at two-mirror laser resonators and determine their resonance frequencies as well.

3.4.1 The Fabry-Perot Interferometer

The Fabry-Perot (F-P) interferometer, named after its inventors Charles Fabry (1867–1945) and Alfred Perot (1863–1925),[17] is a very important instrument used in high precision spectroscopy and accurate spectral measurements. It consists of two parallel, highly reflecting mirrors. In a scanning F-P interferometer configuration, the distance between the mirrors is adjustable. In the case of an F-P etalon, we have a transparent plate of constant thickness with partially reflecting parallel surfaces.

We will follow the standard notation[18] to analyze the transmission of the F-P etalon shown in Fig. 3.33. It consists of a plate of thickness L and refractive index n', surrounded by a medium of refractive index n on both sides.

We assume that light with a field amplitude of E_0 is obliquely incident on the etalon with an angle of incidence of θ_1. The corresponding angle of refraction inside the etalon is θ_2. For the case where

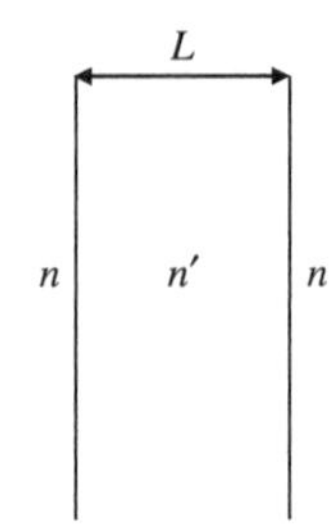

FIGURE 3.33 Sketch of a Fabry-Perot etalon.

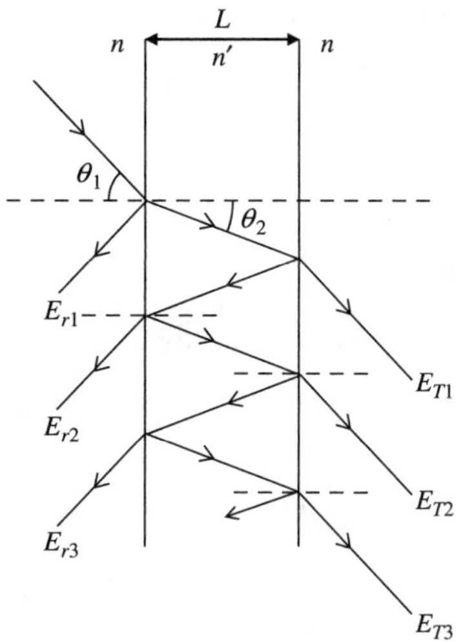

FIGURE 3.34 Sketch of the F-P etalon showing the incident field along with the first three reflected (E_{r1}, E_{r2}, and E_{r3}) and transmitted (E_{T1}, E_{T2}, and E_{T3}) fields after multiple internal reflections.

the light passes from the outer medium (refractive index = n) into the substrate (refractive index = n'), we take the amplitude reflection and transmission coefficients as r and t, respectively. As the field passes from the etalon to the outer medium, the corresponding reflection and transmission coefficients are r' and t'. From our discussion of reflection and transmission coefficients in Sec. 1.10, we see that $r' = -r$ and $t = t'$. We want to calculate how the intensity transmission varies with the frequency ν of the incident light. Since there occurs infinitely many partial reflections within the etalon (see Fig. 3.34 which shows the first three reflected and transmitted field components), we superimpose all of the transmitted field components to determine the total transmitted field E_T.

Looking at Fig. 3.34, we see that the transmitted field components E_{T1}, E_{T2}, and E_{T3} will be given in terms of E_0 by the respective expressions

$$E_{T1} = E_0 t t \tag{3.158}$$

$$E_{T2} = E_0 t r' r' \exp(i\delta) t = E_0 t t (r'^2 \exp(i\delta)) = E_{T1}(r'^2 \exp(i\delta)) \tag{3.159}$$

and

$$E_{T3} = E_0 t r' \exp(i\delta) r' r' r' \exp(i\delta) t = E_{T2}(r'^2 \exp(i\delta)) \tag{3.160}$$

Above, δ is the round-trip phase acquired by the wave along the normal to the interface and is given by

$$\delta = (k_0 n' \cos\theta_2)(2L) = \frac{4\pi}{\lambda} n' L \cos\theta_2 \tag{3.161}$$

where λ and k_0 are the vacuum wavelength and wave vector. As can be seen from Eqs. (3.158) to (3.160), the pth transmitted field E_{Tp} will be given by

$$E_{Tp} = E_0 t^2 R^{p-1} (\exp(i\delta))^{p-1} \tag{3.162}$$

Above, $R = r'^2$ and $t^2 = 1 - R$. To find the total transmitted field E_T, we sum over all of the partially transmitted components, which gives a geometric series that can be readily evaluated:

$$\begin{aligned} E_T &= E_0 t^2 (1 + (R)\exp(i\delta) + ((R)\exp(i\delta))^2 + \ldots) \\ &= E_0 (1 - R) \sum_{p=1}^{\infty} [(R)\exp(i\delta)]^{p-1} = \frac{E_0 (1 - R)}{1 - (R)\exp(i\delta)} \end{aligned} \tag{3.163}$$

The intensity transmission T of the etalon will be given by

$$\begin{aligned} T &= \left|\frac{E_T}{E_0}\right|^2 = \left|\frac{(1-R)}{1-(R)\exp(i\delta)}\right|^2 \\ &= \frac{(1-R)^2}{|(1 - R\cos\delta) - iR\sin\delta|^2} = \frac{(1-R)^2}{(1 - R\cos\delta)^2 + R^2\sin^2\delta} \\ &= \frac{(1-R)^2}{(1-R)^2 + 4R\sin^2\dfrac{\delta}{2}} \end{aligned} \tag{3.164}$$

Equation (3.164) is the key result that summarizes the resonance characteristics of the F-P etalon. First, note that the transmission of the etalon depends on two parameters: the intensity reflectivity R and the round-trip phase shift δ. R can be set to a desired value by coating the surfaces of the etalon. The phase shift δ, on the other hand, depends on the frequency as well as the thickness and the refractive index. Here are some important characteristics of the transmission function:

1. The intensity transmission T will have maxima whenever $\sin\delta/2 = 0$, or when the round-trip phase shift satisfies

$$\delta = 2m\pi \tag{3.165}$$

Here, m is an integer. This is the resonance condition for the Fabry-Perot etalon. The resonance we are talking about refers to the local maxima observed in the intensity transmission whenever the condition in Eq. (3.165) is met. Stated in words, at the transmission resonances, the round-trip phase shift is an integral multiple of 2π.

2. In practice, for an F-P etalon of fixed thickness, transmission resonances will be observed as the frequency of the incident light beam is varied. We assume that ν_m represents the mth resonant frequency. We then have

$$\delta = \frac{4\pi}{\left(\frac{c}{\nu_m}\right)} n' L\cos\theta_2 = 2m\pi \tag{3.166}$$

or

$$\nu_m = m\nu_{FSR} \tag{3.167}$$

where ν_{FSR} is the free spectral range, often abbreviated as FSR, and is given by

$$\nu_{FSR} = \frac{c}{2n'L\cos\theta_2} \tag{3.168}$$

The free spectral range ν_{FSR} corresponds to the frequency spacing between two adjacent resonant frequencies where the transmission is maximum and, in general, depends on the refractive index of the substrate, the thickness of the etalon, and the angle of refraction within the etalon, as can be seen from Eq. (3.168). At normal incidence, θ_2 becomes zero and the expression for the free spectral range simplifies to

$$\nu_{FSR} = \frac{c}{2n'L} \tag{3.169}$$

Table 3.1 gives the free-spectral range for an air-spaced Fabry-Perot interferometer at different mirror separations.

Mirror Separation (L)	Free Spectral Range (ν_{FSR})
0.15 mm	1 THz
1.5 mm	100 GHz
15 mm	10 GHz
15 cm	1 GHz
150 cm	100 MHz
15 m	10 MHz
150 m	1 MHz
1500 m	1 kHz

TABLE 3.1 Free-Spectral Range for an Air-Spaced Fabry-Perot Interferometer at Different Mirror Separations

The lowest two rows may not look realistic for practical examples. However, with multi-pass resonators, it is now possible to attain resonators with mirror spacing at such length scales.

3. The minima $T_{\min}$ in the intensity transmission will occur at those frequencies where $\sin\delta/2 = 1$. This gives

$$T = \frac{(1-R)^2}{(1-R)^2 + 4R} = \frac{(1-R)^2}{(1+R)^2} \tag{3.170}$$

As the reflectivity of the surfaces approaches 100%, the minimum value $T_{\min}$ will tend to 0 and the contrast between the maxima and the minima will become more pronounced. As an example, Fig. 3.35 shows the variation of T as a function of frequency for (a) $R = 0.1$ and (b) $R = 0.8$. For the case where $R = 0.1$, the frequency selectivity of the interferometer is poor in the sense that the transmission peaks are broad and the contrast is low since the transmission varies only over the range from 67% to 100%. When the reflectivity of each surface is increased to 0.8, the width of the transmission band becomes significantly narrower and the contrast between the minimum and maximum transmission becomes much larger. In particular, T varies from 1.2% to 100%.

4. We can use the above results to determine the frequency bandwidth of the Fabry-Perot etalon. We have already seen in Fig. 3.35 that the spectral width of the intensity transmission

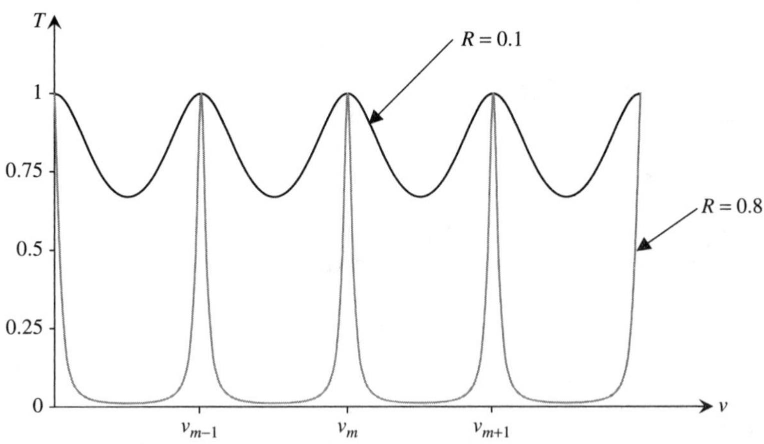

FIGURE 3.35 Calculated variation of the intensity transmission T of a Fabry-Perot etalon as a function of frequency for (*a*) $R = 0.1$ and (*b*) $R = 0.8$.

band depends strongly on the reflectivity of the etalon surfaces. To determine the full width at half maximum (FWHM) of the F-P transmission peak, we first note that local maxima occur at $\delta = 2m\pi$. Let the width (FWHM) in phase units be $2\bar{\Delta}$. Then, at $\delta = 2m\pi + \bar{\Delta}$, we have $T = 1/2$. Since

$$\sin\left(\frac{2m\pi + \bar{\Delta}}{2}\right) = \sin(m\pi)\cos\frac{\bar{\Delta}}{2} + \cos(m\pi)\sin\frac{\bar{\Delta}}{2} = (-1)^m \sin\left(\frac{\bar{\Delta}}{2}\right) \tag{3.171}$$

we get

$$\frac{(1-R)^2}{(1-R)^2 + 4R\sin^2\left(\frac{\delta}{2}\right)} = \frac{(1-R)^2}{(1-R)^2 + 4R\sin^2\left(\frac{\bar{\Delta}}{2}\right)} = \frac{1}{2} \tag{3.172}$$

or

$$(1-R)^2 = 4R\sin^2\left(\frac{\bar{\Delta}}{2}\right) \tag{3.173}$$

Assuming that $\bar{\Delta}$ is small, we further have $\sin\left(\bar{\Delta}/2\right) \approx \bar{\Delta}/2$, resulting in

$$\bar{\Delta} = \sqrt{\frac{(1-R)^2}{R}} = \frac{(1-R)}{\sqrt{R}} \tag{3.174}$$

By using Eq. (3.166), we can determine the frequency bandwidth (FWHM) $\Delta\nu_{1/2}$ from

$$\Delta\nu_{1/2} = \frac{c}{2n'L\cos\theta_2}\frac{2\bar{\Delta}}{2\pi} = \left(\frac{c}{2n'L\cos\theta_2}\right)\frac{1}{\pi}\frac{(1-R)}{\sqrt{R}} = \frac{\nu_{FSR}}{F} \tag{3.175}$$

where F is known as the finesse of the interferometer and is given in terms of the surface reflectivity as

$$F = \frac{\pi\sqrt{R}}{1-R} \tag{3.176}$$

Example Consider a free-space F-P interferometer where the mirrors are separated by 1 mm. The corresponding FSR will be $3 \times 10^{10}/(2 \times 0.1) = 150$ GHz. If the reflectivity of each surface is 90% (i.e., $R = 0.9$), we further have

$$F = \frac{\pi\sqrt{0.9}}{1-0.9} \approx 30$$

giving a spectral bandwidth of 5 GHz (FWHM). If R is increased to 0.99, the finesse becomes 313. We should note with caution that Eq. (3.176) is correct only for the ideal case where the etalon is perfectly parallel and the surfaces are perfectly flat. In practice, F will not of course become infinitely large as R approaches unity due to mirror imperfections such as scattering losses and deviations from perfect parallelism. To summarize, two important design rules for a F-P etalon are worth remembering:

1. Adjust the thickness to control the free-spectral range.
2. Choose the appropriate reflectivity, for a given etalon thickness, to control the transmission bandwidth.

3.4.2 Scanning Fabry-Perot Spectrometer

Scanning F-P spectrometers are frequently used in measuring the spectral intensity distribution of optical signals. In other words, we can use a scanning F-P spectrometer to determine how the intensity varies with frequency. However, the parameters of the F-P interferometer should be carefully chosen in order to avoid ambiguities in the spectral measurements.

Figure 3.36 shows a sketch of a scanning F-P spectrometer used to analyze the spectral content of an optical signal. In its simplest form, the spectrometer consists of two closely spaced mirrors (M_1 and M_2). One of the mirrors is mounted on a piezo-electric transducer (PZT) which can provide small displacements of the order of microns or less when driven by an applied voltage. As the movable mirror (in this case M_2) is scanned, the transmission maxima of the Fabry-Perot interferometer are swept and a replica of the optical spectrum is recorded by a detector which monitors the light transmitted through M_2. To gain further insight about how the spectrometer operates, we need to understand by what distance we should translate the movable mirror, and also decide how to determine the mirror separation and the reflectivity. We are already assuming here that

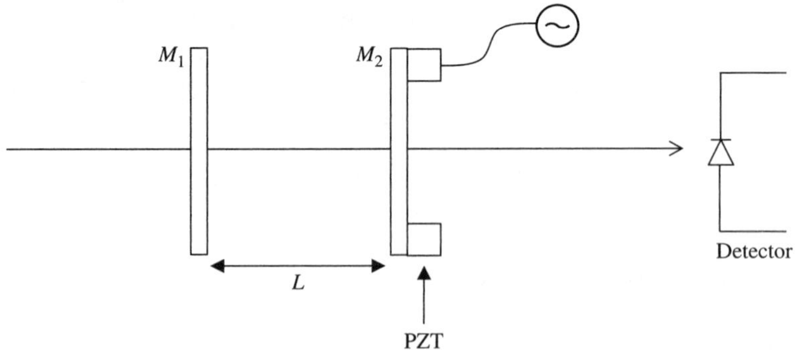

FIGURE 3.36 Sketch of a scanning Fabry-Perot spectrometer.

the distance over which we need to dither M_2 is very small compared to the separation between the mirrors. This turns out to be a very good assumption, as we will see soon.

Let us assume that we have an air-spaced spectrometer with a mirror separation of L. At normal incidence, the mth resonance frequency will be given by

$$\nu_m = m\frac{c}{2L} \tag{3.177}$$

Taking the derivative of both sides, we can determine the shift $\Delta\nu_m$ due to a change ΔL in the mirror separation. $\Delta\nu_m$ is given by

$$|\Delta\nu_m| = m\frac{c}{2L^2}\Delta L = \left(\frac{c}{2L}\right)\frac{m\Delta L}{L} = \nu_{FSR}\left(\frac{1}{c/2\nu_m}\right)\Delta L = \nu_{FSR}\frac{\Delta L}{\lambda_m/2} \tag{3.178}$$

Above, λ_m is the vacuum wavelength corresponding to the mth resonance frequency. Equation (3.178) summarizes a very important property of the scanning F-P spectrometer. When the mirror separation is changed by half a wavelength, the transmission peak of the spectrometer is shifted by one free spectral range. Since the same transmission pattern repeats after one free spectral range, in principle, we do not need to sweep the movable mirror by more than half a wavelength. Consider signals around the carrier wavelength of 1 μm. To sweep over one free spectral range, the mirror needs to be moved by 0.5 μm only. In practice, it is useful to sweep the mirror far enough (at least two free spectral ranges) so that the next Fabry-Perot transmission peak also appears for calibration purposes.

Suppose that we are interested in measuring the spectrum of a hypothetical signal whose spectral intensity distribution is as shown in Fig. 3.37*a*. The full spectral width of the signal is $\Delta\nu_1$. In addition, there is a narrower local peak of width $\Delta\nu_2$ on the spectrum. We are interested in designing the F-P spectrometer in such a way as to give the correct measurement of this spectrum and also enable us to resolve the narrower peak. This is best achieved by using a F-P spectrometer whose intensity transmission as a function of frequency may look something like what is shown in Fig. 3.37*b*. Note that the free spectral range is larger than $\Delta\nu_1$, so that only one of the transmission peaks coincides with the non-zero portion of the spectrum as the mirror is translated. As is shown in Fig. 3.37*b*, for the particular instant depicted, the detector will register the intensity $I(\nu_1)$. Furthermore, the reflectivity of the mirrors should be adjusted so that the finesse is large enough to give a transmission bandwidth $\Delta\nu_{1/2}$, much smaller than $\Delta\nu_2$, say $\Delta\nu_2/10$. This way, we can also resolve the finer structure

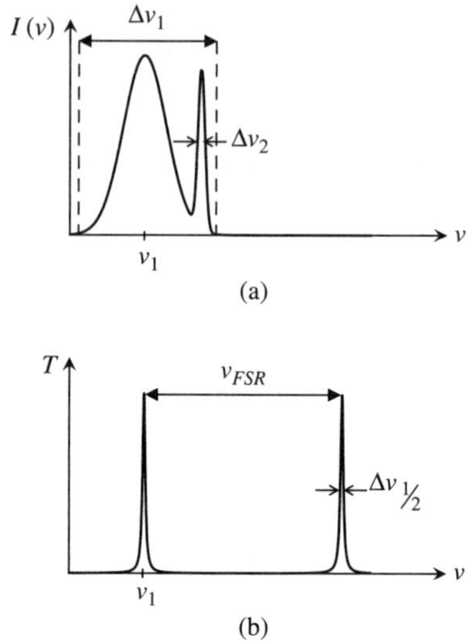

FIGURE 3.37 (*a*) The spectrum of a hypothetical signal and (*b*) the transmission of a suitable Fabry-Perot spectrometer that can be used to measure this spectrum.

on the spectrum. As the movable mirror is translated, the detector will produce a signal whose intensity is proportional to the spectral intensity distribution.

If the mirror spacing is not close enough so that several transmission peaks appear across the spectrum, as is the case depicted in Fig. 3.38*a* and *b*, the measured signal will be ambiguous. For the

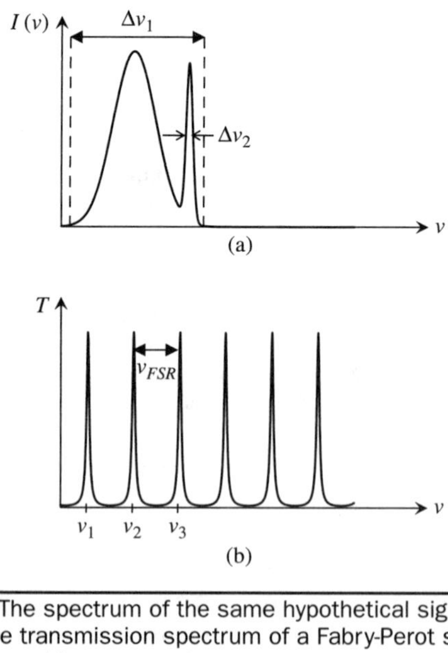

FIGURE 3.38 (*a*) The spectrum of the same hypothetical signal as in Fig. 3.37*a*. (*b*) The transmission spectrum of a Fabry-Perot spectrometer that would result in an ambiguous spectrum measurement.

particular case shown here, the detector will register the sum of intensities at different frequencies and the generated signal will not correspond to the actual spectrum.

Example Let us design an F-P spectrometer to analyze a spectrum whose total width $\Delta \nu_1 = 15$ GHz. The length should be such that the free-spectral range is at least larger than 15 GHz. This gives a mirror separation L of

$$L = \frac{3\times 10^{10}}{2(15\times 10^{9})} = 1 \text{ cm}$$

If the narrowest feature to be resolved has a width of 1.5 GHz, the transmission width can be chosen as 150 MHz. This requires a finesse of 100. By using Eq. (3.176), we find that a reflectivity of 97% gives a finesse of 103, which works out in this case.

3.4.3 Resonance Frequencies of Two-Mirror Cavities

Due to the wave nature of the light trapped inside an optical resonator, the energy of the Hermite-Gaussian modes supported by the cavity builds up only at special frequencies. To find these characteristic resonance frequencies, we use the fact that the round-trip phase shift acquired by the circulating beam should be an integral multiple of 2π in order to have constructive interference. Before looking at the general case of a resonator consisting of two curved mirrors, we first look at the simpler cavity with two plane end mirrors. In this case, a plane wave satisfies the boundary conditions on the two flat mirrors. Recalling that the spatial phase of such a wave is given by

$$\Phi = -kz \tag{3.179}$$

the resonance condition for a cavity of length d becomes

$$2k_q d = q2\pi \tag{3.180}$$

Here, q is an integer (not to be confused with the Gaussian q-parameter) and k_q is the wave vector of the qth mode. Converting to frequency, the qth resonance frequency ν_q will be given by

$$\nu_q = \frac{k_q c}{2\pi} = q\frac{c}{2d} \tag{3.181}$$

The result in Eq. (3.181) is identical to what we obtained in the previous section for the air-spaced ($n = 1$) Fabry-Perot interferometer. We now consider the general two-mirror cavity depicted in Fig. 3.39. Let the waist be located as before at $z = 0$. Now, we need to take into account the additional phase shift of the higher-order modes which, using Eq. (3.114), becomes

$$\Phi = -\left(kz - (1+\ell+m)\tan^{-1}\left(\frac{z}{z_0}\right)\right) \tag{3.182}$$

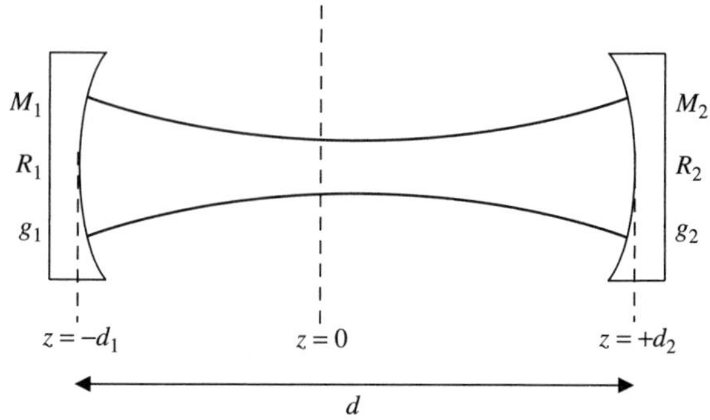

FIGURE 3.39 Sketch of a two-mirror resonator of length d. The resonance frequencies can be determined by requiring that the round-trip phase shift acquired by the circulating beam is an integral multiple of 2π.

Here, z_0 is the Rayleigh range of the beam supported by the resonator, and ℓ and m are integers. For a single pass through the resonator, the resonance condition now becomes

$$\Phi(z = d_2) - \Phi(z = -d_1) = q\pi \tag{3.183}$$

From Eq. (3.182), we further have

$$\Phi(z = d_2) = -\left[kd_2 - (1 + \ell + m)\tan^{-1}\left(\frac{d_2}{z_0}\right)\right] \tag{3.184}$$

and

$$\Phi(z = -d_1) = -\left[-kd_1 + (1 + \ell + m)\tan^{-1}\left(\frac{d_1}{z_0}\right)\right] \tag{3.185}$$

Hence, the single-pass phase difference becomes

$$\Phi(z = d_2) - \Phi(z = d_1)$$
$$= -\left(k(d_1 + d_2) - (1 + \ell + m)\left[\tan^{-1}\left(\frac{d_1}{z_0}\right) + \tan^{-1}\left(\frac{d_2}{z_0}\right)\right]\right) \tag{3.186}$$

It can be shown that[3]

$$\tan^{-1}\left(\frac{d_1}{z_0}\right) + \tan^{-1}\left(\frac{d_2}{z_0}\right) = \cos^{-1}\sqrt{g_1 g_2} \tag{3.187}$$

where g_1, g_2 are the resonator parameters given by Eqs. (3.123) and (3.124). Hence, the wave number of the resonant modes satisfies

$$kd - (1+\ell+m)\cos^{-1}\sqrt{g_1 g_2} = q\pi \tag{3.188}$$

resulting in the following expression for the frequency of the resonant modes:

$$\nu = \frac{c}{2d}\left[q + \frac{(1+\ell+m)}{\pi}\cos^{-1}\sqrt{g_1 g_2}\right] \tag{3.189}$$

Note that now, three integers (q, ℓ, m) specify each resonant mode of the cavity with two characteristic frequency spacings given by $c/2d$ and $(c/2d)(\cos^{-1}\sqrt{g_1 g_2}/\pi)$. This is schematically shown in Fig. 3.40.

Example There is a very important special case referred to as the symmetric confocal resonator. This is shown in Fig. 3.41*a*. Note that each mirror has the same radius *R*, and the separation between the mirrors is also equal to *R*. In this case, $g_1 = g_2 = 0$ and the resonance frequencies are given by

$$\begin{aligned}\nu &= \frac{c}{2d}\left[q + \frac{(1+\ell+m)}{\pi}\cos^{-1}\sqrt{0}\right]\\ &= \frac{c}{2d}\left[q + \frac{(1+\ell+m)}{\pi}\frac{\pi}{2}\right] = \frac{c}{2d}\left[q + \frac{1+\ell+m}{2}\right]\end{aligned} \tag{3.190}$$

For this special case, the resonance frequencies due to higher-order modes fall exactly in the middle of consecutive axial modes, giving an effective mode spacing of $c/4d$ between the resonant modes. This configuration is especially important in the construction of Fabry-Perot spectrometers for high-resolution spectroscopy. For this application, one requires a spectrometer with a large

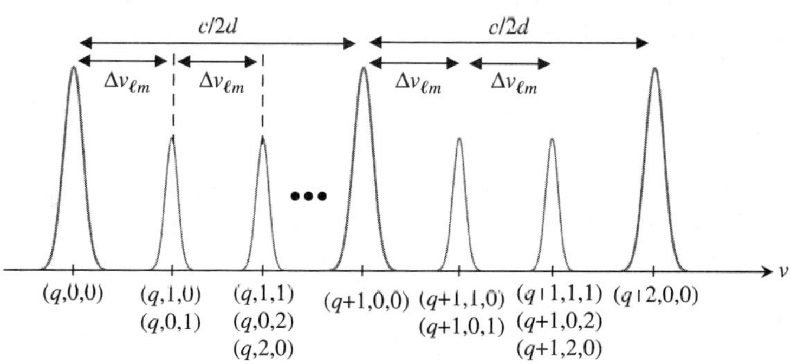

FIGURE 3.40 Resonant frequencies of a two-mirror optical resonator for different values of the mode integers (q, ℓ, m). Higher-order spatial modes are separated by $\Delta\nu_{\ell m}$.

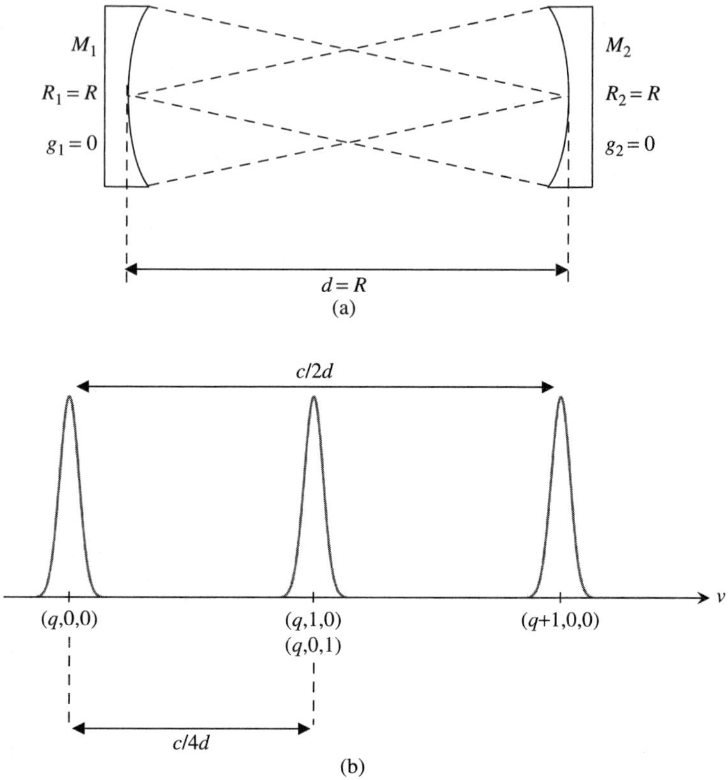

FIGURE 3.41 (*a*) Symmetric, confocal Fabry-Perot resonator. (*b*) The resonance frequencies of a confocal Fabry-Perot resonator.

mirror spacing. However, if flat mirrors are used to achieve this, the alignment of the spectrometer becomes very difficult beyond several millimeters of separation. Instead, it is easier to build this by using a symmetric confocal resonator, with the added advantage that the free-spectral range is $c/4d$, not $c/2d$.

Problems

3.1 Show that under the slowly varying envelope approximation, the transverse amplitude distribution $u(r, z)$ satisfies the wave equation given by Eq. (3.12).

3.2 Show that the functions $q(z)$ and $P(z)$ satisfy the coupled ordinary differential equations given by Eqs. (3.14) and (3.15).

3.3 Based on the solutions for $P(z)$ and $q(z)$ in a homogeneous medium, show that the electric field of the Gaussian beam can be put into the form given by Eqs. (3.33) to (3.36).

3.4 Show by first calculating the cycle-averaged Poynting vector, that the total power carried by a Gaussian beam, crossing any plane is given by Eq. (3.42).

3.5 Show that the expressions for the beam waist location d_f and the corresponding Rayleigh range z_2 are given by Eqs. (3.90) and (3.91).

3.6 Derive Eqs. (3.125) to (3.127) for the parameters d_1, d_2, and z_0 of a two-mirror resonator.

3.7 For a two-mirror resonator, show that the spotsizes on mirrors M_1 and M_2 are given by $w_1^2 = d\lambda/\pi\sqrt{g_2/g_1(1-g_1g_2)}$ and $w_2^2 = d\lambda/\pi\sqrt{g_1/g_2(1-g_1g_2)}$, respectively.

3.8 For the flat-curved cavity with $R = 0.5$ m, plot the variation of the beam waist on the flat mirror as a function of the mirror separation d for (a) $\lambda = 1.25$ μm and (b) $\lambda = 2$ μm. (Hint: the result is shown in Fig. 3.20). Next, take $\lambda = 1.25$ μm and plot the variation of the maximum spotsize as a function of R between $R = 0.01$ and 0.1 m.

3.9 Beam analysis of the four-mirror resonator: Study the algorithm given in Sec. 3.3.3 to analyze the four-mirror optical resonator depicted in Fig. 3.23. Write a numerical routine to reproduce the spot-size graphs given in Figs. 3.25 to 3.28.

3.10 Show that for a two-mirror MPC, (A + D)/2 is given by Eq. (3.156). Furthermore, derive the expression (3.157) for the separation between the end mirrors of a q-preserving MPC consisting of a flat and curved end mirror.

3.11 Consider a GRIN lens rod with the ray transfer matrix

$$\mathbf{M}_I = \begin{bmatrix} \cos\beta z & \frac{1}{\beta}\sin\beta z \\ -\beta\sin\beta z & \cos\beta z \end{bmatrix}$$

(*a*) A collimated Gaussian beam with a beam waist of w_0 is incident at $z = 0$. At what value(s) of z will it be focused?

(*b*) What will be the beam waist of the focused beam?

An optical cavity is formed by putting a reflective coating to each end of a GRIN lens rod of length d as shown in Fig. 3.42.

(*c*) Analyze the stability of the resonator as a function of d.

(*d*) Is there any value of d for which the resonator will be marginally stable? If so, find the smallest non-zero value of d where the cavity is marginally stable.

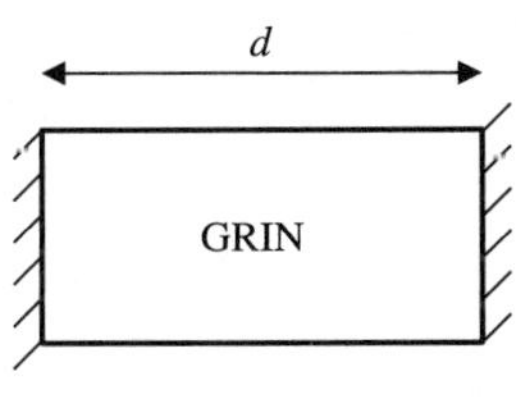

FIGURE 3.42 Sketch of the GRIN-rod resonator for Prob. 3.11.

3.12 By using a scanning Fabry-Perot interferometer, it is desired to analyze the mode structure of a He-Ne laser (center wavelength = 632.8 nm) whose cavity length is 33.3 cm.

(*a*) Neglecting the transverse mode structure, what is the frequency difference between the adjacent longitudinal modes of the He-Ne laser? Take the refractive index of the gas mixture as 1.
(*b*) Suppose that a Fabry-Perot etalon is used to analyze the mode structure of the He-Ne laser. If the separation between the air-spaced mirrors of the interferometer is 6 mm, what should the reflectivity of each mirror be in order to resolve the adjacent longitudinal modes of the He-Ne laser?

3.13 A laser cavity is constructed between a curved mirror (M_2) and a flat mirror (M_1) as shown in Fig. 3.43. The length of the cavity is $d = 8$ cm, and the radius of curvature of the second mirror (M_2) is $R = 10$ cm. The wavelength of the circulating light is 1 μm.

(*a*) At what location(s) inside the cavity is the beam waist located?
(*b*) Calculate the spotsize of the beam, in μm, at the location of M_1.
(*c*) Calculate the spotsize, in μm, at the location of M_2.
(*d*) For what values of *d* will the cavity be stable?

3.14 A collimated Gaussian output of a He-Ne laser ($\lambda = 0.6328$ μm) is incident on a thin converging lens of focal length $f = 10$ cm. Find the location and the size of the focused beam waist after the lens if the incident beam waist has a radius of (a) 1 mm and (b) 100 μm.

3.15 Suppose that a Gaussian beam is incident on a circular aperture whose radius is equal to the spotsize $w(z)$. How much of the power will be transmitted through the aperture if one neglects diffraction effects?

3.16 The symmetric confocal resonator has two identical concave mirrors with the same radius of curvature *R*, separated by a distance $d = R$. As you can see, it is at the edge of stability. Starting with the general expressions for the two-mirror resonator, show that the beam waist at the center of the resonator is given by

$$w_0 = \sqrt{\frac{\lambda d}{2\pi}}$$

What is the beam waist if the radius of curvature of each mirror is 20 cm and output wavelength is 1.34 μm?

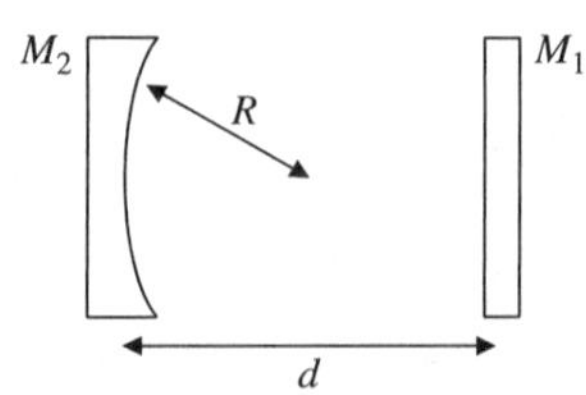

FIGURE 3.43 Sketch of the two-mirror flat-curved resonator for Prob. 3.13.

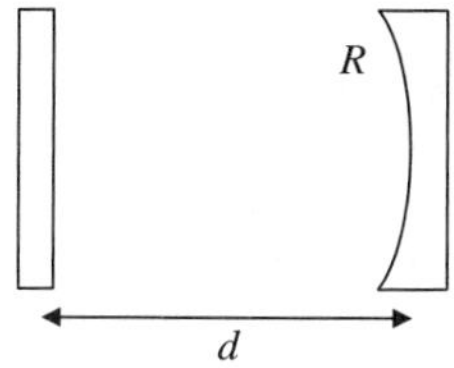

Figure 3.44 Two-mirror resonator for Prob. 3.17.

3.17 What is the frequency separation between the higher-order resonant transverse modes for the cavity shown in Fig. 3.44 with $d = 12$ cm and $R = 15$ cm?

3.18 Astigmatic compensation of four-mirror resonators: In the analysis of the four-mirror resonator discussed in Sec. 3.3.3, we neglected astigmatic effects. In a typical laser resonator, the gain crystal is usually placed at Brewster's angle, as a result of which the effective path length for the saggital and tangential beams becomes different. The effect can be compensated for by tilting the curved mirrors. As such, the effective focal length of the tilted mirror becomes different for the saggital and tangential beams. This method of compensating for astigmatism is described in Ref. 19. Read this reference and develop a computer code to determine the mirror tilts that will compensate for astigmatism. In particular, write down the cavity round-trip matrix for the saggital and tangential beams. If the tilt angle of the curved mirrors is 0, you should find that the stability ranges do not overlap. Now, tilt the curved mirrors to determine the angle where the stability regions line up at the same values of the cavity parameters. In your calculations, you can use the parameters for the cavity shown in Fig. 3.25.

References

1. A. E. Siegman, *Lasers* (University Science Books, Mill Valley, Calif. 1986).
2. J. T. Verdeyen, *Laser Electronics*, 3d ed. (Prentice Hall, Upper Saddle River, N. J., 1995).
3. A. Yariv, *Optical Electronics in Modern Communications*, 5th ed. (Oxford University Press, New York, 1997).
4. C. C. Davis, *Lasers and Electro-Optics* (Cambridge University Press, Cambridge, 1996).
5. P. W. Milonni and J. H. Eberly, *Lasers* (John Wiley & Sons, Inc., New York, 1988).
6. B. E. A. Saleh and M. C. Teich, *Fundamentals of Photonics*, 2d ed. (John Wiley & Sons, Inc, Hoboken, N. J., 2007).
7. H. Kogelnik and T. Li, "Laser beams and resonators," *Applied Optics*, **5**(10), 1550–1567 (1966).
8. H. Kogelnik and T. Li, "Laser beams and resonators," *Proceedings of the Institute of Electrical and Electronics Engineers*, **54**(10), 1312–1329 (1966).
9. A. Yariv and P. Yeh, *Optical Waves in Crystals* (John Wiley & Sons, Inc., New York, 1984).

10. A. E. Siegman, "Unstable optical resonators for laser applications," *Proceedings of the Institute of Electrical and Electronics Engineers*, **53**(3), 277–287 (1965).
11. A. E. Siegman, "Unstable Optical Resonators," *Applied Optics*, **13**(2), 353–367 (1974).
12. D. Herriot, H. Kogelnik, and R. Kompfner, "Off-axis paths in spherical mirror interferometers," *Applied Optics*, **3**(4), 523–526 (1964).
13. A. Sennaroglu and J. G. Fujimoto, "Design criteria for Herriott-type multi-pass cavities for ultrashort pulse lasers," *Optics Express*, **11**, 1106–1113 (2003).
14. A. Sennaroglu, A. M. Kowalevicz, E. P. Ippen, and J. G. Fujimoto, "Compact femtosecond lasers based on novel multi-pass cavities," *IEEE Journal of Quantum Electronics*, **40**, 519–528 (2004).
15. A. Sennaroglu and J. G. Fujimoto, "Multipass-cavity femtosecond solid-state lasers," in *Solid-State Lasers and Applications*, A. Sennaroglu, Ed. (CRC Press, Boca Raton, Fla., 2007).
16. S. H. Cho, B. E. Bouma, E. P. Ippen, and J. G. Fujimoto, "Low-repetition-rate high-peak-power Kerr-lens mode-locked Ti:Al_2O_3 laser with a multiple-pass cavity," *Optics Letters*, **24**(6), 417–419 (1999).
17. J. F. Mulligan, "Who were Fabry and Pérot?," *American Journal of Physics*, **66**, 797–802 (1998).
18. M. Born and E. Wolf, *Principles of Optics*, 6th ed. (Cambridge University Press, Cambridge, 1997).
19. H. W. Kogelnik, C. V. Shank, A. Dienes, and E. P. Ippen, "Astigmatically compensated cavities for Cw dye lasers," *IEEE Journal of Quantum Electronics*, **QE 8**(3), 373–379 (1972).

CHAPTER 4

Light-Matter Interactions

Let us provide some motivation for why we want to understand light-matter interactions in photonics. In the previous chapters, we talked about the electromagnetic nature of light and laser beams. In an optical communication system, for example, such a light beam is launched into a fiber to transmit signals over a large distance. The immediate question that comes to our mind is this: At what wavelength(s) should we send the light down an optical fiber? Or more generally, does it matter what wavelength we choose? It turns out that the wavelength of light does indeed matter for two very important reasons:

1. There is wavelength-dependent attenuation inside the fiber due to absorption and scattering. Hence, we choose the particular wavelength windows in which attenuation is minimized.
2. Another very important phenomenon that affects the propagation of light signals is dispersion. Dispersion arises because of the wavelength dependence of the refractive index and causes signals at different wavelengths to propagate at different speeds. If there is excessive dispersion at the wavelength of choice, the propagating signals will be distorted and the quality as well as the amount of information that can be sent will be reduced. It is therefore important to identify the wavelength(s) for a given material where dispersive pulse distortion is minimized. More fundamentally, we also need to explore the origin of the wavelength dependence of the refractive index.

Wavelength-dependent attenuation and dispersion are important not just in fibers but also in many other fascinating phenomena observed in daily experience. Selective wavelength-dependent absorption is what gives rise to the colors of most objects. Dispersion results in the formation of rainbow colors as refracted rays at different wave lengths separate inside rain drops. Finally, a fascinating outcome of the wavelength dependence of scattering is the color of the sky which is predominantly blue during day time when looking away from the Sun, and orange-red when one looks in the direction of the Sun at sunset.

We can gain a fairly accurate understanding of these observed phenomena by using the classical theory of light-matter interactions. We will first develop the classical electron oscillator model introduced by Lorentz to explain the origin of the wavelength-dependent absorption and refractive index. In fact, one can deduce the wavelength dependence of absorption from a full knowledge of the wavelength dependence of refractive index, or vice versa. This is summarized in Kramers-Kronig relations which we will derive from the causality principle. We will further discuss the mechanisms that lead to the broadening of absorption bands, such as collision broadening and Doppler broadening. We will end the discussion of the classical theory by obtaining an expression for the wavelength-dependent attenuation coefficient for Rayleigh scattering based on the results we obtained earlier in electric dipole radiation.

The classical model has several shortcomings, and a more accurate description of light-matter interactions requires the use of the photon picture of light. In the second part of the chapter, we will develop an elementary quantum theory of light starting with a treatment of Planck's hypothesis and blackbody radiation. The results obtained from blackbody radiation theory, in particular, the expression for the spectral energy density function, is later used in the development of the elementary quantum theory of light-matter interactions. We will discuss Einstein's model of optical processes, Einstein's A and B coefficients, spectral broadening due to spontaneous emission, and the relation between the A and B coefficients. Rate equations and optical amplification will be discussed in Chap. 6. References 1 to 9 provide further discussion on the topics discussed in this chapter. Quantum mechanical derivation of the Einstein's coefficients will be discussed in Chap. 5.

4.1 Classical Theory of Light-Matter Interactions

We will start with a classical treatment of light-matter interactions. In other words, in order to describe the response of atoms to electromagnetic waves, we will use Newton's laws of motion. The model is quite simple and captures some of the most important features of optical resonance phenomena. Here, we discuss the assumptions and key results of the Lorentz model. Historically, the Lorentz model was first systematically formulated by H. A. Lorentz and discussed in his classic book *The Theory of Electrons*.[10] In our analysis, we further derive the functional form of the Sellmeier equation and discuss absorption resonance phenomena.

4.1.1 The Lorentz Model

To capture the essence of the Lorentz model, we first consider an atom with Z electrons as schematically shown in Fig. 4.1. Here, Z is the atomic number of the atom. According to this model, each electron remains bound to the atom under the action of a linear, spring-like restoring force. The spring constant of the binding force on the jth electron is k_j. We further assume that the nucleus is too massive in comparison with the electron and remains at rest.

To simplify our analysis, we consider the one-dimensional motion of the electron only as shown in Fig. 4.2. In particular, we assume that the electron is constrained to move only along the x direction. In static

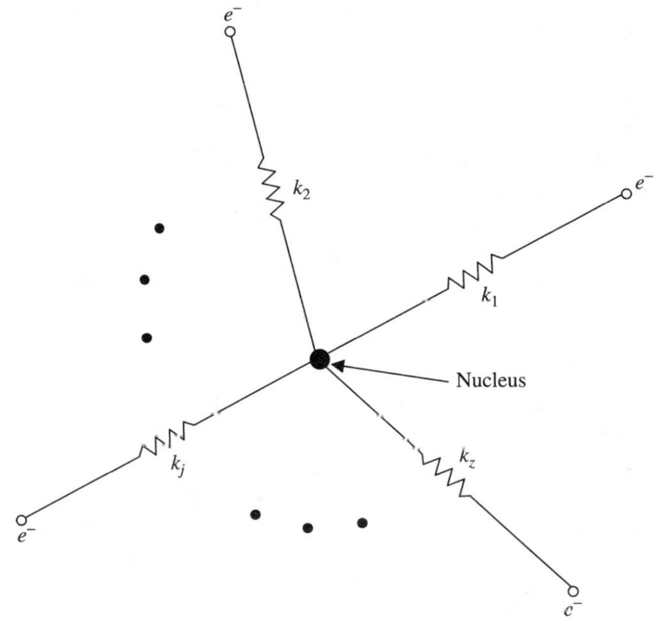

Figure 4.1 Schematic of the classical atom with Z electrons. The jth electron is bound to the atom by a linear restoring force with a spring constant of k_j.

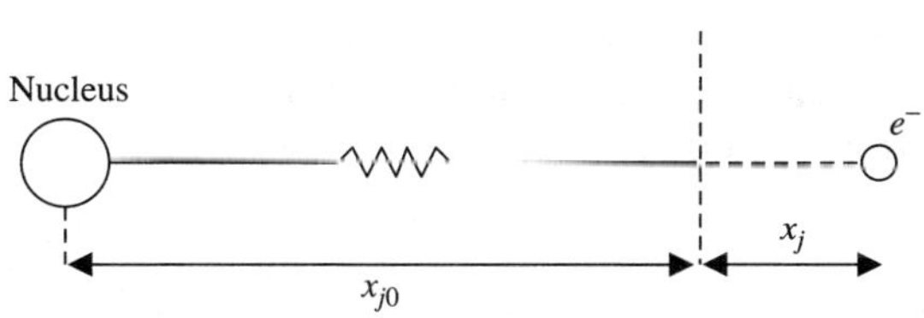

Figure 4.2 Sketch of the jth electron undergoing one-dimensional motion. The deviation from the equilibrium position x_{j0} is given by x_j.

equilibrium, the separation between the jth electron and the nucleus is x_{j0}. Under the action of a net external force, the electron departs from the equilibrium position, the deviation from the equilibrium position given by x_j. To study the motion of the electron under the influence of the external electromagnetic waves, we assume that the atom is interacting with a propagating plane electromagnetic wave whose electric field vector is given by

$$\vec{E}(\vec{r},t) = \hat{e}E_0 \exp(i(\omega t - kz)) \tag{4.1}$$

Above, E_0 is the amplitude of the electric field, ω is the angular frequency, k is the wave number, and $\hat{e}$ is the polarization vector which satisfies $\hat{e} \cdot \hat{z} = 0$. In general, $\hat{e}$ will be in the x-y plane. For the particular case considered here, we choose $\hat{e} = \hat{x}$. In the optical regime, the wavelength $\lambda = 2\pi/k$ is much larger than the atomic dimensions which are of the order of angstroms. Hence, we can completely neglect the spatial variation of the electric field over the extent of the atom and only consider the temporal dependence. Of course, this simplification, also known as the electric dipole approximation, should be critically evaluated as one goes to smaller and smaller wavelengths such as in the deep ultraviolet or x-ray region. For electromagnetic waves with wavelengths above 10 nm, the approximation is quite accurate. Next, we investigate the nature of the electromagnetic force on the electron. This is given by the Lorentz force, which has the general form

$$\vec{F}_L = -e(\vec{E}(\vec{r},t) + \vec{u} \times \vec{B}(\vec{r},t)) \tag{4.2}$$

Above, e is the electron charge, $\vec{u}$ is the electron velocity, and $\vec{E}(\vec{r},t)$ and $\vec{B}(\vec{r},t)$ are the electric and magnetic fields of the electromagnetic wave, respectively. In our analysis, we will stay in the non-relativistic regime, meaning that the electron speeds are at all times much smaller than the speed of light. It can be shown that (see Prob. 4.1) in this regime, the magnetic contribution to the Lorentz force is negligible, leading to the simplified expression

$$\vec{F}_L = -e\vec{E}(\vec{r},t) = -\hat{x}eE_0 \exp(i(\omega t - kz)) \tag{4.3}$$

for the electromagnetic force. The linear restoring force $\vec{F}_{rj}$ on the electron due to the nucleus is assumed to be proportional to the displacement from the equilibrium point and is given by

$$\vec{F}_{rj} = -k_j\vec{r}_j = -\hat{x}k_jx_j \tag{4.4}$$

Later, in Chap. 11, we will investigate the case where the linearity assumption is relaxed to investigate nonlinear optical properties. To account for absorption, we introduce a phenomenological damping force $\vec{F}_{dj}$ which is proportional to the instantaneous velocity of the electron. For the jth electron, $\vec{F}_{dj}$ becomes

$$\vec{F}_{dj} = -b_j \frac{d\vec{r}_j}{dt} = -\hat{x} b_j \frac{dx_j}{dt} \tag{4.5}$$

Above, b_j is a constant. So, altogether, we have three forces acting on the electron in this model. Newton's second law of motion takes the form

$$m_e \frac{d^2\vec{r}_j}{dt^2} = \vec{F}_{\text{NET}} = \vec{F}_L + \vec{F}_{rj} + \vec{F}_{dj} \tag{4.6}$$

Above, m_e is the electron mass and $\vec{F}_{\text{NET}}$ is the vector sum of all of the forces acting on the jth electron. For our particular case, Eq. (4.6) simplifies to

$$\hat{x} m_e \frac{d^2 x_j}{dt^2} = -\hat{x} e E_0 \exp(i(\omega t - kz)) - \hat{x} k_j x_j - \hat{x} b_j \frac{dx_j}{dt} \tag{4.7}$$

Equation (4.7) can be rearranged to obtain the following second-order, linear, inhomogeneous differential equation for the electron displacement x_j:

$$\frac{d^2 x_j}{dt^2} + \gamma_j \frac{dx_j}{dt} + \omega_{0j}^2 x_j = \frac{-eE_0}{m_e} \exp(i(\omega t - kz)) \tag{4.8}$$

Above, the damping parameter γ_j and the resonance frequency ω_{0j} are defined according to

$$\gamma_j = \frac{b_j}{m_e} \tag{4.9}$$

and

$$\omega_{0j} = \sqrt{\frac{k_j}{m_e}} \tag{4.10}$$

respectively. Next, we want to solve Eq. (4.8) by neglecting the position dependence of the driving force on the right-hand side. The solution of a second-order, linear, inhomogeneous differential equation has two parts: the homogeneous solution and the particular solution. The homogeneous solution is obtained by setting the driving force on the right-hand side equal to zero. The corresponding homogeneous solution $x_{jh}(t)$ satisfies

$$\frac{d^2x_{jh}(t)}{dt^2}+\gamma_j\frac{dx_{jh}(t)}{dt}+\omega_{0j}^2x_{jh}(t)=0 \tag{4.11}$$

By assuming a solution of the form

$$x_{jh}(t)=A_j\exp(rt) \tag{4.12}$$

where r is yet an unknown exponent to be determined, and by substituting the general form of the solution into Eq. (4.11), we find that r satisfies the quadratic equation

$$r^2+\gamma_j r+\omega_{0j}^2=0 \tag{4.13}$$

the solution to which is given by the two roots

$$r_{1,2}=\frac{-\gamma_j\pm\sqrt{\gamma_j^2-4\omega_{0j}^2}}{2}=\frac{-\gamma_j\pm i\sqrt{4\omega_{0j}^2-\gamma_j^2}}{2} \tag{4.14}$$

For most cases of practical interest, the resonance frequency is much larger than the damping parameter ($\omega_{0j}>>\gamma_j$) and, as can be seen from Eq. (4.14), the homogeneous solution consists of two linearly independent damped oscillations. Clearly, the homogeneous solutions also depend on the initial conditions. However, due to their decaying nature, they quickly tend to zero after times much larger than $1/\gamma_j$. Hence, in the derivation of the atomic response, and the refractive index function, we will assume that the homogeneous solution $x_{jh}(t)$ can be neglected and that steady-state has been reached. To find the steady-state response, we assume that the solution $x_{jp}(t)$ can be expressed in the form

$$x_{jp}(t)=x_{j0}\exp(i(\omega t-kz)) \tag{4.15}$$

By differentiating Eq. (4.15) and substituting into the differential equation [Eq. (4.8)], we get, after factoring out the exponential factor,

$$x_{j0}\exp(i(\omega t-kz))\left[-\omega^2+i\gamma_j\omega+\omega_{0j}^2\right]=\frac{-eE_0}{m_e}\exp(i(\omega t-kz)) \tag{4.16}$$

Hence, the steady-state displacement vector of the jth electron becomes

$$\vec{r}_j(t)=\hat{x}\frac{\dfrac{eE_0}{m_e}}{\left(\omega^2-\omega_{0j}^2\right)-i\omega\gamma_j} \tag{4.17}$$

To derive an expression for the electric susceptibility and the corresponding refractive index, we first note that the induced dipole moment $\vec{p}_j(t)$ of the jth electron is given by

$$\vec{p}_j(t) = -e\vec{r}_j(t) = \hat{x}\frac{e^2E_0}{m_e}\frac{1}{\left(\omega_{0j}^2 - \omega^2\right) + i\omega\gamma_j} \tag{4.18}$$

This expression is for one of the electrons of the atom. To find the polarization vector $\vec{P}(t)$ of the medium, we sum over all of the electrons and calculate the total electric dipole moment per unit volume. If the volume density of the atoms is N, then $\vec{P}(t)$ becomes

$$\begin{aligned}
\vec{P}(t) &= \sum_{j=1}^{Z} N\vec{p}_j(t) \\
&= \hat{x}\sum_{j=1}^{Z}\left[\frac{Ne^2E_0}{m_e}\frac{1}{\left(\omega_{0j}^2 - \omega^2\right) + i\omega\gamma_j}\right]\exp(i(\omega t - kz)) \\
&= \sum_{j=1}^{Z}\left[\frac{Ne^2}{m_e}\frac{1}{\left(\omega_{0j}^2 - \omega^2\right) + i\omega\gamma_j}\right]\hat{x}E_0\exp(i(\omega t - kz))
\end{aligned} \tag{4.19}$$

Note that $\vec{P}(t)$ is in the form $\vec{P}(t) = \varepsilon_0\chi_e\vec{E}$, where the electric susceptibility χ_e is immediately obtained from Eq. (4.19) as

$$\chi_e = \sum_{j=1}^{Z}\frac{Ne^2}{m_e\varepsilon_0}\frac{1}{\left(\omega_{0j}^2 - \omega^2\right) + i\omega\gamma_j} \tag{4.20}$$

Because of the damping force, the susceptibility and the dielectric constant are both complex-valued. We will soon associate the real part with the dispersive refractive index and the complex part with absorption. The complex dielectric constant $\varepsilon(\omega)$ now becomes

$$\begin{aligned}
\varepsilon = \varepsilon_0(1+\chi_e) &= \varepsilon_0\left(1 + \sum_{j-1}^{Z}\frac{Ne^2}{m_e\varepsilon_0}\frac{1}{\left(\omega_{0j}^2 - \omega^2\right) + i\omega\gamma_j}\right) \\
&= \varepsilon_0\left(1 + \sum_{j=1}^{Z}\frac{Ne^2}{m_e\varepsilon_0}\frac{\left(\left(\omega_{0j}^2 - \omega^2\right) - i\omega\gamma_j\right)}{\left(\omega_{0j}^2 - \omega^2\right)^2 + \gamma_j^2\omega^2}\right)
\end{aligned} \tag{4.21}$$

As we noted above, the dielectric constant is complex of the form

$$\varepsilon = \varepsilon' + i\varepsilon'' \tag{4.22}$$

where the real and imaginary parts are given by the respective expressions

$$\varepsilon' = \varepsilon_0\left(1 + \sum_{j=1}^{Z} \frac{Ne^2}{m_e\varepsilon_0} \frac{\left(\omega_{0j}^2 - \omega^2\right)}{\left(\omega_{0j}^2 - \omega^2\right)^2 + \gamma_j^2\omega^2}\right) \tag{4.23}$$

and

$$\varepsilon'' = -\sum_{j=1}^{Z} \frac{Ne^2}{m_e} \frac{\gamma_j\omega}{\left(\omega_{0j}^2 - \omega^2\right)^2 + \gamma_j^2\omega^2} \tag{4.24}$$

4.1.2 Near-Resonance Approximation for the Dielectric Constant

Let us assume that the frequency ω is close to only one of the resonances, say the mth resonance frequency which we shall denote as $\omega_{0m} = \omega_0$. Let the corresponding damping parameter be $\gamma_m = \gamma$. Furthermore, assume that ω is far enough from all other resonances so that the contributions coming from these have a very weak frequency dependence. We will refer to the non-resonant terms as the background contribution. In this case, we can separate the resonant and background contributions and express ε as

$$\varepsilon = \varepsilon' + i\varepsilon'' = \varepsilon_0\left(1 + \chi_b' + \chi_r' + i\chi_b'' + i\chi_r''\right) \tag{4.25}$$

where the real and imaginary parts of the background and resonant susceptibility terms are given by the expressions

$$\chi_b' = \sum_{j\neq m} \frac{Ne^2}{m_e\varepsilon_0} \frac{\left(\omega_{0j}^2 - \omega^2\right)}{\left(\omega_{0j}^2 - \omega^2\right)^2 + \gamma_j^2\omega^2} \tag{4.26}$$

$$\chi_r' = \frac{Ne^2}{m_e\varepsilon_0} \frac{\left(\omega_0^2 - \omega^2\right)}{\left(\omega_0^2 - \omega^2\right)^2 + \gamma^2\omega^2} \tag{4.27}$$

$$\chi_b'' = -\sum_{j\neq m} \frac{Ne^2}{m_e\varepsilon_0} \frac{\gamma_j\omega}{\left(\omega_{0j}^2 - \omega^2\right)^2 + \gamma_j^2\omega^2} \tag{4.28}$$

and

$$\chi_r'' = -\frac{Ne^2}{m_e\varepsilon_0} \frac{\gamma\omega}{\left(\omega_0^2 - \omega^2\right)^2 + \gamma^2\omega^2} \tag{4.29}$$

In Eqs. (4.25) to (4.29), the subscripts b and r denote the background and resonant contributions, respectively. To proceed, we make the following simplifying assumptions:

1. The background contribution χ_b' to the real part ε' dominates over all other terms and can be treated as a constant over the spectral extent of the resonant absorption band near ω_0 since it has a weak frequency dependence.
2. The term χ_b'' has a weak frequency dependence and can be treated as a constant.
3. The terms χ_r', χ_b'', χ_r'' are all small compared to $(1+\chi_b')$.

We now introduce the complex-valued refractive index $\bar{n}$ defined according to

$$\bar{n} \equiv \sqrt{\frac{\varepsilon}{\varepsilon_0}} \tag{4.30}$$

By using Eq. (4.25), $\bar{n}$ becomes

$$\bar{n} = \sqrt{\frac{\varepsilon}{\varepsilon_0}} = \left(1+\chi_b'+\chi_r'+i\left(\chi_b''+\chi_r''\right)\right)^{1/2}$$

$$= n_0\left(1+\frac{\chi_r'}{1+\chi_b'}+i\frac{\left(\chi_b''+\chi_r''\right)}{1+\chi_b'}\right)^{1/2} \tag{4.31}$$

where the background refractive index n_0 is given by

$$n_0 = \sqrt{1+\chi_b'} \tag{4.32}$$

By making the approximations outlined above, and keeping the first-order correction terms in the Taylor expansion of Eq. (4.31), we obtain

$$\bar{n} \approx n_0\left(1+\frac{\chi_r'}{2n_0^2}+i\frac{\left(\chi_b''+\chi_r''\right)}{2n_0^2}\right) = n - in'' \tag{4.33}$$

where the real and imaginary parts of the complex refractive index are given by

$$n = n_0 + \frac{\chi_r'}{2n_0} \tag{4.34}$$

and

$$n'' = -\frac{\chi_b''+\chi_r''}{2n_0} \tag{4.35}$$

In this book, we adopt the convention of denoting the real part of the complex-valued refractive index as n. This is in agreement with the standard convention and has already been used in the previous chapters. It is a more important issue here, since we need to keep track of the complex refractive index as well as its real and imaginary parts.

By using Eq. (4.27), n can be expressed as

$$n = n_0 + \frac{Ne^2}{2n_0 m_e \varepsilon_0} \frac{(\omega_0^2 - \omega^2)}{(\omega_0^2 - \omega^2)^2 + \gamma^2\omega^2} \tag{4.36}$$

We leave it as an exercise for the reader (see Prob. 4.2) to show that when ω is very close to the resonance frequency, Eq. (4.36) simplifies to

$$n = n_0 + \frac{Ne^2}{4n_0 m_e \varepsilon_0 \omega} \frac{\omega_0 - \omega}{(\omega_0 - \omega)^2 + (\gamma/2)^2} \tag{4.37}$$

Figure 4.3 shows the variation of n as a function of the angular frequency ω around the resonance peak. Note that at resonance, $\omega = \omega_0$ and the refractive index is equal to the background refractive index n_0. Depending on whether n increases or decreases with the frequency, we identify two different regions. In the so-called normal dispersion regime, corresponding to regions I and III in Fig. 4.3, n increases with increasing frequency. However, right around the resonance frequency ω_0 (region II), n decreases with ω. This is known as anomalous dispersion.

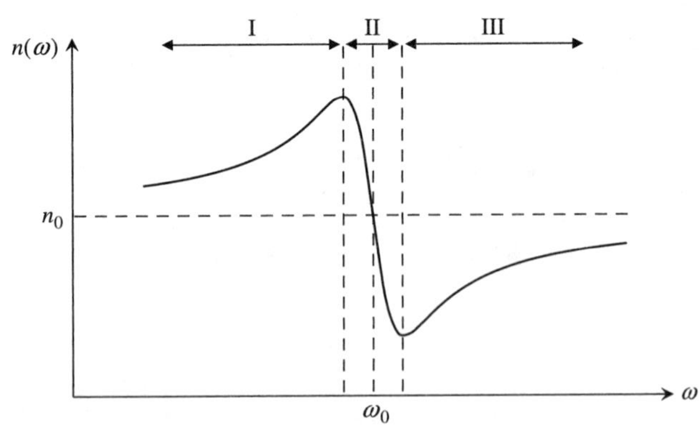

FIGURE 4.3 Variation of the real part n of the complex refractive index around the resonance frequency ω_0. Regions I and III show normal dispersion, whereas region II shows anomalous dispersion.

What about the imaginary part of the complex refractive index? Since the complex refractive index is of the form $\bar{n} = n - in''$, the wave vector of the propagating electromagnetic wave will become

$$k = k_0(n - in'') = k_0 n - ik_0 n'' \tag{4.38}$$

The resulting electric field will have an exponentially decaying amplitude given by

$$\begin{aligned}\vec{E}(\vec{r},t) &= \hat{x}E_0 \exp(i(\omega t - kz)) = \hat{x}E_0 \exp(i(\omega t - (k_0 n - ik_0 n'')z)) \\ &= \hat{x}E_0 \exp(i(\omega t - k_0 nz))\exp\left(-\frac{\alpha(\omega)}{2}z\right)\end{aligned} \tag{4.39}$$

where the absorption coefficient $\alpha(\omega)$ is related to the imaginary part of the refractive index through

$$\alpha(\omega) = 2k_0 n''(\omega) = -2\frac{\omega}{c}\left(\frac{\chi_b''}{2n_0} + \frac{\chi_r''}{2n_0}\right) = \alpha_b + \alpha_r(\omega) \tag{4.40}$$

With this definition of $\alpha(\omega)$, the corresponding intensity of the propagating wave decays as

$$I(z) = I_0 \exp(-\alpha(\omega)z) \tag{4.41}$$

The result in Eq. (4.41) is the well-known Beer's law. Once again, the damping force included in the model is responsible for the dissipation of energy, and hence the attenuation of the propagating electromagnetic wave. By using Eq. (4.29), the resonant contribution to the absorption coefficient can be expressed as

$$\alpha_r(\omega) = -\frac{\omega}{c}\frac{\chi_r''}{n_0} = \frac{\omega}{n_0 c}\frac{Ne^2}{m_e \varepsilon_0}\frac{\gamma\omega}{\left(\omega_0^2 - \omega^2\right)^2 + \gamma^2\omega^2} \tag{4.42}$$

Similar to the refractive index, it can be shown (see Prob. 4.2) that when the frequency is close to the center of the absorption band, the resonant contribution to the absorption coefficient takes the form

$$\alpha_r(\omega) = \frac{Ne^2}{4n_0 cm_e \varepsilon_0}\frac{\gamma}{(\omega_0 - \omega)^2 + (\gamma/2)^2} \tag{4.43}$$

As can be seen from above, the higher the concentration (N) of the absorbing atoms, the larger the absorption coefficient. The absorption coefficient per atom is also an important parameter that describes the strength of absorption due to a particular atom type and for a particular

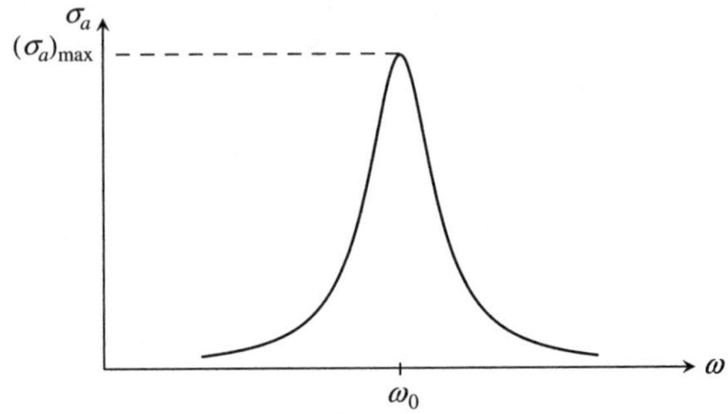

FIGURE 4.4 Variation of the absorption cross section as a function of the angular frequency ω near an absorption resonance. For an electron under the action of a damping force, a linear restoring force and the driving electromagnetic force, the shape of the resonance curve is described by the Lorentzian distribution.

resonance. This quantity is known as the absorption cross section σ_a and is given by

$$\sigma_a = \frac{\alpha_r}{N} = \frac{e^2}{4n_0 c m_e \varepsilon_0} \frac{\gamma}{(\omega_0 - \omega)^2 + (\gamma/2)^2} \tag{4.44}$$

The variation of σ_a as a function of the angular frequency ω is further shown in Fig. 4.4. The particular functional form of σ_a given in Eq. (4.44) is known as the Lorentzian distribution. It is customary to express the spectral dependence of the absorption cross section in terms of a normalized distribution function. This distribution is also known as the spectral lineshape function or simply the lineshape function. For a Lorentzian distribution given by Eq. (4.44) and shown in Fig. 4.4, the corresponding normalized lineshape function expressed in terms of frequency ν will be denoted by $g_L(\nu)$. Later, in the semi-classical treatment of stimulated emission and stimulated absorption, we will make use of the normalized lineshape functions. Since $\omega = 2\pi\nu$, we can express $g_L(\nu)$ as

$$g_L(\nu) = \frac{A}{(\nu_0 - \nu)^2 + (\gamma/4\pi)^2} \tag{4.45}$$

where the constant A is chosen so that the normalization condition

$$\int_{-\infty}^{\infty} d\nu g_L(\nu) = 1 \tag{4.46}$$

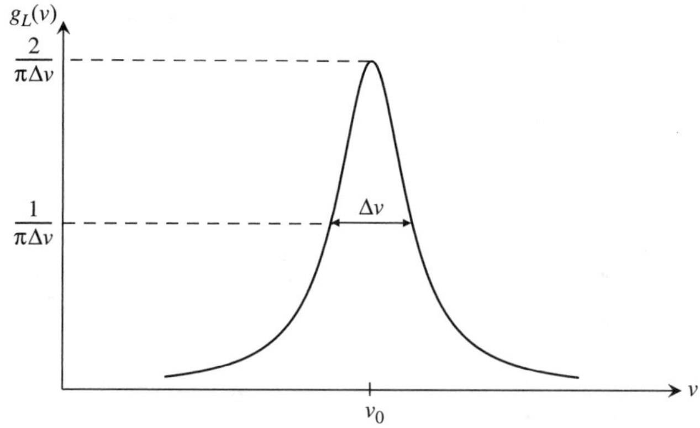

FIGURE 4.5 Sketch of the normalized Lorentzian lineshape function $g_L(\nu)$ with a spectral width (full width at half maximum, FWHM) of $\Delta\nu$.

is satisfied. It can be shown that (see Prob. 4.3) the normalized Lorentzian distribution $g_L(\nu)$ is of the form

$$g_L(\nu) = \frac{\Delta\nu/2\pi}{(\nu-\nu_0)^2 + (\Delta\nu/2)^2} \tag{4.47}$$

where the full width at half maximum (FWHM) is given by $\Delta\nu = \gamma/2\pi$. In other words, it is the energy loss due to the presence of the damping force that leads to the finite width of the absorption line in this case. The width of the resonance curve is directly proportional to the damping parameter. As the losses due to the damping force become smaller, the resonance peak becomes sharper with a narrower spectral width. The normalized Lorentzian distribution $g_L(\nu)$ is sketched in Fig. 4.5.

4.1.3 The Oscillator Strength

The classical Lorentz model has a serious shortcoming which we can see by examining the area under the σ_a–ν curve (Fig. 4.4). To do this, we first express the absorption cross section in terms of the normalized Lorentzian lineshape function. By using Eqs. (4.44) and (4.47), we can further express $\sigma_a(\nu)$ as (see Prob. 4.4)

$$\sigma_a(\nu) = \frac{e^2}{4n_0 m_e \varepsilon_0 c} g_L(\nu) \tag{4.48}$$

By assuming that the other resonance bands are sufficiently far away, we can integrate both sides of Eq. (4.48) to obtain the integrated absorption cross section $(\sigma_a)_T$. Since $g_L(\nu)$ is normalized, we get

$$(\sigma_a)_T = \frac{e^2}{4n_0 m_e \varepsilon_0 c} \tag{4.49}$$

Equation (4.49) gives the area under the $\sigma_a - \nu$ curve for a particular absorption band. Note that according to our classical model, the only adjustable parameter which affects the integrated absorption cross section is the background refractive index n_0. Typically, however, there is a much more pronounced variation in the integrated absorption cross section for different absorption lines, which cannot be accounted for by considering changes in the background refractive index. This deficiency of the classical model can be remedied and the correct form for $(\sigma_a)_T$ can be calculated by using quantum mechanics, which takes into account the different spatial distributions of the wave functions for different energy levels involved in the optical transition. The full quantum mechanical derivation of $(\sigma_a)_T$ is not treated in this text. Instead, we introduce a phenomenological parameter known as the oscillator strength f to account for variations in the strength of the absorption band. We will express the absorption cross section for a general band (even if it is not described by a Lorentzian distribution) in the form

$$\sigma_a(\nu) = \frac{e^2 f}{4n_0 m_e \varepsilon_0 c} g(\nu) \tag{4.50}$$

where f is the oscillator strength, n_0 is the background refractive index, and $g(\nu)$ is the normalized lineshape function, not necessarily Lorentzian, satisfying Eq. (4.46).

How can we experimentally determine the oscillator strength of an absorption line? Well, experimentally, we can measure the resonant absorption coefficient $\alpha_r(\nu)$ of the material after correcting for background losses due to reflection and scattering. Let the sample length be d_s. The intensity transmission T of the sample in the presence of resonant absorption will be given by

$$T(\nu) = \exp(-\alpha_r(\nu) d_s) \tag{4.51}$$

Hence, $\alpha_r(\nu)$ can be deduced from

$$\alpha_r(\nu) = -\frac{1}{d_s} \ln(T(\nu)) \tag{4.52}$$

If the total center density in the sample is N, then we can calculate $\sigma_a(\nu)$ from

$$\sigma_a(\nu) = \frac{\alpha_r(\nu)}{N} \tag{4.53}$$

The area under the σ_a–ν curve will be related to the oscillator strength f through

$$\int d\nu \frac{\alpha_r(\nu)}{N} = f \frac{e^2}{4n_0 c\varepsilon_0 m_e} \tag{4.54}$$

The following example provides an application of these equations.

Example A hypothetical absorbing medium has a peak resonance wavelength of 1250 nm. It is given that the resonance has a Lorentzian lineshape with an oscillator strength of $f = 3\times10^{-4}$ and a spectral width of 20 THz. The background refractive index is 2.

(a) What is the maximum value of the absorption cross section?
(b) Suppose that a tunable monochromatic light is sent through this sample. What is the maximum absorption that can be recorded as the wavelength of the source is tuned across the absorption band? Take the length of the sample as 15 mm and the density of the absorbing atoms as $10^{19}\,\text{cm}^{-3}$.

Solution

(a) By using Eq. (4.50), the resonant absorption cross section can be calculated from

$$\sigma_a(\nu) = \frac{e^2 f}{4n_0 m_e \varepsilon_0 c} g_L(\nu)$$

At $\nu = \nu_0$,

$$g_L(\nu) = \frac{2}{\pi\Delta\nu} = \frac{2}{\pi(20\times10^{12})} = 3.18\times10^{-14}\ \text{Hz}^{-1}$$

By using $f = 3\times10^{-4}$, $n_0 = 2$ and the known values of the fundamental constants, we get

$$(\sigma_a)_{\max} = \sigma_a(\nu = \nu_0) = 1.26\times10^{-19}\,\text{cm}^2$$

(b) Maximum absorption occurs at the wavelength where $g_L(\nu)$ peaks. Hence, $(\sigma_a)_{\max} = 1.26\times10^{-19}\,\text{cm}^2$

$$\alpha_{\max} = N\sigma_{\max} = 10^{19}(1.26\times10^{-19}) = 1.26\ \text{cm}^{-1}$$

Hence maximum absorption $= 1 - \exp[-(1.5)(1.26)] = 85\%$.

4.1.4 The Sellmeier Equation

Another very important approximation to the dielectric constant can be made in the transparency region of the material where ω is away from all of the absorption resonances. This regime is of great practical importance since all optical windows, transmission optics such as substrates or lenses are used in their transparency range. In this case, we can neglect absorption and also assume that $|\omega_{0j} - \omega| >> \gamma_j$ to obtain

$$\frac{\varepsilon}{\varepsilon_0} = \bar{n}^2 = n^2 = 1 + \sum_{j=1}^{Z} \frac{Ne^2}{m_e\varepsilon_0} \frac{1}{\left(\omega_{0j}^2 - \omega^2\right)} \tag{4.55}$$

Conventionally, this relation is expressed in terms of the vacuum wavelength λ given by $\lambda = 2\pi c/\omega$, resulting in

$$n^2 - 1 = \sum_{j=1}^{Z} \frac{Ne^2}{4\pi^2 c^2 m_e \varepsilon_0} \frac{\lambda^2 \lambda_{0j}^2}{\lambda^2 - \lambda_{0j}^2} \tag{4.56}$$

For a realistic material, one typically includes three to five terms expressed in the form

$$n^2 - 1 = \sum_j \frac{A_j \lambda^2}{\lambda^2 - \lambda_j^2} \tag{4.57}$$

The fitting parameters A_j and λ_j are determined by doing least squares fit to the experimentally measured refractive index data of the sample. The empirical relation in Eq. (4.57) was named after W. Sellmeier, who proposed its use in the nineteenth century.[11] The classic example that is used in photonics is the Sellmeier equation for fused silica which is given by[12]

$$n^2 - 1 = \frac{0.6961663\lambda^2}{\lambda^2 - (0.0684043)^2} + \frac{0.4079426\lambda^2}{\lambda^2 - (0.1162414)^2} + \frac{0.8974794\lambda^2}{\lambda^2 - (9.896161)^2} \tag{4.58}$$

Above, λ is in μm and the index data were measured near room temperature ($T = 293\text{K}$). Based on the data given in Eq. (4.58), Fig. 4.6 shows the calculated variation of the refractive index n as a function

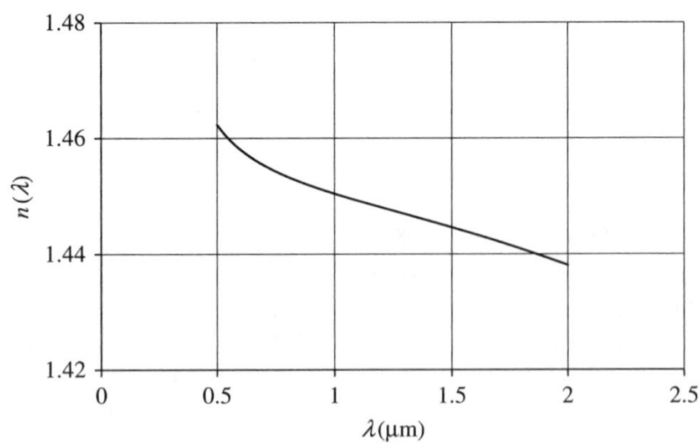

FIGURE 4.6 Calculated variation of the refractive index n as a function of the vacuum wavelength λ in the 0.5–2 μm range for fused silica. The plot was generated by using the index data in Ref. 12.

of the vacuum wavelength λ in the 0.5–2 μm region for fused silica. Note that the refractive index monotonically decreases with increasing wavelength in this wavelength range.

4.1.5 Kramers-Kronig Relations

Going through the previous analysis of absorption and dispersion, one naturally asks the question of whether there is a more general relationship between the real (dispersive) and imaginary (absorption) parts of the complex dielectric constant. The answer is indeed yes! And such relations, known as the Kramers-Kronig relations, may be derived based on the principle of causality.

We start with the general relation between the displacement vector $\vec{D}$ and electric field $\vec{E}$ expressed in the complex angular frequency domain:

$$\vec{D}(\bar{\omega}) = \varepsilon(\bar{\omega})\vec{E}(\bar{\omega}) = \varepsilon_0(1+\chi(\bar{\omega}))\vec{E}(\bar{\omega}) \tag{4.59}$$

In Eq. (4.59), $\bar{\omega}$ is the complex angular frequency, which can be written in terms of the real and imaginary components ω_r and ω_i as

$$\bar{\omega} = \omega_r + i\omega_i \tag{4.60}$$

The product relation in the frequency domain given by Eq. (4.59) implies that there is a nonlocal connection between $\vec{D}(t)$ and $\vec{E}(t)$ in the time domain. To see how $\vec{D}(t)$ and $\vec{E}(t)$ will be related, we introduce the Fourier transform pairs defined as

$$\vec{D}(t) = \int_{-\infty}^{\infty} d\bar{\omega}\vec{D}(\bar{\omega})\exp(i\bar{\omega}t) \tag{4.61}$$

and

$$\vec{D}(\bar{\omega}) = \frac{1}{2\pi}\int_{-\infty}^{\infty} dt\vec{D}(t)\exp(-i\bar{\omega}t) \tag{4.62}$$

By using Eq. (4.59), $\vec{D}(t)$ can be written as

$$\begin{aligned}\vec{D}(t) &= \int_{-\infty}^{+\infty} d\bar{\omega}\varepsilon_0(1+\chi(\bar{\omega}))\vec{E}(\bar{\omega})\exp(i\bar{\omega}t)\\ &= \varepsilon_0\vec{E}(t) + \varepsilon_0\int_{-\infty}^{\infty} d\bar{\omega}\chi(\bar{\omega})\vec{E}(\bar{\omega})\exp(i\bar{\omega}t)\end{aligned} \tag{4.63}$$

Now, let the Fourier transform pair for the susceptibility function χ be defined according to

$$\chi(\bar{\omega}) = \frac{1}{2\pi}\int_{-\infty}^{\infty} dt\chi(t)\exp(-i\bar{\omega}t) \tag{4.64}$$

and

$$\chi(t) = \int_{-\infty}^{\infty} d\bar{\omega}\chi(\bar{\omega})\exp(i\bar{\omega}t) \tag{4.65}$$

By using Eqs. (4.63) and (4.64), the displacement vector can be expressed as

$$\begin{aligned}\vec{D}(t) &= \varepsilon_0\vec{E}(t) + \varepsilon_0\int_{-\infty}^{\infty} d\bar{\omega}\frac{1}{2\pi}\int_{-\infty}^{\infty} d\tau\chi(\tau)\exp(-i\bar{\omega}\tau)\vec{E}(\bar{\omega})\exp(i\bar{\omega}t)\\ &= \varepsilon_0\vec{E}(t) + \frac{\varepsilon_0}{2\pi}\int_{-\infty}^{\infty} d\tau\chi(\tau)\int_{-\infty}^{\infty} d\bar{\omega}\vec{E}(\bar{\omega})\exp[i\bar{\omega}(t-\tau)]\\ &= \varepsilon_0\vec{E}(t) + \frac{\varepsilon_0}{2\pi}\int_{-\infty}^{\infty} d\tau\chi(\tau)\vec{E}(t-\tau)\end{aligned} \tag{4.66}$$

As can be seen, by nonlocality, we mean that the displacement vector at time t depends on the value of the electric field at other times. Due to causality, we require that the displacement vector depends only on the previous values of $\vec{E}$. In other words, the function $\chi(\tau)$ is such that it is zero for $\tau < 0$. Hence we express the causal form of the above equation as

$$\vec{D}(t) = \varepsilon_0\vec{E}(t) + \frac{\varepsilon_0}{2\pi}\int_{0}^{\infty} d\tau\chi(\tau)\vec{E}(t-\tau) \tag{4.67}$$

Let us next investigate the characteristics of the dielectric function $\varepsilon(\bar{\omega})$ in the complex $\bar{\omega}$ plane. By using Eq. (4.64), $\varepsilon(\bar{\omega})$ becomes

$$\begin{aligned}\frac{\varepsilon(\bar{\omega})}{\varepsilon_0} &= 1 + \chi(\bar{\omega}) = 1 + \frac{1}{2\pi}\int_{-\infty}^{\infty} d\tau\chi(\tau)\exp(-i\bar{\omega}\tau)\\ &= 1 + \frac{1}{2\pi}\int_{0}^{\infty} d\tau\chi(\tau)\exp(-i\bar{\omega}\tau)\end{aligned} \tag{4.68}$$

due to causality. From Eq. (4.60), we further have

$$\frac{\varepsilon(\bar{\omega})}{\varepsilon_0} = 1 + \frac{1}{2\pi}\int_0^{\infty} d\tau\chi(\tau)\exp[-i(\omega_r + i\omega_i)\tau]$$

$$= 1 + \frac{1}{2\pi}\int_0^{\infty} d\tau\chi(\tau)\exp(-i\omega_r\tau)\exp(\omega_i\tau) \tag{4.69}$$

Let us assume that $\chi(t)$ remains finite at large times as t tends to ∞. Then $\varepsilon(\bar{\omega})$ and its derivative exist for $\omega_i < 0$. In the jargon of complex analysis, we say that the function $\varepsilon(\bar{\omega})/\varepsilon_0$ is an analytical function in the lower half of the complex $\bar{\omega}$ plane (i.e., $\omega_i < 0$). If $\varepsilon(\bar{\omega})$ is an analytical function, then $\chi(\bar{\omega})$ will also be an analytical function in the lower half of the complex $\bar{\omega}$ plane. To obtain the Kramers-Kronig relations, we integrate $\chi(\bar{\omega})/(\bar{\omega}-\omega)$ around the contour shown in Fig. 4.7. Here, ω is one specific value of the real angular frequency. Since $\chi(\bar{\omega})$ is analytic and no poles are included within the region bounded by the contour, we know from the Cauchy integral theorem that

$$\oint \frac{d\bar{\omega}\chi(\bar{\omega})}{\bar{\omega}-\omega} = 0 \tag{4.70}$$

The contour integral can be broken up into four contributions given by

$$\oint \frac{d\bar{\omega}\chi(\bar{\omega})}{\bar{\omega}-\omega} = \int_{-\infty}^{\omega-\delta} \frac{d\bar{\omega}\chi(\bar{\omega})}{\bar{\omega}-\omega} + \int_{\omega+\delta}^{\infty} \frac{d\bar{\omega}\chi(\bar{\omega})}{\bar{\omega}-\omega} + \int_{C_1} \frac{d\bar{\omega}\chi(\bar{\omega})}{\bar{\omega}-\omega} + \lim_{R\to\infty}\int_{C_2} \frac{d\bar{\omega}\chi(\bar{\omega})}{\bar{\omega}-\omega} \tag{4.71}$$

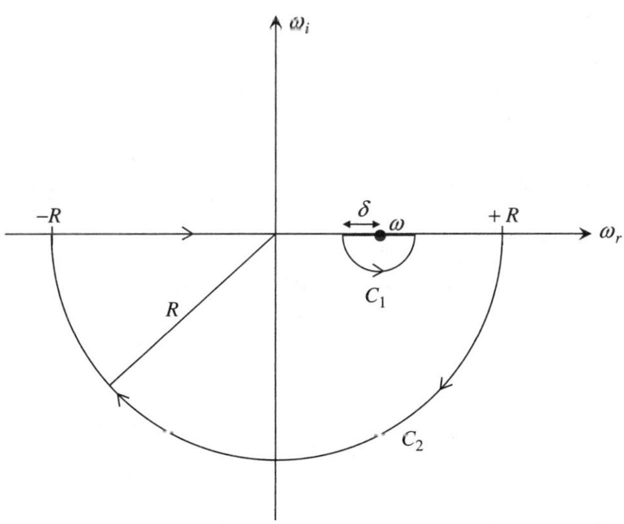

FIGURE 4.7 The contour used in the complex $\bar{\omega}$ plane to derive the Kramers-Kronig relations.

Since $\chi(\bar{\omega})$ is analytic, we have

$$\lim_{R\to\infty} \int_{C_2} \frac{d\bar{\omega}\chi(\bar{\omega})}{\bar{\omega}-\omega} = 0 \tag{4.72}$$

Furthermore, as the radius δ of the semicircle tends to zero, we can write the first two terms on the right-hand side of Eq. (4.71) in terms of a principal-value integral as

$$\lim_{\delta\to 0}\left(\int_{-\infty}^{\omega-\delta} \frac{d\bar{\omega}\chi(\bar{\omega})}{\bar{\omega}-\omega} + \int_{\omega+\delta}^{\infty} \frac{d\bar{\omega}\chi(\bar{\omega})}{\bar{\omega}-\omega}\right) = P\int_{-\infty}^{\infty} \frac{d\bar{\omega}\chi(\bar{\omega})}{\bar{\omega}-\omega} \tag{4.73}$$

where the symbol P in Eq. (4.73) designates the principal-value integral.

Now, what happens to the contour integral around C_1 as $\delta \to 0$? To see this, we make a change of variables according to $\bar{\omega}-\omega = \delta e^{i\phi}$ with $d\bar{\omega} = i\delta e^{i\phi} d\phi$ to obtain

$$\lim_{\delta\to 0} \int_{C_1} \frac{d\bar{\omega}\chi(\bar{\omega})}{\bar{\omega}-\omega} = \lim_{\delta\to 0} \int_{\pi}^{0} d\phi \frac{i\delta e^{i\phi}}{\delta e^{i\phi}} \chi(\omega+\delta e^{i\phi}) = -i\pi\chi(\omega) \tag{4.74}$$

Putting these contributions together in Eq. (4.71), we get

$$P\int_{-\infty}^{\infty} \frac{d\bar{\omega}\chi(\bar{\omega})}{\bar{\omega}-\omega} - i\pi\chi(\omega) = 0 \tag{4.75}$$

or

$$[\chi_r(\omega) + i\chi_i(\omega)] = \frac{1}{i\pi} P\int_{-\infty}^{\infty} d\bar{\omega} \frac{\chi_r(\bar{\omega}) + i\chi_i(\bar{\omega})}{\bar{\omega}-\omega} \tag{4.76}$$

Separating the real and imaginary parts leads to the Kramers-Kronig relations

$$\chi_r(\omega) = \frac{1}{\pi} P\int_{-\infty}^{\infty} d\bar{\omega} \frac{\chi_i(\bar{\omega})}{\bar{\omega}-\omega} \tag{4.77}$$

and

$$\chi_i(\omega) = -\frac{1}{\pi} P\int_{-\infty}^{\infty} d\bar{\omega} \frac{\chi_r(\bar{\omega})}{\bar{\omega}-\omega} \tag{4.78}$$

With no exaggeration, this is one of the most elegant results in physics based on the causality principle. Historically, these relations were first derived independently by Kronig and Kramers in 1926 and 1927, respectively.[1, 13] In fact, in the first article published by R. L. Kronig in 1926, the author mentions in a footnote that similar results had been obtained by H. A. Kramers.[14] Two functions related to each other through the principal-value integrals given by Eqs. (4.77) and (4.78) are said to form a Hilbert transform pair.

A one-sentence summary of the equations would be: You cannot have absorption without dispersion and vice versa. In other words, suppose that we have access to an ideal measuring system with which the absorption of a material can be measured over the full frequency range. Then, based on the knowledge of the absorption spectrum alone, one can then deduce, using Eq. (4.77), the full frequency dependence of the refractive index at any desired frequency ω. In real cases, evaluation of the principal value integrals in Eqs. (4.77) and (4.78) requires numerical computation. Further discussion of Kramers-Kronig relations may be found in Refs. 1, 3, and 15.

4.1.6 Spectral Broadening Mechanisms

Absorption and emission lines are spectrally broadened as a result of several mechanisms. Actually, we have already seen one example of this, namely the broadening of the absorption lines due to the damping of the electron motion, characterization by the γ parameter. As we discussed in the previous section, this leads to a linewidth that is proportional to γ. There are two other important spectral broadening mechanisms which we will investigate in this section:

1. Collision broadening
2. Doppler broadening

These two mechanisms are quite distinct in character. Collision broadening is an example of what we will call homogeneous broadening. In this case, at least ideally, all atoms in the medium have the same spectral response. On the other hand, Doppler broadening is an example of inhomogeneous broadening where different groups of atoms have a different spectral response. Later, after discussing the elementary quantum model for optical transitions, we will also look at another broadening mechanism due to the finite fluorescence lifetime of excited atoms. This will be treated in the section "Spontaneous Emission" in Sec. 4.2.2.

Collision Broadening

Let us consider a gaseous medium consisting of atoms that move randomly in all directions. In such a case, elastic collisions will cause abrupt changes in the oscillation phase of the electrons driven by the Lorentz force. The effect of collisions on the oscillating electron is schematically shown in Fig 4.8.

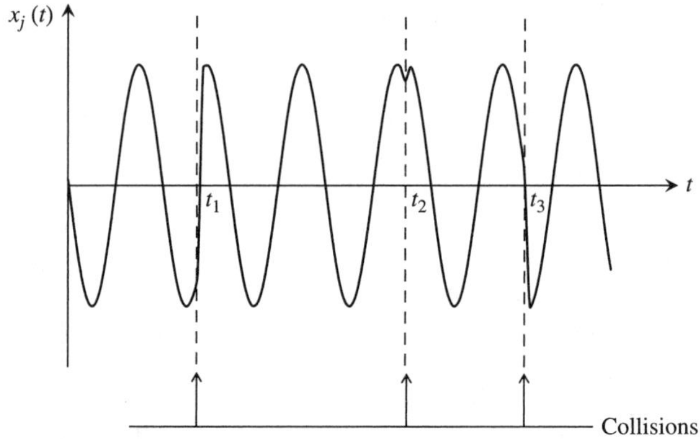

FIGURE 4.8 Effect of collisions on an oscillating electron. Abrupt random shifts occur in the oscillation phase at the times t_1, t_2, and t_3.

The times designated as t_1, t_2, and t_3 in Fig. 4.8 are random instants when the atom undergoes a collision with other atoms moving around in the medium. After each collision, the oscillation of the electron resumes at the same frequency but with a different initial phase. We can show in this case that, for an ensemble of atoms, the effect of collisions is equivalent to having a damping force whose damping constant γ is related to the mean frequency of collisions.

To obtain a quantitative expression for the spectral width of collision broadening, we consider the damped oscillatory motion of one of the electrons after the driving field has been turned off, and we analyze this with and without collisions. Let us again consider the jth electron in the absence of collisions. When the field is turned off, the equation of motion becomes

$$\frac{d^2x_j}{dt^2}+\gamma_j\frac{dx_j}{dt}+\omega_{0j}^2x_j=0 \tag{4.79}$$

To solve for $x_j(t)$ we assume a solution of the form $x_j(t)=x_{j0}\exp(rt)$, where the exponent r satisfies the quadratic equation

$$r^2+\gamma_j r+\omega_{0j}^2=0 \tag{4.80}$$

The two roots are given by

$$r=\frac{-\gamma_j\pm i\sqrt{4\omega_{0j}^2-\gamma_j^2}}{2}=\frac{-\gamma_j\pm i\omega_j}{2} \tag{4.81}$$

where ω_j becomes

$$\omega_j = \sqrt{4\omega_{0j}^2 - \gamma_j^2} \tag{4.82}$$

The resulting dipole moment of the electron becomes

$$\vec{p}_j(t) = -\hat{x}ex_j(t) = -\hat{x}e\exp\left[\frac{-\gamma_j t}{2}\right]\left(x_1\exp\left[\frac{-i\omega_j t}{2}\right] + x_2\exp\left[\frac{+i\omega_j t}{2}\right]\right) \tag{4.83}$$

Above, x_1 and x_2 are constants that depend on the initial conditions. If there are no collisions in the medium, each atom will have the same temporal response (at least ideally), and the medium polarization becomes

$$\vec{P}(t) = N\vec{p}_j(t) \tag{4.84}$$

where N is the total atomic density. On the other hand, if there are collisions, then the total polarization will be smaller because, as time progresses, the number of atoms which have not yet undergone collisions will decrease. Here we assume that since the oscillation phase of the atoms becomes random after collisions, the total polarization due to those atoms which have undergone collisions averages out to zero. Hence, we now need to recalculate the polarization of the medium by using the number of atoms which have not yet undergone a collision. We can calculate this by introducing the parameter T_2 which is the mean free time between collisions. The quantity $1/T_2$ can then be interpreted as the probability per unit time that an atom will undergo a collision. If we denote by dP_C, the probability that the atom will undergo a collision in a small time interval dt, dP_C can be calculated from

$$dP_C = \frac{dt}{T_2} \tag{4.85}$$

Now, let us calculate $P_{NC}(t)$, the probability that the atom will not undergo a collision up to time t. To do this, we subdivide the time interval $[0,t]$ into M equal segments as shown in Fig. 4.9.

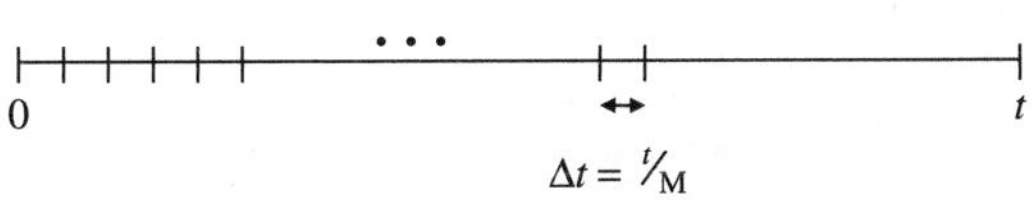

FIGURE 4.9 The time interval $[0,t]$ subdivided into M equal segments. This is used in calculating the probability that the atom will not undergo a collision up to time t.

$P_{NC}(t)$ will be given by

$$P_{NC}(t) = \prod_{j=1}^{M} P_{NCj} \tag{4.86}$$

where P_{NCj} is the probability that the atom does not undergo a collision in the jth interval of duration Δt and is given by

$$P_{NCj} = 1 - \frac{\Delta t}{T_2} = \left(1 - \frac{t}{MT_2}\right) \tag{4.87}$$

Hence, $P_{NC}(t)$ becomes

$$P_{NC}(t) = \left(1 - \frac{t}{MT_2}\right)^M \tag{4.88}$$

To obtain the exact expression for $P_{NC}(t)$, we take M to ∞. We leave it as an exercise (see Prob. 4.5) for the reader to show that this gives

$$P_{NC}(t) = \lim_{M\to\infty}\left(1 - \frac{t}{MT_2}\right)^M = \exp\left(-\frac{t}{T_2}\right) \tag{4.89}$$

As we see from above, the probability that the atom will not undergo a collision will exponentially decrease with time. Hence, the number $N'(t)$ of the atoms which have not yet undergone a collision becomes

$$N'(t) = N\exp\left(-\frac{t}{T_2}\right) \tag{4.90}$$

resulting in the modified medium polarization

$$\vec{P}'(t) = -\hat{x}Ne\exp\left[-\left(\frac{\gamma_j}{2} + \frac{1}{T_2}\right)t\right]\left[x_1\exp\left(-i\frac{\omega_j t}{2}\right) + x_2\exp\left(+i\frac{\omega_j t}{2}\right)\right] \tag{4.91}$$

In other words, the effective polarization decay rate becomes $(\gamma_j/2 + 1/T_2)$. Starting from the equation of motion for the electron, we could have obtained the same result by assuming that the effective damping force on the electron is given by

$$\vec{F}_d = -\hat{x}m_e\left(\gamma_j + \frac{2}{T_2}\right)\frac{dx_j}{dt} \tag{4.92}$$

Although the foregoing analysis is not very rigorous, we conclude, based on Eq. (4.91) that the lineshape function in the presence of collisions still remains as Lorentzian, but its overall width $\Delta\nu$ becomes

$$\Delta\nu = \frac{1}{2\pi}\left(\gamma_j + \frac{2}{T_2}\right) \tag{4.93}$$

In real gaseous systems, T_2 is related to the density of the atoms and hence to the pressure. At one extreme, where the pressure is very low, the mean free time between the collisions will be very long ($T_2 \to \infty$) and $\Delta\nu = \gamma_j/2\pi$. As the pressure is increased, the collision rate will go up and we expect $1/T_2$ to increase with pressure. Empirically, we can say that due to homogeneous broadening mechanisms in gas, the spectral width of the absorption lines will vary with the gas pressure P_g according to

$$\Delta\nu = \Delta\nu_0 + \Delta\nu_c\left(\frac{P_g}{P_{g0}}\right) \tag{4.94}$$

where $\Delta\nu_0$ is the natural width independent of pressure and the term $\Delta\nu_c/P_{g0}$ gives the change in the spectral width per unit change in the gas pressure. $\Delta\nu_c/P_{g0}$ can be empirically determined for different types of gaseous systems. As an example, the collision-induced spectral broadening of the He-Ne transition with a 1:1 He:Ne mixture at the wavelength of 632.8 nm is of the order of $\Delta\nu_c/P_{g0} \approx 100\,\text{MHz/torr}$.[16, 17]

Doppler Broadening

Doppler broadening results because of the fact that the spectral response of atoms is shifted due to the relative motion between the atom and the light source. We know from special relativity that the apparent frequency ν' of a wave source moving away from an observer with speed u is given by (see, for example, Ref. 18)

$$\nu' = \nu\sqrt{\frac{1 - u/c}{1 + u/c}} \tag{4.95}$$

where ν is the frequency measured by an observer at rest with respect to the source. This is schematically shown in Fig. 4.10. If u is much smaller than c, the above relation reduces to

$$\nu' \cong \nu\left(1 - \frac{u}{c}\right) \tag{4.96}$$

This phenomenon, that is, the shift of the apparent frequency due to the relative motion between the source and the observer, is known as

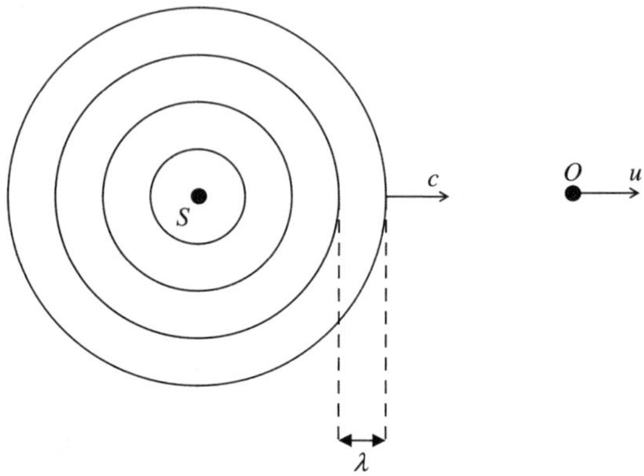

FIGURE 4.10 An observer O moving with a speed u away from a stationary source S of electromagnetic waves. The apparent frequency ν' detected by the observer will be less than the frequency $\nu = c/\lambda$ due to the Doppler effect.

the Doppler effect. In a gaseous medium, atoms move with different velocities. Since the Doppler shift gives rise to variation in the resonance frequencies, and since there is a distribution of velocities, we expect to see an inhomogeneous broadening of the absorption lines. To obtain the Doppler lineshape of a gaseous medium, consider an atom which is moving away from the electromagnetic wave source with a speed of u. Assuming that there are no other broadening mechanisms, the Doppler shifted frequency $\nu(1-u/c)$ of the wave should equal the resonance frequency ν_0 so that absorption can take place.

We therefore have

$$\nu\left(1-\frac{u}{c}\right)=\nu_0 \tag{4.97}$$

or

$$\nu=\nu_0\left(1-\frac{u}{c}\right)^{-1}\cong\nu_0\left(1+\frac{u}{c}\right) \tag{4.98}$$

In other words, we can say that the group of atoms moving with speed u have their resonance frequency shifted to $\nu_0(1+u/c)$. We then ask the following question: What fraction of atoms will absorb radiation with frequency in the range ν to $\nu+d\nu$? Clearly, these are the atoms that move with velocity in the range u to $u + du$ where, from Eq. (4.98), u and ν are related through

$$u=\frac{c}{\nu_0}(\nu-\nu_0) \tag{4.99}$$

From the Maxwell-Boltzmann velocity distribution, the fractional df_{MB} of atoms moving with velocity in the range u to $u + du$ is given by

$$df_{MB} = \left(\frac{m_a}{2\pi k_B T}\right)^{\frac{1}{2}} \exp\left[-\frac{m_a u^2}{2k_B T}\right] du \tag{4.100}$$

Above, the subscript MB stands for "Maxwell-Boltzmann," k_B is the Boltzmann constant ($1.38\times10^{-23}\,J/K$), m_a is the mass of the atom, and T is the absolute temperature measured in degrees Kelvin. Since $du = (c/\nu_0)d\nu$, we see that the Doppler broadened lineshape function $g_D(\nu)$ should be of the form

$$g_D(\nu) = \left(\frac{m_a c^2}{2\pi k_B T \nu_0^2}\right)^{\frac{1}{2}} \exp\left[-\frac{m_a c^2(\nu-\nu_0)^2}{2k_B T \nu_0^2}\right] \tag{4.101}$$

The Doppler lineshape function is shown in Fig. 4.11.

It can be further shown that (see Prob. 4.6) the width (FWHM) of the Doppler lineshape function is given by

$$\Delta\nu_D = \nu_0\left(\frac{8k_B T}{m_a c^2}\ln 2\right)^{\frac{1}{2}} \tag{4.102}$$

Note that the Doppler linewidth depends on the temperature and also on the center frequency of the absorption band. In the limit where homogeneous broadening effects are negligible, the Doppler linewidth is proportional to the square root of the absolute temperature.

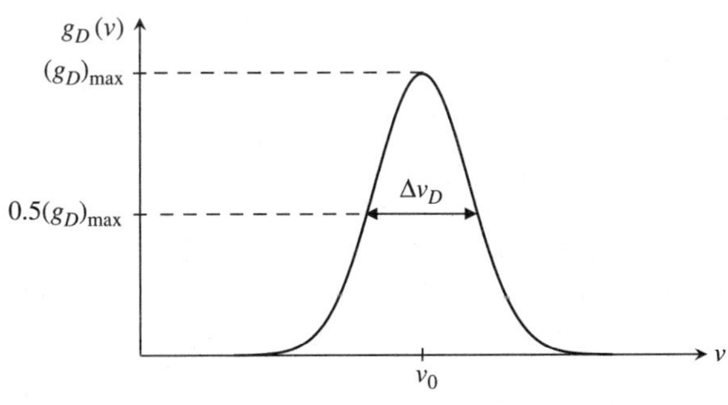

FIGURE 4.11 The Doppler lineshape function. The peak value $(g_D)_{max}$ is given by $\sqrt{m_a c^2 / 2\pi k_B T \nu_0^2}$.

Example Let us calculate the Doppler linewidth of the He-Ne laser transition at 633 nm. Here, the transition is due to neon atoms for which the atomic mass becomes

$$m_a = \frac{20.18\times 10^{-3}}{6.02\times 10^{23}} = 3.352\times 10^{-26}\,\text{kg. At } T = 300\text{ K, we get}$$

$$\Delta\nu_D = \left(\frac{3\times 10^8}{633\times 10^{-9}}\right)\sqrt{\frac{8(1.38\times 10^{-23})(300)\ln 2}{(3.352\times 10^{-26})(3\times 10^8)^2}} = 1.31\times 10^9 \text{ Hz} = 1.31 \text{ GHz}$$

To see the effect of the center frequency and atomic mass on the linewidth, let's repeat the above calculation for the CO_2 transition at $\lambda = 10.6$ μm at 300 K. For $\lambda_0 = 10.6\times 10^{-6}$ m, we have

$$\nu_0 = \frac{3\times 10^8}{10.6\times 10^6} = 2.83\times 10^{13} \text{ Hz}$$

molar mass of $CO_2 = 12 + 2(16) = 44$ g/mole, giving

$$m_{CO_2} = \frac{44\times 10^{-3}}{6.02\times 10^{23}} = 7.3\times 10^{-26} \text{ kg}$$

and

$$\Delta\nu_D = (2.83\times 10^{13})\sqrt{\frac{8(1.38\times 10^{-23})(300)\ln 2}{(7.3\times 10^{-26})(3\times 10^8)^2}} = 52.9 \text{ MHz}$$

In a realistic gaseous system, both collision broadening and Doppler broadening coexist, leading to a lineshape profile known as the Voigt profile. For a discussion of the Voigt profile, we refer the reader to Refs. 4, 5, and 10 listed at the end of the chapter.

4.1.7 Rayleigh Scattering

As part of the classical treatment of light-matter interactions, we next discuss a very important phenomenon known as Rayleigh scattering. Rayleigh scattering explains, among other things, the reason why the sky is blue during daytime and reddish at sunset. It is also a critical factor in determining where the optimum transmission window of a fiber lies, and sets the limits of the lowest possible loss in fiber-optic communication systems.

To gain an intuitive feel for Rayleigh scattering, let us consider the passage of electromagnetic waves through a medium in which resonant absorption is negligible. In other words, we assume that the frequency ω of the wave is away from all of the medium resonance frequencies. As in the case of the Lorentz model, the incident wave will set the electrons into oscillatory motion. As we saw in Sec. 1.12, oscillating dipoles radiate energy. If the medium is perfectly homogeneous, the radiated fields interfere in such a way as to give a wave moving along the same direction as the original wave, with an effective dielectric constant that we calculated in the previous sections. If, on the other hand, there are density fluctuations in the medium, the scattered waves from different atoms will no longer be coherent, redirecting a portion of the electromagnetic energy from the original direction of motion, and hence causing the attenuation of the wave.

Let the electric field of the propagating wave be given by

$$\vec{E}(\vec{r},t) = \hat{x}E_0 \exp[i(\omega t - kz)] \tag{4.103}$$

The resulting polarization of the medium will be

$$\begin{aligned}\vec{P}(\vec{r},t) &= \varepsilon_0\chi_e\vec{E}(\vec{r},t) = \hat{x}\varepsilon_0\chi_e E_0 \exp[i(\omega t - kz)] \\ &= \hat{x}\varepsilon_0(n^2-1)E_0 \exp[i(\omega t - kz)]\end{aligned} \tag{4.104}$$

In other words, the induced electric dipole per atom will be

$$\vec{p} = \hat{x}p_0 \exp[i(\omega t - kz)] \tag{4.105}$$

where

$$p_0 = \frac{\varepsilon_0(n^2-1)E_0}{N} \tag{4.106}$$

In Eq. (4.106), p_0 is the amplitude of the induced dipole moment. As we saw in Chap. 1 [Eq. (1.292)], the total radiated power P_{rad} due to this dipole will be given by

$$P_{\text{rad}} = \frac{p_0^2\omega^4}{4\pi\varepsilon_0(3c^3)} = \frac{1}{4\pi\varepsilon_0}\frac{\omega^4\varepsilon_0^2(n^2-1)^2E_0^2}{N^2(3c^3)} \tag{4.107}$$

Now, consider a thin slab of this scattering medium of length dz and cross-sectional area a as shown in Fig. 4.12.

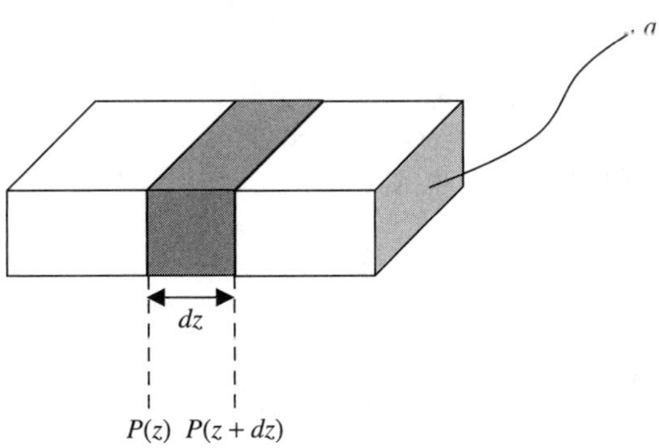

FIGURE 4.12 Thin slab of a scattering medium of length *dz* and cross sectional area *a*.

Since the scattering centers are incoherent, we can add the power scattered from all atoms to determine how much power loss will occur in the wave. Let the wave power at the locations z and $z + dz$ be $P(z)$ and $P(z + dz)$, respectively. If the cross-sectional area of the slab is a then,

$$P(z+dz) - P(z) = \text{change in the wave power}$$
$$= [I(z+dz) - I(z)]a = -Nadz\frac{1}{4\pi\varepsilon_0}\left(\frac{\omega^4\varepsilon_0^2(n^2-1)^2E_0^2}{N^2(3c^3)}\right) \tag{4.108}$$

Above, $I(z)$ is the wave intensity and $(Nadz)$ is the total number of scatters in the thin slab. Also noting that

$$I(z) = \frac{1}{2}\frac{E_0^2}{\eta} = \frac{1}{2}\varepsilon_0 cE_0^2 \tag{4.109}$$

we get

$$\frac{I(z+dz) - I(z)}{dz} = \frac{dI}{dz} = -\left(\frac{1}{2}\varepsilon_0 cE_0^2\right)\frac{\omega^4(n^2-1)^2}{6\pi Nc^4} = -\alpha_R I \tag{4.110}$$

where α_R is the intensity attenuation coefficient due to Rayleigh scattering and is given by

$$\alpha_R = \frac{(n^2-1)^2\omega^4}{6\pi Nc^4} \tag{4.111}$$

Clearly, the wave intensity will decay exponentially due to Rayleigh scattering according to

$$I(z) = I_0 \exp[-\alpha_R z] \tag{4.112}$$

where I_0 is the initial intensity.

We have obtained a very important result that summarizes the key features of Rayleigh scattering. The intensity of a propagating wave decreases exponentially with distance. Note also that the amount of Rayleigh scattering is a strong function of frequency. If the frequency of light is decreased by a factor of 2, the differential attenuation coefficient (α_R) decreases by a factor of 16. This strong frequency dependence also explains the color of the sky. During daytime, when we look away from the sun, we see that the sky is predominantly blue in color since blue scatters most among the rainbow colors. At sunset, on the other hand, we see light rays that come directly from the Sun. These are the wavelength components that undergo the least amount of scattering, and hence are at the larger wavelength end of the spectrum, i.e., the orange-red region.

Another important characteristic of Rayleigh scattering is that the light scattered at right angles is predominantly linearly polarized. To see this, consider the idealized case where unpolarized light is incident on the scattering medium from the left as shown in Fig. 4.13. The original direction of propagation is chosen along the z-axis and the observer looks along the y-axis. The field components polarized along and perpendicular to the x-axis are shown. The field polarized along the y-axis will induce oscillating dipoles aligned along the y-axis ($\vec{p}_1$) which radiate predominantly in the x-direction. The observer looking in the direction perpendicular to the original direction of wave propagation will not see the waves radiated by these dipoles. On the other hand, the field component polarized along the x-axis ($\vec{E}_2$), will induce dipoles aligned parallel to the x-axis ($\vec{p}_2$), and the radiated fields which are linearly polarized along the x-direction will reach the observer. This is because, as we saw in Sec. 1.12, an ideal dipole does not radiate along the dipole axis. Hence, the scattered light seen by the observer is predominantly polarized along the x-axis. For a more rigorous analysis of the polarization of scattered light, see Ref. 1.

Rayleigh scattering also plays a very important role in optical fibers. The differential loss coefficient α_R due to Rayleigh scattering in fused silica glass is given by[19]

$$\alpha_R = \frac{A_0}{\lambda^4} \tag{4.113}$$

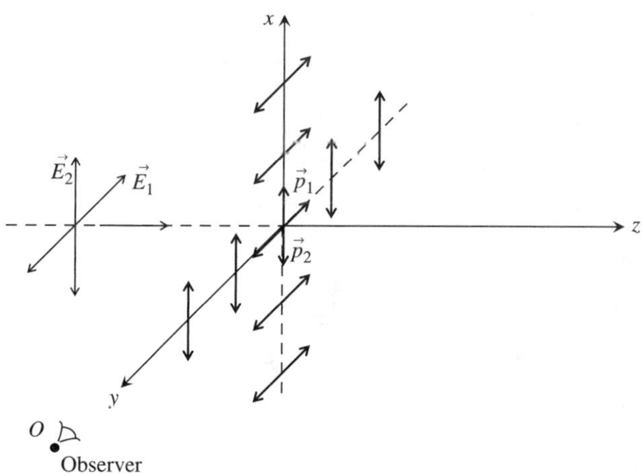

FIGURE 4.13 Rayleigh scattering of unpolarized light. The observer looking in a direction perpendicular to the direction of wave propagation sees scattered light that is predominantly linearly polarized.

where the constant A_0 has the value $A_0 = -(0.73 - 0.78)(\text{dB/km})(\mu\text{m})^4$. Note that the constant A_0 is given in decibel units, which give power ratios on a logarithmic scale. The decibel ratio, abbreviated as dB, is defined as

$$\text{dB} = 10\log\frac{P_2}{P_1} \tag{4.114}$$

where P_2 and P_1 are two power levels. In the case of Rayleigh scattering, P_1 and P_2 designate the launched power into the fiber and the transmitted power, respectively.

Example Consider an optical fiber where $A = -0.75(\text{dB/km})\mu\text{m}^4$. The loss experienced by light at the wavelength of 1 μm over 1 km will be –0.75 dB. Setting

$$-0.75 = 10\log\frac{P_2}{P_1}$$

we find $P_2 = 0.84P_1$. In other words, 16% of the light will be lost due to Rayleigh scattering over 1 km of fiber.

In a real silica fiber, there are other loss mechanisms as well such as absorption due to the hydroxide impurities. The overall loss factor in silica-based communication fibers is a complicated function of wavelength. A typical wavelength-dependent loss curve for silica fibers is sketched in Fig. 4.14. Efforts in purifying silica resulted in loss values as low as about 0.2 dB/km near the wavelength of 1.55 μm.[19] Hence, the wavelength of 1.55 μm has become the preferred wavelength for long-distance optical communication systems based on silica fibers.

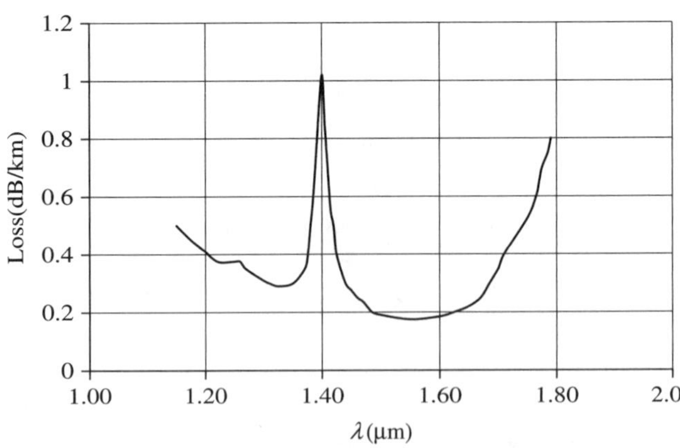

FIGURE 4.14 Typical wavelength-dependent loss curve for silica fibers (adapted from Ref. 19).

4.1.8 Electromagnetic Pulse Propagation in Dispersive Media

In the previous sections we discussed, based on a classical analysis, the origin of absorption and dispersion in materials. By dispersion, we refer to the wavelength dependence of the refractive index. Now, we provide a simple description of how electromagnetic pulse propagation can be analyzed in a dispersive medium. We will revisit this topic and provide a more rigorous treatment in Chap. 11.

Let us assume that a pulse with a spectral width of $\Delta\lambda$ enters the dispersive medium. As we saw earlier in Sec. 1.9.5, the pulse will move at the group velocity u_g given by

$$u_g = \frac{d\omega}{dk} \tag{4.115}$$

We leave it as an exercise for the reader to show that (see Prob. 4.7) u_g can be equivalently expressed as

$$u_g = \frac{c}{n - \lambda \dfrac{dn}{d\lambda}} \tag{4.116}$$

As the pulse propagates, its temporal width will monotonically increase due to dispersion. To estimate the amount of temporal broadening, we can crudely assume that the pulse consists of groups of waves where the widest spectral separation between the groups is on the order of $\Delta\lambda$, as sketched in Fig. 4.15. Since each group will move through the material with a different group velocity, the pulse will spread in time. In the context of fiber optics, the dispersive pulse broadening is characterized by using the group velocity dispersion parameter D_{GV}, which is defined according to

$$D_{GV} = \frac{1}{d}\frac{dT_p}{d\lambda} \tag{4.117}$$

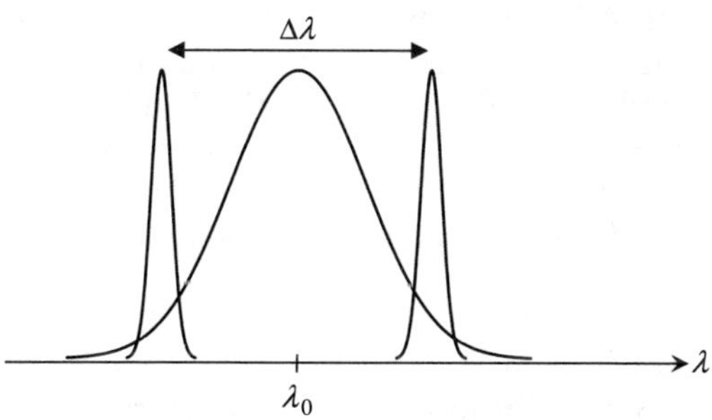

FIGURE 4.15 Spectrum of a pulse with a width of $\Delta\lambda$. Wave groups at the extreme ends of the pulse spectrum are separated by approximately $\Delta\lambda$.

where d is the propagation distance, T_p is the pulse transmission time, and λ is the vacuum center wavelength of the pulse. It can be shown that (see Prob. 4.7) D_{GV} can be expressed in terms of the second derivative of the refractive index n as

$$D_{GV} = -\frac{\lambda}{c}\frac{d^2 n}{d\lambda^2} \tag{4.118}$$

Now, let's go back to the pulse broadening problem. Due to the difference in the center wavelengths, wave groups separated by $\Delta\lambda$ in the wavelength domain will arrive at the location $z = d$ at different times. Assuming that all groups overlap in time at $z = 0$, we can approximately say that the temporal width of the pulse at $z = d$ will become

$$\Delta t = \Delta\left(\frac{d}{u_g}\right) \cong d\frac{d}{d\lambda}\left(\frac{1}{u_g}\right)\Delta\lambda = d\,D_{GV}\,\Delta\lambda \tag{4.119}$$

This phenomenon is commonly referred to as dispersion induced pulse broadening. A more accurate analysis requires knowledge of the initial temporal pulse profile and is treated in Sec. 11.2.4. Based on the result given in Eq. (4.119), D_{GV} can be expressed as

$$D_{GV} = \frac{\Delta t}{d\Delta\lambda} \tag{4.120}$$

Stated in words, the group-velocity dispersion parameter D_{GV} expresses the amount of dispersion induced temporal broadening per unit propagation length per unit spectral width. The commonly used units of D_{GV} in fiber optic systems is picoseconds/km-nm. If the pulses are sent at the special wavelength where $D_{GV} = 0$, minimum pulse broadening occurs. In a real system, pulse distortion will not completely vanish even if $D_{GV} = 0$ due to higher-order dispersion contributions. Figure 4.16 shows the calculated variation of D_{GV} as a function of wavelength for silica in the 1–1.5 μm range. The calculation is based on the refractive index data given in Eq. (4.58).[12] Note that the zero dispersion point is near $\lambda = 1.27$ μm.[20] The zero dispersion wavelength calculated here is determined based on material dispersion only. There is additional dispersion that occurs due to the fact that modes propagating down the fiber are not plane waves, but have spatial confinement which is a characteristic of the guiding geometry. As such, the propagation wave vector of the guided modes is a function of frequency and this leads to what is known as waveguide dispersion. Waveguide dispersion can be engineered to shift the zero

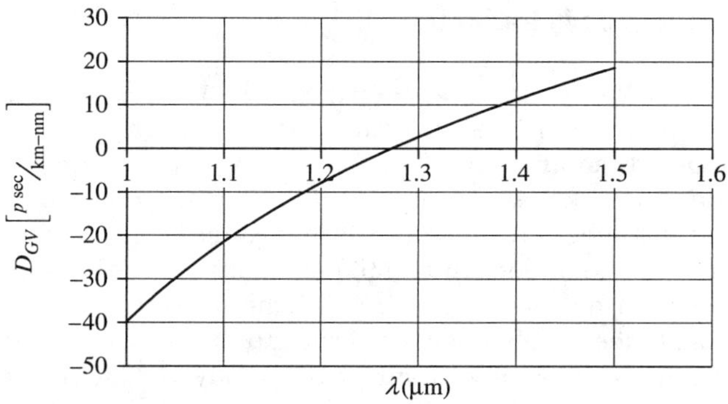

FIGURE 4.16 Calculated variation of the group velocity dispersion parameter D_{GV} as a function of wavelength for silica in the 1–1.5 μm range.

dispersion wavelength to the minimum attenuation wavelength of 1.55 μm.

Finally, we leave it as an exercise (see Prob. 4.9) for the reader to show that the approximate temporal width Δt of a pulse with an initial duration of τ_0 becomes

$$\Delta t = d\frac{D_{GV}}{2\pi c}\frac{\lambda^2}{\tau_0} \tag{4.121}$$

after propagating a distance of d in a dispersive medium with a group velocity dispersion parameter of D_{GV}. Hence the shorter the initial pulse, the broader the dispersion induced temporal broadening.

4.2 Elementary Quantum Theory of Light-Matter Interactions

Electromagnetic radiation also exhibits particle-like properties. Without introducing rigorous quantum mechanical tools, we develop in this section the photon picture of light and apply it to the study of optical processes including absorption and emission. The results for blackbody radiation play a central role in the development of the photon picture. So, we start with a discussion of blackbody radiation. We then present Einstein's model of stimulated and spontaneous optical transitions. An excellent account on the early developments of the quantum theory of radiation may be found in Refs. 18 and 21. Quantum mechanical treatment of optical processes will be discussed in Chap. 5.

4.2.1 Blackbody Radiation

General Characteristics of Blackbody Radiation

Objects at a finite temperature emit electromagnetic radiation, a phenomenon commonly referred to as blackbody radiation or thermal radiation. A well-known example of this is the red-hot coal in a barbeque. In the later part of the nineteenth century, physicists became very interested in understanding the spectral characteristics of blackbody radiation, an effort which, through the work of Max Planck and others, led to the development of the quantum theory of radiation. In this section we give an account of blackbody radiation, since the results are very important in the discussion of the elementary quantum theory of light-matter interactions. In particular, we will derive the expression for the spectral energy density and spectral radiant intensity functions based on Planck's hypothesis, and outline some of the important characteristics of blackbodies.

An ideal blackbody is one which absorbs all incident radiation. In practice, it can be realized by opening a tiny hole on the surface of an otherwise closed cavity as shown in Fig. 4.17.

A beam of light entering through the hole undergoes successive random reflections from the cavity walls and, in principle, has very low probability of escaping. Hence, the tiny hole acts as an ideal blackbody. In thermal equilibrium, the temperature of the cavity is kept constant, in which case, the radiant power emitted by the blackbody is equal to the absorbed power.

To characterize a blackbody experimentally, we can measure the spectral radiant intensity function $R(\lambda)$, which is defined as: $R(\lambda)d\lambda =$ power emitted by a blackbody per unit area in the wavelength range $\lambda \to \lambda + d\lambda$.

Two very important spectral properties of blackbodies were already known from experiments conducted with blackbodies by the end of the nineteenth century:

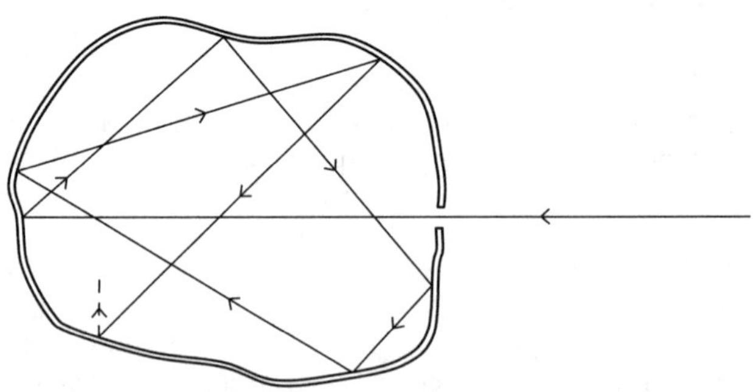

FIGURE 4.17 A tiny hole opened on the surface of an otherwise closed cavity acts as an ideal blackbody.

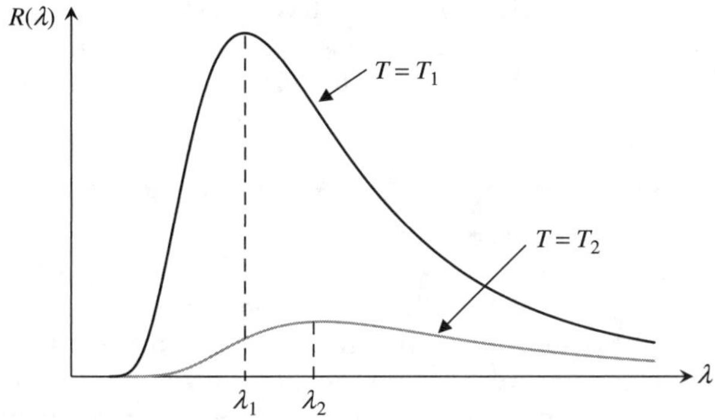

FIGURE 4.18 Typical variation of the spectral radiant intensity function $R(\lambda)$ for an ideal blackbody as a function of wavelength λ at two different temperatures T_1 and T_2. Here, $T_1 > T_2$. Increasing the temperature lowers the wavelength of peak emission according to Wien's displacement law.

1. Wien's displacement law: The measured spectral dependence of $R(\lambda)$ for a typical blackbody was known to be represented by a peaked distribution, as shown in Fig. 4.18. It was further known that when the temperature of the blackbody increases, the area under the $R(\lambda)$ curve also increases and the wavelength of peak emission shifts according to

$$\lambda_1 T_1 - \lambda_2 T_2 = 2.89777 \times 10^{-3} \text{m.K} \tag{4.122}$$

where λ_1 and λ_2 designate the wavelengths of peak emission at the absolute temperatures T_1 and T_2, respectively. This result is known as Wien's displacement law.

2. Stefan-Boltzmann law: As mentioned above, the area under the $R(\lambda)$ curve depends on the temperature. It was also discovered that the integrated radiant intensity R_T defined according to

$$R_T = \int_0^\infty d\lambda \; R(\lambda) \tag{4.123}$$

varies as a strong function of the absolute temperature T, given by the well known Stefan-Boltzmann law

$$R_T = \sigma_{SB} \; T^4 \tag{4.124}$$

Above, σ_{SB} is the Stefan-Boltzmann constant with the experimentally measured value of $5.67040 \times 10^{-8}\,\text{W}/\text{m}^2\text{-K}^4$. Based on thermodynamic arguments, Boltzmann was able to show the T^4 dependence of R_T.

However, there was no theory yet with which to derive an expression for σ_{SB} in terms of fundamental constants. Thus, began the work of Rayleigh, Jeans, Planck, and others to gain a deeper understanding of blackbody radiation.

Relation Between the Spectral Energy Density and Spectral Radiant Intensity

Before we review the classical and early quantum theory of blackbody radiation, we will first derive the relation between the spectral energy density distribution and the spectral radiant intensity. The spectral energy density distribution $\rho(\lambda)$ is defined according to $\rho(\lambda)d\lambda =$ electromagnetic energy per unit volume in the wavelength range $\lambda \to \lambda + d\lambda$.

We will assume that the energy is stored in the electromagnetic standing wave modes supported by the blackbody cavity. $\rho(\lambda)d\lambda$ can then be expressed as

$$\rho(\lambda)d\lambda = N(\lambda)\bar{E}(\lambda)d\lambda \tag{4.125}$$

where $N(\lambda)$ is the number of electromagnetic modes per unit volume in the interval λ to $\lambda + d\lambda$ and $\bar{E}(\lambda)$ represents the average energy per mode. So, our task is reduced to the determination of $N(\lambda)$ and $\bar{E}(\lambda)$. Whether the analysis is classical or quantum mechanical, the expression for $N(\lambda)$ remains the same. The expression for $\bar{E}(\lambda)$, however, will greatly differ for the two cases as we will see soon. Before embarking on these calculations, we note that to go from $\rho(\lambda)$ to the experimentally measured $R(\lambda)$, we can use the relation

$$R(\lambda) = \frac{c}{4}\rho(\lambda) \tag{4.126}$$

where, as usual, c is the vacuum speed of light. To see where this relation comes from, we consider the blackbody geometry depicted in Fig. 4.19.

The energy dE contained in the element of volume d^3r per unit wavelength interval is given by

$$dE = \rho(\lambda)d^3r = \rho(\lambda)r^2 \sin\theta \; dr d\theta d\phi \tag{4.127}$$

The energy will be radiated equally in all directions. The contribution to the Poynting vector dS due to the element of volume d^3r will be

$$dS = \frac{1}{4\pi r^2}\frac{dE}{dt} = \frac{\rho(\lambda)c}{4\pi}\sin\theta \; d\theta d\phi \tag{4.128}$$

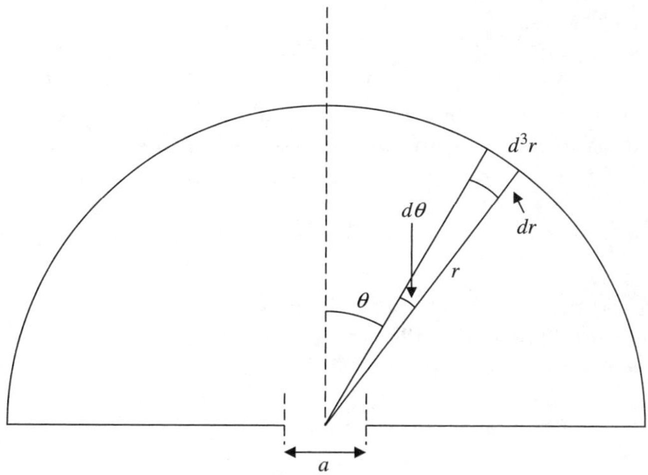

FIGURE 4.19 The blackbody geometry used to derive the relation between the spectral energy density function and the spectral radiant intensity function.

where we used $dr/dt = c$ for electromagnetic waves. In vector form, we can write $d\vec{S}$ as

$$d\vec{S} = -\hat{r}\frac{\rho(\lambda)c}{4\pi}\sin\theta d\theta d\phi \tag{4.129}$$

The power dP falling on the blackbody aperture will be given by

$$dP = \vec{a}\cdot d\vec{S} \tag{4.130}$$

where $\vec{a}$ is the area vector of the aperture and is given by $\vec{a} = -\hat{z}a$. Therefore, dP will become

$$dP = \frac{a\rho(\lambda)c}{4\pi}(\hat{z}\cdot\hat{r})\sin\theta d\theta d\phi = \frac{a\rho(\lambda)c}{4\pi}\cos\theta\sin\theta\, d\theta d\phi \tag{4.131}$$

Now, the infinitesimal contribution dR to the spectral radiant intensity can be calculated from

$$dR = \frac{dP}{a} = \frac{\rho(\lambda)c}{8\pi}\sin 2\theta d\theta d\phi \tag{4.132}$$

To find the total radiant intensity $R(\lambda)$, we integrate over the hemispherical blackbody to obtain

$$R(\lambda) = \int_0^{2\pi} d\phi \int_0^{\pi/2} d\theta \sin 2\theta \left(\frac{\rho(\lambda)c}{8\pi}\right) = \frac{\rho(\lambda)c}{8\pi}(2\pi)\left[-\frac{1}{2}\cos 2\theta\right]_0^{\pi/2} = \frac{c}{4}\rho(\lambda) \tag{4.133}$$

hence recovering the result given in Eq. (4.126).

Density of Modes $N(\lambda)$

Next, we calculate the density of electromagnetic modes $N(\lambda)$. To derive the expression for $N(\lambda)$, we first consider a one-dimensional cavity with perfectly conducting end walls as shown in Fig. 4.20*a*.

The standing wave mode polarized along the $\hat{x}$ direction will consist of counter-propagating waves of the form

$$E_x(z,t) = A\exp[i(\omega t - k_z z)] + B\exp[i(\omega t + k_z z)] \tag{4.134}$$

where k_z is the wave vector, ω is the angular frequency, and A and B are constants. The end walls are perfect conductors so that the field satisfies the boundary condition

$$E_x(z = 0,t) = E_x(z = d,t) = 0 \tag{4.135}$$

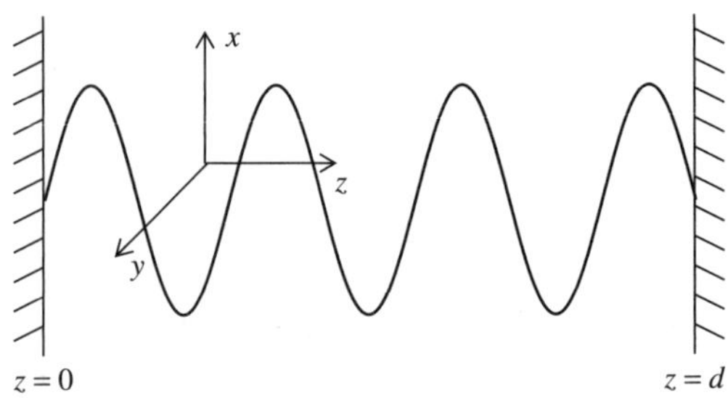

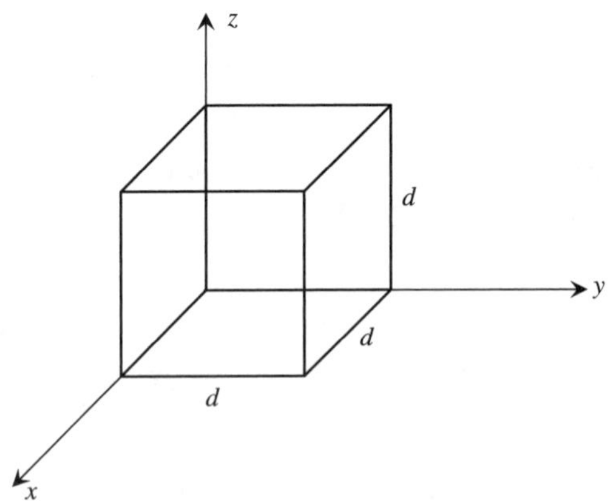

FIGURE 4.20 (*a*) One-dimensional and (*b*) three-dimensional cavity with perfectly conducting walls used in the derivation of the density of modes function $N(\lambda)$.

Applying the above boundary conditions at $z = 0$, we find that $A = -B$, and E_x can be expressed as

$$E_x(z,t) = E_0 \sin k_z z \exp(i\omega t) \tag{4.136}$$

Furthermore, $E_x(z = d,t) = 0$ leading to $\sin k_z d = 0$ or $k_z d = n_z \pi$, where n_z is an integer. Hence, the allowed values of the wave vector k_z are discrete, given by

$$k_z = n_z \pi / d \tag{4.137}$$

Note that in this case, there is a second independent mode polarized along the y-direction with the same k_z value. We will use this fact while counting the total number of modes.

Now, we can extend this result to three dimensions. Without going into the details, we note that in three dimensions each component of the wave vector $\vec{k}$ will satisfy a relation similar to Eq. (4.137). Specifically, we will have, for a cubical resonator with edge length of d (see Fig. 4.20b),

$$\begin{aligned} k_x &= \frac{n_x \pi}{d} \\ k_y &= \frac{n_y \pi}{d} \\ k_z &= \frac{n_z \pi}{d} \end{aligned} \tag{4.138}$$

where n_x, n_y and n_z are integers. The magnitude of the wave vector $|\vec{k}|$ will be given by

$$|\vec{k}| = \sqrt{k_x^2 + k_y^2 + k_z^2} = \frac{\pi}{d}\sqrt{n_x^2 + n_y^2 + n_z^2} \tag{4.139}$$

Since $k = \omega/c = 2\pi\nu/c$, we see that the discrete resonance frequencies ν of the modes can be calculated from

$$\nu = \frac{c}{2d}\sqrt{n_x^2 + n_y^2 + n_z^2} \tag{4.140}$$

The important conclusion we draw from this analysis is that the resonance frequencies of a blackbody cavity are discrete. We only considered a simple cubical cavity to arrive at this result. For a cavity of arbitrary shape, the discreteness of the mode frequencies will still be valid, except it could be more challenging to come up with an analytical expression for the mode frequencies as in Eq. (4.140).

How can we use the above result given by Eq. (4.140) to develop a simple method of calculating the number of modes in a frequency/

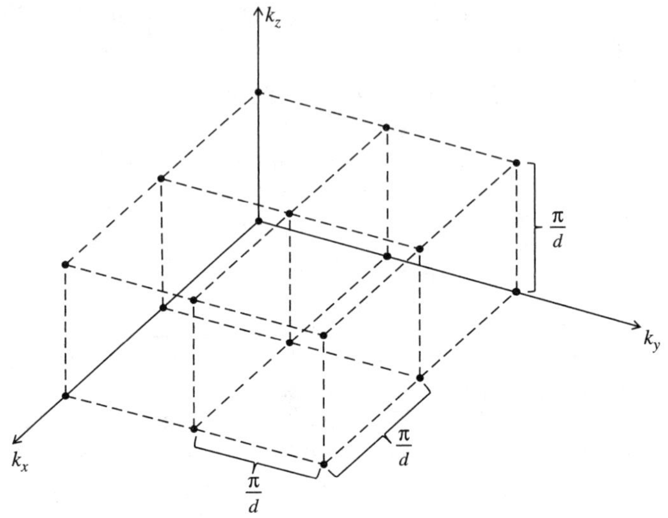

FIGURE 4.21 Allowed values (k_x,k_y,k_z) of the wave vector for the resonant modes of a cubical cavity of edge length *d*.

wavelength interval? It is more convenient to do this enumeration in the k-space. Let's take a look at Fig. 4.21. The points marked in Fig. 4.21 designate the allowed values of (k_x,k_y,k_z) in the k-space. Careful examination shows that the volume occupied by one mode is $(\pi/d)^3$. If the magnitude of k is small, one can draw a spherical shell of radius k in the first octant and manually count the number of allowed modes. However, if k becomes large, so that within a range of k values the number of allowed modes is large, we can, to a good approximation, calculate the number of modes $N(k)$ as follows:

$N(k)dk$ = number of resonant cavity modes in the wave number range from k to $k+dk$ per unit volume.

$$= 2\times\frac{\text{volume in } k\text{-space between } k \text{ and } k+dk}{d^3\left(\pi/d\right)^3} \tag{4.141}$$

The extra factor of 2 accounts for the fact that for each value of k, there are two independent electromagnetic polarizations as was noted for the one-dimensional case. For large k, the region between k and $k+dk$ will be a spherical shell of thickness dk and volume $4\pi k^2 dk$. This is shown in Fig. 4.22. Hence, $N(k)$ becomes

$$N(k)dk = 2\left(\frac{1}{8}\right)\frac{4\pi k^2 dk}{\left(\pi/d\right)^3 d^3} = \frac{k^2}{\pi^2}dk \tag{4.142}$$

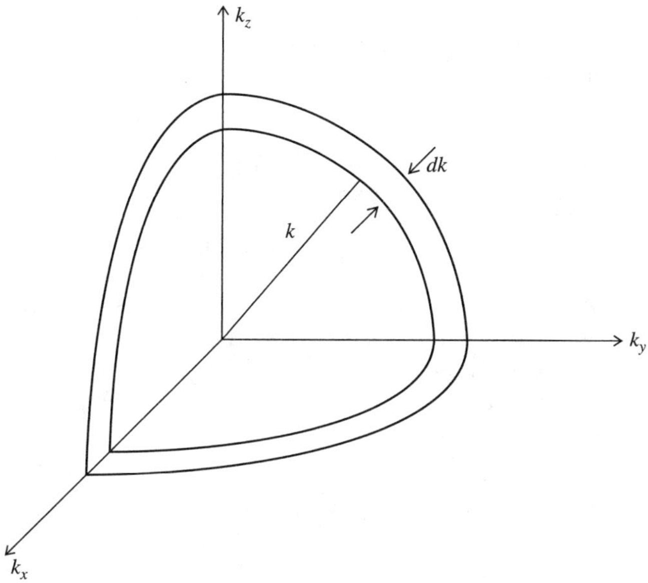

FIGURE 4.22 The spherical shell of thickness *dk* between the wave number *k* and *k + dk*.

To express this in terms of frequency, we note that $\nu = kc/2\pi$ or $dk = (2\pi/c)d\nu$, giving

$$N(\nu)d\nu = \left(\frac{2\pi}{c}\right)\left(\frac{2\pi\nu}{c}\right)^2 \frac{1}{\pi^2}d\nu = \frac{8\pi\nu^2}{c^3}d\nu \tag{4.143}$$

Equivalently, we can express this in terms of wavelength λ, where $\lambda = c/\nu$ and $d\lambda = (-c/\nu^2)d\nu$, giving

$$N(\lambda)d\lambda = \frac{8\pi}{c^3}\left(\frac{c}{\lambda}\right)^2 c\,d\lambda = \frac{8\pi}{\lambda^4}d\lambda \tag{4.144}$$

Average Energy per Mode

We now discuss how the average energy $\bar{E}(\lambda)$ per mode is calculated. Classically, $\bar{E}(\lambda)$ is calculated by using the Boltzmann distribution which states that the probability $P_B(E)dE$ that a system will have an energy in the interval E to $E + dE$ is given by

$$P_B(E)dE = C\exp\left[-\frac{E}{k_B T}\right]dE \tag{4.145}$$

where C is a constant, k_B is the Boltzmann constant, and T is the absolute temperature. In this case, the average energy $\bar{E}(\lambda)$ turns out to be independent of wavelength and comes to

$$\bar{E}(\lambda) = \frac{\int_0^\infty dE E \exp\left[-E/k_B T\right]}{\int_0^\infty dE \exp\left[-E/k_B T\right]} = k_B T \tag{4.146}$$

Equivalently, we could say, based on the equipartition theorem of statistical physics, that the average energy per degree of freedom is $1/2\ k_B T$, and the resulting average energy for two degrees of freedom (due to the two independent polarizations) is $k_B T$. Hence, the resulting expression for $R(\lambda)$ becomes

$$R(\lambda)d\lambda = \frac{c}{4}\rho(\lambda)d\lambda = \frac{c}{4}\left(\frac{8\pi}{\lambda^4}\right)k_B T d\lambda = \frac{2\pi c k_B T}{\lambda^4} d\lambda \tag{4.147}$$

This is the classical result for $R(\lambda)$, known as the Rayleigh-Jeans law. It has good agreement with experimentally measured spectral radiant intensity data at long wavelengths but fails as $\lambda \to 0$, this failure being commonly referred to as the ultraviolet catastrophe of classical physics.

To correctly account for the observed spectral form of $R(\lambda)$, in 1900 Max Planck, put forward his famous quantum hypothesis, according to which the energy associated with each degree of freedom is not continuous but instead can only take an integer multiple of a quantum of energy E_0. E_0, which is further taken to be proportional to the frequency of the radiation mode, can be written as

$$E_0 = h\nu \tag{4.148}$$

where h is the Planck's constant. Hence the energy of the mode can be an integer multiple of E_0 in the form

$$E_m = mh\nu \tag{4.149}$$

Above, m is an integer. To derive the quantum expression for $R(\lambda)$, note that $N(\lambda)$ stays the same. What differs is the average energy per mode, since now only discrete values of energy are allowed. $\bar{E}(\lambda)$ can be calculated from

$$\bar{E}(\lambda) = \frac{\sum_{m=0}^{\infty} E_m \exp[-\beta E_m]}{\sum_{m=0}^{\infty} \exp[-\beta E_m]} \tag{4.150}$$

where $\beta = 1/k_B T$. The sums appearing in Eq. (4.150) can be analytically calculated (see Prob. 4.10) to get

$$\bar{E}(\lambda) = \frac{hc/\lambda}{\exp\left[\frac{hc}{\lambda k_B T}\right] - 1} \tag{4.151}$$

Planck Distribution Function for Blackbody Radiation

Based on the above results, the spectral radiant intensity $R(\lambda)$ becomes

$$R(\lambda)d\lambda = \frac{c}{4}\left(\frac{8\pi}{\lambda^4}\right)\frac{hc}{\lambda}\frac{1}{\exp\left[\frac{hc}{\lambda k_B T}\right] - 1}d\lambda = \frac{2\pi hc^2}{\lambda^5}\frac{1}{\exp\left[\frac{hc}{\lambda k_B T}\right] - 1}d\lambda \tag{4.152}$$

Equation (4.152) is known as Planck's blackbody distribution function and accurately describes the observed wavelength dependence of $R(\lambda)$. In terms of frequency, the spectral energy density $\rho(\nu)$ can further be written as

$$\rho(\nu)d\nu = \frac{8\pi h\nu^3/c^3}{\exp\left[\frac{h\nu}{k_B T}\right] - 1}d\nu \tag{4.153}$$

by using Eq. (4.143). Note that in the long wavelength limit

$$\exp\left[\frac{hc}{\lambda k_B T}\right] \simeq 1 + \frac{hc}{\lambda k_B T} \tag{4.154}$$

and $R(\lambda)$ becomes

$$R(\lambda) \simeq \frac{2\pi hc^2}{\lambda^5}\frac{\lambda k_B T}{hc} = \frac{2\pi c k_B T}{\lambda^4} \tag{4.155}$$

which is identical to the Rayleigh-Jeans result of Eq. (4.147). The quantum and classical expressions for $R(\lambda)$ are plotted in Fig. 4.23.

By using the quantum expression for $R(\lambda)$, we can also prove Stefan-Boltzmann law.

The total radiant intensity R_T will be given by

$$R_T = \int_0^\infty d\lambda R(\lambda) = 2\pi hc^2 \int_0^\infty \frac{d\lambda\,\lambda^{-5}}{\exp\left[\frac{hc}{\lambda k_B T}\right] - 1} \tag{4.156}$$

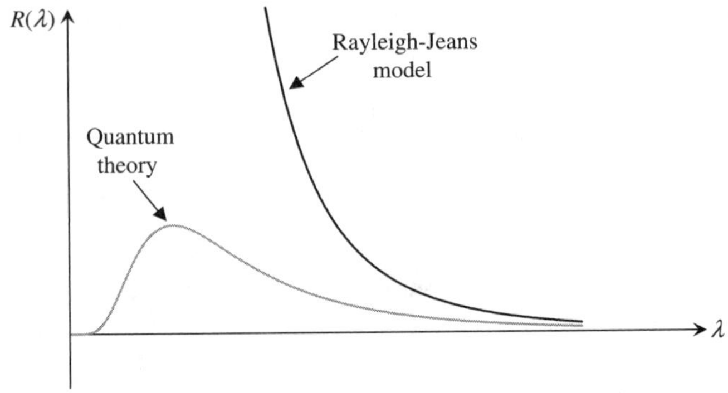

FIGURE 4.23 Wavelength dependence of the spectral radiant intensity $R(\lambda)$ based on Planck's quantum model and the Rayleigh-Jeans classical model.

Let the dimensionless variable x be defined by $x = hc/\lambda k_B T$ with $dx = (-hc/\lambda^2 k_B T)d\lambda$. By substituting for λ and $d\lambda$ in Eq. (4.156), we can express R_T as

$$R_T = \frac{2\pi k_B^4 T^4}{h^3 c^2} \int_0^\infty \frac{dx x^3}{e^x - 1} \tag{4.157}$$

The integral appearing in Eq. (4.157) is in terms of a dimensionless variable and is simply a constant. Hence, we see that R_T is indeed proportional to T^4 as we expect from Stefan-Boltzmann law. Furthermore, by using the result

$$\int_0^\infty dx \frac{x^3}{e^x - 1} = \frac{\pi^4}{15} \tag{4.158}$$

we obtain

$$R_T = \frac{2\pi^5 k_B^4}{15 h^3 c^2} T^4 \tag{4.159}$$

Hence, the Stefan-Boltzmann constant σ_{SB} can be expressed in terms of the fundamental constants as

$$\sigma_{SB} = \frac{2\pi^5 k_B^4}{15 h^3 c^2} \tag{4.160}$$

From the empirically known value of $5.67040 \times 10^{-8}\,\mathrm{W/m^2\text{-}K^4}$ for σ_{SB}, Planck's constant h was further determined to be $h = 6.626 \times 10^{-34}\,\mathrm{J \cdot s}$. The same value of h appeared in other independent experiments such as the photoelectric effect and Compton scattering, verifying that h is indeed a fundamental constant of nature. We leave the derivation of Wien's law from Planck's distribution function for $R(\lambda)$ as an exercise (see Prob. 4.11).

Example For our sun, we can take the blackbody temperature as $T = 5800\,\text{K}$. From Wien's displacement law, the wavelength of maximum emission λ_{max} becomes

$$\lambda_{max} = \frac{2898\ \mu\text{m-K}}{5800} \cong 500\,\text{nm}$$

which is in the green region of the electromagnetic spectrum.

We can also calculate the radiated intensity in the wavelength interval of 500–520 nm. Taking λ as 500 nm, we have

$$\frac{hc}{\lambda k_B T} = \frac{(6.626\times 10^{-34})(3\times 10^{8})}{(500\times 10^{-9})(1.38\times 10^{-23})(5800)} = 4.77$$

and

$$dR_T = R(\lambda)d\lambda = \frac{2\pi hc^2}{\lambda^5}\frac{1}{\exp\left[\dfrac{hc}{\lambda k_B T}\right]-1}d\lambda$$

$$= \frac{2\pi(6.626\times 10^{-34})(3\times 10^{8})^2}{(500\times 10^{-9})^5}\frac{1}{e^{4.97}-1}(20\times 10^{-9})$$

$$= 1.68\frac{\text{MW}}{\text{m}^2} = 168\frac{\text{W}}{\text{cm}^2}$$

This means that if we can use a filter that passes only the colors in the wavelength range of 500–520 nm, we will measure 168 W of emitted power from the blackbody surface of area 1 cm^2.

4.2.2 Elementary Quantum Theory of Optical Processes

We will discuss in detail, the quantum mechanics of light-matter interactions at a more rigorous level in Chap. 5. For the time being, we will borrow two important results from quantum mechanics to build the photon description of light-matter interactions. Here are the two important ingredients:

1. The energy levels of a microscopic system such as atoms, ions, or molecules are discrete. This is schematically shown in Fig. 4.24. For example, $|1\rangle$ is the state designation of the lowest energy level with energy E_1. The next level is $|2\rangle$ with energy E_2 and so on. Actually, as we noted in the discussion of broadening processes, these energy levels are not infinitely sharp but instead have a certain width due, for example, to collisions or the Doppler effect. We will soon discover that spontaneous decay from the excited states also broadens these lines.

FIGURE 4.24 Microscopic systems such as atoms and molecules have discrete energy levels.

2. We will treat light as consisting of particles, called photons, each of which possesses an energy

$$E = h\nu \tag{4.161}$$

where h is Planck's constant and ν is the frequency. There are two processes which we need to investigate in light-matter interactions: absorption and emission. In terms of the photon picture of light, these processes are summarized in the sketches of Fig. 4.25*a* and *b*. For simplicity, we consider a two-level system. In the case of absorption, a photon with energy $h\nu$ approximately equal to the energy difference $E_2 - E_1$ is absorbed by the atom. As a result, the atom which is originally in the lower state $|1\rangle$, is raised to the excited state $|2\rangle$. The reverse process occurs during emission, when the excited atom makes a transition from the upper state $|2\rangle$ to the lower state $|1\rangle$ and emits a photon of energy $h\nu \approx E_2 - E_1$. In the description of light-matter interactions and the operation of lasers, it is crucial to know the rate at which absorption and emission processes occur. For instance, we need to be able to determine how many of the atoms are raised from the lower level to the upper level per unit time during absorption. In 1917,

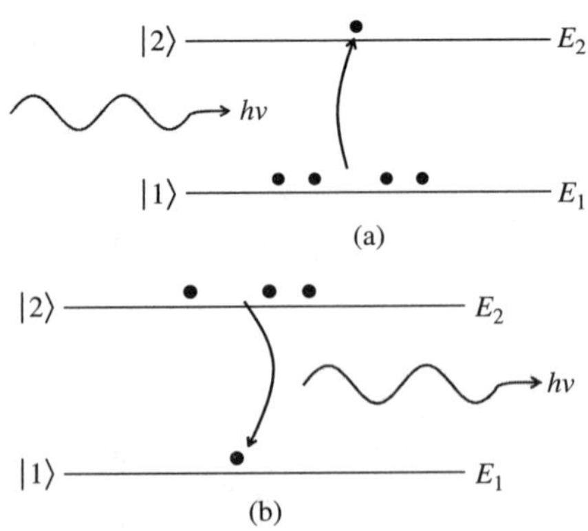

FIGURE 4.25 Photon picture of (*a*) absorption and (*b*) emission.

Einstein gave a phenomenological description of three radiative processes.[22, 23] In the next section, we will go over his theory and discuss the models of transition rates for stimulated absorption, spontaneous emission, and stimulated emission.

Stimulated Absorption

Let us assume that we have a two-level atomic system with broadened energy levels described by the lineshape function $g(\nu)$. If light with frequency $\nu \approx (E_2 - E_1)/h$ is incident on the system, it will be absorbed and will stimulate the transition of atoms from the lower level to the upper level. This process is known as stimulated absorption, or simply, absorption. The rate of transition of the atoms will depend on the number of incident photons arriving per unit time, their frequency, and the intrinsic absorption strength of the atom. In general, light falling on the atom will not be monochromatic and its color distribution can be represented by the spectral energy density $\rho(\nu)$. According to Einstein's phenomenological model, the rate at which the atomic density N_2 will increase in the upper level will be given by

$$\frac{dN_2}{dt} = N_1 B_{12} \int_0^\infty d\nu \rho(\nu) g(\nu) \tag{4.162}$$

where N_1 is the atomic density in the lower level and B_{12} is known as Einstein's B coefficient for stimulated absorption. As we will see in Chap. 5, it is possible to derive a quantum mechanical expression for the B coefficient in terms of the wave functions of the two levels and the electric dipole operator. As before, $g(\nu)$ is normalized and describes the probability that light at frequency ν will interact with the atomic system, the probability being maximum when $\nu = (E_2 - E_1)/h = \nu_0$. Two limiting cases of Eq. (4.162) are important. First, suppose that we excite the atoms with broadband light where the spectral width of $\rho(\nu)$ is much larger than that of $g(\nu)$. In this case, we can express $g(\nu)$ in terms of the Dirac delta function as

$$g(\nu) = \delta(\nu - \nu_0) \tag{4.163}$$

Note that $g(\nu)$ is still normalized. Then, we get

$$\int_0^\infty d\nu \rho(\nu) \delta(\nu - \nu_0) = \rho(\nu_0) \tag{4.164}$$

and Eq. (4.162) becomes

$$\frac{dN_2}{dt} = +B_{12} N_1 \rho(\nu_0) \tag{4.165}$$

The second limiting case, which is very important in the discussion of optical amplification, occurs when the interacting light beam is monochromatic at frequency ν with a spectral width much narrower than that of the lineshape function. In this case, we can express the spectral energy density $g(\nu')$ at frequency ν' as

$$\rho(\nu') = \rho_\nu \delta(\nu' - \nu) \tag{4.166}$$

leading to

$$\frac{dN_2}{dt} = +B_{12}N_1 \int_0^\infty d\nu' \rho_\nu \delta(\nu' - \nu) g(\nu') = B_{12}N_1 \rho_\nu g(\nu) \tag{4.167}$$

Above, ρ_ν is the total energy density (in units of J/cm^3) of the monochromatic wave at the frequency ν.

Spontaneous Emission

Spontaneous emission is induced by the vacuum fluctuations of the quantized electromagnetic field. To treat the effect in a phenomenological way, we consider excited atoms at the upper level $|2\rangle$. It turns out that even in the absence of stimulating photons, these atoms can undergo decay to the lower level and emit photons. The so-called radiative decay rate, leading to the emission of photons, is proportional to the density of atoms in the excited level and is given by

$$\left(\frac{dN_2}{dt}\right)_{\text{rad}} = -A_{21}N_2 \tag{4.168}$$

where N_2 is the density of atoms in level $|2\rangle$ and A_{21} is known as Einstein's A coefficient for spontaneous emission. The subscript rad designates that this is the radiative decay rate. We will derive a quantum mechanical expression for the A coefficient in Chap. 5. Different from stimulated emission which we will discuss next, photons are emitted in random directions and with random phases during spontaneous emission. This is schematically shown in Fig. 4.26.

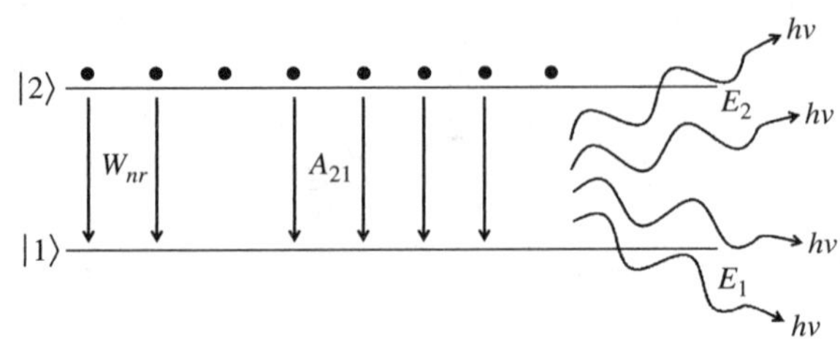

FIGURE 4.26 Spontaneous decay of excited atoms.

In most realistic systems, the decay almost always has a nonradiative component W_{nr} as well. As a result of nonradiative decay, the energy lost by the atom is given in the form of heat to the host and no photon emission results. The overall decay rate of atoms in level $|2\rangle$ can therefore be written in the form

$$\frac{dN_2}{dt} = -\frac{N_2}{\tau_f} \tag{4.169}$$

where τ_f is known as the fluorescence lifetime and is given by

$$\tau_f = \frac{1}{A_{21} + W_{nr}} \tag{4.170}$$

Suppose that we have somehow excited a certain initial density N_0 of atoms into the upper state at $t = 0$, and for $t > 0$ there are no stimulating photons to cause further excitation. From Eq. (4.169), we see that the upper-state population will decay exponentially with time according to

$$N_2(t) = N_0 \exp\left(-\frac{t}{\tau_f}\right) \tag{4.171}$$

If the decay rate has a substantial radiative component, we can in fact use a fast enough photodiode to monitor this decay process by looking at the time dependence of the emitted fluorescence signal near ν_0. The total radiated power $P_{\text{rad}}(t)$ will be given by

$$P_{\text{rad}}(t) = h\nu_0\left(\frac{dN_2}{dt}\right)_{\text{rad}} = A_{21}h\nu_0 N_2(t) = A_{21}h\nu_0 N_0 \exp\left(-\frac{t}{\tau_f}\right) \tag{4.172}$$

Hence, we can determine τ_f by finding the time when the fluorescence signal decays to $1/e$ of its initial value. Note again that although we are measuring the decay curve for the emitted photons, the fluorescence lifetime is related to both the radiative and the non-radiative decay rates. Several factors, including active center concentration and temperature, will influence the rate of nonradiative decay processes. At low concentrations and low temperatures, τ_f typically approaches its maximum possible value τ_r, which is the radiative lifetime given by $1/A_{21}$.

We can look at the Fourier transform of the emitted fluorescence signal to gain some insight about how spontaneous emission affects the lineshape of the transition.[8] From Eq. (4.172), the time-dependent electric field amplitude $E(t)$ of the emitted fluorescence will be of the form

$$E(t) = E_0 \exp\left[-\frac{t}{2\tau_f}\right]\cos(\omega_o t) \tag{4.173}$$

where $\omega_0 = 2\pi\nu_0$. The Fourier transform $E(\omega)$ of $E(t)$ will give the spectral amplitude distribution of the emitted signal. $E(\omega)$ can be calculated from

$$\begin{aligned} E(\omega) &= \frac{1}{\sqrt{2\pi}}\int_{-\infty}^{\infty} dt E(t)\exp(-i\omega t) \\ &= \frac{1}{\sqrt{2\pi}}\int_{0}^{\infty} dt \frac{E_0}{2}\exp\left[\left(\frac{-1}{2\tau_f} - i\omega\right)t\right][\exp(i\omega_o t) + \exp(-i\omega_o t)] \\ &= \frac{E_0}{2\sqrt{2\pi}}\int_{0}^{\infty} dt\left\{\exp\left[\left(\frac{-1}{2\tau_f} + i(\omega - \omega_0)\right)t\right] + \exp\left[\left(\frac{-1}{2\tau_f} + i(\omega + \omega_0)\right)t\right]\right\} \\ &= -\frac{E_0}{4\sqrt{2\pi}}\left[\frac{\frac{1}{2\tau_f} + i\omega}{(\omega_0^2 - \omega^2) - \left(\frac{1}{2\tau_f}\right)^2 + i\frac{\omega}{\tau_f}}\right] \end{aligned} \tag{4.174}$$

By assuming that both ω and ω_0 are much larger than $1/\tau_f$, and by taking ω nearly equal to ω_0, Eq. (4.174) simplifies to

$$E(\omega) \cong \frac{-E_0}{4\sqrt{2\pi}}\left[\frac{i}{2(\omega_0 - \omega) + i(1/\tau_f)}\right] \tag{4.175}$$

Expressing in terms of frequency, we get

$$E(\nu) = \frac{-E_0}{16\pi\sqrt{2\pi}} \frac{i}{(\nu - \nu_0) + i(1/4\pi\tau_f)} \tag{4.176}$$

The fluorescence lineshape function $g(\nu)$ will be proportional to $|E(\nu)|^2$. Up to a normalization constant which we call A, we have, therefore,

$$g(\nu) = \frac{A}{(\nu - \nu_0)^2 + \left(\frac{1}{4\pi\tau_f}\right)^2} \tag{4.177}$$

We saw previously that this is the Lorentzian distribution. To recall, the normalized lineshape function for a Lorentzian was given by

$$g_L(\nu) = \frac{\Delta\nu/2\pi}{(\nu-\nu_0)^2 + (\Delta\nu/2)^2} \tag{4.178}$$

Comparing Eqs. (4.177) and (4.178), we see that spontaneous decay also leads to homogeneous spectral broadening with a width (FWHM) given by

$$\Delta\nu = \frac{1}{2\pi\tau_f} \tag{4.179}$$

In the limit when all broadening processes (such as collisions or Doppler broadening) and nonradiative decay are negligible, we see that the radiative decay rate A_{21} gives the fundamental limit to $\Delta\nu_0$ for the spectral width of an atomic transition, which becomes

$$\Delta\nu_0 = \frac{A_{21}}{2\pi} \tag{4.180}$$

In typical systems, however, broadening effects will play a more dominant role and result in a spectral width much broader than $\Delta\nu_0$.

Stimulated Emission

Again, let's consider excited atoms in the upper state $|2\rangle$. In addition to spontaneous emission, these excited atoms can be stimulated to make downward transition to the lower level by resonant photons with energy $h\nu \cong E_2 - E_1$. This is known as stimulated emission and results in the generation of photons which ideally have the same frequency, phase, direction, and polarization state as the stimulating photons. This is schematically shown in Fig. 4.27. The presence of stimulated emission causes the decay of the atoms from the upper state, the rate of decay being proportional to the spectral energy

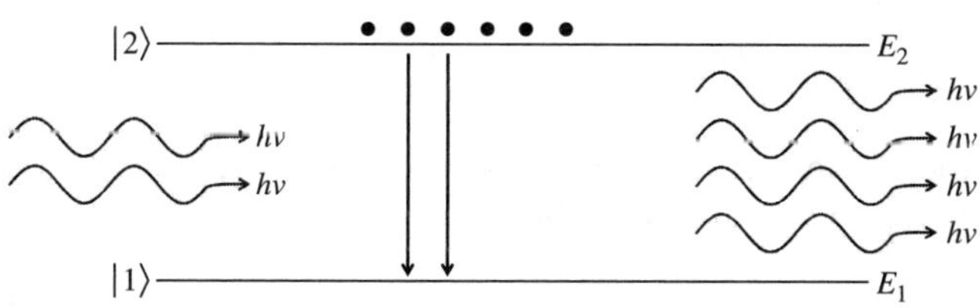

FIGURE 4.27 Stimulated emission in a two-level atomic system.

density of the photons, the lineshape function, and the density of excited atoms. In equation form, we have

$$\frac{dN_2}{dt} = -B_{21}N_2\int_0^\infty d\nu\rho(\nu)g(\nu) \tag{4.181}$$

Above, B_{21} is known as Einstein's B coefficient for stimulated emission. The negative sign in Eq. (4.181) signifies that stimulated emission leads to a decrease of the upper-state population.

Similar to the case of stimulated absorption, the decay rate becomes

$$\frac{dN_2}{dt} = -B_{21}N_2\rho(\nu_0) \tag{4.182}$$

and

$$\frac{dN_2}{dt} = -B_{21}N_2\rho_\nu g(\nu) \tag{4.182}$$

in the case of broadband and monochromatic excitation, respectively.

Relation between *A* and *B* Coefficients

It turns out that A and B coefficients are not independent of each other. We can use an equilibrium analysis and the results of blackbody radiation to derive the relation between the A and B coefficients. Specifically, let us analyze the interaction of the atoms on the blackbody walls with the radiation present inside the blackbody cavity. There are three ongoing optical processes: spontaneous emission, stimulated absorption, and stimulated emission. To simplify the analysis, we assume that each atom is a two-level system and the atomic linewidth function $g(\nu)$ is much narrower than the spectral energy density $\rho(\nu)$. In this case, the net time rate of change of the upper-state population density N_2 will be given by

$$\frac{dN_2}{dt} = -A_{21}N_2 - B_{21}N_2\rho(\nu_0) + B_{12}N_1\rho(\nu_0) \tag{4.184}$$

In equilibrium, the net rate of change will be zero, leading to

$$-A_{21}N_2 - B_{21}N_2\rho(\nu_0) + B_{12}N_1\rho(\nu_0) = 0 \tag{4.185}$$

From the above equation, we can determine the ratio of the populations in the upper and lower states as

$$\frac{N_2}{N_1} = \frac{B_{12}\rho(\nu_0)}{B_{21}\rho(\nu_0) + A_{21}} \tag{4.186}$$

We also know from statistical thermodynamics that N_2/N_1 in equilibrium is given by the well-known Boltzmann statistics

$$\frac{N_2}{N_1} = \frac{g_2}{g_1}\exp\left(-\frac{(E_2 - E_1)}{k_B T}\right) = \frac{g_2}{g_1}\exp\left(-\frac{h\nu_0}{k_B T}\right) \tag{4.187}$$

where k_B is the Boltzmann constant, T is the absolute temperature, and g_1 and g_2 are the degeneracy factors of the states $|1\rangle$ and $|2\rangle$. For example if $g_1 = 3$, we have three linearly independent quantum states, each with energy E_1.

Combining Eqs. (4.186) and (4.187), we can eliminate the population densities to obtain

$$\frac{B_{12}\rho(\nu_0)}{B_{21}\rho(\nu_0) + A_{21}} = \frac{g_2}{g_1}\exp\left(-\frac{h\nu_0}{k_B T}\right) \tag{4.188}$$

which, when solved for the spectral energy density $\rho(\nu_0)$, gives

$$\rho(\nu_0) = \frac{\dfrac{A_{21}}{B_{21}}}{\dfrac{B_{12}}{B_{21}}\dfrac{g_1}{g_2}\exp\left(\dfrac{h\nu_0}{k_B T}\right) - 1} \tag{4.189}$$

However, we know from our earlier discussion of blackbody radiation, that the spectral energy density function is

$$\rho(\nu_0) = \frac{8\pi h\nu_0^3}{c^3}\frac{1}{\exp\left(\dfrac{h\nu_0}{k_B T}\right) - 1} \tag{4.190}$$

Comparing Eqs. (4.189) and (4.190), we see that

$$\frac{A_{21}}{B_{21}} = \frac{8\pi h\nu_0^3}{c^3} \tag{4.191}$$

and

$$g_2 B_{21} = g_1 B_{12} \tag{4.192}$$

A more rigorous quantum mechanical approach involves the use of time-dependent perturbation theory to derive the expressions for the A and B coefficients independently. As we will show in Chap. 5, this gives the same result for the ratio A_{21}/B_{21} as was obtained in

Eq. (4.191). Finally, in a medium with refractive index n, we replace c by (c/n) to get

$$\frac{A_{21}}{B_{21}} = \frac{8\pi h \nu_0^3 n^3}{c^3} \tag{4.193}$$

Problems

4.1 Show that for an atom interacting with a plane electromagnetic wave, the magnitudes of the electric and magnetic contributions of the Lorentz force satisfy the inequality $\left|\vec{u} \times \vec{B}(\vec{r},t)\right| / \left|\vec{E}(\vec{r},t)\right| \leq u/c$, where u is the electron speed and c is the speed of light.

4.2 Show that when the frequency is close to the resonance frequency, the refractive index variation with frequency is given by Eq. (4.37). Furthermore, show that under the same approximation, the resonant contribution to the absorption coefficient is given by Eq. (4.43).

4.3 Show that the normalized Lorentzian distribution is given by Eq. (4.47).

4.4 Show that the absorption cross section can be expressed in terms of the normalized Lorentzian lineshape function as given by Eq. (4.48).

4.5 Use L'Hopital's rule to show that as M tends to infinity, the expression for $P_{NC}(t)$ is given by Eq. (4.89).

4.6 Show that the width (FWHM) of the Doppler lineshape function is given by Eq. (4.102).

4.7 Show that the group velocity of an electromagnetic pulse in a dispersive medium is given by Eq. (4.116). Also show that the group velocity dispersion parameter, as defined by Eq. (4.117), can be calculated by using Eq. (4.118).

4.8 By using the refractive index data for silica given in Eq. (4.58), reproduce the wavelength-dependent dispersion curve for silica shown in Fig. 4.16 in the 1–1.5 μm range.

4.9 Show that the approximate temporal width of a pulse with an initial duration of τ_0 is given by Eq. (4.121) after a propagation distance of d in a dispersive medium characterized by the group velocity dispersion parameter D_{GV}.

4.10 By using Eq. (4.150), show that the average energy $\bar{E}(\lambda)$ per mode is given by Eq. (4.151).

4.11 By using Planck's blackbody distribution for $R(\lambda)$, derive Wien's displacement law in Eq. (4.122).

4.12 A hypothetical absorber has the measured differential absorption coefficient as a function of frequency shown in Fig. 4.28. The active center concentration in the medium is 10^{14} cm^{-3}. Calculate the oscillator strength of the transition. Assume that the refractive index of the medium is approximately 1.

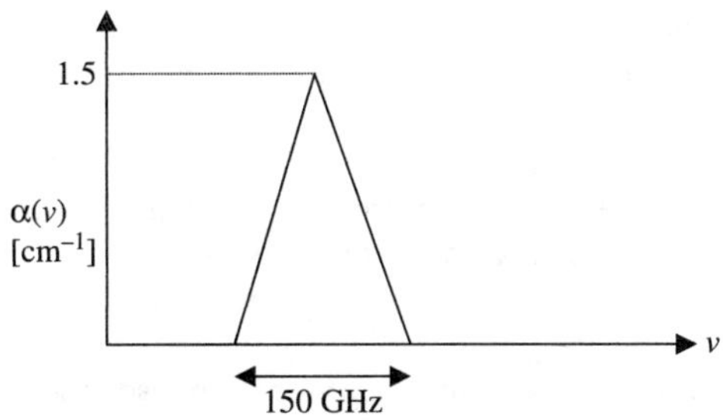

FIGURE 4.28 Problem 4.12.

4.13 At a wavelength of 0.8 μm, the loss coefficient of a new optical glass is 2 dB/km due to Rayleigh scattering. If light with a wavelength of 1.6 μm is launched into a 1-km-long fiber made of this material, calculate the fraction of the incident power that will be lost due to scattering.

4.14 Monochromatic light at a wavelength of 650-nm is incident on a dielectric slab. Measurements indicate that each surface has an intensity reflection of 15% at normal incidence. Furthermore, the absorption (excluding reflective losses) of a 10-mm-long sample is 45%. What is the complex refractive index n at this wavelength? In your calculations, assume that the imaginary part of n is much smaller than the real part. Your answer, however, should give the imaginary as well as the real part of n.

4.15 A certain material has a Sellmeier equation given by

$$n(\lambda) = 1 + \frac{A}{\lambda^2} - \frac{B}{\lambda^3}$$

where $A = 2\ (\mu\text{m})^2$ and $B = 1.5\ \mu\text{m}^3$. At what wavelength could one achieve electromagnetic pulse propagation with minimum broadening?

4.16 Consider Fig. 4.29 where a laser is used to measure the absorption characteristics of a gaseous medium. Assume that the laser gain medium is inhomogeneously broadened. The output will then consist of many longitudinal modes which are equally spaced in frequency.

(*a*) If the laser cavity has a length $L = 100$ cm, what is the frequency spacing between two adjacent output modes (neglect the transverse mode structure of the laser)?

(*b*) The multi-mode output of the laser is sent through a gas cell in which the absorption lines are also predominantly Doppler broadened (i.e., homogeneous

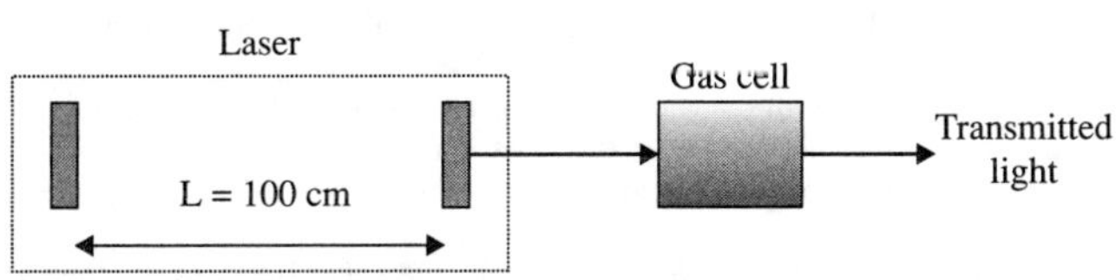

FIGURE 4.29 Problem 4.16.

broadening of the lines is negligible). In the experiments, the laser is adjusted such that one of the longitudinal modes (call it the mth mode) coincides exactly with the center of the absorption band. The Doppler width (FWHM) of this absorption band is 200 MHz. If the cell absorbs 70% of the mth mode, what will the absorption of the $(m + 1)$th mode be? You may make the following assumptions:

1. The spectral width of each laser mode is negligible with respect to the width of the absorption line. Hence, you may treat each mode as a monochromatic wave.
2. The gaseous medium has only one absorption resonance in the frequency range of interest.

References

1. J. D. Jackson, *Classical Electrodynamics,* 3d ed. (John Wiley & Sons, Inc., New York 1998).
2. B. E. A. Saleh and M. C. Teich, *Fundamentals of Photonics,* 2d ed. (John Wiley & Sons, Inc., Hoboken, N.J., 2007).
3. A. Yariv, *Optical Electronics in Modern Communications,* 5th ed. (Oxford University Press, New York, 1997).
4. P. W. Milonni and J. H. Eberly, *Lasers* (John Wiley & Sons, Inc., New York, 1988).
5. A. E. Siegman, *Lasers* (University Science Books, Mill Valley, Calif., 1986).
6. S. G. Lipson and H. Lipson, *Optical Physics,* 2d ed. (Cambridge University Press, Cambridge, 1981).
7. C. R. Pollock, *Fundamentals of Optoelectronics* (Irwin, Chicago, 1995).
8. J. T. Verdeyen, *Laser Electronics,* 3d ed. (Prentice Hall, Upper Saddle River, N.J., 1995).
9. G. R. Fowles, *Introduction to Modern Optics* (Dover, New York, 1975).
10. H. A. Lorentz, *The Theory of Electrons* (Dover, New York, 1952).
11. W. Sellmeier, "Zur Erklärung der abnormen Farbenfolge im Spectrum einiger Substanzen," *Annalen der Physik und Chemie,* **219**, 272–282 (1871).
12. I. H. Malitson, "Interspecimen Comparison of the refractive index of fused silica," *Journal of the Optical Society of America,* **55**, 1205–1209 (1965).
13. G. B. Arfken and H. J. Weber, *Mathematical Methods for Physicists,* 5th ed. (Harcourt Academic Press, Amsterdam, 2001).
14. R. d. L. Kronig, "On the theory of dispersion of x-rays," *Journal of the Optical Society of America and Review of Scientific Instruments,* **12**, 547–557 (1926).
15. C. A. Brau, *Modern Problems in Classical Electrodynamics* (Oxford University Press, Oxford, 2004).
16. W. R. Bennett, V. P. Chebotay, and J. W. Knutson, "Direct observation of collision broadening and effect of resonant interactions on gas-laser transitions," *Physical Review Letters,* **18**(17), 688–692 (1967).
17. J. W. Knutson and W. R. Bennett, "Line broadening of 3s2-2pn laser transitions of neon," *Physical Review A,* **13**(1), 318–325 (1976).
18. K. Krane, *Modern Physics,* 2d ed. (John Wiley & Sons, Inc., New York, 1996).
19. H. Kanamori, H. Yokota, G. Tanaka, M. Watanabe, Y. Ishiguro, I. Yoshida, T. Kakii, S. Itoh, Y. Asano, and S. Tanaka, "Transmission characteristics and reliability of pure-silica-core single-mode fibers," *Journal of Lightwave Technology,* **LT-4**, 1144–1150 (1986).
20. D. N. Payne and W. A. Gambling, "Zero material dispersion in optical fibers," *Electronics Letters,* **11**(8), 176–178 (1975).
21. R. Eisberg and R. Resnick, *Quantum Physics of Atoms, Molecules, Solids, Nuclei, and Particles* (John Wiley & Sons, Inc., New York, 1985).
22. A. Einstein, "Quantum theory of radiation," *Physikalische Zeitschrift,* **18**, 121–128 (1917).
23. A. Pais, *Subtle Is the Lord: The Science and the Life of Albert Einstein* (Oxford University Press, Oxford, 1982).

CHAPTER 5

Quantum Theory of Light-Matter Interactions

In Chap. 4, Einstein's A and B coefficients were introduced by using a phenomenological approach based on the population rate equations. In this chapter, we will go a step further to develop a more rigorous theory of light-matter interactions and to derive quantum mechanical expressions for these coefficients. Needless to say, we need quantum mechanics to do this. So, we start with a condensed review of the principles of quantum mechanics and apply these to the study of light-matter interactions. Of course, the reader will not become a "certified quantum mechanic" just by reading a chapter on these concepts that have far reaching consequences in both science and technology! We simply provide the basics at an elementary level to accommodate readers with little or no prior introduction to quantum mechanics. Further discussion of the principles and methods may be found in many other excellent sources on the subject.[1-19]

We first start with a discussion of de Broglie's hypothesis and introduce the wave function. The probability interpretation and the postulates of quantum mechanics are then discussed. The solution of the Schrödinger equation for several one-dimensional examples is given. In particular, the harmonic oscillator problem is treated by solving the Schrödinger equation and also by using operator methods. Time-dependent perturbation theory is then used to derive the expression for the transition rate in the presence of a harmonic perturbation. By treating the electromagnetic radiation classically, Einstein's B coefficient is first derived in terms of the fundamental constants and the dipole matrix element. Then, the quantization of the electromagnetic field is discussed and the operator representation of the electric field is used to derive the quantum mechanical expression for the Einstein's A coefficient. Finally, we discuss the interaction of an ideal two-level atomic system with electromagnetic radiation and Rabi oscillations.

5.1 From Particle Waves to Operators

5.1.1 de Broglie's Hypothesis and Wave Packets

Earlier, we saw in the discussion of blackbody radiation that electromagnetic radiation shows particle-like properties and introduced the concept of a photon. In 1923, the French physicist de Broglie[20] further proposed that every material particle with a finite momentum p exhibits wave-like properties with a characteristic wavelength λ given by

$$\lambda = \frac{h}{p} \tag{5.1}$$

In Eq. (5.1), h is the Planck's constant introduced in Chap. 4. de Broglie's hypothesis was later experimentally verified by Davisson and Germer, who in 1927 demonstrated the diffraction of an electron beam from a periodic lattice by using a nickel crystal.[21] In this section, our task will therefore be to develop the mathematical formalism for describing objects that exhibit both wave and particle properties. This phenomenon is commonly referred to as wave-particle duality and the subject that studies the properties of particles exhibiting wave-particle duality is known as wave mechanics or quantum mechanics. To start off, let's try to answer two basic questions regarding the wave aspects of particles:

1. What does the wave amplitude physically represent?
2. Can we come up with an equation that describes the evolution of this wave in time and space?

As we will soon find out, the evolution of this particle wave is described by the Schrödinger equation.

By now, we have gained ample experience in dealing with waves. Based on the discussions in Chap. 1, we go ahead and express the wave function of a de Broglie particle as

$$\psi(x,t) = \psi_0 \exp[i(kx - \omega t)] \tag{5.2}$$

This represents a particle moving along the x-axis with a wave number k and angular frequency ω. Since $\lambda = 2\pi/k$, we further see that the particle momentum is related to k through

$$p = \frac{h}{\lambda} = \frac{h}{2\pi}k = \hbar k \tag{5.3}$$

Above, we introduced the constant "h-bar," designated as $\hbar$, which frequently appears in quantum mechanics and is given by $\hbar = h/2\pi$. For the general case, if the particle propagates in an arbitrary direction described by the wave vector

$$\vec{k} = k_x\hat{x} + k_y\hat{y} + k_z\hat{z} \tag{5.4}$$

the wave function becomes

$$\psi(x,t) = \psi_0 \exp\left[i(\vec{k}\cdot\vec{r} - \omega t)\right] \tag{5.5}$$

where the position vector $\vec{r}$ is given by

$$\vec{r} = x\hat{x} + y\hat{y} + z\hat{z} \tag{5.6}$$

Above, $\hat{x}$, $\hat{y}$, and $\hat{z}$ denote the unit vectors. As in the case of light waves, ω will be a function of k and the equation

$$\omega = \omega(k) \tag{5.7}$$

gives the dispersion relation for the particle waves.

To motivate the physical interpretation for the wave amplitude, consider the double-slit experiment performed not with light, but instead with an electron beam. The sketch of the experiment is shown in Fig. 5.1. The incident plane wave describing the electron beam will be split after passage through the slits. Two waves ψ_1 and ψ_2 emanating from the slits will reach the observation screen and form an interference pattern. To find the resultant wave amplitude at the location with position vector $\vec{r}$, we will

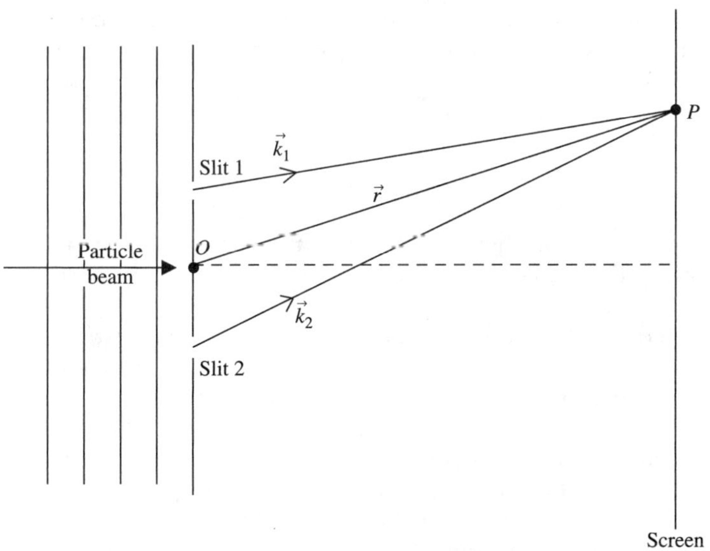

FIGURE 5.1 Double-slit experiment performed with an electron beam that exhibits wave-particle duality. An interference pattern will be observed on the viewing screen.

assume that the superposition principle holds and add the amplitudes of the waves coming from slits 1 and 2. The resultant wave function ψ will be given by

$$\psi(\vec{r},t) = \psi_1(\vec{r},t) + \psi_2(\vec{r},t) = \psi_{10}\exp\left[i\left(\vec{k}_1\cdot\vec{r} - \omega t\right)\right] + \psi_{20}\exp\left[i\left(\vec{k}_2\cdot\vec{r} - \omega t\right)\right] \tag{5.8}$$

Note that since the direction of propagation is different from each slit to the observation point, we assigned a different $\vec{k}$ vector value ($\vec{k}_1$ and $\vec{k}_2$) to the waves ψ_1 and ψ_2.

The wave function describing the particle beams is inherently complex-valued. Hoping to obtain a physically meaningful interpretation of the wave amplitude, we look at the magnitude squared of ψ and obtain

$$\begin{aligned}|\psi(\vec{r},t)|^2 &= |\psi_1(\vec{r},t) + \psi_2(\vec{r},t)|^2 = \left(\psi_1(\vec{r},t) + \psi_2(\vec{r},t)\right)\left(\psi_1^*(\vec{r},t) + \psi_2^*(\vec{r},t)\right)\\ &= |\psi_{10}|^2 + |\psi_{20}|^2 + \psi_{10}\psi_{20}^*\exp\left[i\left(\vec{k}_1 - \vec{k}_2\right)\cdot\vec{r}\right] + \psi_{20}\psi_{10}^*\exp\left[-i\left(\vec{k}_1 - \vec{k}_2\right)\cdot\vec{r}\right]\\ &= |\psi_{10}|^2 + |\psi_{20}|^2 + 2\,\mathrm{Re}\left[\psi_{10}\psi_{20}^*\exp\left[i\left(\vec{k}_1 - \vec{k}_2\right)\cdot\vec{r}\right]\right]\end{aligned} \tag{5.9}$$

First, note that this function is time-independent and forms a persistent periodic pattern on the screen. In particular, the third term on the right hand side of Eq. (5.9) is an interference term and varies periodically with position on the screen. To see this more clearly, consider the specific case where $\psi_{10} = \psi_{20} = \psi_0$ is real. Then $|\psi(\vec{r},t)|^2$ becomes

$$\begin{aligned}|\psi(\vec{r},t)|^2 &= 2\psi_0^2 + 2\psi_0^2\cos\left(\left(\vec{k}_1 - \vec{k}_2\right)\cdot\vec{r}\right)\\ &= 2\psi_0^2\left[1 + \cos\left(\left(\vec{k}_1 - \vec{k}_2\right)\cdot\vec{r}\right)\right]\end{aligned} \tag{5.10}$$

Consider the hypothetical experiment where the screen is a detector array that can register the location of the arriving particles. In fact, such an experiment performed with an electron beam was reported in 1989.[22] Results indicate that the spatial distribution of electrons received on the screen also follows an interference profile similar to what we calculated for $|\psi|^2$ in Eq. (5.10). This suggests that the amplitude squared of the wave is proportional to the probability of observing an electron at that location. Wherever $|\psi|^2$ has a local maximum, the number of arriving electrons is also maximum. Wherever it vanishes, the likelihood of registering electrons becomes very low. We will therefore follow the standard quantum mechanical interpretation and take $|\psi(\vec{r},t)|^2\, d^3r$ as the probability of finding a particle at time t, within a volume of d^3r around the location $\vec{r}$. Since the particle will be located somewhere over all space, we further require that the wave function satisfy the normalization condition

$$\int d^3r \left|\psi(\vec{r},t)\right|^2 = 1 \tag{5.11}$$

As a specific example, let's look at the particle moving with a definite momentum p along the x direction. From Eq. (5.5), we see that $|\psi|^2 = |\psi_{10}|^2$ is a constant as a function of x. In this case, $\psi(x,t)$ cannot be normalized since it extends over all space. We still see that the particle is likely to be found at any location with equal probability. In other words, whereas the uncertainty in the value of the momentum is 0, the uncertainty Δx in the particle position is ∞, and we see here a special case of the Heisenberg uncertainty principle.

So, what should we do to describe a particle which is localized in space? It turns out that in this case, we can use a superposition of plane waves with different k values and a wave number-dependent amplitude. Such plane wave superpositions which describe localized particles are known as wave packets. By assuming that the amplitude distribution $a(k)$ varies continuously with k about a central value k_0 as sketched in Fig. 5.2, the wave function of the packet can be written as

$$\psi(x,t) = \int_{-\infty}^{\infty} dk\, a(k) \exp[i(kx - \omega t)] \tag{5.12}$$

To gain some insight about the characteristics of the wave, we can assume that the dispersion relation about k_0 can be approximated up to second order according to

$$\omega(k) \cong \omega(k_0) + (k - k_0)\left.\frac{d\omega}{dk}\right|_0 + \frac{1}{2}(k - k_0)^2 \left.\frac{d^2\omega}{dk^2}\right|_0 \tag{5.13}$$

where the subscript 0 assigned to the derivatives indicates that the differentiation is evaluated at $k = k_0$. We leave it as an exercise

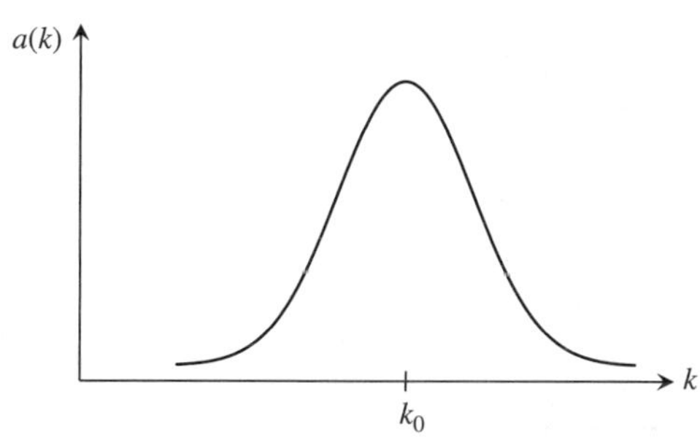

FIGURE 5.2 Spectral amplitude distribution function $a(k)$ for a wave packet.

(see Prob. 5.1) for the reader to show that for a gaussian amplitude distribution given by

$$a(k) = a_0 \exp\left[-\alpha(k - k_0)^2\right] \tag{5.14}$$

the resulting wave function at (x, t) becomes

$$|\psi(x,t)|^2 = \frac{a_0^2 \pi}{\sqrt{(\alpha^2 + \beta^2 t^2)}} \exp\left[-\frac{(x - u_g t)^2 \alpha}{2(\alpha^2 + \beta^2 t^2)}\right] \tag{5.15}$$

Two important observations are in order. First, note that the width of the distribution is increasing as times passes, this is illustrated in Fig. 5.3. Second, we see that the center of the wave packet is moving with a velocity of

$$u_g = \left.\frac{d\omega}{dk}\right|_{k=k_0} \tag{5.16}$$

This is identical to the group velocity we introduced back in Chap. 1 for light waves. In order to reconcile the wave and particle properties, we further assume that the group velocity of the wave packet is equal to the velocity p/m of the particle to obtain[5]

$$\frac{d\omega}{dk} = \frac{\hbar k}{m} \tag{5.17}$$

Integrating Eq. (5.17) and multiplying by $\hbar$, we get, after setting the integration constant to zero,

$$\hbar\omega = \frac{(\hbar k)^2}{2m} = K \tag{5.18}$$

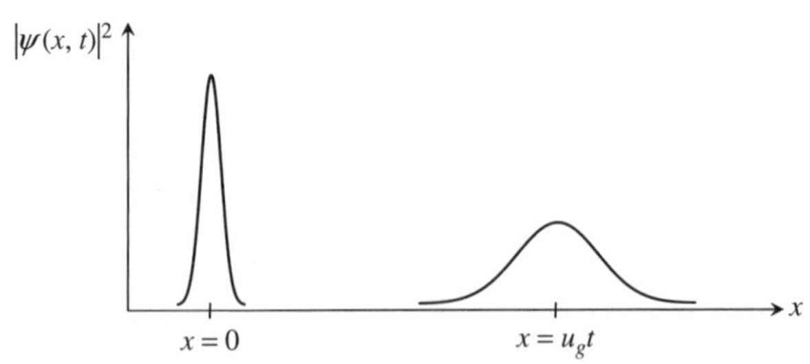

FIGURE 5.3 The probability distribution $|\psi(x,t)|^2$ describing a propagating wave packet.

Hence, we can identify $\hbar\omega$ as the energy of the particle which, in this case, is equal to the kinetic energy K. We did the same thing before for photons in Chap. 4.

5.1.2 Schrödinger Equation

The argument we developed in the earlier section can be used to come up with a differential equation that gives the temporal and spatial evolution of the wave function $\psi(x,t)$ of a wave packet moving in the presence of a potential. Since, we identified $\hbar\omega$ as the total energy (E) and $\hbar k$ as the momentum (p) of a particle, we will now rewrite the expression for the wave function of the localized particle as

$$\psi(x,t) = \int_{-\infty}^{\infty} dk\, a(k) \exp\left[\frac{i(px - Et)}{\hbar}\right] \tag{5.19}$$

Consider the one-dimensional motion of a particle of mass m in a position-dependent potential well with energy $V = V(x)$. From the conservation of energy, we know that

$$E = K + V \tag{5.20}$$

Now, let's apply linear operations to ψ so as to bring out E and K in terms of its derivatives. We note from Eq. (5.19) that taking the derivative of $\psi(x,t)$ with respect to time and rearranging, we get

$$E = i\hbar \frac{1}{\psi} \frac{\partial \psi}{\partial t} \tag{5.21}$$

Furthermore, differentiating twice with respect to x and multiplying by $-\hbar^2/2m$, we obtain

$$K = -\frac{\hbar^2}{2m} \frac{1}{\psi} \frac{\partial^2 \psi}{\partial x^2} \tag{5.22}$$

Now, substituting into Eq. (5.20) and rearranging gives

$$-\frac{\hbar^2}{2m} \frac{\partial^2 \psi}{\partial x^2} + V(x)\psi(x,t) = i\hbar \frac{\partial \psi}{\partial t} \tag{5.23}$$

This is the famous Schrödinger equation.[23-25] What we have provided here is not a rigorous derivation of the Schrödinger equation. The analysis simply shows that a simple application of conservation of energy and use of the assumed functional form for the wave function of the localized particle [Eq. (5.19)] justifies the form of the Schrödinger equation. The Schrödinger equation has no proof. Our firm trust in it stems from the fact that its predictions have been verified by experimental observations in many different cases.

Next, let us seek a solution of Eq. (5.23) in the form

$$\psi(x,t) = u(x)T(t) \tag{5.24}$$

Substituting into Eq. (5.23) and dividing by $u(x)T(t)$, we get

$$-\frac{\hbar^2}{2m}\frac{1}{u(x)}\frac{d^2u}{dx^2}+V(x)=i\hbar\frac{1}{T(t)}\frac{dT}{dt} \tag{5.25}$$

The left-hand side is only a function of x, whereas the right-hand side is only a function of t. Hence, each side of Eq. (5.25) should be a constant, which we choose as E to obtain two separate ordinary differential equations:

$$-\frac{\hbar^2}{2m}\frac{d^2u}{dx^2}+V(x)u(x)=Eu(x) \tag{5.26}$$

and

$$\frac{1}{T(t)}\frac{dT}{dt}=\frac{E}{i\hbar} \tag{5.27}$$

Equation (5.26) is known as the time-independent Schrödinger equation. The separation constant E, which gives the energy eigenvalues, is determined from the solution of Eq. (5.26) for a given potential energy and boundary conditions. To each energy eigenvalue E, there corresponds an eigenfunction $u_E(x)$ that satisfies Eq. (5.26). E can have a continuous or a discrete spectrum. We will explore several examples of potential wells and determine the energy eigenvalues and the corresponding eigenfunctions in the following sections. For a given E, Eq. (5.27) can be readily solved to yield the full solution

$$\psi(x,t)=u_E(x)\exp\left[-i\frac{E}{\hbar}t\right] \tag{5.28}$$

for the time-dependent wave function. For the general case, the solution of the Schrödinger equation will involve a continuous or a discrete sum over all possible solutions of the form given in Eq. (5.28).

As a final remark, we can further generalize the time-dependent Schrödinger equation to three dimensions by noting that in this case, the wave function for the wave vector $\vec{k}$ will be of the form

$$\psi_{\vec{k}}=\exp\left[\frac{i}{\hbar}(\vec{p}\cdot\vec{r}-Et)\right] \tag{5.29}$$

where $\vec{p}=\hbar\vec{k}$. Now, the kinetic energy becomes

$$K=-\frac{\hbar^2|\vec{p}|^2}{2m}=-\frac{\hbar^2}{2m}\frac{1}{\psi}\nabla^2\psi \tag{5.30}$$

Hence, the three-dimensional version of Eq. (5.23) reads

$$-\frac{\hbar^2}{2m}\nabla^2\psi + V(\vec{r})\psi = i\hbar\frac{\partial\psi}{\partial t} \tag{5.31}$$

5.1.3 Expectation Values

Since the wave equation has a probabilistic interpretation, we can use it to calculate the expectation values of physical observables. Consider the particle position as an example. Assuming one-dimensional motion, the expectation value $< x >$ of the particle's position will be given by

$$< x > = \int_{-\infty}^{\infty} dx\, x|\psi(x,t)|^2 = \int_{-\infty}^{\infty} dx\, \psi^*(x,t)\, x\, \psi(x,t) \tag{5.32}$$

Generalizing to three dimensions, we get

$$\langle\vec{r}(t)\rangle = \int_{\text{allspace}} d^3r\, \psi^*(\vec{r},t)\vec{r}\,\psi(\vec{r},t) \tag{5.33}$$

Going back to one dimension, we can, in a similar way, calculate the expectation value of any function of x, say $f(x)$, by using

$$\langle f(x)\rangle = \int_{-\infty}^{+\infty} dx\, \psi^*(x,t)\, f(x)\psi(x,t) \tag{5.34}$$

At first sight, it may appear that this is all there is to expectation values. However, this is not true when we consider the momentum of the particle. Due to the uncertainty principle, we cannot exactly express the particle momentum as a function of position, and hence the above approach summarized by Eq. (5.34) fails in calculating the expectation values of momentum $\vec{p}$. To proceed, note that

$$p_x = m\frac{dx}{dt} \tag{5.35}$$

Following the approach in Ref. 5, we set out to find $\langle p_x\rangle$ from

$$\langle p_x\rangle = m\frac{d\langle x\rangle}{dt} \tag{5.36}$$

By using Eq. (5.32) for $\langle x\rangle$, we obtain

$$\begin{aligned}\langle p_x\rangle &= m\frac{d\langle x\rangle}{dt} = m\frac{d}{dt}\int_{-\infty}^{+\infty} dx\psi^*(x,t)x\psi(x,t)\\ &= m\int_{-\infty}^{+\infty} dx\left[\frac{\partial\psi^*}{\partial t}x\psi(x,t) + \psi^*(x,t)x\frac{\partial\psi}{\partial t}\right]\end{aligned} \tag{5.37}$$

Next, we use the Schrödinger equation [Eq. (5.23)] to express the time derivative of the wave function as

$$\frac{\partial \psi}{\partial t} = \frac{1}{i\hbar}\left(-\frac{\hbar^2}{2m}\frac{\partial^2 \psi}{\partial x^2} + V\psi\right) \tag{5.38}$$

Taking the complex conjugate of Eq. (5.38) gives

$$\frac{\partial \psi^*}{\partial t} = -\frac{1}{i\hbar}\left(-\frac{\hbar^2}{2m}\frac{\partial^2 \psi^*}{\partial x^2} + V\psi^*\right) \tag{5.39}$$

Substituting into the expression for $\langle p_x \rangle$, we then obtain

$$\begin{aligned}\langle p_x \rangle &= m\int_{-\infty}^{+\infty} dx\left\{\left(\frac{\hbar}{2mi}\frac{\partial^2\psi^*}{\partial x^2} - \frac{V}{i\hbar}\psi^*\right)x\psi + \psi^* x\left(-\frac{\hbar}{2mi}\frac{\partial^2\psi}{\partial x^2} + \frac{V}{i\hbar}\psi\right)\right\} \\ &= \frac{\hbar}{2i}\int_{-\infty}^{+\infty} dx\left[\frac{\partial^2\psi^*}{\partial x^2}x\psi - \psi^* x\frac{\partial^2\psi}{\partial x^2}\right]\end{aligned} \tag{5.40}$$

We leave it as an exercise for the reader (see Prob. 5.2) to show by using integration by parts that $\langle p_x \rangle$ simplifies to

$$\langle p_x \rangle = \frac{\hbar}{i}\int_{-\infty}^{+\infty} dx\,\psi^*\frac{\partial \psi}{\partial x} \tag{5.41}$$

5.1.4 Quantum Mechanical Operators

In quantum mechanics, each physical observable is represented by a mathematical operator. To motivate how this is done, let us revisit the expression for the position expectation value. We can equivalently write the expectation value $\langle x \rangle$ in the form

$$\langle x \rangle = \int_{-\infty}^{+\infty} dx\,\psi^*\hat{X}\psi \tag{5.42}$$

where $\hat{X}$ is defined as the operator that represents the position of the particle. Here, the $\hat{X}$ operator simply corresponds to "multiplication by x."

Turning back to the momentum expectation value, we can equivalently express $\langle p_x \rangle$ as

$$\langle p_x \rangle = \int_{-\infty}^{+\infty} dx\,\psi^*\hat{P}_x\psi \tag{5.43}$$

where, from Eq. (5.41), the momentum operator $\hat{P}_x$ can be defined according to

$$\hat{P}_x = \frac{\hbar}{i}\frac{\partial}{\partial x} \tag{5.44}$$

Now that we have introduced the position and momentum operators, we can go one step further and rewrite the Schrödinger equation in terms of these operators. Note that in terms of $\hat{P}_x$, we have

$$-\frac{\hbar^2}{2m}\frac{\partial^2\psi}{\partial x^2} = \frac{\hat{P}_x^{\,2}}{2m}\psi \tag{5.45}$$

where $\hat{P}_x^2$ simply means "apply the differential operator $(\hbar/i)(\partial/\partial x)$ twice." Hence, the Schrödinger equation can be recast into the operator form

$$\hat{H}\psi(x,t) = i\hbar\frac{\partial\psi}{\partial t} \tag{5.46}$$

where $\hat{H}$ is known as the hamiltonian operator (or the total energy operator) and is given by

$$\hat{H} = \frac{\hat{P}_x^2}{2m} + V(x) = -\frac{\hbar^2}{2m}\frac{\partial^2}{\partial x^2} + V(x) \tag{5.47}$$

5.2 Postulates of Quantum Mechanics

In this section, we summarize some of the basic postulates of quantum mechanics and discuss the general properties of operators. A more complete discussion of the principles may be found in Refs. 1 to 6. We also introduce the Dirac notation which provides a convenient, short-hand notation to write down expectation values and matrix elements. Finally, Schrödinger and Heisenberg representations of a quantum system evolving in time are reviewed.

5.2.1 The Wave Function

All information about a quantum mechanical system is contained in the wave function $\psi(\vec{r},t)$. We already introduced the wave function in the previous section. This can be determined by solving the Schrödinger equation, given in operator form by

$$\hat{H}\psi = i\hbar\frac{\partial\psi}{\partial t} \tag{5.48}$$

Knowing $\psi(\vec{r},t)$, we can then calculate probabilities, expectation values, and transition rates. In fact, this will be very important when we discuss stimulated absorption/emission as well as spontaneous emission.

5.2.2 Observables and Operators

Every physical observable is represented by a linear hermitian operator. Here, the two attributes of the operator must be emphasized:

(1) linearity and (2) hermicity. Let us investigate these properties in more detail. If an operator, say $\hat{O}$, is linear, then it satisfies the relation

$$\hat{O}\sum_{i=1}^{n} a_i \psi_i = \sum_{i=1}^{n} a_i \hat{O} \psi_i \tag{5.49}$$

where a_i represents constant expansion coefficients and n is an integer. As a specific example, let us take $\hat{O}$ as d/dx. This is a linear operator since, for example,

$$\frac{d}{dx}(2y_1(x) + 3y_2(x)) = 2\frac{dy_1}{dx} + 3\frac{dy_2}{dx} \tag{5.50}$$

Next, we introduce the hermitian conjugate $\hat{O}^+$ of the operator $\hat{O}$. $\hat{O}$ and $\hat{O}^+$ satisfy the relation

$$\int d^3r \left(\hat{O}\psi_1\right)^* \psi_2 = \int d^3r \psi_1{}^* \hat{O}^+ \psi_2 \tag{5.51}$$

for two arbitrary wave functions ψ_1 and ψ_2. By definition, if $\hat{O}$ is hermitian, then

$$\hat{O}^+ = \hat{O} \tag{5.52}$$

In other words, a hermitian operator satisfies

$$\int d^3r \left(\hat{O}\psi_1\right)^* \psi_2 = \int d^3r \psi_1{}^* \hat{O} \psi_2 \tag{5.53}$$

We leave it as an exercise for the reader (see Prob. 5.3) to show that both the position and momentum operators are hermitian.

5.2.3 Quantum Measurements

If we measure a physical observable represented by the operator $\hat{O}$, the measurement will give one of the eigenvalues of the operator $\hat{O}$ which can be found by solving the eigenvalue equation

$$\hat{O}\psi_n = O_n \psi_n \tag{5.54}$$

Above O_n is the nth eigenvalue of the observable, and ψ_n is the corresponding eigenfunction. The time-independent Schrödinger equation [Eq. (5.26)] discussed before is an example of such an eigenvalue equation. Its solution yields all of the possible energy eigenvalues that can be measured for the system. If the wave function of the system just before the measurement is ψ_n, then we know for sure that the measurement will yield the value O_n. If the wave function is not one of the eigenfunctions, we will not know, a priori, which eigenvalue will be obtained, although the measurement gives one of the eigenvalues. Lastly, suppose that a measurement has been made and the eigenvalue O_n has been measured. It follows that now the wave function has collapsed to ψ_n and repeating the measurement again will produce O_n.

Since different values of O_n correspond to the possible measurement outcomes of O, a physical observable, they should be real. This is always the case for hermitian operators. To see, consider the integral $\int d^3r \psi_n{}^* \hat{O}\psi_n$, where ψ_n is the nth eigenfunction. We have

$$\int d^3r\, \psi_n^* \hat{O}\psi_n = \int d^3r\, \psi_n^* O_n \psi_n = O_n \int d^3r\, \psi_n^* \psi_n \tag{5.55}$$

Since $\hat{O}$ is hermitian, we can alternatively write the above relation as

$$\int d^3r\, \psi_n^* \hat{O}\psi_n = \int d^3r \left(\hat{O}\psi_n\right)^* \psi_n = O_n^* \int d^3r\, \psi_n^* \psi_n \tag{5.56}$$

In other words,

$$\left(O_n - O_n^*\right) \int d^3r\, \psi_n^* \psi_n = 0 \tag{5.57}$$

However, the integral $\int d^3r\, \psi_n^* \psi_n \neq 0$ and hence, we obtain

$$O_n = O_n^* \tag{5.58}$$

which means that the eigenvalues are real.

Here is another very important property of hermitian operators. Suppose that we solve the eigenvalue equation $\hat{O}\,\psi_n = O_n \psi_n$ and determine the eigenvalues and the corresponding eigenfunctions. It turns out that eigenfunctions corresponding to different eigenvalues are orthogonal. By definition, if two functions ψ_m and ψ_n are orthogonal, then their inner product vanishes according to

$$\int d^3r\, \psi_m^* \psi_n = 0 \tag{5.59}$$

To prove this property, let us assume that ψ_m and ψ_n are two eigenfunctions of the operator $\hat{O}$ with eigenvalues O_m and O_n, respectively. We have

$$\hat{O}\,\psi_m = O_m \psi_m \tag{5.60}$$

and

$$\hat{O}\,\psi_n = O_n \psi_n \tag{5.61}$$

We multiply Eq. (5.60) by ψ_n^* to obtain

$$\psi_n^* \hat{O}\psi_m = O_m \psi_n^* \psi_m \tag{5.62}$$

We then take the complex conjugate of Eq. (5.61) and multiply by ψ_m to obtain

$$\psi_m\left(\hat{O}\psi_n\right)^* = O_n\psi_m\psi_n^* \tag{5.63}$$

Above, we used the fact that $O_n = O_n^*$. We now subtract Eq. (5.63) from Eq. (5.62) and integrate over all space to get

$$\int d^3r\,\psi_n^*\hat{O}\psi_m - \int d^3r\,\psi_m\left(\hat{O}\psi_n\right)^* = \int d^3r\left(\hat{O}\psi_n\right)^*\psi_m - \int d^3r\,\psi_m\left(\hat{O}\psi_n\right)^*$$

$$= 0 = (O_m - O_n)\int d^3r\,\psi_m\psi_n^* \tag{5.64}$$

But, O_m is different from O_n, which results in

$$\int d^3r\,\psi_m\psi_n^* = 0 \tag{5.65}$$

Depending on the symmetry of the system, we may end up having degenerate eigenvalues. If an eigenvalue is nth order degenerate, this means that there are n linearly independent eigenfunctions with the same eigenvalue. In this case, one can form suitable linear combinations to obtain n orthogonal wavefunctions. This technique is known as the Schmidt orthogonalization procedure, and we refer the reader to Ref. 2 for a detailed description of the method.

5.2.4 Completeness and Closure Relation

The eigenfunctions of the eigenvalue equation

$$\hat{O}\psi_n = O_n\psi_n \tag{5.66}$$

form a complete set. Consider the case where the eigenvalues are discrete. Completeness means that an arbitrary wave function ψ can be expressed as a linear superposition of these eigenfunctions as

$$\psi = \sum_i a_i\psi_i \tag{5.67}$$

The expansion coefficients can be determined by multiplying each side of Eq. (5.67) by ψ_j^* and integrating over all space. Since eigenfunctions corresponding to different eigenvalues are orthogonal, they satistfy

$$\int d^3r\,\psi_i\psi_j^* = \delta_{ij} \tag{5.68}$$

where δ_{ij} is the Kroenecker delta, which is 1 when $i = j$ and zero otherwise. Hence, we get

$$\int d^3r\,\psi_j^*\psi = \sum_i a_i\int d^3r\,\psi_j^*\psi = \sum_i a_i\delta_{ij} = a_j \tag{5.69}$$

or

$$a_i = \int d^3r\, \psi_i^* \psi \tag{5.70}$$

We refer the reader again to other references (Refs. 2 and 5, for example) for a discussion of completeness and orthogonality for the case where the eigenvalue spectrum is continuous. If ψ is normalized to begin with, satisfying

$$\int d^3r |\psi|^2 = 1 \tag{5.71}$$

then $|a_i|^2$ = probability that a measurement on the system gives the eigenvalue O_i with the corresponding wave function ψ_i. Getting back to Eq. (5.67), we substitute the expression for a_i to obtain

$$\begin{aligned}\psi(\vec{r}) &= \sum_i \left(\int d^3r' \psi_i^*(\vec{r}\,')\psi(\vec{r}\,')\right)\psi_i(\vec{r}) \\ &= \int d^3r' \left(\sum_i \psi_i^*(\vec{r}\,')\psi_i(\vec{r})\right)\psi(\vec{r}\,')\end{aligned} \tag{5.72}$$

Since

$$\int d^3r' \delta^3(\vec{r}\,' - \vec{r})\psi(\vec{r}\,') = \psi(\vec{r}) \tag{5.73}$$

where $\delta^3(\vec{r})$ is the three-dimensional Dirac delta function, we conclude that

$$\sum_i \psi_i^*(r')\psi_i(\vec{r}) = \delta^3(\vec{r}\,' - \vec{r}) \tag{5.74}$$

Equation (5.74) is known as the closure relation.

5.2.5 Expectation Values

The expectation value $\langle O \rangle$ of an observable is calculated from

$$\langle O \rangle = \int d^3r\, \psi^* \hat{O} \psi \tag{5.75}$$

Suppose that ψ can be represented as a linear superposition of the eigenfunctions of $\hat{O}$ as

$$\psi = \sum_i a_i \psi_i \tag{5.76}$$

where a_i is constant and

$$\hat{O}\psi_i = O_i \psi_i \tag{5.77}$$

Above, O_i is the ith eigenvalue of $\hat{O}$ with the corresponding eigenfunction ψ_i. Then,

$$\langle O\rangle = \int d^3r \left(\sum_i a_i^* \psi_i^*\right)\hat{O}\left(\sum_j a_j \psi_j\right) = \int d^3r \sum_i \sum_j a_i^* a_j \psi_i^* O_j \psi_j$$

$$= \sum_i \sum_j a_i^* a_j O_j \delta_{ij} = \sum_i O_i |a_i|^2 \tag{5.78}$$

In other words, the expectation value of an operator is given by the sum of the eigenvalues, each weighted by the magnitude squared of the expansion coefficient.

5.2.6 Commutator of Two Operators

Suppose that we have two operators $\hat{O}$ and $\hat{P}$. In general, the order in which they act on a function is important. In other words $\hat{O}\hat{P}\psi_1$ is in general different from $\hat{P}\hat{O}\psi_1$. To quantify this property, we define the commutator $[\hat{O},\hat{P}]$ of two operators as

$$[\hat{O},\hat{P}] = \hat{O}\hat{P} - \hat{P}\hat{O} \tag{5.79}$$

If $[\hat{O},\hat{P}] = 0$ we say that $\hat{O}$ and $\hat{P}$ commute. If $\hat{O}$ and $\hat{P}$ have a complete orthogonal set of simultaneous eigenfunctions, we say that they represent compatible observables. If this is the case, the observables represented by the operators $\hat{O}$ and $\hat{P}$ can be simultaneously measured. It is also true that if two observables can be simultaneously measured, then the operators representing these observables commute. To prove this, consider an arbitrary wave function ψ and express it as a superposition of simultaneous eigenfunctions of $\hat{O}$ and $\hat{P}$:

$$\psi = \sum_i a_i \psi_i \tag{5.80}$$

Since ψ_i is a simultaneous eigenfunction of $\hat{O}$ and $\hat{P}$, we have

$$\hat{O}\psi_i = O_i \psi_i \tag{5.81}$$

and

$$\hat{P}\psi_i = P_i \psi_i \tag{5.82}$$

Operating on ψ with $\hat{O}\hat{P}$ we get

$$\hat{O}\hat{P}\psi = \sum_i a_i \hat{O}\hat{P}\psi_i = \sum_i a_i \hat{O} P_i \psi_i = \sum_i a_i P_i O_i \psi_i \tag{5.83}$$

Similarly,

$$\hat{P}\hat{O}\psi = \sum_i a_i \hat{P}\hat{O}\psi_i = \sum_i a_i \hat{P}O_i\psi_i = \sum_i a_i P_i O_i \psi_i \tag{5.84}$$

and hence,

$$[\hat{O}, \hat{P}] = 0 \tag{5.85}$$

The converse is also true: if $[\hat{O}, \hat{P}] = 0$, then $\hat{O}$ and $\hat{P}$ represent compatible observables. To prove this, we assume that the eigenvalues of $\hat{O}$ are nondegenerate. If ψ_i is an eigenfunction of $\hat{O}$, then

$$\hat{O}\psi_i = O_i\psi_i \tag{5.86}$$

Since $[\hat{O}, \hat{P}] = 0$, we have

$$\hat{O}\hat{P}\psi_i = \hat{P}\hat{O}\psi_i = \hat{P}O_i\psi_i = O_i(\hat{P}\psi_i) \tag{5.87}$$

In other words, we see that the function $(\hat{P}\psi_i)$ is an eigenfunction of $\hat{O}$ with eigenvalue O_i. If the eigenvalues of $\hat{O}$ are nondegenerate, there should be only one eigenfunction ψ_i with the eigenvalue O_i. Therefore, $(\hat{P}\psi_i)$ can at most be a scalar multiple of ψ_i, and we have

$$\hat{P}\psi_i = P_i\psi_i \tag{5.88}$$

which shows that ψ_i is a simultaneous eigenfunction of both $\hat{O}$ and $\hat{P}$. Depending on the symmetry of the system, there may be more than two compatible observables as in the case of the hydrogen atom, where the set of compatible observers includes the energy, total angular momentum, axial component of the angular momentum, and spin.

The above important result about compatible observers can be used in the determination of the energy eigenfunctions of one-dimensional potential wells which possess even symmetry. To investigate this further, we introduce the parity operator $\hat{\Pi}$ which is defined as

$$\hat{\Pi}\,\psi(x) \equiv \psi(-x) \tag{5.89}$$

Eigenvalues of $\hat{\Pi}$ can be obtained from the equation

$$\hat{\Pi}\,\psi(x) = \lambda\,\psi(x) \tag{5.90}$$

Applying the parity operator twice, we get

$$\hat{\Pi}^2\psi(x) = \hat{\Pi}\,\psi(-x) = \psi(x) = \lambda^2\psi(x) \tag{5.91}$$

giving $\lambda = \pm 1$. The eigenfunction $\psi_e(x)$ with eigenvalue +1 satisfies

$$\psi_e(x) = \hat{\Pi}\,\psi_e(x) = \psi_e(-x) \tag{5.92}$$

Hence, it has even symmetry. The eigenfunction $\psi_o(x)$ with eigenvalue −1 further satisfies

$$\psi_o(-x) = -\psi_o(x) \tag{5.93}$$

and possesses odd symmetry. If the potential is an even function, the hamiltonian will commute with the parity operator. To see this, look at the kinetic energy and potential terms of $\hat{H}$ separately. For the kinetic energy, we have

$$\hat{\Pi}\frac{\hat{p}^2}{2m}\psi(x) = -\frac{\hbar^2}{2m}\hat{\Pi}\frac{d^2\psi}{dx^2} = -\frac{\hbar^2}{2m}\frac{d^2\psi(-x)}{dx^2} \tag{5.94}$$

and

$$\frac{\hat{p}^2}{2m}\hat{\Pi}\,\psi(x) = -\frac{\hbar^2}{2m}\frac{d^2}{dx^2}(\psi(-x)) = -\frac{\hbar^2}{2m}(-1)^2\frac{d^2\psi(-x)}{dx^2} \tag{5.95}$$

Hence,

$$\left[\hat{\Pi}, \frac{\hat{p}^2}{2m}\right] = 0 \tag{5.96}$$

For the potential term, we see that

$$\hat{\Pi}V(x)\psi(x) = V(-x)\psi(-x) = V(x)\psi(-x) \tag{5.97}$$

since $V(-x) = V(x)$ for an even potential. Also,

$$V(x)\hat{\Pi}\psi(x) = V(x)\psi(-x) \tag{5.98}$$

so that

$$[\hat{\Pi}, V(x)] = 0 \tag{5.99}$$

Putting Eqs. (5.96) and (5.99) together, we get

$$[\hat{\Pi}, \hat{H}] = 0 \tag{5.100}$$

for one-dimensional potentials with even symmetry. Hence, the eigenfunctions will be simultaneous eigenfunctions of both the parity and hamiltonian operators.This means that if the potential well has even symmetry, we can group the energy eigenfunctions into two sets with even and odd symmetry.

5.2.7 Dirac Notation

We now introduce a very useful and powerful notation due to P. A. M Dirac. According to this notation, we represent the wave function ψ_2 by a state vector $|\psi_2\rangle$ known as a "ket". Similarly, the complex conjugate ψ_1^* is represented by the "bra" vector $\langle\psi_1|$. When a bra and a ket vector are combined, we obtain the inner product $\langle\psi_1|\psi_2\rangle$ which is called a "braket" defined by

$$\langle\psi_1|\psi_2\rangle \equiv \int d^3r\psi_1^*\psi_2 \tag{5.101}$$

We see clearly from this definition that

$$\langle\psi_1|\psi_2\rangle = \langle\psi_2|\psi_1\rangle^* \tag{5.102}$$

We can express various integrals by using the Dirac notation. For example, consider the defining relation for the hermitian adjoint of an operator $\hat{O}$ which was given earlier:

$$\int d^3r(\hat{O}\psi_1)^*\psi_2 = \int d^3r\psi_1^*\hat{O}^+\psi_2 \tag{5.103}$$

By using the Dirac notation, this can be expressed in the equivalent form

$$\langle\hat{O}\psi_1|\psi_2\rangle = \langle\psi_1|\hat{O}^+|\psi_2\rangle \tag{5.104}$$

Note further that

$$\langle\psi_1|a\psi_2\rangle = a\langle\psi_1|\psi_2\rangle \tag{5.105}$$

whereas

$$\langle a\psi_1|\psi_2\rangle = a^*\langle\psi_1|\psi_2\rangle \tag{5.106}$$

As another example, we can express the closure relation by using the Dirac notation. Suppose that $|\psi_i\rangle$ forms a complete orthonormal set. An arbitrary state $|\psi\rangle$ can be written in terms of the complete set as

$$|\psi\rangle = \sum_i a_i|\psi_i\rangle \tag{5.107}$$

Now, form a braket of each side with $\langle\psi_j|$ to obtain

$$\langle\psi_j|\psi\rangle = \sum_i a_i\langle\psi_j|\psi_i\rangle = a_j \tag{5.108}$$

or, changing the index from i to j, we have

$$a_i = \langle \psi_i | \psi \rangle \tag{5.109}$$

Now, we can write the inner product $\langle \phi | \psi \rangle$ as

$$\langle \phi | \psi \rangle = \sum_i a_i \langle \phi | \psi_i \rangle = \sum_i \langle \psi_i | \psi \rangle \langle \phi | \psi_i \rangle = \langle \phi | \sum_i | \psi_i \rangle \langle \psi_i | \psi \rangle \tag{5.110}$$

Hence, we see that

$$\sum_i | \psi_i \rangle \langle \psi_i | = \hat{1} \tag{5.111}$$

where $\hat{1}$ represents the unity operator. Equation (5.111) is equivalent to the closure relation given earlier in Eq. (5.74).

5.2.8 Schrödinger and Heisenberg Pictures

There are two commonly used equivalent ways to characterize the time evolution of a quantum system. These are referred as the Schrödinger and Heisenberg pictures. Having introduced the notion of ket state vectors in an abstract vector space, we can now go ahead and describe these two approaches. The Schrödinger equation:

$$\hat{H} | \psi(t) \rangle = i\hbar \frac{\partial}{\partial t} | \psi(t) \rangle \tag{5.112}$$

has the formal solution

$$| \psi(t) \rangle = \hat{U}(t) | \psi(0) \rangle \tag{5.113}$$

where $| \psi(0) \rangle$ is the initial state vector, and the evolution operator $\hat{U}(t)$ has the form

$$\hat{U}(t) = \exp\left[\frac{-i\hat{H}t}{\hbar} \right] = \sum_{l=0}^{\infty} \frac{\left(-i\hat{H}t/\hbar \right)^l}{l!} \tag{5.114}$$

Now, suppose that we look at the expectation value $O(t)$ of an operator $\hat{O}$ in the state $| \psi(t) \rangle$ given by

$$O(t) = \langle \psi(t) | \hat{O} | \psi(t) \rangle \tag{5.115}$$

This can be equivalently written as

$$O(t) = \langle \psi(0) | \hat{U}(t)^{+} \hat{O} \hat{U}(t) | \psi(0) \rangle = \langle \psi(0) | \hat{O}(t) | \psi(0) \rangle \tag{5.116}$$

where the time-dependent operator $\hat{O}(t)$ is given by

$$\hat{O}(t) = \hat{U}(t)^{+}\hat{O}\hat{U}(t) = \exp\left[\frac{+i\hat{H}t}{\hbar}\right]\hat{O}\exp\left[\frac{-i\hat{H}t}{\hbar}\right] \tag{5.117}$$

According to the Schrödinger picture, we associate no time dependence with the operator $\hat{O}$, and assume that the state vector $|\psi(t)\rangle$ evolves in time. The Heisenberg picture, on the other hand, assumes that the state vector remains constant in time equal to its initial form $|\psi(0)\rangle$, and the operator $\hat{O}(t)$ is time dependent given by Eq. (5.117). If the Heisenberg picture is adopted, we can determine the time evolution of the operator $\hat{O}(t)$ by differentiating Eq. (5.117) with respect to time. We get

$$\begin{aligned}\frac{d\hat{O}(t)}{dt} &= \frac{d\hat{U}^{+}}{dt}\hat{O}\hat{U}(t) + \hat{U}(t)^{+}\hat{O}\frac{d\hat{U}}{dt} = \frac{i\hat{H}}{\hbar}\left(\hat{U}(t)^{+}\hat{O}\hat{U}(t)\right) - \frac{i}{\hbar}\hat{U}(t)^{+}\hat{O}\hat{U}(t)\hat{H} \\ &= \frac{i}{\hbar}\left[\hat{H},\hat{O}(t)\right]\end{aligned} \tag{5.118}$$

5.3 One-Dimensional Potential Wells

In this section, we will look at two important potential well problems and describe how the Schrödinger equation is solved in each case to determine the energy eigenvalues and eigenstates. These are the infinite potential well and the finite potential well problems.

5.3.1 Particle in an Infinite Potential Well

Let us consider a particle trapped inside an infinite well whose potential energy $V(x)$ is given by

$$V(x) = \begin{cases} \infty & |x| > d/2 \\ 0 & |x| < d/2 \end{cases} \tag{5.119}$$

The potential well is sketched in Fig. 5.4.

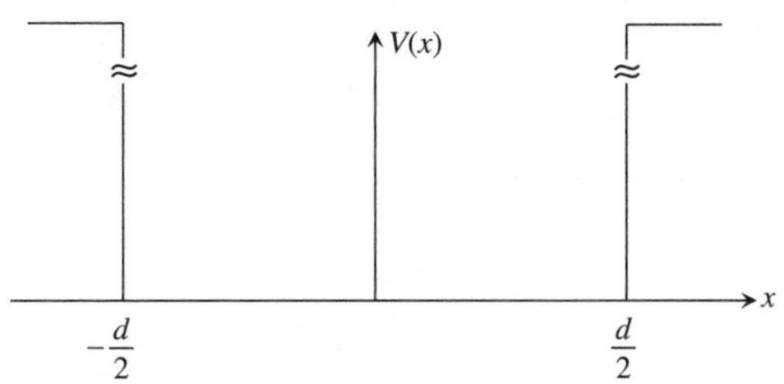

FIGURE 5.4 Potential energy of an infinite well.

As discussed earlier, the spatial part of the wave function $u(x)$ satisfies

$$-\frac{\hbar^2}{2m}\frac{d^2u}{dx^2}+V(x)u(x)=E\,u(x) \tag{5.120}$$

Since $V(x)$ is infinitely large for $|x|>d/2$, $u(x)$ is zero in this region. Inside the well, $u(x)$ will satisfy

$$\frac{d^2u}{dx^2}=\frac{-2mE}{\hbar^2}u(x) \tag{5.121}$$

subject to the boundary condition $u(-d/2)=u(+d/2)=0$. Note that E must be positive; otherwise, $u(x)$ is a combination of hyperbolic sine and hyperbolic cosine functions and meeting the boundary conditions at both ends of the well is not possible. We let

$$k=\sqrt{\frac{2mE}{\hbar^2}} \tag{5.122}$$

Since the potential has even symmetry, the solution of $u(x)$ for $E > 0$ consists of two linearly independent sets, each with a definite symmetry about the midpoint (in this case, it is the origin) of the well. Even solutions are given by

$$u_{\text{even}}(x)=A\cos kx \tag{5.123}$$

To satisfy the boundary conditions we have

$$k_{\text{even}}^{m}\frac{d}{2}=(2m-1)\frac{\pi}{2} \tag{5.124}$$

where $m = 1, 2, 3, \ldots$. For odd solutions,

$$u_{\text{odd}}(x)=B\sin kx \tag{5.125}$$

with

$$k_{\text{odd}}^{m}\frac{d}{2}=m\pi=2m\left(\frac{\pi}{2}\right) \tag{5.126}$$

Above, $m = 1, 2, 3, \ldots$. Combining the above results, we see that the nth eigenvalue of k is given by

$$k_n=\frac{n(\pi/2)}{d/2}=\frac{n\pi}{d} \tag{5.127}$$

where odd values of n correspond to solutions with even symmetry about the origin, and even values of n give the solutions with odd symmetry about the origin. The corresponding nth energy eigenvalue becomes

$$E_n = \frac{\hbar^2 k^2}{2m} = n^2 \frac{\hbar^2 \pi^2}{2md^2} \tag{5.128}$$

Since $|\psi|^2$ gives the probability distribution, we can determine the constants A and B appearing in Eqs. (5.123) and (5.125) by requiring that

$$\int_{-\frac{d}{2}}^{+\frac{d}{2}} dx |u(x)|^2 = 1 \tag{5.129}$$

Since the periodic function goes to zero at the boundaries of the well, we have an integer number of half de Broglie waves within the well for any one of the solutions. Hence, for the even case, for example,

$$\int_{-\frac{d}{2}}^{+\frac{d}{2}} dx |u(x)|^2 = A^2 \int_{-\frac{d}{2}}^{+\frac{d}{2}} dx \cos^2(kx) = A^2 \left(\frac{1}{2}\right)(d) \tag{5.130}$$

or

$$A = \sqrt{\frac{2}{d}} \tag{5.131}$$

The same result can be obtained for the constant B of the odd case. Figure 5.5 shows the four lowest energy levels of the particle trapped in the infinite well. Figure 5.6a through d further shows

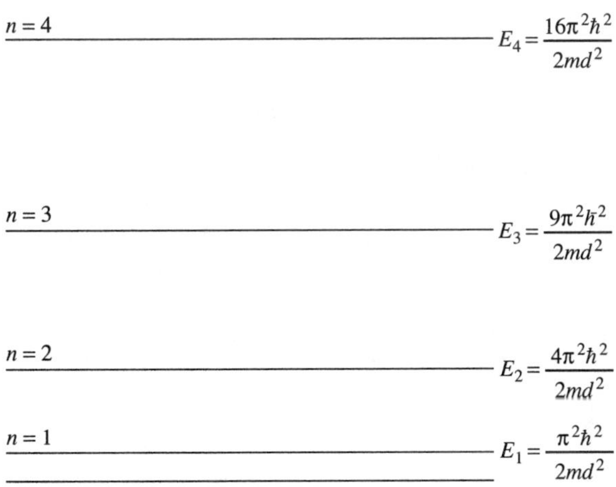

FIGURE 5.5 Four lowest energy levels of the particle trapped in an infinite well of width d.

the corresponding wave functions of the lowest four levels. Note that in this case, we are characterizing the well with a single parameter: the width of the well. Also note that as d becomes smaller, in other words, as the quantum confinement increases, the spacing between the energy eigenvalues also increases. Finally, the nth energy scales as n^2.

The full expression for the time dependent wave function of the nth state can then be written as

$$\psi_n(x,t) = \begin{cases} \sqrt{\frac{2}{d}} \cos\left(\frac{n\pi x}{d}\right) \exp\left[-\frac{iE_n t}{\hbar}\right] & n = 2m-1 = \text{odd} \\ \sqrt{\frac{2}{d}} \sin\left(\frac{n\pi x}{d}\right) \exp\left[-\frac{iE_n t}{\hbar}\right] & n = 2m = \text{even} \end{cases} \tag{5.132}$$

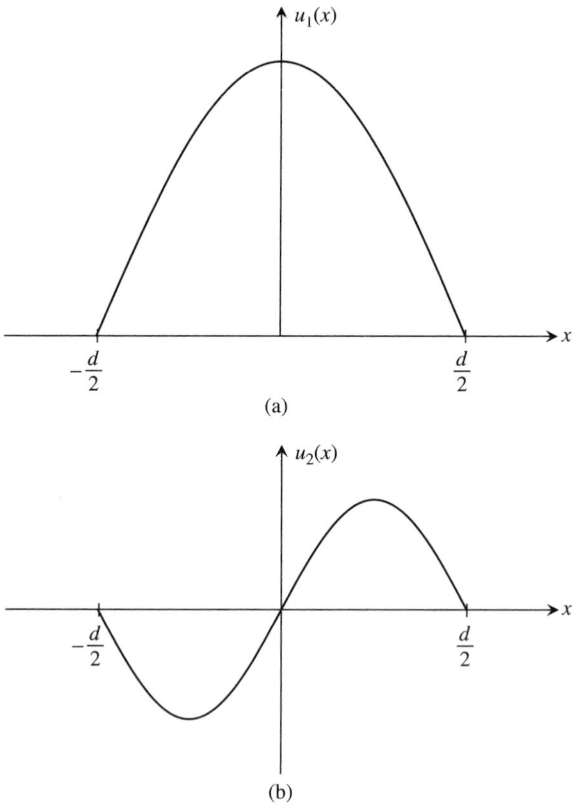

FIGURE 5.6 Wave functions of the four lowest levels of the particle trapped in an infinite well of width d.

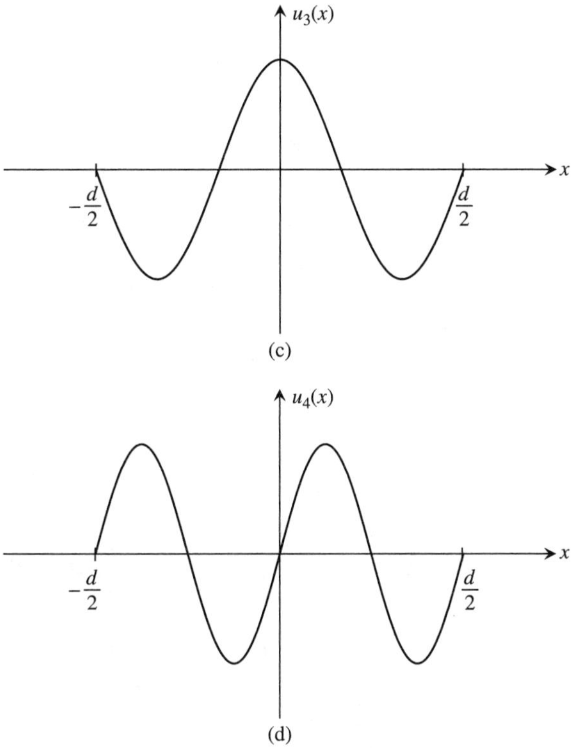

FIGURE 5.6 *(Continued)*

5.3.2 Particle in a Finite Well

Now we consider a particle trapped inside a well whose potential $V(x)$ is given by

$$V(x) = \begin{cases} 0 & |x| \geq \dfrac{d}{2} \\ -V_o & |x| < \dfrac{d}{2} \end{cases} \tag{5.133}$$

as shown in Fig. 5.7. In this case, we can find energy eigenvalue solutions for $E > 0$ and $E < 0$. The spectrum of energy eigenvalues for $E > 0$ is continuous. We will not treat this case here. Discussion on the derivation of the continuous eigenvalue spectrum can be found in Refs. 2, 5, and 6. Here, we will concentrate on the case $E < 0$ to find the bound states of the particle.

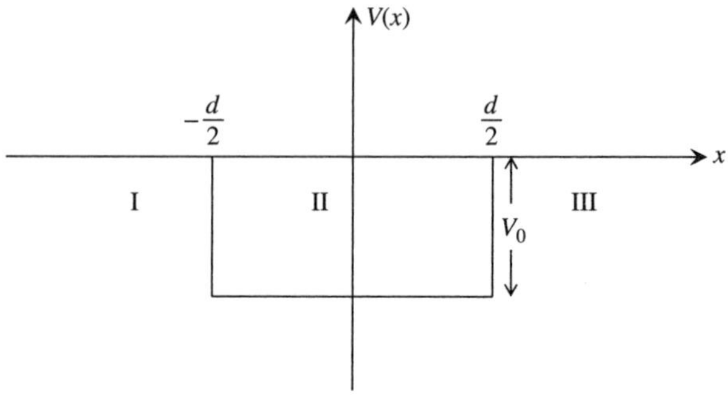

FIGURE 5.7 Potential energy of the finite well.

Referring to Fig. 5.7, the time-independent Schrödinger equation for the regions I, II, and III becomes

$$\frac{d^2u}{dx^2} = \begin{cases} -\frac{2m|E|}{\hbar^2}u(x) & |x| > +\frac{d}{2} \quad \text{(Regions I,III)} \\ -\frac{2m}{\hbar^2}\left(V_o - |E|\right)u(x) & |x| < +\frac{d}{2} \quad \text{(Region II)} \end{cases} \tag{5.134}$$

Letting

$$k^2 = \frac{2m}{\hbar^2}\left(V_o - |E|\right) \tag{5.135}$$

and

$$\kappa^2 = \frac{2m|E|}{\hbar^2} \tag{5.136}$$

the solution $u(x)$ can be written as

$$u(x) = \begin{cases} A_1 \exp(\kappa x) & x < -\frac{d}{2} \\ A_2 \cos kx + A_3 \sin kx & -\frac{d}{2} < x < \frac{d}{2} \\ A_4 \exp(-\kappa x) & x > \frac{d}{2} \end{cases} \tag{5.137}$$

Note that outside the well, we choose the decaying solutions to ensure that the probability distribution remains finite at all values of x.

Again, since the potential energy has even symmetry, we obtain the even and odd solutions separately. For the even solutions, we have

$$u(x) = A_2 \cos kx \tag{5.138}$$

for $-d/2 < x < d/2$. To find the allowed k values, we require the continuity of $u(x)$ and du/dx at $x = \pm d/2$. We leave it as an exercise for the reader (see Prob. 5.4) to show that k and κ satisfy the eigenvalue equation

$$k \tan\left(\frac{kd}{2}\right) = +\kappa \tag{5.139}$$

To express Eq. (5.139) in terms of dimensionless variables, we define

$$\bar{x} = \frac{kd}{2} \tag{5.140}$$

In terms of $\bar{x}$, the eigenvalue equation can be expressed as (see Prob. 5.4)

$$\bar{x}\tan(\bar{x}) = \sqrt{\alpha - \bar{x}^2} \tag{5.141}$$

where the dimensionless potential strength parameter α is given by

$$\alpha = \frac{mV_0 d^2}{2\hbar^2} \tag{5.142}$$

Similarly, the dimensionless eigenvalue equation for odd solutions becomes (see Prob. 5.5)

$$\bar{x}\cot\bar{x} = -\sqrt{\alpha - \bar{x}^2} \tag{5.143}$$

or

$$\bar{x}\tan\left(\bar{x} + \frac{\pi}{2}\right) = \sqrt{\alpha - \bar{x}^2} \tag{5.144}$$

Graphical solutions of the eigenvalue equations are shown in Figs. 5.8 and 5.9. Figure 5.8 shows the graphical solution of Eq. (5.141) for the even states of an electron trapped in a one-dimensional well with $V_0 = 20$ eV and $d = 1$ nm. Note that there are four bound states with even symmetry. Figure 5.9 further shows the graphical solution of Eq. (5.144) for the odd states. The well parameters are the same, giving rise to four bound states with odd symmetry.

Note that the number of bound states decreases as the width or the depth of the well is reduced. Such a case is shown in Fig. 5.10*a* and *b* for a well of the same width as above ($d = 1$ nm) but with lower depth of $V_0 = 2$ eV. Now, there are only two bound even states and one bound odd state. By using Eq. (5.137), the spatial variation of

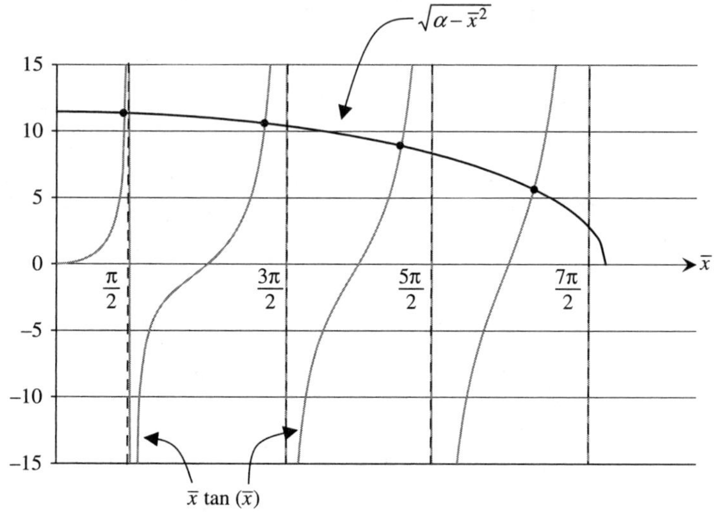

Figure 5.8 Graphical solution of the eigenvalue equation for the even states of an electron trapped in a one-dimensional well with $V_0 = 20$ eV and $d = 1$ nm.

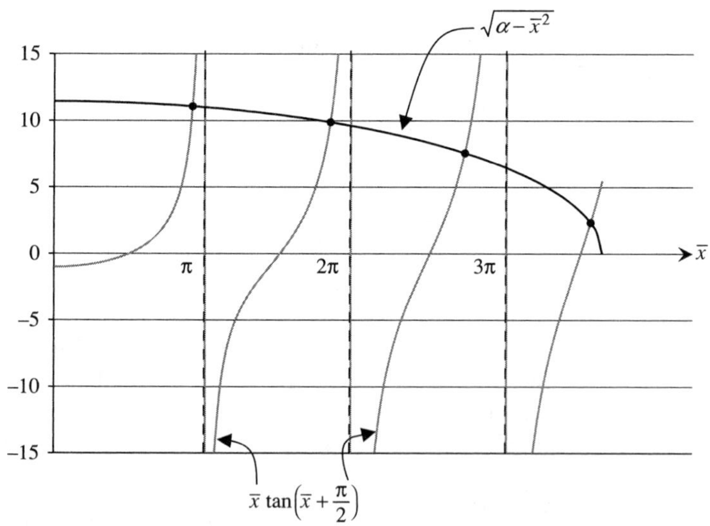

Figure 5.9 Graphical solution of the eigenvalue equation for the odd states of an electron trapped in a one-dimensional well with $V_0 = 20$ eV and $d = 1$ nm.

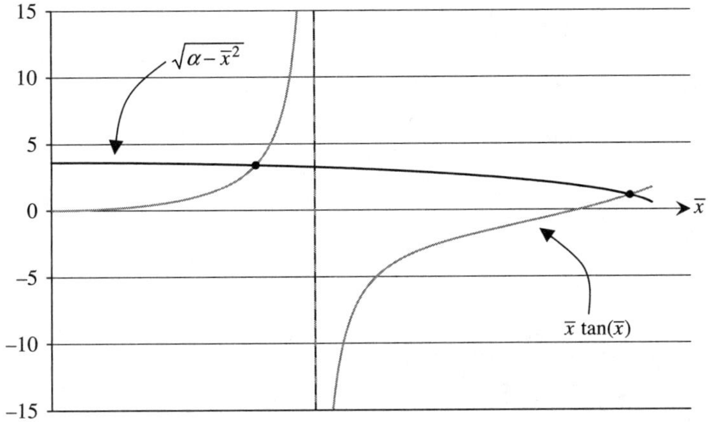

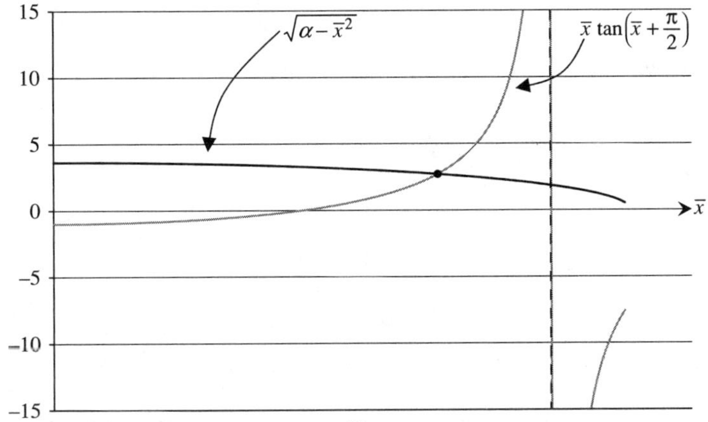

FIGURE 5.10 Graphical solution of the eigenvalue equation for the (a) even and (b) odd states of an electron trapped in a one-dimensional well with $V_0 = 2$ eV and $d = 1$ nm.

the corresponding even and odd wavefunctions is shown in Fig. 5.11*a* to *c*. As can be seen from the graphical solutions, there remains only one bound even state when

$$\sqrt{\alpha} < \pi \tag{5.145}$$

or when

$$\frac{mV_0 d^2}{2\hbar^2} < \pi^2 \tag{5.146}$$

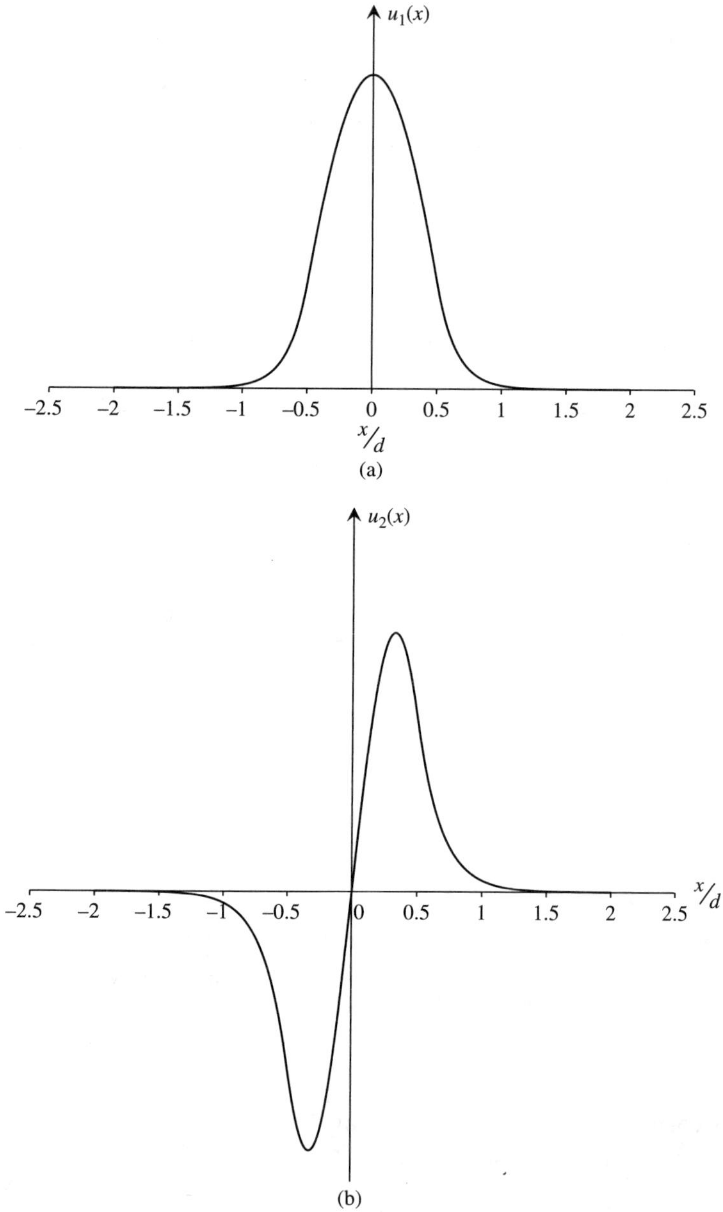

FIGURE 5.11 Spatial variation of the wave functions for the three bound states of an electron trapped in a one-dimensional well with $V_0 = 2$ eV and $d = 1$ nm.

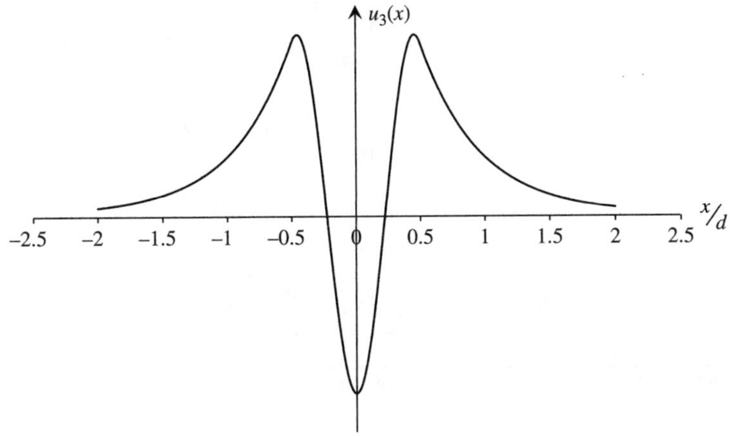

FIGURE 5.11 *(Continued)*

In the case of the states with odd symmetry, a solution will exist if

$$\sqrt{\alpha} > \frac{\pi}{2} \tag{5.147}$$

or

$$\frac{mV_0 d^2}{2\hbar^2} \geq \frac{\pi^2}{4} \tag{5.148}$$

5.4 Quantum Mechanical Harmonic Oscillator

One of the most important cases that comes up in a wide range of applications is the quantum mechanical harmonic oscillator. In this case, the particle moves under the action of a symmetric potential given by

$$V(x) = \frac{1}{2}\kappa_0 x^2 \tag{5.149}$$

where κ_0 is the spring constant. Figure 5.12 shows the variation of the harmonic oscillator potential energy as a function of position. Classically, the particle undergoes oscillatory motion under such a potential with an oscillation frequency of

$$\omega = \sqrt{\frac{\kappa_0}{m}} \tag{5.150}$$

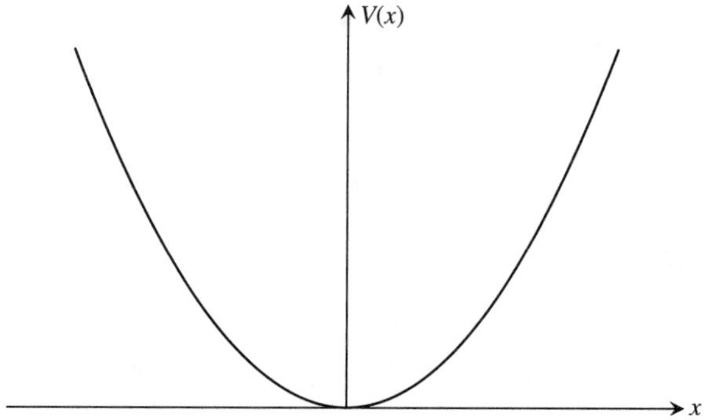

FIGURE 5.12 Variation of the harmonic oscillator potential energy as a function of position.

It is customary for reasons that will become apparent shortly to express the potential energy in terms of ω as

$$V(x) = \frac{1}{2} m\omega^2 x^2 \tag{5.151}$$

The time-independent Schrödinger equation then takes the form

$$-\frac{\hbar^2}{2m}\frac{d^2u}{dx^2} + \frac{1}{2} m\omega^2 x^2 u(x) = Eu(x) \tag{5.152}$$

As before, we are interested in finding the energy eigenvalues E and the corresponding eigenstates $u(x)$. We can express Eq. (5.152) in terms of a dimensionless variable y defined according to

$$y \equiv \beta x \tag{5.153}$$

Equation (5.152) then becomes

$$-\frac{\hbar^2}{2m}\beta^2 \frac{d^2u}{dy^2} + \frac{1}{2} m\omega^2 \frac{y^2}{\beta^2} u(y) = Eu(y) \tag{5.154}$$

or

$$\frac{d^2u}{dy^2} + \frac{2mE}{\hbar^2\beta^2} - \frac{m^2\omega^2}{\hbar^2\beta^4} y^2 u(y) = 0 \tag{5.155}$$

To turn Eq. (5.155) into a dimensionless form, we choose β such that

$$\frac{m^2\omega^2}{\hbar^2\beta^4} = 1 \tag{5.156}$$

Note that with this choice, the scaling parameter β simply becomes

$$\beta = \sqrt{\frac{m\omega}{\hbar}} \tag{5.157}$$

We further let the dimensionless energy eigenvalue $\bar{E}$ be defined according to

$$\bar{E} = \frac{2mE}{\hbar^2\beta^2} = \frac{2E}{\hbar\omega} \tag{5.158}$$

Based on these transformations, the dimensionless form of the Schrödinger equation becomes

$$\frac{d^2u}{dy^2} + (\bar{E} - y^2)u = 0 \tag{5.159}$$

Equation (5.159) can be solved analytically by using different approaches. In this chapter, we will explore the series solution as well as the use of operators. We first attempt a series solution of the form

$$u = \sum_{i=0}^{\infty} a_i y^i \tag{5.160}$$

By substituting Eq. (5.160) into Eq. (5.159), we obtain

$$\sum_{i=2}^{\infty} i(i-1)a_i y^{i-2} + \bar{E}\sum_{i=0}^{\infty} a_i y^i - \sum_{i=0}^{\infty} a_i y^{i+2} = 0 \tag{5.161}$$

By redefining the indices so that the same power of y appears as the general term in each sum, the above equation becomes

$$\sum_{i=0}^{\infty} (i+2)(i+1)a_{i+2} y^i + \bar{E}\sum_{i=0}^{\infty} a_i y^i - \sum_{i=2}^{\infty} a_{i-2} y^i = 0 \tag{5.162}$$

The coefficients a_i will satisfy the following relations:

$$i = 0 : (1)(2)a_2 + \bar{E}a_0 = 0 \tag{5.163}$$

$$i = 1 : (3)(2)a_3 + \bar{E}a_1 = 0 \tag{5.164}$$

and

$$i \geq 2 : (i+2)(i+1)a_{i+2} + \bar{E}a_i - a_{i-2} = 0 \tag{5.165}$$

What we have in Eq. (5.165) is a three-term recurrence relation, from which it is rather difficult, at least to the author's knowledge, to obtain a general expression for a_i. To obtain a two-term recurrence relationship, we first solve the asymptotic form of Eq. (5.159) as $y \to \infty$, and factor out this solution. As $y \to \infty$, Eq. (5.159) approaches

$$\frac{d^2u_a}{dy^2} = y^2 u_a(y) \tag{5.166}$$

where $u_a(y)$ is the asymptotic solution. With an educated guess, we see that the asymptotic solution will be of the form

$$u_a(y) = \exp\left(\pm\frac{1}{2}y^2\right) \tag{5.167}$$

This can be easily verified by substitution and by assuming that y is very large. Furthermore, we rule out the solution $\exp(y^2/2)$, since we are looking for square integrable solutions that decay at large y. Hence, we now seek a solution to Eq. (5.159) by assuming that

$$u(y) = H(y)\exp\left(-\frac{1}{2}y^2\right) \tag{5.168}$$

Above, the function $H(y)$ remains to be determined. We leave it as an exercise for the reader (see Prob. 5.9) to show that $H(y)$ now satisfies the differential equation

$$\frac{d^2H}{dy^2} - 2y\frac{dH}{dy} + (\bar{E} - 1)H(y) = 0 \tag{5.169}$$

We attempt once again a series solution by letting

$$H(y) = \sum_{j=0}^{\infty} a_j y^j \tag{5.170}$$

Substituting into Eq. (5.169), we obtain

$$\sum_{j=2}^{\infty} j(j-1)a_j y^{j-2} - 2y\sum_{j=0}^{\infty} j a_j y^{j-1} + (\bar{E} - 1)\sum_{j=0}^{\infty} a_j y^j = 0 \tag{5.171}$$

Adjusting indices so that each summation starts from $j = 0$, we then get

$$\sum_{j=0}^{\infty}(j+2)(j+1)a_{j+2}y^j - \sum_{j=0}^{\infty} 2j a_j y^j + \sum_{j=0}^{\infty}(\bar{E} - 1)a_j y^j = 0 \tag{5.172}$$

or

$$\sum_{j=0}^{\infty} y^j\left[(j+2)(j+1)a_{j+2} - 2ja_j + (\bar{E} - 1)a_j\right] = 0 \tag{5.173}$$

Now, the recursion relation connecting the coefficients a_{j+2} and a_j can be written as

$$a_{j+2} = \frac{2j+1-\bar{E}}{(j+2)(j+1)} a_j \tag{5.174}$$

We see clearly from the above analysis that there are two linearly independent solutions, each with a definite symmetry, as we expect for a potential with even symmetry. The even and odd solutions are of the form

$$H_{\text{even}}(y) = a_0 + a_2 y^2 + a_4 y^4 + a_6 y^6 + \cdots \tag{5.175}$$

and

$$H_{\text{odd}}(y) = a_1 + a_3 y^3 + a_5 y^5 + \cdots \tag{5.176}$$

It turns out that the function $H(y)$ behaves asymptotically as $\exp(y^2)$. Hence, we see that the product $H(y)\exp(-1/2y^2)$ will vary as $\exp(1/2y^2)$ as $y \to \infty$, and still we do not get a square integrable solution. We can easily fix this problem and get physically meaningful solutions. In order to obtain square integrable solutions, we require that $\bar{E} = 2j+1$ when j is equal to some integer, say n. Then, the series solution has a finite number of terms and the nth dimensionless energy eigenvalue $\bar{E}_n$ becomes

$$\bar{E}_n = 2n+1 \tag{5.177}$$

The corresponding energy eigenvalue E_n and the associated wave function become

$$E_n = \hbar\omega\left(n + \frac{1}{2}\right) \tag{5.178}$$

and

$$\psi_n(y) = A_n H_n(y) \exp\left(-\frac{1}{2}y^2\right) \tag{5.179}$$

Above, A_n is the normalization constant. As we show in App. A, $H_n(y)$ is the nth-order Hermite polynomial which can be computed by using the formula

$$H_n(y) = (-1)^n \exp(y^2) \frac{d^n}{dy^n}[\exp(-y^2)] \tag{5.180}$$

Furthermore, we have the symmetry property

$$H_n(-y) = (-1)^n H_n(y) \tag{5.181}$$

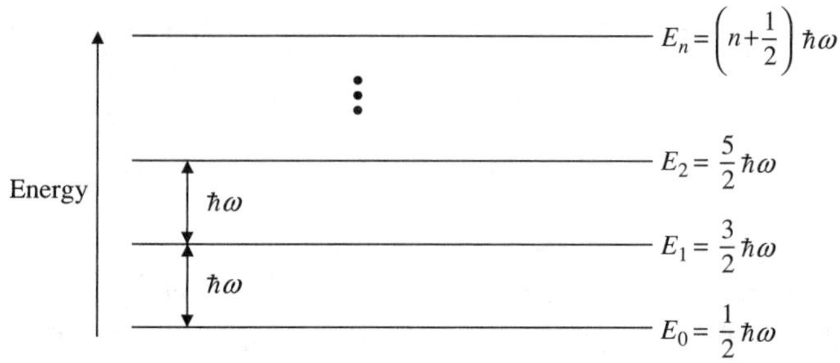

FIGURE 5.13 Energy levels of the quantum harmonic oscillator.

Hence, an $H_n(y)$ with even n has even symmetry, and an $H_n(y)$ with odd n has odd symmetry. Figure 5.13 shows the energy levels of the quantum harmonic oscillator. Note that they are equally spaced and the difference between two successive energy levels is $\hbar\omega$. Furthermore, Fig. 5.14*a* through *c* show sketches of the wavefunctions corresponding to the three lowest energy levels. As expected, the symmetry of the wavefunctions alternates going from one level to the next.

It is also shown in App. A that Hermite polynomials of different orders obey the orthogonality relation

$$\int_{-\infty}^{\infty} dy H_n(y) H_m(y) \exp(-y^2) = \delta_{mn} 2^n n! \sqrt{\pi} \tag{5.182}$$

Hence, the full normalized solution of the Schrödinger equation for the harmonic oscillator becomes

$$\psi_n(x,t) = \sqrt{\frac{1}{2^n n!}\sqrt{\frac{m\omega}{\pi\hbar}}}\, H_n\left(\sqrt{\frac{m\omega}{\hbar}}x\right)\exp\left(-\frac{m\omega x^2}{2\hbar}\right)\exp\left[-i\omega\left(n+\frac{1}{2}\right)t\right] \tag{5.183}$$

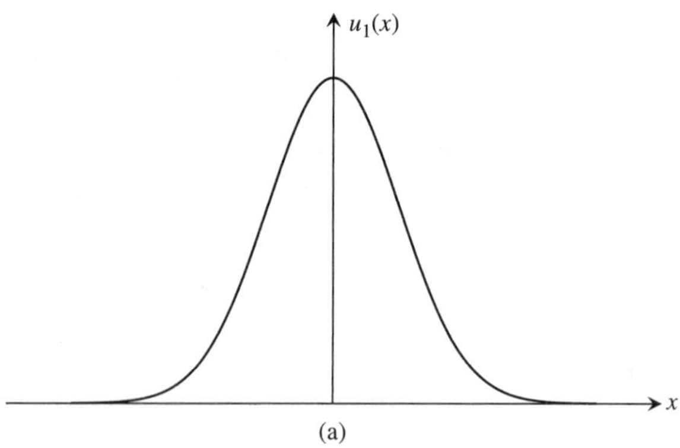

FIGURE 5.14 Wave functions corresponding to the three lowest energy levels of the quantum harmonic oscillator.

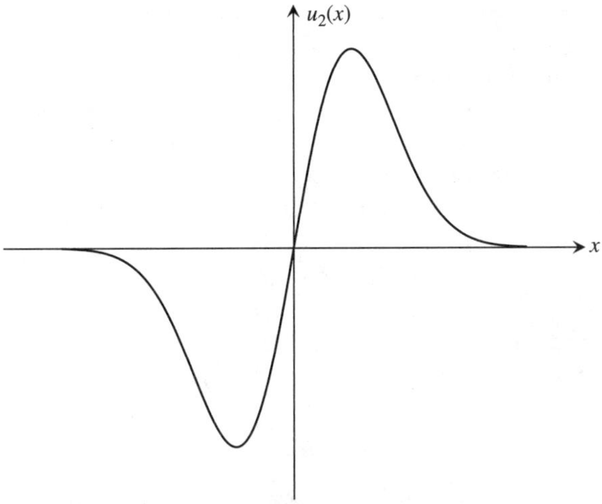

(b)

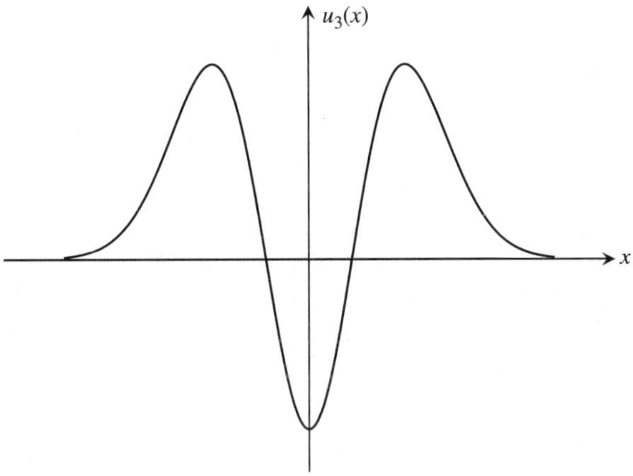

(c)

FIGURE 5.14 *(Continued)*

5.5 Operator Method for the Solution of the Quantum Harmonic Oscillator

In addition to the method we outlined in the previous section, there is an alternative way of obtaining the energy eigenvalue spectrum of the harmonic oscillator. As such, instead of solving a differential equation, we introduce operators and explore their algebraic properties to determine the energy eigenvalues. In this approach, we will not worry about the exact mathematical form of the wave functions in the position space, but simply treat them as state vectors (or kets) which belong to an abstract, infinite dimensional vector space known as the Hilbert space. Providing a rigorous mathematical account of Hilbert spaces is beyond the scope of this book. Instead, we will focus on a practical application of the operator method and use it to obtain the energy eigenvalues. The operator method will again be used later when we discuss the quantization of the electromagnetic fields and spontaneous emission.

The hamiltonian operator of the harmonic oscillator is given by

$$\hat{H} = \frac{\hat{p}_x^2}{2m} + \frac{1}{2} m\omega^2 \hat{x}^2 \tag{5.184}$$

where $\hat{p}_x$ and $\hat{x}$ denote the momentum and position operators which obey the commutation relation

$$\left[\hat{x}, \hat{p}_x\right] = i\hbar \tag{5.185}$$

We now introduce two new operators $\hat{a}$ and $\hat{a}^+$ defined according to

$$\hat{a} \equiv \sqrt{\frac{m\omega}{2\hbar}} \left(\hat{x} + \frac{i}{m\omega} \hat{p}_x\right) \tag{5.186}$$

and

$$\hat{a}^+ \equiv \sqrt{\frac{m\omega}{2\hbar}} \left(\hat{x} - \frac{i}{m\omega} \hat{p}_x\right) \tag{5.187}$$

Note that since $\hat{x}$ and $\hat{p}_x$ are hermitian, $\hat{a}^+$ is the hermitian conjugate of $\hat{a}$, as suggested by the above notation. We leave it as an exercise for the reader (see Prob. 5.10) to show that the hamiltonian can now be written as

$$\hat{H} = \hbar\omega \left(\hat{a}^+ \hat{a} + \frac{1}{2}\right) \tag{5.188}$$

Furthermore, we have

$$[\hat{a}, \hat{a}^+] = 1 \tag{5.189}$$

To find the energy eigenvalues, we introduce the operator $\hat{n}$ defined by

$$\hat{n} \equiv \hat{a}^+ \hat{a} \tag{5.190}$$

In terms of $\hat{n}$, $\hat{H}$ now becomes

$$\hat{H} = \hbar\omega \left(\hat{n} + \frac{1}{2}\right) \tag{5.191}$$

Now, let's try to find the eigenvalues of $\hat{n}$ by solving

$$\hat{n}\left|\alpha\right\rangle = \alpha\left|\alpha\right\rangle \tag{5.192}$$

We can explore the properties of the eigenvalues α by using the operators $\hat{a}$ and $\hat{a}^+$. First, note that if the corresponding eigenstate $|\alpha\rangle$ is normalized, then

$$\alpha = \langle\alpha\,|\,\hat{n}\,|\,\alpha\rangle = \langle\alpha\,|\,\hat{a}^+\hat{a}\,|\,\alpha\rangle = \langle\psi'\,|\,\psi'\rangle \geq 0 \tag{5.193}$$

where

$$|\psi'\rangle = \hat{a}|\alpha\rangle \tag{5.194}$$

Next, let us consider $\hat{n}$ acting on $\hat{a}|\alpha\rangle$. We get

$$\begin{aligned}\hat{n}\,\hat{a}|\alpha\rangle &= \hat{a}^+\hat{a}\,\hat{a}|\alpha\rangle = (\hat{a}\hat{a}^+ - [\hat{a},\hat{a}^+])\hat{a}|\,\alpha\rangle = (\hat{a}\hat{a}^+ - 1)\hat{a}|\,\alpha\rangle = \hat{a}(\hat{a}^+\hat{a} - 1)|\,\alpha\rangle \\ &= \hat{a}(\hat{n} - 1)|\alpha\rangle = (\alpha - 1)\hat{a}|\alpha\rangle\end{aligned} \tag{5.195}$$

Hence, $\hat{a}|\alpha\rangle$ must be an eigenstate of $\hat{n}$ with eigenvalue $(\alpha - 1)$. So, we can write $\hat{a}|\alpha\rangle$ as

$$\hat{a}|\alpha\rangle = c|\alpha - 1\rangle \tag{5.196}$$

where c is a constant. To determine c, we examine the inner product $\langle\hat{a}\alpha|\hat{a}\alpha\rangle$:

$$\langle\hat{a}\alpha\,|\,\hat{a}\alpha\rangle = \langle\alpha\,|\,\hat{a}^+\hat{a}\,|\,\alpha\rangle = \langle\alpha\,|\,\hat{n}\,|\,\alpha\rangle = \alpha = |c|^2\langle\alpha - 1\,|\,\alpha - 1\rangle = |c|^2 \tag{5.197}$$

Hence, we see that

$$\hat{a}|\alpha\rangle = \sqrt{\alpha}|\alpha - 1\rangle \tag{5.198}$$

Note that when $\hat{a}$ acts on $|\alpha\rangle$, it lowers the eigenvalue by 1. Because of this property, $\hat{a}$ is known as the lowering or annihilation operator. If we keep on applying $\hat{a}$, we will get states with successively lower eigenvalues. As such, every time $\hat{a}$ is applied to $|\alpha\rangle$, the value of α will be lowered by 1. However, above we saw that $\alpha \geq 0$. The values of α should be such that the lowest value is zero, so that when we try to lower this further, we get

$$\hat{a}|0\rangle = \sqrt{0}|0 - 1\rangle = 0 \tag{5.199}$$

We conclude from this observation that α should be an integer given by $\alpha = n = 0, 1, 2, \ldots$. It is customary to represent the eigenstates of $\hat{n}$ as $|n\rangle$, where

$$\hat{n}|n\rangle = n|n\rangle \tag{5.200}$$

and

$$\hat{a}|n\rangle = \sqrt{n}\,|n-1\rangle \tag{5.201}$$

We leave it as an exercise for the reader (see Prob. 5.12) to show that when $\hat{a}^{+}$ acts on $|n\rangle$, we get

$$\hat{a}^{+}|n\rangle = \sqrt{n+1}\,|n+1\rangle \tag{5.202}$$

Hence, $\hat{a}^{+}$ is referred to as the raising or creation operator. Finally, since the hamiltonian operator is given by Eq. (5.191), we see that the energy eigenvalues become

$$E_n = \hbar\omega\left(n+\frac{1}{2}\right) \tag{5.203}$$

as expected from our earlier analysis.

5.6 Time-Dependent Perturbation Theory

As we discussed at the beginning of this chapter, we want to develop the formalism for treating light-matter interactions quantum mechanically. This can be done by using the time-dependent perturbation theory. We will not discuss the time-independent perturbation theory in this book, but more information on this method can be found in Refs. 2, 3, 5, and 6.

Let us assume that the atomic system in the absence of the electromagnetic wave is described by the hamiltonian $\hat{H}_0$ whose eigenstates satisfy

$$\hat{H}_0|\phi_n\rangle = E_n|\phi_n\rangle \tag{5.204}$$

Here, $|\phi_n\rangle$ belongs to a complete orthonormal set of eigenstates of $\hat{H}_0$, and E_n represents the corresponding eigen energies. What happens if we now shine light on this system? Quantum mechanically, we now have a different hamiltonian that represents the atom-light system. We express the new hamiltonian $\hat{H}_T$ of the combined system as

$$\hat{H}_T(t) = \hat{H}_0 + \lambda\hat{H}(t) \tag{5.205}$$

where $\hat{H}(t)$ represents the interaction energy of the atom in the presence of the electromagnetic wave and λ is the perturbation parameter. Now, the time-dependent wave function of the system will satisfy

$$\hat{H}_T(t)|\psi(t)\rangle = i\hbar\frac{\partial}{\partial t}|\psi(t)\rangle \tag{5.206}$$

Since $|\phi_n\rangle$ forms a complete orthonormal set, we can expand $|\psi(t)\rangle$ as a superposition in the form

$$|\psi(t)\rangle = \sum_n a_n(t)\exp\left(\frac{-iE_n t}{\hbar}\right)|\phi_n\rangle \tag{5.207}$$

For a given form of $\hat{H}(t)$, we can find the time dependence of the expansion coefficients $a_n(t)$, and hence determine the transition probabilities. Substituting Eq. (5.207) into Eq. (5.206), we obtain

$$\sum_n [E_n + \lambda\hat{H}(t)]a_n(t)\exp\left(\frac{-iE_n t}{\hbar}\right)|\phi_n\rangle = i\hbar\sum_n\left[\dot{a}_n(t) - \frac{iE_n}{\hbar}a_n(t)\right]\exp\left(\frac{-iE_n t}{\hbar}\right)|\phi_n\rangle \tag{5.208}$$

Taking the inner product with $\langle\phi_m|$ and using the orthonormality relation

$$\langle\phi_m|\phi_n\rangle = \delta_{mn} \tag{5.209}$$

we get

$$\sum_n [E_n a_n(t)\delta_{mn} + \lambda H_{mn}(t)a_n(t)]\exp\left(\frac{-iE_n t}{\hbar}\right) = \sum_n [i\hbar\dot{a}_n + E_n a_n(t)]\delta_{mn}\exp\left(\frac{-iE_n t}{\hbar}\right) \tag{5.210}$$

where $H_{mn}(t)$ is the matrix element given by

$$H_{mn}(t) = \langle\phi_m|\hat{H}(t)|\phi_n\rangle \tag{5.211}$$

Simplifying Eq. (5.210), we obtain

$$i\hbar\frac{da_m}{dt} = \sum_n \lambda H_{mn}(t)a_n(t)\exp\left[\frac{i(E_m - E_n)t}{\hbar}\right] \tag{5.212}$$

Now, we assume that the solution for $a_m(t)$ can be expanded as a power series of λ in the form

$$a_m(t) = a_m^{(0)}(t) + \lambda a_m^{(1)}(t) + \lambda^2 a_m^{(2)}(t) + \cdots = \sum_{k=0}^{\infty}\lambda^k a_m^{(k)}(t) \tag{5.213}$$

where $a_m^{(k)}(t)$ is the kth-order contribution to $a_m(t)$. By substituting Eq. (5.213) into Eq. (5.212) and collecting terms belonging to the same power of λ, we can construct a perturbation solution to any order of the parameter λ we wish. Beyond the lowest orders, however, this gets quite cumbersome. For our purposes, looking at the two lowest-order corrections will be sufficient. Of course, this assumes that the perturbing electromagnetic wave is weak. Clearly, this assumption needs to be revisited when one considers the interaction of high-intensity light with the atomic system. From the coefficients of λ^0 and λ, we get

$$\frac{da_m^{(0)}}{dt} = 0 \tag{5.214}$$

and

$$i\hbar \frac{da_m^{(1)}}{dt} = \sum_n H_{mn}(t) a_n^{(0)}(t) \exp\left[\frac{i(E_m - E_n)t}{\hbar}\right] \tag{5.215}$$

Looking at Eq. (5.214), we see that the zeroth-order solution to $a_m(t)$ is $a_m^{(0)}$ = constant for all states. This is the solution we would get if the interaction energy $\hat{H}(t)$ were zero and all of the expansion coefficients remain at their initial values. Next, let us look at the first-order correction given in Eq. (5.125). Specifically, we assume that the system is initially in state $|\phi_\ell\rangle$ with energy E_ℓ as sketched in Fig. 5.15. In this case, $a_\ell(0) = 1$ and all other values of a_n's are zero for $n \neq \ell$. We have

$$i\hbar \frac{da_m^{(1)}}{dt} = H_{m\ell}(t) \exp\left[\frac{i(E_m - E_\ell)t}{\hbar}\right] \tag{5.216}$$

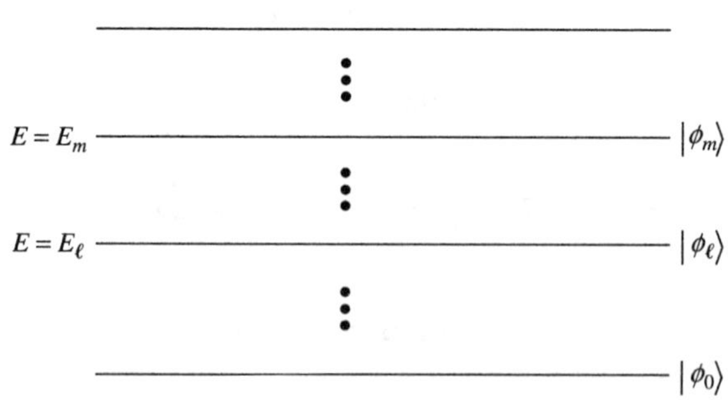

FIGURE 5.15 Two states of an atomic system at energies E_ℓ and E_m. Perturbation theory can be used to calculate the transition rate between these states if a weak external electromagnetic field is present.

This can be readily integrated to give

$$a_m^{(1)}(t) = \frac{-i}{\hbar}\int_0^t dt' H_{m\ell}(t')\exp\left[\frac{i(E_m - E_\ell)t'}{\hbar}\right] \tag{5.217}$$

Above, $\left|a_m^{(1)}(t)\right|^2$ gives the probability, to the first order in the perturbation parameter λ, that as a result of the interaction with the external electromagnetic wave, the system will be in state $\left|\phi_m\right\rangle$ at time t.

5.7 Fermi's Golden Rule and Einstein's *B* Coefficient

Of particular importance is the case where the interaction energy is given by

$$\lambda\hat{H}(t) = \hat{H}\exp(-i\omega t) + \hat{H}^{+}\exp(i\omega t) \tag{5.218}$$

where $\hat{H}^{+}$ is the hermitian adjoint of $\hat{H}$. This particular choice for $\hat{H}(t)$ guarantees that it is hermitian. Further note that here, $\hat{H}$ and $\hat{H}^{+}$ are not themselves functions of time. To proceed and investigate the likelihood of optical transitions between the states $\left|\phi_\ell\right\rangle$ and $\left|\phi_m\right\rangle$, we first define the transition frequency $\omega_{m\ell}$ as

$$\omega_{m\ell} = \frac{E_m - E_\ell}{\hbar} \tag{5.219}$$

$H_{m\ell}(t)$ now becomes

$$H_{m\ell}(t) = \left\langle\phi_m\middle|\hat{H}\middle|\phi_\ell\right\rangle\exp(-i\omega t) + \left\langle\phi_m\middle|\hat{H}^{+}\middle|\phi_\ell\right\rangle\exp(i\omega t) \tag{5.220}$$

Letting

$$\left\langle\phi_m\middle|\hat{H}\middle|\phi_\ell\right\rangle = H_{m\ell} \tag{5.221}$$

we get

$$\left\langle\phi_m\middle|\hat{H}^{+}\middle|\phi_\ell\right\rangle = \left\langle\hat{H}\phi_m\middle|\phi_\ell\right\rangle = \left\langle\phi_\ell\middle|\hat{H}\middle|\phi_m\right\rangle^{*} = H_{\ell m}^{*} \tag{5.222}$$

We leave it as an exercise for the reader (see Prob. 5.13) to show that $a_m^{(1)}(t)$ becomes

$$a_m^{(1)}(t) = \frac{-i}{\hbar}\left[H_{m\ell}\exp\left[\frac{i(\omega_{m\ell}-\omega)t}{2}\right]\frac{\sin[(\omega_{m\ell}-\omega)t/2]}{[(\omega_{m\ell}-\omega)/2]}\right.$$
$$\left. + H_{\ell m}^{*}\exp\left[\frac{i(\omega_{m\ell}+\omega)t}{2}\right]\frac{\sin[(\omega_{m\ell}+\omega)t/2]}{[(\omega_{m\ell}+\omega)/2]}\right] \tag{5.223}$$

Looking at Eq. (5.223), we see that the dominant contribution to $a_m^{(1)}(t)$ will occur when $\omega \cong \omega_{m\ell}$ and $\omega \cong -\omega_{m\ell}$. When $\omega \cong \omega_{m\ell}$, we have

$$E_m \cong E_\ell + \hbar\omega \tag{5.224}$$

which means that we are looking at an upward transition originating from the state $|\phi_\ell\rangle$ to a higher lying level $|\phi_m\rangle$. In other words, we can say that the term $H_{m\ell}\exp(-i\omega t)$ in the interaction potential represents stimulated absorption. Let's see what happens when $\omega \cong \omega_{m\ell}$. In this case, the second term in $\hat{H}(t)$ will be negligible and $a_m^{(1)}(t)$ will become

$$a_m^{(1)}(t) = \frac{-i}{\hbar} H_{m\ell} \exp\left[\frac{i(\omega_{m\ell}-\omega)t}{2}\right] \frac{\sin[(\omega_{m\ell}-\omega)t/2]}{[(\omega_{m\ell}-\omega)/2]} \tag{5.225}$$

Letting

$$\Delta\omega = \omega_{m\ell} - \omega \tag{5.226}$$

the corresponding probability $P_m^{(1)}(t) = |a_m^{(1)}(t)|^2$ of occupying state $|\phi_m\rangle$ becomes

$$P_m^{(1)}(t) = \frac{|H_{ml}|^2}{\hbar^2} \frac{\sin^2(\Delta\omega t/2)}{(\Delta\omega/2)^2} \tag{5.227}$$

The function

$$y(\Delta\omega) = \frac{\sin^2(\Delta\omega t/2)}{(\Delta\omega/2)^2} \tag{5.228}$$

is sketched in Fig. 5.16. Note that as $\Delta\omega$ goes to 0, $y(\Delta\omega)$ tends to t^2. Furthermore, as $t \to \infty$, $y(\Delta\omega)$ will approach a constant multiple of

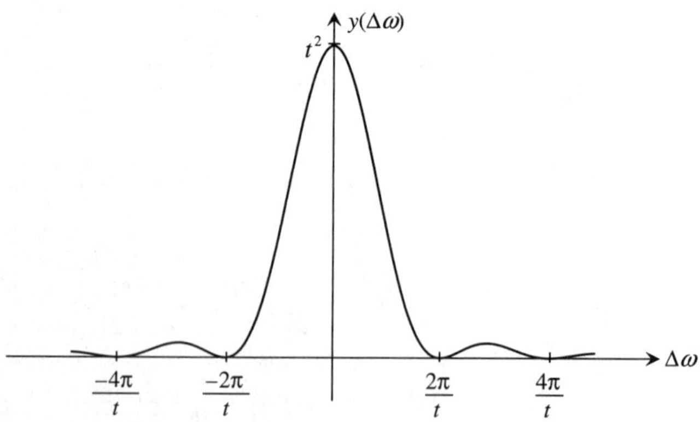

FIGURE 5.16 Variation of the function $y(\Delta\omega)$ with $\Delta\omega$.

the Dirac delta function. We leave it as an exercise for the reader (see Prob. 5.14) to show that in this limit, $y(\Delta\omega)$ becomes

$$\lim_{t\to\infty} y(\Delta\omega) = 2\pi t\delta(\Delta\omega) \tag{5.229}$$

For large t, the corresponding probability $P_m^{(1)}(t)$ of occupying state $|\phi_m\rangle$ becomes

$$P_m^{(1)}(t) = \left|a_m^{(1)}(t)\right|^2 = \frac{\left|H_{m\ell}\right|^2}{\hbar^2} 2\pi t\delta(\omega_{m\ell} - \omega) \tag{5.230}$$

Note that

$$\delta(\omega_{m\ell} - \omega) = \delta\left(\frac{E_m - E_\ell - \hbar\omega}{\hbar}\right) = \hbar\delta(E_m - E_\ell - \hbar\omega) \tag{5.231}$$

giving

$$P_m^{(1)}(t) = \frac{2\pi}{\hbar}\left|H_{m\ell}\right|^2 t\delta(E_m - E_\ell - \hbar\omega) \tag{5.232}$$

It can be shown in a similar way that when $\omega \cong -\omega_{m\ell}$, we have $E_m = E_\ell - \hbar\omega$. This corresponds to stimulated emission and the corresponding probability distribution becomes $(2\pi/\hbar)\left|H_{\ell m}\right|^2 t\delta(E_m - E_\ell + \hbar\omega)$. Combining the two terms together, we get

$$P_m^{(1)}(t) = \frac{2\pi}{\hbar}\left|H_{m\ell}\right|^2 t\delta(E_m - E_\ell - \hbar\omega) + \frac{2\pi}{\hbar}\left|H_{\ell m}\right|^2 t\delta(E_m - E_\ell + \hbar\omega) \tag{5.233}$$

where the first and second terms correspond to absorption and stimulated emission, respectively. The corresponding transition rate $\Gamma_{\ell\to m}$ (also denoted as $W_{\ell\to m}$ in some references) is defined as

$$\Gamma_{\ell\to m} \equiv \frac{dP_m^{(1)}(t)}{dt} \tag{5.234}$$

and is given by

$$\Gamma_{\ell\to m} = \frac{2\pi}{\hbar}\left|H_{m\ell}\right|^2 t\delta(E_m - E_\ell - \hbar\omega) + \frac{2\pi}{\hbar}\left|H_{\ell m}\right|^2 t\delta(E_m - E_\ell + \hbar\omega) \tag{5.235}$$

This is the celebrated Fermi's Golden Rule. Before looking at the derivation of Einstein's B coefficient, we express Fermi's Golden Rule in terms of the linear frequency. Letting

$$\frac{E_m - E_\ell}{h} = \nu_0 \tag{5.236}$$

we get

$$\begin{aligned}\Gamma_{\ell\to m} &= \frac{2\pi}{\hbar}\left|H_{m\ell}\right|^2 \delta((\nu_0-\nu)h) + \frac{2\pi}{\hbar}\left|H_{\ell m}\right|^2 \delta(h(\nu_0+\nu)) \\ &= \frac{1}{\hbar^2}\left|H_{m\ell}\right|^2 \delta(\nu_0-\nu) + \frac{1}{\hbar^2}\left|H_{\ell m}\right|^2 \delta(\nu_0+\nu)\end{aligned} \tag{5.237}$$

At first sight, the expression for the transition rate may look nonphysical since it contains a singularity where the argument of the Dirac delta function vanishes. This arises because in the ideal case that we considered, the energy levels are infinitely sharp. We know already from our classical analysis given in Chap. 4, that this is not the case due to different broadening mechanisms. Even if the broadening processes such as collisions or Doppler broadening are negligible, there is always a fundamental limit to the spectral width of the transitions due to the finite lifetime of the upper state. As discussed in Chap. 4, we characterized this by introducing the lineshape function $g(\nu)$ defined according to

$g(\nu)d\nu$ = Probability that an electromagnetic wave with frequency in the range ν to $\nu + d\nu$ will interact with the system (5.238)

Hence, the total transition rate will be given by

$$\Gamma_{\text{total}} = \int_0^\infty d\nu \Gamma_{\ell\to m}(\nu) g(\nu) \tag{5.239}$$

Let us consider the downward transition only. This leads to induced or stimulated emission characterized by the B_{21} coefficient as sketched in Fig. 5.17.

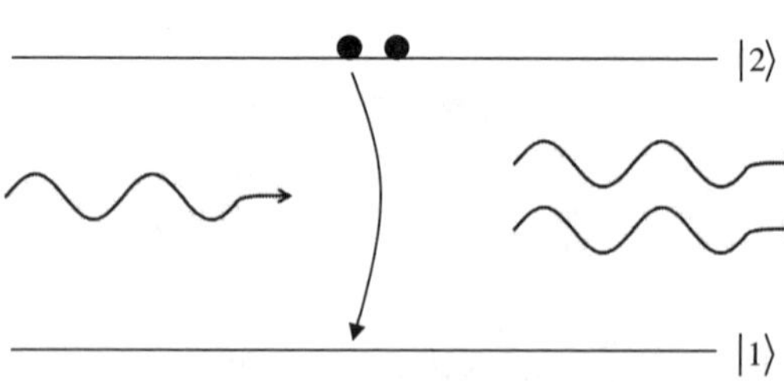

FIGURE 5.17 Sketch showing stimulated emission taking place between states $|2\rangle$ and $|1\rangle$ of a two-level atomic system.

Here, E_m is the energy of level $|1\rangle$ and E_ℓ is the energy of the upper level $|2\rangle$. We define the transition frequency ν_{21} as

$$\nu_{21} = \frac{E_\ell - E_m}{h} \tag{5.240}$$

Γ_{total} now becomes

$$\Gamma_{\text{total}} = \int_0^\infty d\nu \frac{|H_{21}|^2}{\hbar^2} \delta(\nu - \nu_{21}) g(\nu) = \frac{|H_{21}|^2}{\hbar^2} g(\nu_{21}) \tag{5.241}$$

Now, we can obtain the quantum mechanical expression for B_{21} by identifying Γ_{total} as the phenomenalogical transition rate per atom. From the rate equation analysis [see Eq. (4.182)], asssuming a monochromatic wave, the transition rate $(1/N_2)dN_2/dt$ per atom can be written as

$$\frac{1}{N_2}\left(\frac{dN_2}{dt}\right)_{\text{emission}} = -\int d\nu \rho(\nu) g(\nu) B_{21} = -\rho_\nu g(\nu_{21}) B_{21} \tag{5.242}$$

where ρ_ν is the electromagnetic energy density of the monochromatic wave at the frequency ν.

Hence, we get

$$\Gamma_{\text{total}} = \frac{|H_{21}|^2}{\hbar^2} g(\nu_{21}) = \left|\frac{1}{N_2}\frac{dN_2}{dt}\right| = \rho_\nu B_{21} g(\nu_{21}) \tag{5.243}$$

leading to the quantum mechanical expression

$$B_{21} = \frac{|H_{21}|^2}{\hbar^2 \rho_\nu} \tag{5.244}$$

for $B_{21} \cdot B_{21}$ is commonly expressed in terms of the electric dipole matrix element. To derive this expression, note that the interaction hamiltonian can be written as

$$\lambda \hat{H}(t) = -\hat{\vec{d}} \cdot \vec{E}(t) = -\hat{\vec{d}} \cdot \hat{e} E_0 \cos \omega t \tag{5.245}$$

Here, $\hat{e}$ is the unit vector that describes the state of polarization of the electromagnetic wave, E_0 is the peak electric field, $\omega = 2\pi\nu$, and $\hat{\vec{d}}$ is the vector-valued electric dipole operator. The corresponding dipole moment matrix element $\vec{d}_{21}$ is given by

$$\vec{d}_{21} = -\langle 2 | e\hat{\vec{r}} | 1 \rangle \tag{5.246}$$

where $\hat{\vec{r}}$ is the vector-valued electron position operator. Note that H_{21} becomes

$$H_{21} = \frac{E_0}{2}\left(\hat{e}\cdot\vec{d}_{21}\right) \tag{5.247}$$

The energy density ρ_v of the electromagnetic wave can be calculated from

$$\rho_v = \frac{|\vec{S}|}{c} = \frac{1}{c}\left(\frac{|E_0|^2}{2\eta_0}\right) = \frac{1}{c}\frac{|E_0|^2}{2\sqrt{\frac{\mu_0}{\varepsilon_0}}} = \frac{\varepsilon_0}{2}|E_0|^2 \tag{5.248}$$

Above, $\vec{S}$ is the usual Poynting vector, η_0 is the vacuum impedance, μ_0 is the vacuum permeability, and ε_0 is the vacuum permittivity.

Note that B_{21} is proportional to $\left|\hat{e}\cdot\vec{d}_{21}\right|^2$. In an amorphous solid or a gas, we will see the average value of $\left|\hat{e}\cdot\vec{d}_{21}\right|^2$ over all possible orientations. To calculate this average value, let the angle between $\hat{e}$ and $\vec{d}_{21}$ be θ. This gives

$$\left|\hat{e}\cdot\vec{d}_{21}\right|^2 = d_{21}^2\cos^2\theta \tag{5.249}$$

where d_{21}^2 is the magnitude squared of the electric dipole matrix element given by

$$d_{21}^2 = e^2\left(x_{21}^2 + y_{21}^2 + z_{21}^2\right) \tag{5.250}$$

Since the dipole orientation of different electrons will be random, we replace $\cos^2\theta$ by its average over the full solid angle of 4π, which is given by

$$\left\langle\cos^2\theta\right\rangle = \frac{1}{4\pi}\int_0^{2\pi} d\phi \int_0^{\pi} d\theta \cos^2\theta \sin\theta = \frac{1}{3} \tag{5.251}$$

Finally, B_{21} becomes

$$B_{21} = \frac{\frac{|E_0|^2}{4}\frac{|d_{21}|^2}{3}}{\hbar^2\frac{\varepsilon_0|E_0|^2}{2}} = \frac{|d_{21}|^2}{6\hbar^2\varepsilon_0} \tag{5.252}$$

So, this is one of the most important results we wanted to derive in this chapter. It clearly shows how the strength of stimulated emission (represented by B_{21}) depends on the wavefunctions of the atomic states (through d_{21}) and fundamental constants. However, it is not the full story. So far, we treated the electromagnetic field classically, and the atom quantum mechanically. This does not account for spontaneous emission whose correct analysis requires that the electromagnetic field be also quantized. We treat this subject next. Ironically, although spontaneous emission is more readily observed in everyday experince, basically from all ordinary sources of light such as the sun, incandescent lamps, or candles, to name a few, its analysis requires more advanced quantum mechanics in comparison with stimulated emission!

5.8 Quantization of Electromagnetic Fields

We will first employ the potential formulation introduced in Sec. 1.12 to quantize the electromagnetic fields. Our approach is similar to those used in Refs. 8, 9, 11, 15, 16, and 18. First, a note on the gauge choice. This was briefly discussed in Sec. 1.12. Here, we comment on it in more detail to justify the choice of the vector and scalar potentials $\vec{A}(\vec{r},t)$ and $V(\vec{r},t)$. Recall that if the scalar and vector potentials are $V_0(\vec{r},t)$ and $\vec{A}_0(\vec{r},t)$, respectively, the corresponding electric and magnetic fields are given by [see Eqs. (1.40) and (1.274)]

$$\vec{E}_0 = -\nabla V_0 - \frac{\partial \vec{A}_0}{\partial t} \tag{5.253}$$

and

$$\vec{B}_0 = \nabla \times \vec{A}_0 \tag{5.254}$$

In fact, if the original potentials are transformed according to

$$\vec{A} = \vec{A}_0 - \nabla \Psi_g \tag{5.255}$$

and

$$\phi = V_0 + \frac{\partial \Psi_g}{\partial t} \tag{5.256}$$

we find that the resulting fields remain unchanged. Here Ψ_g is any arbitrary "well-behaved" scalar function. To see this, we compute the new electric and magnetic fields $\vec{E}$ and $\vec{B}$ to get

$$\begin{aligned}\vec{E} &= -\nabla V - \frac{\partial \vec{A}}{\partial t} = -\nabla V_0 - \nabla \frac{\partial \Psi_g}{\partial t} - \frac{\partial \vec{A}_0}{\partial t} + \frac{\partial}{\partial t} \nabla \Psi_g \\ &= -\nabla V_0 - \frac{\partial \vec{A}_0}{\partial t} = \vec{E}_0\end{aligned} \tag{5.257}$$

and

$$\vec{B} = \nabla \times \vec{A} = \nabla \times \vec{A}_0 - \nabla \times (\nabla \Psi_g) = \nabla \times \vec{A}_0 = \vec{B}_0 \tag{5.258}$$

This property of the electromagnetic fields, i.e., the fact that $\vec{E}$ and $\vec{B}$ remain invariant under the transformations given by Eqs. (5.255) and (5.256), is referred to as gauge invariance. Furthermore, Eqs. (5.255) and (5.256) are known as gauge transformations. Gauge invariance gives us quite a bit of freedom in choosing the particular form of $\vec{A}(\vec{r},t)$ and $V(\vec{r},t)$. It can be further shown that we can, in fact, equivalently choose a particular form for $\nabla \cdot \vec{A}$ without altering the fields. The specific choice, known as the Lorentz gauge, where

$$\nabla \cdot \vec{A} + \mu_0 \varepsilon_0 \frac{\partial V}{\partial t} = 0 \tag{5.259}$$

was also used earlier in Sec. 1.12. In the quantization of electromagnetic fields, we commonly use the Coulomb gauge where

$$\nabla \cdot \vec{A} = 0 \tag{5.260}$$

which leads to [by using Eq. (1.275)]

$$\nabla^2 V(\vec{r},t) = -\frac{\rho(\vec{r},t)}{\varepsilon_0} \tag{5.261}$$

This was already solved in Sec. 1.3.2, giving

$$V(\vec{r},t) = \frac{1}{4\pi\varepsilon_0} \int d^3r' \frac{\rho(\vec{r}\,',t)}{|\vec{r} - \vec{r}\,'|} \tag{5.262}$$

In the quantization of electromagnetic fields, we will assume that the region of interest is free of charges and currents, and Coulomb gauge is applicable, leading to

$$\nabla \cdot \vec{A} = 0 \tag{5.263}$$

and

$$V = 0 \tag{5.264}$$

Hence, $\vec{E}$ and $\vec{B}$ will be given by

$$\vec{E} = -\frac{\partial \vec{A}}{\partial t} \tag{5.265}$$

and

$$\vec{B} = \nabla \times \vec{A} \tag{5.266}$$

An arbitrary vector magnetic potential $\vec{A}$ can be expressed as an infinite superposition of modes in the form

$$\vec{A}(\vec{r},t) = \frac{1}{\sqrt{\varepsilon_0}} \sum_l x_l(t)\, \vec{U}_l(\vec{r}) = \sum_l \vec{A}_l(\vec{r},t) \tag{5.267}$$

where $\vec{U}_l(\vec{r})$ belongs to a complete set of functions. Since there are no free currents, $\vec{A}_l(\vec{r},t)$ further satisfies, using Eq. (1.281)

$$\nabla^2 \vec{A}_l - \frac{1}{c^2}\frac{\partial^2 \vec{A}_l}{\partial t^2} = 0 \tag{5.268}$$

By using the mode expansion given in Eq. (5.267), we get

$$x_l(t)\nabla^2 \vec{U}_l(\vec{r}) - \frac{1}{c^2}\vec{U}_l(\vec{r})\ddot{x}_l(t) = 0 \tag{5.269}$$

In cartesian coordinates, the x-component of the above equation becomes

$$x_l(t)\nabla^2 U_{lx}(\vec{r}) - \frac{1}{c^2}U_{lx}(\vec{r})\ddot{x}_l(t) = 0 \tag{5.270}$$

Above, $U_{lx}(\vec{r})$ is the x-component of $\vec{U}_l(\vec{r})$. Dividing out by $x_l(t)\ U_{lx}(\vec{r})$ we obtain

$$\frac{c^2\nabla^2 U_{lx}(\vec{r})}{U_{lx}(\vec{r})} = \frac{\ddot{x}_l(t)}{x_l(t)} = -\omega_l^2 \tag{5.271}$$

where ω_l is introduced as a separation constant. This leads to two decoupled differential equations:

$$\ddot{x}_l(t) + \omega_l^2 x_l(t) = 0 \tag{5.272}$$

and

$$\nabla^2 U_{lx}(\vec{r}) + \frac{\omega_l^2}{c^2}U_{lx}(\vec{r}) = 0 \tag{5.273}$$

A similar result can be obtained for the y- and z-components of $\vec{U}_l(\vec{r})$. Adding these components together, one can rewrite Eq. (5.273) in vector form as

$$\nabla^2 \vec{U}_l(\vec{r}) + \frac{\omega_l^2}{c^2}\vec{U}_l(\vec{r}) = 0 \tag{5.274}$$

where $\nabla^2\vec{U}_l(\vec{r})$ is the vector Laplacian of $\vec{U}_l(\vec{r})$. $\vec{E}$ and $\vec{B}$ can now be computed by using

$$\vec{E}(\vec{r},t) = -\frac{1}{\sqrt{\varepsilon_0}}\sum_l \dot{x}_l(t)\vec{U}_l(\vec{r}) \tag{5.275}$$

and

$$\vec{B}(\vec{r},t) = \frac{1}{\sqrt{\varepsilon_0}}\sum_l x_l(t)\nabla\times\vec{U}_l(\vec{r}) \tag{5.276}$$

For reasons that will become clear shortly, we let

$$\dot{x}_l(t) = p_l(t) \tag{5.277}$$

and hence, $\vec{E}$ becomes

$$\vec{E}(\vec{r},t) = -\frac{1}{\sqrt{\varepsilon_0}}\sum_l p_l(t)\vec{U}_l(\vec{r}) \tag{5.278}$$

We can prove that if the region of interest is surrounded by a perfectly conducting surface, the spatial mode functions $U_l(\vec{r})$ form an orthogonal set and obey the orthogonality condition

$$\int_V d^3r\vec{U}_l(\vec{r})\vec{U}_{l'}(\vec{r}) = \delta_{l,l'} \tag{5.279}$$

To prove the above statement, we first note that, since the electromagnetic field will vanish within the perfectly conducting boundary surface S, the tangential component of $\vec{E}$ and the normal component of $\vec{B}$ will also be zero on S. In other words,

$$\left(\vec{U}_l\right)_{\text{tan}} = 0 \tag{5.280}$$

and

$$\left(\nabla\times\vec{U}_l\right)_{\text{norm}} = 0 \tag{5.281}$$

on S, where the subscripts tan and norm refer to the tangential and normal components, respectively. We now use the vector identity

$$\nabla\cdot\left(\vec{U}_1\times\vec{U}_2\right) = \vec{U}_2\cdot\left(\nabla\times\vec{U}_1\right) - \vec{U}_1\cdot\left(\nabla\times\vec{U}_2\right) \tag{5.282}$$

with which $\nabla\cdot\left(\vec{U}_l\times\left(\nabla\times\vec{U}_{l'}\right)\right)$ becomes

$$\nabla\cdot\left(\vec{U}_l\times\left(\nabla\times\vec{U}_{l'}\right)\right) = \left(\nabla\times\vec{U}_{l'}\right)\cdot\left(\nabla\times\vec{U}_l\right) - \vec{U}_l\cdot\left(\nabla\times\nabla\times\vec{U}_{l'}\right) \tag{5.283}$$

Similarly,

$$\nabla\cdot\left(\vec{U}_{l'}\times\left(\nabla\times\vec{U}_l\right)\right)=\left(\nabla\times\vec{U}_l\right)\cdot\left(\nabla\times\vec{U}_{l'}\right)-\vec{U}_{l'}\cdot\left(\nabla\times\nabla\times\vec{U}_l\right) \tag{5.284}$$

Subtracting Eq. (5.283) from Eq. (5.284), we get

$$\begin{aligned}\nabla\cdot&\left(\vec{U}_{l'}\times\left(\nabla\times\vec{U}_l\right)-\vec{U}_l\times\left(\nabla\times\vec{U}_{l'}\right)\right)\\&=\vec{U}_l\cdot\left(\nabla\times\nabla\times\vec{U}_{l'}\right)-\vec{U}_{l'}\cdot\left(\nabla\times\nabla\times\vec{U}_l\right)\end{aligned} \tag{5.285}$$

From Eq. (5.274), we see that

$$\nabla\times\nabla\times\vec{U}_l=-\nabla^2\vec{U}_l=+\frac{\omega_l^{\,2}}{c^2}\vec{U}_l(\vec{r}) \tag{5.286}$$

since $\nabla\cdot\vec{U}_l=0$ in our particular gauge. Hence, we get

$$\nabla\cdot\left(\vec{U}_{l'}\times\left(\nabla\times\vec{U}_l\right)-\vec{U}_l\times\left(\nabla\times\vec{U}_{l'}\right)\right)=+\frac{\left(\omega_{l'}^{\,2}-\omega_l^2\right)}{c^2}\vec{U}_l\cdot\vec{U}_{l'} \tag{5.287}$$

We now integrate Eq. (5.287) over the volume and use the divergence theorem to obtain

$$\oint_S d\vec{a}\cdot\left(\vec{U}_{l'}\times\left(\nabla\times\vec{U}_l\right)-\vec{U}_l\times\left(\nabla\times\vec{U}_{l'}\right)\right)=+\frac{\left(\omega_{l'}^{\,2}-\omega_l^2\right)}{c^2}\int_V d^3r\vec{U}_l\cdot\vec{U}_{l'} \tag{5.288}$$

We need to know the normal component of $\vec{U}_{l'}\times(\nabla\times\vec{U}_l)$ to compute the surface integral. The normal component will be related to the tangential components of $\vec{U}_{l'}$ (electric field) and $\nabla\times\vec{U}_l$ (magnetic field) on the surface. Since the tangential electric field is zero, we conclude that the surface integral should be zero, leading to

$$\int_V d^3r\vec{U}_l\cdot\vec{U}_{l'}=0 \tag{5.289}$$

when $l\neq l'$. We further assume that the normalization constants are chosen in such a way as to give

$$\int_V d^3r\vec{U}_l\cdot\vec{U}_l=1 \tag{5.290}$$

Hence, the result in Eq. (5.279) is obtained.

To arrive at the quantization result, we now need to calculate the total energy of the electromagnetic field. The total electromagnetic energy density u is given by

$$u=u_e+u_m=\frac{1}{2}\varepsilon_0\left|\vec{E}\right|^2+\frac{1}{2\mu_0}\left|\vec{B}\right|^2 \tag{5.291}$$

The resulting total energy H can be calculated from

$$H = \int_V d^3r\, u \tag{5.292}$$

By using the expressions for $\vec{E}(\vec{r},t)$ and $\vec{B}(\vec{r},t)$ given in Eqs. (5.278) and (5.276), the electric and magnetic energy densities become

$$u_e = \frac{1}{2}\varepsilon_0 \left|\vec{E}\right|^2 = \frac{1}{2}\sum_l \sum_{l'} p_l(t) p_{l'}(t) \vec{U}_l(\vec{r}) \cdot \vec{U}_{l'}(\vec{r}) \tag{5.293}$$

and

$$u_m = \frac{1}{2\mu_0}\left|\vec{B}\right|^2 = \frac{1}{2\mu_0\varepsilon_0}\sum_l \sum_{l'} x_l(t) x_{l'}(t) \left(\left(\nabla \times \vec{U}_l\right)\cdot\left(\nabla \times \vec{U}_{l'}\right)\right) \tag{5.294}$$

Before calculating the total energy, we focus on the volume integral involving the integrand $(\nabla \times \vec{U}_l)\cdot(\nabla \times \vec{U}_{l'})$. As before,

$$\nabla \cdot \left(\vec{U}_l \times \left(\nabla \times \vec{U}_{l'}\right)\right) = \left(\nabla \times \vec{U}_{l'}\right)\cdot\left(\nabla \times \vec{U}_l\right) - \vec{U}_l \cdot \left(\nabla \times \nabla \times \vec{U}_{l'}\right) \tag{5.295}$$

giving

$$\begin{aligned}\int_V d^3r \left(\nabla \times \vec{U}_{l'}\right)\cdot\left(\nabla \times \vec{U}_l\right) &= \oint_S d\vec{a} \cdot \left(\vec{U}_l \times \left(\nabla \times \vec{U}_{l'}\right)\right) + \int_V d^3r\, \vec{U}_l \cdot \left(\nabla \times \nabla \times \vec{U}_{l'}\right) \\ &= 0 + \frac{\omega_{l'}{}^2}{c^2}\int_V d^3r \vec{U}_l \cdot \vec{U}_{l'} = \frac{\omega_{l'}{}^2}{c^2}\delta_{l,l'}\end{aligned} \tag{5.296}$$

Hence, the total energy H becomes

$$\begin{aligned}H &= \frac{1}{2}\sum_l \sum_{l'} p_l(t) p_{l'}(t) \int_V d^3r \vec{U}_l \cdot \vec{U}_{l'} + \frac{1}{2\mu_0\varepsilon_0}\sum_l \sum_{l'} x_l(t) x_{l'}(t)\, \frac{\omega_{l'}{}^2}{c^2}\delta_{l,l'} \\ &= \sum_l \left(\frac{p_l^2}{2} + \frac{1}{2}\omega_l{}^2 x_l^2\right)\end{aligned} \tag{5.297}$$

Now, isn't this amazing!? The reader clearly appreciates by now why we were so eager to go over a detailed analysis of the harmonic oscillator in the earlier sections. Also note that the electromagnetism background introduced in Chap. 1 proved to be very useful here. The total energy of each mode looks exactly like the energy of the classical harmonic oscillator of unit mass. Hence, we have arrived at the important result that each mode of the electromagnetic field is equivalent to a harmonic oscillator with a unit mass. Furthermore, the electric field resembles the momentum, whereas the magnetic field looks like the position. Therefore, we can proceed and use the quantization results we already obtained for the harmonic oscillator in Sec. 5.4 and 5.5. In particular, to find the operator representations of the fields, we express the functions $x(t)$ and $p(t)$ by the operators $\hat{x}$ and $\hat{p}_x$. As we saw earlier, $\hat{x}$ and $\hat{p}_x$ can be expressed in terms of the annihilation and creation operators ($\hat{a}$ and $\hat{a}^+$). For a harmonic oscillator with unit mass, we obtain

$$\hat{a}(t) = \frac{1}{\sqrt{2\hbar\omega_l}}\left(\omega_l\hat{x} + i\hat{p}_x\right) \tag{5.298}$$

and

$$\hat{a}^+ = \frac{1}{\sqrt{2\hbar\omega_l}}\left(\omega_l\hat{x} - i\hat{p}_x\right) \tag{5.299}$$

Adding Eqs. (5.298) and (5.299) together, we get

$$\hat{x} = \sqrt{\frac{\hbar}{2\omega}}(\hat{a} + \hat{a}^+) \tag{5.300}$$

Subtracting gives

$$\hat{p} = \sqrt{\frac{h\omega}{2}}(\hat{a} - \hat{a}^+)\frac{1}{i} \tag{5.301}$$

At this point, we choose to proceed with the Heisenberg representation where the operators are time dependent. We leave it as an exercise for the reader (see Prob. 5.15) to show that the time-dependent annihilation operator for the lth mode is given by

$$\hat{a}(t) = \hat{a}(0)\exp(-i\omega_l t) \tag{5.302}$$

Here, we take $\hat{a}(0)$ as the annihilation operator $\hat{a}$ in the Schrödinger representation.

Now, what about the electric field operator? The electric field operator for the lth mode will be

$$\hat{\vec{E}}_l(\vec{r},t) = -\frac{1}{\sqrt{\varepsilon_0}}\hat{p}_l(t)\vec{U}_l(\vec{r}) \tag{5.303}$$

From Eq. (5.274), the solution for $\vec{U}_l(\vec{r})$ can be written as

$$\vec{U}_l(\vec{r}) = C\hat{e}_l \exp(i\vec{k}\cdot\vec{r}) + C\hat{e}_l^* \exp(-i\vec{k}\cdot\vec{r}) \tag{5.304}$$

where C is the normalization constant. A solution for $\hat{\vec{E}}(\vec{r},t)$ can be written in the form

$$\hat{\vec{E}}(\vec{r},t) = i\frac{C}{\sqrt{\varepsilon_0}}\sqrt{\frac{\hbar\omega_l}{2}}(\hat{e}\hat{a}(t)\exp(i\vec{k}\cdot\vec{r}) - \hat{e}^*\hat{a}^+(t)\exp(-i\vec{k}\cdot\vec{r})) \tag{5.305}$$

Above, $\hat{e}$ and $\hat{e}^*$ are unit vectors, whereas $\hat{a}(t)$ and $\hat{a}^+(t)$ are operators. To find the normalization constant C, we set

$$\int_V d^3r \frac{1}{2}\varepsilon_0 |E|^2 = \frac{1}{2}\hbar\omega_l\left(n_l + \frac{1}{2}\right) \tag{5.306}$$

where $|E|$ now is the magnitude of the expectation value of $\hat{\vec{E}}$ in the state $|n_l\rangle$. We obtain

$$\begin{aligned}|E|^2 &= \left\langle n_l \middle| \hat{\vec{E}}\cdot\hat{\vec{E}}^* \middle| n_l \right\rangle = \frac{C^2}{\varepsilon_0}\frac{\hbar\omega}{2}\langle n_l|(\hat{e}\hat{a}(t)\exp(i\vec{k}\cdot\vec{r}) - \hat{e}^*\hat{a}^+(t)\exp(-i\vec{k}\cdot\vec{r}))\\ &\quad\cdot(\hat{e}^*\hat{a}^+(t)\exp(-i\vec{k}\cdot\vec{r}) - \hat{e}\hat{a}(t)\exp(i\vec{k}\cdot\vec{r}))|n_l\rangle\\ &= \frac{C^2\hbar\omega}{2\varepsilon_0}\langle n_l|\hat{a}\hat{a}^+ + \hat{a}^+\hat{a}|n_l\rangle = \frac{C^2\hbar\omega}{2\varepsilon_0}\langle n_l|2\hat{a}^+\hat{a}+1|n_l\rangle = \frac{C^2\hbar\omega}{2\varepsilon_0}(2n_l+1)\end{aligned} \tag{5.307}$$

We then get

$$\frac{1}{2}\varepsilon_0\frac{C^2\hbar\omega_l}{2\varepsilon_0}(2n_l+1)V = \frac{1}{2}\hbar\omega_l\left(n_l+\frac{1}{2}\right) \tag{5.308}$$

or

$$C = \frac{1}{\sqrt{V}} \tag{5.309}$$

where V is the volume of the region. Hence, the quantum mechanical expression for the electric field operator for the lth mode of a plane electromagnetic wave can be written as

$$\hat{\vec{E}}_l(\vec{r},t) = i\sqrt{\frac{\hbar\omega_l}{2\varepsilon_0 V}}(\hat{e}\hat{a}\exp(i(\vec{k}\cdot\vec{r}-\omega_l t)) - \hat{e}^*\hat{a}^+\exp(-i(\vec{k}\cdot\vec{r}-\omega_l t))) \tag{5.310}$$

5.9 Quantum Mechanical Derivation of Einstein's *A* Coefficient

Let us assume that we have an atom interacting with a single radiation mode represented by the electric field operator in Eq. (5.310). This means that we are now considering the interaction between the atom and the quantized electromagnetic field. We further employ the electric dipole approximation and take $\exp(i\vec{k}\cdot\vec{r})\cong 1$. Let the atom be initially in state $|2\rangle$, and let there be n_l photons in the radiation mode. We will represent this state by the composite wave function

$$|\psi_1\rangle = |2\rangle|n_l\rangle \tag{5.311}$$

After the decay of the atom to the lower level $|1\rangle$, the radiation mode will gain one photon, resulting in the final composite wave function

$$|\psi_2\rangle = |1\rangle|n_l+1\rangle \tag{5.312}$$

This is schematically shown Fig. 5.18.

In the electric dipole approximation, the interaction hamiltonian will become

$$\hat{H} = \hat{\vec{E}}\cdot\hat{\vec{d}} = i\sqrt{\frac{\hbar\omega_l}{2\varepsilon_0 V}}[(\hat{e}\cdot\hat{\vec{d}})\hat{a}\exp(-i\omega_l t) - (\hat{e}^*\cdot\hat{\vec{d}})\hat{a}^+\exp(i\omega_l t)] \tag{5.313}$$

By using Fermi's Golden Rule, the decay rate $\Gamma_{2\to1}$ can now be expressed as

$$\Gamma_{2\to1} = +\frac{2\pi}{\hbar}|H_{12}|^2\,\delta(E_2 - E_1 - \hbar\omega) \tag{5.314}$$

where

$$H_{12} = i\sqrt{\frac{\hbar\omega_l}{2\varepsilon_0 V}}\left(\hat{e}\cdot\vec{d}_{21}\right)\langle n_l|\hat{a}|n_l+1\rangle = i\sqrt{\frac{\hbar\omega_l}{2\varepsilon_0 V}}\left(\hat{e}\cdot\vec{d}_{21}\right)\sqrt{n_l+1} \tag{5.315}$$

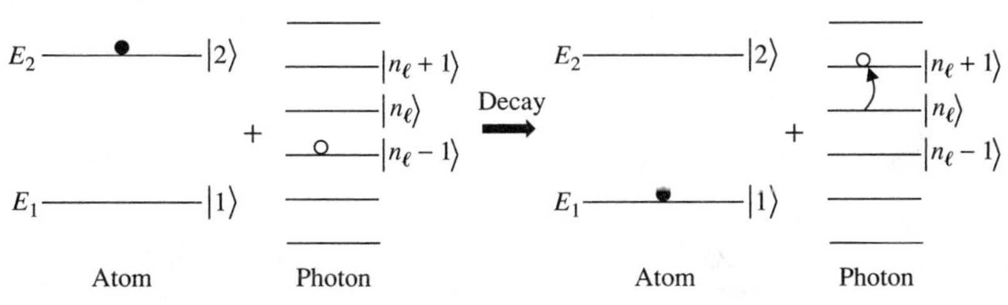

FIGURE 5.18 The composite system consisting of the atom and the quantized radiation field before and after the emission of one photon.

where

$$\vec{d}_{21} = \left\langle 2 \middle| \hat{\vec{d}} \middle| 1 \right\rangle \tag{5.316}$$

Hence, the decay rate becomes

$$\Gamma_{2\to 1} = \frac{2\pi}{\hbar}\frac{\hbar\omega_l}{2\varepsilon_0 V}\left|\hat{e}\cdot\vec{d}_{21}\right|^2 (n_l+1)\delta(E_2 - E_1 - \hbar\omega_l)$$

$$= \frac{\pi e^2\omega_l}{\varepsilon_0 V}\left|\hat{e}\cdot\vec{r}_{21}\right|^2 (n_l+1)\delta(E_2 - E_1 - \hbar\omega_l) \tag{5.317}$$

$\Gamma_{2\to 1}$ is the sum of the stimulated and spontaneous rates. In particular, spontaneous rate is the component of $\Gamma_{2\to 1}$ that does not depend on the photon number and is given by

$$(\Gamma_{2\to 1})_{\text{spontaneous}} = \frac{\pi e^2\omega_l}{\varepsilon_0 V}\left|\hat{e}\cdot\vec{r}_{21}\right|^2 \delta(E_2 - E_1 - \hbar\omega_l) \tag{5.318}$$

Note that this is the decay rate into only one mode, namely the mode at frequency ω_l. The total spontaneous emission rate can be calculated by multiplying the above rate by the density of modes and integrating over all frequencies. We let

$$\omega_0 = \frac{E_2 - E_1}{\hbar} \tag{5.319}$$

giving

$$(\Gamma_{2\to 1})_{\text{spontaneous}} = \frac{\pi e^2\omega_l}{\varepsilon_0 \hbar V}\left|\hat{e}\cdot\vec{r}_{21}\right|^2 \delta(\omega_0 - \omega) \tag{5.320}$$

Back in Chap. 4, we determined the density of modes $N(k)dk$ as [see Eq. (4.142)]

$$N(k)dk = \frac{k^2}{\pi^2}dk \tag{5.321}$$

Since

$$k = \frac{\omega}{c}n \tag{5.322}$$

where n is the refractive index of the medium, $N(\omega)d\omega$ becomes

$$N(\omega)d\omega = \frac{\omega^2 n^3}{\pi^2 c^3}d\omega \tag{5.323}$$

The total spontaneous decay rate over the volume V becomes

$$\sum(\Gamma_{2\to 1})_{\text{spontaneous}} = A_{21} = \int_0^\infty d\omega_l \frac{\pi e^2 \omega_l}{\varepsilon_0 \hbar V} \left|\hat{e}\cdot\vec{r}_{21}\right|^2 \frac{\delta(\omega_0 - \omega_l) V \omega_l^2 n^3}{\pi^2 c^3}$$

$$= \frac{n^3 e^2 \left|\hat{e}\cdot\vec{r}_{21}\right|^2 \omega_l^3}{\pi \varepsilon_0 \hbar c^3} \tag{5.324}$$

Similar to the averaging we did for $\left|\hat{e}\cdot\vec{d}_{lm}\right|^2$ in Sec. 5.7, we get

$$A_{21} = \frac{1}{3\pi} \frac{n^3 e^2 \left|r_{21}\right|^2 \omega_0^3}{\varepsilon_0 \hbar c^3} = \frac{2n^3 e^2 \left|r_{21}\right|^2 \omega_0^3}{3\varepsilon_0 h c^3} \tag{5.325}$$

By using Eq. (5.252), the ratio A_{21}/B_{21} becomes

$$\frac{A_{21}}{B_{21}} = \frac{2n^3 \left|d_{21}\right|^2 \omega_0^3}{3\varepsilon_0 h c^3} \frac{6\hbar^2 \varepsilon_0}{\left|d_{21}\right|^2} = \frac{8\pi h v_0^3 n^3}{c^3} \tag{5.326}$$

as we obtained in Eq. (4.193) back in Chap. 4. This completes the quantum mechanical derivation of Einstein coefficients. See Refs. 9, 11, and 16 for further discussion of spontaneous emission.

5.10 Ideal Two-Level System and Rabi Oscillations

In the presence of strong fields produced by laser beams, it is easily possible to go beyond the regime where perturbation theory is valid and end up in a scenario where considerable variation occurs in the populations of the atomic levels. Such a case can be treated analytically for an ideal two-level sytem, provided that the interaction is fast enough so that spontaneous decay rates are negligible. From our earlier analysis, we derived the following exact equation for the temporal evolution of the probability coefficients:

$$i\hbar \frac{da_m}{dt} = \sum_n \lambda H_{mn}(t) a_n(t) \exp\left[\frac{i(E_m - E_n)t}{\hbar}\right] \tag{5.327}$$

We now look at the specific atomic system whose energy level diagram is as sketched in Fig. 5.19. Here, $h\omega_0 = E_2 - E_1$, ω is the frequency of the monochromatic electromagnetic wave, and $\Delta = \omega_0 - \omega$ is termed the detuning parameter.

For our ideal case involving two energy levels, Eq. (5.327) reduces to two coupled differential equations:

$$i\hbar \frac{da_2}{dt} = \lambda H_{21}(t) a_1(t) \exp[i\omega_0 t] \tag{5.328}$$

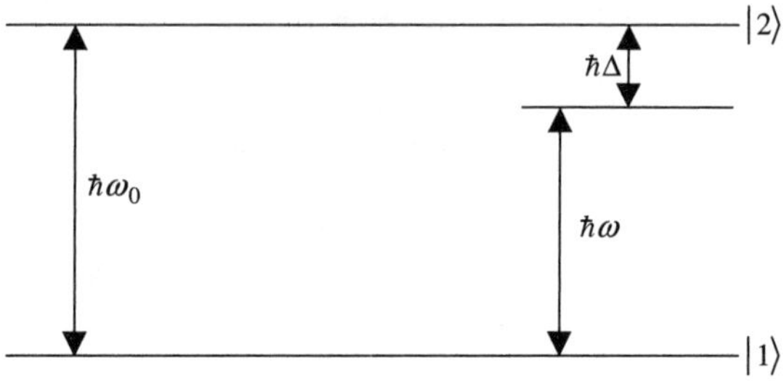

FIGURE 5.19 Two-level atomic system interacting with a monochromatic electromagnetic wave whose frequency is detuned from the atomic resonance by $\Delta = \omega_0 - \omega$.

and

$$i\hbar \frac{da_1}{dt} = \lambda H_{12}(t) a_2(t) \exp[-i\omega_0 t] \tag{5.329}$$

Since there are only two energy levels, probability amplitudes further satisfy the condition

$$|a_1(t)|^2 + |a_2(t)|^2 = 1 \tag{5.330}$$

at all times. The interaction hamiltonian will be

$$\lambda \hat{H}(t) = (\hat{e} \cdot \hat{\vec{d}}) E_0 \cos \omega t \tag{5.331}$$

with the corresponding matrix element

$$\lambda H_{21}(t) = \hbar \chi \cos(\omega t) \tag{5.332}$$

where

$$\chi = \frac{(\hat{e} \cdot \vec{d}_{21}) E_0}{\hbar} \tag{5.333}$$

Note also that

$$\lambda H_{12}(t) = \hbar \chi^* \cos(\omega t) \tag{5.334}$$

Equations of motion for $a_1(t)$ and $a_2(t)$ now become

$$i \frac{da_1}{dt} = \chi^* \cos(\omega t) \exp[-i\omega_0 t] a_2(t) \tag{5.335}$$

and

$$i\frac{da_2}{dt} = \chi\cos(\omega t)\exp[i\omega_0 t]a_1(t) \tag{5.336}$$

Expanding the trigonometric function in terms of complex exponentials, Eqs. (5.335) and (5.336) can be written as

$$i\frac{da_1}{dt} = \frac{\chi^*}{2}(\exp[i(\omega-\omega_0)t]+\exp[-i(\omega+\omega_0)t])a_2(t) \tag{5.337}$$

and

$$i\frac{da_2}{dt} = \frac{\chi}{2}(\exp[i(\omega+\omega_0)t]+\exp[i(\omega_0-\omega)t])a_1(t) \tag{5.338}$$

Let us assume that ω is close to ω_0 so that $\Delta = \omega_0 - \omega$ is small. The terms containing $(\omega+\omega_0)$ in Eqs. (5.337) and (5.338) will oscillate very rapidly and, over sufficiently long time intervals, will average to zero. Hence, we can neglect them. This is commonly referred to as the rotating wave approximation. The above equations then simplify to

$$i\frac{da_1}{dt} = \frac{\chi^*}{2}\exp[i(\omega-\omega_0)t]a_2(t) \tag{5.339}$$

and

$$i\frac{da_2}{dt} = \frac{\chi}{2}\exp[i(\omega_0-\omega)t]a_1(t) \tag{5.340}$$

Let us suppose that all of the atoms are initially in the lower state. We leave it as an exercise for the reader (see Prob. 5.15) to show that $a_1(t)$ and $a_2(t)$ are given by

$$a_1(t) = \exp\left(-i\Delta\frac{t}{2}\right)\left[\cos\left(\frac{\Omega_R t}{2}\right)+i\frac{\Delta}{\Omega_R}\sin\left(\frac{\Omega_R t}{2}\right)\right] \tag{5.341}$$

and

$$a_2(t) = -i\frac{\chi}{\Omega_R}\exp\left(i\Delta\frac{t}{2}\right)\sin\left(\frac{\Omega_R t}{2}\right) \tag{5.342}$$

where Ω_R is referred to as the Rabi frequency[10, 12] (named after Isodor Isaac Rabi) and is given by

$$\Omega_R = \sqrt{\Delta^2 + |\chi|^2} \tag{5.343}$$

Note that the Rabi frequency depends on the electric field strength. The corresponding occupation probabilities of the two states become

$$|a_1(t)|^2 = \cos^2\left(\frac{\Omega_R t}{2}\right) + \frac{\Delta^2}{\Omega_R^2}\sin^2\left(\frac{\Omega_R t}{2}\right) \tag{5.344}$$

and

$$|a_2(t)|^2 = \frac{|\chi|^2}{\Omega_R^2}\sin^2\left(\frac{\Omega_R t}{2}\right) \tag{5.345}$$

Based on the foregoing analysis, we draw the following important conclusions that summarize the interaction of an ideal two-level atom with a monochromatic wave having a frequency ω close to the resonance frequency ω_0:

1. Under the action of the external field, the population in the upper level varies sinusoidally at the Rabi frequency Ω_R. The Rabi frequency is determined by the field strength E_0 and the amount of detuning through Eq. (5.343). Hence, the oscillation frequency of the population will increase when the field strength E_0 or the amount of detuning is increased.
2. When the detuning parameter is non-zero ($\Delta \neq 0$), the maximum value of $|a_2(t)|^2$, which can be interpreted as the fraction of the excited atoms, will be

$$\left[|a_2(t)|^2\right]_{\max} = \frac{|\chi|^2}{\Omega_R^2} = \frac{\Omega_R^2 - \Delta^2}{\Omega_R^2} \tag{5.346}$$

By tuning the frequency of the radiation to exactly ω_0, which means that $\Delta = 0$, we can place all of the atoms of the ideal two-level system into the upper level. In this case, $\Omega_R = |\chi|$. The population in level $|2\rangle$ will become maximum when

$$\frac{\Omega_R t}{2} = \frac{\pi}{2} \tag{5.347}$$

or when $\Omega_R t = \pi$. Such an electromagnetic pulse needed to raise all of the atoms to the upper level is commonly known as a π-pulse. Figure 5.20 shows the variation of $|a_2(t)|^2$ as a function of the normalized time $|\chi|t$ for the $a = \Delta/|\chi|$ values of (a) 0, (b) 0.25, (c) 0.50, and (d) 1. As a increases, the maximum value of $|a_2(t)|^2$ monotonically decreases as shown in Eq. (5.346).

In our discussion of light-matter interactions, we bypassed many important topics and mainly focused on an elementary quantum mechanical analysis of stimulated and spontaneous emission. The references cited at the end of this chapter provide further discussion on a rich variety of quantum phenomena encountered in light-matter interactions.

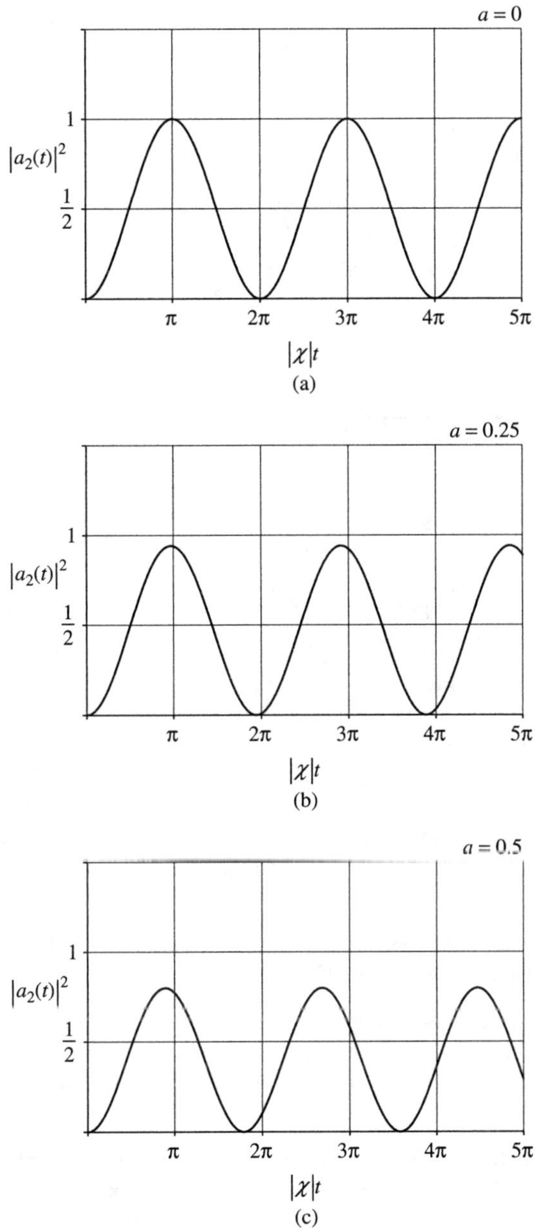

FIGURE 5.20 Variation of $|a_2(t)|^2$ as a function of the normalized time $|\chi|t$ for the $a = \Delta/|\chi|$ values of (*a*) 0, (*b*) 0.25, (*c*) 0.50, and (*d*) 1.

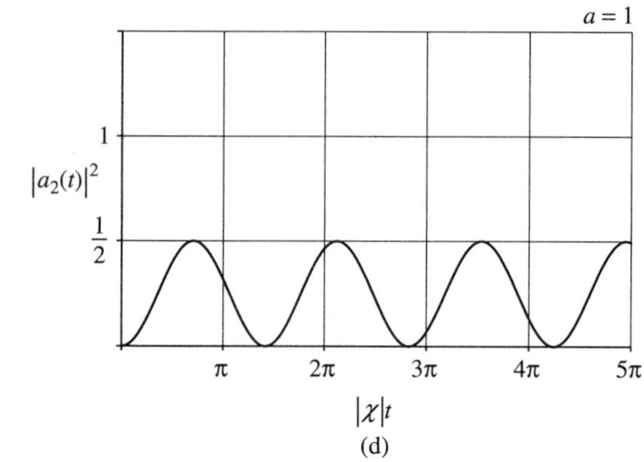

Figure 5.20 (*Continued*)

Problems

5.1 Show that for a gaussian amplitude distribution given by Eq. (5.11), Eq. (5.12) gives the corresponding wavefunction.

5.2 By using integration by parts and by assuming that both ψ as well as its derivative vanish at $x = \pm\infty$, show that the expectation value of the momentum is given by Eq. (5.41).

5.3 Show that the position and momentum operators are hermitian.

5.4 Show that the eigenvalue equation for the even solutions of the finite well is given by Eq. (5.139). Further show that the dimensionless form is given by Eq. (5.141).

5.5 Show that the dimensionless eigenvalue equation for the odd solutions of the finite well is given by Eq. (5.144).

5.6 Consider the one-dimensional well with $V_0 = 20$ eV and $d = 1$ nm. Write a computer code to determine the eigenvalues of the normalized wave number $\bar{x}$ for the even and odd cases. The graphical solutions are displayed in Figs. 5.8 and 5.9.

5.7 Show that, for a given eigenvalue $\bar{x}$, the wave function of the even state can be expressed in terms of normalized coordinates as

$$u(x) = \begin{cases} \exp\left[2\sqrt{\alpha - \bar{x}^2}\, y\right] & y < -\dfrac{1}{2} \\ \dfrac{\exp\left[-\sqrt{\alpha - \bar{x}^2}\right]}{\cos \bar{x}} \cos(2\bar{x}y) & -\dfrac{1}{2} < y < \dfrac{1}{2} \\ \exp\left[-2\sqrt{\alpha - \bar{x}^2}\, y\right] & y > \dfrac{1}{2} \end{cases}$$

where

$$y = \frac{x}{d}$$

and α is given by Eq. (5.142).

5.8 Consider the one-dimensional well with $V_0 = 2$ eV and $d = 1$ nm. Write a computer code to determine the eigenvalues of the normalized wave number x for the even and odd cases. The graphical solutions are displayed in Fig. 5.10*a* and *b*. Also write a program to display the spatial variation of the even and odd eigenfunctions shown in Fig. 5.11*a* through *c*.

5.9 Show that $H(y)$ satisfies the differential equation given by Eq. (5.169).

5.10 Show that the hamiltonian operator of the harmonic oscillator can be written in terms of the annihilation and creation operators, as given in Eq. (5.188).

5.11 Show that the commutation of the annihilation and creation operators is given by Eq. (5.189).

5.12 Show that when the creation operator acts on $|n\rangle$, the result is given by Eq. (5.202).

5.13 Show that $a_m^{(1)}(t)$ is given by Eq. (5.223).

5.14 Show that as $t \to \infty$, the function $y(\Delta\omega)$ is given by Eq. (5.229).

5.15 Show that the solutions for $a_1(t)$ and $a_2(t)$ are given by Eqs. (5.341) and (5.342).

5.16 Analysis of 1-D semiconductor quantum wells
It is possible to form one-dimensional finite potential wells by growing layers of semiconductors with different bandgap energy. We will discuss this further in Chap. 7. To gain some practice in making energy-level calculations, assume that a 5-nm layer of GaAs (bandgap energy = 1.42 eV) is sandwiched between thick layers of $Al_{0.2}Ga_{0.8}As$ (bandgap energy = 1.67 eV). This structure forms a potential well with a thickness of 5 nm for the conduction electrons. Assume that the potential depth is approximately equal to 67% of the band-gap energy difference. Calculate the difference in eV between the lowest bound energy level of the electrons and the bottom of the well. Take the effective mass of the conduction band electrons as 0.067 m_e, where m_e is the rest mass of a free electron.

5.17 Radiative lifetime of an electron trapped in a 1-D infinite potential well. Although this example is not very realistic, one can make an order-of-magnitude estimate for radiative lifetime by considering the idealized case of a particle trapped in an infinite one-dimensional potential well. Use the wavefunctions for the infinite well to estimate the radiative lifetime for an electron making a transition from the first excited state to the ground state. What is the radiative lifetime when the well width is 5 nm and 1 nm? Take the refractive index as 1.

5.18 Repeat the above analysis for a one-dimensional harmonic oscillator with a mass of m and classical resonance frequency of ω. Estimate the radiative decay rate going from state $|n\rangle$ to $|n-1\rangle$.

References

1. P. A. M. Dirac, *The Principles of Quantum Mechanics* (Clarendon Press, Oxford, 1930).
2. R. H. Dicke and J. P. Wittke, *Introduction to Quantum Mechanics* (Addison Wesley, Reading, Mass. 1960).
3. L. Pauling and E. B. Wilson, *Introduction to Quantum Mechanics with Applications to Chemistry* (Dover, New York, 1963).
4. C. Cohen-Tannoudji, B. Diu, and F. Laloe, *Quantum Mechanics I and II* (John Wiley & Sons, Inc., New York, 1977).
5. S. Gasiorowicz, *Quantum Physics,* 2d ed. (John Wiley & Sons, Inc., New York, 1996).
6. R. L. Liboff, *Introductory Quantum Mechanics,* 4th ed. (Addison-Wesley, San Fransisco, 2003).
7. B. D. Bartolo, *Optical Interactions in Solids* (John Wiley & Sons, Inc., New York, 1968).
8. A. Yariv, *Quantum Electronics* (Wiley, New York, 1975).
9. A. Yariv, *An Introduction to Theory and Applications of Quantum Mechanics* (John Wiley & Sons, Inc., New York, 1982).
10. L. Allen and J. H. Eberly, *Optical Resonance and Two-Level Atoms* (Dover, New York, 1987).
11. R. Loudon, *The Quantum Theory of Light,* 2d ed. (Oxford Clarendon Press, Oxford, 1988).
12. P. W. Milonni and J. H. Eberly, *Lasers* (John Wiley & Sons, Inc., New York, 1988).
13. B. Henderson and G. F. Imbusch, *Optical Spectroscopy of Inorganic Solids* (Clarendon Press, Oxford, 1989).
14. A. Imamoglu and Y. Yamamoto, *Mesoscopic Quantum Optics* (John Wiley & Sons, Inc., New York, 1999).
15. M. Orszag, *Quantum Optics* (Springer, Berlin, 2000).
16. C. C. Gerry and P. L. Knight, *Introductory Quantum Optics* (Cambridge University Press, Cambridge, 2005).
17. D. A. B. Miller, *Quantum Mechanics for Scientists and Engineers* (Cambridge University Press, Cambridge, 2008).
18. M. O. Scully and M. S. Zubairy, *Quantum Optics* (Cambridge University Press, Cambridge, 1997).
19. P. L. Hagelstein, S. D. Senturia, and T. P. Orlando, *Introductory Applied Quantum and Statistical Mechanics* (John Wiley & Sons, Inc., Hoboken, N.J., 2004).
20. I. James, *Remarkable Physicists: From Galileo to Yukawa* (Cambridge University Press, Cambridge, 2004).
21. C. Davisson and L. H. Germer, "Diffraction of electrons by a crystal of nickel," *Physical Review,* **30**(6), 705–740 (1927).
22. A. Tonomura, J. Endo, T. Matsuda, T. Kawasaki, and H. Ezawa, "Demonstration of Single-Electron Buildup of an Interference Pattern," *American Journal of Physics,* **57**(2), 117–120 (1989).
23. E. Schrodinger, "Quantisation as an eigen value problem," *Annalen Der Physik,* **79**(4), 361–368 (1926).
24. E. Schrodinger, „Quantification of the eigen value problem," *Annalen Der Physik,* **80**(13), 437–490 (1926).
25. E. Schrodinger, "Quantification of the eigenvalue problem," *Annalen Der Physik,* **79**(6), 489–527 (1926).

CHAPTER 6

Lasers

We are now ready to describe how lasers operate. We first provide an overview of the general characteristics of laser light. This includes a discussion of some of the important concepts such as monochromaticity, spectral brightness, and temporal coherence, which are among the distinguishing characteristics of laser light. We emphasize that lasers are not just amplifiers but oscillators. We then discuss the conditions under which optical gain can be achieved, and explore the pumping requirements of two-, three-, and four-level systems. Oscillation condition and spectral characteristics of the laser output are then discussed. Frequency pulling is briefly explained. In this chapter, we have not provided a thorough review of different types of lasers, since this has been done in many excellent references listed at the end of the chapter.[1-4] Instead, as a specific type of lasers, we focus on solid-state lasers and develop useful models that can be used to characterize the continuous-wave as well as the pulsed operation of these lasers. In particular, we focus on four-level lasers and discuss the plane-wave analysis of continuous-wave operation, relaxation oscillations, Q-switching, and active mode locking. We end the chapter with a description of how ultrashort laser pulses are measured. Semiconductor lasers form another important class of widely used laser systems. We discuss the electronic and photonic properties of semiconductors and describe the operation of semiconductor lasers in Chap. 7. For further discussion on the physics of lasers, see Refs. 1 to 8.

6.1 General Characteristics of Lasers

6.1.1 Lasers Are Oscillators at Optical Frequencies

First and foremost, a laser is an oscillator. As such, it combines optical amplification with positive feedback to generate coherent light oscillation at a specific wavelength. Although the acroynm "loser" (light oscillation by stimulated emission of radiation) more accurately describes the process of light generation in a laser, the acroynm "laser" (light amplification by stimulated emission of radiation) was historically adopted for obvious reasons! The schematic of a generic oscillator is shown in Fig. 6.1.

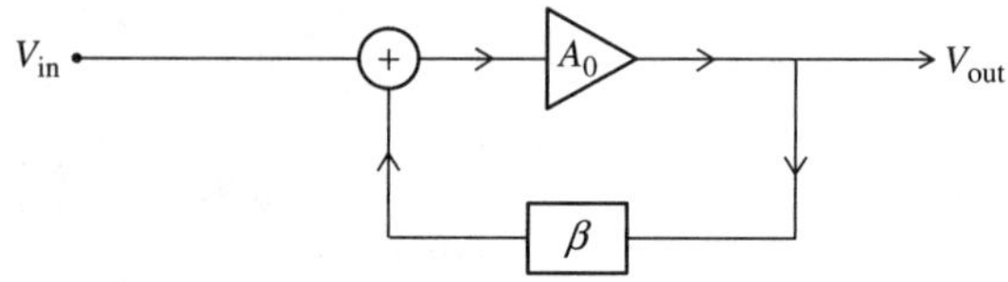

FIGURE 6.1 Schematic of a general oscillator.

The signal after the additive mixer is amplified by an amount A_0 and a portion βV_{out} of the output signal is returned back and combined with the input by using a feedback network. Hence, the output signal can be written as

$$V_{out} = A_0(V_{in} + \beta V_{out}) \tag{6.1}$$

or, solving for V_{out}, we get

$$V_{out} = \frac{A_0}{1-\beta A_0} V \tag{6.2}$$

In Eq. (6.2), βA_0 is referred to as the loop gain. When βA_0 is unity, the circulating signal has enough gain to overcome losses and a finite output signal is produced even when the input is zero. In a real system, the background noise initiates the finite output of the oscillator. This is clearly seen from Eq. (6.2), where the denominator blows up when $\beta A_0 = 1$. The condition $\beta A_0 = 1$ is typically met over a narrow range of frequencies, especially when frequency selective filters are used in the feedback network. Then, the output is in the form of a sinusoidal signal at a well-defined frequency, hence the name "oscillator." In the case of a laser, the feedback is provided by a resonator consisting of highly reflecting mirrors as shown in the sketch in Fig. 6.2.

The frequency selectivity comes from the spectroscopic characteristics of the gain medium, which has absorption and emission bands at well-defined frequencies as discussed in Chap. 4 on light-matter

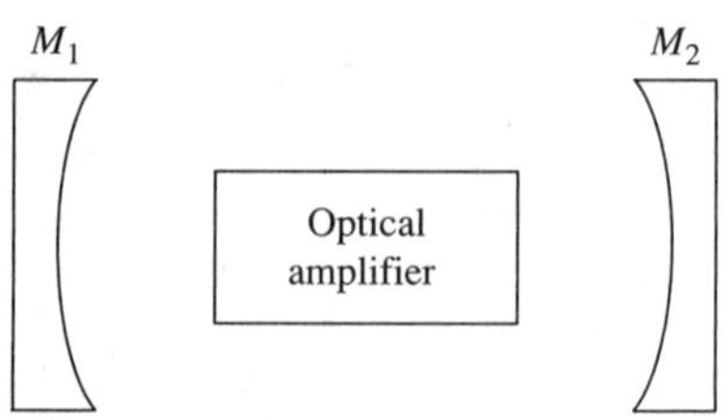

FIGURE 6.2 Sketch of a laser resonator containing a gain medium and highly reflecting mirrors for feedback.

interaction. Wavelength-dependent reflectivity of the end mirrors or use of wavelength-selective intracavity elements such as prisms, diffraction gratings, or etalons can also provide finer frequency selectivity.

6.1.2 Lasers Produce Directional Light

The output beam coming out of a laser resonator is directional. To say the least, the beam does not spread as much as, for example, what emanates from conventional light sources such as incandescent lamps. However, we should note that even in the most ideal case, beam spreading takes place due to diffraction, as was discussed in Chap. 3. When a laser resonator is properly aligned and beam distortion effects such as thermal lensing are minimized, the output beam can be purely TEM_{00}, in which case the spotsize of the beam is given by [see Eq. (3.35)]

$$w(z) = w_0\sqrt{1+\left(\frac{z-z_f}{z_0}\right)^2} \tag{6.3}$$

Here, z_0 is the Rayleigh range, which depends on the beam waist w_0 and the wavelength λ through

$$z_0 = \frac{n\pi w_0^2}{M^2\lambda} \tag{6.4}$$

Here, n is the refractive index and M^2 is the M-square parameter of the beam introduced in Sec. 3.1.10. For a diffraction-limited TEM_{00} laser output, $M^2 = 1$. Again, note that the smaller the wavelength is, the larger the Rayleigh range. In this case, less spreading results for the same initial beam waist. Is the output of a practical laser always diffraction limited? The answer is clearly no. For example, if the cross-sectional area of the gain medium is large enough to support the oscillation of higher-order modes, the output will be multimode, in which case $M^2 > 1$ and the corresponding Rayleigh range will be smaller than that of the ideal Gaussian beam at the same wavelength and with the same beam waist. In such a case, to improve the quality of the output beam, intracavity apertures can be placed to prevent the oscillation of the higher order modes, since these occupy a larger volume than that of the TEM_{00} mode.

6.1.3 Lasers Can Produce Monochromatic Light

Lasers can produce very sharp colors! Stated in more quantitative terms, the spectral bandwidth $\Delta\nu$ of the laser output can be made very small in relation to the center frequency ν. Numerous novel techniques have been proposed and demonstrated for the stabilization of the frequency and phase of lasers,[9, 10] and fractional linewidths

($\Delta\nu/\nu$) of less than 10^{-15} have been achieved.[11] Note however, that typically, spectrally selective elements such as birefringent filters, prisms, or diffraction gratings, as well as active stabilization techniques are necessary to achieve narrow-line output.

6.1.4 Lasers Produce Light with High Spectral Brightness

One of the most important features that distinguishes laser output from other sources of incoherent light such as the sun or light bulbs is what is termed "spectral brightness." The spectral brightness B of a light source is defined as

$$B \equiv \frac{P}{a\Omega\Delta\nu} \tag{6.5}$$

where P is the output power emitted from the source of area a, $\Delta\nu$ is the spectral width of the source, and Ω is the solid-angle subtended by the emitted radiation. To gain an intuitive feel for what type of spectral brightness levels are possible with incoherent and laser sources, we look at two examples: the sun and a He-Ne laser.

For a blackbody emitting over a frequency interval $\Delta\nu$, it can be shown that the spectral brightness can be calculated from (see Prob. 6.1)

$$B = \frac{hc}{2\lambda^3}\frac{1}{\exp(hc/\lambda k_B T)-1} \tag{6.6}$$

where h is Planck's constant, c is the vacuum speed of light, k_B is the Boltzmann constant, λ is the wavelength, and T is the absolute temperature in degrees Kelvin. For the sun, we can take $\lambda = 500$ nm and $T = 5800$ K. This gives

$$\frac{hc}{\lambda k_B T} = \frac{(6.626\times10^{-34})(3\times10^{8})}{(500\times10^{-9})(1.38\times10^{-23})(5800)} = 4.97 \text{ and we get}$$

$$B = \frac{(6.626\times10^{-34})(3\times10^{8})}{2(500\times10^{-9})}\frac{1}{\exp(4.97)-1}$$

$$= 5.58\times10^{-9}\frac{\text{W}}{\text{m}^2-\text{sr}-\text{Hz}} \cong 5.6\times10^{-13}\frac{\text{W}}{\text{cm}^2-\text{sr}-\text{Hz}}$$

To remind the reader, steradian (sr) is the unit of solid angle. Let us next contrast this with the spectral brightness of a modest He-Ne laser with an output power of 0.5 mW and a typical spectral width of $\Delta\nu = 1$ MHz $= 1\times10^{6}$ Hz. Again, we leave it as an exercise for the reader to show that (see Prob. 6.2), at large distances from the waist location, the product of the beam cross-sectional area and solid angle for a Gaussian beam is given by

$$a\Omega \approx \lambda^2 \tag{6.7}$$

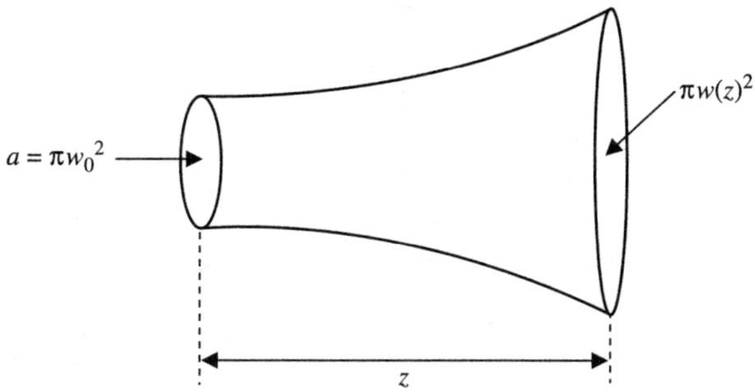

FIGURE 6.3 Sketch of a Gaussian beam, showing the initial beam cross-sectional area *a*.

where $\lambda = 632.8$ nm for the He-Ne laser output. Figure 6.3 shows a sketch of a Gaussian beam that shows the initial beam cross-sectional area *a*. Hence, the spectral brightness factor for the He-Ne laser becomes

$$B = \frac{P}{(a\Omega)\Delta\nu} = \frac{0.5 \times 10^{-3}}{(6328 \times 10^{-10})^2(10^6)} = 1249 \frac{\text{W}}{\text{m}^2 - \text{sr} - \text{Hz}}$$

$$= 0.125 \frac{\text{W}}{\text{cm}^2 - \text{sr} - \text{Hz}}$$

Notice that even though the output power of a laser may be low, the narrow spectral linewidth and directionality result in much higher spectral brightness factors than for incoherent light sources.

6.1.5 Lasers Produce Highly Coherent Light

Again, due in principle to the narrow spectral linewidth of the output, lasers can be designed to produce highly coherent light. Qualitatively, the term "coherence" describes how closely a wave source resembles a perfect sinusoidal wave. If the light source is perfectly coherent, it can be split into beams, which, when superimposed, can form a distinct interference pattern even after a very long distance from the light source. In practice, no wave source is perfectly coherent due to its finite spectral bandwidth. Having a finite spectral width simply means that the source consists of sinusoids with different frequencies over a finite spectral range. Even if these sinusoids are in phase with each other at the source, they will eventually lose the fixed phase relationship since each component acquires a different amount of phase over a given propagation distance. This is irrespective of how narrow the spectral width of the source is. The distance after the source where the fixed phase relationship is lost is known as the coherence length l_c of the source. This is related to the coherence time τ_c through

$$l_c = \tau_c c \tag{6.8}$$

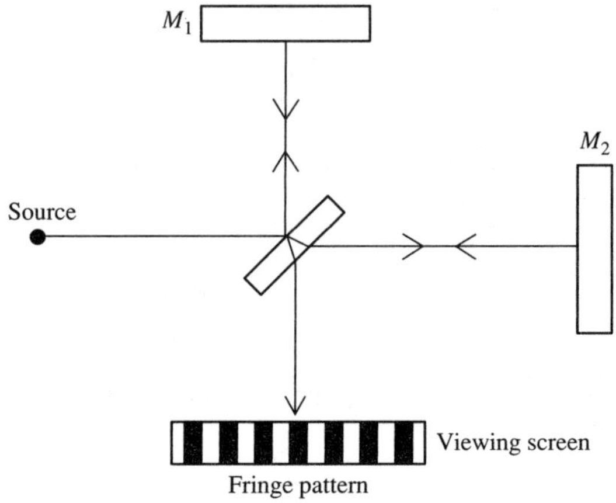

FIGURE 6.4 A Michelson interferometer which can be used to investigate the coherence properties of a light source.

We can roughly define τ_c as the time interval over which the source remains approximately as a sinusoid with frequency equal to the central frequency of the spectrum.

Theory of coherence is among the most challenging areas of photonics. The interested reader can refer to Ref. 12 for a thorough treatment of the subject. To gain an elementary quantitative understanding of coherence, we consider a Michelson interferometer sketched in Fig. 6.4.

In this setup, the beam coming from the light source is split by the beam splitter (B. S.) into two beams which propagate with different delays before being combined by the same beam splitter. The interference of the beams is observed on the viewing screen. Let us assume that the degree of attenuation in each arm is different so that, in general, the amplitudes of the interfering waves are different. Let the electric fields of the waves returning from arms 1 and 2 be $\vec{E}_1$ and $\vec{E}_2$, respectively. The total intensity I_{screen} on the viewing screen can be calculated as

$$I_{\text{screen}} = \frac{1}{2\eta_0}\left\langle \left|\vec{E}_T\right|^2 \right\rangle = \frac{1}{2\eta_0}\left\langle \left(\vec{E}_1 + \vec{E}_2\right)\cdot\left(\vec{E}_1^* + \vec{E}_2^*\right)\right\rangle$$

$$= \frac{\left\langle \left|\vec{E}_1\right|^2 \right\rangle}{2\eta_0} + \frac{\left\langle \left|\vec{E}_2\right|^2 \right\rangle}{2\eta_0} + \frac{1}{2\eta_0}\left\langle 2\,\mathrm{Re}\left[\vec{E}_1\cdot\vec{E}_2^*\right]\right\rangle \tag{6.9}$$

Above, η_0 is the impedance of free space, $\vec{E}_T$ is the total electric field $\left(\vec{E}_1 + \vec{E}_2\right)$, the superscript $*$ denotes the complex conjugate, and $\langle\ \rangle$ denotes the time average of each quantity. To proceed, we assume that our light source is stationary, which means that the choice of the initial time does not affect the average quantities. In other words, a shift in the time origin will leave the time averages invariant. The first two quantities in Eq. (6.9) above, therefore, correspond to the intensities I_1 and I_2 of the individual beams, respectively. If we look at the third term, we note that although $\vec{E}_1$ and $\vec{E}_2$ originate from the same source, they follow different paths before reaching the screen, and hence will undergo different amounts of delay. Let us assume that the time needed for the wave represented by $\vec{E}_1$ to reach the screen is t. If the temporal delay between $\vec{E}_1$ and $\vec{E}_2$ is τ, we can express the intensity on the screen as

$$I_{\text{screen}} = I_1 + I_2 + \frac{1}{2\eta_0} 2\,\text{Re}(\Gamma_{12}(\tau)) \tag{6.10}$$

where $\Gamma_{12}(\tau)$, known as the mutual coherence function, is given by

$$\Gamma_{12}(\tau) = \left\langle \vec{E}_1(t) \cdot \vec{E}_2^*(t+\tau) \right\rangle = \lim_{T \to \infty} \frac{1}{T} \int_0^T dt\, \vec{E}_1(t) \cdot \vec{E}_2^*(t+\tau) \tag{6.11}$$

In Eq. (6.11), T is the time window over which the signal measurement was performed. It is customary to express I_{screen} in the form

$$I_{\text{screen}} = I_1 + I_2 + 2\sqrt{I_1 I_2}\, \text{Re}(\gamma_{12}(\tau)) \tag{6.12}$$

where $\gamma_{12}(\tau)$ is the normalized mutual coherence function defined by

$$\gamma_{12}(\tau) = \frac{\Gamma_{12}(\tau)/2\eta_0}{\sqrt{I_1 I_2}} \tag{6.13}$$

Since $I_i = (1/2\eta_0)\Gamma_{ii}(0)$, $\gamma_{12}(\tau)$ can also be expressed as

$$\gamma_{12}(\tau) = \frac{\Gamma_{12}(\tau)}{\sqrt{\Gamma_{11}(0)\Gamma_{22}(0)}} \tag{6.14}$$

From the Schwartz inequality,[13] we have

$$\left|\Gamma_{12}(\tau)\right| \le \sqrt{\left|\Gamma_{11}(0)\right|\left|\Gamma_{22}(0)\right|} \tag{6.15}$$

which gives us the important result

$$0 \le \left|\gamma_{12}(\tau)\right| \le 1 \tag{6.16}$$

Depending on the value of $\left|\gamma_{12}(\tau)\right|$, the degree of coherence of a light source can be categorized into three cases:

1. Complete coherence: When a source is completely coherent, in other words, when we have a perfectly sinusoidal output

waveform, the normalized mutual coherence function becomes $|\gamma_{12}(\tau)| = 1$.

2. Partial coherence: In this case, the magnitude of the normalized coherence function lies between 0 and 1, satisfying the inequality $0 < |\gamma_{12}(\tau)| < 1$.
3. Complete incoherence: In this case, we are not able to observe any interference fringe and $|\gamma_{12}(\tau)| = 0$.

As we will soon see with a particular example, $\gamma_{12}(\tau)$ is a complex periodic function of τ that varies between the extreme values $|\gamma_{12}(\tau)|$ and $-|\gamma_{12}(\tau)|$. Hence, if we use a detector to register the fringe pattern on the viewing screen, as the detector moves, the intensity of the interference pattern varies from the minimum value I_{min} to the maximum value I_{max} given respectively as

$$I_{\text{min}} = I_1 + I_2 - 2\sqrt{I_1 I_2}\,|\gamma_{12}| \tag{6.17}$$

and

$$I_{\text{max}} = I_1 + I_2 + 2\sqrt{I_1 I_2}\,|\gamma_{12}| \tag{6.18}$$

Hence, we can actually measure the value of $|\gamma_{12}|$ by using a Michelson interferometer, which is balanced so that $I_1 = I_2$. If we define the fringe visibility (also known as fringe contrast) V according to

$$V = \frac{I_{\text{max}} - I_{\text{min}}}{I_{\text{max}} + I_{\text{min}}} \tag{6.19}$$

we find that $V = |\gamma_{12}|$ for the case where $I_1 = I_2$. In words, the experimentally measured fringe visibility in a balanced Michelson interferometer corresponds to the magnitude of the normalized mutual coherence function. This is a very important result that gives us a practical recipe for experimentally determining the degree of coherence of a light source. Also note that if $|\gamma_{12}| = 0$, all coherence between the interfering beams is lost and no fringe pattern is visible. When we say that lasers can be designed to produce coherent light, we mean in a more accurate sense that the output can have a normalized mutual coherence value whose magnitude is close to unity.

Following Ref. 14, we discuss a specific case to show how $\gamma_{12}(\tau)$ is calculated. Suppose that we have a nearly monochromatic light source whose output consists of a train of pure sine waves. The oscillations are interrupted by periodic dephasing events of repetition frequency $1/\tau_c$. In other words, for each time window of duration

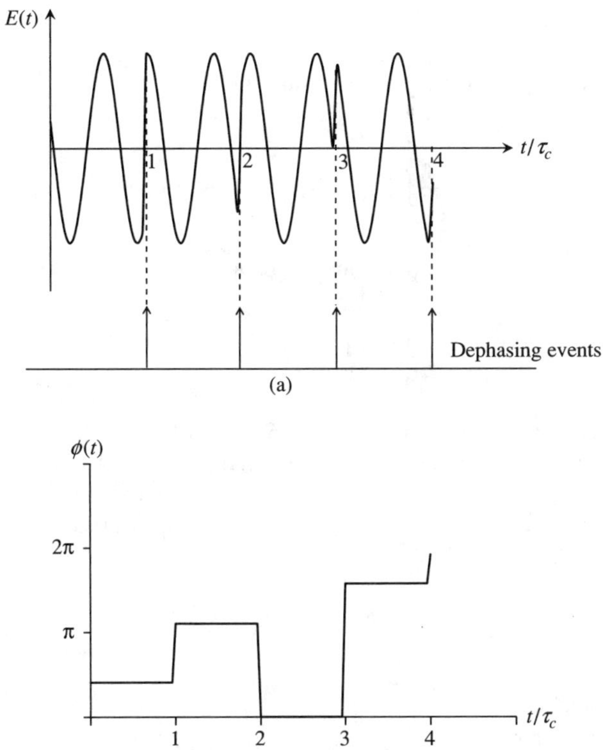

FIGURE 6.5 (*a*) Output of a light source producing sinusoidal waves that are periodically interrupted by dephasing events separated in time by τ_c. (*b*) The corresponding time variation of the phase. Abrupt changes occur at the dephasing events.

τ_c, the output is a pure sine wave. At the end of each time segment of duration τ_c, the oscillation is abruptly interrupted and another pure wave with a random initial phase starts. A sample output of the source is shown in Fig. 6.5*a*.

We can calculate the normalized self-coherence function $\gamma(\tau)$ of the source by taking

$$\vec{E}_1 = \vec{E}_2 = \vec{E}_0 \exp(i\omega t)\exp(i\phi(t)) \tag{6.20}$$

Then, we get

$$\Gamma_{11}(0) = \Gamma_{22}(0) = |E_0|^2 \lim_{T\to\infty} \frac{1}{T}\int_0^T dt(1) = |E_0|^2 \tag{6.21}$$

and

$$\gamma(\tau) = \frac{1}{|E_0|^2} \lim_{T\to\infty} \frac{1}{T} \int_0^T dt \vec{E}_1(t) \cdot \vec{E}_2^*(t+\tau)$$

$$= \frac{1}{|E_0|^2} \lim_{T\to\infty} \frac{1}{T} \int_0^T dt \vec{E}_0 \cdot \vec{E}_0^* \exp(i\omega t) \exp(i\phi(t))$$

$$\times \exp(-i\omega(t+\tau)) \exp(-i\phi(t+\tau))$$

$$= \exp(-i\omega\tau) \lim_{T\to\infty} \frac{1}{T} \int_0^T dt \exp[i(\phi(t) - \phi(t+\tau))] \qquad (6.22)$$

For the case plotted in Fig. 6.5, Fig. 6.6*a* and *b* show the variation of $\phi(t+\tau)$, and $\phi(t) - \phi(t+\tau)$ for the case where $\tau < \tau_c$. In particular, we chose the specific value of $\tau = 0.2\tau_c$. Note that for the first interval $0 < t < \tau_c$, the phase difference is zero over the interval $0 < t < \tau_c - \tau$,

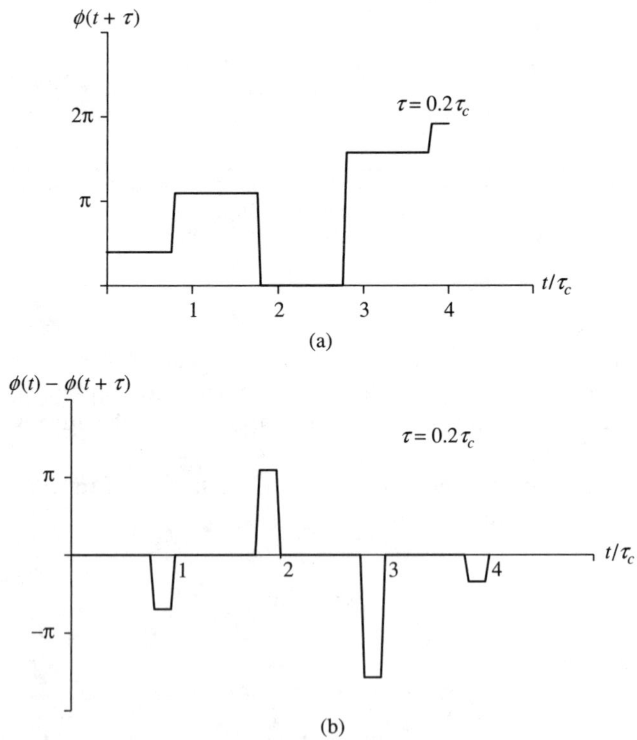

FIGURE 6.6 Calculated variation of (*a*) $\phi(t+\tau)$ and (*b*) $\phi(t) - \phi(t+\tau)$ for the electric field shown in Fig. 6.5*a* and *b*. The delay is taken as $\tau = 0.2\tau_c$.

and then acquires a random value, say $\Delta\phi_1$. This pattern repeats in each of the following segments with a different random phase value. Contrast this with the case $\tau > \tau_c$, where the situation is quite different. A particular case with $\tau = 1.2\tau_c$ is sketched in Fig. 6.7*a* and *b*. We leave it as an exercise for the reader (see Prob. 6.3) to show, by using Eq. (6.22) and based on the observations in Figs. 6.6 and 6.7, that the normalized mutual coherence function $\gamma(\tau)$ becomes

$$\gamma(\tau) = \begin{cases} \left(1 - \tau/\tau_c\right)\exp(-i\omega\tau) & \tau < \tau_c \\ 0 & \tau > \tau_c \end{cases} \tag{6.23}$$

Variation of the magnitude of $\gamma(\tau)$ as a function of delay is further shown in Fig. 6.8. Hence, we obtain the important result that for a light source whose perfectly sinusoidal output is periodically interrupted

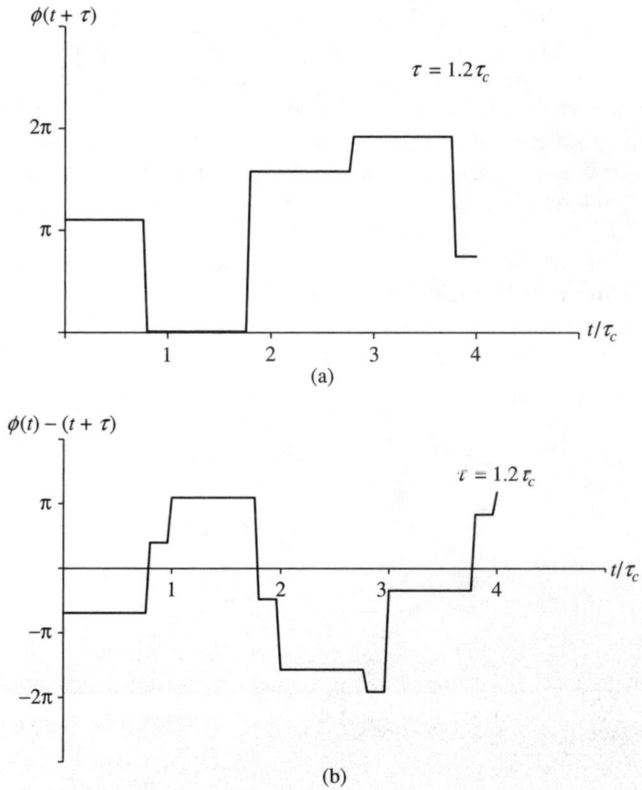

FIGURE 6.7 Calculated variation of (*a*) $\phi(t+\tau)$ and (*b*) $\phi(t) - \phi(t+\tau)$ for the electric field shown in Fig. 6.5*a* and *b*. The delay is taken as $\tau = 1.2\tau_c$.

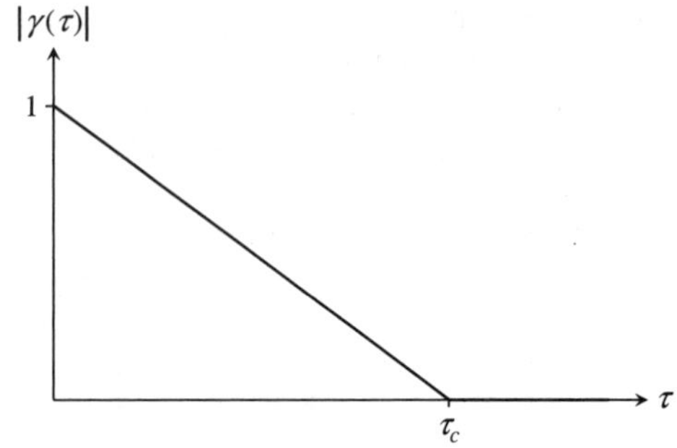

FIGURE 6.8 Variation of the magnitude of $\gamma(\tau)$ as a function of delay for a light source whose perfectly sinusoidal output is periodically interrupted at the frequency $1/\tau_c$.

at the frequency $1/\tau_c$, the mutual coherence becomes zero for delays larger than τ_c. This also gives us a simple intuitive understanding of what is meant by the coherence time. For this particular example, the temporal duration τ_c of an uninterrupted pure sine wave also corresponds to the maximum delay that can be introduced before losing the fringe visibility in the Michelson interference experiment. In terms of distance, the path difference between the interfering beams should not exceed the maximum value of $l_c = c\tau_c$, where l_c is the coherence length defined in Eq. (6.8).

Lastly, to see how the coherence time is related to the spectral distribution of the light source, we continue with the same example and calculate the Fourier transform of a single uninterrupted sine wave of duration τ_c and carrier frequency ω_0 as shown in Fig. 6.9.

The spectral amplitude distribution $E(\omega)$, which can be calculated from

$$E(\omega) = \frac{1}{\sqrt{2\pi}} \int_0^{\tau_c} dt \exp[i(\omega_0 - \omega)t] \tag{6.24}$$

has the spectral intensity distribution $|E(\omega)|^2$ given by (see Prob. 6.4)

$$|E(\omega)|^2 = \frac{2}{\pi} \frac{\sin^2((\omega_0 - \omega)\tau/2)}{(\omega_0 - \omega)^2} \tag{6.25}$$

Normalized variation of $|E(\omega)|^2$ is shown in Fig. 6.9*b*. The corresponding spectral width $\Delta\nu$ (FWHM) becomes

$$\Delta\nu = \left(\frac{4}{2\pi}\right)\frac{1.39}{\tau_c} = \frac{0.88}{\tau_c} \approx \frac{1}{\tau_c} \tag{6.26}$$

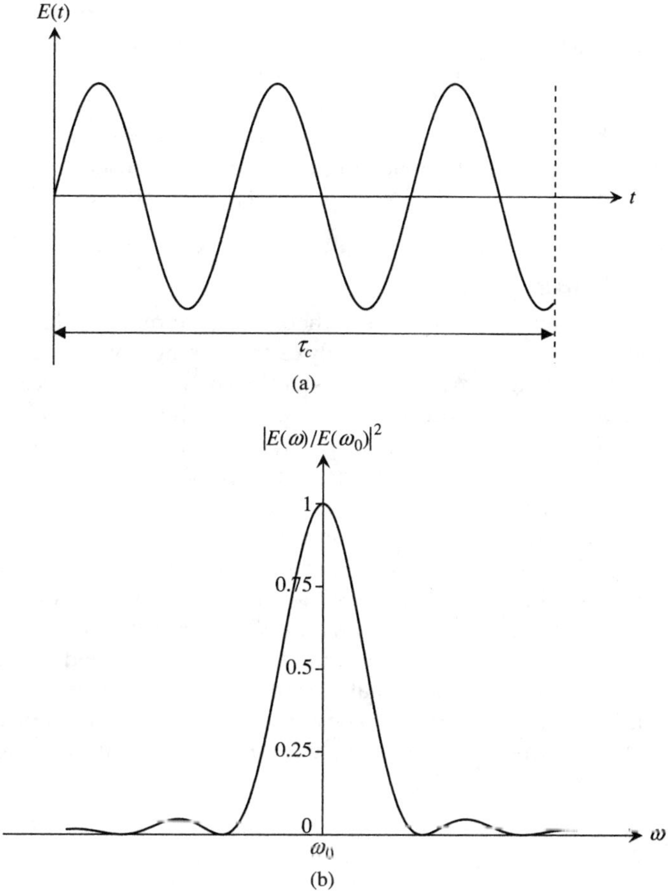

FIGURE 6.9 (*a*) A single sine wave of duration τ_c and (*b*) its normalized spectral intensity distribution.

Hence, based on this particular example, we see that the coherence time of a light source with a spectral width of $\Delta\nu$ is of the order of $1/\Delta\nu$. If the linewidths are specified in wavelength units, the corresponding coherence length l_c can be estimated from

$$l_c = \frac{c}{\Delta\nu} = \frac{c}{\Delta\lambda}\frac{\lambda}{\nu} = \frac{\lambda^2}{\Delta\lambda} \tag{6.27}$$

What we have discussed so far is known as temporal coherence. In other words, this describes the phase relationship between different points of a wave located along the direction parallel to the direction of wave motion. We can also look at how the phase varies for different points located on a plane which is perpendicular to the direction of wave motion. If these points possess a fixed phase relationship, we say that the beam is also spatially coherent. Laser light coming out of resonators typically has a high degree of spatial coherence as well. If higher-order modes oscillate within the gain medium, spatial coherence may be reduced. This can be remedied by introducing pinholes within the cavity to eliminate higher-order mode oscillations and allow the production of the lowest-order TEM_{00} mode.

6.2 Optical Gain

Based on the elementary quantum theory of light matter interactions discussed in Chap. 4, we next analyze the passage of a monochromatic beam through a two-level system and show that optical gain (also called optical amplification) occurs when the upper level population exceeds that of the lower level. We derive an expression for the differential optical gain coefficient and introduce the emission and absorption cross sections.

Let us consider a nearly monochromatic light beam passing through an atomic medium as sketched in Fig. 6.10. For simplicity, the atom is again modeled as a two-level system with N_2 atoms per unit volume in the upper state $|2\rangle$, and N_1 atoms per unit volume in the lower state $|1\rangle$. We further assume that the atomic medium has a length of Δz and a cross-sectional area of a. The light interacting with the atomic system is assumed to be a plane wave. If the light beam is resonant with the atomic system with $h\nu \approx E_2 - E_1$, it will induce

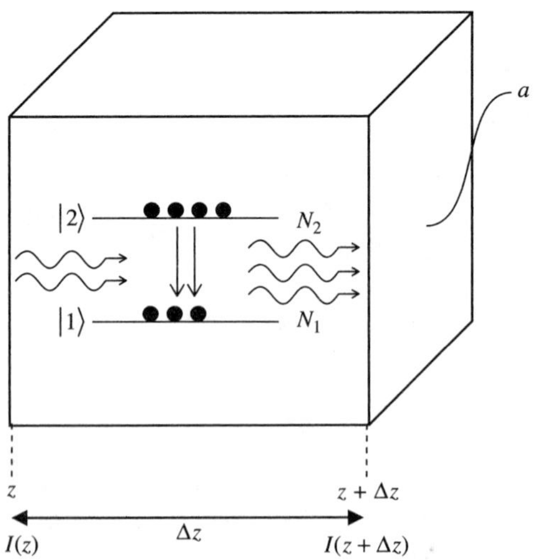

FIGURE 6.10 Passage of a nearly monochromatic light beam through a two-level atomic medium.

downward transitions via stimulated emission, and also upward transitions because of absorption. The photons emitted by stimulated emission will be added to the beam, resulting in an increase of the intensity while absorption will cause a decrease. The net change in the intensity will therefore depend on the atomic populations in the two levels as well as on the strength of stimulated emission and absorption rates. It turns out, we can come up with a simple rule that summarizes whether or not optical gain is possible. In addition to absorption and stimulated emission, we also have spontaneous emission. However, since photons emitted as a result of spontaneous emission go in all directions in a random manner, the contribution to the beam intensity which is propagating in a definite direction will be assumed negligible. In a real system, spontaneous emission will add noise to the emerging beam. We neglect this effect in the present analysis. Let us first look at stimulated emission for the limiting case where the light beam is nearly monochromatic. From Eq. (4.182), we have

$$\frac{dN_2}{dt} = -B_{21}N_2\rho_\nu g(\nu) \tag{6.28}$$

As we also saw earlier [Eq. (4.193)], Einsten's A and B coefficients are further related through

$$\frac{A_{21}}{B_{21}} = \frac{8\pi h\nu^3 n^3}{c^3} \tag{6.29}$$

where n is the refractive index of the medium. Furthermore, since the beam intensity can be expressed as

$$I = \rho_\nu \frac{c}{n} \tag{6.30}$$

Eq. (6.28) becomes

$$\frac{dN_2}{dt} = -\frac{c^3 n I N_2 A_{21}}{n^3(8\pi h\nu^3)c} g(\nu) = \frac{-A_{21}\lambda^2}{8\pi n^2} g(\nu)\left(\frac{I}{h\nu}\right)N_2 \tag{6.31}$$

Hence, the rate of change of the upper-level population can be written in the form

$$\frac{dN_2}{dt} = -\sigma_e\left(\frac{I}{h\nu}\right)N_2 \tag{6.32}$$

where σ_e is known as the stimulated emission cross section and is given by

$$\sigma_e = \frac{A_{21}\lambda^2}{8\pi n^2} g(\nu) \tag{6.33}$$

When expressed in this form, σ_e is in terms of spectroscopic parameters which can be experimentally determined for the system: the radiative decay rate (A_{21}), wavelength of the optical transition, refractive index (n), and the lineshape function $g(\nu)$. ($I/h\nu$) tells us the number of photons incident on the atomic system per unit time per unit area and can be conveniently thought of as the photon flux. Repeating the above analysis for the time rate of change of the population density in the lower level, we get, by using $g_2 B_{21} = g_1 B_{12}$ [Eq. (4.192)]

$$\frac{dN_2}{dt} = B_{12}N_1\rho_\nu g(\nu) = \frac{g_2}{g_1}B_{21}N_1\frac{In}{c}g(\nu)$$

$$= \frac{g_2}{g_1}\frac{cA_{21}\lambda^2}{8\pi h\nu n^3}\frac{In}{c}g(\nu)N_1 = \sigma_a\left(\frac{I}{h\nu}\right)N_1 \tag{6.34}$$

Above, σ_a is the absorption cross section given by

$$\sigma_a = \frac{g_2}{g_1}\frac{A_{21}\lambda^2}{8\pi n^2}g(\nu) \tag{6.35}$$

Now, we can write down the net rate of change of the upper state population density in the absence of spontaneous emission as

$$\frac{dN_2}{dt} = -\sigma_e\left(\frac{I}{h\nu}\right)N_2 + \sigma_a\left(\frac{I}{h\nu}\right)N_1 = -\sigma_e\frac{I}{h\nu}\left(N_2 - \frac{g_2}{g_1}N_1\right) \tag{6.36}$$

From the conservation energy, every atom decaying by stimulated emission will contribute one photon of energy $h\nu$ to the propagating beam. Hence, multiplying the negative of Eq. (6.32) by ($a\Delta z h\nu$) will tell us the increase in the power of the propagating beam due to stimulated emission. A sketch of the geometry is shown once again in Fig. 6.11. Denoting by $P(z)$, the power at the location z, we have

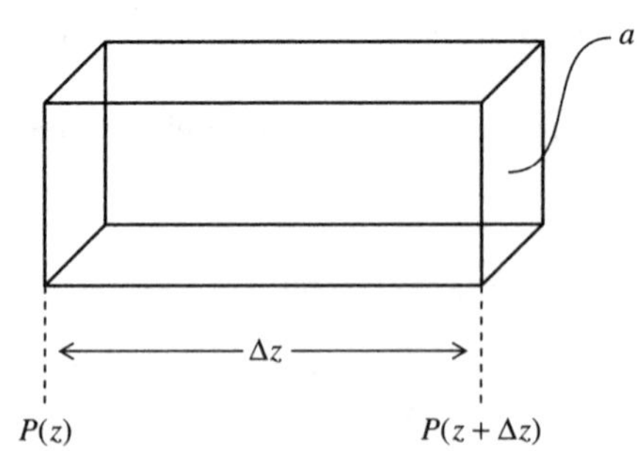

FIGURE 6.11 Sketch of a thin slab of an atomic gain medium showing the incident and transmitted powers.

$$P(z+dz)-P(z)=a(I(z+dz)-I(z))$$
$$=-\left(\frac{dN_2}{dt}\right)(a\Delta z h\nu)=\sigma_e\frac{I}{h\nu}\left(N_2-\frac{g_2}{g_1}N_1\right)a\Delta z h\nu \qquad (6.37)$$

Rearranging Eq. (6.37), we see that the difference in the intensities can be expressed as a derivative in the limit of vanishing thickness, resulting in

$$\frac{I(z+dz)-I(z)}{dz}=\frac{dI}{dz}=\gamma(\nu)I \qquad (6.38)$$

where $\gamma(\nu)$ is the differential gain coefficient given by

$$\gamma(\nu)=\sigma_e\left(N_2-\frac{g_2}{g_1}N_1\right)=\frac{A_{21}\lambda^2}{8\pi n^2}g(\nu)\left(N_2-\frac{g_2}{g_1}N_1\right)$$
$$=\sigma_e\Delta N \qquad (6.39)$$

Above, ΔN is known as the population inversion and is given by

$$\Delta N=\left(N_2-\frac{g_2}{g_1}N_1\right) \qquad (6.40)$$

We have obtained a very important result. Whenever the population inversion is positive ($\Delta N > 0$), the resonant light beam propagating through the atomic system will experience optical amplification, since dI/dz in Eq. (6.38) will also be positive in this case. When the degeneracies of the upper and lower levels are equal ($g_1 = g_2$), population inversion (i.e., $\Delta N > 0$) is achieved by having a larger number of atoms in the upper level than in the lower level. In thermal equilibrium this is not possible, since we know from Boltzmann statistics that the occupation numbers of energy states decrease exponentially with increasing energy according to

$$N_2=N_1\exp\left(-h\nu/k_BT\right) \qquad (6.41)$$

We therefore need a mechanism that raises a sufficiently large number of atoms into the upper state so that the condition $\Delta N > 0$ can be met. Clearly, this is a nonequilibrium case and requires an external energy source to maintain a larger population in the upper state. Methods aimed at achieving population inversion are commonly referred to as optical pumping. In the next section we will investigate optical pumping of ideal two-, three- and four-level systems, and develop guidelines about which energy-level structure is most suitable for obtaining optical gain.

6.3 Optical Pumping

6.3.1 Rate Equations

In the previous section we developed the expressions for the time rate of change of the population densities due to stimulated emission and absorption in terms of cross sections. We can now write down the net rate of change of the upper-state population N_2 in the presence of all three optical processes as follows:

$$\frac{dN_2}{dt} = \sigma_a\left(\frac{I}{h\nu}\right)N_1 - \frac{N_2}{\tau_f} - \sigma_e\left(\frac{I}{h\nu}\right)N_2 \tag{6.42}$$

Above, I is the intensity of the electromagnetic wave which is nearly resonant with this transition, h is Planck's constant, ν is the frequency, σ_a is the absorption cross section, σ_e is the stimulated emission cross section, τ_f is the fluorescence lifetime of the upper state, and N_1 is the population density of the lower level. This form for dN_2/dt shown in Eq. (6.42) is frequently used to model population dynamics and is referred to as the rate equation for upper-state population density. Although we have not shown it explicitly in Eq. (6.42), keep in mind that the cross sections σ_a and σ_e both have frequency dependence as can be seen from Eqs. (6.33) and (6.35).

By using the form of the rate equation in Eq. (6.42), we now examine the optical pumping requirements for ideal two-, three-, and four-level atomic systems.

6.3.2 Ideal Two-Level System

In an ideal two-level system we can neglect light-matter interactions with all other levels except for two, as shown in Fig. 6.12. Here, the system consists of a lower state $|1\rangle$ with energy E_1, and an upper state $|2\rangle$ with energy E_2. The atomic system further interacts with a nearly monochromatic electromagnetic wave of frequency

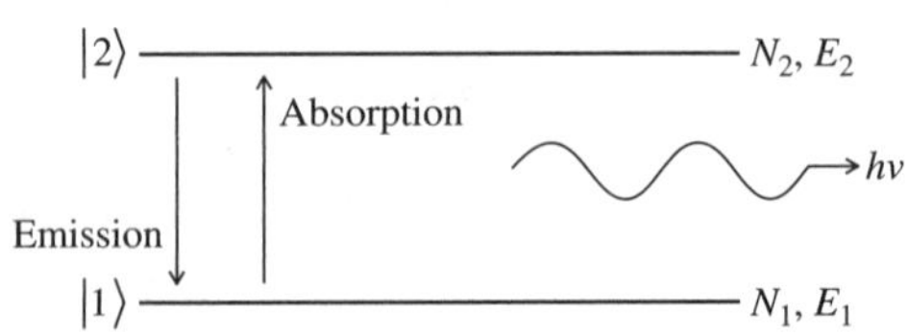

FIGURE 6.12 An ideal two-level atomic system interacting with an electromagnetic wave of frequency $\nu \approx E_2 - E_1/h$.

$\nu \approx E_2 - E_1/h$, which is assumed to overlap with the lineshape function. The population densities of the lower and upper levels are N_1 and N_2, respectively. We further assume that the degeneracy factors of the two levels are equal ($g_1 = g_2$) and the absorption cross section is equal to the stimulated emission cross section ($\sigma_e = \sigma_a$), both designated as σ.

Since only two levels are involved, the rate equation for the upper- and lower-level populations becomes

$$\frac{dN_2}{dt} = -\frac{dN_1}{dt} = -\sigma\frac{I}{h\nu}(N_2 - N_1) - \frac{N_2}{\tau_f} \tag{6.43}$$

with

$$N_1 + N_2 = N_t \tag{6.44}$$

where N_t is the total population density of atoms. Since there are only two levels involved, the same beam with intensity I causes stimulated absorption and emission. We leave it as an exercise for the reader (see Prob. 6.5) to show that the steady-state population inversion ΔN for the two-level atom can be written as

$$\Delta N = \frac{-N_t}{1 + \dfrac{I}{I_s}} \tag{6.45}$$

where I_s is the saturation intensity which can be expressed in terms of the atomic parameters as

$$I_s = \frac{h\nu}{2\sigma\tau_f} \tag{6.46}$$

We can make several important conclusions about the ideal two-level system based on the results of Eqs. (6.45) and (6.46):

1. In an ideal two-level system, the population inversion is always negative. Hence, it is never possible to achieve optical amplification in this case. In more realistic systems, we can have what are termed as quasi-three-level systems in which there are two manifolds of closely spaced energy levels. It turns out that one can obtain amplification between different levels of such a two-manifold atomic system. Many examples based on solid-state gain media activated with ytterbium or thulium ions, for example, exist. Further information on the modeling of population dynamics in quasi three-level lasers may be found in Refs. 15 to 17.

2. Note that as the intensity I increases, saturation occurs and the magnitude of population inversion decreases, asymptotically approaching zero at large intensities. Saturation of population inversion, and similarly gain, are very important phenomena which are essential in describing the steady-state operation of lasers. This is an effect which we could not account for in our classical analysis of light-matter interactions in Chap. 4. Physically, we have nearly all of the atoms in the ground state at very low intensities. As the intensity increases, more atoms are raised to the upper state due to absorption and this causes a decrease in the magnitude of inversion. However, this cannot go on indefinitely. At very large intensities, stimulated emission and absorption rates dominate over the spontaneous decay rate and equalize the populations of the two levels since we assumed that the cross sections and hence, the decay rates, for absorption and emission are equal.
3. We have identified I_s as the saturation intensity. Physically, this is the intensity level where a noticeable departure begins to occur from the small-signal value ($-N_t$) of the population inversion. We will see later that saturation intensity also plays an important role in the determination of the output power of a laser operated in the so-called continuous-wave mode with constant output power as a function of time.

6.3.3 Ideal Three-Level Systems

We need the interaction of more than two levels with the electromagnetic radiation in order to achieve population inversion. The simplest energy-level scheme that enables the creation of population inversion is the ideal three-level system depicted in Fig. 6.13.

Looking at Fig. 6.13, it is apparent that a three-level system consists of three energy levels with some special characteristics. The pump source, be it a flashlamp, an electrical discharge process, or similar source, first excites atoms from the ground state $|1\rangle$ to the upper pump level $|3\rangle$. The spontaneous decay rate from level $|3\rangle$ to level $|2\rangle$ occurs fast enough so that we assume the atomic populations reside either in the upper level $|2\rangle$ or the ground state $|1\rangle$. Hence, the

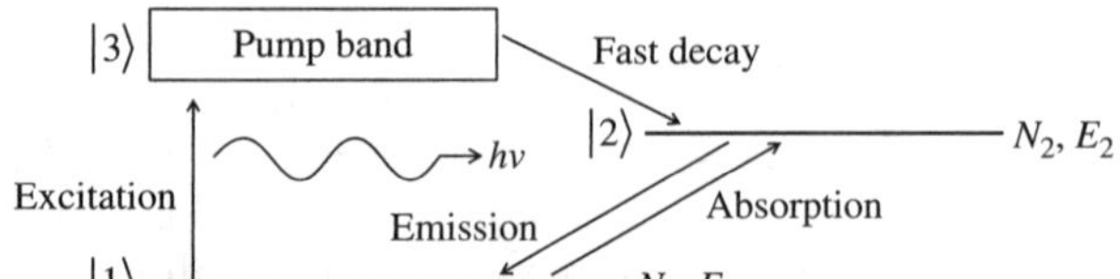

FIGURE 6.13 Energy-level diagram of an ideal three-level atomic system.

population density N_3 is negligible. Again, a nearly monochromatic electromagnetic wave with frequency $\nu \approx E_2 - E_1/h$ interacts with the two levels as sketched in Fig. 6.13. The rate equation for the upper-level population density has an additional pumping term and can be written in the form

$$\frac{dN_2}{dt} = -\sigma\frac{I}{h\nu}(N_2 - N_1) - \frac{N_2}{\tau_f} + W_p N_1 \tag{6.47}$$

In Eq. (6.47), the pumping rate per atom W_p has the units of inverse time. It can take on different forms depending on the nature of the excitation. For example, if we are using optical excitation with another light source such as a flashlamp or another laser with an average optical frequency of ν_{31}, W_p becomes

$$W_p = \frac{I_p}{h\nu_{31}}\sigma_a \tag{6.48}$$

where σ_a is now the absorption cross section for excitation from the ground state $|1\rangle$ to the upper pump level $|3\rangle$, and I_p is the pump intensity at the frequency ν_{31}. For the sake of generality, we leave the pumping rate as W_p. Later, when we revisit the rate equations for four-level solid-state systems, we will express the pumping rate in terms of the absorption cross section and pump light intensity as shown in Eq. (6.48).

Again, we leave it as an exercise for the reader to show that (see Prob. 6.6) the steady-state population inversion is given by

$$\Delta N = \frac{\Delta N_0}{1 + \frac{I}{I_s}} \tag{6.49}$$

where the small-signal inversion ΔN_0 and the saturation intensity I_s are given by the respective terms

$$\Delta N_0 = \frac{N_t[W_p\tau_f - 1]}{1 + W_p\tau_f} \tag{6.50}$$

and

$$I_s = \frac{h\nu(1 + W_p\tau_f)}{2\sigma\tau_f} \tag{6.51}$$

We note that in contrast to two-level systems, it is now possible to achieve population inversion in a three-level system, provided that the condition $W_p\tau_2 > 1$ is satisfied. This means that, as can be expected intuitively, the rate at which atoms are excited by the pump to the upper level $|2\rangle$ should exceed the spontaneous decay rate from this level. Also, similar to the previous case, we again observe saturation as the intensity increases. However, this time, there is additional dependence of the saturation intensity on the pumping rate as well, as can be seen from Eq. (6.51).

6.3.4 Four-Level Atomic Systems

Four-level atomic systems have more favorable features for obtaining population inversion and low-threshold laser operation. A schematic of an ideal four-level atomic system is shown in Fig. 6.14. A pumping mechanism first raises the atoms from the ground state $|g\rangle$ to the pump band $|3\rangle$. For an ideal four-level system, the decay rates from the pump band to the upper level $|2\rangle$, and from the lower level $|1\rangle$ to the ground state, are very fast. As a result, we can neglect the populations in levels $|3\rangle$ and $|1\rangle$ (i.e., $N_3 = N_1 \approx 0$). Hence, any number of atoms placed in level $|2\rangle$ will readily provide population inversion between states $|2\rangle$ and $|1\rangle$, since $\Delta N = N_2$ in this case. The nearly monochromatic electromagnetic wave with frequency $\nu \approx E_2 - E_1/h$ further interacts with the system. The rate equation for the population density of level $|2\rangle$ becomes

$$\frac{dN_2}{dt} = W_p N_g - \sigma \frac{I}{h\nu}(N_2 - N_1) - \frac{N_2}{\tau_f} \tag{6.52}$$

where N_g is the ground-state population and the other parameters are as defined in the previous sections. Since the atoms are either in the ground state $|g\rangle$ or the upper level $|2\rangle$, we further have

$$N_2 + N_g = N_t \tag{6.53}$$

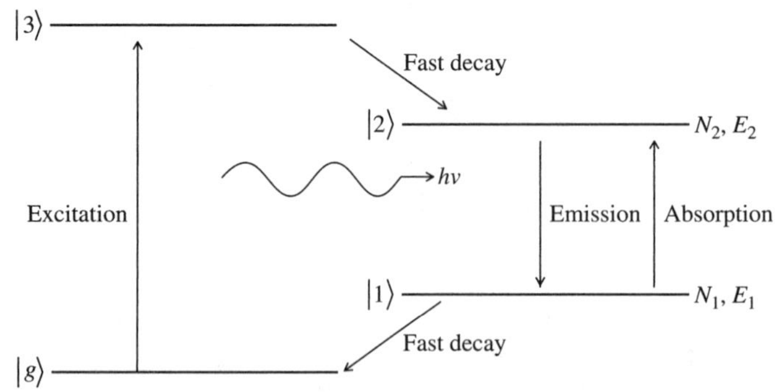

FIGURE 6.14 Energy-level diagram of an ideal four-level atomic system.

Here, N_t is the total atomic density. Again, it can be shown that (see Prob. 6.6) in steady state, the population inversion can be expressed in the form

$$\Delta N = \frac{\Delta N_0}{1 + \frac{I}{I_s}} \tag{6.54}$$

where the small-signal inversion ΔN_0 and the saturation intensity I_s are given by

$$\Delta N_0 = \frac{N_t \quad W_p \tau_f}{1 + W_p \tau_f} \tag{6.55}$$

and

$$I_s = \frac{h\nu}{\sigma \tau_f}(1 + W_p \tau_f) \tag{6.56}$$

Hence, we conclude from this rate-equation analysis that, in the presence of pumping, an ideal four-level atomic system always has population inversion. This does not mean that a four-level atomic system, when pumped, always has optical amplification, because the signal typically experiences attenuation due to scattering or other competing absorption processes in a real situation. To obtain net amplification, the pumping level (given by the magnitude of W_p) should be above a certain threshold value. We will investigate this further in the discussion on laser oscillation. We also note that when the pumping level is low (i.e., $W_p \tau_f << 1$), the expression for the saturation intensity in Eq. (6.56) simplifies to

$$I_s = \frac{h\nu}{\sigma \tau_f} \tag{6.57}$$

We will provide examples of three-level and four-level laser systems in Sec. 6.5.2.

6.4 Laser Oscillation

Let us suppose that a suitable optical gain medium and a pumping mechanism are available to produce amplification at a desired wavelength. The gain medium can be solid, liquid, or gas. In this section, we want to provide a qualitative discussion of how laser oscillation can be obtained by answering the following questions in a systematic way:

1. What is the condition that must be satisfied in order to obtain laser oscillation? Related to this, how much pump power is required to obtain lasing?

2. When the pump power is increased above the threshold, how do we determine the output power of the laser?
3. Does the presence of optical gain influence in any way the oscillation frequency of a laser?
4. Does the nature of spectral broadening (homogeneous or inhomogeneous) influence the frequency of the laser output?

6.4.1 Condition for Lasing Threshold

As we discussed before, the differential gain coefficient γ can be written in the form

$$\gamma = \sigma_e \Delta N \tag{6.58}$$

where the stimulated emission cross section σ_e is a material parameter, and ΔN is the population inversion. For both three-level and four-level systems, we saw in the earlier sections that ΔN can be written in the form

$$\Delta N = \frac{\Delta N_0}{1 + I/I_s} \tag{6.59}$$

Above, ΔN_0 is the inversion that can be achieved at very low levels of laser intensity, and I_s is the saturation intensity. Combining Eqs. (6.58) and (6.59), the differential gain coefficient becomes

$$\gamma = \frac{\gamma_0}{1 + I/I_s} \tag{6.60}$$

where γ_0 $(\gamma_0 = \sigma_e \Delta N_0)$ is the small-signal differential gain coefficient. Due to its intensity dependence, γ is commonly referred to as the saturated gain coefficient since, as a result of increased stimulated emission rate at high signal intensities, the differential gain coefficient decreases with increasing intensity. From Eq. (6.60), we see that when $I = I_s$, the saturated gain coefficient becomes half as large as the small-signal value. The intensity dependence of γ is shown in Fig. 6.15.

So, let's address the first question. To obtain light oscillation, we need to combine the optical amplification mechanism with positive feedback. In the case of lasers, this is done by using an optical resonator consisting of highly reflecting mirrors. The simplest resonator configuration consisting of only two mirrors is schematically shown in Fig. 6.16*a*.

Note that here, R_1 and R_2 represent the intensity reflection coefficients of the mirrors M_1 and M_2, respectively. The small signal gain coefficient γ_0 of the gain medium depends on the pumping level.

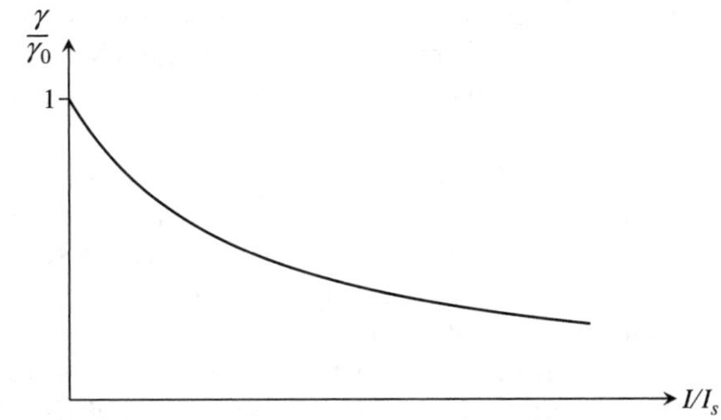

FIGURE 6.15 Variation of the differential saturated gain coefficient γ as a function of intensity.

Typically, there will be loss mechanisms within the gain medium due to absorption by impurities, scattering, or surface imperfections. We can use a differential loss coefficient α_L to represent all sources of loss inside the gain medium. One way to reduce losses due to surface reflections involves putting an antireflection coating on the two surfaces of the gain medium. Another possibility is to use the gain medium at Brewster incidence, as shown in Fig. 6.16*b*. In this configuration, the circulating beam hits the surfaces of the gain medium (assumed to be a parallel slab) at Brewster incidence, in which case the TM polarization with the electric field oriented parallel

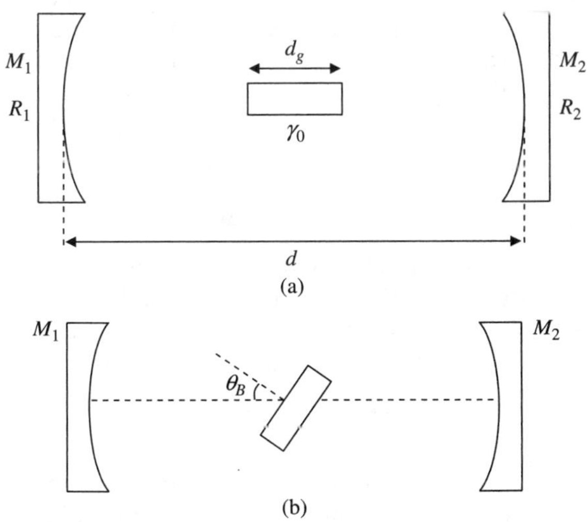

FIGURE 6.16 (*a*) Sketch of a two-mirror laser resonator. (*b*) A two-mirror resonator with a gain medium positioned at Brewster incidence.

to the plane of incidence suffers only minimal losses. Since the TE polarization has higher reflective losses than the TM polarization, this resonator configuration is ideally suited for generating polarized light output from the laser as well.

The end mirrors of the resonator must be well aligned so that photons can complete a round trip and come back to the same point. Suppose that some number of photons emitted by spontaneous emission and moving close to the optical axis of the resonator begin circulating inside the resonator. The small-signal gain coefficient γ_0 needs to have a minimum value before the total losses during one round trip are overcome and oscillation can start. Let us follow the intensity of the circulating beam once around the cavity. At low intensities, the small-signal differential gain coefficient has no intensity dependence and the intensity of the passing beam will grow exponentially with distance according to

$$I(z) = I(0)\exp[(\gamma_0 - \alpha_L)z] \tag{6.61}$$

We have obtained Eq. (6.61) by using Eq. (6.38), and also by noting that in the presence of attenuation, the net small-signal gain will be $\gamma_0 - \alpha_L$. Denoting by I_m and I_{m+1} the intensities at some point within the cavity after the mth and $(m+1)$th round trips, we have

$$I_{m+1} = I_m R_1 R_2 \exp(2d_g(\gamma_0 - \alpha_L)) \tag{6.62}$$

where d_g is the length of the gain medium. In addition to the exponential growth through the gain medium, Eq. (6.62) also takes into account the resonator losses due to the finite transmission of the end mirrors. At the threshold of oscillation, $I_{m+1} = I_m$. In other words, the round-trip optical gain is just large enough to overcome the losses in the gain medium and losses due to output coupling from the partially reflecting end mirrors. Hence, at threshold, we have

$$R_1 R_2 \exp(2d_g((\gamma_0)_{th} - \alpha_L)) = 1 \qquad \text{or} \qquad (\gamma_0)_{th} = \alpha_L + \frac{1}{2d_g}\ln\left[\frac{1}{R_1 R_2}\right] \tag{6.63}$$

Equation (6.63) tells us the intuitively obvious result that the lower the losses and the higher the end mirror reflectivities, the lower will be the threshold small-signal gain. Having determined $(\gamma_0)_{th}$, we can find the threshold inversion from

$$(\Delta N_0)_{th} = \frac{(\gamma_0)_{th}}{\sigma_e} \tag{6.64}$$

Then, depending on whether we have a three-level or a four-level laser system, we can determine the threshold pumping rate $(W_p)_{th}$ needed to achieve this population inversion by using the rate-equation results of Sec. 6.3.

There are many possible resonator geometries that can be used in order to obtain laser oscillation, in addition to the two-mirror resonator shown in Fig. 6.16*a* and *b*. Figure 6.17 shows three other possibilities. Figure 6.17*a* and *b* shows a three-mirror and a z-type four-mirror resonator, respectively. These, as well as the two-mirror resonator shown in Fig. 6.16, are standing-wave resonators in which the circulating beam makes two passes through the gain

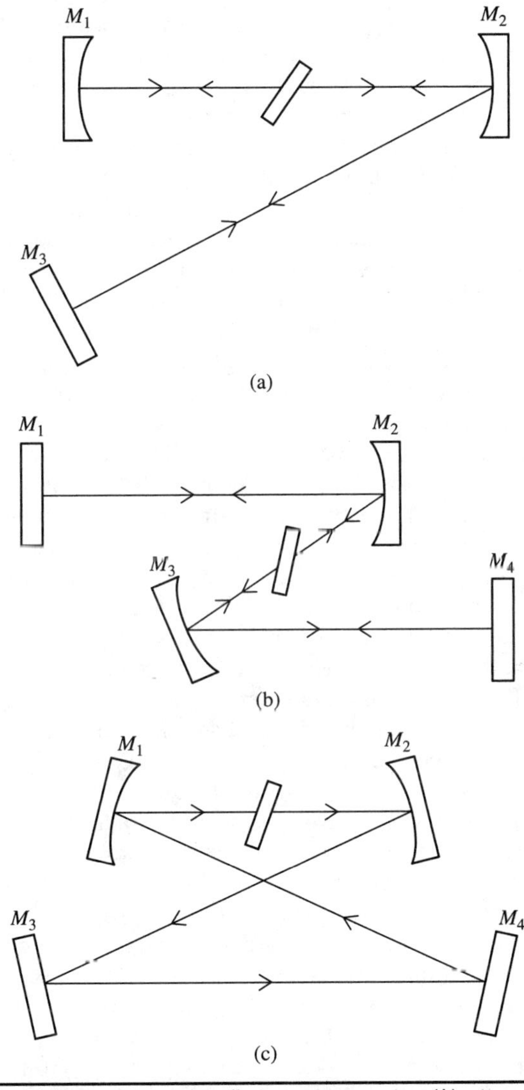

FIGURE 6.17 (*a*) Three-mirror standing-wave resonator, (*b*) z-type four-mirror standing-wave resonator, and (*c*) x-type four-mirror traveling-wave resonator.

medium during one round trip. The x-type four-mirror resonator shown in Fig. 6.17*c* is a traveling-wave type. Here, the circulating beam passes through the gain medium only once during each round trip. The three- and four-mirror resonators provide more freedom in beam size control in the gain medium by adjusting the arm lengths and radii of curvature of the curved mirrors.

6.4.2 Output Power of the Laser

Regarding our third question, gain saturation plays a very important role in determining the output power coming from the laser. Once oscillation threshold is passed with $\gamma_0 > (\gamma_0)_{th}$, the circulating beam inside the cavity sees exponential gain every time it passes through the gain medium. After a sufficient number of round trips, the intensity grows to a level where it becomes comparable to the saturation intensity I_s. When this happens, the differential gain coefficient begins to decrease from its small-signal value, due to increased rate of stimulated emission, until the saturated gain coefficient is again equal to the total losses. So, in steady state, assuming a constant output intensity as a function of time, we now have $\gamma = (\gamma_0)_{th}$. This condition is satisfied at some finite level of intracavity intensity, say I_0. Hence, the output intensity, say going through the second mirror, will be $I_0(1-R_2)$. From a knowledge of the effective beam area inside the gain medium, we can then determine the output power from the laser. Later, we will provide a more quantitative analysis of how the output power is calculated.

Above the lasing threshold, one can obtain different regimes of laser oscillation in terms of the time dependence of the output power. During continuous-wave (cw) operation, the output power of the laser remains constant (or nearly constant given various sources of fluctuation in a real system) as a function of time as depicted in Fig. 6.18*a*. In Fig. 6.18*a*, P_0 denotes the constant output power of the laser. If a time-dependent loss or gain mechanism is present, it is possible to operate the laser in pulsed mode where the output comes out as a series of pulses as shown in Fig. 6.18*b*. If a pulsed pumping source is used to generate the optical gain, we say that the laser is gain switched. If a time-dependent loss is introduced, the laser is said to operate in the Q-switched mode. The pulse duration (shown as τ_p in Fig. 6.18*b*) and the repetition frequency (this corresponds to $1/T_0$, where T_0 is the period of the pulse train shown in Fig. 6.18*b*) are typically variable and can be controlled by changing the pump energy, loss level, cavity length, and so on. Pulsewidths as short as tens of picoseconds have been generated from Q-switched lasers[18]. More commonly, they are in the nanosecond or microsecond range. A different type of pulsed operation results when the laser is operated in the mode-locked regime. In this case, a periodic pulse train is produced with a repetition frequency (this corresponds to $1/T_R$, where T_R is the period of the pulse train shown in Fig. 6.18*c*) equal to the free spectral range ν_{FSR} of the resonator ($\nu_{FSR} = c/2d$, d = cavity length) (see Fig. 6.18*c*). Here, typical pulse durations (shown as τ_p in Fig. 6.18*c*) are in the picosecond to

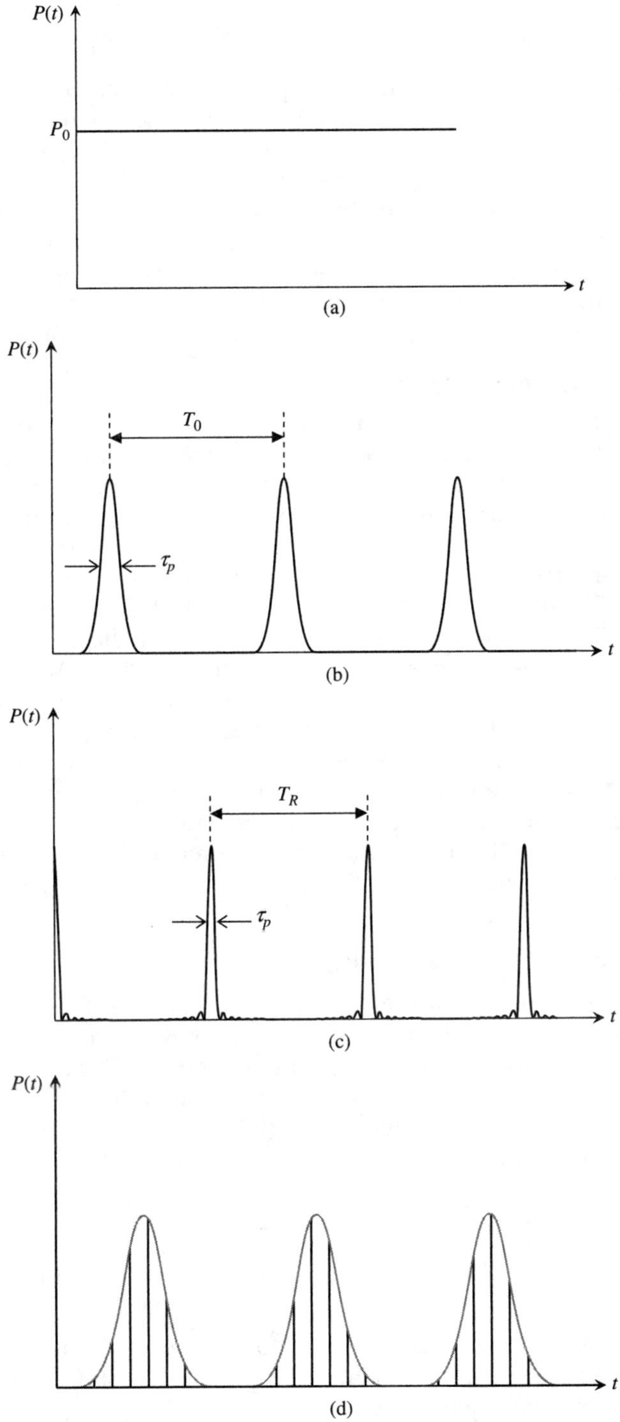

Figure 6.18 Output power of a laser during (*a*) continuous-wave (cw) operation, (*b*) pulsed (Q-switched or gain-switched) laser oscillation, (*c*) mode-locked operation, and (*d*) Q-switched mode-locked operation.

femtosecond range. In fact, more complicated temporal power variation is also possible, as in the case of Q-switched mode-locked operation, where the periodic mode-locked train at the repetition frequency of $\nu_{FSR} = c/2d$ appears under a Q-switched envelope (Fig. 6.18*d*). We will investigate these regimes of operation in the upcoming sections. We note in passing that by using what have become known as semiconductor saturable absorber mirrors (SESAM)[19] or saturable Bragg reflectors (SBR)[20] inside a laser resonator, it is possible to operate the same laser in these different pulsed regimes by varying the level of pumping.

6.4.3 Frequency Pulling in Laser Oscillators

We would naively expect at the first sight that laser oscillation would occur at the frequencies equal to the empty-cavity resonance frequencies discussed in Chap. 3. Actually, this is not a bad approximation for certain types of lasers, especially those with broad gain bandwidths such as tunable solid-state gain media. However, this assumption is not in general correct due to an effect commonly referred to as frequency pulling. As such, the oscillation frequencies of a laser containing an amplifying medium are shifted from those of the empty-cavity resonant modes due to the dispersion of the amplifier. Actually, this should come as no surprise, since we saw exactly the same effect in Chap. 4 when we discussed the classical theory of absorption resonances.

To find an expression for the shifted lasing resonant frequencies, we first revisit the results of Chap. 4 and assume a homogeneously broadened medium. In particular, we found that the resonant contribution to the differential absorption coefficient $\alpha_r(\nu)$ could be expressed in the form [see Eqs. (4.43) and (4.48)]

$$\alpha_r(\nu) = \frac{Ne^2}{4n_0 m_e \varepsilon_0 c} g_L(\nu) \tag{6.65}$$

where $g_L(\nu)$ is the Lorentzian lineshape function, N is the atomic density, e is the electron charge, n_0 is the background refractive index, m_e is the electron mass, ε_0 is the dielectric permittivity of vacuum, and c is the speed of light. We also found that the deviation of the frequency-dependent refractive index $n(\nu)$ from the background value n_0 could be calculated from [see Eq. (4.37)]

$$\Delta n(z) = n(\nu) - n_0 = \frac{Ne^2}{4n_0 m_e \varepsilon_0}\left(\frac{\nu_0 - \nu}{\nu}\right)\frac{1}{4\pi^2\left[(\nu_0 - \nu)^2 + \left(\frac{\Delta\nu}{2}\right)^2\right]} \tag{6.66}$$

By recognizing the functional form for the normalized Lorentzian lineshape function $g_L(\nu)$ from Eq. (4.47), we see that $\Delta n(\nu)$ in Eq. (6.66) can be expressed in terms of $\alpha_r(\nu)$ as

$$\Delta n = \frac{c}{2\pi\nu}\left(\frac{\nu_0 - \nu}{\Delta\nu}\right)\alpha_r(\nu) \tag{6.67}$$

Here, $\Delta\nu$ is the spectral width (FWHM) of the resonance. In the case of an amplifier with a differential gain coefficient of $\gamma(\nu)$, all we have to do is replace $\alpha_r(\nu)$ in Eq. (6.67) by $-\gamma(\nu)$, resulting in

$$\Delta n = \frac{-c}{2\pi\nu}\left(\frac{\nu_0 - \nu}{\Delta\nu}\right)\gamma(\nu) \tag{6.68}$$

Taking the length of the gain medium as d_g, the corresponding phase shift acquired by the passing beam will be given by

$$\begin{aligned}\Delta\phi &= k\,(n_0 + \Delta n)\,d_g = \frac{2\pi\nu}{c}\,(n_0 + \Delta n)\,d_g \\ &= \frac{2\pi\nu}{c}n_0 d_g - d_g\left(\frac{\nu_0 - \nu}{\Delta\nu}\right)\gamma(\nu)\end{aligned} \tag{6.69}$$

As we did in Chap. 3, we can calculate the frequencies of the axial modes by assuming a TEM_{00} spatial mode, and by requiring that the round-trip phase shift due to free-space propagation of the beam, as well as due to gain dispersion, will be an integral multiple of 2π. This leads to

$$2(d + (n_0 - 1)d_g)k_m + 2\frac{\nu_m - \nu_0}{\Delta\nu}\,\gamma(\nu_m)d_g = m2\pi \tag{6.70}$$

where m is an integer, $k_m = (2\pi\nu_m)/c$ is the wave vector corresponding to the mth resonance frequency (ν_m), and d is the total length of the laser resonator. Note that we have obtained a nonlinear implicit equation for ν_m, since $\gamma(\nu_m)$ also has frequency dependence. By rearranging Eq. (6.70) and by neglecting d_g in relation to d, we obtain

$$\nu_m = m\frac{c}{2d} + \frac{\nu_0 - \nu_m}{\Delta\nu}\,\gamma(\nu_m)\frac{cd_g}{2\pi d} \tag{6.71}$$

Suppose that the width $\Delta\nu$ is large compared to $(\nu_0 - \nu_m)$ so that the shift produced in ν_m is small. Without exactly solving Eq. (6.71), we can still gain a qualitative understanding of what happens to the oscillation frequency in the presence of gain dispersion. If $\nu_m > \nu_0$, the shift will be negative, meaning that ν_m will become smaller than $m(c/2d)$ and approach ν_0 (since ν_0 is less than ν). If ν_m was smaller than ν_0 to begin with, the shift would be positive, again pushing ν_m toward ν_0. We see that in either case, the presence of gain dispersion tends to pull the cavity oscillation frequency toward the center of the gain profile. This is known as frequency pulling in the jargon of laser physics. The effect is most pronounced in gas lasers where the gain linewidths $\Delta\nu$ are narrow. This is clearly seen from Eq. (6.71). In laser oscillators where the departure of the oscillation frequency from the line center is very small in comparison with the gain bandwidth $\Delta\nu$, we can neglect frequency pulling effects.

6.4.4 Laser Output Frequency

Now we explore the influence of homogeneous and inhomogeneous broadening on the spectral characteristics of a laser oscillator. Based on the analysis of the previous section, let us assume that the gain bandwidth is much larger than the separation of the oscillation frequencies from the line center so that we can neglect frequency pulling effects. As we discussed earlier in Chap. 3, the resonator supports the growth of electromagnetic modes at the resonant cavity frequencies given by Eq. (3.189). For the sake of simplicity, let us assume that only the TEM_{00} spatial mode is allowed to oscillate. In this case, the axial modes of the resonators will be equally spaced in frequency with a separation of $c/2d$ between two adjacent modes. Once the pumping rate is raised above the threshold level, the number of oscillating modes will depend on whether the gain medium is homogeneously or inhomogeneously broadened.

First, consider a homogeneously broadened gain medium with a Lorentzian lineshape. This is schematically shown in Fig. 6.19*a*. In an ideal homogeneously broadened gain medium, all atoms have the same spectral response. This means that the cavity mode closest to the gain peak begins to oscillate and uses up all of the gain at the expense of all the other modes. As saturation increases, the whole gain curve shifts down such that at the resonant frequency closest to the maximum of the curve, the saturated gain equals the net cavity losses (see Fig. 6.19*b*). No other mode has enough gain to oscillate, and we have a single oscillating mode. In reality, however, many factors cause deviations from an ideal homogeneous response, and

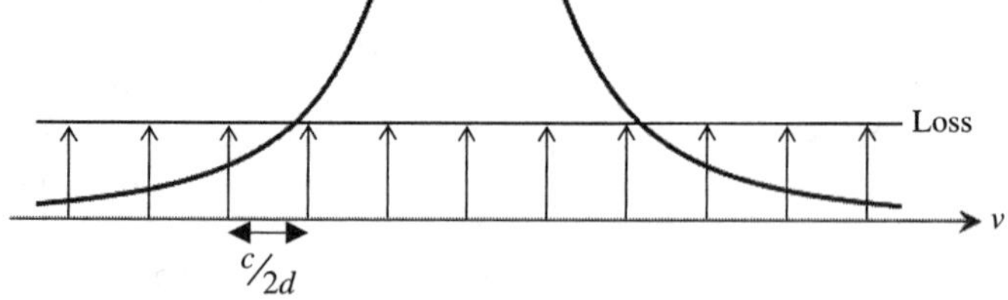

FIGURE 6.19 (*a*) Small-signal, homogeneously broadened gain profile with axial modes separated by $c/2d$. (*b*) Saturated gain profile of a homogeneously broadened gain medium in steady state showing the oscillation of only one mode.

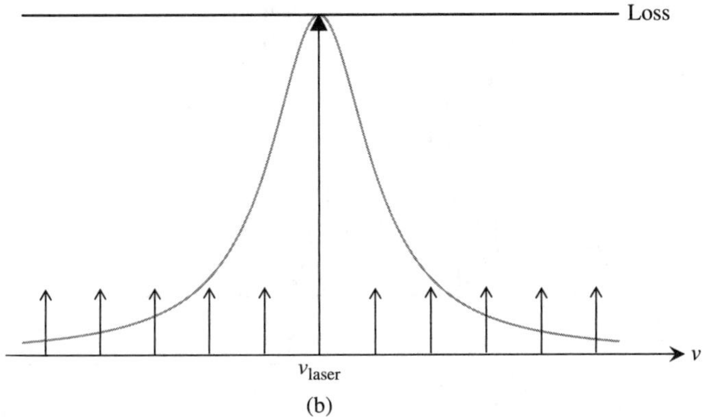

FIGURE 6.19 (*Continued*)

forcing the laser to operate in a single longitudinal mode requires quite a bit of careful design!

In an inhomogeneously broadened gain medium, the situation is quite different. Different groups of atoms will have a different spectral response and will experience saturation effects independently. Looking at Fig. 6.20*a*, the modes at frequencies $\nu_1 - \nu_7$, each with gain above the loss level, can all oscillate simultaneously and form the steady-state saturated gain profile depicted in Fig. 6.20*b*. Hence, inhomogeneously broadened laser gain media will support multi-longitudinal-mode operation. If the spatial mode is purely TEM_{00}, then the spacing between two adjacent oscillating modes will be $c/2d$ when frequency pulling effects are negligible.

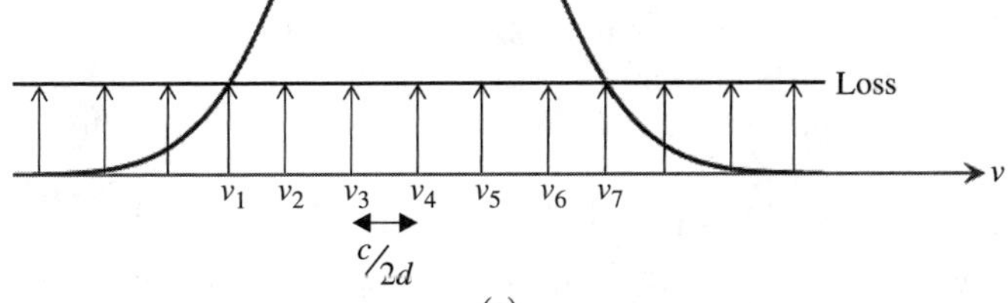

FIGURE 6.20 (*a*) Small-signal, inhomogeneously broadened gain profile with axial modes separated by $c/2d$. (*b*) Saturated gain profile of an inhomogeneously broadened gain medium in steady state. Here, modes $\nu_1 - \nu_7$ will simultaneously oscillate.

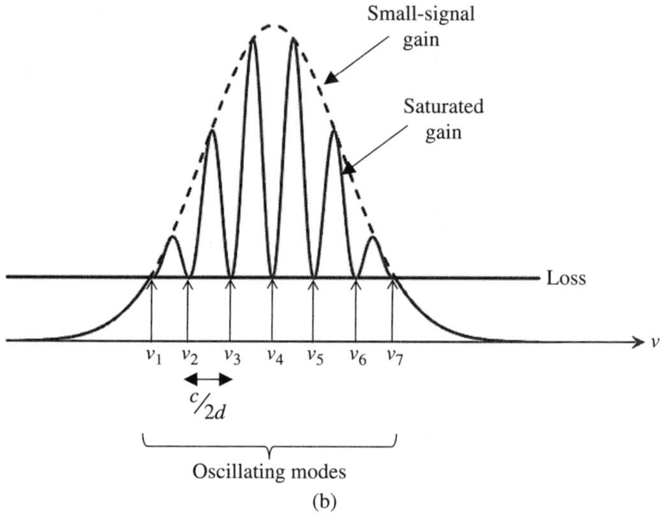

FIGURE 6.20 (*Continued*)

6.5 Solid-State Lasers

To date, lasing has been successfully achieved in gaseous, liquid, and solid media. For a thorough treatment of different types of lasers and the underlying mechanisms which result in optical amplification, we refer the reader to Refs. 3, 4, and 21. In this section, we will specifically focus on solid-state lasers. We will first discuss the general characteristics of solid-state lasers. Then, based on the rate equations developed earlier, we will revisit the analysis of an optically pumped four-level system to derive the coupled differential equations satisfied by the pump and laser beam intensities. These are useful in the modeling of solid-state gain, as well as saturable absorber media. We will then make the thin-length approximation and obtain the coupled differential equations satisfied by the gain and instantaneous power. With this set of equations, we will then analyze the cw and pulsed operation of lasers. Finally, we will provide an elementary treatment of mode locking. The rate equations which we will specifically develop for the analysis of solid-state lasers can in fact be used to model the behavior of a wide variety of laser systems. Later, in Chap. 7, we will provide a detailed treatment of semiconductor lasers.

6.5.1 General Characteristics of Solid-State Lasers

In solid-state lasers, optical amplification is produced by using insulating single-crystals, ceramics, or glasses doped with special laser-active ions. In fact, the ruby laser, which was the very first laser invented in 1960, belongs to this category.[22] Because solid-state lasers offer the advantages of durability and long operational lifetimes, they have long been among the preferred candidates for many applications in science and technology. We will only provide a descriptive picture of solid-state laser materials here without going into a detailed quantum mechanical analysis. For a thorough review of atomic spectroscopy and physics of solid-state laser materials, see Refs. 23 through 27.

The medium into which the active ions are doped is known as the host. Several families of crystals including garnets, fluorides, silicates, tungstates, vanadates, among others, as well as different types of glasses (silicates, phosphates, borates, etc.) have been used as hosts.[28] Naturally occurring crystals typically suffer from the presence of imperfections as well as impurities which cause excessive losses. Therefore, laser-grade host crystals are prepared by using synthetic growth methods. The host must possess many favorable physical properties in order to qualify as a robust medium for light amplification. These include high optical transparency, i.e., very low background absorption and scattering, in the lasing region of the active ion; high mechanical strength; resistance to fracture; resistance to moisture; chemical stability; high heat conductivity; and so on. Of course, it is always a great challenge to find all of these virtues in the same host! We note in passing that although semiconductors are also solids, in the generally adopted classification of lasers, the term "solid-state" refers only to ion-doped solids. One important distinguishing characteristic is that whereas solid-state lasers are optically excited by using flashlamps, arc lamps, or other lasers, semiconductor lasers typically operate via electrical excitation.

What about the types of laser-active ions? Typically, ions from the rare earth or the transition metal series of the periodic table are used in solid-state lasers. Under most circumstances, doping concentrations are low, hardly exceeding a few percent. When optically excited with flashlamps or other lasers, strong absorption results between the electronic levels of the active ions, producing population inversion and enabling the amplification of light. By using different ion-host combinations, it is possible to access different wavelength regions of the electromagnetic spectrum, from the ultraviolet to mid-infrared. The spectral characteristics of solid-state lasers vary depending on whether rare-earth or transition metal ions are used.

In the case of rare-earth ion-doped solid-state lasers (examples of ions include Nd, Ho, Yb, Pr, Tm, and Er), the laser-active 4f electron is electrostatically shielded from the surrounding lattice by the electrons in the outer closed xenon-type shell. This screening minimizes the coupling of the laser-active electron with the surrounding lattice vibrations. As a result, optical emission/absorption bands are relatively sharp and/or the resonance wavelengths of the ion do not vary significantly from one host to another. Let us look at the specific example of the yttrium aluminum garnet crystal (abbreviated as YAG, chemical formula $Y_3Al_5O_{12}$) doped with trivalent neodymium (Nd^{3+}) ions, commonly designated as Nd^{3+}:YAG or simply Nd:YAG. This is one of the most widely used solid-state gain media, first reported in

1964.[29] Incorporation of Nd^{3+} ions into the YAG host causes the formation of several absorption bands which correspond to different optical transitions originating from the ground state to higher energy levels of the ion. Figure 6.21*a* and *b* shows the absorption bands of the Nd:YAG crystal (doped with 1 mol.% neodymium) in the 300 to 900 nm and 2300 to 2400 nm range. The data were gathered in the author's laboratory. Note the sharpness of the absorption resonances. Closely spaced lines that are visible in the absorption spectrum of Nd:YAG occur because of the splitting of the otherwise degenerate quantum

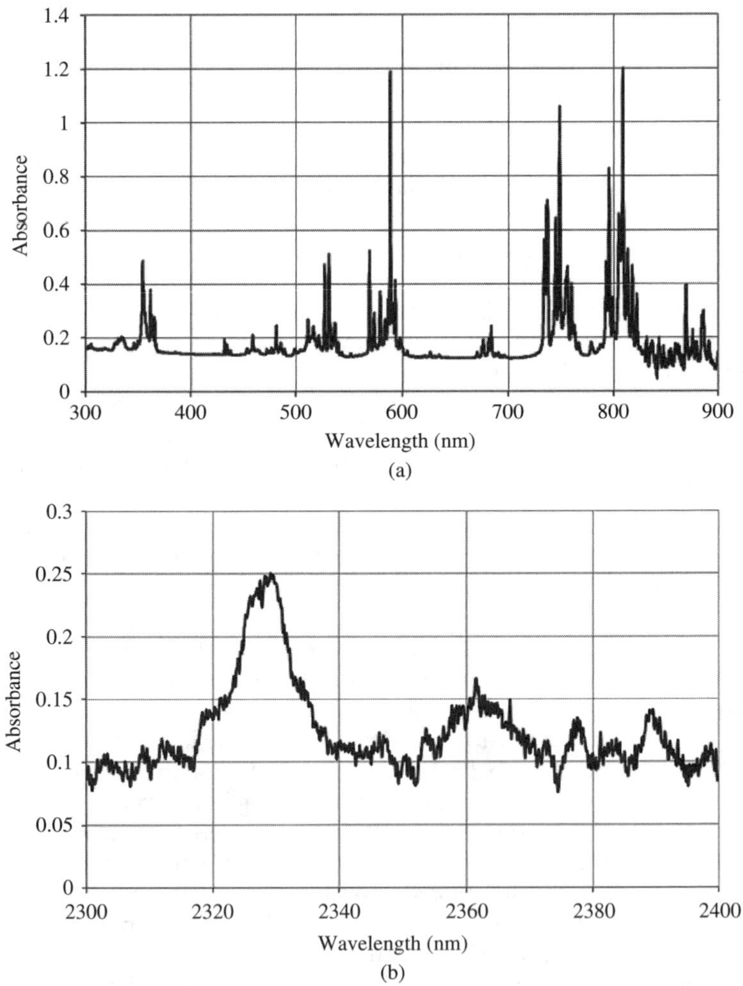

FIGURE 6.21 Measured optical density of Nd:YAG over the (*a*) 300–900 nm and (*b*) 2300–2400 nm range. (*c*) Measured optical density of Nd:tellurite glass over the 300–900 nm range.

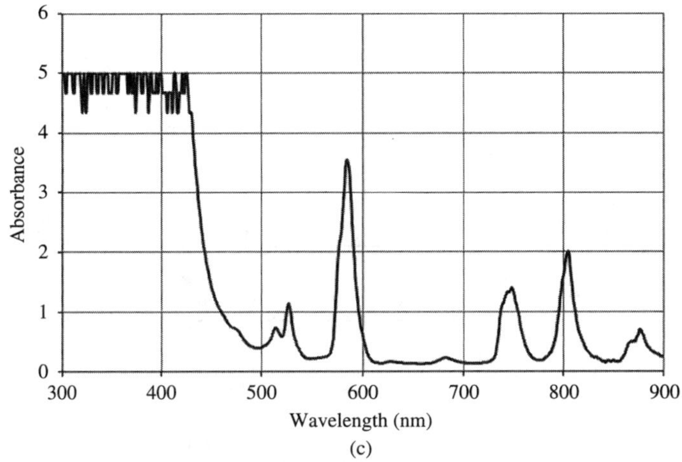

FIGURE 6.21 (*Continued*)

states; the splitting originating from the interaction of the ion with the neighboring host ions. In both cases, we have plotted the variation of the measured optical density (OD), defined as Log(1/T), where T is the transmission of the samples. Strictly speaking, the optical density here does not fully represent the losses due to resonant absorptions, since the data were not corrected for attenuation resulting from surface reflections and scattering. However, the location and general spectral features of the absorption bands are still clearly visible. In the case of Fig. 6.21*b*, the data were more noisy since the absorption strength of this band is rather weak. Contrast the absorption spectrum of Nd:YAG in Fig. 6.21*a* with the absorption spectrum of neodymium-doped tellurite glass over the same spectral range from 300 to 900 nm shown in Fig. 6.21*c*. Due to the amorphous nature of the glass host, there is additional spectral broadening in comparison with that in a perfectly ordered single crystal, resulting in wider bands. However, due to weak electron-phonon coupling, the wavelengths of the band centers remain nearly the same in both hosts. This is an important characteristic of rare-earth ion-doped solid-state lasers, as mentioned earlier. We note in passing that other rare earth ions such as ytterbium or thulium possess broader absorption and emission bands in relation to those of neodymium. However, again due to the weak ion-lattice coupling mentioned above, the center wavelengths of these transitions do not vary significantly from one host to another.

From a full knowledge of the absorption spectrum, it is also possible to construct the energy-level diagram of the ion. The data shown in Fig. 6.21*a* and *b* are not complete since absorption bands exist in other wavelength ranges as well. The energy-level diagram of Nd: YAG showing the low-lying quantum states is depicted in Fig. 6.22. Splitting of the levels is not shown in Fig. 6.22. Each level is designated by using the standard spectroscopic notation $^{2S+1}L_J$, where L is

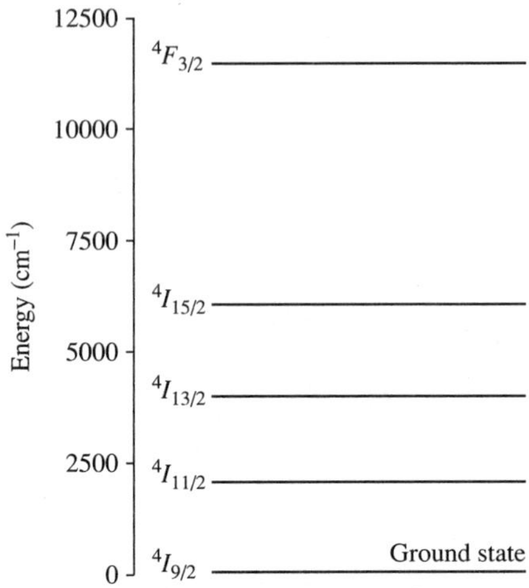

FIGURE 6.22 Energy level diagram of the neodymium ion in YAG showing the approximate energy levels of the low-lying quantum states (adapted from Ref. 28).

the total orbital angular momentum number, S is the total spin, $(2S+1)$ is the multiplicity of the atomic term, and J is the total angular momentum number (sum of L and S). L can only take integer values 0, 1, 2, 3, 4, 5, 6, and so on, represented by the respective symbols S, P, D, F, G, H, I. As an example, the ground state of Nd:YAG is $^4I_{9/2}$, meaning that $L = 6$, $S = 3/2$, and $J = 9/2$. Note that the absorption band near 2330 nm in Fig. 6.21*b* represents the transition from the ground state $^4I_{9/2}$ to the upper level $^4I_{13/2}$. Also note that the energy-level diagram of Fig. 6.22 does not show the levels for which the transition wavelengths are below 800 nm. Suppose that we excite Nd: YAG with light near the wavelength of 808 nm. This will raise the ions from the ground state to the pump level that populates the $^4F_{3/2}$ upper state. Ions decaying from the $^4F_{3/2}$ level to the lower levels shown in Fig. 6.22 will give rise to various transitions in the near infrared wavelengths, the strongest emission being near 1064 nm. The corresponding emission spectrum in the 1050 to 1090 nm wavelength range is shown in Fig. 6.23. Downward transitions originating from the $^4F_{3/2}$ level to the $^4I_{9/2}$, $^4I_{11/2}$, $^4I_{13/2}$ levels give rise to the emission bands near 946, 1064, and 1319 nm, respectively.

Let us now turn our attention to solid-state hosts doped with transition metal ions such as Ni^{2+}, V^{2+}, Co^{2+}, Ti^{3+}, Cr^{3+}, and so on. Lasers based on these active ions are referred to as tunable solid-state lasers.[30] Many examples exist, including Ti^{3+}: sapphire,[31] Co^{2+}:MgF_2,[32] Cr^{4+}:forsterite,[33] and so on. In particular, we will look at the spectroscopic characteristics of ZnSe crystal doped with divalent chromium ions, designated as Cr^{2+}:ZnSe or Cr:ZnSe.[34] Here, lasing occurs as a result of the transitions between the electronic levels of the 3D electrons.

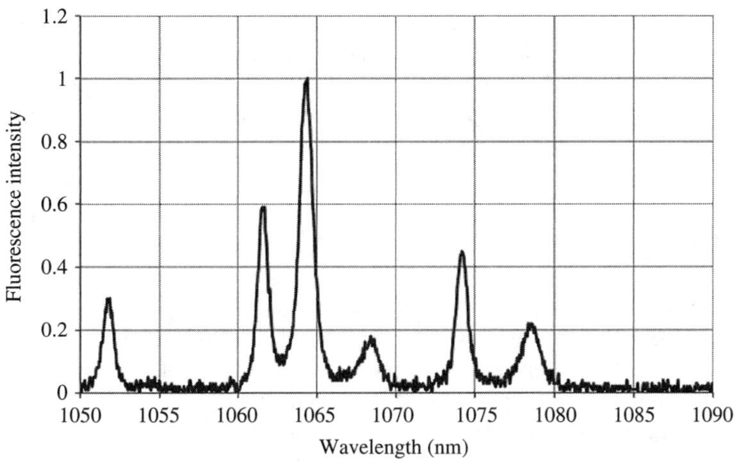

FIGURE 6.23 Emission spectrum of Nd:YAG from 1050 to 1090 nm. The excitation wavelength is 808 nm.

Because the laser active ion is not shielded from the surrounding vibrating lattice, strong electron-phonon coupling takes place, giving rise to broadly tunable output. To provide a simplified quantum mechanical picture of electron-phonon coupling, let us assume that $\hat{H}_0$ is the hamiltonian operator representing the total energy of the ion in the absence of lattice vibrations. Of course, $\hat{H}_0$ consists of several interaction terms due to the nucleus, electron-electron repulsion, spin-orbit coupling, the neighboring crystal field, and so on. Without going into the details, $\hat{H}_0$ will have a set of electronic energy eigenvalues E_i and eigenstates $|\psi_i\rangle$ which satisfy

$$\hat{H}_0|\psi_i\rangle = E_0|\psi_i\rangle \tag{6.72}$$

If we now take into account the lattice vibrations as well, the total hamiltonian $\hat{H}$ of the ion-lattice system can be written in the form

$$\hat{H} = \hat{H}_0 + \hat{H}_L \tag{6.73}$$

where $\hat{H}_L$ represents the energy due to the vibrating crystal. $\hat{H}_L$ can be written as

$$\hat{H}_L = V_L + \sum_i \frac{\hat{P}_i^2}{2M_i} \tag{6.74}$$

where V_L is the interionic potential energy, and $\hat{P}_i$ is the momentum operator of the ith ion of mass M_i. Under the Born-Oppenheimer approximation, which neglects ionic motions in comparison with electronic motion, the two hamiltonian operators $\hat{H}_0$ and $\hat{H}_L$ are not coupled, meaning that the total energy of the electron-lattice system will be the sum of the eigen energies of $\hat{H}_0$ and $\hat{H}_L$. For small oscillations of the lattice ions around their equilibrium locations, V_L will have a quadratic dependence on the displacement. Hence, the energy eigenvalues E_L of the lattice hamiltonian will be given by the familiar harmonic oscillator expression

$$E_L = h\nu_L\left(n+\frac{1}{2}\right) \tag{6.75}$$

where h is Plank's constant and n is an integer that gives the number of lattice quanta of energy, each of which is called a "phonon." ν_L is the classical frequency of the ionic vibration modes, also known as the phonon frequency. Hence, the total energy E of the ion-lattice system can be approximated as

$$E = E_i + h\nu_L\left(n+\frac{1}{2}\right) \tag{6.76}$$

Equation (6.76) states that the total energy of the ion-lattice system has equally spaced energy levels due to quantized lattice vibrations on top of the electronic energy levels. We can use what is known as a "configuration coordinate diagram" to show the energy levels in the presence of electron-phonon coupling. Figure 6.24 shows an example of a single configuration-coordinate diagram of an ion coupled to a vibrating lattice. Here, we used the parameter x to designate the configuration coordinate, which simply refers to the distance between the laser-active ion and a neighboring ion. When the eigen energy of the static lattice is E_1, the equilibrium distance is x_{01}. The corresponding lowest energy of the ion-lattice system is then $E = E_1 + 1/2h\nu_L$.

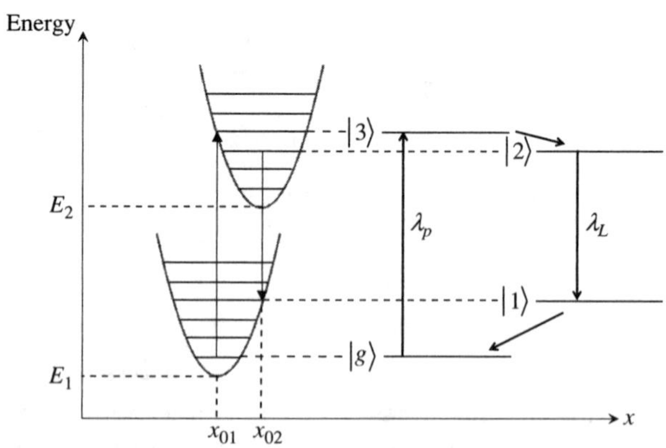

FIGURE 6.24 Single configuration-coordinate diagram showing the coupling of a laser-active ion with the vibrating lattice.

Now, suppose that a resonant photon is absorbed and the ion is excited to a higher state, say the next electronic eigen energy E_2 of the static lattice. In this case, the strength of the lattice coupling also changes and a new equilibrium position x_{02} is reached.

To see the effect of electron-phonon coupling on the emission spectrum, let us assume that the ion originally in the ground state $|g\rangle$ absorbs a pump photon at the wavelength of λ_p and goes up to the excited state $|3\rangle$. From state $|3\rangle$, very rapid nonradiative decay occurs to the lowest vibrational level $|2\rangle$ by the emission of phonons. This causes heat generation within the host and no photons are emitted. It turns out that the state $|2\rangle$ is the upper laser level. From here, either spontaneous or stimulated emission can take the ion to the lower laser level $|1\rangle$ to generate photons at the lasing wavelength of λ_L. From the lower level $|1\rangle$, rapid nonradiative decay again takes place, bringing the ion back to the ground state. Hence, we see another example of a four-level system with the additional effect of lattice vibrations. Because the number of emitted phonons varies during emission, and because there is a distribution of phonon frequencies in a real crystal, the absorption and emission spectra are broadened. Furthermore, the location of the central emission wavelength as well as the amount of phonon-induced broadening will vary greatly from one host to another, even for the same type of laser-active ion. As an example, Fig. 6.25*a* and *b* shows the absorption and emission bands of Cr:ZnSe. Note the significantly wider widths of these bands in comparison with those of Nd:YAG shown in Figs. 6.21 and 6.23.

Once excited to the upper level, the chromium ions will decay by spontaneous emission to the lower level in the absence of stimulated

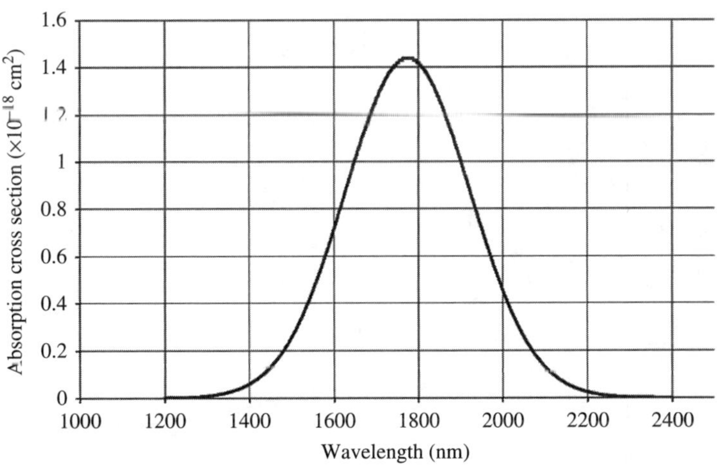

(a)

FIGURE 6.25 (*a*) Absorption and (*b*) emission spectra of Cr:ZnSe.

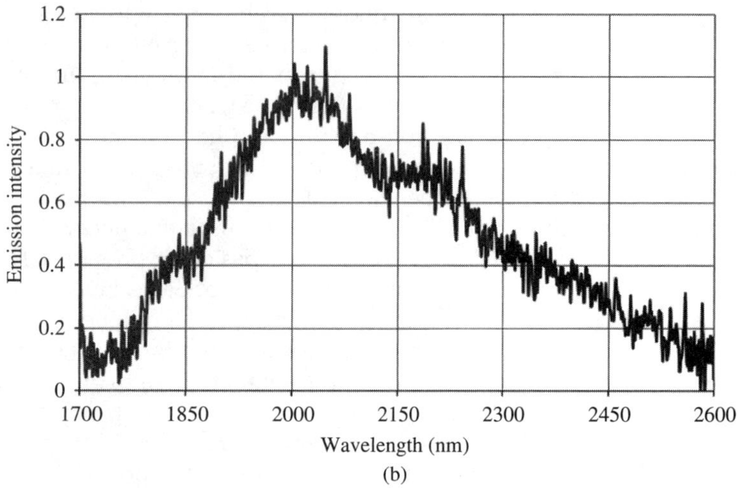

Figure 6.25 (*Continued*)

processes. Figure 6.26 shows the measured fluorescence signal for a sample of Cr:ZnSe with a chromium ion concentration of $15 \times 10^{18}/\text{cm}^3$. The samples were excited with very short pulses at the wavelength of 1570 nm. Note that the fluorescence intensity decays exponentially, as expected for a spontaneous decay process. The exact value of the fluorescence lifetime turns out to depend on the crystal temperature as well as on the active ion concentration. For the particular sample shown in Fig. 6.26, the fluorescence lifetime is about 4 μs. Figure 6.27*a* and *b* shows how the fluorescence lifetime of Cr:ZnSe varies as a function of the concentration of chromium ions and the

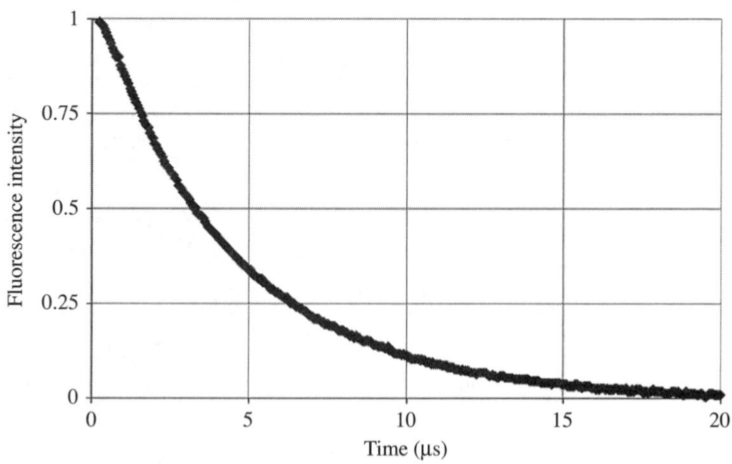

Figure 6.26 Measured fluorescence intensity as a function of time for Cr:ZnSe.

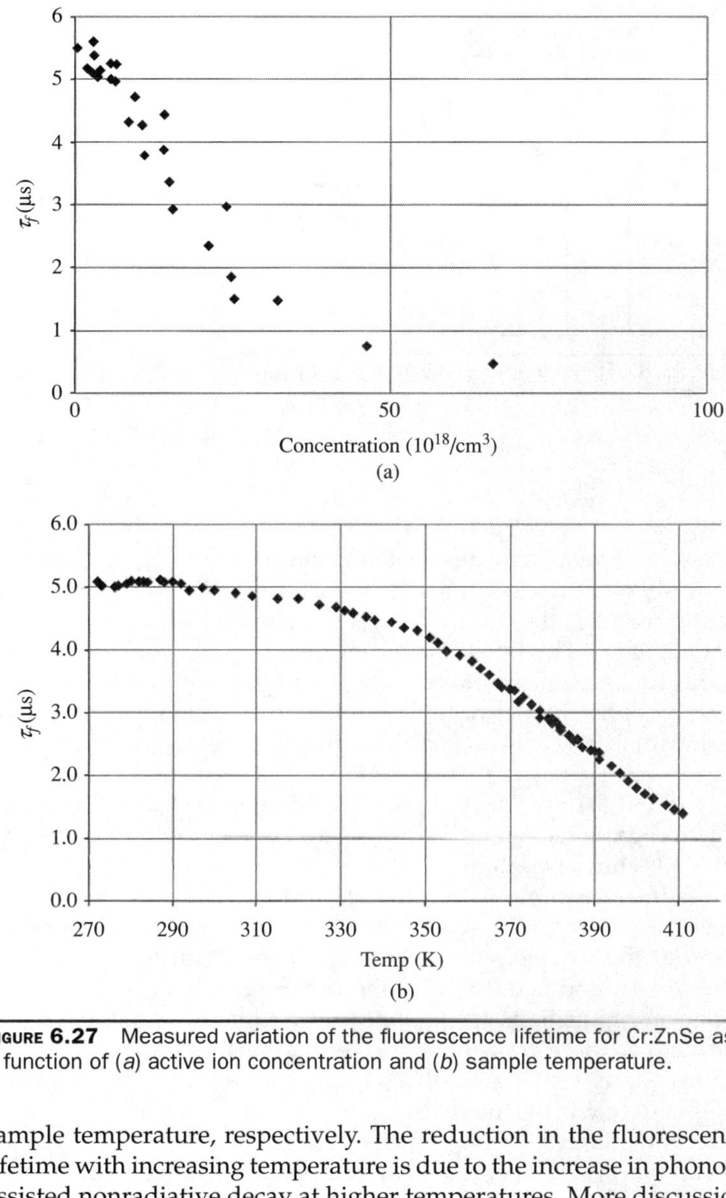

FIGURE 6.27 Measured variation of the fluorescence lifetime for Cr:ZnSe as a function of (*a*) active ion concentration and (*b*) sample temperature.

sample temperature, respectively. The reduction in the fluorescence lifetime with increasing temperature is due to the increase in phonon-assisted nonradiative decay at higher temperatures. More discussion on the spectroscopic properties of Cr:ZnSe may be found in Refs. 34 through 36.

6.5.2 Examples of Three-Level and Four-Level Transitions

The classic example of a three-level solid-state laser system is ruby, which is sapphire (Al_2O_3) crystal doped with trivalent chromium, abbreviated as Cr^{3+}:Al_2O_3. Figure 6.28 shows the relevant energy

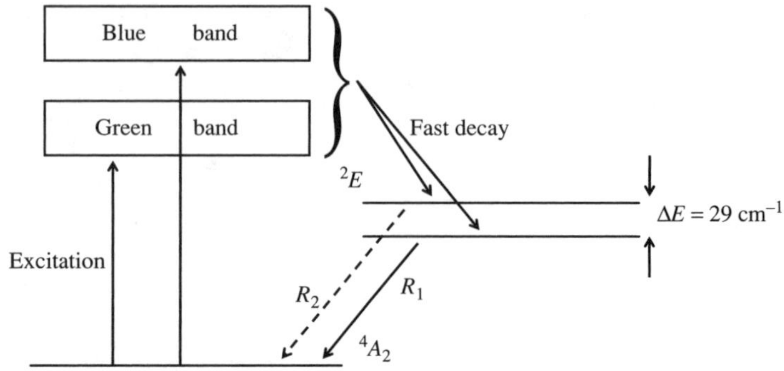

Figure 6.28 The energy-level diagram of ruby (Cr^{3+}:Al_2O_3). (Adapted from Ref. 28).

levels that are involved in the 694.3-nm laser emission. After excitation to the upper pump bands (designated as blue pump and green pump in Fig. 6.28), rapid decay brings the ions to the ^2E level. This level has a slight splitting, the energy difference between the two levels being about 29 cm^{-1}. The transitions from the lower and upper levels of the ^2E band are referred to as the R_1 and R_2 transitions. Because the lower level has higher occupation due to Boltzmann statistics, the R_1 transition always has higher population inversion, and hence, lasing occurs for the R_1 transition only. The output of the ruby laser is 694.3 nm. The fluorescence lifetime and the stimulated emission cross section of ruby are 3 ms (at room temperature) and $2.5 \times 10^{-20}\,\text{cm}^2$, respectively.[28]

Another example of a three-level laser system is the 946-nm transition of Nd:YAG, shown schematically in Fig. 6.29*a*. After excitation at the wavelength of 808 nm, neodymium ions are raised from the ground state $^4I_{9/2}$ to the upper laser level $^4F_{3/2}$. Various downward transitions occur to the lower levels as we discussed earlier. One of these transitions brings the ions directly back to the ground state, leading to photon emission at the wavelength of 946 nm. In contrast, the emission at 1064 nm originating from the same upper laser level $^4F_{3/2}$ to an intermediate lower level ($^4I_{11/2}$) is an example of a four-level laser system, since four levels are involved and the lower laser level is nearly empty (see Fig. 6.29*b*). Hence, the 1064-nm transition of the Nd:YAG laser can be operated with a much lower pumping threshold than that for the 946-nm line. The energy levels involved in the 1319-nm emission of Nd:YAG, with the downward transition from $^4F_{3/2}$ to $^4I_{13/2}$, also form a four-level laser system. The fluorescence lifetime of the $^4F_{3/2}$ level and the stimulated emission cross section of the 1064-nm transition in Nd:YAG are 230 μs and $6.5 \times 10^{-19}\,\text{cm}^2$, respectively.[28]

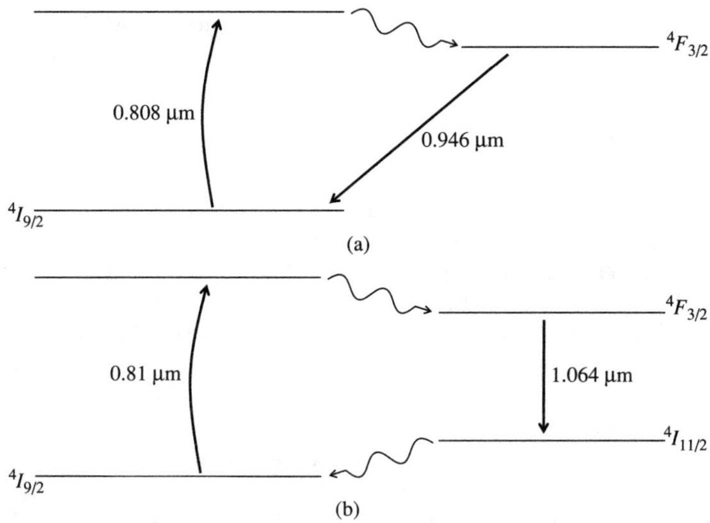

Figure 6.29 Simplified energy-level diagrams of Nd:YAG showing the energy levels involved in the (*a*) 946-nm and (*b*) 1064-nm emission.

6.5.3 Rate Equation Analysis of a Four-Level Solid-State Laser System

Now, we take an ideal, optically pumped four-level laser system and systematically develop the differential equations that describe the spatial evolution of the pump and laser beams through the medium. A schematic of the energy-level diagram is shown in Fig. 6.30.

Here is a summary of the optical transitions involved. Absorption at the pump wavelength of λ_p first excites the ions from the ground-state $|g\rangle$ to the pump band $|3\rangle$. Rapid, nonradiative decay then takes the ions to the upper laser level $|2\rangle$. Both spontaneous and stimulated emission can occur between the upper level $|2\rangle$ and the lower level $|1\rangle$. Finally, ions in level $|1\rangle$ rapidly decay to the ground state $|g\rangle$. Due to the rapid decays, we assume that the populations N_1 and N_3 of the

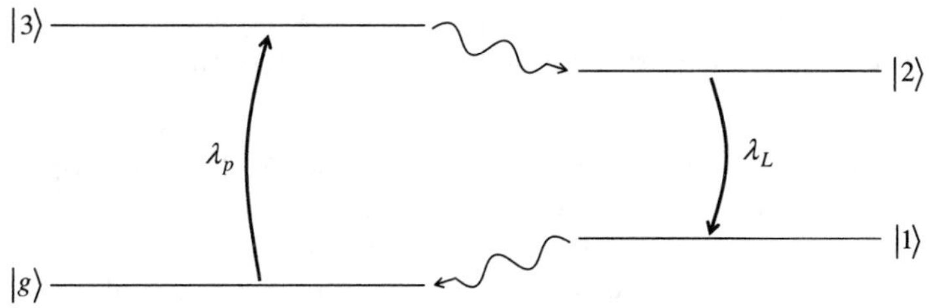

Figure 6.30 Energy-level diagram of an ideal, optically pumped four-level laser system.

lower laser level $|1\rangle$ and the pump band $|3\rangle$ are negligible. As we saw earlier, the population density N_2 of level $|2\rangle$ will satisfy the rate equation

$$\frac{dN_2}{dt} = \frac{\sigma_a I_p}{h\nu_p} N_g - \frac{N_2}{\tau_f} - \frac{\sigma_e I_L}{h\nu_L} N_2 \tag{6.77}$$

where I_p is the pump intensity at the pump wavelength of λ_p, I_L is the laser intensity at the lasing wavelength of λ_L, σ_a is the stimulated absorption cross section, σ_e is the stimulated emission cross section, τ_f is the fluorescence lifetime, ν_p is the pump frequency, ν_L is the laser frequency, h is Planck's constant, and N_g is the ground-state population density. Furthermore,

$$N_g + N_2 = N_T \tag{6.78}$$

where N_T is the total ion population density. In the cw case, the pump and the laser intensities at each point inside the medium are constant as a function of time. In steady state, it can be shown that (see Prob. 6.8) the upper-state population density becomes

$$N_2 = N_T \frac{I_p}{I_{sa}} \frac{1}{1 + \frac{I_p}{I_{sa}} + \frac{I_L}{I_{se}}} \tag{6.79}$$

where I_{sa} and I_{se} are the saturation intensities at the pump and laser wavelengths, given respectively by

$$I_{sa} = \frac{h\nu_p}{\sigma_a \tau_f} \tag{6.80}$$

and

$$I_{se} = \frac{h\nu_L}{\sigma_e \tau_f} \tag{6.81}$$

Absorption at the Pump Wavelength

Based on the result given in Eq. (6.79), we can now investigate the passage of the co-propagating pump and laser beams through a thin section of the medium of length dz and cross-sectional area a. This is shown in Fig. 6.31. First, we look at the pump beam. We note that

$\frac{(I_p(z+dz) - I_p(z))a}{h\nu_p}$ = the number of pump photons lost within the medium of volume adz per unit time
= the total number of ions that undergo absorption at the wavelength λ_p per unit time

$$= -\frac{\sigma_a I_p}{h\nu_p} N_g(adz) \tag{6.82}$$

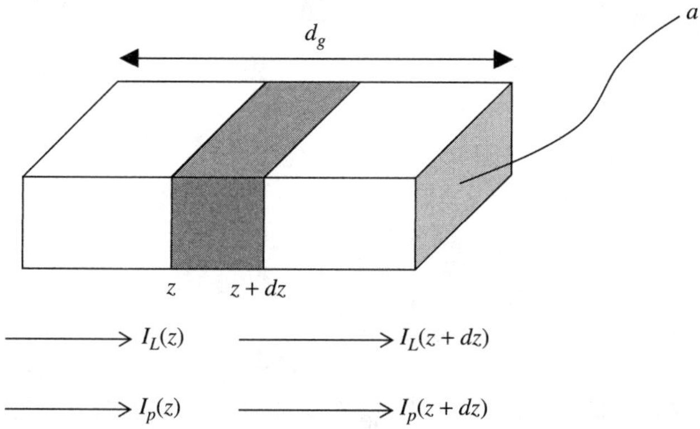

FIGURE 6.31 Passage of the co-propagating pump and laser beams through a thin section of the medium of length *dz* and cross-sectional area *a*.

Rearranging and making use of the limit definition of the derivative, we obtain

$$\frac{1}{I_p}\frac{\partial I_p}{\partial z} = -\sigma_a N_g = -\sigma_a (N_T - N_2) = -\alpha_{p0}\frac{1+\dfrac{I_L}{I_{se}}}{1+\dfrac{I_p}{I_{sa}}+\dfrac{I_L}{I_{se}}} \tag{6.83}$$

where α_{p0} is the small-signal pump absorption coefficient given by

$$\alpha_{p0} - \sigma_a N_T \tag{6.84}$$

Equation (6.83) is a key result that summarizes the absorption characteristics of the pump beam. Let us look at some important limiting cases:

1. No stimulated emission and no pump saturation: In this case, we can neglect the laser intensity and also assume that the absorption saturation intensity is very large ($I_L \cong 0$ and $I_{sa} >> I_p$).

   Intuitively, this case should be as close as we can get to the classical theory of absorption treated in Chap. 4. Equation (6.83) simplifies to

$$\frac{1}{I_p}\frac{dI_p}{dz} = -\alpha_{p0} \tag{6.85}$$

   which has the familiar exponential solution

$$I_p(z) = I_{p0}\exp(-\alpha_{p0} z) \tag{6.86}$$

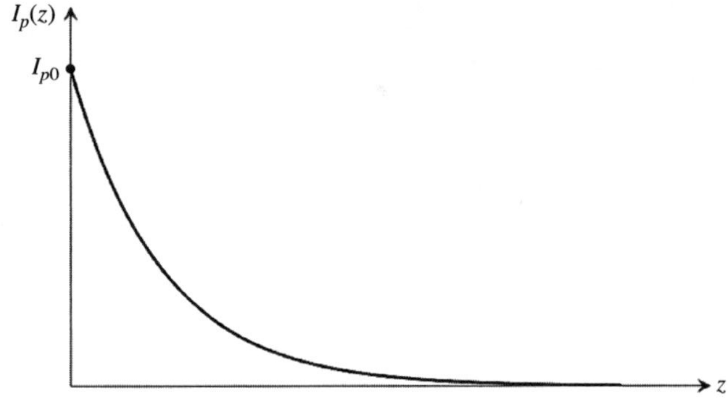

FIGURE 6.32 Exponential decay of the pump intensity in a four-level absorber where stimulated emission and absorption saturation are negligible.

where I_{p0} is the incident intensity at $z = 0$. Indeed, this is exactly what we had for the classical case [see Eq. (4.41)]. The spatial variation of the pump intensity is shown in Fig. 6.32. Equation (6.86) applies to a wide range of simple absorbing media. Furthermore, we see that the differential absorption coefficient α_{p0} depends both on the concentration (N_T) of the ions as well as on the strength of absorption (σ_a).

2. Pump saturation effects: Keeping pump saturation effects and still neglecting stimulated emission, Eq. (6.83) becomes

$$\frac{1}{I_p}\frac{dI_p}{dz} = \frac{-\alpha_{p0}}{1+\frac{I_p}{I_{sa}}} \tag{6.87}$$

Now, the absorption coefficient decreases with increasing intensity, an effect which is routinely observed in saturable absorbers[37, 38] and which we could not account for with the classical theory. Absorption saturation originates from the fact that as more and more ions are excited from the ground state with increasing pump intensity, less remain to absorb the incoming light. When does saturation become important? This is determined by the saturation intensity. As can be seen from Eq. (6.87), saturation will become observable when the pump intensity is comparable to the saturation intensity, which depends inversely on the fluorescence lifetime and the absorption cross section. Both make physical sense. If the absorption cross section is large, even a small pump intensity can excite a large number of ions to the upper laser level.

Hence, departure from the small-signal absorption coefficient will occur at low intensities. If the fluorescence lifetime is long, again, the atoms raised to the upper level will take a long time to come back, leading to a departure of the ground-state population from N_T even at low intensities. All this dependence is summarized in the expression for I_{sa} given in Eq. (6.80).

Finally, what is the role of I_L on absorption saturation? When non-zero, the laser intensity I_L competes to restore the original population in the ground state by increasing the rate of stimulated emission. Hence, for the same level of pumping intensity, an increase in I_L leads to an increase of the differential absorption coefficient. At extremely large values of I_L, the absorption coefficient will asymptotically tend to its small-signal value α_{p0}, as can be seen from Eq. (6.83).

Gain at the Lasing Wavelength

Now, we study the passage of the laser intensity I_L. Using a similar logic as to that for the pump intensity, we note that

$\left(\frac{(I_L(z+dz)-I_L(z))}{h\nu_L}\right)a =$ the total number of photons added to the laser beam within the volume adz per unit time

= the total number of ions which decay by stimulated emission within the same volume per unit time

$$= +\left(\frac{\sigma_e I_L}{h\nu_L}N_2\right)adz \tag{6.88}$$

After using the same procedure to simplify, we obtain

$$\frac{1}{I_L}\frac{dI_L}{dz} = \sigma_e N_2 = g_T \frac{I_p}{I_{sa}} \frac{1}{1+\frac{I_p}{I_{sa}}+\frac{I_L}{I_{se}}} \tag{6.89}$$

where g_T can be defined as the maximum extractable small-signal gain given by

$$g_T = \sigma_e N_T \tag{6.90}$$

So, how can we interpret this result? From Eq. (6.89), in the presence of a non-zero pump intensity and negligible attenuation, we see that $(1/I_L)(dI_L/dz) > 0$, meaning that the laser intensity will grow with propagation distance. As the pump intensity is increased, the differential gain will also increase, but not indefinitely, because we have a finite population density in the ground state which we can excite. At very large pumping intensities, the laser intensity will satisfy the equation

$$\frac{1}{I_L}\frac{dI_L}{dz} = g_T \tag{6.91}$$

Therefore, g_T tells us what the differential gain coefficient will be at very high pumping levels, and this is why it is called the maximum extractable small-signal gain coefficient. At very high pumping levels, the laser intensity will therefore grow exponentially according to

$$I_L(z, I_p >> I_{sa}) = I_L(0)\exp(g_T z) \tag{6.92}$$

To be more realistic, we should also include a phenomenological differential loss coefficient α_L to account for scattering and parasitic absorption at the lasing wavelength. Hence, the differential equation for the laser beam intensity can then be cast into the form

$$\frac{1}{I_L}\frac{dI_L}{dz} = g_T \frac{I_p}{I_{sa}} \frac{1}{1 + \frac{I_p}{I_{sa}} + \frac{I_L}{I_{se}}} - \alpha_L \tag{6.93}$$

As the laser intensity grows and becomes comparable to the saturation intensity I_{se} at the lasing wavelength, the rate of stimulated emission will be large enough to cause a substantial decrease in the population inversion, and decrease the differential gain coefficient. This is the gain saturation phenomenon which we saw earlier as well. It can also be seen from Eq. (6.93) that, as the laser intensity increases, the differential gain coefficient will decrease.

6.5.4 Laser-Matter Interaction under the Thin-Length Approximation

In this section, we will derive a very useful differential equation for the instantaneous power of the laser beam and the fractional power gain by using the rate equations. Our analysis follows the approach given in Refs. 39, 40. Let us assume that the 4-level optical amplifier can be treated as a lumped element with a thin length of $dz = d_g$. The fractional power gain G per pass through the amplifier can be defined as

$$G \equiv \frac{\Delta P_L}{P_L} \tag{6.94}$$

where ΔP_L is the change in the laser power after one single pass through the amplifier. From Eq. (6.88), ΔP_L can be written as

$$\Delta P_L = \Delta I_L a = \sigma_e I_L N_2 a d_g \tag{6.95}$$

Since $P_L = aI_L$, we get, by using Eq. (6.95),

$$G = \sigma_e N_2 d_g \tag{6.96}$$

By multiplying both sides of Eq. (6.77) by $\sigma_e d_g$, and by assuming that pump absorption saturation is negligible with $N_g \approx N_T$, it can be shown that (see Prob. 6.9) the equation satisfied by G can be written in the form

$$\frac{dG}{dt} = -\frac{G(t) - G_0}{\tau_f} - \frac{P_L(t)G(t)}{E_{\text{sat}}} \tag{6.97}$$

Above, E_{sat} is the saturation energy given by

$$E_{\text{sat}} = \frac{h\nu_L}{\sigma_e} a \tag{6.98}$$

and G_0 is the small-signal fractional power gain given by

$$G_0 = \frac{\sigma_a I_p N_g}{h\nu_p} \sigma_e d_g \tau_f \approx \frac{\sigma_a I_p N_T}{h\nu_p} \sigma_e d_g \tau_f \tag{6.99}$$

We will soon use Eq. (6.97) together with the cavity equation for $P_L(t)$ to investigate the cw and pulsed behavior of lasers. Note that in the cw case, we set the time derivative to zero and obtain

$$G_s = \frac{G_0}{1 + \dfrac{P_L}{P_{\text{sat}}}} \tag{6.100}$$

where G_s is the steady-state value of the fractional power gain, and P_{sat} is the saturation power which can be written in the form

$$P_{\text{sat}} = \frac{E_{\text{sat}}}{\tau_f} = \frac{h\nu_L}{\sigma_e \tau_f} a \tag{6.101}$$

We can see from Eq. (6.100) once again that due to stimulated emission, which is represented by the second term on the right-hand side of Eq. (6.97), the power gain will decrease with increasing signal power as given by Eq. (6.100).

6.5.5 Plane-Wave Analysis of a Solid-State Laser

Consider the solid-state laser sketched in Fig. 6.33. Here, R_1 and R_2 are the intensity reflectivities of the end mirrors (not to be confused with the radius of curvature), d is the effective length of the resonator, and d_g is the length of the gain medium. Let us further assume that the differential loss coefficient at the lasing wavelength is α_L.

First, we analyze the evolution of the beam power in the presence of passive resonator losses only. Denoting the beam power at a particular location after the mth round trip as P_m, the power after $(m + 1)$ round trips will be related to P_m through

$$P_{m+1} = P_m R_1 R_2 \exp(-2\alpha_L d_g) = P_m \exp(-2\alpha_T d_g) \qquad (6.102)$$

where we introduced the total differential loss coefficient α_T to account for all losses due to attenuation in the medium (represented by α_L) as well as the finite transmission of the end mirrors. From Eq. (6.102) α_T is given by

$$\alpha_T = \alpha_L + \frac{1}{2d_g}\ln\left(\frac{1}{R_1 R_2}\right) \qquad (6.103)$$

In other words, we are assuming that all losses of the resonator are distributed over the gain medium with an effective differential loss coefficient α_T. If we now employ the thin-length approximation, the decrease in the laser power after one round trip will be given by

$$\frac{\Delta P_L}{P_L} = -2\alpha_T d_g \qquad (6.104)$$

If the fractional change in power is small over one round trip so that $dP_L/dt \approx \Delta P_L/\Delta t$, with $\Delta t = 2d/c$, we get

$$\frac{dP_L}{dt} \approx \frac{\Delta P_L}{\Delta t} = \frac{\Delta P_L}{2d/c} = -\frac{P_L}{\tau_p} \qquad (6.105)$$

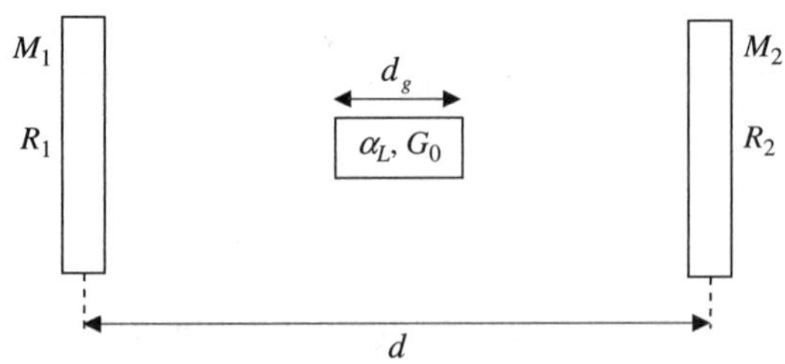

FIGURE 6.33. Sketch of the solid-state laser resonator used in the plane-wave analysis of laser operation.

where τ_p is known as the cavity photon lifetime and is given by

$$\tau_p = \frac{2d/c}{2\alpha_L d_g + \ln\left(1/R_1R_2\right)} \tag{6.106}$$

To gain some insight about why we call τ_p the cavity photon lifetime, we consider the case where the external pump is shut off at time $t = 0$. From Eq. (6.105), the laser beam power inside the resonator will exponentially decay according to

$$P_L(t) = P_L(0)\exp\left(-t/\tau_p\right) \tag{6.107}$$

Hence, we may view τ_p as the characteristic time after which the circulating photon intensity decays to $1/e$ of its initial value after the pump is shut off. The cavity photon lifetime gives us a measure of how much loss there exists inside a resonator. The longer the cavity lifetime, the lower the cavity losses will be, and the longer it takes for the intracavity laser power to decay after the pump is shut off.

The quality factor, often referred to as the Q-factor of a resonator, is another very useful measure of the level of losses inside a resonator. The Q-factor is defined as

$$Q \equiv 2\pi \frac{\text{(energy stored in the resonator)}}{\text{(energy lost in one period of oscillation)}} \tag{6.108}$$

Denoting the frequency of the stored light as ν_0, we get

$$Q = 2\pi \frac{E}{-\Delta E} \approx 2\pi \frac{E}{-\frac{dE}{dt}\frac{1}{\nu_0}} = \omega_0 \frac{E}{-\frac{dE}{dt}} \tag{6.109}$$

by assuming that the change in energy over one period is very small. Above, ω_0 is the angular frequency of the light. The differential equation satisfied by the stored energy becomes

$$\frac{dE}{dt} = -\frac{\omega_0}{Q}E \tag{6.110}$$

giving

$$E(t) = E(0)\exp\left(-\frac{\omega_0}{Q}t\right) \tag{6.111}$$

or

$$P_L(t) = \left|\frac{dE}{dt}\right| = E(0)\frac{\omega_0}{Q}\exp\left(-\frac{\omega_0}{Q}t\right) \tag{6.112}$$

Comparing Eqs. (6.107) and (6.112), we see that the Q-factor and cavity photon lifetime τ_p are related through

$$Q = \omega_0 \tau_p \tag{6.113}$$

Both the cavity photon lifetime τ_p and the cavity quality factor Q are different quantitative measures of how much loss a light beam experiences per round trip as it circulates inside an optical resonator.

Example What is the photon lifetime of a resonator with $d = 1$ m, $R_1 = 100\%$, $R_2 =$ 95%, and $\alpha_L = 0$?

$$\tau_p = \frac{2d/c}{\ln\left(1/R_2\right)} = \frac{2(100)/3\times 10^{10}}{-\ln(0.95)} = 130 \text{ ns}$$

Equation (6.105) is not in its final form because we have not yet included the optical gain. Based on the definition of the fractional power gain G, defined in Eq. (6.94), the change in the laser power over one round trip due to optical gain will become

$$\Delta P_L = 2GP_L \tag{6.114}$$

Again, this change occurs over one round-trip time $T_R = 2d/c$, resulting in the following differential equation for P_L:

$$\frac{dP_L}{dt} = -\frac{P_L(t)}{\tau_p} + \frac{2G(t)}{T_R} P_L(t) \tag{6.115}$$

The two coupled equations [Eq. (6.97) and Eq. (6.115)] for $G(t)$ and $P_L(t)$ can be used to analyze the cw as well as the pulsed operation of solid-state lasers.

6.5.6 Continuous-Wave Analysis of Laser Operation

To analyze the cw behavior of the laser, we set the time derivatives equal to zero in Eqs. (6.97) and (6.115). At the oscillation threshold, the laser power P_L will further be zero and the fractional power gain will be at its threshold value G_{th}. From Eq. (6.115), with $dP_L/dt = 0$, we have

$$\frac{2G_{th}}{T_R} = \frac{1}{\tau_p} \tag{6.116}$$

or

$$G_{th} = \frac{T_R}{2\tau_p} = \left[\alpha_L d_g + \frac{1}{2}\ln\left(\frac{1}{R_1R_2}\right)\right] \tag{6.117}$$

This result is basically equivalent to what we obtained earlier in Eq. (6.63). Once we go above the threshold gain by increasing the pump power, gain will saturate and reach its steady-state value given by Eq. (6.100). In steady state, the saturated gain will be equal to the total losses (or the threshold gain), giving

$$G_s = \frac{G_0}{1+\frac{P_L}{P_{\text{sat}}}} = G_{th} \tag{6.118}$$

In a standing-wave cavity, there are two beams propagating in opposite directions. We assume that the power in each beam is approximately the same. If P^+ denotes the power of the forward propagating beam, then $P_L \approx 2P^+$ and

$$P_{\text{sat}}\left(\frac{G_0}{G_{th}} - 1\right) = 2P^+ \tag{6.119}$$

Hence, the output power, coming through the output coupler with reflectivity of R_2 will be given by

$$P_{\text{out}} = (1-R_2)\frac{P_{\text{sat}}}{2}\left(\frac{G_0}{G_{th}} - 1\right) \tag{6.120}$$

As the pump power increases, the small-signal gain G_0 will also increase and we expect a linear relationship between the output power and the pump power at pumping levels far above the threshold. Of course, this assumes that nonlinear effects such as thermal loading or pump absorption saturation are negligible.

Now, we focus on a specific but very important case where the first mirror has nearly 100% reflectivity ($R_1 = 1$) and the reflectivity of the second mirror, commonly referred to as the output coupler is R ($R = 1–T$, where T is the intensity transmission of the output coupler). The fractional threshold gain G_{th} can now be expressed as

$$G_{th} = \left[\alpha_L d_g + \frac{1}{2}\ln\left(\frac{1}{1-T}\right)\right] \approx \frac{1}{2}[L+T] \tag{6.121}$$

where we have assumed that the transmission of the output coupler is small so that $\ln(1/(1-T)) \approx T$ and L is the total round-trip material loss given by

$$L = 2d_g\alpha_L \tag{6.122}$$

In a 4-level laser, the threshold fractional gain is directly proportional to the upper-level population and also to the pump power P_p. Hence, the incident threshold pump power required for lasing can be expressed in terms of L and T as

$$(P_p)_{th} = A(L+T) \tag{6.123}$$

where A is a constant. Equation (6.123) gives us a useful result that can be used in the determination of the round-trip cavity loss L. In particular, if we can experimentally measure the threshold pump power $(P_p)_{th}$ of a laser for output couplers with different transmission and plot the variation of $(P_p)_{th}$ as a function of T, we expect to get a straight line as shown in Fig. 6.34. The intercept and slope of the line are AL and A, respectively. Therefore, the loss L is simply obtained from the ratio L = intercept/slope. For a discussion of different methods to determine the cavity loss, see Refs. 41 and 42.

Based on the foregoing analysis, we can also determine how much output power will be obtained from the resonator at a pump power above $(P_p)_{th}$. Specifically, we derive the expression for the slope efficiency η which gives the change in the output laser power per unit change in the pump power, defined according to

$$\eta \equiv \frac{\Delta P_{\text{out}}}{\Delta P_p} \tag{6.124}$$

From Eq. (6.99), we can express the small-signal fractional power gain G_0 in terms of the pump power as

$$G_0 = \sigma_a \frac{P_p}{a} \frac{N_g \sigma_e d_g \tau_f}{h\nu_p} \tag{6.125}$$

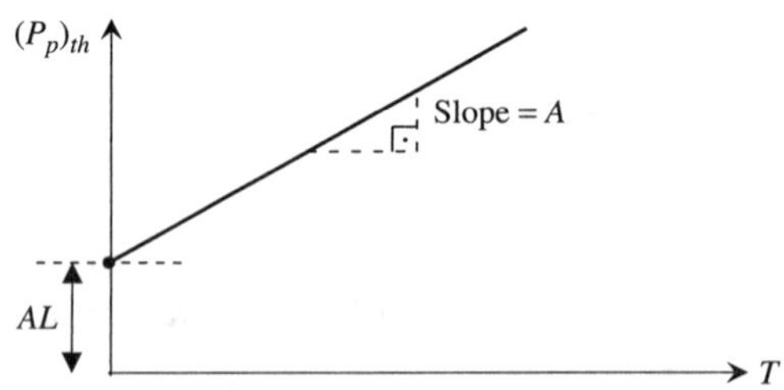

FIGURE 6.34 Variation of the incident threshold pump power $(P_p)_{th}$ for a four-level solid-state laser as a function of the output coupler transmission T.

By using Eq. (6.101) for the saturation power P_{sat} the slope efficiency becomes

$$\eta = \frac{\Delta P_{\text{out}}}{\Delta P_p} = \frac{T\left(\frac{1}{2}\right)\frac{h\nu_L}{\sigma_e\tau_f}a\sigma_a n_g \sigma_e d_g \tau_f}{\frac{1}{2}(L+T)ah\nu_p} = \frac{\lambda_p}{\lambda_L}\frac{T}{T+L}\eta_a \tag{6.126}$$

where η_a is the fractional pump absorption given by

$$\eta_a = \sigma_a N_g d_g \approx \sigma_a N_T d_g \tag{6.127}$$

Equation (6.126) is a very important general result which states that the maximum slope efficiency, when all of the pump power is absorbed (i.e., $\eta_a \approx 1$) and losses are low, is given by λ_p/λ_L, which is known as the quantum efficiency. This makes physical sense because when losses are minimal, the highest efficiency will be obtained when every pump photon is converted to a laser photon. Since the wavelengths are different, the resulting maximum efficiency becomes λ_p/λ_L from Planck's energy formula. The presence of loss reduces the slope efficiency as given by Eq. (6.126). Putting all this information together, the output power of the laser can be expressed in terms of the pump power as

$$P_{\text{out}} = \eta(P_p - (P_p)_{th}) \tag{6.128}$$

as shown in Fig. 6.35.

6.5.7 Relaxation Oscillations

In the earlier section, we analyzed the steady-state solution of the coupled equations for gain and the laser power. What can we say

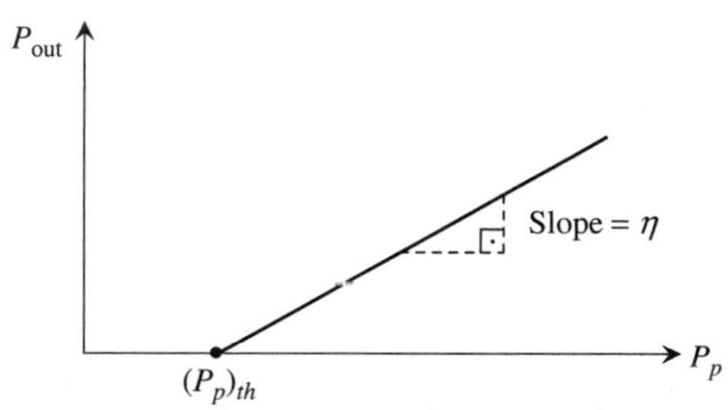

FIGURE 6.35 Variation of the output power as a function of the pump power for a laser.

about the temporal stability of this solution? In other words, if we were to perturb the fractional gain and the laser power, how would they evolve in time? To answer this question, we analyze the effect of perturbations on the steady-state solution. Let G_s and P_s be the steady-state values for $G(t)$ and $P(t)$, respectively. We assume that a perturbation occurs in both quantities according to

$$G(t) = G_s + \Delta G(t) \tag{6.129}$$

and

$$P_L(t) = P_{Ls} + \Delta P_L(t) \tag{6.130}$$

where $\Delta G(t)$ and $\Delta P_L(t)$ represent the deviations in gain and laser power from the steady-state values. We leave it as an exercise for the reader (Prob. 6.11) to show that when only the linear terms are kept, $\Delta G(t)$ and $\Delta P_L(t)$ satisfy the linear coupled differential equations[40]

$$\frac{d\Delta G}{dt} = -\frac{G_s}{E_{\text{sat}}}\Delta P_L - \frac{\Delta G}{\tau_{\text{stim}}} \tag{6.131}$$

and

$$\frac{d\Delta P_L}{dt} = 2\frac{P_{Ls}}{T_R}\Delta G \tag{6.132}$$

where the stimulated lifetime τ_{stim} is defined by

$$\frac{1}{\tau_{\text{stim}}} \equiv \frac{1}{\tau_f}\left(1 + \frac{P_{Ls}}{P_{\text{sat}}}\right) \tag{6.133}$$

By using the fact that $G_{th} = T_R/2\tau_p$ [see Eq. (6.116)], Eqs. (6.131) and (6.132) can be written in matrix form as

$$\frac{d}{dt}\begin{bmatrix} \Delta P_L \\ \Delta G \end{bmatrix} = \begin{bmatrix} 0 & 2\,{}^{P_s}\!\big/_{T_R} \\ -\dfrac{T_R}{2\tau_p E_{\text{sat}}} & -\dfrac{1}{\tau_{\text{stim}}} \end{bmatrix} \begin{bmatrix} \Delta P_L \\ \Delta G \end{bmatrix} \tag{6.134}$$

We leave it as an exercise for the reader (Prob. 6.12) to show, by substitution, that a solution of the form

$$\begin{bmatrix} \Delta P_L \\ \Delta G \end{bmatrix} = \begin{bmatrix} \Delta P_{L0} \\ \Delta G_0 \end{bmatrix} \exp(st) \tag{6.135}$$

satisfies Eq. (6.134), where the exponent s is given by

$$s = -\frac{r}{2\tau_f} \pm i\sqrt{\frac{(r-1)}{\tau_p \tau_f} - \left(\frac{r}{2\tau_f}\right)^2} \tag{6.136}$$

Above, r is the normalized pumping parameter defined according to

$$r \equiv 1 + \frac{P_{Ls}}{P_{\text{sat}}} \tag{6.137}$$

The most important conclusion drawn from this analysis is that because the real part of the exponent s in Eq. (6.136) is negative, the steady-state solution with $G = G_s$ and $P_L = P_{Ls}$ is stable and perturbations exponentially decay back to zero. The decay may have different characteristics depending on the magnitudes of the parameters.

Example As an example, we consider the Nd^{+3}:YVO_4 gain medium with λ_L = 1064 nm, $\sigma_e \approx 10 \times 10^{-19}$ cm^2 and $\tau_f = 100$ μ sec. The laser resonator has a total length of $d = 15$ cm, and mirror reflectivities of $R_1 = 100\%$ and $R_2 = 98\%$. The round-trip cavity loss is $2\alpha_L d_g = 0.02$ (2%). The laser is pumped to the level where $P_{Ls}/P_{sat} = 4$. By using Eq. (6.106), the cavity photon lifetime comes to

$$\tau_p = \frac{2(15)/3 \times 10^{10}}{0.02 + \ln\left(\dfrac{1}{0.98}\right)} = 24.87 \text{ ns}$$

Furthermore,

$$\frac{1}{\tau_{\text{stim}}} = \frac{1}{\tau_f}\left(1 + \frac{P_s}{P_{\text{sat}}}\right) = \frac{5}{\tau_f} = \frac{5}{100 \times 10^{-6}} \text{s}^{-1}$$

Here, we are in the long fluorescence lifetime regime where the cavity photon lifetime is much smaller than the stimulated lifetime, as well as the fluorescence lifetime with $1/\tau_p >> 1/\tau_f$ or $1/\tau_{\text{stim}}$. The exponent s simplifies to (when r>>1)

$$s = -\frac{r}{2\tau_f} \pm i\sqrt{\frac{1}{\tau_{\text{stim}}\,\tau_p}} = -\frac{r}{2\tau_f} + i2\pi\nu_R \tag{6.138}$$

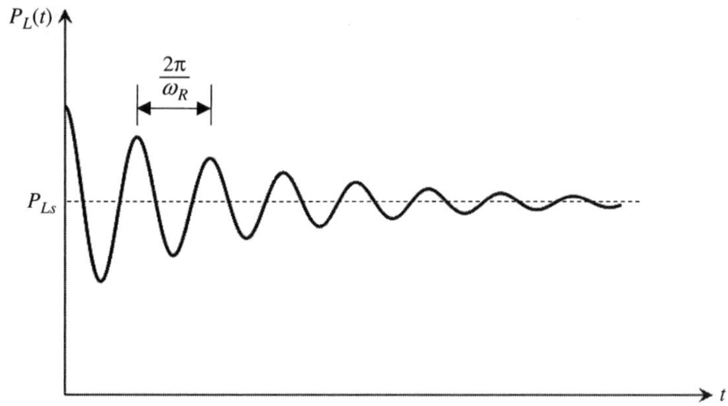

FIGURE 6.36 Relaxation oscillation of a laser perturbed from its steady-state operating point.

where the relaxation oscillation frequency ν_R is given by

$$\nu_R = \frac{1}{2\pi}\sqrt{\frac{1}{\tau_{\text{stim}}\tau_p}} \tag{6.139}$$

In this particular example, ν_R comes to 226 kHz. A typical variation of the laser power exhibiting relaxation oscillations is sketched in Fig. 6.36. This behavior is typical of most solid-state lasers where the cavity photon lifetime is much shorter than the fluorescence lifetime.

If the gain medium is excited at a very high pumping level with $r >> 1$, or if the fluorescence lifetime τ_f is short so that the inequality

$$\frac{r-1}{\tau_p} < \frac{r^2}{4\tau_f} \tag{6.140}$$

is satisfied, then no oscillations occur and the perturbations in gain and laser power decay monotonically to zero.

6.6 Active Q-Switching of Lasers

Q-switching is a very useful and widely used technique for generating high-peak-power, short pulses from lasers. We already discussed the meaning of the cavity Q-factor in Sec. 6.4.5. A high-Q cavity implies low passive losses, which enable low-threshold lasing. In contrast, a lower-Q cavity requires a relatively larger pumping level before oscillation can be achieved. In this section, we will discuss active Q-switching where the loss level inside the cavity is externally controlled in order to obtain pulsed output. Another method, known as passive Q-switching can also be employed to obtain pulsed

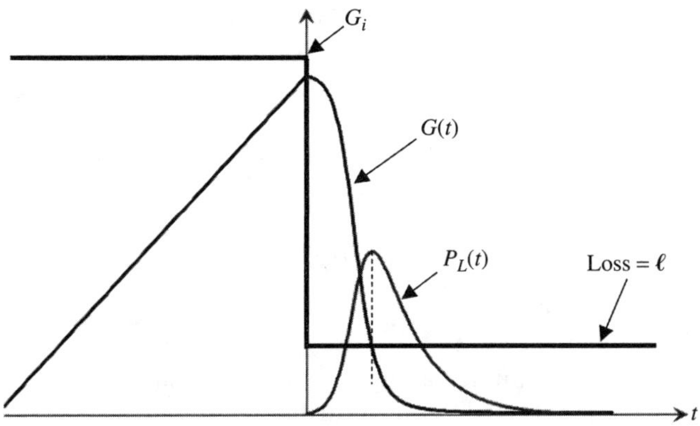

FIGURE 6.37 Variation of the fractional power gain, fractional loss, and the laser power during active Q-switching.

output. In this case, a saturable absorber is placed inside the cavity. The transmission of the saturable absorber increases with increasing light intensity, favoring the oscillation of high-intensity light pulses. Passive saturable absorbers can be designed to produce, Q-switched, mode-locked, or Q-switched and mode-locked output.[19, 39]

Various stages during active Q-switching are shown in Fig. 6.37. In the low-Q stage ($t < 0$ in Fig. 6.37), a large level of loss is first introduced into the resonator to prevent lasing. This can be done in several different ways by using polarization-sensitive optics or simply by misaligning one of the end mirrors. During this time, pumping continues on time scales comparable to or shorter than the fluorescence lifetime, so that population inversion and gain keep building up with minimal decay of excited ions. Then, the high-Q state of the cavity is abruptly restored ($t = 0$ in Fig. 6.37). The pumping scheme is designed in such a way that at this instant, the initial fractional power gain G_i is far above the single-pass fractional loss ℓ. This leads to a rapid exponential growth of the laser power [shown as $P_L(t)$ in Fig. 6.37]. At the same time, stimulated decay of excited ions occurs at a high rate, causing a rapid depletion of the gain $G(t)$ with time. At the instant where $G(t) = \ell$, the instantaneous rate of change of the pulse power vanishes and the pulse reaches its maximum power. From here on, the loss exceeds the gain and the power quickly decays, leading to the generation of a pulse from the resonator. This process can be periodically repeated at a frequency that depends on the fluorescence lifetime of the gain medium, the charging time of the power supplies used for pumping, and the rate of cooling of the gain medium. Reference 28 provides a more detailed discussion of the practical aspects of the design of actively Q-switched lasers.

Here, we provide a simple quantitative analysis of active Q-switching based on the coupled laser-gain equations.

During the pulse build-up period of active Q-switching, pulse dynamics typically occur on time scales much shorter than the fluorescence lifetime. Hence, the spontaneous decay of the ions can be neglected and Eq. (6.97) for $G(t)$ simplifies to

$$\frac{dG}{dt} = -\frac{P_L(t)G(t)}{E_{\text{sat}}} \tag{6.141}$$

We assume that the initial value of the fractional single-pass gain is G_i at $t = 0$. The equation for $P_L(t)$ can further be written as

$$\begin{aligned}\frac{dP_L}{dt} &= -\frac{P_L(t)}{\tau_p} + \frac{2G(t)}{T_R}P_L(t) = \frac{P_L(t)}{T_R}\left[2G(t) - 2\frac{T_R}{2\tau_p}\right] \\ &= \frac{2(G(t)-\ell)}{T_R}P_L(t)\end{aligned} \tag{6.142}$$

where ℓ is the fractional single-pass loss given by

$$\ell = \frac{L}{2} = \frac{2\alpha_L d_g + \ln\left(1/R_1R_2 \right)}{2} \tag{6.143}$$

Dividing Eq. (6.142) by (6.141), we can eliminate the time variable and obtain

$$\frac{dP_L}{dG} = \frac{2E_{\text{sat}}}{T_R}\left(\frac{\ell}{G(t)} - 1\right) \tag{6.144}$$

As initial conditions, we take $P_L(t=0) = 0$ and $G(t=0) = G_i = r\ell$, where r is a dimensionless number that tells us the factor by which the initial fractional gain exceeds the single-pass fractional loss ℓ during the high-Q phase. Integrating Eq. (6.144) with respect to G, we obtain

$$\begin{aligned}P_L(t) - P_L(0) = P_L(t) &= \frac{2E_{sat}}{T_R}\left[\ell \ln\frac{G(t)}{G(0)} - (G(t) - G(0))\right] \\ &= \frac{2E_{\text{sat}}}{T_R}\left[G_i - G(t) + \ell \ln\frac{G(t)}{G_i}\right]\end{aligned} \tag{6.145}$$

At the instant when the pulse power is maximum (i.e., at the level where $P_L(t) = P_{max}$), we have $G(t) = \ell$, since this gives $dP_L/dG = 0$. By using $G(t=0) = G_i = r\ell$, we get for the maximum power

$$P_{max} = \frac{E_{sat}}{\tau_p}[r - 1 - \ln r] \tag{6.146}$$

since $T_R/2\ell = \tau_p$. To find the total energy contained in the pulse, we now need to determine the final gain value G_f where the pulse power reaches zero. From Eq. (6.145), we have

$$G_i - G_f + \ell \ln \frac{G_f}{G_i} = 0 \tag{6.147}$$

Dividing Eq. (6.147) by G_i, we can put Eq. (6.147) into dimensionless form for the unknown $x = G_f/G_i$ as

$$1 - x + \frac{1}{r}\ln(x) = 0 \tag{6.148}$$

Note that the final residual gain can be calculated just by knowing r. On the downside, Eq. (6.148) cannot be explicitly solved for x and a numerical scheme or trial and error needs to be used. Suppose that the final inversion has been determined. For a four-level system, $G(t)$ is directly proportional to the upper-level population inversion $N_2(t)$. Hence, we can determine the output energy E_p of the pulse from

$$E_p = (N_{2i} - N_{2f})ad_g h\nu_L = N_{2i}(1-x)ad_g h\nu_L \tag{6.149}$$

where N_{2i} and N_{2f} are the initial and final upper-state population densities, a is the effective cross-sectional area, d_g is the length of the gain medium, h is Planck's constant, and ν_L is the laser output frequency.

Finally, we can estimate the output pulsewidth. For this, we need to first determine the energy coupled out of the resonator. Due to material losses, not all of the circulating photons will make it out of the cavity. The pulse energy we calculated in Eq. (6.149) is the intracavity pulse energy. To find the output energy, we first separate the passive losses contained in ℓ into two categories:

1. internal material losses, represented by $\ell_i = \alpha_L d_g$
2. out-coupling loss, represented by $\ell_m = \frac{1}{2}\ln 1/R$

Again we are assuming here that output coupling occurs only through the second mirror with reflectivity $R_2 = R$. All of the generated photons, except those lost internally, will eventually come out of the resonator. Hence, the output pulse energy E_{out} can be calculated from

$$E_{out} = N_{2i}(1-x)ad_g h\nu_L \frac{\ell_m}{\ell_m + \ell_i} \approx N_{2i}(1-x)\, ad_g h\nu_L \tag{6.150}$$

The last approximate expression for E_{out} is valid when $\ell_m >> \ell_i$. This is typically the case, because in most Q-switched laser systems we have a few percent of internal losses, whereas the output coupling can be much longer in the 10% to 50% range. Since the output power P_{out} is

$$P_{\text{out}} = \left(\frac{P_L}{2}\right) T \cong 2\ell_m \frac{P_L}{2} \tag{6.151}$$

in a standing-wave resonator, the pulsewidth τ_{pulse} becomes

$$\begin{aligned}\tau_{\text{pulse}} &\approx \frac{E_{\text{out}}}{\ell_m P_{\text{max}}} \approx \frac{E_{\text{out}}}{\ell P_{\text{max}}} = \frac{N_{2i}(1-x)ad_g h\nu_L}{\ell \dfrac{E_{\text{sat}}}{\tau_p}(r-1-\ln r)} \\ &= \frac{N_{2i}(1-x)ad_g h\nu_L}{\ell \dfrac{h\nu_L}{\sigma_e \tau_p} a(r-1-\ln r)} = \frac{\sigma_e N_{2i} d_g (1-x)}{\ell \dfrac{1}{\tau_p}(r-1-\ln r)} \\ &= \tau_p \frac{r(1-x)}{(r-1-\ln r)}\end{aligned} \tag{6.152}$$

Above, we used the fact that the initial fractional power gain G_i is given by $G_i = \sigma_e N_{2i} d_g$ and $r = G_i / \ell$.

Example Actively Q-switched Nd:YAG laser
As an example, let us consider 1 atomic % Nd:YAG laser rod. This means that in the host lattice with the chemical formula $Y_3\,Al_5\,O_{12}$, 1% of the Y^{3+} ions are replaced by Nd^{3+} ions. We leave it as an exercise (Prob. 6.13) for the reader to show that this corresponds to a total neodymium ion concentration of $N_t = 1.39 \times 10^{20}$ cm^{-3}. We take the length and the radius of the rod as 4 cm and 3 mm, respectively. Furthermore, let the internal round-trip loss and output coupler reflectivity be 4% ($0.04 = 2\alpha_L d_g$) and 0.85, respectively. We take the stimulated emission cross section for Nd:YAG as $\sigma_e = 8.8 \times 10^{-19}$ cm^2. The total resonator length is 20 cm.

First, we calculate the threshold inversion needed to achieve lasing. At threshold,

$$G_{th} = \ell = \alpha_L d_g + \frac{1}{2}\ln\left(1/R\right) = \sigma_e (N_2)_{th} d_g$$

or

$$(N_2)_{th} = \frac{1}{\sigma_e}\left[\alpha_L + \frac{1}{2d_g}\ln\left(\frac{1}{R}\right)\right] = \frac{1}{8.8\times 10^{-19}}\left[\frac{0.04}{2(4)} + \frac{1}{2(4)}\ln\left(\frac{1}{0.85}\right)\right]$$

$$= 2.88\times 10^{16}\ \text{cm}^{-3}$$

Note that the total ion density in this case is far more than what is needed to achieve lasing threshold. During Q-switching, we assume that $G_i = 3\ell$ (or $r = 3$). This means that $N_{2i} = 3(N_2)_{th}$. Solving Eq. (6.148) numerically for $r = 3$, we obtain $x = 0.06$. Substituting into Eq. (6.150) finally gives $E_{out} = 4.3$ mJ.

To determine the output pulsewidth, we need the cavity photon lifetime τ_p given by

$$\tau_p = \frac{2d/c}{2\alpha_L d_g + \ln\left(\frac{1}{R_1R_2}\right)} = \frac{2(20)/3\times 10^{10}}{0.04 + \ln(1/0.85)} = 6.58\ \text{ns}$$

From Eq. (6.152), the output pulsewidth becomes

$$\tau_{\text{pulse}} = \tau_p \frac{r(1-x)}{(r-1-\ln r)} = \frac{6.58(1-0.06)(3)}{2-\ln 3} = 21\ \text{ns}$$

The saturation energy of the gain medium is further given by Eq. (6.98) as

$$E_{\text{sat}} = \frac{h\nu_L}{\sigma_e}a = \frac{(6.626\times 10^{-34})(3\times 10^{10})}{(1.06\times 10^{-4})(8.8\times 10^{-19})}\pi(0.15)^2 = 15\ \text{mJ}$$

The maximum power generated by the Q-switched laser hence becomes [from Eq. (6.146)]

$$P_{\max} = \frac{E_{\text{sat}}}{\tau_p}[r-1-\ln r] = \frac{15\times 10^{-3}}{6.58\times 10^{-9}}(3-1-\ln 3) = 2.05\ \text{MW}$$

clearly demonstrating with this realistic example the versatility of Q-switched lasers in producing light with high peak powers.

6.7 Passive Q-Switching with Saturable Absorbers

In passive Q-switching, a saturable absorber is placed inside the laser cavity to introduce large losses at low light intensities and low loss at high intensities. This is schematically shown in Fig. 6.38. Since high-intensity pulses experience relatively lower loss, this arrangement favors the generation of repetitively Q-switched pulses when certain

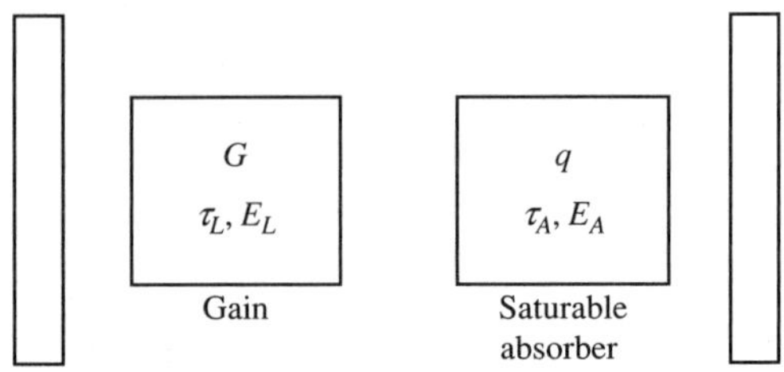

FIGURE 6.38 Schematic of a passively Q-switched laser containing a gain medium and a saturable absorber.

conditions are met. In recent years, passively Q-switched solid-state lasers based on semiconductor saturable absorber mirrors (referred to as SESAM or SBR) have become widely used.[19,20]

We will just outline the theory of passive Q-switching as discussed in Refs. 18 and 39. The saturable absorber will introduce a fractional power loss, represented by Q which will decrease with increasing light intensity. Similar to the gain medium dynamics we discussed back in Sec. 6.5.4, the saturable loss can be modeled by using the equation

$$\frac{dQ}{dt} = -\frac{Q(t) - Q_0}{\tau_A} - \frac{Q(t)P_L(t)}{E_A} \tag{6.153}$$

where $Q(t)$ is the time-dependent, saturable loss which gives the fractional loss in the propagating power, τ_A is the decay time of the absorber atoms, Q_0 is the small-signal loss of the absorber, $P_L(t)$ is the instantaneous intracavity laser power, and E_A is the saturation energy of the absorber given by

$$E_A = \frac{h\nu_L}{\sigma_a} a_{\text{abs}} \tag{6.154}$$

Above, a_{abs} is the effective beam area in the absorber, and σ_a is the absorption cross section of the absorber. Hence, the coupled laser-gain-absorber equations become, together with Eq. (6.153):

$$\frac{dG}{dt} = -\frac{G(t) - G_0}{\tau_f} - \frac{G(t)P_L(t)}{E_{\text{sat}}} \tag{6.155}$$

and

$$\frac{dP_L}{dt} = \frac{2}{T_R}(G(t) - \ell - Q(t))P_L(t) \tag{6.156}$$

Above, ℓ is the single-pass, non-saturable, fractional power loss as given by Eq. (6.143). The other parameters in Eqs. (6.155) and (6.156) are as defined before. It can be shown, based on the analysis in Refs. 39 and 40, that if the gain medium is excited with a cw pump whose power is constant over time, repetitive Q-switching will be obtained if the condition

$$\frac{2q_0 T_f}{\chi} > 1 \tag{6.157}$$

is satisfied. Here, $T_f = \tau_f / T_R$ is the dimensionless fluorescence lifetime, normalized with respect to the cavity round-trip time, and the parameter χ is given in terms of the absorber saturation power P_A ($P_A = E_A / \tau_A$) and laser saturation power P_{sat} ($P_{\text{sat}} = E_{\text{sat}} / \tau_f$), [same as in Eq. (6.101)] as

$$\chi = \frac{P_A}{P_{\text{sat}}} \tag{6.158}$$

Furthermore, the frequency f_{rep} of the repetitive pulse train is given by

$$f_{\text{rep}} = \frac{1}{\tau_f} \frac{G_0}{2Q_0} \tag{6.159}$$

The result in Eq. (6.159) shows that, as expected, as the pump power is increased resulting in larger small-signal gain (G_0), the repetition rate will also increase.

Example Suppose that we have a passively Q-switched Nd:YVO$_4$ laser at 1064 nm ($\tau_f = 100$ μs, $T_R = 1$ns, and $G_0 = 50\%$), containing a Cr:YAG saturable absorber with $Q_0 = 10\% = 0.1$. From Eq. (6.157), passive Q-switching will be obtained when the condition $\chi < 10^4$ is satisfied. Furthermore, the repetition frequency of the laser will be around 25 kHz from Eq. (6.159).

6.8 Mode Locking of Lasers

6.8.1 General Characteristics of Mode Locking

As we saw in the earlier sections, pulse shaping in the case of Q-switching occurs over time scales comparable to the cavity round-trip delay. As such, the pulse duration can be varied by changing the length of the cavity and will range anywhere from picoseconds to microseconds depending on the values of the gain parameters and cavity length.

There is another very important technique that is commonly used in generating pulsed output from lasers. In this case, pulse durations can be significantly shorter than the cavity round-trip delay time and typically fall in the picosecond ($\times 10^{-12}$ sec) and femtosecond ($\times 10^{-15}$ sec) time scales. This occurs as a result of the coherent superposition of the otherwise randomly oscillating modes supported by the optical resonator. Mode locking was first experimentally demonstrated by Hargrove et al. in 1964[43] and has since been applied to many different lasers. The field of ultrafast optics is concerned with the generation of short laser pulses by employing the technique of mode locking and the study of light-matter interactions at short time scales covering the picosecond, femtosecond, and more recently, attosecond ($\times 10^{-18}$ sec) ranges.[44]

To understand how mode locking takes place, consider a laser resonator containing a gain medium and a modulator or a saturable absorber as sketched in Fig. 6.39.

Let us suppose that the spatial distribution of the mode is TEM_{00}. As discussed in Chap. 3, the resonator can support infinitely many longitudinal modes at discrete frequencies given by

$$\nu_q = q\frac{c}{2d} \tag{6.160}$$

where q is an integer, c is the speed of light in vacuum, and d is the effective optical length of the resonator. Furthermore, the frequency spacing between two adjacent modes, also referred to as the free spectral range ν_{FSR} is given by

$$\nu_{FSR} = \nu_{q+1} - \nu_q = \frac{c}{2d} \tag{6.161}$$

Typically, solid-state gain media, which we discussed in Sec. 6.4, have broad enough gain bandwidths to support the oscillation of many longitudinal modes. Suppose that several such modes are simultaneously oscillating. If the difference in the angular frequency of two adjacent modes is ω_F, where

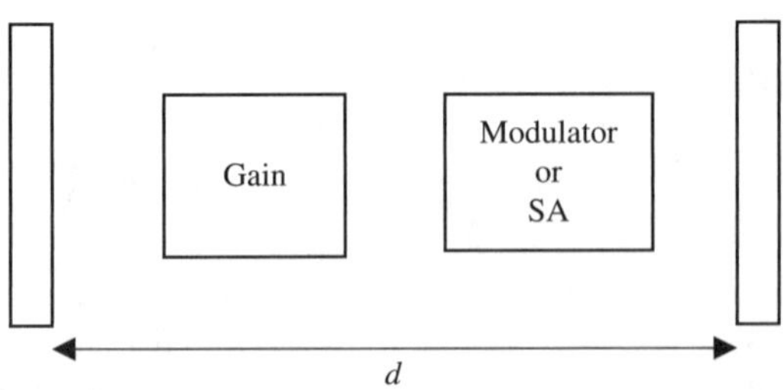

FIGURE 6.39 A laser resonator containing a gain medium and a modulator for mode locking.

$$\omega_F = \omega_{q+1} - \omega_q = 2\pi\nu_{FSR} = \frac{\pi c}{d} \tag{6.162}$$

then the net electric field due to M such oscillating modes will be given, based on the principle of superposition, by

$$E(t) = \sum_{q=0}^{M-1} E_q \exp(i(\omega_0 + q\omega_F)t + i\phi_q(t)) \tag{6.163}$$

Above, E_q is the amplitude of the qth mode, ω_0 is some reference angular frequency, and $\phi_q(t)$ is the phase of the qth mode. Furthermore, we have assumed that the electric field is linearly polarized and focus on only one of its components at a particular location in order to illustrate the essential features of mode locking. If $\phi_q(t)$ varies in a random fashion, then $E(t)$ also varies randomly over time scales comparable to the cavity round-trip time, and when averaged over long times, gives rise to a fairly constant output power. If, on the other hand, the modulator is operated in such a way as to force all of the modes to oscillate with a constant phase, then we have the regime of what is known as mode locking and a periodic train of ultrashort optical pulses can be obtained. To see this, we set all of the phase factors to zero and assume the same amplitude (E_0) for each mode to get

$$E(t) = E_0 \sum_{q=0}^{M-1} \exp(i(\omega_0 + q\omega_F)t) = E_0 \exp(i\omega_0 t)\frac{1-\exp(iM\omega_F t)}{1-\exp(i\omega_F t)}$$

$$= E_0 \exp(i\omega_0 t)\frac{\sin\left(\dfrac{M\omega_F t}{2}\right)}{\sin\left(\dfrac{\omega_F t}{2}\right)} \tag{6.164}$$

The instantaneous output power which is proportional to $|E(t)|^2$ is shown for the cases of $M = 2, 4, 8$, and 16 in Fig. 6.40*a* through *d*. From Fig. 6.40, we can draw several important conclusions about the characteristics of the mode-locked output:

1. The electric field has a local maximum at $t = 0$, given by $M \times E_0$. This repeats after a time interval of T_R which satisfies $\omega_F T_R/2 = \pi$ or $T_R = 2\pi/\omega_F = 2d/c$. In other words, T_R is the familiar cavity round-trip time we saw earlier. Another way of saying this is that the periodic pulse train has a repetition frequency equal to the free spectral range ν_{FSR} of the resonator.

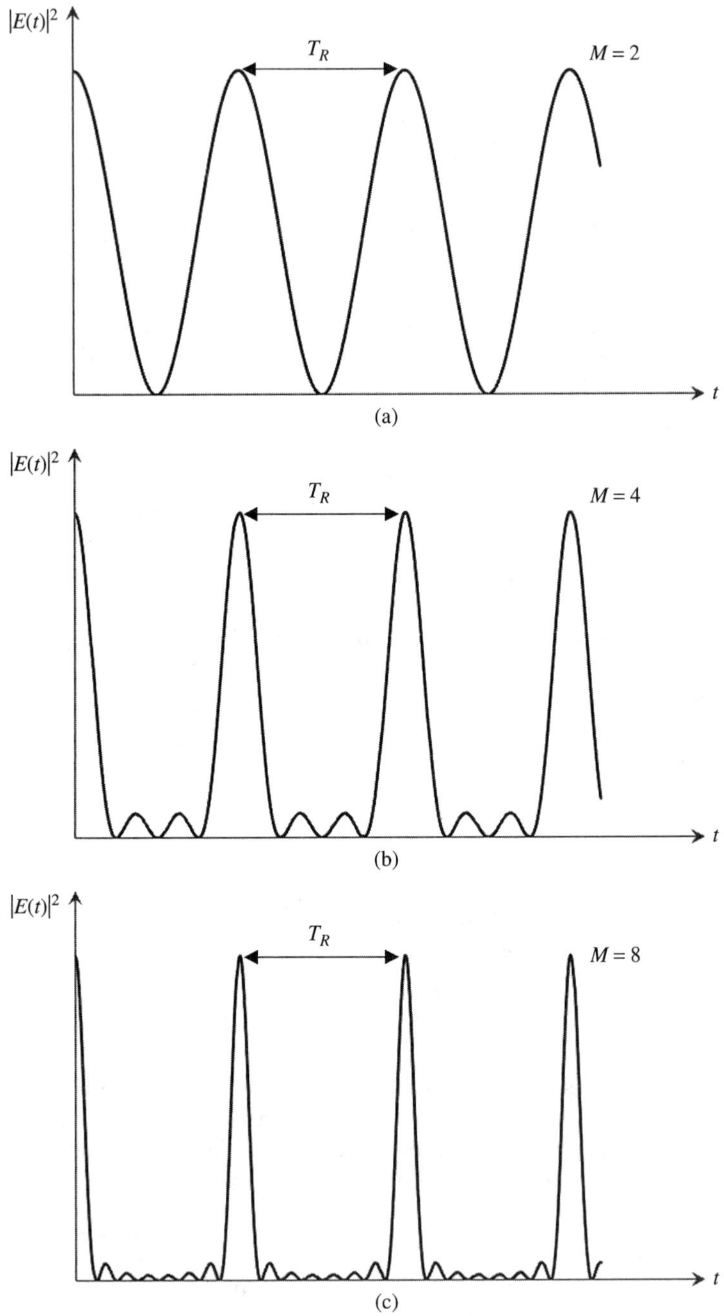

FIGURE 6.40 The output power of a mode-locked laser as a function of time for the cases where the number (M) of locked modes is (*a*) 2, (*b*) 4, (*c*) 8, and (*d*) 16.

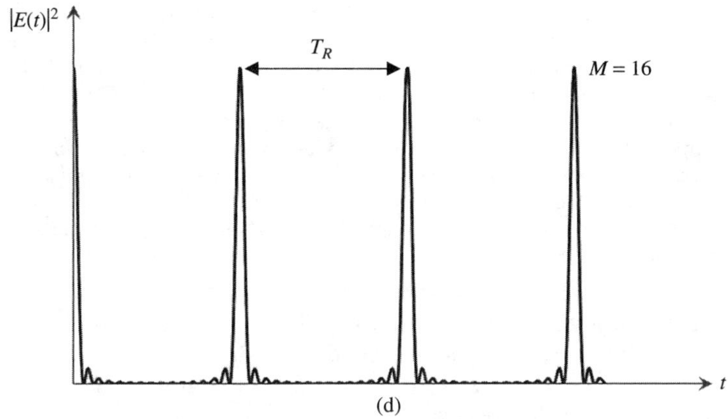

Figure 6.40 (*Continued*)

2. Since the average power is proportional to $M \times |E_0|^2$, the peak power of the pulses is M times larger than the average power. Hence, the larger the number of locked modes, the larger is the corresponding peak power of the mode-locked laser.
3. The pulse width τ_{pulse}, which can be roughly defined as the time difference between the pulse peak and the first zero, can be estimated from

$$\frac{M\omega_F \tau_{\text{pulse}}}{2} = \pi \tag{6.165}$$

resulting in

$$\tau_{\text{pulse}} = \frac{1}{M}\left(\frac{2\pi}{\omega_F}\right) = \frac{T_R}{M} \tag{6.166}$$

Equation (6.166) clearly shows that as the number of locked modes increases, the pulse width decreases. If all of the modes within the spectral gain bandwidth $\Delta\nu$ are locked, M is given by $M \approx \Delta\nu/(c/2d)$ and the pulse width becomes

$$\tau_{\text{pulse}} \approx \frac{1}{\Delta\nu} \tag{6.167}$$

Example To appreciate the effectiveness of mode locking in producing ultrashort optical pulses with high peak powers, let us consider the Cr^{4+}:YAG tunable solid-state gain medium as an example. The Cr^{4+}:YAG medium, when pumped near the wavelength of 1 μm, provides optical gain near the wavelength of 1.5 μm with a spectral bandwidth of approximately 40 THz (40×10^{12} Hz).[38] Suppose that the Cr^{4+}:YAG laser resonator has a length of 1.5 m. The corresponding free spectral range will be 100 MHz from Eq. (6.161). If the full gain bandwidth is used during mode locking, we will have up to 400,000 longitudinal modes locked together. The resulting pulsewidth will be of the order of 25 fs (25×10^{-15} s). Another striking outcome is the peak power of the mode-locked pulses. Suppose that the resonator outputs a modest average power of 100 mW. The peak power of the pulses will then be (400,000 × 100 mW) or 40 kW. So, the laser will then produce 25-fs pulses with a peak power of 40 kW at a repetition rate of 100 MHz, meaning that 100 million of such pulses are produced per second from the resonator.

As the above example illustrates, the number of locked modes can be quite large in a mode-locked laser. In such a case, we can replace the summation in Eq. (6.163) with an integral and $E(t)$ will be given in terms of the spectral amplitude distribution $E_p(\omega)$ as

$$E(t) = \int_{-\infty}^{+\infty} d\omega [E_p(\omega)\exp(i\phi(\omega))\exp(i\omega t)] \tag{6.168}$$

Above, ω is the angular frequency and $\phi(\omega)$ is the phase. For effective mode locking, $\phi(\omega)$ must be constant as a function of ω as was assumed in the case involving the discrete sum of M modes. In a practical laser, however, as the locked bandwidth becomes large, dispersive and nonlinear effects distort the phase, and compensation techniques must be employed to minimize these distortions across the gain bandwidth. When all of the unwanted phase distortions are absent, one can then produce the shortest possible pulse for a given spectral bandwidth. In this case, we say that the pulse is transform-limited. For a transform-limited pulse, the available spectral bandwidth $\Delta\nu$ (FWHM) and the output pulsewidth τ_{pulse} (FWHM) satisfy a reciprocal relation similar to Eq. (6.167). This can be expressed as

$$\Delta\nu\tau_{\text{pulse}} = K \tag{6.169}$$

where K is a constant of the order of unity and is called the time-bandwidth product. The value of K depends on the particular pulse profile generated. For example, for a Gaussian and hyperbolic secant-squared (sech^2) pulse, K = 0.441 and 0.315, respectively.[45, 46]

Several mode locking techniques exist. These are categorized into two groups:

1. Active mode locking: Here, a periodic loss mechanism is introduced into the cavity and the resulting modulation is synchronized with the cavity round-trip time. In other words, we can think of this periodic modulation as a shutter that opens and closes at the frequency equal to the free spectral range of the cavity. The shutter will be open every time a pulse completes a round trip and hence enables the generation

of a periodic pulse train. Either the phase or the amplitude can be modulated during active mode locking. In the next section, we will analyze the case of amplitude modulation (AM) mode locking.

2. Passive mode locking: Here, a saturable absorber is introduced into the resonator. If the absorber parameters are properly optimized, its transmission will increase with increasing light intensity and the resonator will favor the generation of high-intensity mode-locked pulses.

Mode-locked lasers have many applications in science and technology including time-resolved analysis of ultrafast processes,[44, 47, 48] biomedical imaging,[49,50] high harmonic generation,[51-53] and material processing,[54, 55] among others. More recently, mode-locked lasers are gaining importance in metrology applications as well. Since the output of a mode-locked laser consists of equally spaced frequency components (now commonly referred to as a frequency comb), it is possible to stabilize the repetition frequency and the carrier envelope phase and use the output as a high-precision frequency standard over a large spectral range. Many methods have been developed and demonstrated for the generation of stabilized frequency combs from mode-locked lasers.[56-59]

6.8.2 Analysis of AM Mode Locking

Here, we follow Ref. 60 to analyze AM mode locking. An alternative derivation can be found in Refs. 1, 61, and 62. As we discussed above, two adjacent axial modes are separated in frequency by $\omega_F = 2\pi/T_R$. To achieve AM mode locking, a modulator is placed inside the laser cavity to periodically vary the loss. This can be done by using, for example, an acousto-optic deflector, which is discussed in Chap. 8. The amplitude transmission function $T(t)$ of the modulator can be expressed as

$$T(t) = (1 - M) + M\cos\omega_m t \tag{6.170}$$

where M is the modulation parameter and ω_m is the modulation frequency. Variation of $T(t)$ as a function of time is shown in Fig. 6.41.

Let the electric field component incident on the modulator be $E_i(t)$, given by

$$E_i(t) = E_0 \exp(i\omega t) \tag{6.171}$$

Here, E_0 is the field amplitude and ω is the angular frequency. The emerging field $E_{\text{out}}(t)$ after the modulator becomes

$$\begin{aligned} E_{\text{out}}(t) = T(t)E_i(t) = {} & (1-M)E_0\exp(i\omega t) + \frac{M}{2}[\exp(i\omega_m t) \\ & + \exp(-i\omega_m t)](E_0\exp(i\omega t)) = E_0\exp(i\omega t) \\ & + \frac{M}{2}[E_0\exp(i(\omega+\omega_m)t) + E_0\exp(i(\omega-\omega_m)t) - 2E_0\exp(i\omega t)] \end{aligned} \tag{6.172}$$

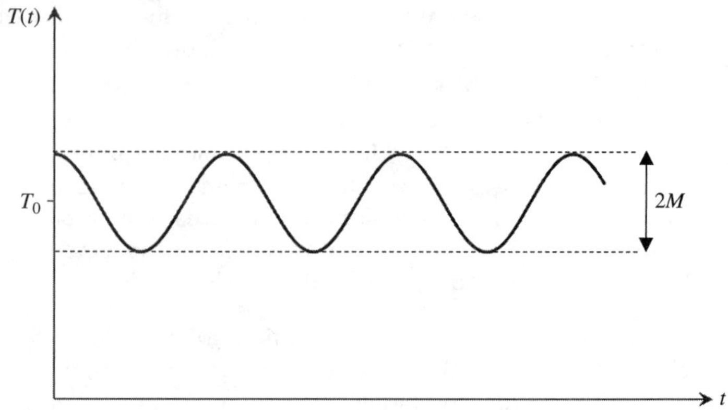

Figure 6.41 Variation of the amplitude modulation function $T(t)$ as a function of time.

The change ΔE in the electric field can be expressed as

$$\begin{aligned}\Delta E &= E_{out}(t) - E_i(t)\\ &= \frac{M}{2}[E_0 \exp(i(\omega-\omega_m)t) - 2E_0 \exp(i\omega t)\\ &\quad + E_0 \exp(i(\omega+\omega_m)t)]\end{aligned} \tag{6.173}$$

We see, therefore, that the periodic modulator at ω_m creates two sidebands at $(\omega-\omega_m)$ and $(\omega+\omega_m)$. We now let E_q be the amplitude of the axial mode at frequency ω_q given by

$$\omega_q = \omega_0 + q\omega_F \tag{6.174}$$

If the modulator drive frequency ω_m is set to ω_F, then the sidebands generated by the modulator exactly overlap with the adjacent cavity modes. Hence, the modulator will coherently feed energy from E_q to the neighboring cavity modes. We see from Eq. (6.173) that during each passage through the modulator, the change in the amplitude of the qth mode becomes

$$\Delta E_q = \frac{M}{2}(E_{q-1} - 2E_q + E_{q+1}) \tag{6.175}$$

To see the effect of the gain medium on the electric field, we assume that we have a homogeneously broadened gain medium with a Lorentzian lineshape. The frequency-dependent amplitude gain $g(\omega)$ can be expressed as

$$g(\omega) = \frac{g_0}{1+\left(\dfrac{\omega_q-\omega_0}{\Omega_g}\right)^2} \tag{6.176}$$

where Ω_g is the gain bandwidth, g_0 is the amplitude gain at the line center, and $\omega_q = \omega_0 + q\omega_F$. If the mode frequencies are close to the center of the gain curve, we can further approximate $g(\omega)$ as

$$g(\omega_q) \approx g_0\left(1-\left(\frac{q\omega_F}{\Omega_g}\right)^2+\cdots\right) \tag{6.177}$$

If the single-pass fractional amplitude loss is ℓ_A, then the net change in the amplitude over a single pass will become

$$\Delta E_q = g_0\left(1-\left(\frac{q\omega_F}{\Omega_g}\right)^2\right)E_q - \ell_A E_q + \frac{M}{2}(E_{q-1} - 2E_q + E_{q+1}) \tag{6.178}$$

Now, we assume that the axial mode frequencies are very closely spaced so that the quantity $\Omega = q\omega_F$ can be treated as a continuous variable. From the limit definition of the derivative function, it can be shown that

$$E_{q-1} - 2E_q + E_{q+1} = \omega_F^2 \frac{d^2E}{d\Omega^2} = \omega_m^2 \frac{d^2E}{d\Omega^2} \tag{6.179}$$

Hence, the change in the electric field during a single pass becomes

$$\Delta E(\Omega) = (g_0 - \ell_A)E(\Omega) - g_0\left(\frac{\Omega}{\Omega_g}\right)^2 E(\Omega) + \frac{1}{2}M\omega_m^2 \frac{d^2E(\Omega)}{d\Omega^2} \tag{6.180}$$

In steady state, we have $\Delta E(\Omega) = 0$, leading to

$$(g_0 - \ell_A)E(\Omega) - g_0\left(\frac{\Omega}{\Omega_g}\right)^2 E(\Omega) + \frac{1}{2}M\omega_m^2 \frac{d^2E(\Omega)}{d\Omega^2} = 0 \tag{6.181}$$

A little bit of pondering shows that Eq. (6.182) looks just like the Schrödinger equation for the harmonic oscillator which we saw back in Chap. 5:

$$-\frac{\hbar^2}{2m}\frac{d^2\psi}{dx^2} + \frac{1}{2}m\omega^2 x^2 \psi = E\psi \tag{6.182}$$

We can therefore solve Eq. (6.181) by making a one-to-one correspondence with Eq. (6.182). The ground-state eigen solution and the corresponding eigenvalue $(g_0 - \ell_A)$ become

$$E(\Omega) = \exp\left(-\frac{\alpha^2\Omega^2}{2}\right) \tag{6.183}$$

and

$$g_0 - \ell_A = \frac{M}{2}(\omega_m)^2\alpha^2 \tag{6.184}$$

respectively, where α is given by

$$\alpha = \left(\frac{2g_0}{M\omega_m^2\Omega_g^2}\right)^{1/4} \tag{6.185}$$

Note that, interestingly, the same differential equation for Hermite polynomials has come up in three different contexts up to now: Gaussian beams, harmonic oscillator, and AM mode locking. The reader may consult App. A for a further discussion of these polynomials.

Above, we have carried out the analysis of AM mode locking in the frequency domain. The solution $E(\Omega)$ given in Eq. (6.183) is therefore the Fourier transform of the temporal solution. The temporal pulse profile $E(t)$ can be calculated from

$$E(t) = \int_{-\infty}^{+\infty} d\Omega \exp(i\Omega t)E(\Omega) \tag{6.186}$$

which yields a Gaussian proportional to $\exp(-t^2/2\alpha^2)$. The corresponding temporal pulsewidth (FWHM) τ_{pulse} becomes

$$\tau_{\text{pulse}} = \frac{\sqrt{2\ln 2}}{\pi}\left(\frac{2g_0}{M}\right)^{1/4}\frac{1}{\sqrt{f_m \Delta\nu_g}} \tag{6.187}$$

which is the well-known Siegman-Kuizenga relation for the pulsewidth in AM mode locking.[61] In Eq. (6.187), f_m is the modulation frequency (in Hz) related to ω_m through $\omega_m = 2\pi f_m$ and $\Delta\nu_g$ is the spectral gain band width (in Hz) related to Ω_g via $\Omega_g = 2\pi\Delta\nu_g$.

The reader may consult Refs. 1, 19, 20, and 62 through 73 for a further discussion on the theory and experimental demonstration of other mode-locking techniques in lasers.

6.8.3 Measurement of Mode-Locked Laser Pulses

Due to the bandwidth limitation of current electronic measurement systems, it is not possible to directly measure the duration of subpicosecond optical pulses. Instead, correlation methods are typically used. In this section, we describe the operation of an autocorrelator for pulsewidth characterization. This method, first reported in Ref. 74, works well if the pulsewidth is long compared to the temporal duration of one optical cycle at the carrier wavelength. As the pulse duration approaches few optical cycles, more sophisticated methods are needed to fully characterize the pulses. For a

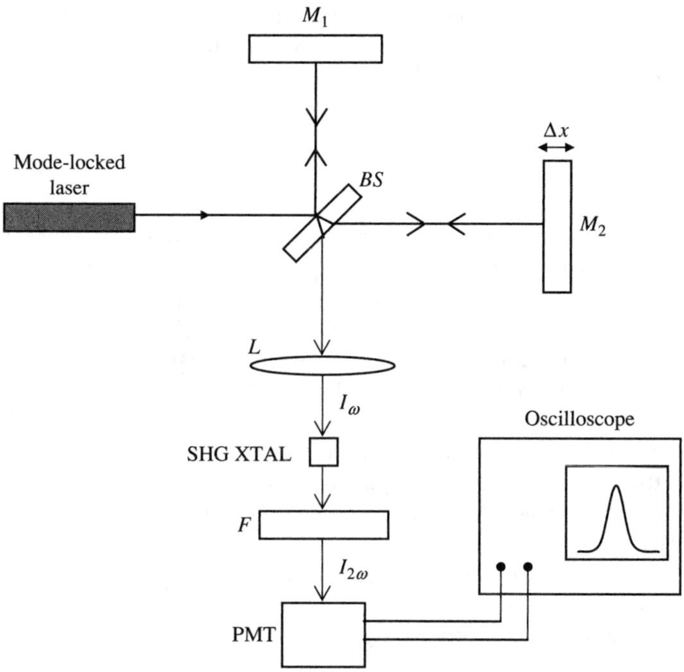

FIGURE 6.42 Schematic of an autocorrelator for measuring the pulse duration of mode-locked lasers.

detailed overview of different approaches, we recommend the reader consult Refs. 45, 46, 75, and 76.

Figure 6.42 shows a schematic of an autocorrelator. In its simplest form, it consists of a Michelson interferometer and a second harmonic generation setup. The output of a mode-locked laser is first split into two equal-intensity beams by a beam splitter (BS in Fig. 6.42). Each beam travels a different optical path length in the arms of a Michelson interferometer. When the two beams are returned by the retroreflecting mirrors M_1 and M_2, they are recombined by the beamplitter and focused with a lens (L) inside a second harmonic crystal (SHG XTAL). We will discuss nonlinear optics and second harmonic generation later in Chap. 11. For the time being, it is sufficient to just mention that the second harmonic crystal, when properly oriented, converts the light at the fundamental frequency of ω into its second harmonic at 2ω with a conversion efficiency that is proportional to the light intensity at ω. The filter (F) placed after the second harmonic crystal separates the second harmonic signal, which is monitored with a sensitive detector such as a photo-multiplier tube (PMT in Fig. 6.42). The operation of PMTs is further discussed in Chap. 9. The two arms of the interferometer are made nearly equal, and one of the arms (terminated with mirror M_2 in Fig. 6.42) is periodically varied to change the temporal delay between the two pulse trains. The temporal delay τ between the pulses can be calculated in terms of the displacement Δx of the variable arm from

$$\tau = \frac{2\Delta x}{c} \tag{6.188}$$

where c is the vacuum speed of light. Note the factor of 2 in Eq. (6.188), since the pulses complete a round trip in the arm before coming back to the beam splitter. This way, we are able to map easily measurable distances (from microns to millimeters) to time delays in the picoseond-femtosecond range. For example, a displacement of 10 μm corresponds to a delay of 66.7 fs (1 fs = 10^{-15} s). Such displacements can be readily obtained with high-precision translation stages. Since the second-harmonic conversion efficiency depends on the intensity at ω, the second-harmonic signal will vary strongly with the delay between the pulses. As a specific case, suppose that the beams are collinear going into the second harmonic crystal. In other words, they spatially overlap. In this case, if the detector response is slow, the measured second harmonic signal intensity $I_{2\omega}$ will be proportional to the function $A(\tau)$ given by[45]

$$A(\tau) = (1 + 2G^{(2)}(\tau)) \tag{6.189}$$

where $G^{(2)}(\tau)$ is the normalized second-order autocorrelation function which can be expressed in terms of the pulse intensity $I_\omega(t)$ and its delayed replica $I_\omega(t + \tau)$ at the fundamental frequency as

$$G^{(2)}(\tau) = \frac{\int_{-\infty}^{+\infty} dt I_\omega(t) I_\omega(t+\tau)}{\int_{-\infty}^{+\infty} dt I_\omega^2(t)} \tag{6.190}$$

Above, τ is the delay given by Eq. (6.188). In the case of a mode-locked laser, the autocorrelation function $A(\tau)$ has a peak at $\tau = 0$ with a width (FWHM) of τ_{AC} as shown in Fig. 6.43. In addition, if the split beams are collinear, then Eqs. (6.189) and (6.190) show that the ratio of the peak second harmonic signal to the background is 3:1 as shown in Fig. 6.43. Actually, seeing a 3:1 contrast ratio in the collinear autocorrelation also verifies that the laser is indeed mode-locked. If we were instead looking at the autocorrelation of a noise burst, the corresponding ratio would be 2:1. To determine the actual pulse duration τ_p from the width of the autocorrelation signal, we need to assume a specific pulse profile. For example τ_{AC}/τ_p is 1.55 and 1.44 for sech^2 and Gaussian pulses, respectively.[45]

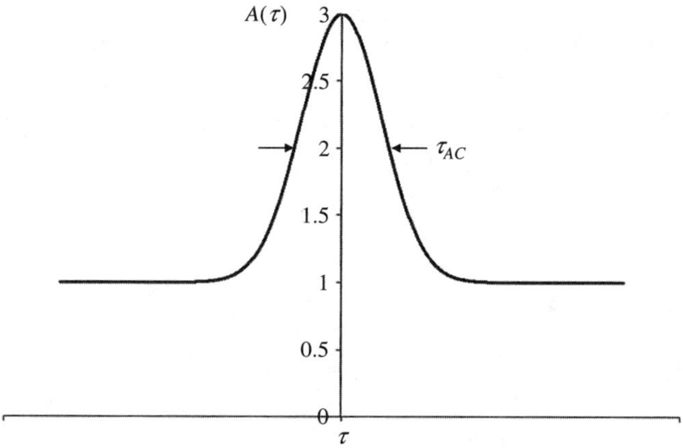

Figure 6.43 Variation of the autocorrelation signal $A(\tau)$ with the temporal delay τ between the pulses.

Problems

6.1 Show that for a blackbody at the absolute temperature T and emitting in the frequency interval $\Delta\nu$, the spectral brightness B is given by Eq. (6.5).

6.2 Show that at large distances from the waist location, the product of beam cross-sectional area and solid angle for a Gaussian beam is given by Eq. (6.7).

6.3 Show, by using Eq. (6.22) and based on the observations in Figs. 6.6 and 6.7, that the normalized mutual coherence function $\gamma(\tau)$ is given by Eq. (6.23).

6.4 Show that the spectral intensity distribution and the corresponding spectral width (FWHM) of a sine wave of finite duration τ_c is given by Eqs. (6.25) and (6.26), respectively.

6.5 Show that for a two-level ideal atomic system interacting with a nearly monochromatic wave resonant with the transition, the steady-state population inversion and the saturation intensity are given by Eqs. (6.45) and (6.46).

6.6 Show that the steady-state population inversion for an ideal three-level system is given by Eq. (6.49) and obtain the expressions (6.50) and (6.51) for the small-signal inversion and saturation intensity.

6.7 Show that the steady-state population inversion for an ideal four-level system is given by Eq. (6.54) and obtain the expressions (6.55) and (6.56) for the small-signal inversion and saturation intensity.

6.8 Show that in steady state, the upper-state population density of an ideal, optically pumped four-level system is given by Eq. (6.79) and derive the expressions (6.80) and (6.81) for the saturation intensities at the pump and laser wavelengths.

6.9 Show, by using Eq. (6.77), that the single-pass fractional power gain $G(t)$ satisfies the differential equation given in Eq. (6.97). Also obtain the steady-state solution G_{ss} for $G(t)$ given in Eq. (6.100).

6.10 The following table gives the measured variation of the threshold pump power for output couplers with different transmission for a Cr:forsterite laser operating at 1270 nm. By doing a linear, least-squares fit to the data, determine the cavity round-trip loss L.

T	$(P_p)_{th}$(mW)
0.001	311
0.025	707
0.047	1100

6.11 Show that the perturbations in the fractional power gain and laser power satisfy the linearized differential equations given in Eqs. (6.131) and (6.132).

6.12 Find the solution of the linear system of equations given in Eq. (6.134) and show that the exponent s is given by Eq. (6.136).

6.13 Consider 1 at. % Nd:YAG crystal. The chemical formula of YAG is $Y_3Al_5O_{12}$. Nd^{3+} ions substitute for the Y^{3+} ions in the lattice. The density of the crystal is 4.56 gm/cm^3. Show that the total density N_T of the Nd^{3+} ions is 1.39×10^{20} cm^{-3}.

6.14 The absorption cross section (at the wavelength of 1064 nm) and upper-state lifetime of a four-level solid-state absorber are 2×10^{-19} cm^2 and 3 μs. The total ion density within the material is 4×10^{18} cm^{-3}.

(*a*) What is the energy of each photon at 1064 nm?

(*b*) How much of the incident power (in %) is absorbed in a 2 cm-long sample at low intensities?

(*c*) What will the absorption coefficient be at the pump intensity of 400 kW/cm^2?

6.15 Consider a solid-state laser with a stimulated emission cross section of 4×10^{-19} cm^2, fluorescence lifetime of 4 μs, output wavelength of 1250 nm, and an effective beam area of 3000 μm^2 in the gain medium. The length of the gain medium is 2 cm with a differential loss coefficient of 0.01 cm^{-1}. The transmission of the output coupler is 5%.

(*a*) If the threshold pump power is 2W, determine the single-pass small-signal power gain G_0 when the pump power is 4W.

(*b*) Calculate the output power when the pump power is 4W.

6.16 Consider a Q-switched laser with an effective cavity length of 15 cm. The reflectivity of the output coupler is 90%. The differential loss

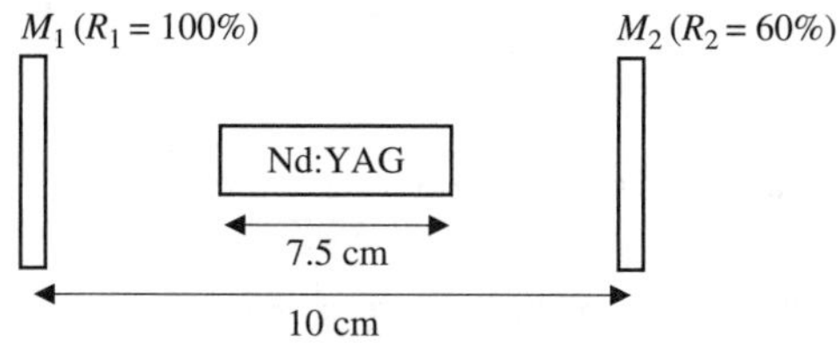

FIGURE 6.44

in the 5-cm-long gain rod is 0.02 cm^{-1}. If the initial gain builds up to ten times the cavity loss, calculate the width of the output pulse.

6.17 Analyze the active Q-switched operation of a 1 at. % Nd:YAG laser with the following parameters (see Fig. 6.44): Rod diameter = 6.2 mm, refractive index of Nd:YAG = 1.82, stimulated emission cross section = 8.8×10^{-19} cm^2, output wavelength = 1064 nm, optical pump wavelength = 808 nm, round-trip internal loss = 7.5%.

(*a*) Assuming that every absorbed photon creates on excited ion, calculate the threshold pump energy needed to reach lasing threshold.

(*b*) Assume that the initial inversion just before the cavity Q is increased is 10 times above the total loss. Estimate the peak output power.

(*c*) Calculate the output energy.

6.18 Consider the active Q-switching of a solid-state laser. The relevant cavity parameters are: effective length = 25 cm, round-trip internal loss = 5%, output coupler reflectivity = 80%, initial gain = 4 times loss, effective beam radius = 3 mm, output wavelength = 1 μ, fluorescence lifetime = 4 ms, stimulated emission cross section = 5×10^{-19} cm^2. Use the above equations to calculate the peak output power in MW obtained from the oscillator. Assume that pulse shaping occurs much faster than spontaneous decay of the inverted atoms.

6.19 A solid-state laser with a center frequency of 1250 nm and a gain bandwidth of 150 nm is actively mode-locked. Assume that the modulation frequency is 80 MHz with a modulation parameter of 10%. Furthermore, the single-pass amplitude gain is 5%.

(*a*) Determine the pulsewidth in picoseconds.

(*b*) What should the length of the cavity be?

6.20 In a solid-state laser operating in the cw regime, the threshold pump power (in W) varies as a function of the fractional output coupler transmission T (i.e., T is between 0 and 1) according to

$$P_{th} = 2 + 40T$$

Determine the total round-trip cavity loss L.

References

1. A. E. Siegman, *Lasers* (University Science Books, Mill Valley, Calif., 1986).
2. P. W. Milonni and J. H. Eberly, *Lasers* (Wiley, New York, 1988).
3. J. T. Verdeyen, *Laser Electronics,* 3d ed. (Prentice Hall, Upper Saddle River, N.J., 1995).
4. A. Yariv, *Optical Electronics in Modern Communications,* 5th ed. (Oxford University Press, New York, 1997).
5. O. Svelto, *Principles of Lasers,* 2d ed. (Plenum, New York, 1982).
6. A. Yariv and P. Yeh, *Photonics: Optical Electronics in Modern Communications,* 6th ed. (Oxford University Press, New York, 2007).
7. C. R. Pollock, *Fundamentals of Optoelectronics* (Irwin, Chicago, 1995).
8. M. Sargent, M. O. Scully, and W. E. Lamb, *Laser Physics* (Addison-Wesley, Reading, Mass. 1974).
9. T. W. Hansch and B. Couillaud, "Laser frequency stabilization by polarization spectroscopy of a reflecting reference cavity," *Optics Communications,* **35**(3), 441–444 (1980).
10. R. W. P. Drever, J. L. Hall, F. V. Kowalski, J. Hough, G. M. Ford, A. J. Munley, and H. Ward, "Laser phase and frequency stabilization using an optical-resonator," *Applied Physics B-Photophysics and Laser Chemistry,* **31**(2), 97–105 (1983).
11. B. C. Young, F. C. Cruz, W. M. Itano, and J. C. Bergquist, "Visible lasers with subhertz linewidths," *Physical Review Letters,* **82**(19), 3799–3802 (1999).
12. L. Mandel and E. Wolf, *Optical Coherence and Quantum Optics* (Cambridge University Press, Cambridge, 1995).
13. G. B. Arfken and H. J. Weber, *Mathematical Methods for Physicists,* 5th ed. (Harcourt Academic Press, Amsterdam, 2001).
14. G. R. Fowles, *Introduction to Modern Optics* (Dover, New York, 1975).
15. T. Y. Fan, "Optimizing the efficiency and stored energy in quasi-3-level lasers," *IEEE Journal of Quantum Electronics,* **28**(12), 2692–2697 (1992).
16. T. Taira, W. M. Tulloch, and R. L. Byer, "Modeling of quasi-three-level lasers and operation of cw Yb:YAG lasers," *Applied Optics,* 36(9), 1867–1874 (1997).
17. W. F. Krupke, "Ytterbium solid-state lasers: The first decade," *IEEE Journal of Selected Topics in Quantum Electronics,* **6**(6), 1287–1296 (2000).
18. G. J. Spuhler, R. Paschotta, R. Fluck, B. Braun, M. Moser, G. Zhang, E. Gini, and U. Keller, "Experimentally confirmed design guidelines for passively Q-switched microchip lasers using semiconductor saturable absorbers," *Journal of the Optical Society of America B-Optical Physics,* **16**(3), 376–388 (1999).
19. U. Keller, K. J. Weingarten, F. X. Kartner, D. Kopf, B. Braun, I. D. Jung, R. Fluck, C. Honninger, N. Matuschek, and J. A. der Au, "Semiconductor saturable absorber mirrors (SESAMs) for femtosecond to nanosecond pulse generation in solid-state lasers," *IEEE Journal of Selected Topics in Quantum Electronics,* **2**(3), 435–453 (1996).
20. S. Tsuda, W. H. Knox, S. T. Cundiff, W. Y. Jan, and J. E. Cunningham, "Mode-locking ultrafast solid-state lasers with saturable Bragg reflectors," *IEEE Journal of Selected Topics in Quantum Electronics,* **2**(3), 454–464 (1996).
21. C. C. Davis, *Lasers and Electro-Optics* (Cambridge University Press, Cambridge, 1996).
22. T. H. Maiman, "Stimulated optical radiation in ruby," *Nature,* **187**(4736), 493–494 (1960).
23. E. U. Condon and G. H. Shortly, *The Theory of Atomic Spectra* (Cambridge University Press, Cambridge, 1964).
24. B. R. Judd, *Operator Techniques in Atomic Spectroscopy* (Princeton University Press, Princeton, 1998).
25. A. A. Kaminskii, *Crystalline Lasers: Physical Processes and Operating Schemes* (CRC Press, Boca Raton, 1996).
26. B. Henderson and G. F. Imbusch, *Optical Spectroscopy of Inorganic Solids* (Clarendon Press, Oxford, 1989).
27. R. C. Powell, *Physics of Solid-State Laser Materials* (Springer Verlag, New York, 1998).

28. W. Koechner, *Solid-State Laser Engineering,* 2d ed. (Springer-Verlag, Berlin, 1988).
29. J. E. Geusic, H. M. Marcos, and L. G. Van Uitert, "Laser oscillations in Nd-doped yttrium aluminum, yttrium gallium and gadolinium garnets, *Applied Physics Letters,* **4**(10), 182–184 (1964).
30. P. F. Moulton, "Tunable solid-state lasers," *Proceedings of the IEEE,* **80**(3), 348–364 (1992).
31. P. F. Moulton, "Spectroscopic and laser characteristics of Ti:Al_2O_3," *Journal of the Optical Society of America B-Optical Physics,* **3**(1), 125–133 (1986).
32. P. F. Moulton, "An investigation of the Co-M_gf_2 Laser system," *IEEE Journal of Quantum Electronics,* **21**(10), 1582–1595 (1985).
33. V. Petricevic, S. K. Gayen, R. R. Alfano, K. Yamagishi, H. Anzai, and Y. Yamaguchi, "Laser action in chromium-doped forsterite," *Applied Physics Letters,* **52**(13), 1040–1042 (1988).
34. L. D. Deloach, R. H. Page, G. D. Wilke, S. A. Payne, and W. F. Krupke, "Transition metal-doped zinc chalcogenides: Spectroscopy and laser demonstration of a new class of gain media," *IEEE Journal of Quantum Electronics* **32**(6), 885–895 (1996).
35. T. J. Carrig, "Transition-metal-doped chalcogenide lasers," *Journal of Electronic Materials* **31**(7), 759–769 (2002).
36. A. Sennaroglu, U. Demirbas, A. Kurt, and M. Somer, "Direct experimental determination of the optimum chromium concentration in continuous-wave Cr^{2+}: ZnSe lasers," *IEEE Journal of Selected Topics in Quantum Electronics,* **13**(3), 823–830 (2007).
37. A. Sennaroglu, "Continuous wave thermal loading in saturable absorbers: Theory and experiment," *Appl. Opt.*, **36**, 9528–9535 (1997).
38. A. Sennaroglu, "Broadly tunable Cr^{4+} doped solid-state lasers in the near infrared and visible," *Progress in Quantum Electronics,* **26**(6), 287–352 (2002).
39. F. X. Kartner, L. R. Brovelli, D. Kopf, M. Kamp, I. Calasso, and U. Keller, "Control of solid-state laser dynamics by semiconductor-devices," *Optical Engineering,* **34**(7), 2024–2036 (1995).
40. F. X. Kaertner, *Ultrafast Optics, Lecture Notes,* http://ocw.mit.edu/OcwWeb/Electrical-Engineering-and-Computer-Science/6-977Spring-2005/LectureNotes/index.htm (MIT, 2005).
41. D. Findlay and R. A. Clay, "Measurement of internal losses in 4-level lasers," *Physics Letters,* **20**(3), 277–278 (1966).
42. J. A. Caird, S. A. Payne, P. R. Staver, A. J. Ramponi, L. L. Chase, and W. F. Krupke, "Quantum electronic-properties of the $Na_3Ga_2Li_3F_{12}$:Cr^{3+} laser," *IEEE Journal of Quantum Electronics,* **24**(6), 1077–1099 (1988).
43. L. E. Hargrove, R. L. Fork, and M. A. Pollack, "Locking of He-Ne laser modes induced by synchronous intracavity modulation," *Applied Physics Letters,* **5**(1), 4–5 (1964).
44. M. Hentschel, R. Kienberger, C. Spielmann, G. A. Reider, N. Milosevic, T. Brabec, P. Corkum, U. Heinzmann, M. Drescher, and F. Krausz, "Attosecond metrology," *Nature,* **414**(6863), 509–513 (2001).
45. E. P. Ippen and C. V. Shank, in *Ultrashort Light Pulses: Springer Topics in Applied Physics,* S. L. Shapiro, ed. (Springer-Verlag, Berlin, 1977), p. 83.
46. D. J. Bradley and G. H. C. New, "Ultrashort pulse measurements," *Proceedings of the IEEE,* **62**(3), 313–345 (1974).
47. A. H. Zewail, "Laser femtochemistry," *Science,* **242**, 1645–1653 (1988).
48. J. G. Fujimoto, J. M. Liu, E. P. Ippen, and N. Bloembergen, "Femtosecond laser interaction with metallic tungsten and nonequilibrium electron and lattice temperatures," *Physical Review Letters,* **53**(19), 1837–1840 (1984).
49. D. Huang, E. A. Swanson, C. P. Lin, J. S. Schuman, W. G. Stinson, W. Chang, M. R. Hee, T. Flotte, K. Gregory, C. A. Puliafito, and J. G. Fujimoto, "Optical coherence tomography," *Science,* **254**, 1178–1181 (1991).
50. W. Denk, J. H. Strickler, and W. W. Webb, "2-Photon laser scanning fluorescence microscopy," *Science,* **248**(4951), 73–76 (1990).

51. T. Brabec and F. Krausz, "Intense few-cycle laser fields: Frontiers of nonlinear optics," *Reviews of Modern Physics*, **72**(2), 545–591 (2000).
52. P. B. Corkum, "Plasma perspective on strong-field multiphoton ionization," *Physical Review Letters*, **71**(13), 1994–1997 (1993).
53. A. Rundquist, C. G. Durfee, Z. H. Chang, C. Herne, S. Backus, M. M. Murnane, and H. C. Kapteyn, "Phase-matched generation of coherent soft X-rays," *Science*, **280**(5368), 1412–1415 (1998).
54. B. N. Chichkov, C. Momma, S. Nolte, F. von Alvensleben, and A. Tunnermann, "Femtosecond, picosecond and nanosecond laser ablation of solids," *Applied Physics a-Materials Science & Processing*, **63**(2), 109–115 (1996).
55. C. B. Schaffer, A. Brodeur, J. F. Garcia, and E. Mazur, "Micromachining bulk glass by use of femtosecond laser pulses with nanojoule energy," *Optics Letters* **26**(2), 93-95 (2001).
56. D. J. Jones, S. A. Diddams, J. K. Ranka, A. Stentz, R. S. Windeler, J. L. Hall, and S. T. Cundiff, "Carrier-envelope phase control of femtosecond mode-locked lasers and direct optical frequency synthesis," *Science*, **288**(5466), 635–639 (2000).
57. H. R. Telle, G. Steinmeyer, A. E. Dunlop, J. Stenger, D. H. Sutter, and U. Keller, "Carrier-envelope offset phase control: A novel concept for absolute optical frequency measurement and ultrashort pulse generation," *Applied Physics B-Lasers and Optics*, **69**(4), 327–332 (1999).
58. T. Udem, R. Holzwarth, and T. W. Hansch, "Optical frequency metrology," *Nature*, **416**(6877), 233–237 (2002).
59. R. Holzwarth, T. Udem, T. W. Hansch, J. C. Knight, W. J. Wadsworth, and P. S. J. Russell, "Optical frequency synthesizer for precision spectroscopy," *Physical Review Letters*, **85**(11), 2264–2267 (2000).
60. H. A. Haus, "Mode-locking of lasers," *IEEE Journal of Selected Topics in Quantum Electronics*, **6**(6), 1173–1185 (2000).
61. A. E. Siegman and D. J. Kuizenga, "Active mode coupling phenomena in pulsed and continuous lasers," *Opto-electronics*, **6**, 43–66 (1974).
62. D. J. Kuizenga and A. E. Siegman, "FM and AM mode locking of the homogeneous laser-part 1: Theory," *IEEE J. Quantum Electron*, **6**, 694–708 (1970).
63. H. A. Haus, J. G. Fujimoto, and E. P. Ippen, "Analytic theory of additive pulse and kerr lens mode-locking," *IEEE Journal of Quantum Electronics*, **28**(10), 2086–2096 (1992).
64. H. A. Haus, J. G. Fujimoto, and E. P. Ippen, "Structures for additive pulse mode-locking," *Journal of the Optical Society of America B-Optical Physics*, **8**(10), 2068–2076 (1991).
65. F. X. Kartner, I. D. Jung, and U. Keller, "Soliton mode-locking with saturable absorbers," *IEEE Journal of Selected Topics in Quantum Electronics*, **2**(3), 540–556 (1996).
66. J. F. Pinto, C. P. Yakymyshyn, and C. R. Pollock, "Acoustooptic mode-locked soliton laser," *Optics Letters*, **13**(5), 383–385 (1988).
67. C. P. Yakymyshyn, J. F. Pinto, and C. R. Pollock, "Additive-pulse mode-locked Nacl-Oh-laser," *Optics Letters*, **14**(12), 621–623 (1989).
68. A. Sennaroglu, C. R. Pollock, H. Nathel, "Continuous-wave self-mode-locked operation of a femtosecond Cr^{4+}:YAG laser," *Optics Letters*, **19**(6)390–392(1994).
69. L. F. Mollenauer and R. H. Stolen, "The *Soliton Laser*," *Optics Letters*, **9**(1), 13–15 (1984).
70. H. A. Haus, "Theory of mode-locking with a slow saturable absorber," *IEEE Journal of Quantum Electronics*, **11**(9), 736–746 (1975).
71. U. Morgner, F. X. Kartner, S. H. Cho, Y. Chen, H. A. Haus, J. G. Fujimoto, E. P. Ippen, V. Scheuer, G. Angelow, and T. Tschudi, "Sub-two-cycle pulses from a Kerr-lens mode-locked Ti:sapphire laser," *Optics Letters*, **24**(6), 411–413 (1999).
72. D. E. Spence, P. N. Kean, W. Sibbett, "60-fsec pulse generation from a self-mode-locked Ti:sapphire laser," *Optics Letters*, **16**(1), 42–44 (1991).
73. H. A. Haus, "Theory of mode-locking with a fast saturable absorber," *Journal of Applied Physics*, **46**(7), 3049–3058 (1975).

74. J. A. Armstron, "Measurement of picosecond laser pulse widths," *Applied Physics Letters*, **10**(1), 16–18 (1967).
75. R. Trebino, K. W. DeLong, D. N. Fittinghoff, J. N. Sweetser, M. A. Krumbugel, B. A. Richman, and D. J. Kane, "Measuring ultrashort laser pulses in the time-frequency domain using frequency-resolved optical gating," *Review of Scientific Instruments*, **68**(9), 3277–3295 (1997).
76. R. Trebino and D. J. Kane, "Using phase retrieval to measure the intensity and phase of ultrashort pulses: Frequency-resolved optical gating," *Journal of the Optical Society of America a-Optics Image Science and Vision*, **10**(5), 1101–1111 (1993).

CHAPTER 7

Semiconductor Lasers

We have devoted a separate chapter to the discussion of semiconductor lasers since they occupy a central place in today's photonics technology. In this class of lasers, optical amplification and oscillation take place inside semiconducting crystalline solids. They posses many favorable features including compactness, long operational lifetime, and chemical stability. In addition, they can be mass produced in large quantities. Ever since their demonstration in the early 1960s,[1-3] they have been used in diverse applications ranging from DVD players to fiber optic communications systems and optical pumping of other lasers, in particular solid-state lasers. Different from solid-state lasers discussed in the previous chapter, semiconductor lasers are electrically excited and can be modulated at high speeds by modulating the driving current.

In this chapter, we start with a condensed review of semiconductor fundamentals. We introduce the concepts of energy bands, electrons/holes, and occupation probabilities. The density of states is derived for the conduction and valence bands to obtain expressions for the electron and hole densities. Electrical properties of doped semiconductors and current-voltage characteristics of the p-n junction are then discussed. Getting a good grasp of the electronic properties of semiconductors helps in better understanding their photonic properties, which make them important as optical detectors as well as coherent light sources. Semiconductor detectors will be discussed in Chap. 9. In the remainder of this chapter, we will derive the condition that must be satisfied for optical amplification to take place in a semiconductor. We then qualitatively explain the operation of semiconductor heterojunction lasers. We will build a fairly simple model of how semiconductor lasers operate by borrowing very basic ideas from solid-state physics. References 4 to 11 provide a more detailed treatment of solid-state physics and semiconductor devices. For further discussion on the optical and lasing characteristics of semiconductors, the reader may consult Refs. 12 to 19.

7.1 Basics of Semiconductors

7.1.1 Energy Bands and Charge Carriers in Semiconductors

Semiconductors are crystalline solids whose electrical conductivities are between those of metals and insulators. One of the most important distinguishing characteristics of semiconductors is that their electrical conductivity can be varied by a large amount by varying the temperature, doping level of impurities, or light illumination. This final characteristic enables their use as efficient optical detectors, as we will see in Chap. 9. In the fabrication of electronic devices such as diodes or transistors, silicon (Si), which is an elemental semiconductor, is widely used. In the case of photonic light sources such as LEDs or semiconductor lasers, gallium arsenide (GaAs) or other compound semiconductors are often used. Optical detectors can be made from all types of semiconductors.

The concept of energy bands is very important in the discussion of semiconductors. In a semiconductor crystalline solid such as silicon, we have a periodic array of closely spaced atoms whose wave functions overlap in space. As such, the electrons of individual atoms can no longer be treated independently. The solution of Schrödinger's equation for a periodic lattice shows that the individual atomic energy levels combine to form energy bands. We can think of these energy bands as collections of closely spaced atomic energy levels that can be approximated as a continuum. As far as the electronic and optical properties of semiconductors are concerned, we need to focus on the outermost, so-called valence electrons, around the atom. According to the Pauli exclusion principle, no two electrons can occupy the same quantum state. Taking into account the spin degeneracy of the electron, this means that we can have at most two electrons in each energy eigenstate. It turns out that at $T = 0$ K (absolute zero), the highest fully occupied band and the lowest empty band are the valence and conduction bands, schematically shown in Fig. 7.1. The vertical axis in Fig. 7.1 shows the energy. There is a forbidden energy gap of width E_g between the valence and the conduction bands. In other words, there is no solution of Schrödinger's equation with an eigen energy that falls within this energy interval. E_g is known as the band-gap energy of the semiconductor and is typically expressed in units of electron volts, abbreviated as eV. 1 eV = 1.602×10^{-19} J and corresponds to the kinetic energy acquired by an electron accelerated across a potential difference of 1V. As far as the electrical conduction characteristics of crystalline solids are concerned, we can classify them into three categories depending on the value of E_g. In insulators, the band-gap energy is typically large with $E_g > 3$ eV. In contrast, we have a very small E_g or no band gap at all in the case of electrical conductors. In the case of semiconductors, E_g values typically fall in the range of 0.1 to 3 eV.

FIGURE 7.1 Energy band diagram of a semiconductor. E_g is the band-gap energy.

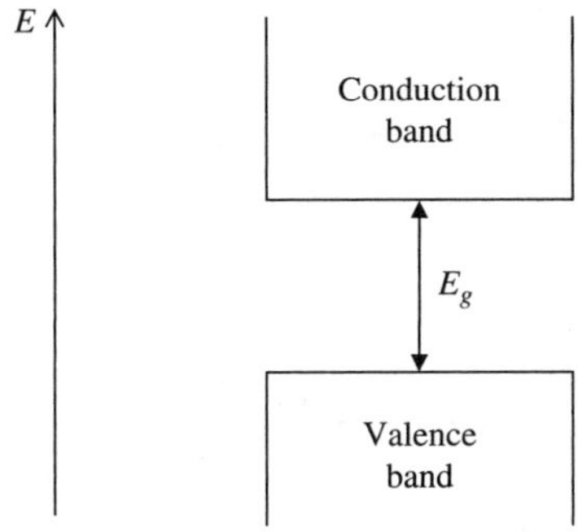

Let us first consider a semiconductor such as silicon at $T = 0$ K (zero kelvin), where we can neglect band-to-band thermal excitations. The valence band has the right number of energy levels to accommodate all of the valence electrons in the crystal. Hence, the valence band will be completely filled with electrons and the conduction band will be completely empty. In this case, we will have no electrical conduction. We can schematically show this by using the simplified two-dimensional bonding picture or the energy band picture which are depicted in Fig. 7.2*a* and *b*, respectively. In crystalline form, each silicon atom forms covalent bonds with its four nearest neighbors in a three-dimensional tetrahedral arrangement. The bonding picture in Fig. 7.2*a* is a simplified sketch of the actual structure and serves the purpose of showing the bound valence electrons around the silicon atoms. In Fig. 7.2*a* we showed each silicon atom with eight solid lines around it, where each solid line corresponds to one shared, bound electron. At $T = 0$ K, there are ideally no broken bonds and all of the valence electrons in the solid are bound to atoms and are not free to contribute to electrical conduction.

As the crystal temperature is increased, an increasing number of electrons will gain sufficient kinetic energy to overcome the binding force of the nuclei and become free carriers. We say, in this case, that a bond is broken, meaning that the electron is freed. This is schematically shown in Fig. 7.3*a*. In the energy band picture, the breaking of the bond corresponds to the excitation of the electron from the valence band to the conduction band (see Fig. 7.3*b*). Now, the electron is free and can contribute to electrical conduction. Note that in addition to the electron, we have a second type of charge carrier in a semiconductor. Under the action of an applied bias, the missing bond, which is referred to as a hole, drifts in a direction opposite to the direction of motion of the electrons and behaves as a positively charged particle. In the band picture shown in Fig. 7.3*b*, the hole corresponds to the empty state left behind in the valence band. We need to keep track of both electrons and holes in the discussion of electrical and optical properties of semiconductors.

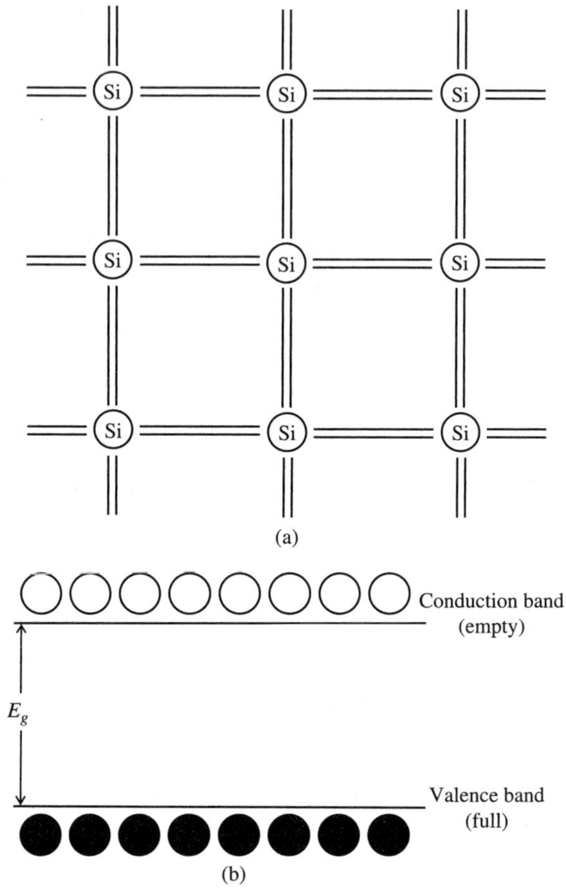

FIGURE 7.2 (*a*) Simplified two-dimensional bonding picture and (*b*) the energy band picture of tetrahedrally coordinated silicon atoms at $T = 0$ K.

7.1.2 Energy-Momentum Relations in Semiconductors

Energy-momentum relations, also referred to as E–k diagrams are very important in the discussion of the photonic properties of semiconductors. Before looking at the effect of the lattice on the energy, let us first consider a free electron. Since the binding potential is zero everywhere in this case, the electron energy is purely kinetic and will be given by

$$E = \frac{p^2}{2m_e} \tag{7.1}$$

where E is the energy, p is the momentum, and $m_e = 9.1 \times 10^{-31}$ kg for the electron. By using de Broglie's hypothesis ($p = h/\lambda$, h = Planck's

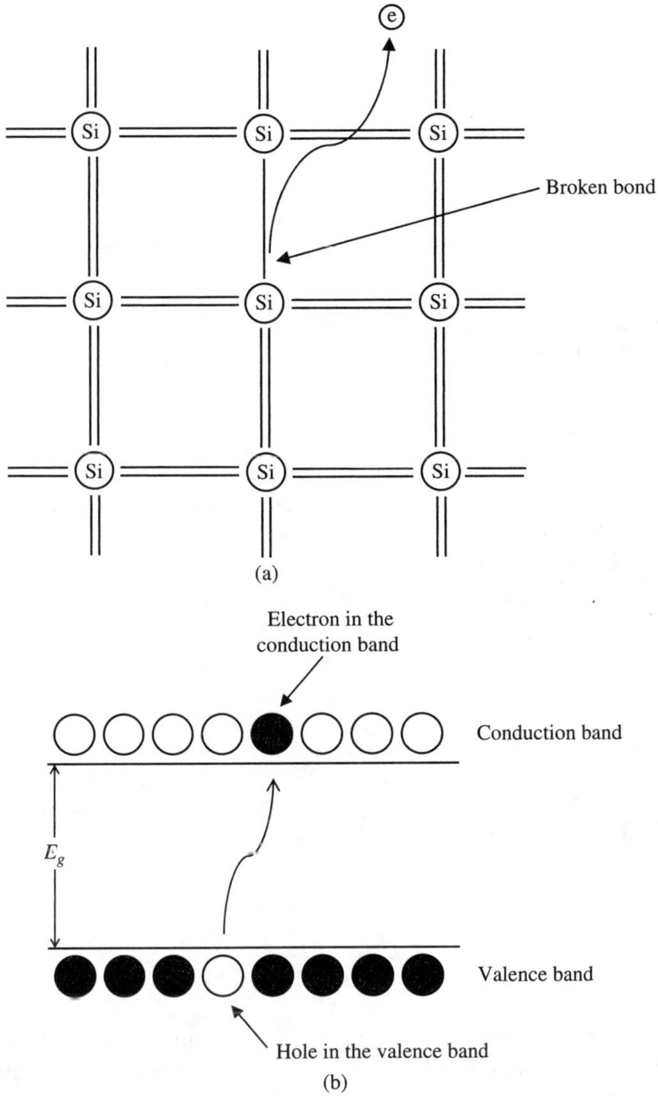

FIGURE 7.3 (*a*) Simplified two-dimensional bonding picture and (*b*) the energy-band picture of tetrahedrally coordinated silicon atoms at a finite temperature.

constant, λ = de Broglie wavelength), introduced in Chap. 5, the energy can be expressed in terms of the wave vector k as

$$E = \frac{\hbar^2 k^2}{2m_e} \tag{7.2}$$

where $\hbar = h/2\pi$. Variation of E as a function of k, known as the E–k diagram, becomes a simple parabola for a free electron as shown in Fig. 7.4.

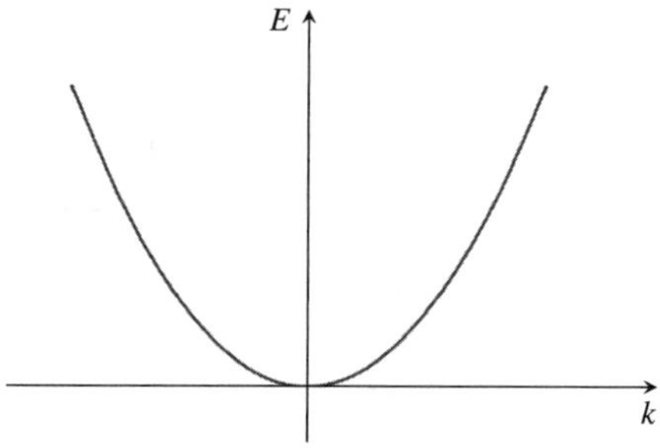

FIGURE 7.4 Variation of energy (E) as a function of the wave vector magnitude (k) for a free electron.

In a crystal, the situation is far more complex, and numerical computation schemes are typically necessary to calculate the variation of the energy with wave number. In general, one finds that the energy E is a periodic function of the components (k_x, k_y, k_z) of the wave vector $\vec{k}$ with periodicity $(\pi/a_1, \pi/a_2, \pi/a_3)$, where a_1, a_2, and a_3 are the lattice constants. The electron energies depend on the direction as well as on the magnitude of $\vec{k}$. Here, we will not be concerned with the techniques of how E–k diagrams are generated. Instead, we will look at certain qualitative features that are important in the discussion of the photonic properties. As an example, let us examine the simplified E–k diagram for silicon near the band edges, shown in Fig. 7.5*a*. Near the bottom of the conduction band, the energy E can be approximated as a quadratic function of k in the form

$$E = E_c + \frac{\hbar^2 k^2}{2m_e^*} \tag{7.3}$$

where E_c is the energy at the bottom of the conduction band, and m_e^* is known as the effective mass of the electron in the conduction band. We can say that near the bottom of the conduction band, the electron behaves nearly as a free particle with an effective mass m_e^*, which is typically different than the free-electron mass. Hence, the overall effect of the periodic lattice is taken into account by assigning an effective mass to the electron or the hole. The smaller the effective mass is, the larger the curvature of the E–k curve. Similarly, the E–k diagram near the top of the valence band has a parabolic shape and the energy is given by

$$E = E_v - \frac{\hbar^2 k^2}{2m_h^*} \tag{7.4}$$

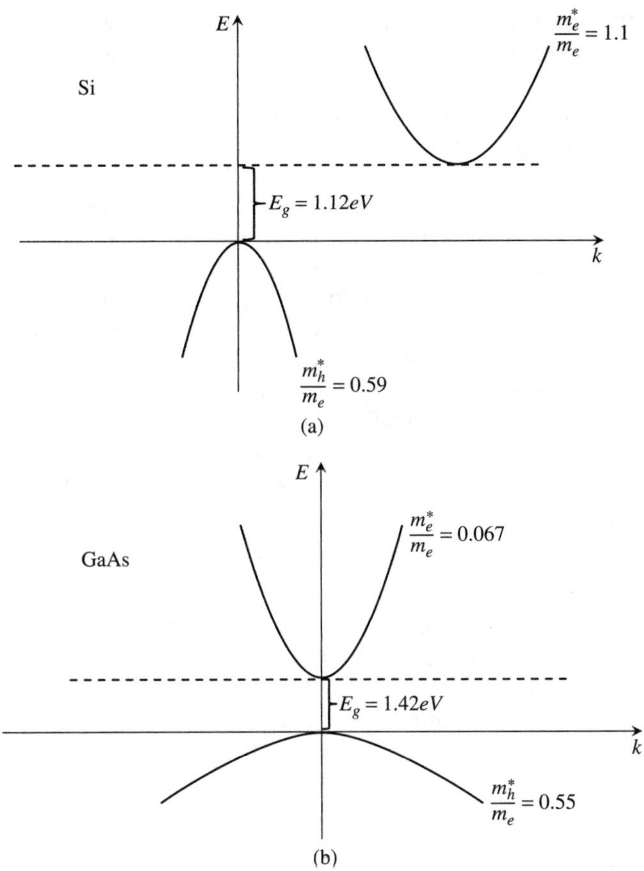

FIGURE 7.5 Sketch of the simplified energy-momentum (E–k) diagrams for (a) silicon and (b) GaAs near the band edges.

Here, m_h^* is the effective hole mass in the valence band, and E_v is the energy at the top of the valence band given by

$$E_v = E_c - E_g \tag{7.5}$$

Note from Fig. 7.5a that silicon is an example of what is known as an indirect-gap semiconductor. In an indirect-gap semiconductor, the valence band maximum and the conduction band minimum occur at different k values. For silicon, $E_g = 1.12$ eV.

In contrast, in a direct-gap semiconductor such as GaAs, the valence band maximum and conduction band minimum occur at the same value of k, as sketched in Fig. 7.5b. $E_g = 1.42$ eV for GaAs. The parabola near the conduction band is steeper since the effective electron mass is significantly smaller than the effective hole mass for GaAs. Note once again that the *E–k* diagrams sketched in Fig. 7.5 are greatly simplified and do not show the exact variation of the energy with wave vector or other details such as the heavy hole/light hole bands. The reader may consult Refs. 4, 5 and 11 for a detailed discussion of band-structure calculations and *E–k* diagrams.

Now, let us discuss some of the qualitative features of interband optical transitions in a semiconductor. We can have absorption or emission of photons. During absorption, an electron which is originally in the valence band absorbs a photon and goes to an empty state in the conduction band. The reverse occurs during emission where an electron originally in the conduction band makes a transition to an empty state in the valence band and emits a photon. In either case, the energy of the photon should be equal or greater than the band-gap energy E_g. Hence, we can say that interband optical transitions in a semiconductor will take place below a maximum wavelength λ_g given by

$$\lambda_g = \frac{c}{\nu_g} = \frac{hc}{E_g} \tag{7.6}$$

Conservation of momentum gives an additional constraint that must be satisfied during optical transitions. Specifically, let us consider photon emission as sketched in the *E–k* diagram of Fig. 7.6*a*. The initial and final wave numbers of the electron are k_1 and k_2, respectively. From the conservation of momentum, we have

$$\begin{aligned} \hbar(k_2 - k_1) &= \text{change in the momentum of the electron} \\ &= 2\pi\frac{\hbar}{\lambda} \end{aligned} \tag{7.7}$$

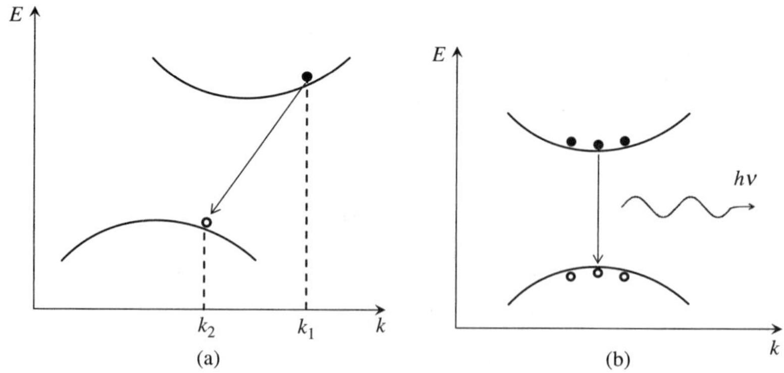

FIGURE 7.6 (*a*) Transition of an electron from the conduction band to the valence band. (*b*) Photon emission is more efficient in direct-gap semiconductors.

or

$$(k_2 - k_1) = \frac{2\pi}{\lambda} \tag{7.8}$$

Above, λ is the wavelength of the emitted photon. Typical k values can be as large as $2\pi/a$, where a is the lattice constant of the order of angstroms. On the other hand, in optical transitions, photon wavelengths are in the range of $0.1\ \mu\text{m} \to 10\ \mu\text{m}$. Therefore, $1/\lambda << 1/a$ and $k_2 \cong k_1$. In other words, conservation of momentum requires that optical transitions as represented in the E–k diagram be nearly vertical. Exceptions occur when an additional phonon is also generated. However, this leads to heat generation and lowers the luminescence efficiency. Figure 7.6*b* shows the case where efficient light generation can occur in a semiconductor. To generate light efficiently, we need electrons near the bottom of the conduction band and holes at the top of the valence band at nearly the same k value. In other words, direct-gap semiconductors such as GaAs are far more efficient as photon emitters than indirect-gap semiconductors. In the case of indirect-gap semiconductors such as silicon, additional phonons are needed to conserve momentum and the emission process is not efficient. Whether the semiconductor is direct-gap or indirect-gap is not an issue during absorption because electrons in the valence band can make vertical transitions to any one of the upper states, which are typically unoccupied in thermal equilibrium.

7.1.3 Types of Semiconductors

We can group inorganic semiconductors into several different groups depending on their chemical composition:

1. Elemental semiconductors: The most important members of this class are silicon and germanium from group IV of the periodic table. Both are indirect-gap semiconductors and cannot be used in ordinary crystalline form as photon emitters. There is ongoing research in developing nanostructured silicon for light generation (see, for example, Refs. 20 to 22). Both silicon and germanium are widely used as optical detectors.
2. Binary semiconductors: These are compound semiconductors consisting of two different elements. As an example, III-V semiconductors consist of one element from group III (such as Al, Ga, or In) and one from group V (P, As, or Sb). The best known binary semiconductor in photonic applications is GaAs, which we introduced in earlier sections. One can also have II-VI compound semiconductors such as ZnS.
3. Ternary compounds: As the name implies, these compound semiconductors consist of three different elements. One widely used example is $Al_xGa_{1-x}As$. Here, a varying amount

of aluminum or gallium is used as the group III element, where x shows the stoichiometric fraction of Al in relation to Ga and varies between 0 and 1. For example, if $x = 0.2$, 20% of the group-III atoms are aluminum and the rest are gallium. One nice feature of ternary semiconductors is that, the band-gap energy of the semiconductor can be varied by changing x during growth. For example, as x goes from 0 to 1, we go from the compound GaAs with band-gap energy E_g of 1.42 eV to AlAs with $E_g = 2.15$ eV.[17]

4. Quaternary semiconductors: In this class of semiconductors, one mixes a varying amount of two group III elements and two group V elements. A wide range of band-gap energies can be obtained by changing the fraction of the different elements. One important example is $In_{1-x}Ga_xAs_{1-y}P_y$. The band-gap energy varies from 0.354 eV ($\lambda_g = 3.5$ μm in the mid-infrared) to 2.27 eV ($\lambda_g = 0.55$ μm in the visible) between the two extreme members InAs and GaP.[23] By varying the composition, and hence the band-gap energy, it is possible to make detectors and light sources over a wide range of wavelengths. This technique is referred to as band-gap engineering.[24]

7.2 Electronic Properties of Semiconductors

A pure, undoped semiconductor is referred to as an intrinsic semiconductor. Here, every time a free electron is generated, a hole is also formed and the density of electrons n is always equal to the density of holes p. In this case, we have the relation

$$n = p = n_i \tag{7.9}$$

where n_i is known as the intrinsic carrier concentration. The electrical and optical properties of semiconductors can be greatly enhanced by doping with controlled amounts of special impurities. In an n-type silicon, as an example, a group V element such as As or P is added. The impurity atom needs four of its valence electrons to form covalent bonds with the neighboring silicon atoms. The fifth electron is very loosely bound and becomes a free charge carrier at room temperature. In this case, the concentration of free electrons can be much more than the concentration of holes, depending on the donor concentration. The reverse occurs in p-type silicon where a group III atom such as Al or B is doped into the crystal. Now, each dopant atom is one short of the right number of valence electrons to form covalent bonds with the neighboring atoms, and this increases the concentration of holes in comparison with electron concentration.

In the next section, we will first develop expressions for the density of states and occupation probabilities to derive the formulas for electron and hole concentrations. We will then look at charge transport in semiconductors, and finally discuss the current-voltage characteristics of p-n junctions.

7.2.1 Density of States

The density of states $\rho(k)$ in a crystalline semiconductor can be defined as

$$\rho(k)dk = \text{the number of allowed electron states with electron wave numbers in the range } k \to k + dk \text{ per unit volume} \tag{7.10}$$

The one-dimensional electron wave functions will have the functional form $\exp(ik_x x)$, where applying periodic boundary conditions over a box of side L, leads to the relation

$$\exp(ik_x x) = \exp[ik_x(x + L)] \tag{7.11}$$

or

$$k_x L = m2\pi \tag{7.12}$$

Above, m is an integer. This means that the separation Δk between two allowed k values will be given by

$$\Delta k = \left(\frac{2\pi}{L}\right) \tag{7.13}$$

Figure 7.7 shows the allowed values of the electron wave vector in a semiconductor crystal in the shape of a cube with edge length of L. The volume occupied by one allowed solution will be given by $(2\pi/L)^3$.

Hence, the number of states per unit volume between k and $k + dk$ will be given by

$$\Rightarrow \rho(k)dk = 2 \times \frac{4\pi k^2 dk}{8\pi^3 / L^3} \frac{1}{L^3} = \frac{k^2}{\pi^2} dk \tag{7.14}$$

The factor of 2 included in Eq. (7.14) accounts for the spin degeneracy of the electron. Now, by using the E–k relations derived above, we can express ρ in terms of the electron energy. We leave it as an exercise (see Prob. 7.1) for the reader to show that the density of states $\rho_c(E)$ near the conduction band minimum can be expressed in the form

$$\rho_c(E)dE = \frac{(2m_e^*)^{3/2}}{2\pi^2\hbar^3}(E - E_c)^{1/2} dE \tag{7.15}$$

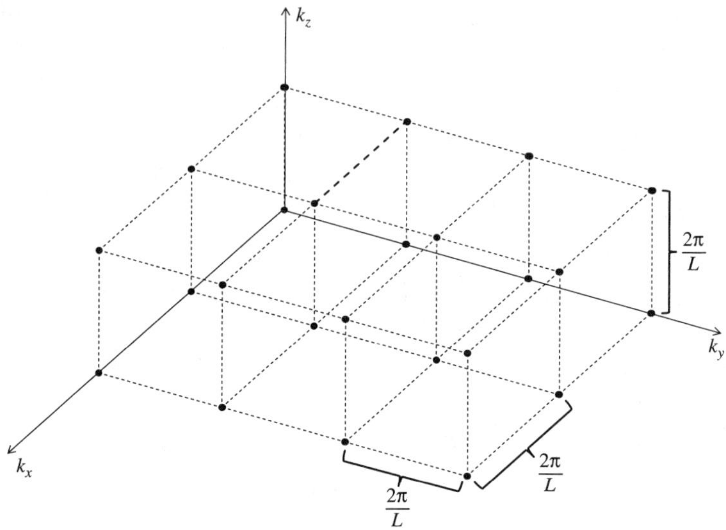

FIGURE 7.7 Allowed values of the electron wave vector in a semiconductor crystal.

Similarly, the density of states $\rho_v(E)$ near the top of the valence band becomes

$$\rho_v(E)dE = \frac{\left(2m_h^*\right)^{3/2}}{2\pi^2\hbar^3}(E_v - E)^{1/2}dE \tag{7.16}$$

The density of states functions for the conduction and valence bands are sketched in Fig. 7.8 as a function of energy.

7.2.2 Occupation Probabilities

At $T = 0$ K, the valence band is completely filled with electrons and the conduction band is completely empty. At higher temperatures, some electrons are raised to higher levels as a result of thermal

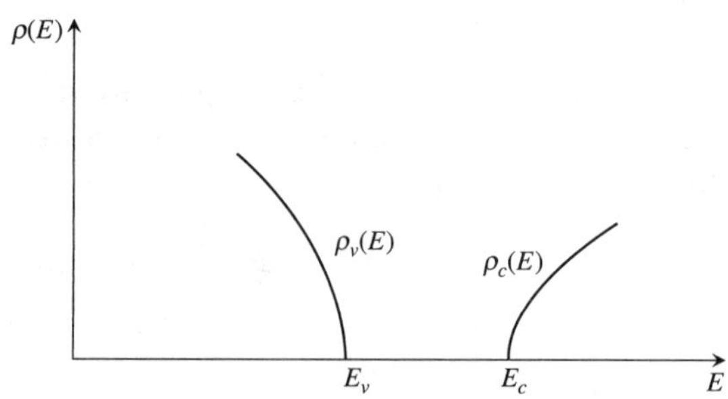

FIGURE 7.8 Sketch of the density of states functions $\rho_c(E)$ and $\rho_v(E)$ for the valence and conduction bands.

excitation. Here, we need statistical mechanics to find the occupation probability of a state at a finite temperature T. This is given by the well-known Fermi-Dirac distribution function

$$f(E) = \frac{1}{\exp[(E - E_f)/k_B T] + 1} \tag{7.17}$$

where $f(E)$ is the probability that a given state at energy E is occupied by an electron, k_B is the Boltzmann constant ($k_B = 1.38 \times 10^{-23}$ J/K), and E_f is the Fermi level energy. At $T = 300$ K, $k_B T = 26$ meV. Clearly, $f(E)$ should be between 0 and 1 since it corresponds to the probability of finding an electron at a certain energy level. Note further that $f(E_f) = 1/2$ at any temperature. In other words, E_f corresponds to the energy level for which the occupation probability is 1/2, provided that an allowed eigenstate of the electron exists at E_f.

The variation of $f(E)$ is shown as a function of energy at $T = 0$ K and $T > 0$ K in Fig. 7.9. At $T = 0$ K, all of the energy levels are occupied up to E_f. Note further that, since $f(E)$ gives the probability of

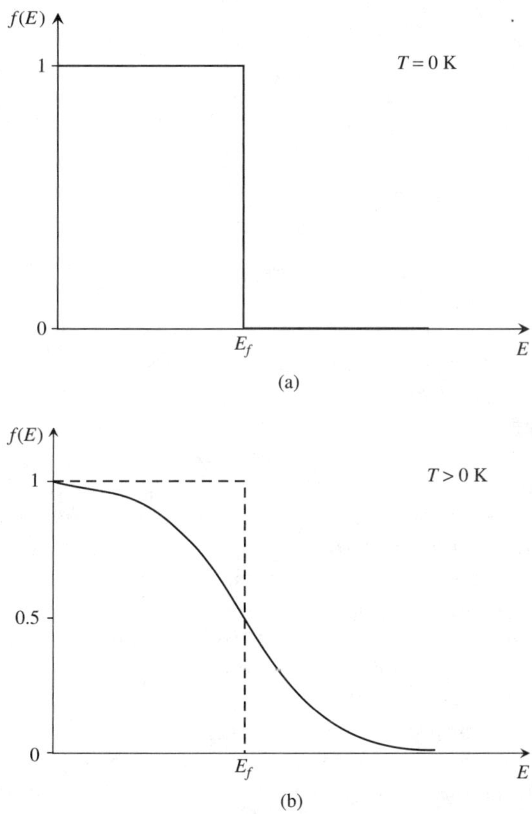

FIGURE 7.9 Fermi-Dirac distribution function $f(E)$ at (*a*) $T = 0$ K and (*b*) $T > 0$ K.

occupancy by an electron, $1-f(E)$ corresponds to the probability that the state is empty or the probability of occupancy by a hole. We leave it as an exercise (see Prob. 7.2) for the reader to show that $f(E)$ can be approximated by

$$f(E) \approx \exp\left[\frac{-(E-E_f)}{k_B T}\right] \tag{7.18}$$

when $E - E_f >> k_B T$. In other words, the high-energy tail of the Fermi function decreases nearly exponentially in the conduction band. Similarly, when $E_f - E >> k_B T$ in the valence band, $1-f(E)$ can be approximated as

$$1-f(E) \approx \exp\left[\frac{-(E_f-E)}{k_B T}\right] \tag{7.19}$$

7.2.3 Carrier Concentrations in Thermal Equilibrium

In thermal equilibrium, the total electron density n in the conduction band will be given by the weighted continuous sum of the density of states multiplied by the occupation probability:

$$n = \int_{E_c}^{\infty} dE\, \rho_c(E) f(E) \tag{7.20}$$

Similarly, the total hole density p becomes

$$p = \int_{-\infty}^{E_v} dE\, \rho_v(E)(1-f(E)) \tag{7.21}$$

Intuitively, it is easy to see that if the effective electron and hole masses were equal, the Fermi level E_f would be in the middle of the band gap for an intrinsic semiconductor so that $n = p = n_i$. In an n-type semiconductor, however, this is no longer the case. The donor atoms introduce an energy level E_D slightly below the conduction band and electrons at this level are easily promoted to the conduction band at room temperature. Since the total electron concentration is larger than the hole concentration, the Fermi level E_{fn} now goes above the middle of the band gap and gets closer to the conduction band edge as shown in Fig. 7.10a. In a p-type semiconductor, the acceptor atoms introduce an acceptor level E_A close to the valence band edge. Electrons can easily make transitions to this level at room temperature resulting in a higher concentration of holes in relation to electrons. Since the electrons in level E_A are still below the conduction band, they are bound and cannot aid in charge transport. In this case, the Fermi level E_{fp} moves below the middle of the band gap.

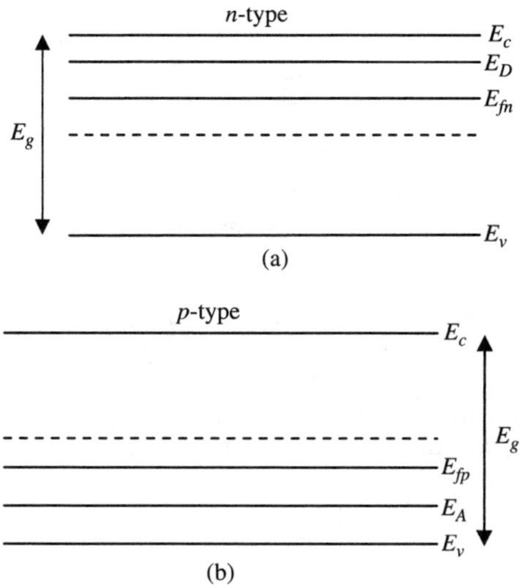

FIGURE 7.10 Energy band diagram of (*a*) *n*-type and (*b*) *p*-type semiconductors.

The exact positions of E_{fn} and E_{fp} depend on the doping level of the impurities.

To calculate the total carrier densities inside the semiconductor, we will first make the approximation that $E - E_f >> k_B T$ in the conduction band and $E_f - E >> k_B T$ in the valence band. A semiconductor in which these approximations are valid is said to nondegenerate. When E_f lies near the middle of the band gap, the semiconductor will be nearly nondegenerate. In more quantitative terms, we will require that $E_c - E_f > 3k_B T$ for electrons and $E_f - E_v > 3k_B T$ for holes in a nondegenerate semiconductor. Under this approximation, it can be shown that (see Prob. 7.3) the total electron and hole concentrations can be expressed in the form

$$n = N_c \exp\left[-\frac{E_c - E_f}{k_B T}\right] \tag{7.22}$$

and

$$p = N_v \exp\left[-\frac{E_f - E_v}{k_B T}\right] \tag{7.23}$$

where the densities N_c and N_v are given respectively by

$$N_c = 2\left(\frac{2\pi m_e^* k_B T}{h^2}\right)^{3/2} \tag{7.24}$$

and

$$N_v = 2\left(\frac{2\pi m_h^* k_B T}{h^2}\right)^{3/2} \tag{7.25}$$

When the above exponential approximations to $f(E)$ hold, the product np is independent of the Fermi level and is given by

$$np = N_c N_v \exp\left(-\frac{E_g}{k_B T}\right) \tag{7.26}$$

Equation (7.26) summarizes what is commonly referred to as the law of mass action in semiconductors. If the semiconductor is further intrinsic, we have $n = p = n_i$ where, based on Eq. (7.26), the intrinsic carrier concentration n_i becomes

$$n_i = (N_c N_v)^{1/2} \exp\left(-\frac{E_g}{2k_B T}\right) \tag{7.27}$$

Note that as the absolute temperature goes up, the intrinsic carrier concentration also increases since higher thermal energies enable the generation of more charge carriers. Based on Eqs. (7.26) and (7.27), the law of mass action can be written in the equivalent form

$$np = n_i^2 \tag{7.28}$$

for both intrinsic (undoped) or extrinsic (doped), nondegenerate semiconductors.

Example As an example, we estimate the intrinsic carrier concentration in Si and GaAs at room temperature. We take the density-of-states effective masses and the band-gap energy as $m_e^* = 1.1m_e$, $m_h^* = 0.59m_e$, $E_g = 1.12$ eV for Si and $m_e^* = 0.068m_e$, $m_h^* = 0.54m_e$, $E_g = 1.42$ eV for GaAs.[7] By using Eqs. (7.24), (7.25), and (7.27), this gives, at $T = 300$ K, $n_i = 7.2 \times 10^9\,\text{cm}^{-3}$ for Si and $n_i = 2.5 \times 10^6\ \text{cm}^{-3}$ for GaAs. The actual intrinsic carrier concentrations in Si and GaAs at room temperature are of the order of $10^{10}\ \text{cm}^{-3}$ and $10^7\ \text{cm}^{-3}$ (Ref. 10). Any discrepancy between the formulas developed here and the actual values is due to the fact that the actual band structures for the conduction and valence bands are more complicated for both Si and GaAs. In particular, the density-of-state effective mass of the actual crystal depends in a more complicated way on the values of the effective masses along the different principal axes of the ellipsoidal energy surface.[10] We did not take this fact into account in the derivation of the formulas for the total carrier concentrations. Also, there is always some variation in the reported values of the band structure parameters.

Despite the fact that it is not very accurate, Eq. (7.27) is still very useful in showing the general trends and in giving an idea about the order of magnitude of the intrinsic carrier concentrations. For example,

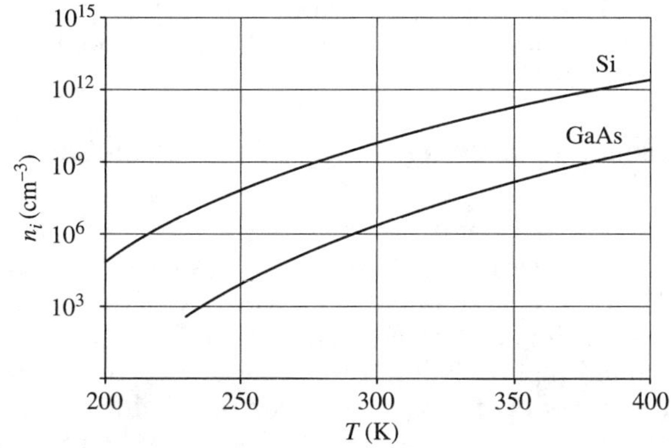

FIGURE 7.11 Estimated variation of the intrinsic carrier concentration n_i as a function of the absolute temperature for Si and GaAs. Material parameters are $m_e^* = 1.1m_e$, $m_h^* = 0.59m_e$, $E_g = 1.12$ eV for Si and $m_e^* = 0.068m_e$, $m_h^* = 0.54m_e$, $E_g = 1.42$ eV for GaAs.

Fig. 7.11 shows the calculated variation of the intrinsic carrier concentration n_i as a function of the absolute temperature for Si and GaAs, based on Eq. (7.27) and the fixed parameters given above. We note that, as expected, n_i increases with temperature. Also, n_i is larger for Si than for GaAs. This can be attributed to the larger band-gap energy of GaAs.

If the Fermi level is within $3k_BT$ of the band edges or if it falls within the bands, as in the case of semiconductor lasers, we have a degenerate semiconductor. In a degenerate semiconductor, we have to use the exact expressions given by Eqs. (7.20) and (7.21), without making any approximation in the form of the Fermi-Dirac distribution function to calculate the electron/hole concentrations and in this case $np \neq n_i^2$.

Now, let us consider a nondegenerate n-type semiconductor. If all of the donor atoms are fully ionized, we have, from charge neutrality,

$$p - n + N_D = 0 \tag{7.29}$$

Since $np = n_i^2$, we can rewrite Eq. (7.29) as a quadratic equation for n in the form

$$n^2 - N_D n - n_i^2 = 0 \tag{7.30}$$

Solving for n and keeping only the positive root, we get

$$n = \frac{N_D + \sqrt{N_D^2 + 4n_i^2}}{2} \tag{7.31}$$

For typical doping levels, we have $N_D >> n_i$, giving $n \approx N_D$ and $p \approx n_i^2/N_D$. Similarly, we have $p \approx N_A$ and $n = n_i^2/N_A$ in a nondegenerate p-type semiconductor. Let us consider Si at room temperature as an example. Typical doping concentrations will be above 10^{14} cm^{-3}. Hence the approximations $N_D >> n_i$ and $N_A >> n_i$ hold under most circumstances.

7.2.4 Modeling of Charge Flow in Semiconductors

Electric current flow inside a semiconductor is influenced by two major processes: drift and diffusion. Let us first consider the drift motion which takes place under the action of an applied electric field $\vec{E}$. Positively charged holes are accelerated in the direction of $\vec{E}$ while electrons are accelerated in the opposite direction. However, this accelerated motion is interrupted by randomly occurring collisions. On a macroscopic scale, charges end up moving with an average drift velocity $\vec{u}$ which is linearly proportional to the applied electric field $\vec{E}$. In the case of electrons, this linear dependence is expressed in equation form as

$$\vec{u} = \mu_n \vec{E} \tag{7.32}$$

where $\vec{u}$ is the drift velocity, $\vec{E}$ is the applied electric field, and the constant of proportionality μ_n is known as the electron mobility. For holes, we have a similar expression where μ_n is replaced by the hole mobility μ_p. To find the resulting current density due to the drift of electrons and holes, consider a semiconductor slab of thickness dz as shown in Fig. 7.12.

The total charge due to electrons in this volume is $neadz$, where e is the electron charge. The electrons will be accelerated to the left. The net current density $\vec{J}_n$ due to electrons will be to the right and its magnitude is given by

$$\begin{aligned} J_n &= \frac{\text{total charge within the volume}}{\text{time (cross-sectional area)}} \\ &= \frac{neadz}{dt(a)} = ne\frac{dz}{dt} = ne\mu_n E \end{aligned} \tag{7.33}$$

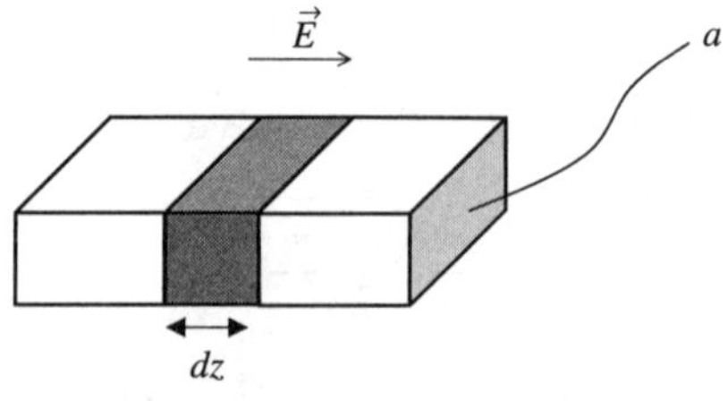

Figure 7.12 A semiconductor slab of thickness dz and cross-sectional area a under an applied external electric field $\vec{E}$.

The drift current due to holes will be in the same direction, given by

$$J_p = pe\mu_p E \tag{7.34}$$

Hence, the total drift current density due to both electrons and holes becomes

$$\vec{J} = \vec{J}_n + \vec{J}_p = e(\mu_n n + \mu_p p)\vec{E} \tag{7.35}$$

For a conducting medium, the current density $\vec{J}$ is related to $\vec{E}$ via

$$\vec{J} = \sigma\vec{E} \tag{7.36}$$

where σ is the electrical conductivity. Hence, for a semiconductor with both electrons and holes, the conductivity can be expressed in terms of the electron and hole mobilities as

$$\sigma = e(\mu_n n + \mu_p p) \tag{7.37}$$

Now, we turn our attention to diffusion. In this case, charge flow occurs as a result of the spatial variation of concentration. Clearly, charges will flow from regions of higher concentration to regions of lower concentration. Empirically, it is known that the current density is proportional to the concentration gradient. Consider the sketches shown in Fig. 7.13 to determine the direction of the diffusion current due to holes and electrons. Looking at Fig. 7.13*a*, if p decreases with increasing z, the hole current will be in the positive z-direction. Since each charge carrier is positively charged, this will give rise to a current density given by

$$\vec{J}_p = -eD_p \nabla p \tag{7.38}$$

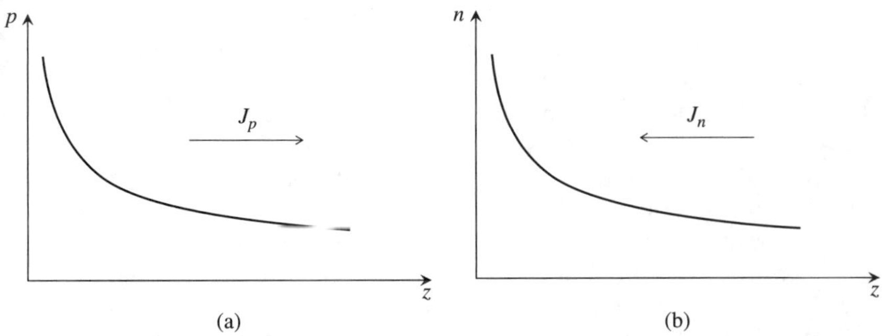

FIGURE 7.13 Direction of the diffusion current in a semiconductor due to a spatial gradient in the (*a*) hole density and (*b*) electron density.

Above, the proportionality constant D_p is known as the hole diffusion coefficient and has the units of cm^2/s. If, instead, we had a concentration gradient for electrons, as shown in Fig. 7.13*b*, we would have electrons diffusing in the positive *z*-direction, giving rise to a current in the negative *z*-direction. Hence, $\vec{J}_n$ is given by

$$\vec{J}_n = eD_n \nabla n \tag{7.39}$$

where D_n is the electron diffusion coefficient. The overall electron current density due to both drift and diffusion becomes

$$\vec{J}_n = e\mu_n n\vec{E} + eD_n \nabla n \tag{7.40}$$

For holes, we have

$$\vec{J}_p = e\mu_p p\vec{E} - eD_p \nabla p \tag{7.41}$$

7.2.5 Charge Continuity Equation for Semiconductors

In this section, we develop, based on the principle of conservation of electric charge, the continuity equation which describes the spatial and temporal evolution of the carrier concentrations in the presence of drift, diffusion, and generation/recombination inside a semiconductor. Consider a region with a finite volume V within a semiconductor. Due to the conservation of electric charge, an outward current flow due to drift and/or diffusion will cause a decrease in the total charge stored within the region. Let us consider the electron density first. The instantaneous electron density n can be expressed as

$$n = n_0 + \Delta n \tag{7.42}$$

where n_0 is the equilibrium carrier concentration and Δn is the deviation in the electron concentration from equilibrium. As an example, Δn may be caused by current injection or illumination with light whose photon energy exceeds the band-gap energy of the semiconductor. Let us assume that G gives the rate of generation of electrons within this volume due to illumination or current injection. When an electron is in the vicinity of a hole, they can annihilate each other releasing heat or photons. This process causes a decrease in the carrier density. On average, an electron will recombine with a hole after a characteristic time τ_n known as the electron recombination time. τ_n depends on the carrier concentration within the region. Typically τ_n may be different from the hole recombination time τ_p. The average rate of decrease of electrons due to recombination will be proportional to Δn and is given by $-\Delta n/\tau_n$. Within a volume V of the semiconductor, we can write the continuity equation for electrons as follows:

$$\frac{d}{dt}\int_V d^3r\rho_V = -\int_S d\vec{a}\cdot\vec{J}_n + \left(\int_V d^3rG - \int_V d^3r\frac{\Delta n}{\tau_n}\right)(-e) \tag{7.43}$$

Here, $\rho_V = -en$ is the volume charge density due to electrons and S is the surface enclosing the volume V. By applying the divergence theorem, we get

$$\frac{\partial}{\partial t}(-en) + \nabla\cdot(e\mu_n n\vec{E} + eD_n\nabla n) = \left(G - \frac{\Delta n}{\tau_n}\right)(-e) \tag{7.44}$$

Since n_0 is not a function of time or position in a uniform semiconductor, we have

$$\frac{\partial}{\partial t}n = \frac{\partial \Delta n}{\partial t} \tag{7.45}$$

and

$$\nabla n = \nabla(\Delta n) \tag{7.46}$$

Hence, the continuity equation can be written for the excess electron concentration Δn as

$$\frac{\partial}{\partial t}(\Delta n) = \mu_n\nabla\cdot[(n_0 + \Delta n)\vec{E}] + D_N\nabla^2(\Delta n) + G - \frac{\Delta n}{\tau_n} \tag{7.47}$$

An important special case occurs outside the depletion region of a p-n junction where $\vec{E}$ is negligible. We will soon discuss the p-n junction in the next section. Furthermore, assuming that changes occur along one dimension only, say the z-axis, Eq. (7.47) simplifies to

$$\frac{\partial \Delta n}{\partial t} = D_n\frac{\partial^2 \Delta n}{\partial z^2} + G - \frac{\Delta n}{\tau_n} \tag{7.48}$$

For holes, $\rho_V = pe$, giving

$$\frac{\partial}{\partial t}(pe) + \nabla\cdot(e\mu_p p\vec{E} - eD_p\nabla p) = \left(G - \frac{\Delta p}{\tau_p}\right)(e) \tag{7.49}$$

Now, the total hole concentration p is related to the equilibrium concentration p_0 and the deviation Δp from the equilibrium concentration through

$$p = p_0 + \Delta p \tag{7.50}$$

Hence, we obtain

$$\frac{\partial \Delta p}{\partial t} = D_p \frac{\partial^2 \Delta p}{\partial z^2} + G - \frac{\Delta p}{\tau_p} \tag{7.51}$$

for the deviation in the hole concentration when the drift current is negligible and the concentration gradient is assumed to be only in the z-direction.

7.2.6 Semiconductor p-n Junctions

Semiconductor p-n junctions which are the building blocks of diodes, transistors and other electronic devices, are among the most important components of the electronic industry. The same thing is also true in photonics. Here, p-n junctions of direct-gap semiconductors are used for light generation, either as laser diodes or light emitting diodes. We have two types of semiconductor junctions, depending on the chemical composition of the host material on either side of the junction:

1. Homojunction: In this case, the host semiconductor on both sides of the junction is the same. For example, we can have a silicon p-n junction.
2. Heterojunction: Here, the semiconductor host on each side of the junction is different. For example, this can be a p-n junction between GaAs and $Al_xGa_{1-x}As$.

We will look at the optical characteristics of p-n junctions in later sections of the chapter. Here, we will first analyze the electrostatics of a homojunction formed between the n-type and p-type regions of the same semiconductor and derive the current-voltage characteristics under forward and reverse bias. For a detailed discussion of how semiconductor junctions are fabricated, see Refs. 9 and 25.

In the analysis, we will assume that the doping concentration abruptly changes at some location taken to be $z = 0$. For simplicity, we will further assume that the device is one dimensional, i.e., changes in doping concentration occur only along the z-direction. This is sketched in Fig. 7.14.

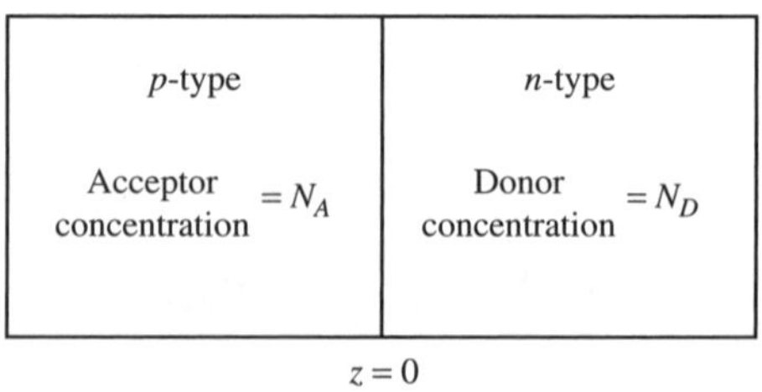

Figure 7.14 A semiconductor p-n junction.

Let us suppose that $n = N_D$ on the n-side and $p = N_A$ on the p-side. Let us look at typical numbers to see what will happen. Take a silicon p-n junction as an example. At $T = 300$ K, n_i is of the order of $10^{10}/\text{cm}^3$. Typical donor and acceptor doping concentrations are in the $10^{15} - 10^{17}/\text{cm}^3$ range. Suppose that the n-side has a donor concentration of $N_D = 10^{15}/\text{cm}^3$. If all of the donor atoms are ionized, this gives a free electron density n of $10^{15}/\text{cm}^3$ and a hole concentration of $p = (10^{10})^2/10^{15} = 10^5/\text{cm}^3$ from the law of mass action given in Eq. (7.28). If the p-side has the same acceptor concentration of $10^{15}/\text{cm}^3$, the electron concentration here will be $10^5/\text{cm}^3$. So, in going from the p-side to the n-side at $z = 0$, we see a huge increase in the electron concentration from $10^5/\text{cm}^3$ to $10^{15}/\text{cm}^3$. Due to this high-concentration gradient, electrons on the n-side of the junction will diffuse to the p-side to equalize the electron concentration, leaving behind a region of positively charged donor ions which are fixed in position. Similarly, holes on the p-side will diffuse to the n-side leaving behind a region of negatively charged acceptor ions. In equilibrium, a region consisting mainly of the immobile acceptor/donor ions will be formed around the junction as sketched in Fig. 7.15. This region, which contains a negligible concentration of free charge carriers, is known as the depletion region. We assume that the edges of the depletion region on the p- and n-side of the junction are at $z = -d_p$ and $z = +d_n$, respectively.

The diffusion of free charge carriers leads to the formation of an electric field directed from the positively charged donor ions toward the negatively charged acceptor ions. In equilibrium, the drift current due to the presence of this electric field is cancelled by the diffusion current due to the concentration gradient of the charges. Since $\vec{E} = -\nabla V$ where V is the electrostatic potential, we further see that a

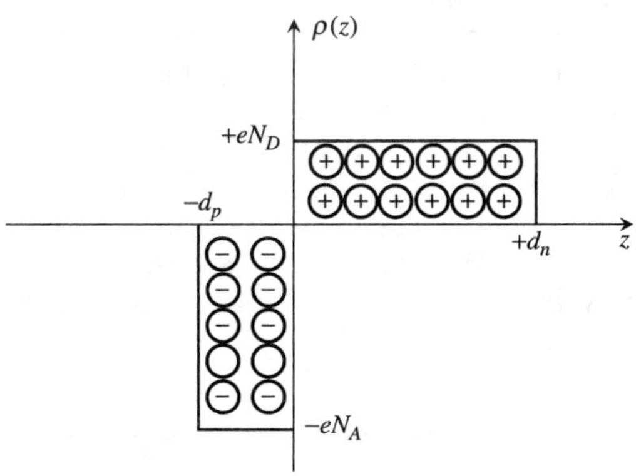

FIGURE 7.15 Variation of the volume charge density $\rho(z)$ across the depletion region of a p-n junction.

potential difference will be set up across the junction with the n-side being at a higher potential relative to the p-side. Now, let us calculate the variation of the electric field and the electrostatic potential across the junction. We will assume that the electric field for $z < -d_p$ and $z > +d_n$ is zero, so that the drop in potential occurs between $z = +d_n$ and $-d_p$. The potential difference

$$V_b = V(z = +d_n) - V(z = -d_p) \tag{7.52}$$

is known as the built-in junction potential and depends on the doping level on either side of the junction. We can calculate the variation of the electrostatic potential across the junction by using Poisson's equation. In one dimension, we have

$$\frac{d^2V}{dz^2} = \frac{-\rho(z)}{\varepsilon} \tag{7.53}$$

where $\rho(z)$ is the volume charge density and ε is the dielectric constant of the semiconductor. We assume a homojunction where the semiconductor host on either side of the junction is the same material, so that ε is the same for $z < 0$ and $z > 0$. For the charge distribution shown in Fig. 7.15, the Poisson equation becomes

$$\frac{d^2V}{dz^2} = \frac{eN_A}{\varepsilon} \tag{7.54}$$

for $-d_p \le z \le 0$ and

$$\frac{d^2V}{dz^2} = \frac{-eN_D}{\varepsilon} \tag{7.55}$$

for $0 \le z \le d_n$. Since the potential drop occurs between $z = -d_p$ and $z = +d_n$, we assume that the electric field $\vec{E} = 0$ for $z \le -d_p$ and $z \ge +d_n$. We also need to apply the electrostatic boundary conditions derived in Chap. 1, to join the solutions for $\vec{E}$ and $V(z)$ on either side of the junction. $V(z)$ will be continuous at $z = 0$ as discussed before. Since there is no surface charge density at $z = 0$, we further see that the displacement vector and hence the electric field will also be continuous for a homojunction. Therefore, we require that both $V(z)$ and dV/dz be continuous at $z = 0$. We leave it as an exercise (see Prob. 7.7) for the reader to show that the resulting expressions for the electrostatic potential and the electric field become

$$V(z) = \begin{cases} \dfrac{1}{2}\dfrac{eN_A}{\varepsilon}(z + d_p)^2 & -d_p \le z \le 0 \\[2ex] V_b - \dfrac{1}{2}\dfrac{eN_D}{\varepsilon}(d_n - z)^2 & 0 \le z \le d_n \end{cases} \tag{7.56}$$

and

$$E_z(z) = \begin{cases} -\frac{eN_A}{\varepsilon}(z + d_p) & -d_p \le z \le 0 \\ -\frac{eN_D}{\varepsilon}(d_n - z) & 0 \le z \le d_n \end{cases} \tag{7.57}$$

By using the fact that the total net charge within the depletion region is zero or by using the boundary conditions, we further have

$$N_A d_p = N_D d_n \tag{7.58}$$

The variation of $V(z)$ and $E_z(z)$ are sketched in Fig. 7.16*a* and *b*, respectively.

From Eq. (7.56), we see that continuity of the electrostatic potential at $z = 0$ gives

$$V(0^-) = \frac{1}{2}\frac{eN_A}{\varepsilon}d_p^2 = V(0^+) = V_b - \frac{1}{2}\frac{eN_D}{\varepsilon}d_n^2 \tag{7.59}$$

or

$$V_b = \frac{e}{2\varepsilon}(N_A d_p^2 + N_D d_n^2) \tag{7.60}$$

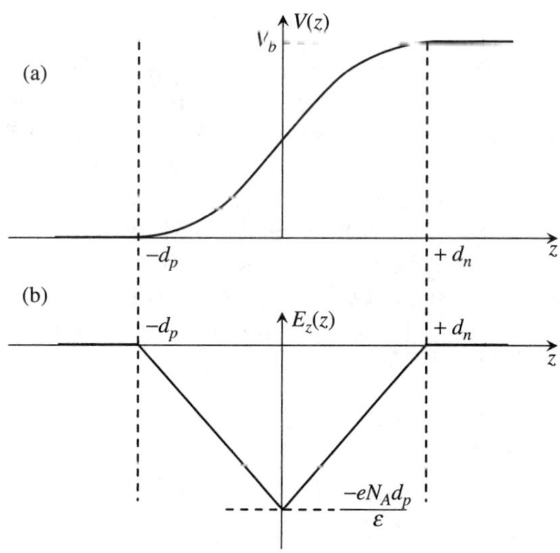

FIGURE 7.16 Variation of the (*a*) electrostatic potential $V(z)$ and (*b*) the electric field $E_z(z)$ across the depletion region of a p-n junction.

Since $N_A d_p = N_D d_n$, we can express the built-in junction potential V_b only in terms of d_p or d_n. In terms of d_p, V_b becomes

$$V_b = \frac{e}{2\varepsilon}\left[N_A d_p^2 + N_D\left(\frac{N_A d_p}{N_D}\right)^2\right] = \frac{e}{2\varepsilon} d_p^2 \left[N_A + \frac{N_A^2}{N_D}\right]$$

$$= \frac{e}{2\varepsilon}\frac{d_p^2}{N_D}[N_A(N_D + N_A)] \tag{7.61}$$

Equation (7.61) tells us that, by knowing the built-in junction potential, we can determine the actual width of the depletion region for the case of an abrupt junction. This is a very useful relationship because, as we will show shortly, it is possible to determine V_b from a knowledge of the doping concentrations on the two sides of the junction. By using Eqs. (7.58) and (7.61), the edge locations d_p and d_n can also be expressed in terms of V_b as

$$d_p = \sqrt{\frac{2\varepsilon V_b}{e}\frac{N_D/N_A}{[N_D + N_A]}} \tag{7.62}$$

and

$$d_n = \frac{N_A}{N_D} d_p = \sqrt{\frac{2\varepsilon V_b}{e}\frac{N_A/N_D}{[N_D + N_A]}} \tag{7.63}$$

Now, we address the issue of how we can determine the built-in potential V_b from the doping concentrations. To do this, we go back to the band picture of the semiconductor. The energy band diagrams of the *p*-type and *n*-type sections are shown in Fig. 7.17*a* and *b*, respectively. As we discussed before, the Fermi level shifts toward the valence band edge in the *p*-type region and toward the conduction band edge in the *n*-type region. It is a known fact from statistical thermodynamics, that in equilibrium the Fermi levels of the *p*- and *n*-sides of a semiconductor p-n junction line up as shown in Fig. 7.17*c*.

From Fig. 7.17, it is clear that

$$eV_b = E_g - e_p - e_n \tag{7.64}$$

For the *n*-type semiconductor,

$$n = N_c \exp\frac{[E_f - E_c]}{k_B T} = N_c \exp\left[\frac{-e_n}{k_B T}\right] = N_D \tag{7.65}$$

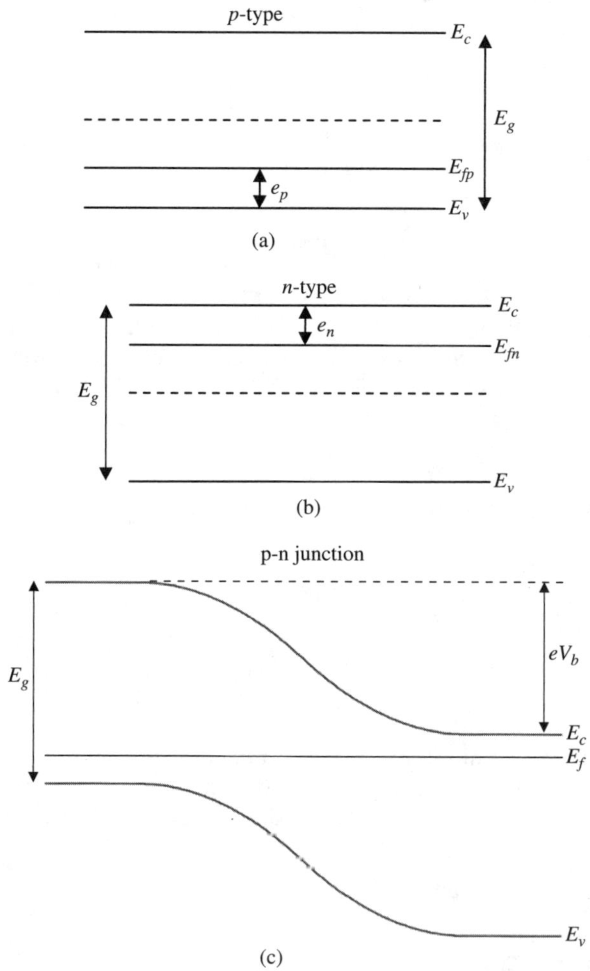

FIGURE 7.17 Energy band diagrams of the (*a*) *p*-type region, (*b*) *n*-type region, and (*c*) p-n junction.

whereas for the *p*-type semiconductor, we have

$$N_A = p = N_v \exp\left(\frac{E_v - E_f}{k_B T}\right) = N_v \exp\left[\frac{-e_p}{k_B T}\right] \tag{7.66}$$

Solving for e_n and e_p from Eqs. (7.65) and (7.66), we obtain

$$e_n = k_B T \ln\left(\frac{N_c}{N_D}\right) \tag{7.67}$$

and

$$e_p = k_B T \ln\left(\frac{N_v}{N_A}\right) \tag{7.68}$$

Adding Eqs. (7.67) and (7.68) gives

$$e_p + e_n = k_B T \ln\left(\frac{N_v N_c}{N_A N_D}\right) \tag{7.69}$$

By using the law of mass action given in Eq. (7.27), we can express the product $N_v N_c$ in terms of the intrinsic carrier concentration and the band-gap energy to obtain

$$e_p + e_n = k_B T \ln\left(\frac{n_i^2 \exp\left(\frac{E_g}{k_B T}\right)}{N_A N_D}\right) = k_B T \ln\left(\frac{n_i^2}{N_A N_D}\right) + E_g \tag{7.70}$$

Hence, the expression for the built-in junction potential becomes

$$V_b = \frac{1}{e}\left[E_g - E_g - k_B T \ln\left(\frac{n_i^2}{N_A N_D}\right)\right] = \frac{k_B T}{e} \ln\left(\frac{N_A N_D}{n_i^2}\right) \tag{7.71}$$

So, note that the built-in junction potential depends on the band structure parameters of the semiconductor (through the intrinsic carrier concentration), the temperature, and the doping levels of the two sides.

Example Silicon p-n junction
Consider a Si p-n junction at $T = 300$ K. We take the band structure parameters of Si as $m_e^* = 1.1m_e$, $m_h^* = 0.59m_e$, $E_g = 1.12$ eV. At $T = 300$ K, by using Eqs. (7.24), (7.25), and (7.27), n_i comes to 7.2×10^9 cm^{-3}. We further assume that the p-n junction has a symmetric doping profile on either side with $N_A = N_D = 10^{15}\,\text{cm}^{-3}$ and that $\varepsilon/\varepsilon_0 = 11.8$ for Si. By using Eq. (7.71), V_b becomes 0.61 V. Since the doping concentrations on either side of the junction are equal, we have, from Eqs. (7.62) and (7.63) $d_p = d_n = 0.63\ \mu\text{m}$, giving a total depletion region width of 1.26 μm. Figure 7.18*a* and *b* further shows the calculated variation of the electrostatic potential $V(z)$ and the corresponding electric field $E_z(z)$ around the junction. The peak electric field at the center of the depletion region is 9700 V/cm.

We can derive the expression for the built-in potential V_b and the general relationship between the diffusion coefficient and the mobility by noting that the net current density due to drift and diffusion should be zero across the p-n junction in equilibrium. Consider the current density for holes. Working in one dimension, we get

$$J_p = e\mu_p p E - eD_p \frac{dp}{dz} = 0 \tag{7.72}$$

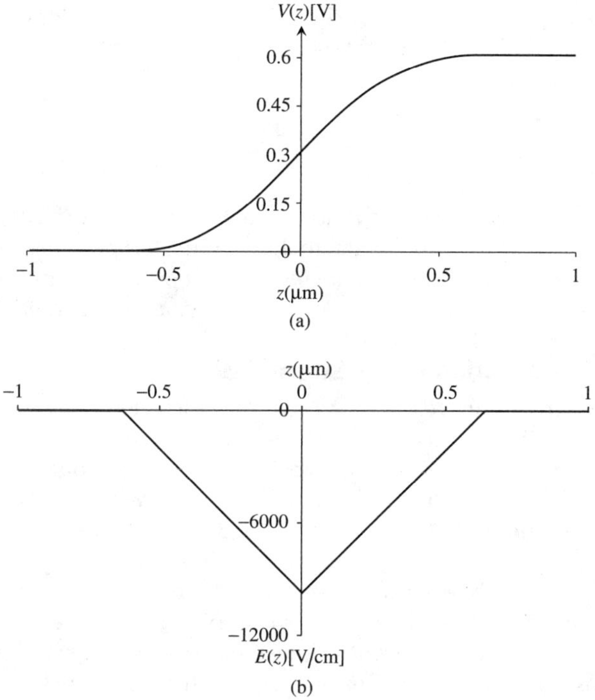

FIGURE 7.18 Calculated variation of (*a*) the electrostatic potential $V(z)$ and (*b*) the electric field $E_z(z)$ for a silicon p-n junction with parameters $m_e^* = 1.1m_e$, $m_h^* = 0.59m_e$, $E_g = 1.12$ eV, $T = 300$ K, $\varepsilon/\varepsilon_0 = 11.8$, and $N_A = N_D = 10^{15}\,\text{cm}^{-3}$.

From above, the electric field in the depletion region is given by

$$E = \frac{D_p}{\mu_p}\frac{1}{p}\frac{dp}{dz} \tag{7.73}$$

Since $dV = -E\,dz$, we have

$$\int_{z=-d_p}^{z=+d_n} dV = V_b = -\frac{D_p}{\mu_p}\int_{p(-d_p)}^{p(+d_n)}\frac{dp}{p} = -\frac{D_p}{\mu_p}\ln\left[\frac{p(+d_n)}{p(-d_p)}\right]$$

$$= \frac{D_p}{\mu_p}\ln\left[\frac{p(-d_p)}{p(+d_n)}\right] \tag{7.74}$$

At $z = -d_p$ (p-side), $p(-d_p) = N_A$ and at $z = +d_p$ (n-side), $p(+d_n) = n_i^2/N_D$ giving

$$V_b = \frac{D_p}{\mu_p}\ln\left[\frac{N_A N_D}{n_i^2}\right] \tag{7.75}$$

Since the expression for V_b derived here should be identical to that given in Eq. (7.71), we see that the diffusion constant D_p and mobility μ_p should be related through

$$\frac{D_p}{\mu_p} = \frac{k_B T}{e} \tag{7.76}$$

This important relation, which shows that the ratio of the diffusion coefficient to mobility is proportional to the absolute temperature, is known as the Einstein relation. The same relation can be derived by following a more rigorous approach based on thermodynamic arguments.[26]

7.2.7 p-n Junction under External Bias

Let us assume that we apply an external forward bias across a p-n junction. In other words, the p-side is connected to the positive terminal of the battery as shown in Fig. 7.19. If the bias voltage is V, the potential difference across the depletion region will now be reduced to $V_b - V$, as shown in Fig. 7.20*a* and *b*. Consider the majority carriers (electrons) on the n-side. After the bias is applied, they will see a smaller potential barrier, and hence diffusion toward the p-side will increase. As a result, at the edge of the depletion region on the p-side, there will be a larger concentration of electrons than n_i^2/N_A. This is known as carrier injection. Similarly, the potential barrier seen by the majority carriers (holes) on the p-side will be reduced, resulting in a larger number of injected holes on the n-side. In the derivation of the junction current, we will assume that electron-hole recombination is negligible in the depletion region. Furthermore, if there is no light incident on the junction, we can also neglect the generation of carriers. Hence, under these assumptions, the total current in the depletion region due to holes and electrons should remain constant. Note that we can alternatively determine the total current from a knowledge of the minority diffusion current at the edges of the depletion region. To see why this is the case, consider the hole current density J_p. In general, this has both a drift and diffusion component. At $z = d_n^+$

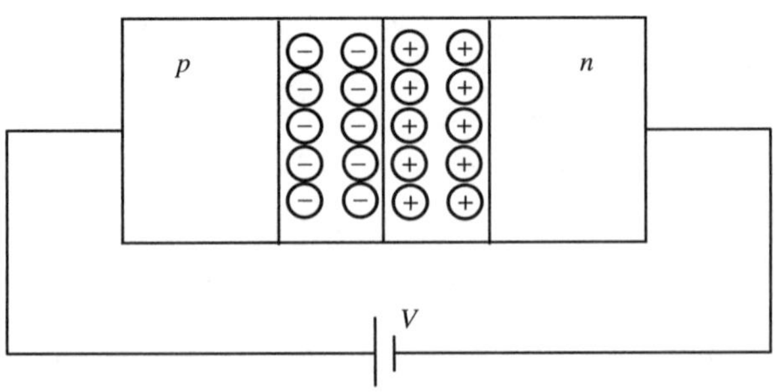

FIGURE 7.19 Forward-biased p-n junction.

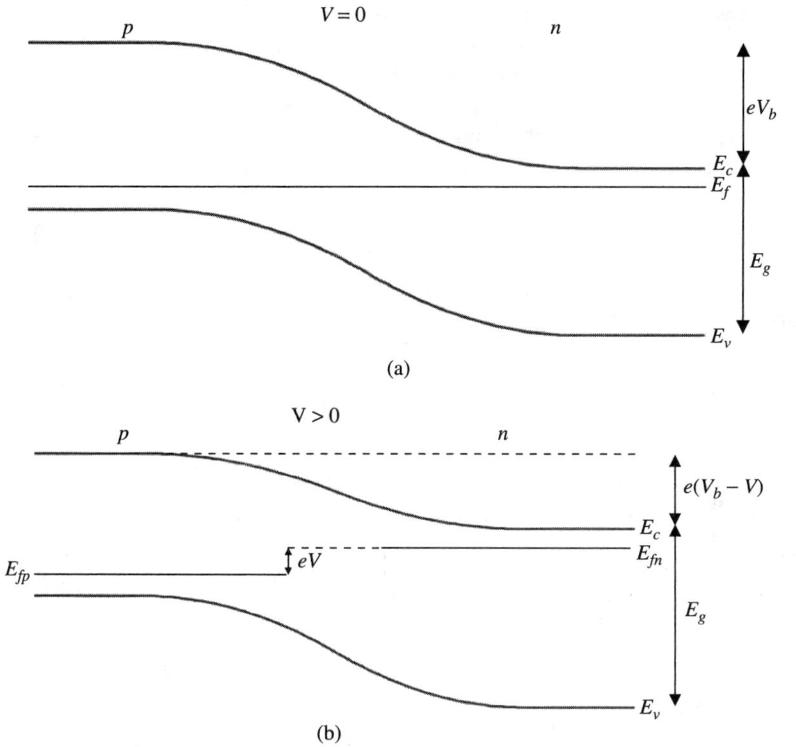

Figure 7.20 Energy band diagram of a p-n junction under (*a*) no bias and (*b*) forward bias.

(i.e., *n*-side of the depletion region edge; see Fig. 7.16), the electric field is zero and the total hole current in the depletion region will be given by the minority hole diffusion current evaluated at $z = d_n^+$. In other words,

$$J_p = -eD_p \left.\frac{dp}{dz}\right|_{z=d_n^+} \tag{7.77}$$

Similarly, the electron component of the current density in the depletion region will be given by

$$J_n = eD_n \left.\frac{dn}{dz}\right|_{z=-d_p^-} \tag{7.78}$$

Above, $z = -d_p^-$ designates the left-hand side of the depletion region edge on the *p*-side. The total current anywhere should then be equal to

$$J = J_n + J_p \tag{7.79}$$

Now, we need to determine the position dependence of the minority charge concentrations in order to calculate the derivatives dp/dz and dn/dz appearing in Eqs. (7.77) and (7.78). To obtain the expressions for p and n near the depletion edge, we assume a nondegenerate semiconductor where n and p are given by the respective relations

$$n = N_c \exp\left[\frac{E_{fn} - E_c}{k_B T}\right] \tag{7.80}$$

and

$$p = N_v \exp\left[\frac{E_v - E_{fp}}{k_B T}\right] \tag{7.81}$$

Earlier, we showed that for a nondegenerate semiconductor, the product np becomes [see Eqs. (7.22)–(7.27)]

$$np = n_i^2 = N_c N_v \exp\left[-\frac{E_c - E_v}{k_B T}\right] \tag{7.82}$$

First, consider the intrinsic semiconductor where $n = n_i$. From Eq. (7.80), n_i becomes

$$n_i = N_c \exp\left[\frac{E_i - E_c}{k_B T}\right] \tag{7.83}$$

where E_i is the Fermi level of the intrinsic semiconductor. Hence n becomes

$$n = n_i \exp\left[\frac{E_{fn} - E_i}{k_B T}\right] \tag{7.84}$$

Similarly, we get

$$p = n_i \exp\left[\frac{E_i - E_{fp}}{k_B T}\right] \tag{7.85}$$

for holes. When the p-n junction is in equilibrium, $E_{fn} = E_{fp}$, and we get $np = n_i^2$, as expected, using Eqs. (7.84) and (7.85). When a forward bias is applied, E_{fn} and E_{fp} are no longer equal and we have

$$np = n_i^2 \exp\left[\frac{E_{fn} - E_{fp}}{k_B T}\right] \tag{7.86}$$

The resulting energy band diagram under forward bias is sketched in Fig. 7.20*b*. Since $E_{fn} - E_{fp} = eV$, we get

$$np = n_i^2 \exp\left[\frac{eV}{k_B T}\right] \tag{7.87}$$

Now, let us look at the edge of the depletion region on the *n*-side. Here $n = N_D$ and p is given by

$$p = \frac{n_i^2}{N_D} \exp\left[\frac{eV}{k_B T}\right] \tag{7.88}$$

On the *p*-side of the depletion region edge, we get

$$n = \frac{n_i^2}{N_A} \exp\left[\frac{eV}{k_B T}\right] \tag{7.89}$$

for the electron concentration n. Hence, the deviations Δn and Δp from the equilibrium concentration levels are given by

$$\Delta p(z = +d_n) = \frac{n_i^2}{N_D}\left(\exp\left[\frac{eV}{k_B T}\right] - 1\right) \tag{7.90}$$

and

$$\Delta n(z = -d_p) = \frac{n_i^2}{N_A}\left(\exp\left[\frac{eV}{k_B T}\right] - 1\right) \tag{7.91}$$

Figure 7.21 shows a rough sketch of how n and p vary around the depletion region of a silicon p-n junction with $n_i = 10^{10}\,\text{cm}^{-3}$, $N_D = 10^{16}\,\text{cm}^{-3}$ and $N_A = 10^{16}\,\text{cm}^{-3}$ for $V = 0$ and $V > 0$. Similar plots and further discussion on the carrier statistics of p-n junctions can be found in Refs. 8 and 10.

By using the continuity equations derived earlier for excess carrier concentrations, we can solve for the position dependence of the injected minority carriers. Let us consider the injected holes on the *n*-side. The excess carrier density Δp satisfies [see Eq. (7.51)]

$$\frac{\partial \Delta p}{\partial t} = G + D_p \frac{\partial^2 \Delta p}{\partial z^2} - \frac{\Delta p}{\tau_p} \tag{7.92}$$

where G is the generation rate, D_p is the hole diffusion coefficient, and τ_p is the hole recombination time. In steady state and with no illumination, we have $\partial/\partial t = 0$ and $G = 0$, giving

$$\frac{\partial^2 \Delta p}{\partial z^2} - \frac{1}{D_p \tau_p} \Delta p = 0 \tag{7.93}$$

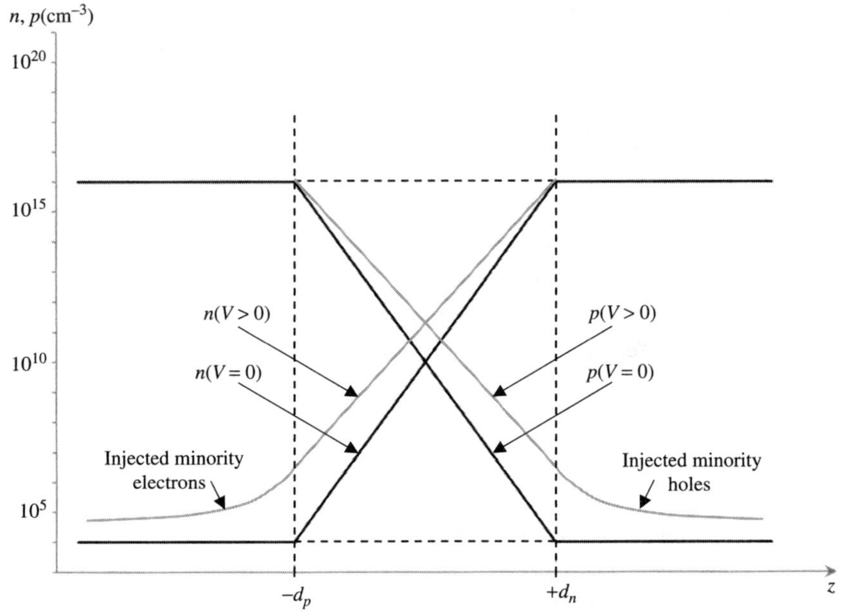

FIGURE 7.21 Sketch of the electron (n) and hole (p) concentrations across the depletion region of a silicon p-n junction under no bias and forward bias ($n_i = 10^{10}\,\text{cm}^{-3}$, $N_D = 10^{16}\,\text{cm}^{-3}$, and $N_A = 10^{16}\,\text{cm}^{-3}$).

By defining the characteristic diffusion length L_P as

$$L_P = \sqrt{D_p \tau_p} \tag{7.94}$$

we rewrite Eq. (7.93) in the form

$$\frac{\partial^2 \Delta p}{\partial z^2} - \frac{\Delta p}{L_P^2} = 0 \tag{7.95}$$

The general solution of Eq. (7.95) has the form

$$\Delta p(z) = A \exp\left[-\frac{z}{L_P}\right] + B \exp\left[+\frac{z}{L_P}\right] \tag{7.96}$$

where A and B are constants. As $z \to \infty$, the excess hole density tends to zero. Furthermore, Δp at $z = +d_n$ is given by Eq. (7.90). Hence, $B = 0$ and $\Delta p(z)$ can be written as

$$\Delta p(z) = \frac{n_i^2}{N_D}\left[\exp\left(\frac{eV}{k_B T}\right) - 1\right] \exp\left(-\frac{(z - d_n)}{L_P}\right) \tag{7.97}$$

The hole diffusion current at $z = d_n$ becomes

$$J_p = -eD_p \left.\frac{dp}{dz}\right|_{z=d_n} = \frac{eD_p n_i^2}{N_D L_P}\left[\exp\left(\frac{eV}{k_B T}\right) - 1\right] \tag{7.98}$$

A similar calculation for J_n at $z = -d_p$ yields

$$J_n = \frac{eD_n n_i^2}{N_A L_N}\left[\exp\left(\frac{eV}{k_B T}\right) - 1\right] \tag{7.99}$$

where $L_N = \sqrt{D_n \tau_n}$ is the characteristic diffusion length for minority electrons on the p-side. Assuming that the cross-sectional area of the p-n junction is a, the total current can be expressed as

$$I = I_0\left[\exp\left(\frac{eV}{k_B T}\right) - 1\right] \tag{7.100}$$

where I_0 is given by

$$I_0 = ean_i^2\left(\frac{D_n}{N_A L_N} + \frac{D_p}{N_D L_P}\right) \tag{7.101}$$

We can do a similar analysis for the reverse bias case. The corresponding energy band diagrams and the carrier concentrations are shown in Figs. 7.22 and 7.23. In fact, the expression for the current derived in

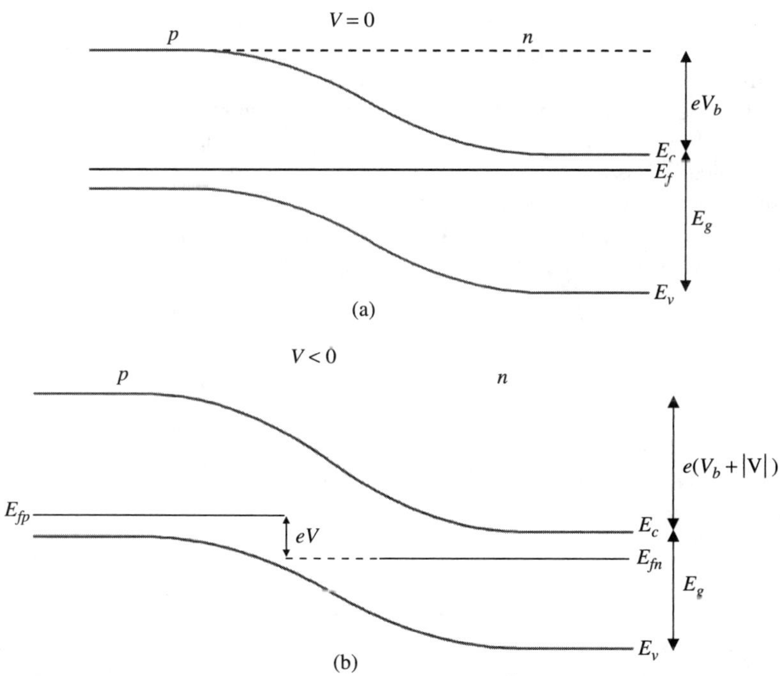

FIGURE 7.22 Energy band diagram of a p-n junction under (*a*) no bias and (*b*) reverse bias.

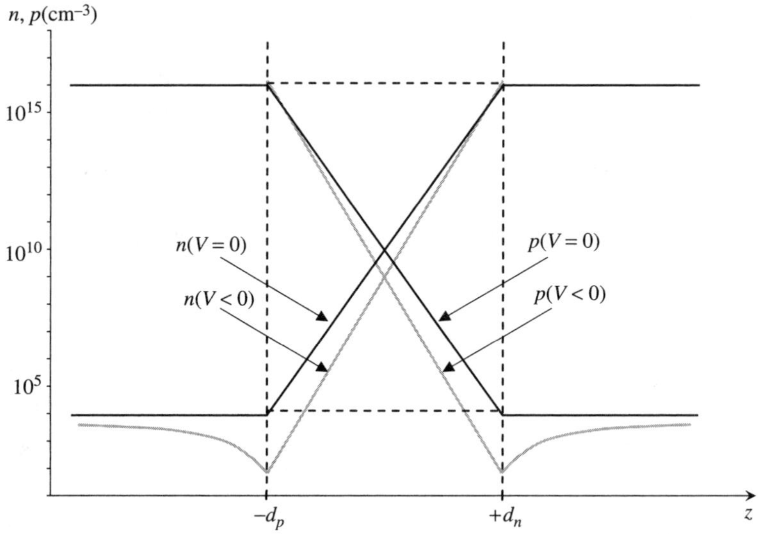

FIGURE 7.23 Variation of the electron (n) and hole (p) concentrations across the depletion region of a p-n junction under no bias and reverse bias ($n_i = 10^{10}\text{cm}^{-3}$, $N_D = 10^{16}\text{cm}^{-3}$, and $N_A = 10^{16}\text{cm}^{-3}$).

Eqs. (7.100) and (7.101) also hold for the case $V < 0$. Qualitatively, because the potential barrier across the junction is increased, very small current flows when the p-n junction is reverse-biased. From Eq. (7.100), we get $I \simeq -I_0$. I_0 is known as the reverse leakage current and typically has a very small value. To summarize, the p-n junction is highly conducting when forward-biased, and hardly any current flows when it is reverse-biased. A typical current-voltage (I-V) characteristic of a p-n-junction is shown in Fig. 7.24.

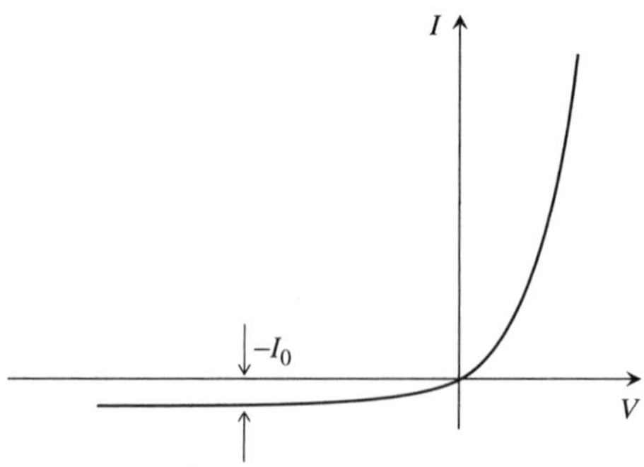

FIGURE 7.24 Typical current-voltage (I-V) characteristic of an ideal p-n junction. I_0 is known as the reverse leakage current.

7.3 Interaction of Light with Semiconductors

We now turn our attention to how photons interact with electrons and holes in semiconductors and explain the processes of absorption, emission, and laser operation. We will specifically focus on GaAs, since this is a direct-gap semiconductor and an efficient photon emitter. We will discuss the concepts of quasi-Fermi levels and joint density of states, and derive expressions for the gain coefficient and the condition for oscillation in a semiconductor laser.

7.3.1 Density of States for GaAs

Let us recall the expressions for the density of states near the edges of the conduction and valance bands:

$$\rho_c(E)dE = \frac{\left(2m_e^*\right)^{3/2}}{2\pi^2\hbar^3}(E-E_c)^{1/2}dE \tag{7.102}$$

and

$$\rho_v(E)dE = \frac{\left(2m_h^*\right)^{3/2}}{2\pi^2\hbar^3}(E_v-E)^{1/2}dE \tag{7.103}$$

In the case of GaAs, the effective electron mass in the conduction band is much smaller than the effective hole mass with $m_e^* = 0.067m_0$ and $m_h^* = 0.55m_0$. The resulting variation of $\rho_c(E)$ and $\rho_v(E)$ are sketched in Fig. 7.25. Note that due to the difference in the effective masses, $\rho_v(E)$ has a much higher slope in comparison with $\rho_c(E)$.

7.3.2 Quasi-Fermi Levels

The Fermi-Dirac distribution $f(E)$ introduced in Sec. 7.2.2 gives the probability that a state at energy E is occupied by an electron. The probability of occupation of a level at $E = E_f$ is further equal to 1/2. At $T = 0$ K, all states up to the Fermi level E_f are occupied, while the probability of occupation of states above $E = E_f$ is zero.

What we have discussed above is applicable to semiconductors in thermal equilibrium. When an external bias is applied to inject electrons and holes, or when the semiconductor is exposed to intense pump light with frequency ν such that $h\nu > E_g$, we get a nonequilibrium situation where the electron and hole concentrations are above their equilibrium values. Typically, in such a case, intraband relaxation occurs at a much faster rate (typically 10^{-12} s) than interband transitions (typically 10^{-9} s; this is also known as electron hole recombination time), and charge carriers quickly reach a state of quasi-equilibrium, filling the lowest available levels in each band. Once quasi-equilibrium conditions prevail, we can still use the Fermi-Dirac function to describe the occupation probability of states in each band. However, the Fermi

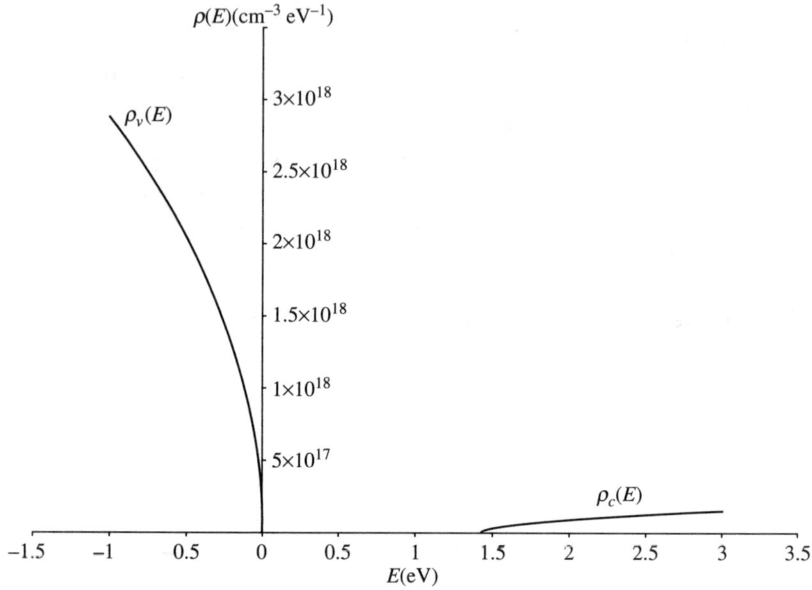

FIGURE 7.25 Variation of the density of states $\rho_c(E)$ and $\rho_v(E)$ as a function of energy near the edges of the conduction and valence bands for GaAs.

level E_{fn} for electrons is now different from the Fermi level E_{fp} for holes. E_{fn} and E_{fp} are referred to as the quasi-Fermi levels. In fact, we have already used the concept of quasi-Fermi levels in the previous section to discuss the current-voltage characteristics of the biased p-n junction. This concept is further useful while analyzing the interaction of light with electrons and holes occupying the same location, which is the case in semiconductor lasers or light emitting diodes.

In quasi-equilibrium, the probability of a state occupied in the conduction band is given by

$$f_c(E) = \frac{1}{\exp\left(\frac{E - E_{fn}}{k_B T}\right) + 1} \tag{7.104}$$

Similarly, the probability that a state in the valance band is occupied becomes

$$f_v(E) = \frac{1}{\exp\left(\frac{E - E_{fp}}{k_B T}\right) + 1} \tag{7.105}$$

It is often the case in the heavily doped *n*- or *p*-type regions of a p-n junction, that the quasi-Fermi levels are actually within the bands.

Then, the high-energy exponential approximation to $f(E)$ [as given in Eq. (7.18), for example] is no longer valid. The total electron and hole densities are then given by

$$n = \frac{\left(2m_e^*\right)^{3/2}}{2\pi^2\hbar^3} \int_{E_C}^{\infty} dE \frac{(E - E_c)^{1/2}}{\exp\left(\frac{E - E_{fn}}{k_B T}\right) + 1} \tag{7.106}$$

and

$$p = \frac{\left(2m_h^*\right)^{3/2}}{2\pi^2\hbar^3} \int_{-\infty}^{E_V} dE (E_v - E)^{1/2} \left[1 - \frac{1}{\exp\left(\frac{E - E_{fp}}{k_B T}\right) + 1}\right] \tag{7.107}$$

respectively. From a knowledge of n and p, the quasi-Fermi levels E_{fn} and E_{fp} can be numerically determined by using the above integral relations. This will be important when we discuss the condition for optical amplification in a semiconductor because the relation between E_{fn} and E_{fp} will be critical. At $T = 0$ K, the quasi-Fermi levels can be determined analytically. In the case of electrons in the conduction band, all states will be occupied up to $E = E_{fn}$, yielding

$$n = \frac{\left(2m_e^*\right)^{3/2}}{2\pi^2\hbar^3} \int_{E_C}^{E_{fn}} dE (E - E_c)^{1/2} \tag{7.108}$$

By making a change of variables to $u = E - E_c$, n can be readily calculated as

$$n = \frac{\left(2m_e^*\right)^{3/2}}{2\pi^2\hbar^3} \frac{2}{3} (E_{fn} - E_c)^{3/2} \tag{7.109}$$

giving

$$E_{fn} = E_c + \frac{\hbar^2}{2m_e^*} (3\pi^2 n)^{2/3} \tag{7.110}$$

At higher temperatures, the quasi-Fermi levels can be determined numerically as discussed later.

7.3.3 Exact Carrier Density Calculations in Degenerate Semiconductors

Starting with Eq. (7.106) and by introducing the dimensionless integration variable x as

$$x = \frac{E - E_c}{k_B T} \tag{7.111}$$

it can be shown (see Prob. 7.11) that the electron concentration n can be written in the form

$$n = \frac{\left(2m_e^* k_B T\right)^{3/2}}{2\pi^2 \hbar^3} \int_0^\infty dx \frac{x^{1/2}}{1+\exp(x-x_0)} \tag{7.112}$$

where x_0 is given by

$$x_0 = \frac{E_{fn} - E_c}{k_B T} \tag{7.113}$$

Similarly, we have, for the hole concentration

$$p = \frac{\left(2m_h^* k_B T\right)^{3/2}}{2\pi^2 \hbar^3} \int_0^\infty dy \frac{y^{1/2}}{1+\exp(y-y_0)} \tag{7.114}$$

where y is given by

$$y = \frac{E_v - E}{k_B T} \tag{7.115}$$

and $y_0 = (E_v - E_{fp})/k_B T$.

7.3.4 Joint Optical Density of States Function

In our discussion of optical interactions in semiconductors, we will focus on band-to-band (also known as interband) transitions. Here, the energy of the absorbed or emitted photon is above the band-gap energy, i.e., $hc/\lambda > E_g$, where λ is the photon wavelength. This, by no means, is the only possible process that can take place. It is also possible to observe other transitions between impurity level(s) and a band, between energy levels of a free carrier within a band, and between different phonon levels, as well as excitonic transitions. The latter is the formation of a bound electron-hole pair which is similar to a hydrogen atom. When the bound electron and hole recombine, a photon may be emitted. We will not analyze these processes here.

To characterize the strength of the optical transitions, we will first derive the expression for what is known as the joint optical density of states $\rho_j(\nu)$ defined according to:

$\rho_j(\nu)d\nu$ = number of states per unit volume that can interact with a photon of frequency between ν and $\nu + d\nu$ (7.116)

To derive the expression for $\rho_j(\nu)$, let us examine a direct-gap semiconductor in the k-space as shown in Fig. 7.26. Suppose that an electron in an occupied state at energy E_2 makes a transition down to

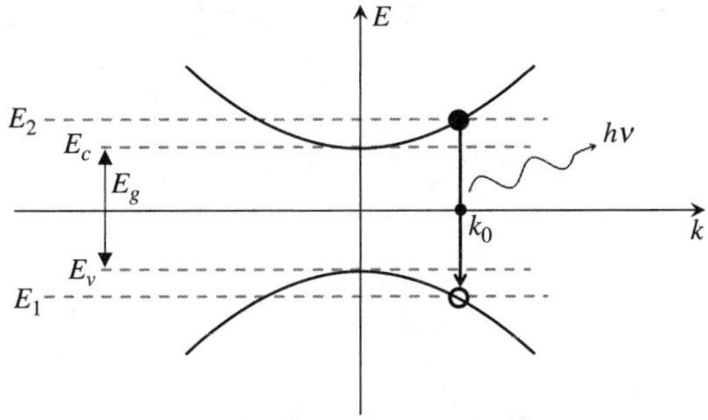

FIGURE 7.26 Energy band diagram showing the energy levels involved in the derivation of the joint optical density of states.

an empty state at E_1. Both the electron and the hole have approximately the same wave vector of k_0. In the effective mass approximation, the energy of the emitted photon will be given by

$$h\nu = E_2 - E_1 = E_c + \frac{\hbar^2 k_0{}^2}{2m_e^*} - \left(E_v - \frac{\hbar^2 k_0{}^2}{2m_h^*}\right)$$

$$= E_g + \frac{\hbar^2 k_0{}^2}{2m_e^*} + \frac{\hbar^2 k_0{}^2}{2m_h^*} \tag{7.117}$$

Introducing the reduced effective mass $\overline{m}$ defined according to

$$\frac{1}{\overline{m}} = \frac{1}{m_e^*} + \frac{1}{m_h^*} \tag{7.118}$$

we obtain

$$k_0^2 = \frac{2\overline{m}}{\hbar^2}(h\nu - E_g) \tag{7.119}$$

From before, the density-of-states function was derived to be

$$\rho(k_0)dk_0 = \frac{k_0^2}{\pi^2}dk_0 \tag{7.120}$$

The joint optical density of states can be expressed in terms of $\rho(k_0)$ by using

$$\rho_j(\nu)\,d\nu = \rho(k_0(\nu))\left(\frac{dk_0}{d\nu}\right)d\nu \tag{7.121}$$

where k_0 is given by Eq. (7.119). We leave it as an exercise (see Prob. 7.12) to show that $\rho_j(\nu)$ becomes

$$\rho_j(\nu)\,d\nu = \frac{(2\overline{m})^{3/2}}{\pi\hbar^2}(h\nu - E_g)^{1/2} \tag{7.122}$$

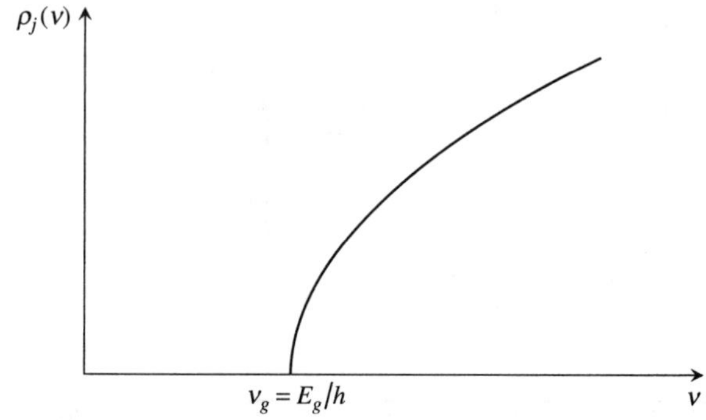

FIGURE 7.27 Variation of the joint optical density-of-states function $\rho_j(\nu)$ with photon frequency.

for $h\nu > E_g$. Variation of the joint optical density of states $\rho_j(\nu)$ as a function of photon frequency is sketched in Fig. 7.27. Note that $\rho_j(\nu)$ increases as the square root of the energy difference from the band edge.

7.3.5 Condition for Optical Amplification in Semiconductors

In Chap. 6, we considered the interaction of a monochromatic beam with an atomic system and derived the following equation [see Eq. (6.39)] for the differential gain coefficient $\gamma(\nu)$:

$$\gamma(\nu) = A_{21}\frac{\lambda^2}{8\pi n_s^2}g(\nu)\left(N_2 - \frac{g_2}{g_1}N_1\right) \tag{7.123}$$

To remind ourselves, A_{21} is Einstein's A coefficient which gives the radiative decay rate of the excited atoms, λ is the wavelength of the monochromatic light passing through the atomic system, n_s is the refractive index of the semiconductor, $g(\nu)$ is the lineshape function, N_1 and N_2 are the lower- and upper-level population densities, and g_1 and g_2 are the degeneracy factors of the two levels. Now, in the case of a semiconductor, we do not have discrete energy levels, rather, we have a distribution of energy levels. The density of states corresponding to the optical transition at the wavelength $\lambda = c/\nu$ is described by the joint optical density-of-states function $\rho_j(\nu)$ given by Eq. (7.122). $\rho_j(\nu)d\nu$ tells us the number of states per unit volume that can undergo a transition to emit photons in the frequency range ν to $\nu + d\nu$, provided that $h\nu > E_g$. Therefore, we can calculate the contribution of $d\gamma$ to the gain coefficient arising from the levels that emit in the range of ν to $\nu + d\nu$. As in an atomic system, both absorption and emission will occur simultaneously. The rate of absorption R_a

will be proportional to the probability that the lower level E_1 is filled and the upper level E_2 is empty. In other words,

$$R_a \propto f_v(E_1)(1 - f_c(E_2)) \tag{7.124}$$

where

$$f_v(E_1) = \frac{1}{\exp\left[\dfrac{E_1 - E_{fp}}{k_B T}\right] + 1} \tag{7.125}$$

and

$$f_c(E_2) = \frac{1}{\exp\left[\dfrac{E_2 - E_{fn}}{k_B T}\right] + 1} \tag{7.126}$$

Similarly, we have the proportionality

$$R_e \propto f_c(E_2)(1 - f_v(E_1)) \tag{7.127}$$

for the rate of emission R_e, with $f_v(E_1)$ and $f_c(E_2)$ given by Eqs. (7.125) and (7.126), respectively. Note that since we have a nonequilibrium case here, different Fermi levels have been assumed for electrons and holes. The net downward emission rate will be proportional to $R_e - R_a$, which satisfies the proportionality relation

$$(R_e - R_a) \propto f_c(E_2)(1 - f_v(E_1)) - f_v(E_1)(1 - f_c(E_2)) = f_c(E_2) - f_v(E_1) \tag{7.128}$$

Hence, $d\gamma$ becomes

$$d\gamma = A_{21} \frac{\lambda^2}{8\pi n_s^2} g(\nu) \rho_j(\nu)(f_c(E_2) - f_v(E_1)) d\nu \tag{7.129}$$

Typically, the line shape function $g(\nu)$ is very narrow compared to the other distribution functions and can be approximated as a Dirac delta function in the form

$$g(\nu, \nu_0) = \delta(\nu - \nu_0) \tag{7.130}$$

where now ν_0 represents the frequency of the monochromatic wave. Hence, the total differential gain coefficient at the frequency ν_0 becomes

$$\gamma(\nu_0) = \int_0^\infty d\nu A_{21} \frac{\lambda^2}{8\pi n_s^2} \delta(\nu - \nu_0) \rho_j(\nu)[f_c(E_2) - f_v(E_1)] \tag{7.131}$$

Einstein's coefficient A_{21} is given by $1/\tau_r$, where τ_r is the radiative recombination time. Carrying out the integration in Eq. (7.131) by using Eq. (7.130), we obtain for $\gamma(\nu_0)$

$$\gamma(\nu_0) = \frac{\lambda_0^2}{8\pi n_s^2 \tau_r} \frac{(2\bar{m})^{3/2}}{\pi\hbar^2}(h\nu_0 - E_g)^{1/2}(f_c(E_2) - f_v(E_1)) \tag{7.132}$$

In order to obtain gain, $\gamma(\nu_0)$ must be positive, which implies that $f_c(E_2) - f_v(E_1) > 0$. In other words,

$$\frac{1}{\exp\left[\dfrac{E_2 - E_{fn}}{k_B T}\right] + 1} > \frac{1}{\exp\left[\dfrac{E_1 - E_{fp}}{k_B T}\right] + 1} \tag{7.133}$$

or

$$E_{fn} - E_{fp} > E_2 - E_1 \tag{7.134}$$

This means that the semiconductor must bc pumped sufficiently strongly so that carrier injection leads to a separation between the quasi-Fermi levels larger than the photon energy $h\nu = E_2 - E_1$. Whenever this condition is satisfied, the semiconductor will act as an amplifier. We will provide a detailed example based on the parameters of GaAs to illustrate this.

Before ending this section, we also comment on the electron-hole recombination time in more detail. In general, the recombination time depends on the doping concentration in the semiconductor. To see this, first consider a semiconductor in thermal equilibrium. Let the equilibrium concentrations of electrons and holes be n_0 and p_0, respectively. In equilibrium, the thermal generation rate of electron-hole pairs will be equal to the rate of recombination. The rate of recombination is proportional to the product np, since increasing either the hole or electron concentration should lead to larger recombination rate. The overall rate can therefore be written as

$$\frac{dn}{dt} = G_T - \beta n_0 p_0 = 0 \tag{7.135}$$

Above, G_T is the thermal generation rate, and β is a constant which is a characteristic of the particular semiconductor. Let us now suppose that we have an additional generation rate G due to, say, light illumination. The new concentrations will become $n_0 + \Delta n$ and $p_0 + \Delta p$, where $\Delta n = \Delta p$. Now, the rate of change of the electron concentration will be given by

$$\frac{dn}{dt} = \frac{d}{dt}(n_0 + \Delta n) = G + G_T - \beta(n_0 + \Delta n)(p_0 + \Delta p) \tag{7.136}$$

To first order in Δn, we get

$$\frac{d\Delta n}{dt} = G - \beta(n_0 + p_0)\Delta n = G - \frac{\Delta n}{\tau_r} \tag{7.137}$$

where the radiative recombination time τ_r appearing in Eq. (7.137) is given by

$$\tau_r = \frac{1}{\beta(n_0 + p_0)} \tag{7.138}$$

So, note that the recombination time depends on the equilibrium concentrations under low injection where $\Delta n \ll n_0, p_0$. In fact, in the analysis of the I-V characteristics of the biased p-n junction, we did assume that the recombination time would be different on different sides of the junction. In an n-type semiconductor with $n_0 \cong N_D$ and $p_0 \ll n_0$, we have

$$\tau_r = \frac{1}{\beta N_D} \tag{7.139}$$

Similarly,

$$\tau_r = \frac{1}{\beta N_A} \tag{7.140}$$

for a p-type semiconductor. For GaAs, β is about $2 \times 10^{-10}\text{cm}^3/\text{s}$.[12]

7.3.6 Concentration and Optical Gain Calculations in GaAs

In this section, we present several examples to show how to calculate the carrier concentrations and the optical gain coefficient $\gamma(\nu_0)$ in GaAs. We assume the following parameters for GaAs: $m_e^* = 0.067m_e$, $m_h^* = 0.55m_e$, $E_g = 1.42$ eV, $n_s = 3.6$, $\tau_r = 2 \times 10^{-9}$ s and $T = 300$ K. Since the particular choice of the parameter values can cause a significant variation in the results compared to what one would see in a realistic system, the results should be seen as estimates rather than exact calculations. Accurate analysis of specific semiconductor lasers may be found in other sources (see for example, Refs.14 and 17).

Weighted Density of States for GaAs

First, let us calculate the weighted density of states $\bar{\rho}_c(E)$ and $\bar{\rho}_v(E)$ for electrons and holes which can be defined as the density of states weighted by the occupation probabilities according to

$$\bar{\rho}_c(E)dE = \rho_c(E)f_c(E)dE \tag{7.141}$$

and

$$\bar{\rho}_v(E)dE = \rho_v(E)(1 - f_v(E))dE \tag{7.142}$$

where $f_c(E)$ and $f_v(E)$ are as defined in Eqs. (7.104) and (7.105). Figure 7.28*a* and *b* shows the variation of $\bar{\rho}_v(E)$ and $\bar{\rho}_c(E)$ for GaAs for two different values of the quasi-Fermi levels. In this calculation, we took $E_v = 0$ eV (the edge of the valence band). Note that there is no contribution to the weighted density of states for energies falling in the band gap, in other words, for energy values between 0 and 1.42 eV for GaAs. Furthermore, in both cases, although the density-of-states function monotonically increases with energy away from the band edges, the weighted densities reach a maximum and then decay, since the occupation probabilities diminish as we go further and further from the band edges. In Fig. 7.28*a*, we chose $E_{fp} = E_v$ and $E_{fn} = E_c$. Note that in this case, the total hole concentration ($7.8 \times 10^{18}/\text{cm}^3$) is nearly 23 times larger than the total electron density ($3.3 \times 10^{17}/\text{cm}^3$). The asymmetry arises from the vast difference in the effective masses of the holes and electrons in GaAs. In Fig. 7.28*b*, we plot the weighted

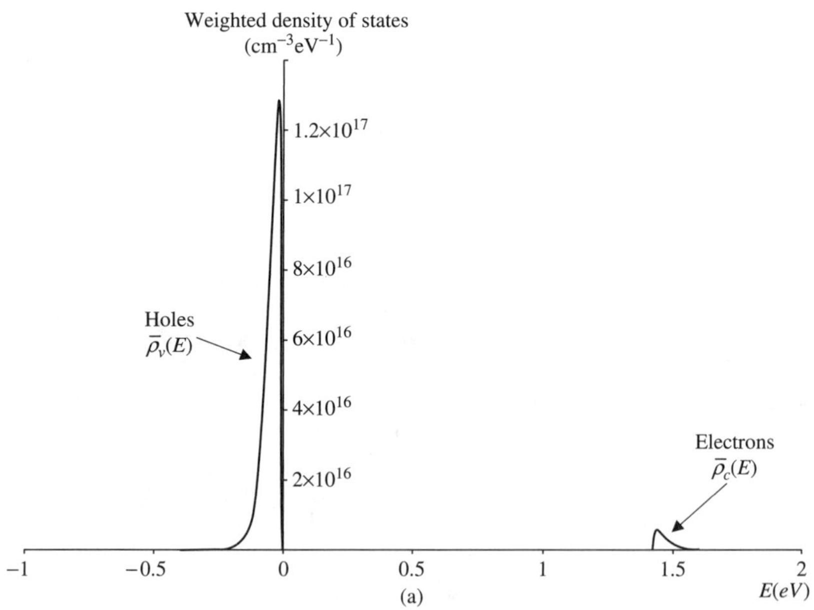

FIGURE 7.28 (*a*) Calculated variation of the weighted density-of-states functions $\bar{\rho}_v(E)$ and $\bar{\rho}_c(E)$ for GaAs for the case where $E_{fp} = E_v$ and $E_{fn} = E_c$. (*b*) Calculated variation of the weighted density-of-states functions $\bar{\rho}_v(E)$ and $\bar{\rho}_c(E)$ for GaAs for the case where $E_{fp} = E_v$ and E_{fn} is adjusted to make the total electron concentration equal to the total hole concentration.

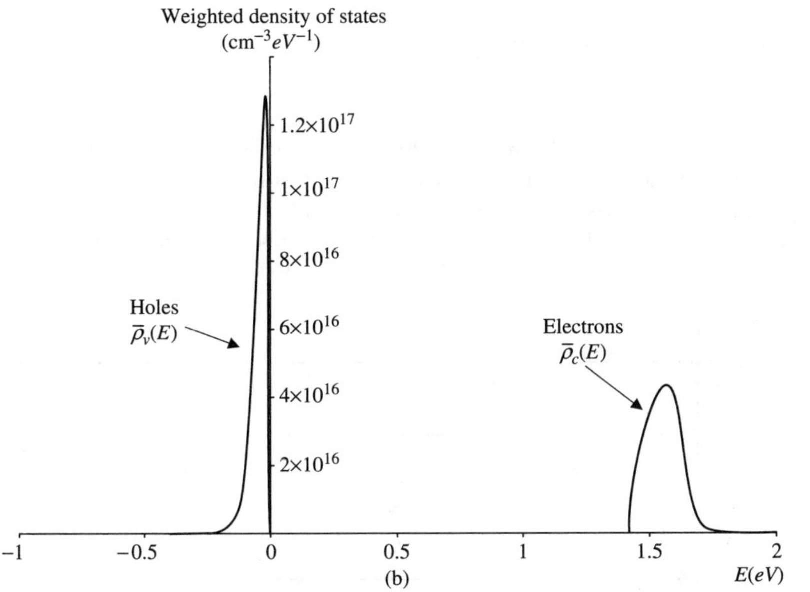

FIGURE 7.28 (*Continued*)

density of states for the case where $E_{fp} = E_v$, and where the electron and hole concentrations are equal to each other. Looking at Fig. 7.28*b*, the area under each curve is equal. We leave it as an exercise for the reader to numerically determine the E_{fn} value relative to the conduction band edge where the total electron and hole concentrations are both equal to $7.8 \times 10^{10}/\text{cm}^3$ (see Probs. 7.13–7.15).

Total Electron and Hole Concentrations

As was discussed earlier, the total electron and hole concentrations n and p can be calculated numerically by using

$$n = \frac{(2m_e^* k_B T)^{3/2}}{2\pi^2\hbar^3} \int_0^\infty dx \frac{x^{1/2}}{1+\exp[x - x_0]} \tag{7.143}$$

and

$$p = \frac{(2m_h^* k_B T)^{3/2}}{2\pi^2\hbar^3} \int_0^\infty dy \frac{y^{1/2}}{1+\exp[y - y_0]} \tag{7.144}$$

where the quasi-Fermi levels E_{fn} and E_{fp} are related to the normalized Fermi level offsets x_0 and y_0 through

$$E_{fn} = E_c + k_B T x_0 \tag{7.145}$$

and

$$E_{fp} = E_v - k_B T y_0 \tag{7.146}$$

To illustrate this, we plot in Fig. 7.29*a* and *b* the variation of the total electron and hole concentrations as a function of the Fermi level position. Again, we took $E_v = 0$ eV. First, note that the vertical axis in each plot is logarithmic. In Fig. 7.29*a*, as E_{fn} approaches and goes inside the conduction band ($E_{fn} > 1.42$ eV), the electron density increases by

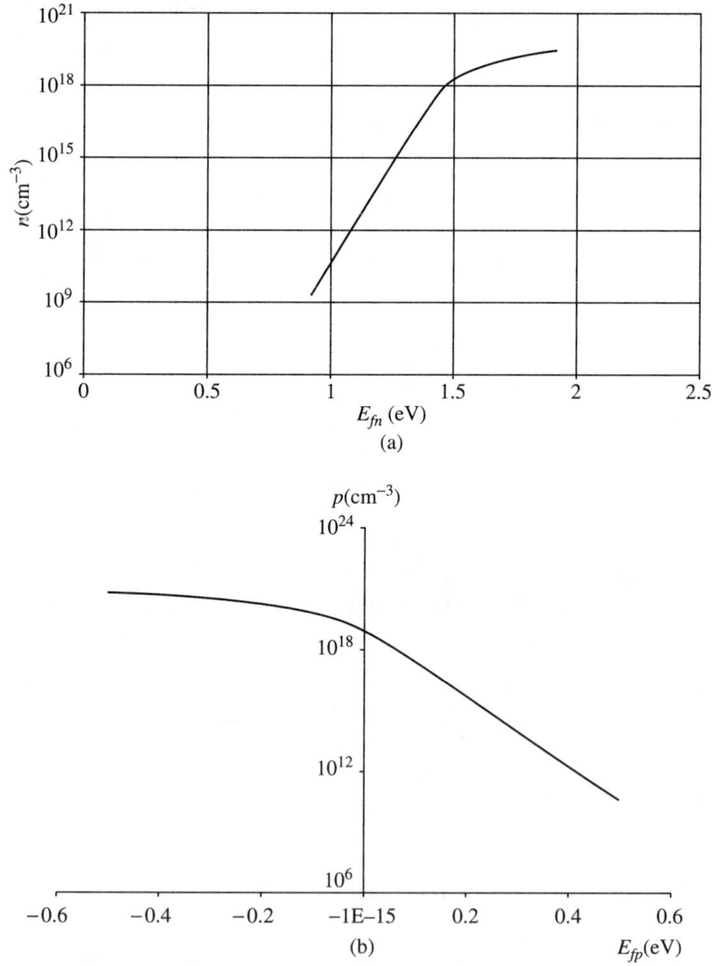

FIGURE 7.29 (*a*) Calculated variation of the electron density n as a function of the quasi-Fermi level E_{fn} for GaAs at $T = 300$ K. (*b*) Calculated variation of the hole density p as a function of the quasi-Fermi level E_{fp} for GaAs at $T = 300$ K ($m_e^* = 0.067m_e$, $m_h^* = 0.55m_e$, $E_g = 1.42$ eV).

orders of magnitude. For example, when E_{fn} goes from 0.9 eV to 1.9 eV, n increases by 10 orders of magnitude from 1.7×10^9 cm^{-3} to 2.7×10^{19} cm^{-3}. In practice, this is done by carrier injection when, for example, a GaAs p-n junction is forward-biased, as we will discuss in the next section. A similar trend is seen in Fig. 7.29*b* for the hole concentration. In this case, as the quasi-Fermi level goes into the valence band (in other words $E_{fp} < 0$ eV), the hole concentration exponentially increases.

Now, it is clear from the foregoing examples that if we want to increase the separation between the quasi-Fermi levels and achieve optical amplification, we need to increase the total carrier concentrations. In the next example, we work backwards, and numerically determine the quasi-Fermi levels E_{fn} and E_{fp} from a knowledge of the electron and hole concentrations. To do this, the normalized Fermi level offsets x_0 and y_0 are first numerically determined from Eqs. (7.143) and (7.144) for a particular value of n and p. Then E_{fn} and E_{fp} are readily determined from Eqs. (7.145) and (7.146). We leave the details of the numerical calculation as an exercise (see Prob. 7.17). The results are shown in Fig. 7.30. Notice that when $n = p = 1.26 \times 10^{18}$cm^{-3}, $E_{fn} - E_{fp} = 1.42$ eV, shown by the dotted arrows. This is the critical concentration level beyond which we will begin to see amplification of light, since interband optical transitions will have photon energies given by $E_2 - E_1 > E_g$ and, as we discussed in Sec. 7.3.5, $E_{fn} - E_{fp} > E_2 - E_1$ for optical amplification to take place. Of course, more carrier injection will be required in a practical laser in order to overcome passive

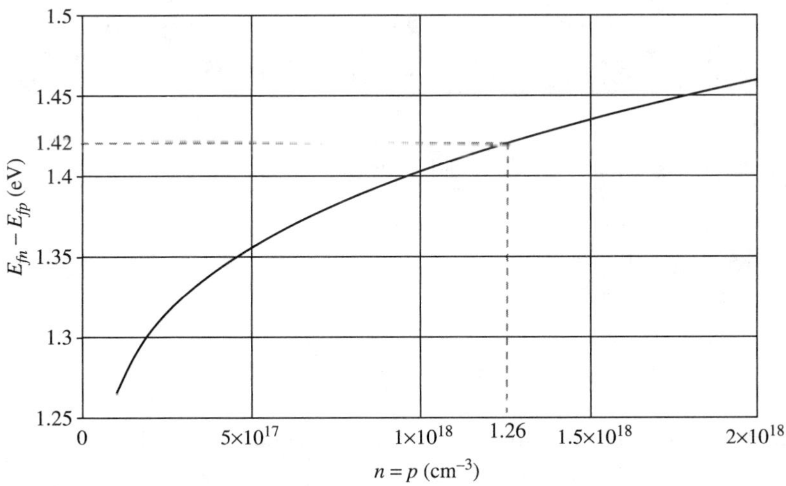

FIGURE 7.30 Calculated variation of $E_{fn} - E_{fp}$ as a function of $n = p$ for GaAs at $T = 300$ K.

losses arising from scattering, absorption or finite transmission of the facets to reach optical oscillation.

Calculation of the Differential Gain Coefficient

Having seen how the carrier concentrations and quasi-Fermi levels are determined, we now describe the numerical calculation of the frequency-dependent differential gain coefficient $\gamma(\nu_0)$. First, recall that $h\nu_0 > E_g$ for interband transitions. For a particular choice of ν_0, we first determine the corresponding wave vector k_0 from

$$k_0^2 = \frac{2\overline{m}}{\hbar^2}(h\nu_0 - E_g) \tag{7.147}$$

The upper and lower energy levels E_2 and E_1, which are involved in the emission of the photon with frequency ν_0, are then determined from

$$E_2 = E_c + \frac{\hbar^2 k_0^2}{2m_e^*} = E_c + \frac{\overline{m}}{m_e^*}(h\nu_0 - E_g) \tag{7.148}$$

and

$$E_1 = E_v - \frac{\hbar^2 k_0^2}{2m_h^*} = E_v - \frac{\overline{m}}{m_h^*}(h\nu_0 - E_g) \tag{7.149}$$

Above, $\overline{m}$ is the electron-hole reduced mass given by Eq. (7.118). Remember that we are calculating the differential gain coefficient for a certain level of carrier injection, in other words, for a given value of n and p. So, the corresponding quasi-Fermi levels E_{fn} and E_{fp} can be determined from the numerical solution of Eqs. (7.143) and (7.144) as described in the previous section. Once, E_2, E_1, E_{fn}, and E_{fp} are determined, the occupation probabilities $f_c(E_2)$ and $f_v(E_1)$ for the upper and lower levels can be calculated using

$$f_c(E_2) = \frac{1}{\exp\left[\dfrac{E_2 - E_{fn}}{k_B T}\right] + 1} \tag{7.150}$$

and

$$f_v(E_1) = \frac{1}{\exp\left[\dfrac{E_1 - E_{fp}}{k_B T}\right] + 1} \tag{7.151}$$

Finally, $\gamma(\nu_0)$ is determined as a function of frequency by using

$$\gamma(\nu_0) = \frac{\lambda_0^2}{2h^2 n_s^2 \tau_r}(2\overline{m})^{3/2}(f_c(E_2) - f_v(E_1))(h\nu_0 - E_g)^{1/2} \tag{7.152}$$

Note once again that the above relation for $\gamma(\nu_0)$ is valid for $h\nu_0 > E_g$ only. Figure 7.31 *a–c* shows the variation of $(f_c(E_2) - f_v(E_1))$, $(h\nu_0 - E_g)^{1/2}$, and $\gamma(\nu_0)$ as a function of the photon energy $E = h\nu_0$ for the injected carrier densities of $n = p = 2.5 \times 10^{18}\,\text{cm}^{-3}$ ($m_e^* = 0.067m_e$, $m_h^* = 0.55m_e$, $E_g = 1.42$ eV, $n_s = 3.6$, $\tau_r = 2 \times 10^{-9}$ s and $T = 300$ K). Since the semiconductor becomes transparent for $n = p = 1.26 \times 10^{18}$ cm^{-3}, we chose a larger value for the carrier concentrations to make $\gamma(\nu_0)$ greater than 0. As we noted earlier, the actual measured values of gain may differ from what we calculate here due to the uncertainty in the material parameters. In particular, the recombination time has concentration dependence. We assumed a constant value of 2 ns in all

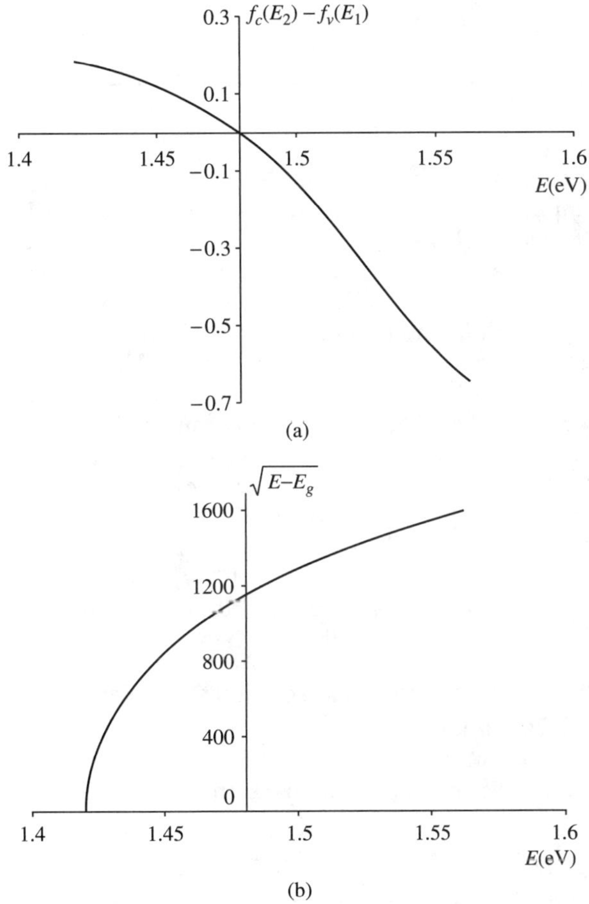

FIGURE 7.31 Calculated variation of (*a*) $(f_c(E_2) - f_v(E_1))$, (*b*) $(h\nu_0 - E_g)^{1/2}$, and (*c*) $\gamma(\nu_0)$ as a function of the photon energy $E = h\nu_0$ for the injected carrier densities of $n = p = 2.5 \times 10^{18}$ cm^{-3} ($m_e^* = 0.067m_e$, $m_h^* = 0.55m_e$, $E_g = 1.42$ eV, $n_s = 3.6$, $\tau_r = 2 \times 10^{-9}$ s, and $T = 300$ K).

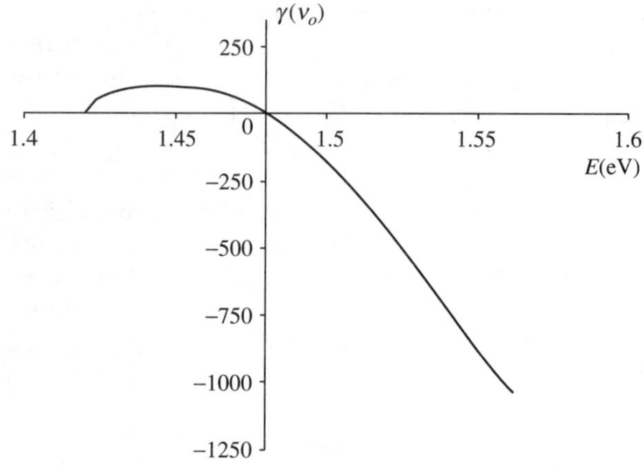

FIGURE 7.31 *(Continued)*

of the calculations. The results still show the important features of optical amplification in a semiconductor. First of all, note that as soon as the difference $(f_c(E_2) - f_v(E_1))$ in the occupation probabilities becomes negative, the material ceases to act as an amplifier. In the particular case analyzed here, this happens at the photon energy of 1.48 eV. Second, note that differential gain coefficients can be tens of cm^{-1} or higher, depending on the carrier injection level. This is far larger than what is typically obtained in solid-state lasers. Hence, short active lengths of the order of tens of microns can be sufficient to enable amplification in semiconductors.

Figure 7.32 further shows the variation of $\gamma(\nu_0)$ as a function of the photon energy $E = h\nu_0$ for the injected carrier densities of (*a*) $n = p = 2\times10^{18}\,cm^{-3}$ and (*b*) $n = p = 3\times10^{18}\,cm^{-3}$ ($m_e^* = 0.067m_e$, $m_h^* = 0.55m_e$, $E_g = 1.42$ eV, $n_s = 3.6$, $\tau_r = 2\times10^{-9}$ s, and $T = 300$ K). Note that as the carrier densities are increased, the maximum differential gain and the spectral width of the gain band also increase.

Calculation of the Differential Absorption Coefficient in Thermal Equilibrium

Consider a semiconductor in thermal equilibrium. In this case, there will be a single Fermi level for both electrons and holes. Hence,

$$f_c(E_2) - f_v(E_1) = \frac{1}{\exp\left[\dfrac{E_2 - E_f}{k_B T}\right] + 1} - \frac{1}{\exp\left[\dfrac{E_1 - E_f}{k_B T}\right] + 1} \tag{7.153}$$

Recall that E_2 lies within the conduction band and E_1 lies in the valence band. If the Fermi level lies within the band gap sufficiently away

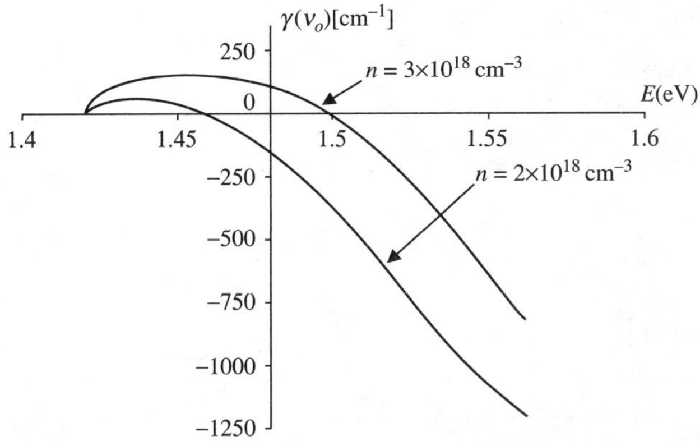

FIGURE 7.32 Calculated variation of $\gamma(\nu_0)$ as a function of the photon energy $E = h\nu_0$ for the injected carrier densities of (*a*) $n = p = 2\times10^{18}$ cm^{-3} and (*b*) $n = p = 3\times10^{18}$ cm^{-3} ($m_e^* = 0.067m_e$, $m_h^* = 0.55m_e$, $E_g = 1.42$ eV, $n_s = 3.6$, $\tau_r = 2\times10^{-9}$ s, and $T = 300$ K).

from the band edges (say more than $3k_BT$ where $k_BT = 0.026$ eV at $T = 300$ K), then

$$f_c(E_2) - f_v(E_1) \cong 0 - 1 = -1 \tag{7.154}$$

and the semiconductor acts as an absorber for photon energies $h\nu > E_g$. By using Eq. (7.152), the absorption coefficient $\alpha(\nu_0)$ is then given by

$$\alpha(\nu_0) = -\gamma(\nu_0) = \frac{\lambda_0^2}{2h^2 n_s^2 \tau_r}(2\overline{m})^{3/2}(h\nu_0 - E_g)^{1/2} \tag{7.155}$$

The variation of $\alpha(\nu_0)$ is shown in Fig. 7.33 as a function of the photon energy for GaAs for the following set of parameters: $m_e^* = 0.067m_e$, $m_h^* = 0.55m_e$, $E_g = 1.42$ eV, $n_s = 3.6$, and $\tau_r = 1\times10^{-9}$ns at $T = 300$ K. Again, note that differential absorption coefficient values are very large, making the material essentially opaque at photon energies larger than the band-gap energy.

7.3.7 Types of Semiconductor Lasers

The Homojunction Semiconductor Laser

The homojunction semiconductor laser consists of a p-n junction between a heavily doped *p* region and a heavily doped *n* region of the same type of semiconductor host as sketched in Fig. 7.34. Direct-gap semiconductors such as GaAs are used to enable efficient light generation by electron-hole recombination.

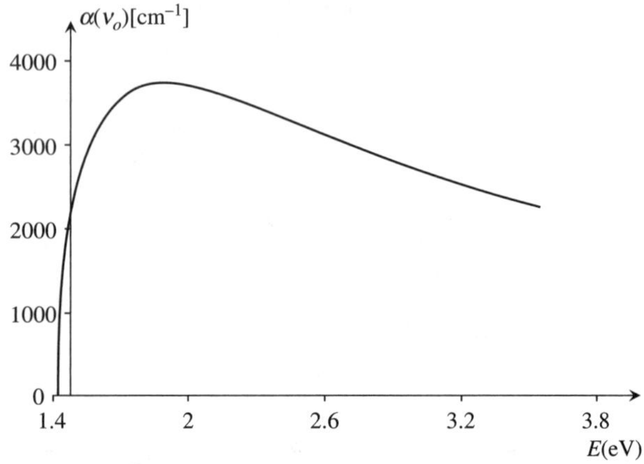

FIGURE 7.33 Calculated variation of $\alpha(\nu_0)$ as a function of the photon energy for GaAs ($m_e^* = 0.067m_e$, $m_h^* = 0.55m_e$, $E_g = 1.42$ eV, $n_s = 3.6$, and $\tau_r = 1 \times 10^{-9}$ ns at $T = 300$ K).

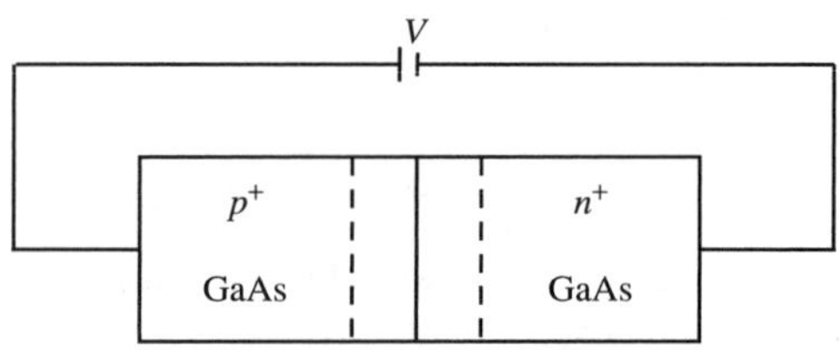

FIGURE 7.34 Sketch of a forward-biased homojunction semiconductor laser.

Application of a forward bias reduces the built-in potential barrier and causes injection of electrons and holes into the junction where radiative electron-hole recombination leads to light amplification. This is schematically shown in Fig. 7.35*a* and *b* for the no-bias and forward-bias cases. When the round-trip gain exceeds the losses, light oscillation can be achieved. The homojunction architecture suffers from the drawback that no mechanism exists to control the confinement of carriers or the laser mode within the active junction region.

The Heterojunction Semiconductor Laser

Both carrier and light confinement can be achieved by using a p-n junction made from different semiconductor materials. This is known as the heterojunction laser.[27–30] A commonly used architecture consisting of regions of GaAs and $Al_xGa_{1-x}As$ is shown in Fig. 7.36. The actual device architectures that are currently used can be much more sophisticated, but in essence, the basic structure of a heterojunction has a thin layer of GaAs (shown with thickness *d* in Fig. 7.36)

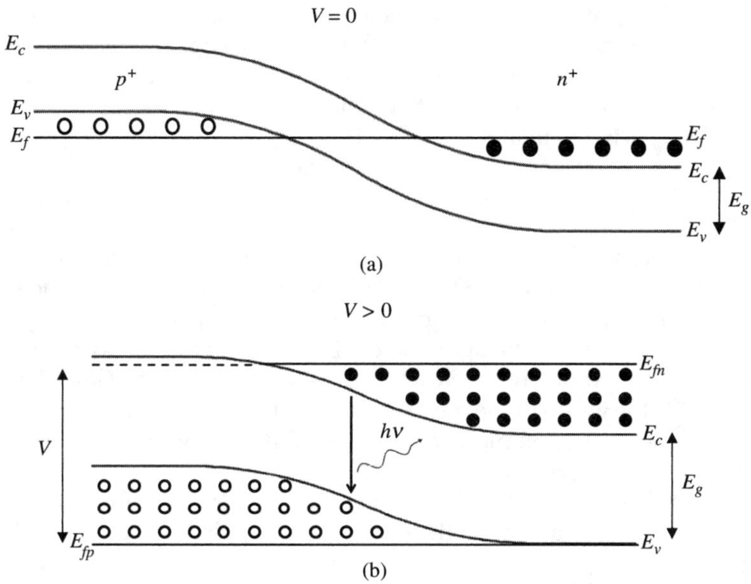

FIGURE 7.35 Heavily doped homojunction under (*a*) no bias and (*b*) forward bias.

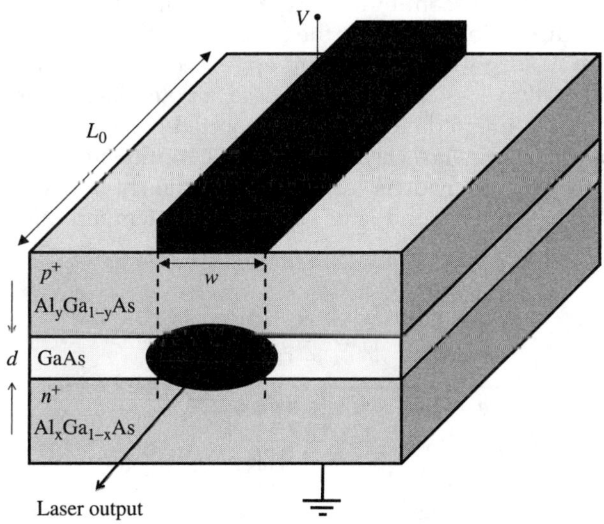

FIGURE 7.36 Sketch of a GaAs- $Al_xGa_{1-x}As$ heterojunction semiconductor laser.

sandwiched between a heavily p-doped $Al_yGa_{1-y}As$ and a heavily n-doped $Al_xGa_{1-x}As$ layer. In general, the mole fractions x and y of the surrounding layers are different. These layers can be grown by various methods, including metallo-organic chemical vapor deposition (MOCVD), molecular beam epitaxy (MBE), and others.[31] Here are the basic characteristics of the heterojunction laser: The GaAs layer serves as the active region. Under forward bias, large numbers of electrons and holes are injected into the active region from the n^+ and p^+ layers, respectively. The refractive index of the ternary semiconductor $Al_xGa_{1-x}As$ is less than that of GaAs and is given by[15, 32, 33]

$$n(x) = 3.590 - 0.710x + 0.091x^2 \tag{7.156}$$

As x increases from 0 to 1 (i.e., from GaAs to AlAs), the refractive index decreases from 3.59 to 2.97. The higher index of the active region in relation to the neighboring heavily doped AlGaAs layers provides guiding of the light and confines the laser mode to a dimension of approximately $w \times d$, where d is the width of the GaAs layer and w is the width of the metallic contact. The actual spatial distribution of the supported modes can be calculated by approximating the structure as a planar waveguide.

Furthermore, the band-gap energy of the ternary semiconductor $Al_xGa_{1-x}As$ increases with the molar fraction x of the Al atoms and is given by[15, 32]

$$E_g = (1.424 + 1.247x)\text{eV} \tag{7.157}$$

for $x < 0.37$. The discontinuity in the bandgap energy ΔE_g is split between a discontinuity ΔE_c for the conduction band edge, and a discontinuity ΔE_v for the valence band edge (see Fig. 7.37). The split is not equal, with $\Delta E_c \cong 1.5\Delta E_v$. In any case, a potential barrier is introduced for both injected electrons and injected holes, confining them mainly to the active GaAs layer. The double confinement of light and injected carriers has made possible the development of low-threshold, continuous-wave semiconductor lasers at room temperature.

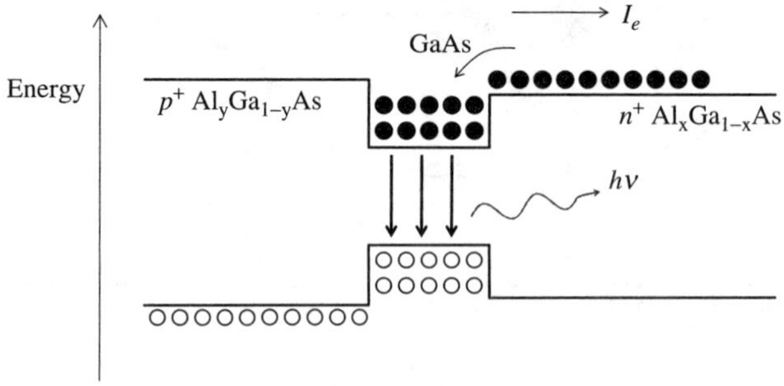

FIGURE 7.37 Simplified energy band sketch of a heterojunction semiconductor laser with an active GaAs layer surrounded by heavily *p*-doped $Al_yGa_{1-y}As$ and heavily *n*-doped $Al_xGa_{1-x}As$ layers. Solid and open circles represent electrons and holes, respectively.

Threshold Current Estimation in a Semiconductor Laser

We can use the gain analysis discussed earlier to estimate the threshold current densities needed to reach lasing oscillation in a semiconductor laser. Typically, in a real structure, a portion of the confined electromagnetic mode leaks into the AlGaAs regions and experiences losses. In this crude analysis, we will neglect the leakage of the mode and assume that most of the mode is confined to the active region. The primary source of loss will be due to Fresnel reflection from the end facets of GaAs. Taking $n = 3.59$ for GaAs, reflection R per facet becomes

$$R = \left(\frac{3.59-1}{3.59+1}\right)^2 = 0.32 \tag{7.158}$$

Let γ be the small-signal gain. Then, at threshold, we require that

$$R^2 \exp[2\gamma_{th} L_0] = 1 \tag{7.159}$$

or

$$\gamma_{th} = \frac{-1}{2L_0}\ln(R^2) = \frac{1}{L_0}\ln\left(\frac{1}{R}\right) \tag{7.160}$$

Let the length L_0 of the laser be 300 μm. γ_{th} becomes

$$\gamma_{th} = \frac{1}{300\times 10^{-4}}\ln\left(\frac{1}{0.32}\right) = 38\ \text{cm}^{-1}$$

From our calculations for GaAs, we find, by using $\tau_r = 2$ ns and the relevant parameters for GaAs, that the peak gain becomes 38 cm^{-1} for $n = p \cong 1.81\times 10^{18}\,\text{cm}^{-3}$ (see Prob. 7.21). Now, to find the threshold current density, consider the active region sketched in Fig. 7.38.

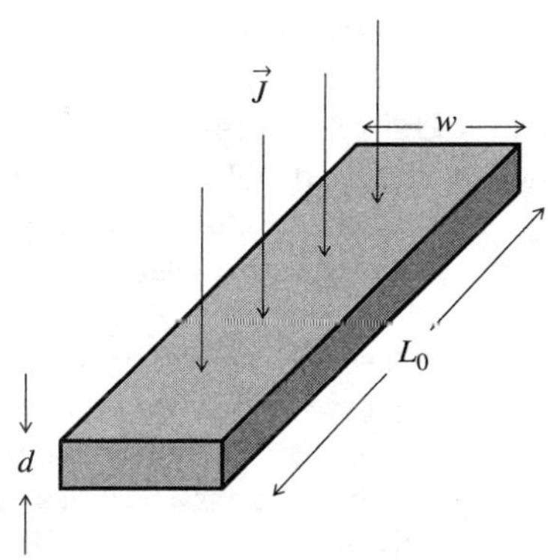

FIGURE 7.38 Sketch of the active region used in the calculation of the threshold current density.

The cross-sectional area a of the junction is $a = L_0 w$. Let the generation rate of e-h pairs per unit volume be G. The steady state carrier concentration will be $G\tau_r$, where τ_r is the recombination time. Hence,

$$\left(\frac{J_{th}a}{e}\right)\frac{1}{ad}\tau_r = n_{th} \tag{7.161}$$

or

$$J_{th} = ed\left(\frac{n_{th}}{\tau_r}\right) \tag{7.162}$$

Taking $d = 0.1\ \mu\text{m}$. J_{th} becomes

$$(1.6\times 10^{-19})(0.1\times 10^{-4})\left(\frac{1.81\times 10^{18}}{2\times 10^{-9}}\right) = 1448\frac{\text{A}}{\text{cm}^2}$$

Suppose that the contact area is $a = L_0 \times W = 300\ \mu\text{m} \times 50\ \mu\text{m}$. Then the threshold current will be $I_{th} = (300\times 10^{-4})(50\times 10^{-4})(1448) =$ 217 mA.

In a more realistic device, the threshold current will be somewhat higher than what we calculated due to losses experienced by the mode in the surrounding AlGaAs layers and surface imperfections. Also note that the threshold current is proportional to d and will decrease as d is made narrower. In typical devices, there is an optimum d value below which the threshold current begins to increase again.[14] There are two classes of commonly used heterostructure semiconductor lasers. One is based on the GaAs/AlGaAs system which we discussed above. Depending on the value of the mole fraction x, these lasers can provide light emission over the wavelength range 0.75–0.88 μm. Another commonly used laser system based on the quaternary compound $Ga_xIn_{1-x}As_yP_{1-y}$ emits in the 1.0–1.6 μm wavelength range.[17] The latter is particularly useful in fiber-optic communication systems operating near the zero-dispersion (1300 nm) or minimum-attenuation (1550 nm) wavelengths.

Other Semiconductor Light Sources

Besides the homojunction and heterojunction semiconductor lasers discussed in the previous sections, there are many other architectures of layered semiconductors that are used for light generation. These include light-emitting diodes (LEDs), quantum well lasers, vertical-cavity surface-emitting lasers, and quantum cascade lasers, among others. In this section, we briefly discuss some of these systems.

Light-emitting diodes (LEDs)[34] generate incoherent light as a result of electroluminescence, where electrons and holes in a forward-biased p-n junction recombine radiatively to generate photons. Electroluminescence is an example of spontaneous emission and occurs before the laser oscillation threshold is reached. The color of an LED depends on the choice of the direct-gap semiconductor material, and LEDs with output from ultraviolet to infrared are currently available. The brightness of LEDs has been greatly improved over the years, and at present they are being increasingly used in lighting applications. Advantages of LEDs over conventional light sources include lower power consumption, more precise color control, and faster modulation capability. There is also an increasing effort aimed at the development of LEDs based on electroluminescence in organic semiconductors.[35,36]

In quantum well lasers, active layer thicknesses are so small that quantum confinement effects begin to dominate.[37,38] As a result of quantum confinement, the density-of-states functions do not vary continuously, as in a three-dimensional system, but in steps. Furthermore, the output wavelength of the laser can be tailored by adjusting the thickness of the active layer, and it is possible to attain wavelengths that are shorter than what one would obtain with a bulk active region made with the same material. Quantum well lasers can be operated with very low threshold current density.

Quantum cascade lasers operate in a way fundamentally different than heterojunction lasers.[39] In a nutshell, they are "unipolar" devices where only one type of carrier is involved in laser action. Injected electrons moving in a potential staircase emit photons at the steps consisting of coupled quantum wells. The emission wavelength depends on quantum confinement and can be varied from mid-infrared to far infrared by using the same active material.

In vertical-cavity surface-emitting lasers, referred to as VCSELs,[40] the optical cavities are at right angles in relation to those of conventional edge-emitting diode lasers that we discussed earlier. This gives them several superior characteristics, including the ability to produce a circularly symmetric output beam and smaller active region volume. More importantly, they can be fabricated in large numbers in the form of addressable arrays for applications in optical interconnects.[41]

Problems

7.1 Show that the density-of-states functions $\rho_c(E)$ and $\rho_v(E)$ for the valence and conduction bands are given by Eqs. (7.15) and (7.16), respectively.

7.2 Show that the approximate expressions for the electron and hole occupation probabilities are given by Eqs. (7.18) and (7.19).

7.3 Show that when $E - E_f >> k_B T$ in the conduction band and $E_f - E >> k_B T$ in the valence band, the total electron and hole concentrations are given by Eqs. (7.22) and (7.23).

7.4 Based on the results for the carrier concentrations, reproduce the $n_i - T$ plots for Si and GaAs shown in Fig. 7.11.

7.5 Find the Fermi level E_f in an intrinsic semiconductor with respect to the maximum valence band energy E_v and minimum conduction band energy E_c.

7.6 Consider crystalline silicon at $T = 300$ K, with $m_e^*/m_e = 1.08$ and $m_h^*/m_e = 0.81$. The band-gap energy of silicon is 1.12 eV.

(*a*) Calculate the intrinsic carrier concentration n_i.

(*b*) If the semiconductor is doped with phosphorous with a density of $5 \times 10^{15}/\text{cm}^3$, calculate the new position of the Fermi level with respect to the minimum energy E_c of the conduction band.

7.7 Show, by solving Poisson's equation, that the electrostatic potential and the electric field for the p-n junction are given by Eqs. (7.56) and (7.57).

7.8 Consider a silicon p-n junction with $m_e^* = 1.1m_e, m_h^* = 0.59m_e, E_g = 1.12$ eV, $T = 300$ K, $N_A = 5\times10^{15}$ cm^{-3}, $N_D = 10^{16}$ cm^{-3}, and $\varepsilon/\varepsilon_0 = 11.8$. Calculate

(*a*) The built-in junction potential
(*b*) The width of the depletion region
(*c*) The maximum value of the built-in electric field

7.9 Write a computer code to plot the variation of the electrostatic potential $V(z)$ and the corresponding electric field $E_z(z)$ around the junction. Generate plots of $V(z)$ and $E_z(z)$ for a silicon p-n junction with $m_e^* = 1.1m_e$, $m_h^* = 0.59m_e$, $E_g = 1.12$ eV, $T = 300$ K, $N_A = 5\times10^{15}$cm^{-3}, $N_D = 10^{16}$cm^{-3}, and $\varepsilon/\varepsilon_0 = 11.8$.

7.10 Write a computer code to plot the density-of-states functions $\rho_c(E)$ and $\rho_v(E)$ as a function of energy for GaAs similar to the plots in Fig. 7.25. Use $m_e^* = 0.067m_0$, $m_h^* = 0.55m_0$, and $E_g = 1.42$ eV.

7.11 Show that the electron and hole concentrations in quasi-equilibrium are given by Eqs. (7.112) and (7.114).

7.12 Show that the expression for the joint density of states $\rho_j(\nu)$ is given by Eq. (7.122).

7.13 Write a computer code to plot the weighted density-of-states functions $\bar{\rho}_v(E)$ and $\bar{\rho}_c(E)$ for GaAs shown in Fig. 7.28*a*. Use the parameter values $m_e^* = 0.067m_e$, $m_h^* = 0.55m_e$, $E_g = 1.42$ eV, for GaAs with $T = 300$ K, $E_{fp} = E_v$ and $E_{fn} = E_c$.

7.14 Write a computer code to calculate the total electron and hole concentrations n and p without making any approximations. For the parameters given in Prob. 7.13 for GaAs, determine n and p. What is the ratio p/n in this case?

7.15 By keeping $E_{fp} = E_v$, determine numerically, for the parameters given in Prob. 7.13 for GaAs, the value of E_{fn} for which $n = p$ at $T = 300$ K. Plot the corresponding variation of $\bar{\rho}_v(E)$ and $\bar{\rho}_c(E)$ as a function of energy for the valence and conduction bands to reproduce Fig. 7.28*b*.

7.16 Reproduce the numerical calculations for the hole and electron concentrations as a function of the quasi-Fermi levels shown in Fig. 7.29*a* and *b*.

7.17 Write a computer code to determine the quasi-Fermi levels from a knowledge of the electron and hole concentrations. Then, keeping $n = p$, plot the variation of $E_{fn} - E_{fp}$ as a function of n as shown in Fig. 7.30 for GaAs. Use the parameter values $m_e^* = 0.067m_e$, $m_h^* = 0.55m_e$, $E_g = 1.42$ eV, for GaAs with $T = 300$ K.

7.18 Using the parameter values in Prob. 7.17, determine the electron/hole concentration (taken equal to each other) where $E_{fn} - E_{fp} = E_g$ for GaAs.

7.19 Write a computer code to reproduce the plots of $(f_c(E_2) - f_v(E_1))$, $(h\nu_0 - E_g)^{1/2}$, and $\gamma(\nu_0)$ as a function of the photon energy $E = h\nu_0$ shown in Fig. 7.31 *a*–*c*. Take the injected carrier densities as $n = p = 2.5 \times 10^{18}\,\text{cm}^{-3}$ ($m_e^* = 0.067m_e$, $m_h^* = 0.55m_e$, $E_g = 1.42$ eV, $n_s = 3.6$, $\tau_r = 2 \times 10^{-9}$ s, and $T =$ 300 K). Numerically determine the energy where $(f_c(E_2) - f_v(E_1))$ becomes zero.

7.20 Write a computer code to reproduce the plots given in Fig. 7.32. In particular, determine the maximum differential gain coefficient and the spectral width of the gain band for (*a*) $n = p = 2 \times 10^{18}$ cm^{-3} and (*b*) $n = p = 3 \times 10^{18}$ cm^{-3}.

7.21 Determine the maximum differential gain coefficient for GaAs when the injected carrier densities are $n = p = 1.8 \times 10^{18}\,\text{cm}^{-3}$. Use the parameters $m_e^* = 0.067m_e$, $m_h^* = 0.55m_e$, $E_g = 1.42$ eV, $n_s = 3.6$, $\tau_r = 2 \times 10^{-9}$ s, and $T = 300$ K.

7.22 Determination of the absorption band edge in semiconductor quantum wells.

We can use the tools developed in Chap. 5 to investigate the effect of quantum confinement on the absorption band edge in one-dimensional semiconductor quantum wells. Suppose that the quantum well is formed by sandwiching a thin layer of GaAs between relatively thick layers of $Al_xGa_{1-x}As$.

(*a*) What is the wavelength corresponding to the absorption edge of bulk GaAs?

(*b*) Suppose that a 5-nm layer of GaAs is grown between thick layers of $Al_{0.25}Ga_{0.75}As$. By using Eq. (7.157), what will be the band-gap energy of $Al_{0.25}Ga_{0.75}As$?

(*c*) Assume that the potential well depths formed in the conduction and valence bands are 2/3 and 1/3 of the band-gap discontinuity, respectively. Make a sketch showing the potential energy as a function of position for the conduction and valence bands.

(*d*) Taking the well width as 5 nm, determine the position of the lowest bound energy level for the conduction-band electrons and valence-band holes. Take $m_e^* = 0.067m_e$ and $m_h^* = 0.55m_e$.

(*e*) By assuming that optical transitions will take place between the bound energy states of the confined electrons and holes, determine the wavelength corresponding to the absorption edge of this quantum well structure.

(*f*) Repeat part (*e*) by taking the well width as 10 nm. What do you expect to see as the width of the quantum well is further increased?

References

1. N. G. Basov, O. N. Krokhin, and Y. M. Popov, "Production of negative temperature states in p-n junctions of degenerate semiconductors," *Soviet Physics-JETP*, **40**, 1320–1321 (1961).
2. R. N. Hall, G. E. Fenner, J. D. Kingsley, T. J. Soltys, and R. O. Carlson, "Coherent light emission from GaAs p-n junctions," *Phys. Rev. Lett.*, **9**, 366 (1962).
3. M. I. Nathan, W. P. Dumke, G. Burns, F. H. Dill, and G. Lasher, "Stimulated emission of radiation from gaAs p-n junctions," *Applied Physics Letters*, **1**(3), 62–64 (1962).
4. N. W. Ashcroft and N. D. Mermin, *Solid State Physics* (Saunders College, Philadelphia, 1976).
5. C. Kittel, *Introduction to Solid State Physics*, 7th ed. (John Wiley & Sons, Inc., New York, 1996).
6. J. S. Blakemore, *Semiconductor Statistics* (Dover, New York, 1987).
7. R. F. Pierret, *Semiconductor Fundamentals*, Modular Series on Solid State Devices Volume I (Addison-Wesley, Reading, Mass., 1983).
8. G. W. Neudeck, *The PN Junction Diode*, Modular Series on Solid-State Devices Volume II (Addison-Wesley, Reading, Mass., 1983).
9. B. G. Streetman, *Solid State Electronic Devices* (Prentice-Hall, Englewood Cliffs, N.J., 1980).
10. S. M. Sze, *Physics of Semiconductor Devices* (John Wiley & Sons, Inc., New York, 1969).
11. S. L. Chuang, *Physics of Optoelectronic Devices* (John Wiley & Sons, Inc., New York, 1995).
12. C. R. Pollock, *Fundamentals of Optoelectronics* (Irwin, Chicago, 1995).
13. B. E. A. Saleh and M. C. Teich, *Fundamentals of Photonics*, 2d ed. (John Wiley & Sons, Inc., Hoboken, N.J., 2007).
14. A. Yariv, *Optical Electronics in Modern Communications*, 5th ed. (Oxford University Press, New York, 1997).
15. J. T. Verdeyen, *Laser Electronics*, 3d ed. (Prentice-Hall, Upper Saddle River, N.J., 1995).
16. C. C. Davis, *Lasers and Electro-Optics* (Cambridge University Press, Cambridge, 1996).
17. L. A. Coldren and S. W. Corzine, *Diode Lasers and Photonic Integrated Circuits* (John Wiley & Sons, Inc., New York, 1995).
18. G. P. Agrawal and N. K. Dutta, *Long-Wavelength Semiconductor Lasers* (Van Nostrand Reinhold, New York, 1986).
19. K. F. Brennan, *The Physics of Semiconductors* (Cambridge University Press, Cambridge, 1999).
20. L. Pavesi, L. Dal Negro, C. Mazzoleni, G. Franzo, and F. Priolo, "Optical gain in silicon nanocrystals," *Nature*, **408**(6811), 440–444 (2000).
21. O. Bisi, S. Ossicini, and L. Pavesi, "Porous silicon: A quantum sponge structure for silicon based optoelectronics," *Surface Science Reports*, **38**(1-3), 1–126 (2000).
22. M. L. Brongersma, A. Polman, K. S. Min, E. Boer, T. Tambo, and H. A. Atwater, "Tuning the emission wavelength of Si nanocrystals in SiO_2 by oxidation," *Applied Physics Letters*, **72**(20), 2577–2579 (1998).
23. O. Madelung, *Semiconductors-Basic Data* (Springer, Berlin, 1996).
24. F. Capasso, "Band-gap engineering: From physics and materials to new semiconductor devices," *Science*, **235**(4785), 172–176 (1987).
25. S. Wolf and R. N. Tauber, *Silicon Processing for the VLSI Era* (Lattice Press, Sunset Beach, Calif., 1986).
26. F. Reif, *Fundamentals of Statistical and Thermal Physics*, International ed. (McGraw-Hill, Auckland, 1985).
27. H. Kressel and H. Nelson, "Close-confinement gallium arsenide pn junction lasers with reduced optical loss at room temperature," *RCA Review*, **30**(1), 106–113 (1969).

28. Z. I. Alferov, V. M. Andreev, V. I. Korolkov, E. L. Portnoi, and D. N. Tretyako, "Coherent radiation of epitaxial heterojunction structures in AlAs-GaAs system," *Soviet Physics Semiconductors-USSR*, **2**(10), 1289–1291 (1969).
29. H. Kroemer, "A proposed class of heterojunction injection lasers," *Proceedings of the IEEE*, **51**(12), 1782–1783 (1963).
30. Z. Alferov, "Double heterostructure lasers: Early days and future perspectives," *IEEE Journal of Selected Topics in Quantum Electronics*, **6**(6), 832–840 (2000).
31. G. B. Stringfellow, *Organometallic Vapor Phase Epitaxy* (Academic Press, San Diego, 1999).
32. H. C. Casey and M. B. Panish, *Heterostructure Lasers Part B: Materials and Operating Characteristics* (Academic Press, New York, 1978).
33. H. C. Casey, D. D. Sell, and M. B. Panish, "Refractive index of $Al_xGa_{1-x}As$ between 1.2 and 1.8 eV," *Appl. Phys. Lett.*, **24**, 63–65 (1974).
34. N. Holonyak, "From transistors to light emitters," *IEEE Journal of Selected Topics in Quantum Electronics*, **6**(6), 1190–1200 (2000).
35. J. H. Burroughes, D. D. C. Bradley, A. R. Brown, R. N. Marks, K. Mackay, R. H. Friend, P. L. Burns, and A. B. Holmes, "Light-emitting-diodes based on conjugated polymers," *Nature*, **347**(6293), 539–541 (1990).
36. D. Braun and A. J. Heeger, "Visible-light emission from semiconducting polymer diodes," *Applied Physics Letters*, **58**(18), 1982–1984 (1991).
37. N. Holonyak, R. M. Kolbas, R. D. Dupuis, and P. D. Dapkus, "Quantum-well heterostructure lasers," *IEEE Journal of Quantum Electronics*, **16**(2), 170–186 (1980).
38. Y. Arakawa and A. Yariv, "Quantum-well lasers gain, spectra, dynamics," *IEEE Journal of Quantum Electronics*, **22**(9), 1887–1899 (1986).
39. J. Faist, F. Capasso, D. L. Sivco, C. Sirtori, A. L. Hutchinson, and A. Y. Cho, "Quantum cascade laser," *Science*, **264**(5158), 553–556 (1994).
40. J. L. Jewell, J. P. Harbison, A. Scherer, Y. H. Lee, and L. T. Florez, "Vertical-cavity surface-emitting lasers: Design, Growth, Fabrication, Characterization," *IEEE Journal of Quantum Electronics*, **27**(6), 1332–1346 (1991).
41. D. A. B. Miller, "Rationale and challenges for optical interconnects to electronic chips," *Proceedings of the IEEE*, **88**(6), 728–749 (2000).

CHAPTER 8

Anisotropic Media and Modulation of Light

In this chapter, we investigate the propagation of light waves in anisotropic media. This subject matter also goes by the name of "electromagnetics of crystals" or "crystal optics." The most important characteristic of anisotropic media which distinguishes them from the homogeneous, isotropic media we discussed earlier in Chap. 1, is the variation of the physical properties with direction due to the complex crystal structure. Hence, for example, the induced polarization vector of the medium is no longer parallel to the applied external electric field. In this chapter, we will be mainly concerned with how the optical properties of crystals vary with direction, and how this affects the propagation characteristics of electromagnetic waves. In their transparency range, anisotropic media find many important applications in photonics. Examples include birefringent optical elements for polarization control and modulators. Both of these topics will be discussed in this chapter. Anistropic media play a very important role in nonlinear optics as well. In particular, the birefringence of nonlinear crystals is employed in finding the phase-matching directions during second harmonic generation. These are the directions in which the efficiency of second harmonic generation is optimum. We will describe the determination of phase-matching directions for a uniaxial crystal in Chap. 11.

We start with the theory of electromagnetic wave propagation in anisotropic media, based on Maxwell's equations. We derive the Fresnel equation and introduce the index ellipsoid method, both of which are used in the calculation of the effective indices of refraction. We then introduce Jones calculus, which is a very useful and practical technique in analyzing the effect of anisotropic media on the polarization state of light, and describe the operation of half-wave and quarter-wave plates. We then discuss the electro-optic and acousto-optic effects and briefly review the principle of operation of modulators. For further discussion of anisotropic media and modulators, the reader is referred to Refs. 1 to 8.

8.1 Propagation of Light in an Anisotropic Medium

8.1.1 Dielectric Permittivity Tensor and Principal Indices

In Chap. 1, we discussed the optical properties of isotropic media where the induced polarization vector $\vec{P}$ is proportional to the applied electric field $\vec{E}$. In this case, the electric susceptibility χ_e is a scalar. This is no longer the case in crystals, where each component of the polarization vector is coupled to the other three components of the electric field according to

$$P_x = \varepsilon_0(\chi_{11}E_x + \chi_{12}E_y + \chi_{13}E_z) \tag{8.1}$$

$$P_y = \varepsilon_0(\chi_{21}E_x + \chi_{22}E_y + \chi_{23}E_z) \tag{8.2}$$

and

$$P_z = \varepsilon_0(\chi_{31}E_x + \chi_{32}E_y + \chi_{33}E_z) \tag{8.3}$$

The nine coefficients χ_{ij} appearing in Eqs. (8.1)–(8.3) make up a 3×3 matrix known as the electric susceptibility tensor. It is clear from above that $\vec{P}$ is, in general, not parallel to $\vec{E}$ unless the susceptibility coefficients have special values (for example, if all are identical, in which case we return back to the case of a homogeneous, isotropic dielectric medium). We can express the above three equations [Eqs. (8.1)–(8.3)] in a compact way by using the Einstein summation convention. In this notation, we write the ith component P_i of the polarization vector as

$$P_i = \varepsilon_0\chi_{ij}E_j \tag{8.4}$$

and follow the convention of summing over repeated indices. The summation sign is not explicitly shown. As an example, if we let $i = x$, j can become x, y, or z, and since this index is repeated, Eq. (8.4) simply becomes equivalent to Eq. (8.1).

Furthermore, we can express the displacement vector $\vec{D}$ in terms of the electric field by introducing the dielectric permittivity tensor ε_{ij}. As we discussed before in Sec. 1.3.4, $\vec{D}$ is given in terms of the electric field and the polarization vector by [see Eq. (1.31)]

$$\vec{D} = \varepsilon_0\vec{E} + \vec{P} \tag{8.5}$$

Hence, the ith component D_i of the displacement vector can be written as (again by using the Einstein summation convention)

$$D_i = \varepsilon_{ij}E_j \tag{8.6}$$

where ε_{ij} is the dielectric permittivity tensor given by

$$\varepsilon_{ij} = \varepsilon_0(1 + \chi_{ij}) \tag{8.7}$$

As an example, by setting $i = x$, and by using Eq. (8.6), we can express D_x as

$$D_x = \varepsilon_{xx}E_x + \varepsilon_{xy}E_y + \varepsilon_{xz}E_z \tag{8.8}$$

Now we present a very important property of the dielectric permittivity tensor ε_{ij}: Irrespective of the particular dielectric crystal, ε_{ij} is always symmetric. In other words,

$$\varepsilon_{ij} = \varepsilon_{ji} \tag{8.9}$$

This property is essential, as it says that there will be at most six independent elements in the 3×3 matrix ε_{ij}. As an example, if we know the values of the three diagonal elements and the three elements in the upper part of the matrix, the remaining three elements below the diagonal can be determined from Eq. (8.9).

We can prove the general result given in Eq. (8.9) by using Maxwell's equations. First, note that the electrostatic energy density u_e of the dielectric medium given by Eq. (1.99) can be calculated from

$$u_e = \frac{1}{2}\vec{E}\cdot\vec{D} = \frac{1}{2}E_i\,\varepsilon_{ij}E_j \tag{8.10}$$

In Eq. (8.10) above, we again used the Einstein summation convention to show the explicit dependence of u_e on the components of the electric field. Next, we differentiate Eq. (8.10) with respect to time to find the time rate of change of u_e:

$$\frac{\partial u_e}{\partial t} = \frac{1}{2}\varepsilon_{ij}\left[\frac{\partial E_i}{\partial t}E_j + E_i\frac{\partial E_j}{\partial t}\right] \tag{8.11}$$

Now, let's recall our earlier discussion on the conservation of energy for electromagnetic fields, summarized by Eq. (1.116) in Chap. 1. If we assume that there are no free currents in the medium, Eq. (1.116) reduces to

$$-\nabla\cdot(\vec{E}\times\vec{H}) = \vec{E}\cdot\frac{\partial\vec{D}}{\partial t} + \vec{H}\cdot\frac{\partial\vec{B}}{\partial t} = E_i\varepsilon_{ij}\frac{\partial E_j}{\partial t} + \vec{H}\cdot\frac{\partial\vec{B}}{\partial t} \tag{8.12}$$

Since the term $\vec{E}\cdot\partial\vec{D}/\partial t$ corresponds to the time rate of change of the electric energy density, we obtain, by using Eqs. (8.11) and (8.12),

$$\frac{1}{2}\varepsilon_{ij}\left[\frac{\partial E_i}{\partial t}E_j + E_i\frac{\partial E_j}{\partial t}\right] = \varepsilon_{ij}E_i\frac{\partial E_j}{\partial t} \tag{8.13}$$

or

$$\frac{1}{2}\varepsilon_{ij}\frac{\partial E_i}{\partial t}E_j = \frac{1}{2}\varepsilon_{ij}E_i\frac{\partial E_j}{\partial t} \tag{8.14}$$

By interchanging the indices i and j on the left-hand side of Eq. (8.14), we arrive at the important result $\varepsilon_{ij} = \varepsilon_{ji}$ stated in Eq. (8.9).

Now, let us go back to the susceptibility tensor χ_{ij}. Clearly, the nine coefficients of χ_{ij} will depend on the particular choice of the coordinate system. If the coordinate axes are transformed according to a certain rule, such as a rotation or a reflection, the values will become different. We know from linear algebra that it is possible to find a special set of coordinate axes such that χ_{ij} becomes diagonal. These coordinate axes, in which the susceptibility and hence the dielectric permittivity tensors become diagonal, are referred to as the principal dielectric axes and make up the principal coordinate system of the crystal. In what follows, we will always be working in the principal coordinate system of the crystal, where Eqs. (8.1) to (8.3) simplify to

$$\begin{aligned} P_x &= \varepsilon_0\chi_{11}E_x \\ P_y &= \varepsilon_0\chi_{22}E_y \\ P_z &= \varepsilon_0\chi_{33}E_z \end{aligned} \tag{8.15}$$

In a similar way, the relations between the components of the displacement vector and the electric field become

$$\begin{aligned} D_x &= \varepsilon_{11}E_x \\ D_y &= \varepsilon_{22}E_y \\ D_z &= \varepsilon_{33}E_z \end{aligned} \tag{8.16}$$

In Eq. (8.16), $\varepsilon_{11} = \varepsilon_0(1+\chi_{11})$, $\varepsilon_{22} = \varepsilon_0(1+\chi_{22})$, and $\varepsilon_{33} = \varepsilon_0(1+\chi_{33})$. In the analysis of crystals, we often prefer to work with the refractive indices rather than the permittivities or the susceptibilities. The principal refractive indices n_x, n_y, and n_z of the crystal can further be determined from

$$\begin{aligned} n_x &= \sqrt{\frac{\varepsilon_{11}}{\varepsilon_0}} = \sqrt{1+\chi_{11}} \\ n_y &= \sqrt{\frac{\varepsilon_{22}}{\varepsilon_0}} = \sqrt{1+\chi_{22}} \\ n_z &= \sqrt{\frac{\varepsilon_{33}}{\varepsilon_0}} = \sqrt{1+\chi_{33}} \end{aligned} \tag{8.17}$$

8.1.2 Derivation of the Fresnel Equation for the Effective Index

Now, here is the important question: When a plane electromagnetic wave propagates in some direction inside an anisotropic medium, what value will we take for the index of refraction? Working in the principal coordinate system, we see that there are, in general, three different index values for waves polarized along the x-, y-, and z-directions. It turns out that a systematic application of Maxwell's equations will show us how to determine the effective index for the wave moving in an arbitrary direction. In this section, we will derive the equation satisfied by the effective index. We will be following the notation of Ref. 1 and designate the direction of propagation of the electromagnetic wave by the unit vector $\hat{s}$. The corresponding wave vector $\vec{k}$ and the electric field will be given by

$$\vec{k} = \frac{\omega}{c} n\hat{s} \tag{8.18}$$

and

$$\vec{E}(\vec{r},t) = \vec{E}_0 \exp[i(\omega t - \vec{k}\cdot\vec{r})] \tag{8.19}$$

In Eq. (8.18), n is the effective index of refraction for the wave propagating along the direction given by the unit vector $\hat{s}$. Hence, the corresponding phase velocity will be given by $\vec{u}_p = (c/n)\hat{s}$. Furthermore, the associated magnetic field of the wave can be expressed as

$$\vec{H}(\vec{r},t) = \vec{H}_0 \exp[i(\omega t - \vec{k}\cdot\vec{r})] \tag{8.20}$$

where $\vec{k}$ is given by Eq. (8.18). Above, the complex vectors $\vec{E}_0$ and $\vec{H}_0$ contain information about the amplitude and state of polarization of the wave.

To derive the Fresnel equation, we first start with Faraday's equation

$$\nabla \times \vec{E} = -\mu \frac{\partial \vec{H}}{\partial t} \tag{8.21}$$

By assuming the plane-wave solutions given in Eqs. (8.19) and (8.20), we get

$$\begin{aligned}\nabla \times \vec{E} &= -\frac{i\omega n}{c}\hat{s}\times \vec{E}_0 \exp[i(\omega t - \vec{k}\cdot\vec{r})] \\ &= -i\mu\omega \vec{H}_0 \exp[i(\omega t - \vec{k}\cdot\vec{r})]\end{aligned} \tag{8.22}$$

Dividing out by the common factors, Eq. (8.22) simplifies to

$$\vec{H}_0 = \frac{n}{\mu c}\hat{s}\times \vec{E}_0 \tag{8.23}$$

Similarly, by using the Ampere-Maxwell equation

$$\nabla \times \vec{H} = \frac{\partial \vec{D}}{\partial t} \tag{8.24}$$

we get

$$\vec{D}_0 = -\frac{n}{c}\hat{s} \times \vec{H}_0 \tag{8.25}$$

Above, $\vec{D}_0$ is the amplitude of the displacement vector. We note from Eqs. (8.23) and (8.25) that $\vec{D}_0$ is perpendicular to $\vec{H}_0$. In addition, both $\vec{D}_0$ and $\vec{H}_0$ are perpendicular to the direction of propagation $\hat{s}$. However, different from an isotropic medium, $\vec{E}_0 \times \vec{H}_0$ is no longer parallel to $\hat{s}$. Finally, it can be further shown that the vectors $\vec{D}_0$, $\vec{E}_0$, $\hat{s}$, and $\vec{E}_0 \times \vec{H}_0$ all lie in a single plane (see Prob. 8.1).

By using Eqs. (8.23) and (8.25), and the identity

$$\vec{A} \times (\vec{B} \times \vec{C}) = \vec{B}(\vec{A} \cdot \vec{C}) - \vec{C}(\vec{A} \cdot \vec{B}) \tag{8.26}$$

for any three vectors $\vec{A}$, $\vec{B}$, and $\vec{C}$, we can further express $\vec{D}_0$ in terms of $\vec{E}_0$ and $\hat{s}$ as

$$\begin{aligned}\vec{D}_0 &= -\frac{n^2}{\mu c^2}\hat{s} \times (\hat{s} \times \vec{E}_0) = -n^2\varepsilon_0[\hat{s}(\hat{s} \cdot \vec{E}_0) - \vec{E}_0(\hat{s} \cdot \hat{s})] \\ &= n^2\varepsilon_0[\vec{E}_0 - \hat{s}(\hat{s} \cdot \vec{E}_0)]\end{aligned} \tag{8.27}$$

In Eq. (8.27), the factor $1/\mu c^2$ was further replaced by ε_0. Now, let us examine the ith component of $\vec{D}_0$. From Eq. (8.27), we obtain

$$D_{0i} = n^2\varepsilon_0\left[E_{0i} - s_i\left(\hat{s} \cdot \vec{E}_0\right)\right] = n_i^2\varepsilon_0 E_{0i} \tag{8.28}$$

where n_i can be n_x, n_y, or n_z as defined in Eq. (8.17). The right-hand side of Eq. (8.28) follows from the fact that we are working in the principal coordinate system. Rearranging Eq. (8.27) gives E_{0i} as

$$E_{0i} = \frac{n^2 s_i(\hat{s} \cdot \vec{E}_0)}{n^2 - n_i^2} \tag{8.29}$$

Now, we multiply E_{0i} by s_i and sum over $i = x, y$, and z to obtain

$$s_x E_{0x} + s_y E_{0y} + s_z E_{0z} = \hat{s} \cdot \vec{E}_0 = \sum_{i=x,y,z} \frac{n^2 s_i^2}{n^2 - n_i^2}(\hat{s} \cdot \vec{E}_0) \tag{8.30}$$

Dividing by the common term $\hat{s}\cdot\vec{E}_0$ and rearranging gives the equation satisfied by the effective index n:

$$\frac{1}{n^2}=\frac{s_x^2}{n^2-n_x^2}+\frac{s_y^2}{n^2-n_y^2}+\frac{s_z^2}{n^2-n_z^2} \tag{8.31}$$

Equation (8.31) is known as the Fresnel equation for the effective index (not to be confused with the Fresnel equations for reflection coefficients discussed in Chap. 1). To summarize, the Fresnel equation gives the mathematical relation satisfied by the effective index n when the electromagnetic wave moves in the direction specified by $\hat{s}$. Here is an important property of the Fresnel equation: It can be shown that the Fresnel equation (8.31) is quadratic in n^2 (see Prob. 8.2). In other words, for a given value of $\hat{s}$, one obtains two solutions $\pm n_1$ and $\pm n_2$. Thus, for a given direction of propagation, the wave can have two distinct effective indices given by $\pm n_1$ and $\pm n_2$. The $\pm$ sign indicates that the solutions are valid for propagation along $\hat{s}$ or $-\hat{s}$. To illustrate the use of the Fresnel equation and calculation of the associated fields, we next investigate a numerical example.

Example Let us illustrate the use of the Fresnel equation and the properties of the eigenwaves with a numerical example. Suppose that the principal indices of a crystal are $n_x = 1.5$, $n_y = 1.6$, and $n_z = 1.7$. We further assume that the wave propagates along $\hat{s} = 1/\sqrt{3}(\hat{x}+\hat{y}+\hat{z})$. By writing a simple code, the Fresnel equation [Eq. (8.31)] can be solved numerically. However, some caution is necessary. As we noted above, Eq. (8.31) can be reduced to a quadratic equation in n^2. This results because the sum of the squares of the components of $\hat{s}$ equals 1. Hence, if the components of $\hat{s}$ are not specified with a sufficient number of significant figures, then the numerical round-off error introduces a cubic term and leads to three roots. By inspection, one can often identify the physically meaningful solutions. As an example, by using Mathematica, we find that if the components of $\hat{s}$ are specified up to three significant figures, the roots are ± 46, ± 1.655, and ± 1.539. Clearly, the correct roots are ± 1.655 and ± 1.539. Hence, it is important to maintain enough significant figures to guarantee that only two distinct solutions are obtained. Or, even better, one can analytically reduce Eq. (8.31) to a quadratic equation for n^2, in which case only two solutions are always guaranteed without introducing ambiguous roots.

Having determined n_1 and n_2, we can now find the directions of the $\vec{E}$, $\vec{H}$, and $\vec{D}$ fields by using the following steps:

1. For a given n, go back to Eq. (8.29) for E_{0i} to determine the relative values of E_{0x}, E_{0y}, and E_{0z}.
2. Use Eqs. (8.23) and (8.25) to determine the corresponding $\vec{H}$ and D.

Example We continue the earlier example and apply the above steps to find the directions of the fields. Looking at Eq. (8.29), we see that the electric field components corresponding to the effective index n_1 will be given by

$$(E_{0i})_1=\frac{n_1^2 s_i(\hat{s}\cdot\vec{E}_{01})}{n_1^2-n_i^2} \tag{8.32}$$

The factors n_1, s_i, and $\hat{s}\cdot\vec{E}_{01}$ are the same for each component of $\vec{E}_{01}$ in this example. Hence, up to a multiplicative constant, the components of $\vec{E}_{01}$ will be given by

$$(E_{0x})_1 \propto \frac{1}{1.65466^2 - 1.5^2} = 2.0496$$

$$(E_{0y})_1 \propto 5.6211$$

$$(E_{0z})_1 \propto -6.5746$$

Hence, we can express $\vec{E}_{01}$ as

$$\vec{E}_{01} = a_1(2.0496\hat{x} + 5.6211\hat{y} - 6.5746\hat{z})$$

where a_1 is a constant. $\vec{H}_{01}$ will be proportional to $\hat{s}\times\vec{E}_{01}$ and is given by

$$\vec{H}_{01} = b_1\begin{vmatrix} \hat{x} & \hat{y} & \hat{z} \\ 1 & 1 & 1 \\ 2.0496 & 5.6211 & -6.5746 \end{vmatrix}$$

$$= b_1(-\hat{x}12.195 + \hat{y}8.6242 + \hat{z}3.5715)$$

Finally, $\vec{D}_{01}$ will be proportional to

$$\vec{D}_{01} = \begin{vmatrix} \hat{x} & \hat{y} & \hat{z} \\ 1 & 1 & 1 \\ -12.195 & 8.6242 & 3.5715 \end{vmatrix}$$

$$= c_1(-\hat{x}5.05267 - \hat{y}15.76729 + \hat{z}20.820)$$

Above a_1, b_1, and c_1 are constants. Repeating the above analysis for $n_2 = 1.5391$ we obtain (see Prob. 8.4)

$$\vec{E}_{02} = a_2(8.4155\hat{x} - 5.2309\hat{y} - 1.9188\hat{z})$$

$$\vec{H}_{02} = b_2(3.3122\hat{x} + 10.3342\hat{y} - 13.646\hat{z})$$

$$\vec{D}_{02} = c_2(-23.9806\hat{x} + 16.9585\hat{y} + 7.0221\hat{z})$$

where, again, a_2, b_2, and c_2 are constants. We note that

$$\vec{D}_{01}\cdot\vec{D}_{02} = c_1c_2(-\hat{x}5.05267 - \hat{y}15.76729 + \hat{z}20.820)$$

$$\times(-23.9806\hat{x} + 16.9585\hat{y} + 7.0221\hat{z})$$

$$= 0$$

As can be seen from the above numerical example, the displacement vectors $\vec{D}_{01}$ and $\vec{D}_{02}$ determined for the effective indices n_1 and n_2 are orthogonal to each other. This turns out to be a general property of the wave solutions in the anisotropic media. We give a proof of the orthogonality relation in the next section.

8.1.3 Proof of the Orthogonality Property $\vec{D}_{01} \cdot \vec{D}_{02} = 0$

In this section, we prove that the displacement vectors corresponding to the effective indices n_1 and n_2 are orthogonal. Before deriving this general result, we first investigate $\nabla \cdot \vec{D}$. Since the anisotropic medium is assumed to be devoid of free charges, $\nabla \cdot \vec{D} = 0$. $\vec{D}$ is further given by

$$\vec{D} = \vec{D}_0 \exp[i(\omega t - \vec{k} \cdot \vec{r})] \tag{8.33}$$

Hence, it follows that

$$\nabla \cdot \vec{D} = -i\frac{\omega n}{c}\hat{s} \cdot \vec{D}_0 \exp[i(\omega t - \vec{k} \cdot \vec{r})] \tag{8.34}$$

For the divergence to vanish, we therefore require that

$$\hat{s} \cdot \vec{D}_0 = 0 \tag{8.35}$$

Actually, we already knew this result from Eq. (8.25), which states that $\vec{D}_0$ and $\hat{s}$ are perpendicular. By using Eq. (8.29) for E_{0i}, we further obtain

$$D_{0i} = \frac{n^2 \varepsilon_0 n_i^2 s_i (\hat{s} \cdot \vec{E}_0)}{n^2 - n_i^2} \tag{8.36}$$

Writing $\hat{s} \cdot \vec{D}_0$ in component form, we get

$$s_x D_{0x} + s_y D_{0y} + s_z D_{0z} = 0 \tag{8.37}$$

By combining the above result with Eq. (8.36) and dividing by the common factors, we arrive at

$$\frac{n_x^2 s_x^2}{n^2 - n_x^2} + \frac{n_y^2 s_y^2}{n^2 - n_y^2} + \frac{n_z^2 s_z^2}{n^2 - n_z^2} = 0 \tag{8.38}$$

Now, we return back to $\vec{D}_{01} \cdot \vec{D}_{02}$. From Eq. (8.36), we see that

$$D_{0i1} = \frac{\varepsilon_0 n_1^2 s_i n_i^2 (\hat{s} \cdot \vec{E}_{01})}{n_1^2 - n_i^2} \tag{8.39}$$

and

$$D_{0i2} = \frac{\varepsilon_0 n_2^2 s_i n_i^2 (\hat{s} \cdot \vec{E}_{02})}{n_2^2 - n_i^2} \tag{8.40}$$

Taking the inner product of $\vec{D}_{01}$ and $\vec{D}_{02}$, we obtain

$$\vec{D}_{01}\cdot\vec{D}_{02}=n_1^2 n_2^2 \varepsilon_0^2(\hat{s}\cdot\vec{E}_{01})(\hat{s}\cdot\vec{E}_{02})\sum_i \frac{s_i^2 n_i^4}{\left(n_1^2-n_i^2\right)\left(n_2^2-n_i^2\right)} \tag{8.41}$$

With a little bit of algebra, it can be shown that

$$\frac{n_i^4}{\left(n_1^2-n_i^2\right)\left(n_2^2-n_i^2\right)}=\frac{n_i^2}{n_1^2-n_2^2}\left[\frac{n_2^2}{n_2^2-n_i^2}-\frac{n_1^2}{n_1^2-n_i^2}\right] \tag{8.42}$$

Hence $\vec{D}_{01}\cdot\vec{D}_{02}$ becomes

$$\vec{D}_{01}\cdot\vec{D}_{02}=\frac{\varepsilon_0^2 n_1^2 n_2^2(\hat{s}\cdot\vec{E}_{10})(\hat{s}\cdot\vec{E}_{20})}{n_1^2-n_2^2}\sum_i\left(\frac{n_2^2 s_i^2 n_i^2}{n_2^2-n_i^2}-\frac{n_1^2 s_i^2 n_i^2}{n_1^2-n_i^2}\right)=0 \tag{8.43}$$

due to the fact that each term in the summation adds up to 0, as can be seen from Eq. (8.38). Part of the author's handwritten version of this proof appears on the cover of the book! In practice, n, $\vec{E}$, $\vec{D}$, and $\vec{H}$ are determined by using the method of the index ellipsoid which we discuss next.

8.1.4 Method of the Index Ellipsoid

There is an alternative graphical method with which we can determine the effective indices of propagation and the polarization of the corresponding propagating waves. This is referred to as the method of the index ellipsoid. In this section, we will provide an outline of the method. For a rigorous proof of the fact that the graphical solution corresponds to the actual solutions obtained from the Fresnel equation, see Ref. 1.

To introduce the method of the index ellipsoid, we first calculate the electric energy density u_e of the wave in the principal coordinate system. As was discussed earlier, u_e is given by

$$u_e=\frac{1}{2}\vec{E}\cdot\vec{D}=\frac{1}{2}\frac{D_x^2}{n_x^2\varepsilon_0}+\frac{D_y^2}{n_y^2\varepsilon_0}+\frac{D_z^2}{n_z^2\varepsilon_0} \tag{8.44}$$

giving

$$2\,\varepsilon_0 u_e=\frac{D_x^2}{n_x^2}+\frac{D_y^2}{n_y^2}+\frac{D_z^2}{n_z^2} \tag{8.45}$$

According to Eq. (8.45), constant-energy surfaces are ellipsoids in the $\vec{D}$-space. By making a change of variables and defining the dimesionless vector $\vec{r}$ as

$$\vec{r} = \frac{\vec{D}}{\sqrt{2\varepsilon_0 u_e}} \tag{8.46}$$

we can rewrite Eq. (8.45) in the form

$$\frac{x^2}{n_x^2} + \frac{y^2}{n_y^2} + \frac{z^2}{n_z^2} = 1 \tag{8.47}$$

Equation (8.47) describes what is known as the index ellipsoid or the optical indicatrix. The dimensionless coordinate x, for example, is related the the x component of the displacement vector through

$$x = \frac{D_x}{\sqrt{2\varepsilon_0 u_e}} \tag{8.48}$$

In other words, x, y, and z are normalized coordinates parallel to components of $\vec{D}$. Figure 8.1 shows a sketch of the the ellipsoid described by Eq. (8.47). The major axis lengths are $2n_x$, $2n_y$, and $2n_z$.

We can use the index ellipsoid to determine the two indices of refraction and the two corresponding directions of the $\vec{D}$ wave for the two propagating plane waves. Suppose that the direction of propagation $\hat{s}$ is specified with respect to the principal dielectric axes.

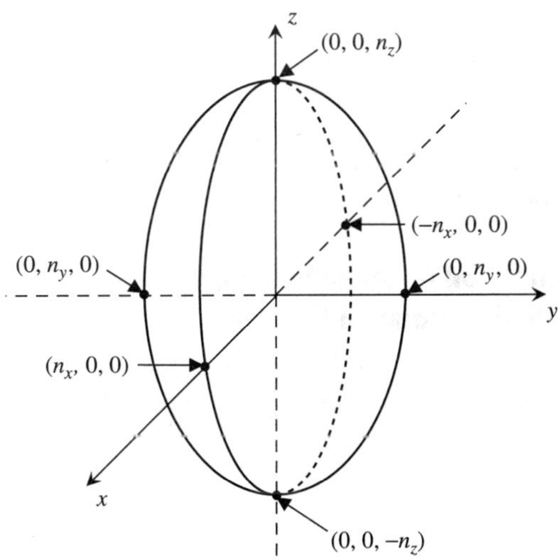

FIGURE 8.1 The index ellipsoid with major axis lengths of $2n_x$, $2n_y$, and $2n_z$.

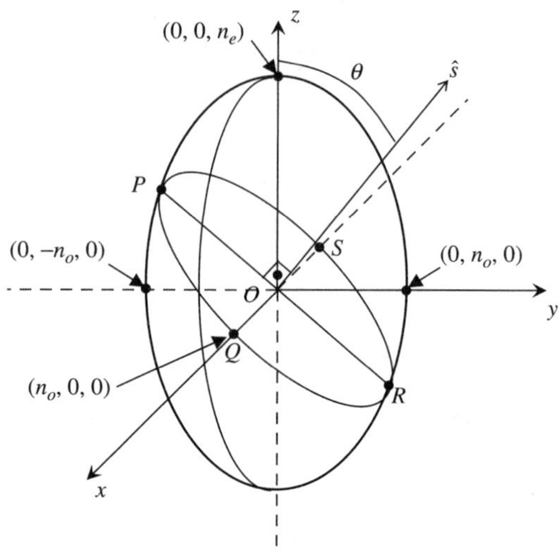

FIGURE 8.2 Sketch showing the use of the index ellipsoid method in the determination of the effective indices of refraction.

Referring to Fig. 8.2, the method of the index ellipsoid consists of the following steps:

1. Find the ellipse formed by the intersection of the index ellipsoid and the plane perpendicular to $\hat{s}$ and going through the origin.
2. The directions of the major and minor axes of the ellipse are the two allowed polarizations of the $\vec{D}$ vector. In general, $\vec{E}$ makes some angle with $\vec{D}$ due to the anisotropic nature of the medium.
3. The lengths of the major and minor axes (OP and OQ in Fig. 8.2) are $2n_1$ and $2n_2$, where n_1 and n_2 are the effective indices of the allowed propagating modes. Hence, the two allowed waves travel with phase velocities of $u_{p1} = c/n_1$ and $u_{p2} = c/n_2$, where c is the speed of light in vacuum.

8.1.5 Application of the Index Ellipsoid Method to Uniaxial Crystals

Crystals, which have a single axis of 3-, 4-, or 6-fold symmetry, are uniaxial. In this case, two of the principal indices of refraction are equal. Conventionally, we take

$$n_x = n_y = n_o \tag{8.49}$$

and

$$n_z = n_e \tag{8.50}$$

Above, n_o and n_e are known as the ordinary and extraordinary indices of refraction, respectively. The z-axis, which is known as the optic axis, becomes the axis of symmetry of the ellipsoid. Uniaxial crystals are divided into two categories, depending on the relative magnitude of n_o and n_e. In positive uniaxial crystals, we have $n_e > n_o$, whereas $n_e < n_o$ in a negative uniaxial crystal.

The solution of the Fresnel equation and the expressions for the effective indices are greatly simplified in this case. To determine the effective indices, we examine the index ellipsoid given by

$$\frac{x^2}{n_0^2}+\frac{y^2}{n_0^2}+\frac{z^2}{n_e^2}=1 \tag{8.51}$$

Due to the circular symmetry about the z-axis, we can take the y-axis to coincide with the projection of the unit vector $\hat{s}$ onto the x-y plane. The cross-sectional view of the ellipsoid is shown in Fig. 8.3. The two allowed polarization directions are parallel to the axes of the ellipse in Fig. 8.2 and are given by the vectors $\overrightarrow{OP}$ and $\overrightarrow{OQ}$. Due to the special symmetry of the uniaxial crystal, note that the length of the vector $\overrightarrow{OQ}$ is always equal to n_o. We refer to this solution as the ordinary wave. The ordinary wave is hence polarized along $\overrightarrow{OQ}$ and always has the same effective index (equal to n_o) and the same direction of polarization ($\overrightarrow{OQ}$), irrespective of the direction of propagation. The other solution which corresponds to the wave polarized along $\overrightarrow{OP}$ gives the extraordinary wave whose effective index is $n_e(\theta)=\left|\overrightarrow{OP}\right|$.

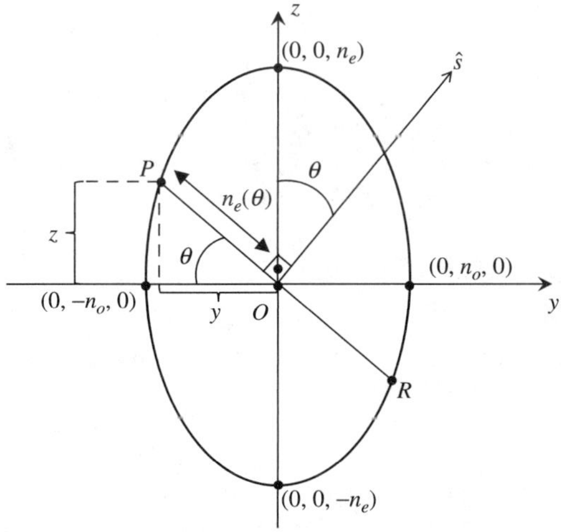

FIGURE 8.3 Cross section of the index ellipsoid for a uniaxial crystal. The extraordinary wave is polarized along $\overrightarrow{OP}$ with an effective index $n_e(\theta)=\left|\overrightarrow{OP}\right|$.

Clearly, as the angle θ between the optic axis and the direction of propagation varies, both the polarization direction and the effective index of the extraordinary wave will change, as can be seen more clearly from Fig. 8.3

We can use the simple geometry shown in Fig. 8.3 to determine $n_e(\theta)$. Note that

$$n_e(\theta)^2 = y^2 + z^2 \tag{8.52}$$

where

$$z = n_e(\theta)\sin\theta \tag{8.53}$$

and

$$y = n_e(\theta)\cos\theta \tag{8.54}$$

Since

$$\frac{z^2}{n_e^2} + \frac{y^2}{n_o^2} = 1 \tag{8.55}$$

we obtain

$$\frac{1}{n_e^2(\theta)} = \frac{\cos^2\theta}{n_o^2} + \frac{\sin^2\theta}{n_e^2} \tag{8.56}$$

Once again, θ is the angle between the direction of propagation and the optic axis. A few special cases are very important. If $\theta = 0$, in other words, when the electromagnetic wave moves along the optic axis, $n_e(0) = n_o$, i.e., both the ordinary and the extraordinary waves have the same effective index in this case. As the name implies, there is only one optic axis in a uniaxial crystal. In a biaxial crystal, on the other hand, n_x, n_y, and n_z are all different and there are two optical axes. Coming back to uniaxial crystals, when $\theta = \pi/2$, in other words, when the direction of propagation is parallel to the x-y plane, $n_e(\pi/2)$ becomes n_e.

8.1.6 Alternative Derivation of the Effective Indices from the Fresnel Equation

We can also obtain the expression for the extraordinary index given by Eq. (8.56) directly from the Fresnel equation. Consider a uniaxial crystal where $n_x = n_y = n_o$ and $n_z = n_e$. Let the direction of the propagation be $\hat{s}$. Note that $\hat{s}$ makes an angle of θ with the z-axis and its projection onto the (x-y) plane is at an angle of ϕ with respect to the x-axis. In this geometry $\hat{s}$ will be given by

$$\hat{s} = \sin\theta\cos\phi\hat{x} + \sin\theta\sin\phi\hat{y} + \cos\theta\hat{z} \tag{8.57}$$

From the Fresnel Equation [Eq. (8.31)], we have

$$\frac{\sin^2\theta\cos^2\phi}{n^2-n_o^2}+\frac{\sin^2\theta\sin^2\phi}{n^2-n_o^2}+\frac{\cos^2\theta}{n^2-n_e^2}=\frac{1}{n^2} \tag{8.58}$$

Multiplying by $\left(n^2-n_o^2\right)^2\left(n^2-n_e^2\right)n^2$ we get

$$\begin{aligned}&\sin^2\theta\cos^2\phi\left(n^2-n_o^2\right)\left(n^2-n_e^2\right)n^2+\sin^2\theta\sin^2\phi\left(n^2-n_o^2\right)\\&\quad\times\left(n^2-n_e^2\right)n^2+\cos^2\theta\left(n^2-n_o^2\right)^2n^2\\&=\left(n^2-n_o^2\right)^2\left(n^2-n_e^2\right)\end{aligned} \tag{8.59}$$

Noting that there is a common factor of $\left(n^2-n_0^2\right)$ with each term and by simplifying, we obtain

$$\begin{aligned}&\left(n^2-n_o^2\right)\left[\sin^2\theta\left(n^2-n_e^2\right)n^2+\cos^2\theta\left(n^2-n_o^2\right)n^2\right.\\&\left.-\left(n^2-n_o^2\right)\left(n^2-n_e^2\right)\right]=0\end{aligned} \tag{8.60}$$

Clearly, one of the solutions is $n=\pm n_o$. To get the second solution, we require

$$\sin^2\theta\left(n^4-n^2n_e^2\right)+\cos^2\theta\left(n^4-n_o^2n^2\right)=n^4-n_e^2n^2-n_o^2n^2+n_o^2n_e^2 \tag{8.61}$$

which simplifies to

$$n^2n_e^2(1-\sin^2\theta)+n^2n_o^2(1-\cos^2\theta)=n_o^2n_e^2 \tag{8.62}$$

giving

$$\frac{1}{n^2}=\frac{1}{n_e^2(\theta)}=\frac{\cos^2\theta}{n_o^{\,2}}+\frac{\sin^2\theta}{n_e^{\,2}} \tag{8.63}$$

as in Eq. (8.56).

8.1.7 Normal Surfaces

The concept of the normal surface is very useful in the graphical analysis of optical effects arising from the anisotropy of a medium. The normal surface, also known as the index surface, can be defined by referring to Fig. 8.4*a* and *b*. Here, the optic axis is taken as the *z*-axis. The direction of propagation, shown by the unit vector $\hat{s}$, makes an angle of θ with the optic axis. To draw the normal surface, we go to a particular point, say *P*, on the surface. For the normal surface, the distance *OP* from the origin to the point *P* is taken equal to the

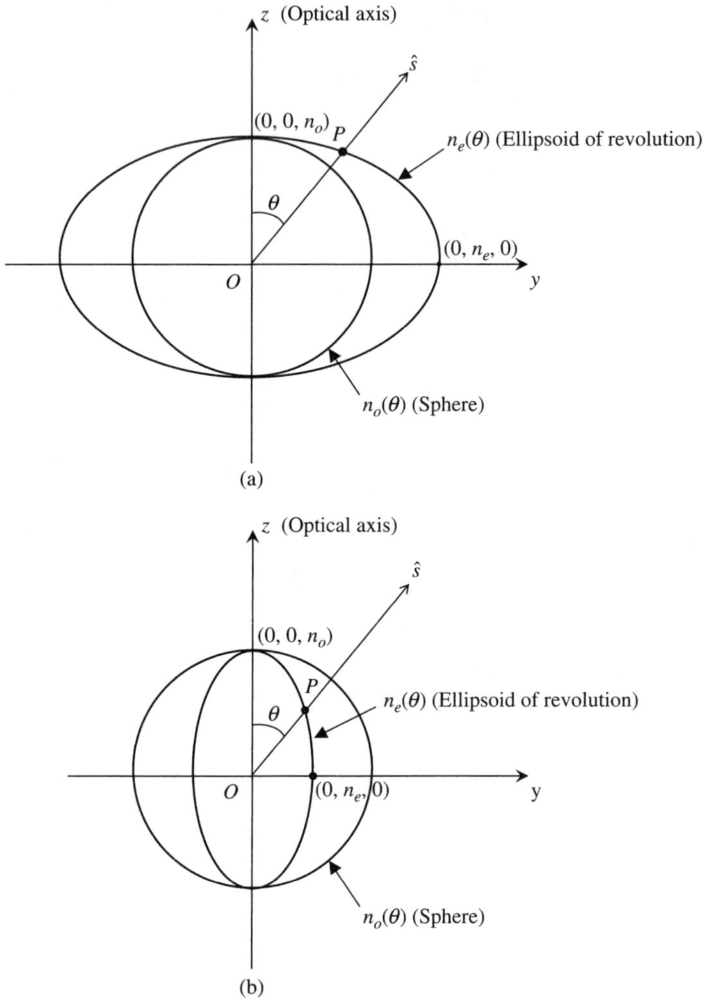

FIGURE 8.4 Normal surfaces for the ordinary and extraordinary waves for a (*a*) positive uniaxial and (*b*) negative uniaxial crystal.

refractive index of the wave moving in this direction. Hence, there will be two normal surfaces, one for the ordinary wave and one for the extraordinary wave. So first, suppose that P coincides with the z-axis. In this case, $\theta = 0$, and as we saw earlier, both the ordinary and the extraordinary waves have the same index n_o. In other words, the normal surfaces for the ordinary and extraordinary waves coincide at $\theta = 0$. For the ordinary wave, the refractive index is always equal to n_o. Hence, the normal surface describing the ordinary wave is a sphere of radius n_o. Now, let's go to $\theta = \pi/2$. Here, the extraordinary wave has the effective index of n_e, so that the normal surface for the extraordinary wave intercepts the y-axis at $y = n_e$. If we repeat this for different values of θ, the locus of points falls on an ellipsoid of revolution for the extraordinary wave as can be seen from Fig. 8.4*a* and *b*. In the particular case shown in Fig. 8.4*a*, the uniaxial crystal is positive with $n_e > n_o$. Fig. 8.4*b* shows the normal surfaces for a negative uniaxial crystal with $n_e < n_o$. See Prob. 8.5 for an application of the normal surfaces in the analysis of double refraction. References 2, 4, and 5 further provide an in-depth discussion of normal surfaces and double refraction.

8.2 Jones Calculus

Jones calculus, originally introduced by R. C. Jones in 1941,[9] provides a systematic method for analyzing the effect of anisotropic, birefringent media on the state of polarization of a light wave. In this section, we first describe the method. As an application, we then explain the operation of polarization optics such as half-wave and quarter-wave plates.

Consider a plane electromagnetic wave propagating along the z-direction. We can express the electric field vector as

$$\vec{E}(\vec{r},t) = \left(\hat{x}E_{x_0} + \hat{y}E_{y_0}e^{i\phi}\right)\exp(i(\omega t - kz)) \tag{8.64}$$

where E_{x_0} and E_{y_0}, assumed to be real, are the components along the x-and y-directions, and ϕ is the relative phase difference between the x-and y-components. In order to keep track of the polarization, we will introduce a normalized 2×1 column vector $\vec{J}$, known as the Jones vector, to represent the relative amplitudes of the x- and y-components of the electric field.

We write the Jones vector $\vec{J}$ as

$$\vec{J} = \begin{bmatrix} J_x \\ J_y \end{bmatrix} \tag{8.65}$$

where

$$J_y / J_x = \frac{E_{y_0}}{E_{x_0}} e^{i\phi} \tag{8.66}$$

and

$$|J_x|^2 + |J_y|^2 = 1 \tag{8.67}$$

As an example, if $\vec{J} = \begin{bmatrix} 1 \\ 0 \end{bmatrix}$, this represents a wave linearly polarized along the x-axis. If $\vec{J} = \begin{bmatrix} 1 \\ 1 \end{bmatrix}$, then we have a linearly polarized wave making an angle of 45° with respect to the x-axis. Going further, based on the convention we used back in Chap. 1, a left-hand (clockwise) circularly polarized wave will be represented by $\vec{J} = \begin{bmatrix} 1 \\ +i \end{bmatrix}$. Let us next assume that a plane wave represented by the Jones vector $\vec{J}$ is incident on a birefringent crystal. The crystal orientation is sketched in Fig. 8.5.

Here is an important note to avoid confusion. The x-, y-, z-axes in Fig. 8.5 do not correspond to the x-, y-, z-axes of the index ellipsoid discussed in the previous section. Rather, they represent a fixed, right-handed coordinate system with respect to which we describe the propagation of the plane wave. Furthermore, the Jones vector $\vec{J}$ gives the relative amplitude information of the wave components along x- and y-directions. The wave is assumed to propagate along the z-axis. In the following treatment, we assume that the birefringent crystal is uniaxial. Furthermore, we assume that the e-axis corresponds to the optic axis (denoted as the z-axis in Figs. 8.1–8.4). The direction of propagation (z-axis in the present discussion) coincides with the x- or y-axis of the index ellipsoid. As such, the two eigenwaves that can propagate are polarized along the e- and o-axis with effective indices of n_e and n_o, respectively. Needless to say, if we chose the direction of propagation to coincide with the optic axis (z-axis of the index ellipsoid), both ordinary and extraordinary waves would have the same effective index and we would not be able to observe or exploit the birefringence effects. Hence, the optic axis (e-axis here) is in the plane of the birefringent plate. More sophisticated birefringent filters, where the optic axis makes an angle with

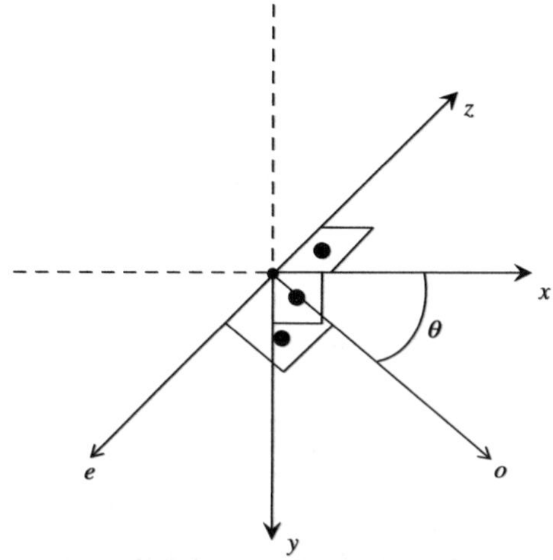

Figure 8.5 Orientation of the birefringent crystal used in the derivation of the Jones matrices.

the plane of the birefringent plate, have been designed and used for broadband tuning of femtosecond lasers.[10]

In general, the components of the incident wave may not be exactly aligned with the *e*-axis and *o*-axis, as is the case shown in Fig. 8.6. Hence, we first need to find the components of the incident wave along the *o*- and *e*-axis. From Fig. 8.6, we see that the component J_o of the Jones vector along the *o*-axis is given by

$$\begin{aligned} J_o &= J\cos(\beta-\alpha) = J\cos\beta\cos\alpha + J\sin\beta\sin\alpha \\ &= J_x\cos\alpha + J_y\sin\alpha \end{aligned} \tag{8.68}$$

Similarly, the component J_e along the *e*-axis becomes

$$\begin{aligned} J_e &= J\sin(\beta-\alpha) = J\sin\beta\cos\alpha - J\cos\beta\sin\alpha \\ &= -J_x\sin\alpha + J_y\cos\alpha \end{aligned} \tag{8.69}$$

Above, α and β are the angles made by the *o*-axis and the Jones vector with the *x*-axis. We can express Eqs. (8.68) and (8.69) in matrix form as

$$\begin{bmatrix} J_o \\ J_e \end{bmatrix} = \begin{bmatrix} \cos\alpha & \sin\alpha \\ -\sin\alpha & \cos\alpha \end{bmatrix} \begin{bmatrix} J_x \\ J_y \end{bmatrix} = \mathbf{R}(\alpha) \begin{bmatrix} J_x \\ J_y \end{bmatrix} \tag{8.70}$$

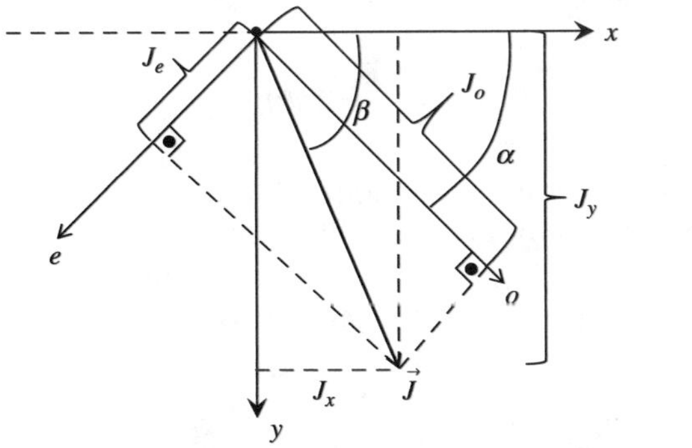

FIGURE 8.6 Sketch of the *x-y* and *o-e* coordinate systems used in the derivation of the rotation matrix $\mathbf{R}(\alpha)$.

where $\mathbf{R}(\alpha)$ is the 2×2 matrix that represents a rotation of α about the z-axis in the clockwise sense, given by

$$\mathbf{R}(\alpha)=\begin{bmatrix}\cos\alpha & \sin\alpha\\ -\sin\alpha & \cos\alpha\end{bmatrix} \tag{8.71}$$

As the wave propagates through the crystal, the ordinary and extraordinary waves will move with different phase velocities, and hence acquire different amounts of phase. The output polarization vector will be

$$\begin{bmatrix}J_o'\\ J_e'\end{bmatrix}=\begin{bmatrix}\exp\left(-in_o\dfrac{\omega}{c}d\right) & 0\\ 0 & \exp\left(-in_e\dfrac{\omega}{c}d\right)\end{bmatrix}\begin{bmatrix}J_o\\ J_e\end{bmatrix} \tag{8.72}$$

In Eq. (8.72), d is the thickness of the birefringent plate, c is the speed of light, and ω is the angular frequency of the wave. Factoring out and neglecting the common phase factor $\exp[-in_o(\omega/c)d]$, the above equation can be written as

$$\begin{bmatrix}J_o'\\ J_e'\end{bmatrix}=\mathbf{M}_{jo}\begin{bmatrix}J_o\\ J_e\end{bmatrix} \tag{8.73}$$

where the Jones matrix $\mathbf{M}_{jo}$ of the birefringent plate becomes

$$\mathbf{M}_{jo}=\begin{bmatrix}1 & 0\\ 0 & \exp(-i\Gamma)\end{bmatrix} \tag{8.74}$$

Above, the relative phase difference Γ, also referred to as the relative retardation, is given by

$$\Gamma=(n_e-n_o)\frac{\omega d}{c}=2\pi(n_e-n_o)\frac{d}{\lambda} \tag{8.75}$$

where λ is the vacuum wavelength. We are interested in finding the polarization state of the emerging field relative to the original coordinate system (x- and y-axes). To obtain this, we apply a counterclockwise rotation of α to the Jones vector expressed in the o-e axes. Hence, the output Jones vector $\vec{J}'$ becomes

$$\vec{J}'=\begin{bmatrix}J_x'\\ J_y'\end{bmatrix}=\begin{bmatrix}\cos\alpha & -\sin\alpha\\ \sin\alpha & \cos\alpha\end{bmatrix}\begin{bmatrix}J_o'\\ J_e'\end{bmatrix}=\mathbf{M}_j(\alpha,\Gamma)\vec{J}_o \tag{8.76}$$

where

$$\mathbf{M}_j(\alpha,\Gamma) = \mathbf{R}(-\alpha)\,\mathbf{M}_{jo}\mathbf{R}(\alpha)$$
$$= \begin{bmatrix} \cos\alpha & -\sin\alpha \\ \sin\alpha & \cos\alpha \end{bmatrix}\begin{bmatrix} 1 & 0 \\ 0 & \exp(-i\Gamma) \end{bmatrix}\begin{bmatrix} \cos\alpha & \sin\alpha \\ -\sin\alpha & \cos\alpha \end{bmatrix}\begin{bmatrix} J_x \\ J_y \end{bmatrix} \tag{8.77}$$

In Eq. (8.77), $\mathbf{M}_j(\alpha,\Gamma)$ is the overall Jones transformation matrix, $\mathbf{R}(\alpha)$ represents a clockwise rotation of α about the z-axis and $\mathbf{M}_{jo}$ is the matrix describing the birefringent plate. Some commonly used matrices in Jones calculus are listed in Table 8.1.

Example Half-wave plate
Here, the relative retardation Γ is π. In other words,

$$(n_e - n_o)2\pi\frac{d}{\lambda} = \pi \tag{8.78}$$

giving

$$d = \frac{\lambda}{2(n_e - n_o)} \tag{8.79}$$

Hence, by knowing the wavelength of the light and the difference in the ordinary and extraordinary indices of refraction, the thickness of the plate can be adjusted to give half-wave retardation. To see the effect of half-wave retardation on the state of polarization of light, we assume that the incident wave is linearly polarized along the x-axis (See Fig. 8.7). It can be shown that (see Prob. 8.7) the overall Jones matrix $\mathbf{M}_j(\alpha,\pi)$ for the half wave plate becomes

$$\mathbf{M}_j(\alpha,\pi) = \begin{pmatrix} \cos 2\alpha & \sin 2\alpha \\ \sin 2\alpha & -\cos 2\alpha \end{pmatrix} \tag{8.80}$$

Description	$\mathbf{M}_{jo}$
Relative retardation of Γ	$\begin{bmatrix} 1 & 0 \\ 0 & \exp(-i\Gamma) \end{bmatrix}$
Polarizer oriented along the x-axis	$\begin{bmatrix} 1 & 0 \\ 0 & 0 \end{bmatrix}$
Polarizer oriented along the y-axis	$\begin{bmatrix} 0 & 0 \\ 0 & 1 \end{bmatrix}$

Table 8.1 Some Commonly Used Jones Matrices

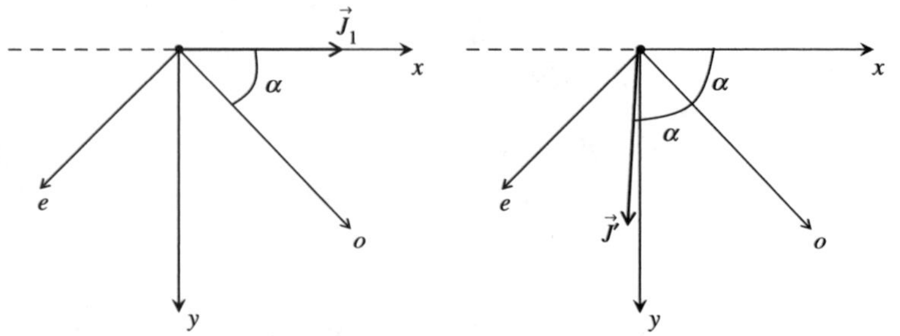

FIGURE 8.7 Effect of a half-wave plate on an incident wave linearly polarized along the *x*-axis.

Taking the initial Jones vector as

$$\vec{J}_1 = \begin{pmatrix} 1 \\ 0 \end{pmatrix} \tag{8.81}$$

we obtain

$$\vec{J}' = \begin{pmatrix} \cos 2\alpha & \sin 2\alpha \\ \sin 2\alpha & -\cos 2\alpha \end{pmatrix} \begin{pmatrix} 1 \\ 0 \end{pmatrix} = \begin{pmatrix} \cos 2\alpha \\ \sin 2\alpha \end{pmatrix} \tag{8.82}$$

Note that, according to Eq. (8.82), the output beam still remains linearly polarized but the direction of polarization is rotated by an angle of 2α. In other words, by rotating the birefringent plate so that the relative orientation of the *o*-axis is changed with respect to the *x*-axis, we can vary the angle α, and hence the amount of rotation of the direction of polarization. Half-wave plates are very useful devices widely used for polarization control in photonic systems. Also note that the magnitude of the output Jones vector remained as unity, since $\cos^2 2\alpha + \sin^2 2\alpha = 1$. This is a general property of the Jones matrix transformations. Since we assumed that the birefringent systems are lossless, the overall Jones matrix $\mathbf{M}_j(\boldsymbol{\alpha}, \boldsymbol{\Gamma})$ is always unitary and preserves the norm of the Jones vector.

Example Quarter-wave plate
In this case, the thickness *d* of the plate is adjusted to

$$d = \frac{\lambda}{4(n_e - n_o)} \tag{8.83}$$

to make the relative retardation Γ equal to $\pi/2$. To investigate the effect of the quarter-wave plate, let us assume that the input wave is again polarized along the *x*-axis and the corresponding Jones vector $\vec{J}_1$ is given by

$$\vec{J}_1 = \begin{pmatrix} 1 \\ 0 \end{pmatrix} \tag{8.84}$$

We leave it as an exercise (see Prob. 8.9) for the reader to show that when $\alpha = \pi/4$, the overall Jones matrix $\mathbf{M}_j(\pi/4, \pi/2)$ becomes

$$\mathbf{M}_j(\pi/4,\pi/2) = \frac{\sqrt{2}}{2}\begin{pmatrix} \exp\left(-i\frac{\pi}{4}\right) & \exp\left(+i\frac{\pi}{4}\right) \\ \exp\left(+i\frac{\pi}{4}\right) & \exp\left(-i\frac{\pi}{4}\right) \end{pmatrix} \tag{8.85}$$

giving

$$\vec{J}' = \mathbf{M}_j\vec{J}_1 = \frac{1}{\sqrt{2}}\begin{pmatrix} \exp\left(-i\frac{\pi}{4}\right) & \exp\left(+i\frac{\pi}{4}\right) \\ \exp\left(+i\frac{\pi}{4}\right) & \exp\left(-i\frac{\pi}{4}\right) \end{pmatrix}\begin{pmatrix} 1 \\ 0 \end{pmatrix}$$

$$= \frac{\exp\left(-i\frac{\pi}{4}\right)}{\sqrt{2}}\begin{pmatrix} 1 \\ i \end{pmatrix} \tag{8.86}$$

According to our convention, the output beam will be left-hand circularly polarized. Hence, the quarter-wave plate is a useful device that can covert linearly polarized light into circularly polarized light. It can be further shown (see Prob. 8.10) that if the input is polarized along the y-axis, the sense of rotation of the circularly polarized output will be reversed.

8.3 Pockels Effect and Electro-Optic Modulators

8.3.1 Pockels Effect

In the previous sections, we discussed the general formulation of wave propagation in anisotropic media. In certain types of crystals, application of an additional external electric field can cause a change in the refractive indices and this effect can be exploited in the design of light modulators. If the change in the refractive index is linearly proportional to the components of the applied electric field, the effect is known as the linear electro-optic effect or the Pockels effect. We will see shortly that devices based on electro-optic crystals can be used to control the intensity or phase of the light. In addition to light modulation, electro-optic crystals can also be used as fast shutters in laser Q-switching or laser cavity dumping.

The reader will recall that the index ellipsoid can be expressed in the form

$$\frac{x^2}{n_x^2} + \frac{y^2}{n_y^2} + \frac{z^2}{n_z^2} = 1 \tag{8.87}$$

where x, y, and z designate the principal dielectric axes of the crystal when no electric field is applied. When an external electric field is applied, the index ellipsoid will contain mixed coordinates, and following the common notation used in photonics (see, for example, Ref. 3), can be written in the form

$$\left(\frac{1}{n^2}\right)_1 x^2 + \left(\frac{1}{n^2}\right)_2 y^2 + \left(\frac{1}{n^2}\right)_3 z^2 + 2\left(\frac{1}{n^2}\right)_4 yz + 2\left(\frac{1}{n^2}\right)_5 xz + 2\left(\frac{1}{n^2}\right)_6 xy = 1 \tag{8.88}$$

Let the applied electric field be of the form

$$\vec{E} = E_x\hat{x} + E_y\hat{y} + E_z\hat{z} \tag{8.89}$$

in Cartesian coordinates. In general, the linear change in the ith component $(1/n^2)_i$ will depend on all three components of $\vec{E}$. To quantify this dependence, we introduce the electro-optic tensor r_{ij} which is defined according to

$$\Delta\left(\frac{1}{n^2}\right)_i = \sum_{j=1}^{3} r_{ij}E_j \tag{8.90}$$

As before, we use the convention that $x = 1$, $y = 2$, and $z = 3$, and i goes from 1 to 6, since there are six terms that describe the index ellipsoid in Eq. (8.88).

As an example, if we want to express the field dependence of $(1/n^2)_3$, we can write this as

$$\Delta\left(\frac{1}{n^2}\right)_3 = r_{31}E_x + r_{32}E_y + r_{33}E_z \tag{8.91}$$

It is clear from the above analysis that the electro-optic tensor can be represented by a 3×6 matrix. Eq. (8.90) can therefore be written in matrix form as

$$\begin{bmatrix} \Delta\left(\frac{1}{n^2}\right)_1 \\ \Delta\left(\frac{1}{n^2}\right)_2 \\ \Delta\left(\frac{1}{n^2}\right)_3 \\ \Delta\left(\frac{1}{n^2}\right)_4 \\ \Delta\left(\frac{1}{n^2}\right)_5 \\ \Delta\left(\frac{1}{n^2}\right)_6 \end{bmatrix} = \begin{bmatrix} r_{11} & r_{12} & r_{13} \\ r_{21} & r_{22} & r_{23} \\ r_{31} & r_{32} & r_{33} \\ r_{41} & r_{42} & r_{43} \\ r_{51} & r_{52} & r_{53} \\ r_{61} & r_{62} & r_{63} \end{bmatrix} \begin{bmatrix} E_x \\ E_y \\ E_z \end{bmatrix} \tag{8.92}$$

The form of the $\mathbf{r}_{ij}$ matrix depends on crystal symmetry. For example, it can be proven that no linear electro-optic effect will occur in crystals with inversion symmetry. For different symmetry classes of crystals, the general form of the $\mathbf{r}_{ij}$ matrices has been tabulated. See, for example Refs. 2 and 3 for a complete overview of different crystal systems and the forms of the corresponding $\mathbf{r}_{ij}$ matrices. For example, for a triclinic crystal, no simplifying relation exists between the 18 elements of the matrix. For a cubic crystal belonging to the symmetry class $\bar{4}3m$ (GaAs being an example of such a crystal), the $\mathbf{r}_{ij}$ matrix is given by[2, 4, 6]

$$\mathbf{r}_{ij} = \begin{bmatrix} 0 & 0 & 0 \\ 0 & 0 & 0 \\ 0 & 0 & 0 \\ r_{41} & 0 & 0 \\ 0 & r_{41} & 0 \\ 0 & 0 & r_{41} \end{bmatrix} \tag{8.93}$$

Example Electro-optic effect in GaAs

Let us use an example to illustrate how the specific symmetry properties of a certain crystal are used to analyze the electro-optic effect. We first choose the GaAs crystal which, as noted above, belongs to the symmetry class $\bar{4}3m$ and has the $\mathbf{r}_{ij}$ matrix given by Eq. (8.93). By using Eq. (8.92), we then get

$$\Delta\left(\frac{1}{n^2}\right)_4 = r_{41}E_x \tag{8.94}$$

$$\Delta\left(\frac{1}{n^2}\right)_5 = r_{41}E_y \tag{8.95}$$

and

$$\Delta\left(\frac{1}{n^2}\right)_6 = r_{41}E_z \tag{8.96}$$

Let us assume that only E_z is non-zero. In other words, the external electric field is applied along the original optical axis of the crystal. In this case, the equation for the index ellipsoid becomes

$$\left(\frac{1}{n^2}\right)_1 x^2 + \left(\frac{1}{n^2}\right)_2 y^2 + \left(\frac{1}{n^2}\right)_3 z^2 + 2r_{41}E_z xy = 1 \tag{8.97}$$

Since GaAs is a cubic crystal, the principal dielectric constants are all equal, giving $n_x = n_y = n_z = n$ and we get

$$\frac{x^2 + y^2 + z^2}{n^2} + 2r_{41}E_z xy = 1 \tag{8.98}$$

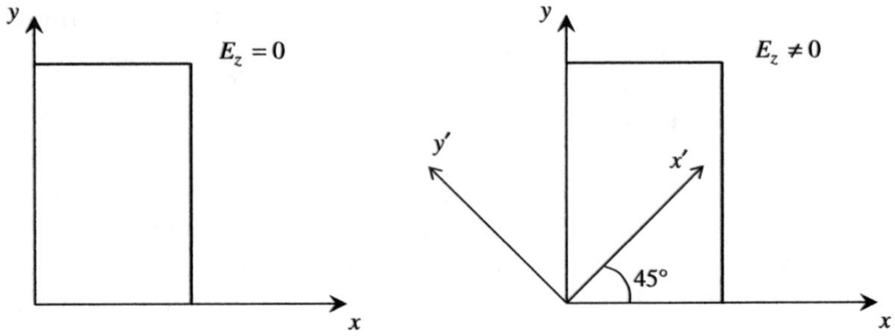

FIGURE 8.8 Orientation of the rotated principal axes x' and y' in the presence of an applied electric field relative to the original principal axes x and y.

Notice that due to the mixed term containing x-y, the principal axes of the ellipse are no longer parallel to the x- and y-axes. Instead, the principal axes are rotated by 45° about an axis perpendicular to the x-y plane as shown in Fig. 8.8. If we denote the transformed principal axes by x', y', and z' we find that

$$z = z' \tag{8.99}$$

$$x = \frac{1}{\sqrt{2}}(x' + y') \tag{8.100}$$

and

$$y = \frac{1}{\sqrt{2}}(-x' + y') \tag{8.101}$$

To check that this transformation actually works, we substitute back into Eq. (8.98) to obtain

$$\frac{1}{2n^2}[(x'+y')^2 + (y'-x')^2 + z^{12}] + 2r_{41}E_z\left(\frac{1}{2}\right)(x'+y')(y'-x') = 1 \tag{8.102}$$

or

$$\left(\frac{1}{n^2} - r_{41}E_z\right)x'^2 + \left(\frac{1}{n^2} + r_{41}E_z\right)y'^2 + z'^2 = 1 \tag{8.103}$$

For example, the principal index n'_x along the x' axis now becomes

$$n'_x = \left(\frac{1}{n^2} - r_{41}E_z\right)^{-1/2} = n\left(1 - n^2 r_{41}E_z\right)^{-1/2} \cong n + \frac{n^3}{2}r_{41}E_z \tag{8.104}$$

where we have assumed that the field-dependent change is very small compared to n. Similarly, n'_y becomes

$$n'_y \cong n - \frac{n^3}{2} r_{41} E_z \tag{8.105}$$

and n'_z remains unchanged as n.

Example Electro-optic effect in $LiNbO_3$

This is another popular electro-optic crystal which belongs to the trigonal crystal symmetry class $3m$ with the electro optic tensor matrix $\mathbf{r}_{ij}$ of the form[2–4]

$$\mathbf{r}_{ij} = \begin{bmatrix} 0 & -r_{22} & r_{13} \\ 0 & r_{22} & r_{13} \\ 0 & 0 & r_{33} \\ 0 & r_{51} & 0 \\ r_{51} & 0 & 0 \\ -r_{22} & 0 & 0 \end{bmatrix} \tag{8.106}$$

Without an external field, $LiNbO_3$ is a uniaxial crystal with $n_x = n_y = n_o$ and $n_z = n_e$. Again, let us assume that the applied e-field $\vec{E} = E\hat{z}$ is along the optical axis. Then, the change in the ellipsoid coefficients will be given by

$$\Delta\left(\frac{1}{n^2}\right)_1 = r_{13} E_z \tag{8.107}$$

$$\Delta\left(\frac{1}{n^2}\right)_2 = r_{13} E_z \tag{8.108}$$

and

$$\Delta\left(\frac{1}{n^2}\right)_3 = r_{33} E_z \tag{8.109}$$

resulting in the index ellipsoid

$$\left(\frac{1}{n_o^2} + r_{13} E_z\right) x^2 + \left(\frac{1}{n_o^2} + r_{13} E_z\right) y^2 + \left(\frac{1}{n_e^2} + r_{33} E_z\right) z^2 = 1 \tag{8.110}$$

Note that the above index ellipsoid can be written in the form

$$\frac{x^2 + y^2}{n_o'^2} + \frac{z^2}{n_e'^2} = 1 \tag{8.111}$$

where

$$\frac{1}{n_o'^2} = \frac{1}{n_o^2} + r_{13} E_z = \frac{1}{n_o^2}\left(1 + n_o^2 r_{13} E_z\right) \tag{8.112}$$

or

$$n_o' \cong n_o - \frac{1}{2} n_o^3 r_{13} E_z \tag{8.113}$$

if we assume that the field contribution to the refractive index is very small and include only the first term of the Taylor expansion in Eq. (8.112). Similarly

$$n_e' \cong n_e - \frac{1}{2} n_o^3 r_{33} E_z \tag{8.114}$$

Note that in this case, the principal axes of the crystal remain unchanged. However, the relative retardation between the ordinary and extraordinary waves can be adjusted by varying the strength of the electric field provided that $r_{13} \neq r_{33}$. For $LiNbO_3$, $r_{13} = 8.5\times10^{-12}\,\text{m/V}$ and $r_{33} = 28.5\times10^{-12}\,\text{m/V}$ near the wavelength of 1550 nm.[11]

8.3.2 Electro-Optic Retardation in GaAs

Let us consider an electro-optic crystal such as GaAs in which a light beam propagates along the axis parallel to the applied electric field. Let this axis of propagation be the z-axis. Once the electric field is applied, the principal axes will be rotated to x' and y' as sketched in Fig. 8.8. Initially, suppose that we launch a wave parallel to the y-axis with

$$\vec{E}(z=0,t) = \hat{y} E_0 \exp[i\omega t] \tag{8.115}$$

The corresponding components along the x' and y' axes will be given by

$$E_{x'}(z=0,t) = \frac{E_0}{\sqrt{2}} \exp[i\omega t] \tag{8.116}$$

and

$$E_{y'}(z=0,t) = \frac{E_0}{\sqrt{2}} \exp[i\omega t] \tag{8.117}$$

After propagating by a distance z, the field components will become

$$E_{x'}(z,t) = \frac{E_0}{\sqrt{2}} \exp\left[i\left(\omega t - \frac{\omega}{c} n_x' z\right)\right] \tag{8.118}$$

and

$$E_{y'}(z,t) = E_{x'}(z,t) \exp(-i\Gamma) \tag{8.119}$$

where the relative phase difference Γ will be given by

$$\Gamma = \frac{\omega}{c} z\left(n'_y - n'_x\right) \tag{8.120}$$

If the crystal length is d and the applied longitudinal voltage is V, E_z becomes V/d and for GaAs we get

$$\Gamma = -\frac{\omega}{c} d\left(2\frac{n^3}{2} r_{41} E_z\right) = \frac{-n^3 \omega r_{41}}{c} V \tag{8.121}$$

Hence, the retardation increases linearly with the applied voltage V. It is customary to express Γ in terms of the voltage required to obtain a retardation of π. This voltage is designated as V_π and can be calculated from

$$V_\pi = \frac{c\pi}{n^3 \omega r_{41}} = \frac{\lambda}{2n^3 r_{41}} \tag{8.122}$$

where λ is the free-space wavelength of the light. Hence, the larger the electro-optic coefficient of the crystal, the smaller is the π-phase voltage. Note also that if all other parameters are kept fixed, the π-phase voltage increases linearly with wavelength. Hence, if modulators are designed for longer wavelengths, proportionally higher bias voltage is needed to obtain the same phase shift. In terms of V_π, the relative retardation Γ can be expressed as

$$\Gamma = -\pi \frac{V}{V_\pi} \tag{8.123}$$

For GaAs, $r_{41} = 1.5x10^{12}$ m/V near the wavelength of 1.52 μm.[12]

8.3.3 Electro-Optic Amplitude Modulators

A typical setup of an electro-optic amplitude modulator is shown in Fig. 8.9. The electro-optic crystal under a bias voltage is placed between two crossed polarizers (P_1 and P_2) whose transmission axes are perpendicular.

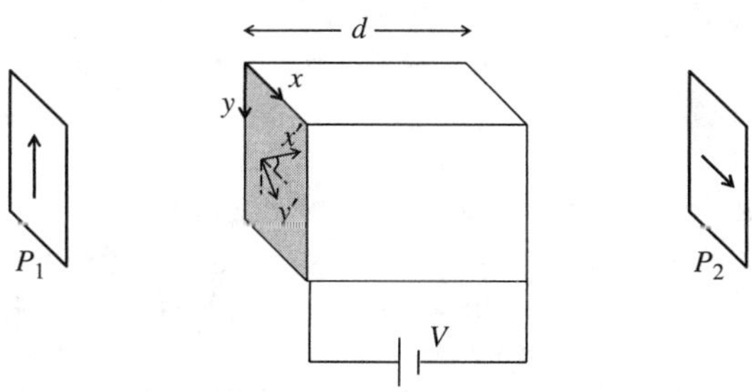

Figure 8.9 Electro-optic amplitude modulator.

We can use Jones matrices to analyze the modulator. Let us consider a crystal such as GaAs. Before, we noted that the original principal coordinate axes are related to the transformed principal axes by Eqs. (8.99) to (8.101). Transforming back, we obtain

$$z' = z \tag{8.124}$$

$$x' = \frac{1}{\sqrt{2}}(x - y) \tag{8.125}$$

and

$$y' = \frac{1}{\sqrt{2}}(x + y) \tag{8.126}$$

So, to go from the laboratory axes (x, y) to the new principal axes (x', y'), we need to apply a rotation of $-45°$ about the z-axis. The input to the system is described by the Jones vector $\begin{bmatrix} 0 \\ 1 \end{bmatrix}$ since the polarizer P_1 passes light linearly polarized along the y-axis. The output Jones vector will be given by (see Prob. 8.11)

$$\vec{J}' = \begin{bmatrix} 1 & 0 \\ 0 & 0 \end{bmatrix} \begin{bmatrix} 1/\sqrt{2} & 1/\sqrt{2} \\ -1/\sqrt{2} & 1/\sqrt{2} \end{bmatrix} \begin{bmatrix} 1 & 0 \\ 0 & e^{-i\Gamma} \end{bmatrix} \begin{bmatrix} 1/\sqrt{2} & -1/\sqrt{2} \\ 1/\sqrt{2} & 1/\sqrt{2} \end{bmatrix} \begin{bmatrix} 0 \\ 1 \end{bmatrix}$$

$$= \begin{bmatrix} -\frac{1}{2} + \frac{1}{2} e^{-i\Gamma} \\ 0 \end{bmatrix} \tag{8.127}$$

The corresponding intensity transmission after the polarizer P_2 will be given by

$$T = \sin^2 \Gamma/2 \tag{8.128}$$

Variation of T with the applied voltage V is shown in Fig. 8.10. As V goes from 0 to V_π, T varies from 0 to 100%. This method allows for the intensity modulation of the beam by modulating the bias voltage V.

8.3.4 Electro-Optic Q-Switching of Lasers

An electro-optic crystal can be used as a fast shutter to Q-switch a laser. When the shutter is closed, no feedback is present to allow the oscillation of the circulating beam. During this period, the population inversion inside the atomic gain medium builds up to levels larger than what would be needed to overcome the losses of the cavity when

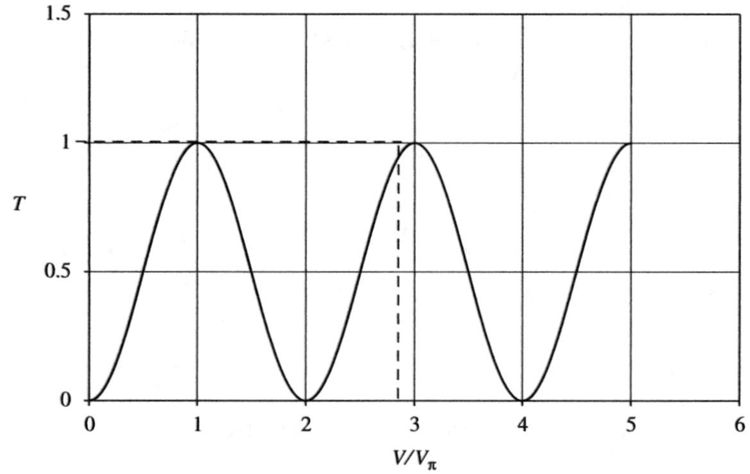

FIGURE 8.10 Variation of the transmission T with the applied voltage V for the electro-optic modulator shown in Fig. 8.9.

the shutter is open. When the shutter is opened, the large inversion stored in the gain medium leads to a rapid growth of the laser intensity and the energy is out-coupled in the form of a short, high-energy pulse. Detailed analysis of Q-switching was given earlier in Sec. 6.6. The general schematic of an electro-optically Q-switched laser is shown in Fig. 8.11.

When no voltage is applied to the electro-optic crystal, the losses in the cavity are low, the Q-factor of the cavity is high, and the vertically polarized circulating beam oscillates. To lower the Q-factor, a quarter-wave voltage is applied to the crystal to cause a phase retardation of $\pi/2$ per pass. The beam retroreflected by M_2 and returning to the polarizer undergoes a total retardation of π and becomes horizontally polarized. The polarizer then blocks the passage of this beam back into the gain medium and no lasing occurs during

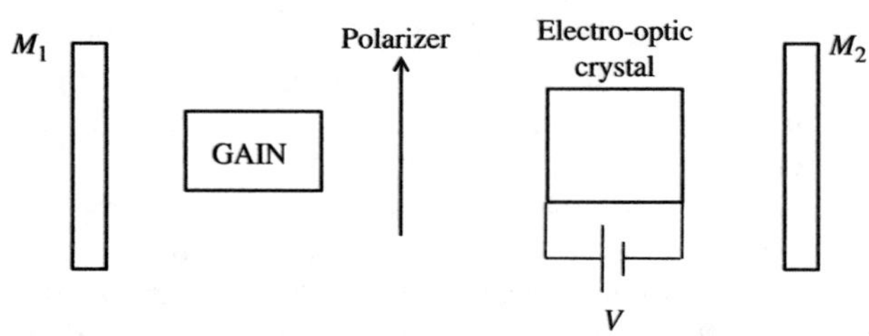

FIGURE 8.11 Electro-optically Q-switched laser.

this phase. If we again assume that the principal axes of the biased electro-optic crystal are rotated by 45° with respect to the laboratory axes, the Jones vector for the beam going back through the vertical polarizer becomes

$$\vec{J}' = \begin{bmatrix} 0 & 0 \\ 0 & 1 \end{bmatrix} \begin{bmatrix} 1/\sqrt{2} & 1/\sqrt{2} \\ -1/\sqrt{2} & 1/\sqrt{2} \end{bmatrix} \begin{bmatrix} 1 & 0 \\ 0 & e^{-i\Gamma} \end{bmatrix} \begin{bmatrix} 1/\sqrt{2} & -1/\sqrt{2} \\ 1/\sqrt{2} & 1/\sqrt{2} \end{bmatrix} \begin{bmatrix} 0 \\ 1 \end{bmatrix}$$

$$= \begin{bmatrix} 0 \\ \frac{1}{2} + \frac{1}{2} e^{-i\Gamma} \end{bmatrix} \tag{8.129}$$

The corresponding transmission T can be calculated from (see Prob. 8.12)

$$T = \cos^2\left(\Gamma/2\right) \tag{8.130}$$

So, if the full retardation is π (i.e., when a quarter-wave voltage is applied to the electro-optic crystal), the beam returning from M_2 will be negligible since $T = 0$. See Refs. 3, 13, and 14 for a further discussion of electro-optic Q-switching and alternative configurations.

8.3.5 Electro-Optic Regenerative Amplifiers

In regenerative amplification, seed femtosecond laser pulses, typically stretched to picosecond durations, are amplified by sending them into a second resonator containing an inverted atomic medium.[15–17] Pulse injection into and extraction from the regenerative amplifier can be achieved by using electro-optical crystals. A simplified schematic of a regenerative amplifier adapted from Ref. 16 is shown in Fig. 8.12.

Let us assume that the incoming seed pulses are vertically polarized. The polarizing beam splitter reflects the incident beam toward the quarter-wave plate (QWP) and the electro-optic crystal, commonly referred to as a Pockels cell (PC). In the injection phase, the applied voltage is zero, so that the beam reflected from the end mirror M_1 becomes horizontally polarized, passes through the polarizing beamsplitter, and enters the cavity containing the amplifying medium. At this instant, a quarter-wave voltage is applied to the Pockels cell so that the circulating beam passing through the Pockels cell and the quarter-wave plate remains horizontally polarized during the amplification phase. After a sufficient number of round trips, the gain gets depleted and the applied voltage is turned back to zero. At this phase, the pulse retroreflected from M_1 becomes vertically polarized and is ejected by the polarizing beamsplitter (PBS) from the cavity. Cavity

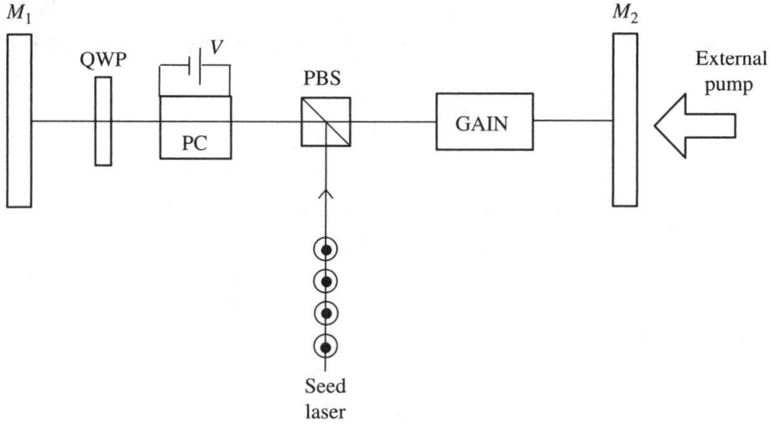

FIGURE 8.12 Schematic of a regenerative amplifier operated with a Pockels cell (PC).

dumping provides a means of extracting the optical energy stored within the cavity. This is typically much higher than what one could obtain from a standard laser operated with an output coupler.

8.3.6 Electro-Optic Phase Modulation

To modulate the phase of an optical signal, a linearly polarized beam is sent onto an electro-optic crystal with its direction of polarization parallel to one of the rotated principal axes (x' or y') of the crystal. This way, the state of polarization of the beam remains invariant but a voltage-dependent phase is acquired. Suppose that the beam is polarized along the x' axis of a GaAs-type phase modulator under longitudinal bias as discussed before. The emerging electric field will be of the form

$$\begin{aligned} E_{x'}(z,t) &= E_0 \exp\left[i\left(\omega t - \frac{\omega}{c} n'_x d\right)\right] \\ &= E_0 \exp\left[i\left(\omega t - \frac{\omega}{c}\left(n + \frac{n^3}{2} r_{41} E_z\right) d\right)\right] \\ &= E_0 \exp\left[-i\frac{\omega}{c} nd\right] \exp\left[i(\omega t - \phi(t))\right] \end{aligned} \tag{8.131}$$

where $\phi(t)$ is given by

$$\phi(t) = \frac{\omega}{c}\frac{n^3}{2} r_{41} V(t) = \frac{2\pi}{\lambda}\frac{n^3}{2} r_{41} V(t) = \frac{\pi}{2}\frac{V(t)}{V_\pi} \tag{8.132}$$

Above, d is the length of the crystal and V_π is the π-phase bias as defined before in Eq. (8.122). Let $V(t)$ be a sinusoidal voltage of the form

$$V(t) = V_0 \sin(\omega_m t) \tag{8.133}$$

where V_0 is the amplitude and ω_m is the modulation frequency. Then $\phi(t)$ can be written as

$$\phi(t) = \phi_0 \sin(\omega_m t) \tag{8.134}$$

with

$$\phi_0 = \frac{\pi}{2}\frac{V_0}{V_\pi} \tag{8.135}$$

and the output electric field takes the form

$$E_{x'}(z,t) = E_0 \exp\left[-i\frac{\omega}{c}nd\right]\exp[i(\omega t - \phi_0 \sin(\omega_m t))] \tag{8.136}$$

As can be seen from the above equation, the dynamic frequency of the wave will be modulated as the applied voltage is varied.

8.4 Acousto-Optic Modulation of Light

Interaction of light with sound waves propagating in a medium is known as the acousto-optic effect. The local density variations due to the propagating sound wave lead to local variations in refractive index and cause diffraction of light. Acousto-optic effect has been exploited in the development of many important photonic devices including scanners and modulators. A detailed description of acousto-optic effect and its photonic applications can be found in Refs. 2, 3, 5, 6, and 18. Here, we provide a simplified discussion to introduce the reader to the basic concepts and describe the operation of acousto-optic modulators.

A sound wave can be simply thought of as a propagating disturbance consisting of local density variations. The regions of high and low density are often referred to as regions of compression and rarefaction. The density disturbance propagates at the sound velocity which varies depending on the rigidity of the medium. In the particular case of solids, sound waves (also referred to as seismic waves in the case of sound waves moving within the Earth's crust) have velocities in the range of several kilometer/second.

Local density and strain variations in a solid further lead to changes in the refractive index. This is known as the photoacoustic effect. A propagating sound wave in a solid will set up a periodic refractive index variation which can be expressed as

$$\delta n(\vec{r},t) = \delta n_0 \exp\left[i\left(\omega_s t - \vec{k}_s \cdot \vec{r}\right)\right] \tag{8.137}$$

where δn_0 is the maximum change in the refractive index induced by the sound wave, ω_s is the angular frequency of the sound wave, and $\vec{k}_s$ is the corresponding wave vector.

Now, consider a plane electromagnetic wave incident on a medium in which a sound wave is propagating. This is illustrated in Fig. 8.13.

The density variation set up by the moving sound wave can be approximately modeled as a set of partially reflecting plane fronts, two consecutive fronts being separated by λ_s. Here, λ_s is the effective sound wavelength in the medium. Because the sound wave moves at velocities much smaller than the speed of light, we can safely assume that the index grating is fixed in space during the interaction with light. Doppler shift in the frequency of the scattered light does in fact take place as we will discuss soon. To find the direction of the scattered wave, we consider a portion of the plane wave which intersects the wave front (shown as 2 in Fig. 8.13) at the points O and P. The incidence angle made between the light wave and the density front is θ_i. For the directions that lead to constructive interference, the optical phase difference between the light phase fronts PR and OS is given by

$$\Delta\phi = k|OP|(\cos\theta_i - \cos\theta_r) = 2m\pi \tag{8.138}$$

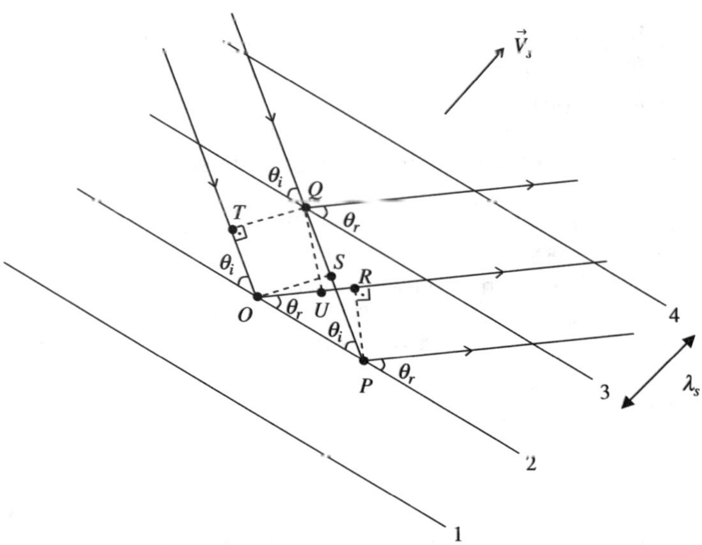

FIGURE 8.13 Diffraction of a plane electromagnetic wave from periodic density variations in a solid set up by acoustic waves.

where $k = (2\pi/\lambda)n$, and n is the refractive index of the medium at the optical wavelength λ. Furthermore, m is an integer that can take the values 0, ±1, ±2 and so on. Simplifying Eq. (8.138) gives

$$|OP|(\cos\theta_i - \cos\theta_r) = m\frac{\lambda}{n} \tag{8.139}$$

The distance $|OP|$ was chosen arbitrarily and can have any value within the transverse extent of the incident beam. The only way Eq. (8.139) will hold for any value of $|OP|$ is if $m = 0$ and $\theta_i = \theta_r$. Second, we need to look at the optical path difference between the wave fronts scattered from adjacent fronts. Such a scenario is shown in Fig. 8.13 where the path difference between the scattered and incident waves is given by

$$\Delta\phi = k(|OT| + |OU|) = 2k|OQ|\sin\theta_i = 2k\lambda_s \sin\theta_i \tag{8.140}$$

In the Bragg regime of diffraction, the acoustic plane wave is assumed to have infinite extent, in which case only the condition $\Delta\phi = 2\pi$ leads to enhanced scattering. This gives

$$2\pi = 2\frac{2\pi}{\lambda}n\lambda_s \sin\theta_i \tag{8.141}$$

or

$$\sin\theta_i = \frac{1}{2n}\frac{\lambda}{\lambda_s} \tag{8.142}$$

So, in the Bragg regime of diffraction, we have only the zeroth order and the first-order diffracted beam. The zeroth-order beam corresponds to the beam that continues along the original direction. We can gain further insight about Bragg diffraction by considering the particle model for the interaction of light with sound. If we treat the light and sound waves as particles (light particles are the familiar photons, whereas sound waves in solids would be phonons), the scattering process can be analyzed by using the conservation of energy and momentum. Conservation of energy leads to

$$h\nu_2 = h\nu_1 + h\nu_s \tag{8.143}$$

or simply

$$\nu_2 = \nu_1 + \nu_s \tag{8.144}$$

Above ν_s is the sound frequency, ν_1 is the frequency of the incident light beam, and ν_2 is the frequency of the scattered light beam. Equivalently, this relation could have been derived by considering the Doppler shift experienced by the scattered beam. Assuming that the sound front moves toward the incoming beam with a speed of u_0, the shifted frequency of the diffracted beam will be given by

$$\nu_2 = \nu_1 \sqrt{\frac{1+u_0/(c/n)}{1-u_0/(c/n)}} \cong \nu_1 \left(1+\frac{nu_0}{c}\right) \tag{8.145}$$

giving

$$\nu_2 - \nu_1 = \nu_1 \frac{u_0}{(c/n)} \tag{8.146}$$

Here, u_0 is the component of the sound velocity in the direction of the incident beam and is given in terms of the magnitude u_s of the sound speed by

$$u_0 = 2u_s \cos\left(\frac{\pi}{2} - \theta_i\right) = 2u_s \sin\theta_i \tag{8.147}$$

Above, the factor of 2 comes from the fact that the image moves at $2u_s$ as the sound front propagates at u_s. From before [Eq. (8.142)], $\sin\theta_i = \lambda/2n\lambda_s$, giving

$$\nu_2 - \nu_1 = \nu_1(2u_s)\frac{\lambda}{2n\lambda_s}\frac{1}{c/n} = \frac{u_s}{\lambda_s} = \nu_s \tag{8.148}$$

in agreement with Eq. (8.144).

Now, what about the conservation of momentum? Conservation of momentum for the light and sound waves gives

$$\hbar\vec{k}_2 = \hbar\vec{k}_1 + \hbar\vec{k}_s \tag{8.149}$$

where $\vec{k}_2$ is the wave vector of the scattered light, and $\vec{k}_1$ is the wave vector of the incident light. The sound wave vector $\vec{k}_s$ is assumed to be unidirectional and also vertical for a plane sound wave of infinite extent. The graphical representation is shown in Fig. 8.14. Typically, since k_s is much smaller than k_1 or k_2 (see Prob. 8.13), the lengths of k_1 and k_2 are nearly the same.

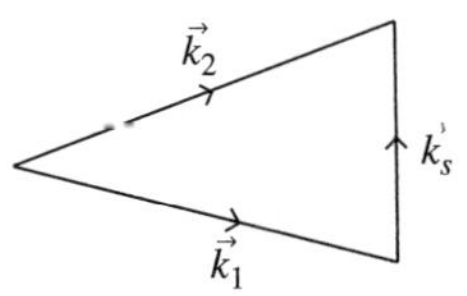

Figure 8.14 Wave vectors of the incident light wave, scattered light wave, and the sound wave, satisfying the conservation of momentum for the case of first-order diffraction.

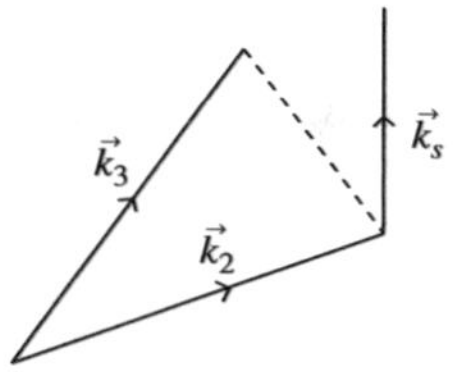

FIGURE 8.15 Orientation of the wave vectors for the incident light wave, scattered light wave, and the sound wave for second-order diffraction. Conservation of momentum cannot be satisfied in this case.

Second order diffraction would require a diffracted wave vector $\vec{k}_3$ with a magnitude nearly equal to that of $\vec{k}_2$ but oriented at an angle such that $\vec{k}_3 = \vec{k}_2 + \vec{k}_s$. Since $\vec{k}_3$ is now at an angle with respect to $\vec{k}_2$, and since $\vec{k}_s$ is fixed in direction (vertical in our case), it is no longer possible to satisfy the conservation of momentum and energy at the same time. This can be seen in the sketch of Fig. 8.15.

To satisfy the conservation of momentum here, we therefore need a sound wave vector $\vec{k}_s'$ that is slightly tilted in direction with respect to $\vec{k}_s$. Such a wave vector component is not present if the sound wave is of infinite extent. However, if we instead had a spatially confined sound wave, that is, a sound wave which has a finite extent in the transverse direction, there will be other wave vector components with directions within, roughly, an angular spread of λ_s/d where λ_s is the sound wave length and d is the width of the beam. This is due to diffraction, an example of which we saw earlier in Chap. 3, when we discussed the propagation of Gaussian beams. In this case, it becomes possible to satisfy both of the conservation laws at the same time for higher-order diffracted beams. This case, where higher-order diffracted beams can be observed, is known as the Raman-Nath regime of acousto-optic interactions. Raman-Nath diffraction will occur when the acoustic beam spreading [given approximately by (λ_s/d)] is large in comparison with the Bragg diffraction angle. Figure 8.16 shows a sketch of the Raman-Nath regime of diffraction where higher-order diffracted light beams can be observed by confining the spatial extent of the sound wave. The dimensionless diffraction parameter Q defined according to

$$Q = \frac{2\pi\lambda d}{n\lambda_s^2} \tag{8.150}$$

is further used to identify the two regimes of diffraction (see, for example, Ref. 2). When $Q < 1$, we are in the Raman-Nath regime, since this corresponds to the case where the acoustic spreading angle (λ_s/d) is large compared to the Bragg angle, which, from Eq. (8.142), is approximately given as $\theta_i \cong \lambda/2n\lambda_s$. Bragg diffraction corresponds to the regime where $Q > 1$.

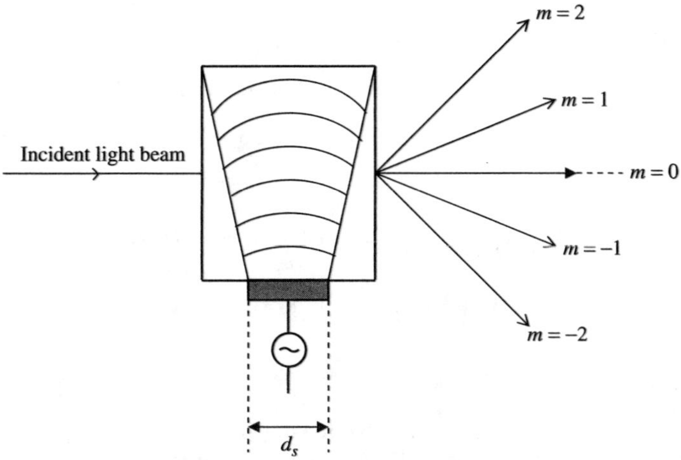

FIGURE 8.16 Acousto-optic diffraction of a light beam in the Raman-Nath regime.

As an example of acousto-optic devices, we briefly discuss the operation of an acousto-optic modulator which is used in amplitude modulation (AM) mode locking[19-21] of a laser. The theory of AM mode locking was discussed earlier in Sec. 6.8.2. A schematic of an AM mode-locked laser is shown in Fig. 8.17.

An acousto-optic Bragg cell driven at some acoustic frequency f_m is placed inside the laser resonator. The acoustic wave set up inside the Bragg crystal causes the diffraction of the passing light and leads to a periodic loss modulation. The circulating laser field suffers no loss only when the acoustic field is instantaneously zero inside the crystal. This happens twice per period. If the cavity length is adjusted to make the pulse repetition frequency $c/2d$ (c = speed of light, d = resonator length) equal to $2f_m$, then a steady-state pulse train is produced from the laser. The temporal separation between the pulses is given by $T_R = 2L/c = 1/2f_m$.

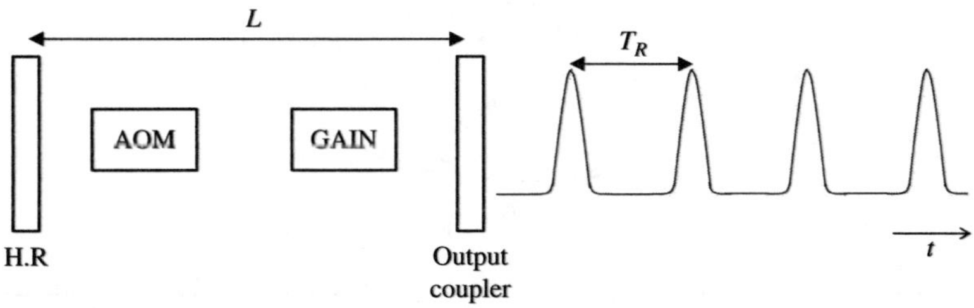

FIGURE 8.17 Schematic of a laser mode-locked with an acousto-optic modulator.

Problems

8.1 Show that the vectors $\vec{D}_0$, $\vec{E}_0$, $\hat{s}$, and $\vec{E}_0 \times \vec{H}_0$ all lie in a single plane.

8.2 Show that the Fresnel Equation given by Eq. (8.31) is quadratic in n^2 and can be written in the form

$$\left(s_x^2 n_x^2 + s_y^2 n_y^2 + s_z^2 n_z^2\right) n^4 - \left(n_x^2 n_y^2 \left(s_x^2 + s_y^2\right) + n_x^2 n_z^2 \left(s_x^2 + s_z^2\right)\right.$$
$$\left. + n_y^2 n_z^2 \left(s_y^2 + s_z^2\right)\right) n^2 + n_x^2 n_y^2 n_z^2 = 0$$

where n is the effective index.

8.3 Write a computer code to solve for the effective indices by using the Fresnel equation given by Eq. (8.31). Assume that the principal indices of refraction are $n_x = 1.5$, $n_y = 1.6$, and $n_z = 1.7$, and that propagation occurs along the direction $\hat{s} = 1/\sqrt{3}(\hat{x} + \hat{y} + \hat{z})$. Investigate what happens to the roots as the number of significant digits in the components of $\hat{s}$ is varied.

8.4 Obtain the fields $\vec{E}_{02}$, $\vec{H}_{02}$, and $\vec{D}_{02}$ corresponding to the effective index $n_2 = 1.5391$ in the numerical example of Sec. 8.1.2.

8.5 Double refraction: Consider a plane wave obliquely incident on a uniaxial crystal. The corresponding normal surfaces are shown in Fig. 8.18 for the case where the optic axis is perpendicular to the surface. The refracted wave will be a mixture of the ordinary and extraordinary waves which move in different directions, hence the name double refraction. The boundary conditions require that the tangential components of the wave vectors in both media remain the same along the interface. Use this idea to find the angle of refraction

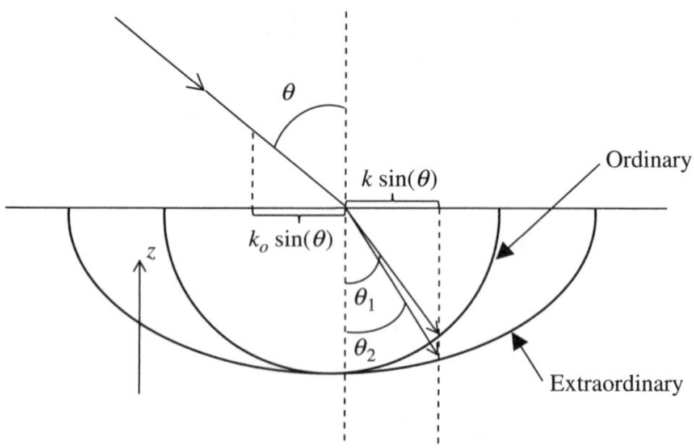

FIGURE 8.18 Problem 8.5.

for the ordinary and extraordinary waves. In particular, show that the angle of refraction θ_2 for the extraordinary wave is given by

$$\tan\theta_2 = \frac{\left(\frac{n_e}{n_o}\right)\sin\theta}{\sqrt{n_e^2 - \sin^2\theta}}$$

8.6 Consider a plane electromagnetic wave which is obliquely incident on a positive uniaxial crystal. The angle of incidence is 50°. The optic axis is perpendicular to the surface. Let $n_o = 1.5$ and $n_e = 1.7$. Calculate the two angles of refraction in the crystal.

8.7 Show that the overall Jones matrix for a half-wave plate is given by Eq. (8.80).

8.8 Show that the overall Jones matrix $\mathbf{M}_j(\alpha, \Gamma)$ is unitary. In other words, show that

$\mathbf{M}_j^{\dagger}(\alpha,\Gamma)\mathbf{M}_j(\alpha,\Gamma) = \mathbf{I}$, where $\mathbf{M}_j^{\dagger}$ is the hermitian conjugate of $\mathbf{M}_j$ defined according to

$(\mathbf{M}_j^{\dagger})_{ik} = (\mathbf{M}_j)_{ki}^{*}$ and $\mathbf{I}$ is the unity matrix. The subscripts k and i refer to the row and column numbers.

8.9 Show that when $\alpha = \pi/4$, the overall Jones matrix $\mathbf{M}_j(\pi/4, \pi/2)$ for the quarter-wave plate is given by Eq. (8.85).

8.10 Analyze the effect of a quarter-wave plate with $\alpha = \pi/4$ on a light beam which is linearly polarized along the y-axis and show that the output is now right-hand circularly polarized.

8.11 For the electro-optic amplitude modulator shown in Fig. 8.9, show that the output Jones vector and the intensity transmission are given by Eqs. (8.127) and (8.128).

8.12 For the electro-optically Q-switched laser shown in Fig. 8.11, show that the output Jones vector and the intensity transmission are given by Eqs. (8.129) and (8.130).

8.13 By taking the velocity and frequency of an acoustic wave as 5 km/s and 50 MHz, compare the wave vector magnitude with that of 1-μm light by assuming that the index of refraction is 1.5.

References

1. M. Born and E. Wolf, *Principles of Optics*, 6th ed. (Cambridge University Press, Cambridge, 1997).
2. A. Yariv and P. Yeh, *Optical Waves in Crystals* (John Wiley & Sons, Inc., New York, 1984).
3. A. Yariv, *Optical Electronics in Modern Communications*, 5th ed. (Oxford University Press, New York, 1997).

4. B. E. A. Saleh and M. C. Teich, *Fundamentals of Photonics*, 2d ed. (John Wiley & Sons, Inc., Hoboken, N.J., 2007).
5. C. C. Davis, *Lasers and Electro-Optics* (Cambridge University Press, Cambridge, 1996).
6. C. R. Pollock, *Fundamentals of Optoelectronics* (Irwin, Chicago, 1995).
7. H. A. Haus, *Waves and Fields in Optoelectronics* (Prentice-Hall, Englewood Cliffs, N.J., 1984).
8. A. Yariv, *Quantum Electronics* (John Wiley & Sons, Inc., New York, 1975).
9. R. C. Jones, "A new calculus for the treatment of optical systems I: Description and discussion of the calculus," *JOSA*, **31**, 488–493 (1941).
10. K. Naganuma, G. Lenz, and E. P. Ippen, "Variable bandwidth birefringent filter for tunable femtosecond lasers," *IEEE Journal of Quantum Electronics*, **28**(10), 2142–2150 (1992).
11. S. H. Lee, S. H. Kim, K. H. Kim, M. H. Lee, and E. H. Lee, "A novel method for measuring continuous dispersion spectrum of electro-optic coefficients of nonlinear materials," *Optics Express*, **17**(12), 9828–9833 (2009).
12. C. A. Berseth, C. Wuethrich, and F. K. Reinhart, "The electrooptic coefficients of GaAs-measurements at 1.32-μm and 1.52-μm and study of their dispersion between 0.9-μm and 10-μm," *Journal of Applied Physics*, **71**(6), 2821–2825 (1992).
13. W. Koechner, *Solid-State Laser Engineering*, 2d ed. (Springer-Verlag, Berlin, 1988).
14. J. T. Verdeyen, *Laser Electronics*, 3d ed. (Prentice Hall, Upper Saddle River, N.J., 1995).
15. G. Vaillancourt, T. B. Norris, J. S. Coe, P. Bado, and G. A. Mourou, "Operation of a 1-khz pulse-pumped Ti-sapphire regenerative amplifier," *Optics Letters*, **15**(6), 317–319 (1990).
16. C. Horvath, A. Braun, H. Liu, T. Juhasz, and G. Mourou, "Compact directly diode-pumped femtosecond Nd:glass chirped-pulse-amplification laser system," *Optics Letters*, **22**(23), 1790–1792 (1997).
17. J. Kawanaka, K. Yamakawa, H. Nishioka, and K. Ueda, "30-mJ, diode-pumped, chirped-pulse Yb:YLF regenerative amplifier," *Optics Letters*, **28**(21), 2121–2123 (2003).
18. E. I. Gordon, "A review of acoustooptical deflection and modulation devices," *Proc. IEEE*, **54**, 1391–1401 (1966).
19. D. J. Kuizenga and A. E. Siegman, "FM and AM mode locking of the homogeneous laser-Part 1: Theory," *IEEE Journal of Quantum Electronics*, **6**, 694–708 (1970).
20. A. E. Siegman and D. J. Kuizenga, "Active mode coupling phenomena in pulsed and continuous lasers," *Opto-Electronics*, **6**, 43–66 (1974).
21. A. E. Siegman, *Lasers* (University Science Books, Mill Valley, Calif., 1986).

CHAPTER 9

Noise and Optical Detection

In Chap. 1, we already touched upon the importance of detectors in photonic systems. Simply put, a detector is a device that converts optical signals to electrical signals. Many types of detectors exist and we will discuss some specific examples in the later sections of this chapter. Before looking at specific detectors, however, we want to discuss the subject of noise. In a detection system, randomly fluctuating signals which overlap with the same spectral range as that of the signal of interest, are referred to as noise and set the limits to how low a signal level we can detect. In this chapter, we first discuss the general aspects of noise and introduce the spectral noise power density which gives a quantitative measure of the amount of noise in a given spectral window. The spectral noise power density is typically expressed in terms of the mean-squared noise current, denoted as $\overline{i_n^2}$. We will then derive the expression for the mean-squared noise current for three commonly encountered types of noise: shot noise, Johnson noise, and background noise. Finally, we will give a concise description of different types of detectors and comment on their noise performance. See Refs. 1 through 7 for a general discussion on noise and optical detectors. Recently, investigation of noise characteristics of lasers has also attracted a lot of attention, with the aim of developing high-precision frequency standards. We will not be discussing this topic here. See, for example, Refs. 8 through 14, and those cited therein for a review of recent activity in this area.

9.1 Different Types of Noise

9.1.1 Spectral Noise Power Density

Suppose that we attempt to detect an optical signal with a semiconductor detector. We assume that the incident light is nearly monochromatic at the frequency v, and hv is larger than the bandgap E_g of the semiconductor. In this case, the incident photons will be energetic enough to cause band-to-band transitions, and the generated carriers will produce a signal current. To determine the average signal current $\overline{I}_s$, let the incident signal power of the light beam be P_s. The number of photons incident on the detector per second will then be given by P_s/hv, since hv is the energy per photon. Let us also assume that the fraction of the incident photons generating electrons is η. η is commonly referred to as the quantum efficiency of the detector. The average signal current $\overline{I}_s$, in other words, the amount of generated charge per second, will be given by

$$\overline{I}_s = \eta \frac{P_s}{hv} e = \eta \frac{eP_s}{hc} \lambda \tag{9.1}$$

The responsivity R, defined as the average generated signal current per unit incident power becomes

$$R = \frac{\overline{I}_s}{P_s} = \frac{\eta e}{hc} \lambda \tag{9.2}$$

A plot of R as a function of λ for what is referred to as an ideal quantum detector is shown in Fig. 9.1.

Two features of the graph shown in Fig. 9.1 are important. First, the linear dependence of R on λ stems directly from the quantized nature of radiation, and hence the name "quantum detector." This is

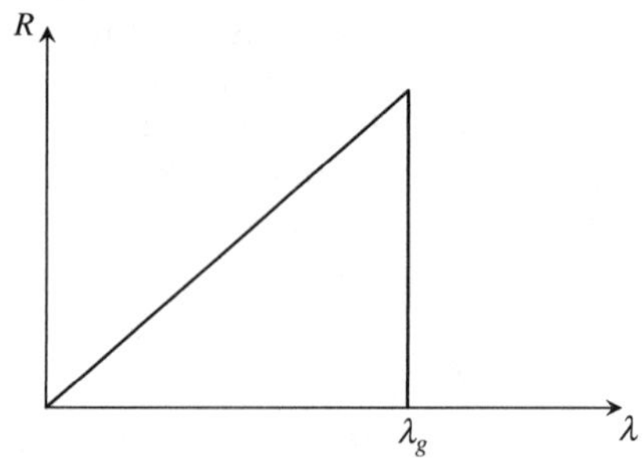

FIGURE 9.1 Variation of the responsivity R as a function of the signal wavelength λ for an ideal quantum detector.

in contrast to what are known as thermal detectors, where the responsivity is constant as a function of wavelength over a broad spectral range. Second, note that R increases linearly with wavelength up to a maximum value λ_g and then drops abruptly to zero. This is because beyond λ_g, photons are no longer energetic enough to cause interband transitions. Hence, electrons cannot be transferred to the conduction band and the signal current goes to zero.

In a practical detection system, the actual measured signal current $i_s(t)$ will further show random fluctuations over time about its average value as sketched in Fig. 9.2.

The cause of the fluctuation may be due to the randomness in the generation of charge carriers or due to random thermal motion of charges. These are referred to as shot noise and Johnson noise, respectively. We will discuss their characteristics in the following sections. Before focusing on specific types of noise, we define the spectral power density function and outline its derivation.

Consider an arbitrary time-dependent current signal $i_s(t)$. This can be expressed as a continuous summation of periodic functions according to

$$i_s(t) = \int_{-\infty}^{+\infty} d\omega I_s(\omega)\exp[i\omega t] \tag{9.3}$$

where $I_s(\omega)$ is the Fourier transform and is given by

$$I_s(\omega) = \frac{1}{2\pi}\int_{\infty}^{+\infty} dt i_s(t)\exp[-i\omega t] \tag{9.4}$$

In Eqs. (9.3) and (9.4), the imaginary number $i = \sqrt{-1}$ should be distinguished from the signal current, which is denoted as $i_s(t)$. Typically,

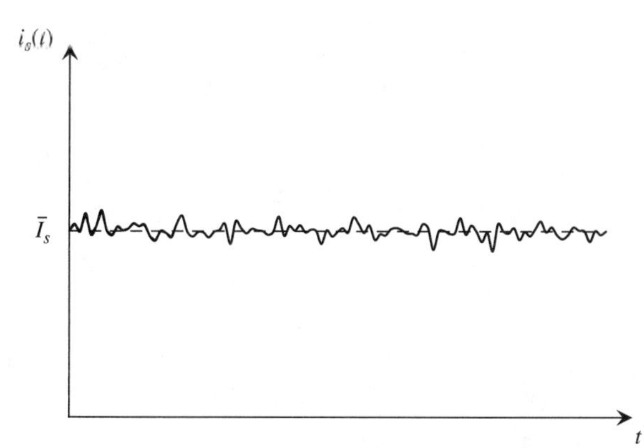

FIGURE 9.2 Actual fluctuating signal current in a real detection system.

we measure the signal over a finite time window of duration, say, T where T is known as the integration time. If $i_s(t)$ is measured over the interval $[-T/2, T/2]$ then the corresponding Fourier transform $I_{sT}(\omega)$ will be given by

$$I_{sT}(\omega) = \frac{1}{2\pi} \int_{-T/2}^{+T/2} dt\, i_s(t) \exp[-i\omega t] \tag{9.5}$$

We note further that if $i_s(t)$ is real, satisfying $i_s(t) = i_s{}^*(t)$, then

$$I_{sT}(-\omega) = \frac{1}{2\pi} \int_{-T/2}^{T/2} dt\, i_s(t) \exp[+i\omega t] = \frac{1}{2\pi} \int_{-T/2}^{T/2} dt (i_s(t) \exp[-i\omega t])^* = I^*_{sT}(\omega) \tag{9.6}$$

Now, let's calculate the average power P_{av} over the integration time T. Assuming a unit resistance, P_{av} will be given by

$$\begin{aligned}
P_{\text{av}} &= \frac{1}{T} \int_{-T/2}^{T/2} dt\, i_s^2(t) = \frac{1}{T} \int_{-T/2}^{T/2} dt\, i_s(t) \int_{-\infty}^{+\infty} d\omega I_{sT}(\omega) \exp[i\omega t] \\
&= \frac{1}{T} \int_{-\infty}^{+\infty} d\omega I_{sT}(\omega) \int_{-T/2}^{T/2} dt\, i_s(t) \exp[i\omega t] = \frac{2\pi}{T} \int_{-\infty}^{+\infty} d\omega I_{sT}(\omega) I_{sT}(-\omega) \\
&= \frac{2\pi}{T} \int_{-\infty}^{+\infty} d\omega I_{sT}(\omega)\, I^*_{sT}(\omega)
\end{aligned} \tag{9.7}$$

It can be shown (see Prob. 9.1) that $I_{sT}(\omega) I^*_{sT}(\omega)$ is an even function of frequency, yielding

$$P_{\text{av}} = \frac{4\pi}{T} \int_0^{\infty} d\omega \left| I_{sT}(\omega) \right|^2 \tag{9.8}$$

P_{av} can be further expressed as

$$P_{\text{av}} = \int_0^{\infty} d\omega \rho(\omega) \tag{9.9}$$

where $\rho(\omega)$ is known as the spectral power density. $\rho(\omega) d\omega$ corresponds to the average signal power due to the frequency components in the range ω to $\omega + d\omega$. In terms of $I_{sT}(\omega)$, $\rho(\omega)$ is given by

$$\rho(\omega) = \frac{4\pi}{T} \left| I_{sT}(\omega) \right|^2 \tag{9.10}$$

In the next sections, we will derive the expressions for $\rho(\omega)$ in the cases of shot noise[15] and Johnson noise.[16]

9.1.2 Shot Noise

Shot noise arises due to the random generation of charge carriers and is one of the most important sources of signal fluctuation. In this section, we will derive the formula for the spectral power density for shot noise.

Consider the parallel-plate capacitor sketched in Fig. 9.3. Every time an electron is released from the cathode it will be attracted to the anode and will give rise to a current impulse $i_s(t)$ such that

$$\int_{-\infty}^{+\infty} dt i_s(t) = +e \tag{9.11}$$

where e is the electron charge. To obtain the expression for $i_s(t)$, we assume that at some intermediate time, the electron is located at a distance x from the cathode as shown in Fig. 9.4.

The electron will induce charges on the two electrodes. Let the induced charge on the cathode and anode be q_c and q_a, respectively. We further assume that the voltage between the plates is nearly zero. If we crudely assume that the electron charge is uniformly distributed on a large plane of area a, we see that the electric field on either side of the electron has a discontinuity given by

$$E^+ - E^- = \frac{\sigma}{\varepsilon_0} = \frac{-e}{a\varepsilon_0} \tag{9.12}$$

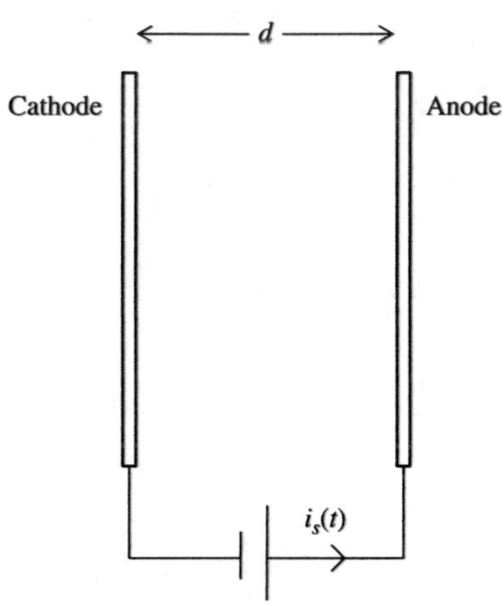

FIGURE 9.3 Parallel-plate capacitor used in the derivation of the shot noise formula.

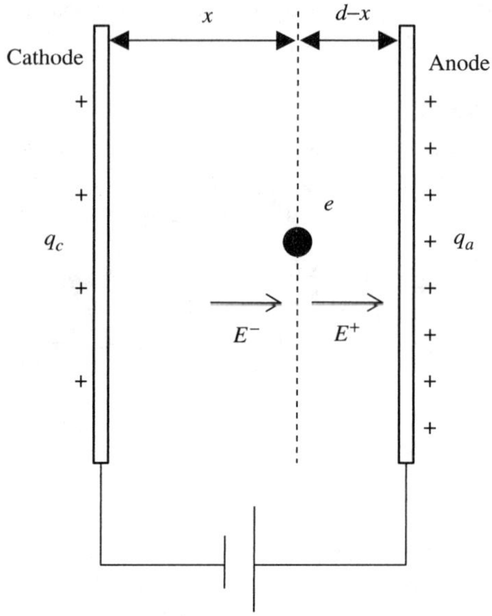

FIGURE 9.4 Induced charges on the plates of the capacitor due to a single electron which is momentarily located at a distance *x* from the cathode.

Above, ε_0 is the vacuum permittivity and σ is the surface charge density. Since the potential difference is nearly zero, we further have

$$E^- x + E^+(d - x) = 0 \tag{9.13}$$

We leave it as an exercise for the reader (Prob. 9.2) to show that the expressions for the induced charges q_a and q_c become

$$q_a = e\frac{x}{d} \tag{9.14}$$

and

$$q_c = e\left(1 - \frac{x}{d}\right) \tag{9.15}$$

The external current due to one electron thus becomes

$$i_s(t) = \frac{dq_a}{dt} = \frac{e}{d}\frac{dx}{dt} \tag{9.16}$$

The corresponding Fourier transform can be calculated from

$$I_s(\omega) = \frac{e}{2\pi d}\int_{t_c}^{t_a} dt \frac{dx}{dt} \exp[-i\omega t] \tag{9.17}$$

where t_c is the time when the electron is generated at the cathode, and t_a is the time when it is received at the anode. Let's take $t_c = 0$ with no loss of generality. We further assume that the electron moves so fast that t_a is very small, in which case the approximation $\exp(-i\omega t) \approx 1$ holds during the transit motion of the electron. Hence,

$$I_s(\omega) = \frac{e}{2\pi d}\int_0^{t_a} dx = \frac{e}{2\pi d}x(t_a) = \frac{e}{2\pi d}d = \frac{e}{2\pi} \tag{9.18}$$

What we calculated in Eq. (9.18) above is for a single electron making a transit from the cathode to the anode. Now, we consider a random train of such events as sketched in Fig. 9.5.

Suppose that a total of N electrons are received over the integration window T. The corresponding average current $\bar{I}_s$ will be given by

$$\bar{I}_s = \frac{Ne}{T} \tag{9.19}$$

The time dependence of the current $i_s(t)$ can be written as

$$i_s(t) = \sum_{j=1}^{N} i_j(t - t_j) \tag{9.20}$$

where we assume that the envelope function $i_j(t - t_j)$ describing each isolated emission event is the same, and t_j is a random variable between 0 and T. So, for example, if we have nine electrons arriving between $t = 0$ and $t = T$, then we have nine random time values in the above summation. The corresponding Fourier transform $I_{sT}(\omega)$ will be given by

$$I_{sT}(\omega) = \sum_{j=1}^{N} I_j(\omega) \tag{9.21}$$

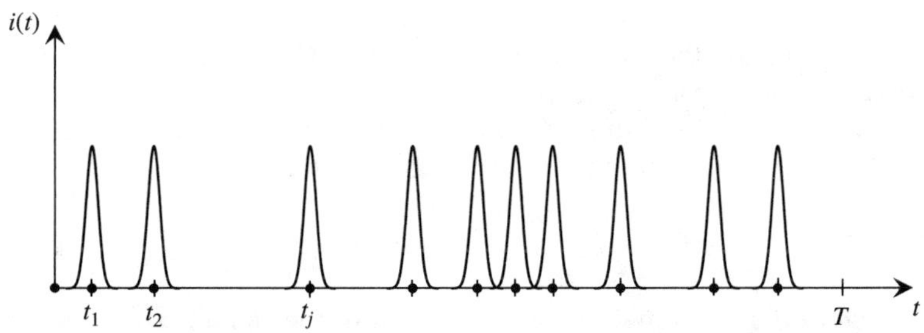

FIGURE 9.5 Random train of current pulses.

where

$$I_j(\omega) = \frac{1}{2\pi}\int_{-\infty}^{+\infty} dt\, i_j(t-t_j)\exp[-i\omega t]$$
$$= \frac{\exp[-i\omega t_j]}{2\pi}\int_{-\infty}^{+\infty} du\, i_s(u)\exp[-i\omega u] = \exp[-i\omega t_j] I_s(\omega) \tag{9.22}$$

Above, the functions $i_s(u)$ and $I_s(\omega)$ describe the time and frequency dependence of one of the current bursts and, for the case of an electron between the capacitor plates, are given by Eqs. (9.16) and (9.17), respectively. Furthermore, we made a change of variables to $u = t - t_j$ to obtain the result in Eq. (9.22). The corresponding spectral power density for the train of random current bursts becomes

$$\rho(\omega) = \frac{4\pi}{T} I_{sT}(\omega) I_{sT}^*(\omega) = \frac{4\pi}{T}\left(\sum_{j=1}^{N} I_s(\omega)\exp\left[-i\omega t_j\right]\right)\left(\sum_{k=1}^{N} I_s^*(\omega)\exp\left[i\omega t_k\right]\right)$$
$$= \frac{4\pi}{T}\left|I_s(\omega)\right|^2 \sum_{j=1}^{N}\sum_{k=1}^{N}\exp\left[-i\omega(t_j - t_k)\right]$$
$$= \frac{4\pi}{T}\left|I_s(\omega)\right|^2\left(N + \sum_{j=1}^{N}\sum_{\substack{k=1\\k\neq j}}^{N}\exp\left[-i\omega(t_j - t_k)\right]\right) \tag{9.23}$$

If N is sufficiently large, the second term, which consists of random complex numbers with unity magnitude, averages to zero, giving

$$\rho(\omega) = 4\pi\left|I_s(\omega)\right|^2 \frac{N}{T} \tag{9.24}$$

In the case of shot noise, $I_s(\omega)$ and $\overline{I}_s$ are given by Eqs. (9.18) and (9.19), resulting in

$$\rho(\omega) = 4\pi\left(\frac{e^2}{4\pi^2}\right)\left(\frac{\overline{I}_s}{e}\right) = \frac{e}{\pi}\overline{I}_s \tag{9.25}$$

Conventionally, the noise power is expressed in terms of a mean-squared noise current $< i_n^2 >$ over the frequency interval from ν to $\nu + d\nu$ (ν is measured in Hz). Since $\omega = 2\pi\nu$, the mean-squared noise current $< i_n^2 >$ for shot noise (assuming unity resistor) becomes

$$< i_n^2 > = \rho(\omega)d\omega = 2\pi\rho(\nu)d\nu = 2e\overline{I}_s d\nu \tag{9.26}$$

Note that the fractional noise component in the signal, which can be defined as $\sqrt{< i_n^2 >}/\overline{I}_s = (i_n)_{\text{rms}}/\overline{I}_s$, decreases as $1/\sqrt{\overline{I}_s}$ for shot noise,

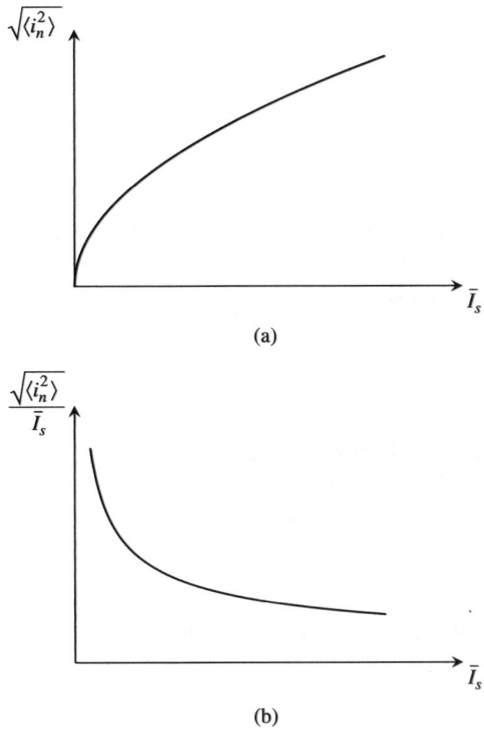

FIGURE 9.6 Variation of (*a*) the root-mean-squared noise current $\sqrt{< i_n^2 >}$, and (*b*) $\sqrt{< i_n^2 >}/\bar{I}_s$ as a function of the average signal current $\bar{I}_s$ for shot noise.

as shown in Fig. 9.6. In other words, the fluctuation due to shot noise will decrease as the average signal level goes up, a very important result that is observed in real detection systems.

9.1.3 Alternative Derivation of the Shot Noise Formula

Let us assume that the mean temporal separation between the randomly generated electrons is τ. In other words, although the electrons are randomly generated, on average, one electron is emitted over a period of τ. Then, the probability $P(1, dt)$ of seeing the generation of one electron over a short time dt will be given by

$$P(1,dt) = \frac{dt}{\tau} \tag{9.27}$$

Similarly, the probability of not seeing an electron over the same interval [i.e., $P(0,dt)$] will be $1-(dt/\tau)$. Let $P(0,t)$ be the probability of not seeing an electron emission over the time interval $[0, t]$.

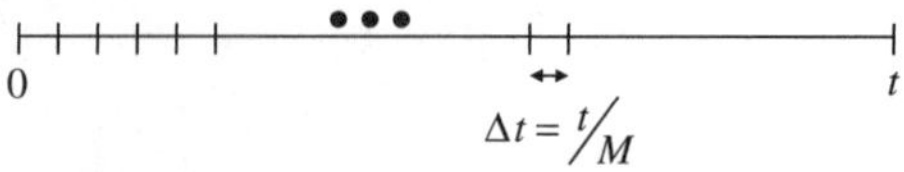

FIGURE 9.7 The time interval [0, *t*], subdivided into *M* equal segments.

Intuitively, as t increases, the probability that no electron has been emitted yet should become vanishingly small. Dividing the interval into M segments as shown in Fig. 9.7, $P(0, t)$ becomes

$$P(0,t) = \prod_{i=1}^{M} P_i \tag{9.28}$$

where P_i, which is the probability of not seeing an electron in the interval $[t_i, t_i + \Delta t]$, is given by

$$P_i = 1 - \frac{\Delta t}{\tau} = 1 - \frac{t}{M\tau} \tag{9.29}$$

Hence, as the width of each time segment tends to zero, $P(0, t)$ can be calculated from

$$P(0,t) = \lim_{M\to\infty}\left(1 - \frac{t}{M\tau}\right)^M = \exp\left[-\frac{t}{\tau}\right] \tag{9.30}$$

Hence, the probability of not seeing a generated electron will decrease exponentially with time as we expect intuitively. Now, we can calculate the probability $P(N, t)$ of receiving N electrons over the time interval $[0, t]$. To determine $P(N, t)$, we note that the probability of receiving N electrons over the time $[0, t + dt]$ is given by

$$\begin{aligned} P(N, t+dt) = {} & \text{probability of receiving } (N-1) \text{ electrons over } [0, t] \\ & \times \text{probability of receiving one more over } dt \\ & + \text{probability of receiving } N \text{ electrons over } [0, t] \\ & \times \text{probability of receiving no more over } dt \end{aligned} \tag{9.31}$$

In equation form, we have

$$P(N,t+dt) = P(N-1,t)\frac{dt}{\tau} + P(N,t)\left(1 - \frac{dt}{\tau}\right) \tag{9.32}$$

Rearranging, Eq. (9.32) becomes

$$\frac{P(N,t+dt) - P(N,t)}{dt} = \frac{P(N-1,t)}{\tau} - \frac{P(N,t)}{\tau} \tag{9.33}$$

or

$$\frac{dP(N,t)}{dt}+\frac{1}{\tau}P(N,t)=\frac{1}{\tau}P(N-1,t) \tag{9.34}$$

The relation in Eq. (9.34) is recursive and, starting with $P(0, t)$ given earlier, it can be solved to yield[1]

$$P(N,t)=\frac{\left(\frac{t}{\tau}\right)^N}{N!}\exp\left[-\frac{t}{\tau}\right] \tag{9.35}$$

The average number of electrons arriving over the interval $[0, t]$ will be $<N>=t/\tau$. Hence $P(N, t)$, which can now be expressed as $P(N,\langle N\rangle)$, becomes

$$P(N,\langle N\rangle)=\frac{<N>^N}{N!}\exp[-<N>] \tag{9.36}$$

This is the well-known Poisson probability distribution. Variation of $P(N,\langle N\rangle)$ as a function of N for different values of $<N>$ is shown in Fig. 9.8.

We leave it as an exercise for the reader (see Prob. 9.3) to show that, starting with $P(N,\langle N\rangle)$ given by Eq. (9.36), the mean value of N is indeed given by $<N>$. Furthermore, the mean value of N^2, denoted as $\langle N^2\rangle$ can be calculated from

$$\langle N^2\rangle = <N>^2 + <N> \tag{9.37}$$

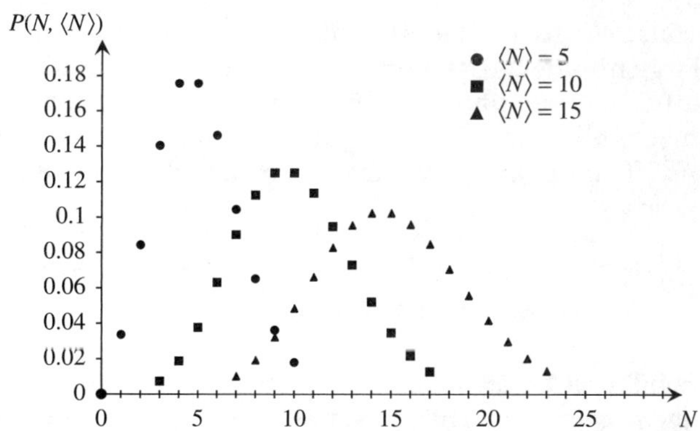

FIGURE 9.8 Variation of $P(N,\langle N\rangle)$ as a function of N for different values of $<N>$.

By using these results, the variance in N, which is defined as

$$\Delta N^2 = \langle (N - \langle N \rangle)^2 \rangle \tag{9.38}$$

is given by (see Prob. 9.3) $\langle N \rangle$. Hence, the root-mean-squared deviation becomes

$$\Delta N = \sqrt{\Delta N^2} = \sqrt{\langle N \rangle} \tag{9.39}$$

Now, let's link these results with shot noise. If, on average, $\langle N \rangle$ electrons are received over the time window $[0, T]$, the corresponding average current will be given by $\bar{I}_s = \langle N \rangle e/T$. Assuming that the charge generation process obeys the Poisson statistics discussed above, the mean-squared noise fluctuation in the current becomes

$$\langle i_n^2 \rangle = \frac{e^2}{T^2}\Delta N^2 = \frac{e^2}{T^2}\langle N \rangle = \frac{e^2}{T^2}\frac{T\bar{I}_s}{e} = \frac{e\bar{I}_s}{T} \tag{9.40}$$

The approximate bandwidth $\Delta\nu$ of the detector will be given by

$$\Delta\nu = \frac{1}{2T} \tag{9.41}$$

leading to the shot noise formula

$$\langle i_n^2 \rangle = 2e\bar{I}_s\Delta\nu \tag{9.42}$$

obtained before in Eq. (9.26).

The Poisson distribution can be derived without solving a recursive differential equation. Following Ref. 7, we focus on the time interval $[0, T]$ and want to find the probability of detecting N electrons. We divide the interval into M small segments where M is a large number. To obtain the exact formula, we will take M to infinity. In each time segment, probability of detecting an electron will be $\Delta t/\tau$, where τ is the average time between two electrons. If the average number of received electrons is $\langle N \rangle$ then

$$\frac{\Delta t}{\tau} = \frac{T/M}{T/\langle N \rangle} = \frac{\langle N \rangle}{M} \tag{9.43}$$

The probability we are after will be the probability of detecting electrons in N of the M segments. The probability of not getting an electron in the remaining $(M - N)$ segments is $[1-(\langle N \rangle/M)]^{M-N}$. Also note that we have to take into account all possible ways in which we can choose N of the M segments without paying attention to the order. This is simply given by the number of N-combinations from a set of M members, which is

$$C_N^M = \binom{M}{N} = \frac{M!}{N!(M-N)!} \tag{9.44}$$

Therefore, the overall probability will be

$$\begin{aligned} P(N,\langle N\rangle,M) &= \frac{M!}{N!(M-N)!}\left(\frac{<N>}{M}\right)^N \left(1-\frac{<N>}{M}\right)^{M-N} \\ &= \frac{<N>^N}{N!}\left(\frac{M!}{M^N(M-N)!}\right)\left(1-\frac{<N>}{M}\right)^{M-N} \end{aligned} \tag{9.45}$$

As $M \to \infty$, we see that the second and third terms in Eq. (9.45) tend to 1 and $\exp[-<N>]$, respectively, giving

$$P(N,\langle N\rangle) = \lim_{M\to\infty} P(N,\langle N\rangle,M) = \frac{<N>^N}{N!}\exp\left[-<N>\right] \tag{9.46}$$

as we obtained in Eq. (9.36).

9.1.4 Johnson Noise

The classic derivation of Johnson noise considers two equal resistors each of resistance R connected via a matched transmission line of length d and characteristic impedance $Z_0 = R$. This is shown in Fig. 9.9.

Current fluctuations due to random thermal motions of electric charge carriers in resistor R_1 will be transmitted by the allowed propagation modes of the transmission line to resistor R_2. Since resistors are equal, no back reflections will take place. Because the noise power is carried by the modes of the transmission line, we need to find the number of modes in a given frequency interval $[\nu, \nu+\Delta\nu]$. The current signal traveling along the transmission line will have a waveform given by

$$i_s(t) = I_0 \exp\left[i(\omega t - kz)\right] \tag{9.47}$$

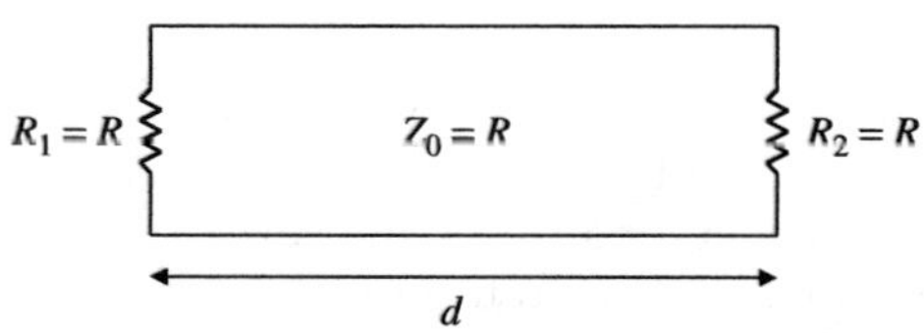

FIGURE 9.9 Matched transmission line of length d used in the analysis of Johnson noise.

Employing periodic boundary conditions, we require that the waveform remains the same after a displacement of d, giving the condition

$$\exp[-ikz] = \exp[-ik(z+d)] \tag{9.48}$$

or

$$kd = 2\pi m \tag{9.49}$$

where m is an integer. This means that the wave vectors of the allowed propagation modes are equally spaced. Since $\nu = kc/2\pi$, we further have

$$\nu_m = m\frac{c}{d} \tag{9.50}$$

where ν_m is the frequency of the mth propagating mode. Hence, the number of modes between ν and $\nu + d\nu$ will be given by $(d/c)d\nu$. Now, let the average energy per mode be $\bar{E}$. The expression for $\bar{E}$ was derived earlier in Eq. (4.151) when we discussed blackbody radiation. Since the mode propagates at the speed c, it will take a time of d/c to transfer the energy $\bar{E}$ to the resistor R_2. Hence, the noise power P_1 delivered to the resistor by one mode will be given by

$$P_1 = \bar{E}\frac{c}{d} = \frac{h\nu}{\exp\left[\frac{h\nu}{k_B T}\right] - 1}\frac{c}{d} \tag{9.51}$$

To remind ourselves, T is the absolute temperature and k_B is the Boltzmann constant. The total noise power ΔP_{noise} delivered to the resistor over the frequency interval $[\nu, \nu + \Delta\nu]$ will be

$$\begin{aligned}\Delta P_{\text{noise}} &= P_1(\text{number of modes in the frequency interval } [\nu, \nu+\Delta\nu]) \\ &= P_1\frac{d}{c}\Delta\nu = \frac{h\nu\Delta\nu}{\exp\left[\frac{h\nu}{k_B T}\right] - 1}\end{aligned} \tag{9.52}$$

If we are in the high-temperature or the low-frequency limit with $k_B T >> h\nu$, the expression for the noise power simplifies to

$$\Delta P_{\text{noise}} \cong k_B T \Delta\nu \tag{9.53}$$

Now, how can we represent this noise power as an effective current source in a detection circuit? This is schematically shown in Fig. 9.10. We know that for a sinusoidal current source of amplitude I_N, the average power delivered to a resistor R will be

$$P_{\text{av}} = \frac{1}{2}I_N^2 R \tag{9.54}$$

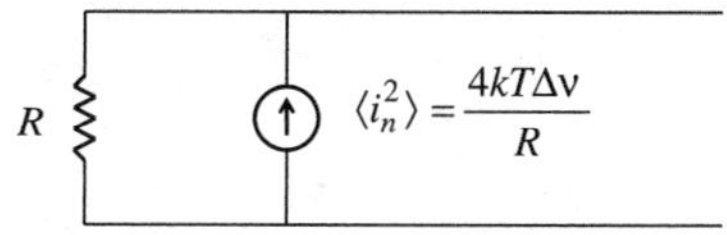

FIGURE 9.10 Equivalent circuit representation of the effective current source due to Johnson noise.

Hence, if we take the mean-squared noise current $< i_n^2 >$ as

$$< i_n^2 > = \frac{4k_B T \Delta \nu}{R} \tag{9.55}$$

we see that it will deliver the average power given in Eq. (9.53) to a matched load of resistance R connected across the terminals of the current source shown in Fig. 9.10. Note that $< i_n^2 >$ is directly proportional to the absolute temperature and can be reduced by cooling the detection system.

9.2 Types of Detectors

9.2.1 Photomultiplier Tubes

A simple schematic of a photomultiplier tube (PMT) is shown in Fig. 9.11. First developed in the early 1930s,[17] this type of detector has low noise and high signal amplification. Radiation at the frequency ν, incident on the photocathode, generates electrons via the photoelectric effect. First discovered by Hertz in 1887 and later explained by Einstein in 1905, the photoelectric effect is the emission of electrons from a surface illuminated by light.[18, 19] Typically, the electrons

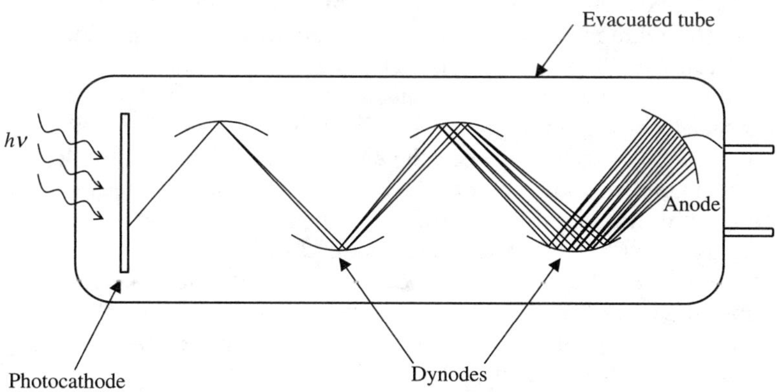

FIGURE 9.11 Schematic of a photomultiplier tube.

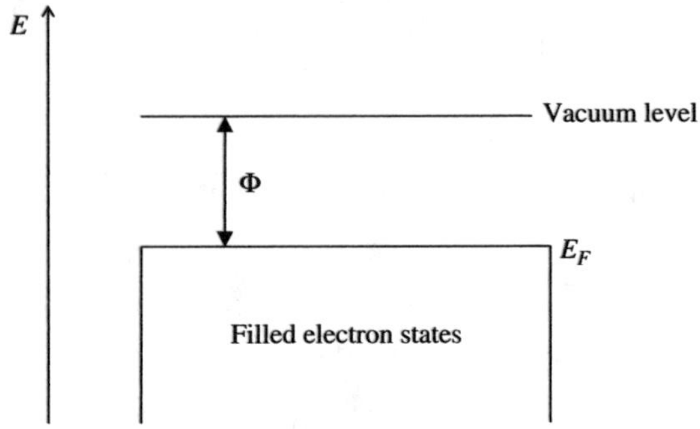

FIGURE 9.12 Energy level diagram for a photocathode. Electrons fill the levels up to the Fermi energy level E_F. An additional energy barrier of Φ (Φ = work function) must be overcome before the electrons can be emitted from the surface.

fill the lowest energy levels up to the Fermi level E_F. To escape the binding force of the host nuclei, electrons require an additional energy Φ which is known as the work function. This is schematically shown in Fig. 9.12.

When $h\nu > \Phi$, electrons can overcome the potential barrier and escape into vacuum, with a maximum kinetic energy of

$$E_{\max} = h\nu - \Phi \tag{9.56}$$

Once released, the electrons are accelerated toward plates, known as dynodes, which are held at increasing positive bias. Upon hitting a dynode, the energetic electrons cause the release of additional electrons. This process is known as secondary electron emission and is characterized by the dimensionless multiplication factor δ, which gives the average number of secondary electrons emitted per incident electron. Typical potential difference between adjacent dynodes is around 100 V. Secondary electron emission leads to the generation of a large electron current which is eventually collected at the anode. If there are N dynodes between the anode and the photo cathode, the current amplification factor is given by δ^N. To minimize the collisions between the accelerating electrons and the gas atoms, the photo cathode, dynodes, and the anode are enclosed inside an evacuated tube. Depending on the type of photocathode material used, which determines the work function Φ, the spectral sensitivity range of the PMT varies. The state-of-the-art commercial PMTs can be operated from ultraviolet to near infrared.[20]

The main sources of noise in a PMT are shot noise and Johnson noise. It can be shown (see Prob. 9.5) that the mean-squared current fluctuation due to shot noise in a PMT, including the noise contributions arising from the intermediate dynodes, is given by[3]

$$\langle i^2 \rangle_{\text{shot}} = 2G^2 e(I_c + I_d)\frac{1-\delta^{-(N+1)}}{1-\delta^{-1}}d\nu \tag{9.57}$$

Above, G is the total current amplification given by δ^N, δ is the secondary electron emission factor, N is the number of dynodes, e is the electron charge, and $d\nu$ is the spectral bandwidth of the detection system. In addition, I_c is the average cathode photocurrent generated by the incoming light and is given by

$$I_c = \frac{P_s}{h\nu}e\eta \tag{9.58}$$

where P_s is the signal power and η is the quantum efficiency introduced earlier in Eq. (9.1). I_d in Eq. (9.57) represents the dark current which exists even when there is no signal and results from the random generation of electrons due to the presence of background radiation such as cosmic rays.

Mean-squared current fluctuation due to Johnson noise is given by Eq. (9.55). The overall signal-to-noise ratio at the output, denoted as (S/N), can be calculated in terms of the mean-squared currents according to

$$\frac{S}{N} = \frac{\langle I_s^2 \rangle}{\langle i_n^2 \rangle_{\text{shot}} + \langle i_n^2 \rangle_{\text{Johnson}}} = \frac{\left(\frac{P_s e\eta}{h\nu}G\right)^2}{2G^2 e(I_c + I_d)\frac{1-\delta^{-(N+1)}}{1-\delta^{-1}}d\nu + 4k_B T \, d\nu/R} \tag{9.59}$$

9.2.2 Photoconductive Detectors

In photoconductive detectors, which can be made from intrinsic or doped semiconductors, light falling on the detector increases the carrier density and decreases the resistance. Changes in the resistance can then be used to determine the amount of light incident on the detector. A possible schematic of a photoconductive detector is shown in Fig. 9.13.

When light falls on the detector (shown as PD in Fig. 9.13), the resistance decreases and the signal voltage increases. The capacitor blocks the DC signals so that only AC variation in the light signals can be detected.

We can estimate the induced current in a photoconductive detector by using the rate equations derived in Chap. 7 in the presence of

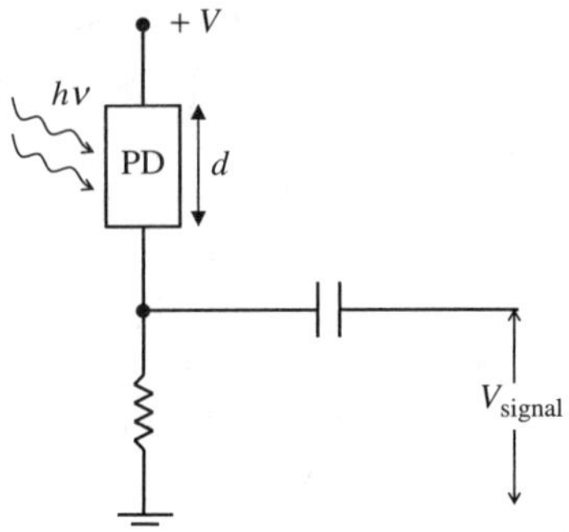

FIGURE 9.13 Schematic of a circuit using a photoconductive detector (PD) for light detection.

electron-hole recombination. Let us assume that τ_r is the average electron-hole recombination time. The steady-state number N_S of the induced charge carriers will be given by

$$N_S = G\tau_r \tag{9.60}$$

where G is the generation rate which can be calculated from

$$G = \frac{P_s}{h\nu}\eta \tag{9.61}$$

Above, P_s is the incident power, $h\nu$ is the photon energy, and η is the quantum efficiency that gives the fraction of the incident photons which results in the generation of charge carriers. As discussed before, each carrier produces a current of

$$i = \frac{eu}{d} \tag{9.62}$$

where d is the length of the semiconductor crystal and u is the average drift velocity of the charge carriers. The total generated photocurrent will be

$$I_s = G\tau_r i = \left(\frac{e\eta}{hc}\right)\left(\frac{\tau_r}{\tau_T}\right)\lambda P_s \tag{9.63}$$

In Eq. (9.63), τ_T is the transit time given by d/u. As can be seen from the above equation, the signal current I_s increases linearly with the incident power and also with the wavelength λ of the light. The linear wavelength dependence is up to a limiting wavelength λ_g, beyond which the photons are not energetic enough to generate charge carriers.

Photoconductive detectors are grouped under two categories. The first group consists of intrinsic semiconductors such as PbS and

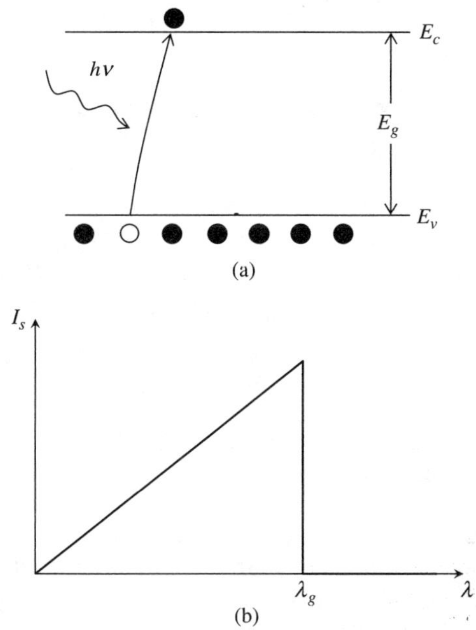

FIGURE 9.14 (*a*) Energy band diagram of a photoconductive detector made of an intrinsic semiconductor and (*b*) the corresponding variation of the generated photocurrent in an ideal intrinsic photoconductive detector as a function of the light wavelength.

PbSe, where induced transitions from the valence band to the conduction band create the charge carriers as shown in Fig. 9.14*a*. In this case, the detector is sensitive to radiation up to the wavelength λ_g given by

$$\lambda_g = \frac{hc}{E_g} \tag{9.64}$$

where E_g is the band-gap energy of the semiconductor. Figure 9.14*b* shows the variation of the photocurrent in an ideal intrinsic photoconductive detector as a function of the light wavelength. For example, E_g is approximately 0.4 eV for PbS at room temperature, giving $\lambda_g = 3.1\ \mu\text{m}$. In the second group, we have doped photoconductive detectors such as Hg-doped Ge (mercury-doped germanium) or Cu-doped Ge (copper-doped germanium). Here, incoming light causes transitions to or from the additional acceptor or donor levels introduced into the forbidden energy band. Figure 9.15*a* shows the case of Cu-doped Ge, where the acceptor level E_A due to copper is 0.04 eV above the valence band edge.

In a donor-type photoconductive detector such as silicon doped with arsenic, the donor level is very close to the conduction band.

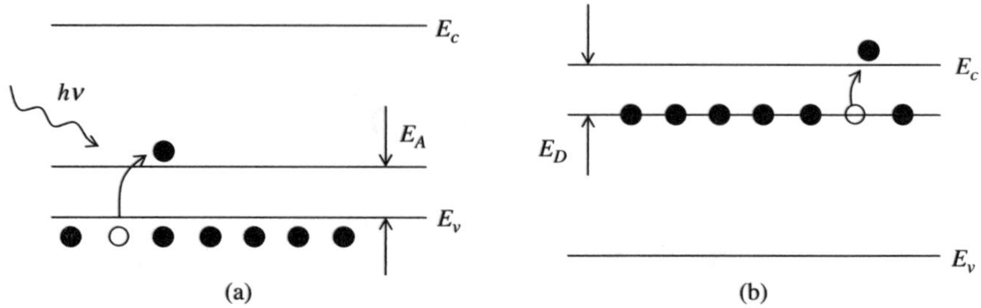

FIGURE 9.15 Energy band diagrams of photoconductive detectors made of doped (*a*) acceptor-type and (*b*) donor-type semiconductors.

In either case, the energy level spacing between the acceptor/donor level and the band edge is very small and this requires the detectors to be cooled to cryogenic temperatures in order to minimize thermal excitations. Again, coming back to the Cu-doped Ge detector, when cooled to below 10 K, this detector is sensitive up to the wavelength of 30 μm. In general, the doped photoconductive detector is sensitive to radiation up to the wavelength λ_0 given by $\lambda_0 = hc/E_A$ (acceptor type) or hc/E_D (donor type).

9.2.3 Semiconductor p-n Junction Detectors

We looked at semiconductor p-n junctions at length in Chap. 7 when we investigated the electrical properties of semiconductors. p-n junctions are also commonly used as detectors, provided that the wavelength of the incident radiation is less than the band edge wavelength $\lambda_g = hc/E_g$, where E_g is the band-gap energy. They are also referred to as semiconductor photodiodes. To understand the basic operating mechanism of a photodiode, let us consider the schematic of a p-n junction shown in Fig 9.16.

Suppose that light consisting of photons with $h\nu > E_g$ is incident on region *I* which is *p*-type (see Fig. 9.16). Absorbed photons will create electron-hole pairs. In this region, electrons are the minority carriers and will face a high probability of recombination with holes.

FIGURE 9.16 Schematic of a p-n junction used as a photodiode.

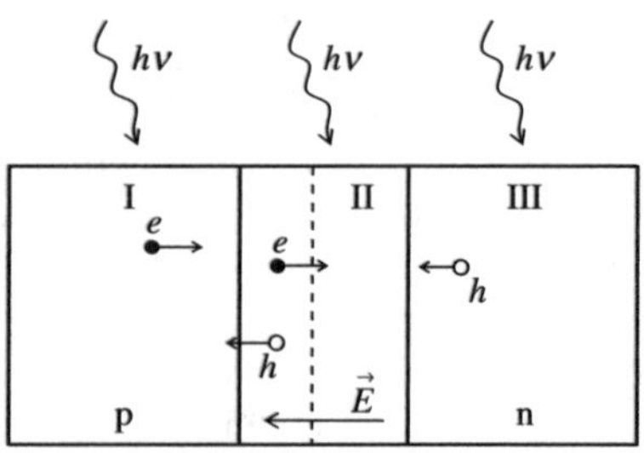

Those electrons which are close to the depletion region (with a small separation compared to the diffusion length) will reach the junction and once within the junction, will drift under the action of the built-in electric field to produce an electrical current in the external circuit. In region III the situation is quite similar, except now the generated holes within the diffusion length reach the junction and contribute to the external current. The best scenario occurs in region II where both the electron and the hole generated by the photon drift under the action of the electric field. In this case, we can show that each electron-hole pair created by the incoming photon contributes a charge of size e (e = charge of one electron) in the external circuit. To see this, let us assume that the depletion region may be modeled as a simple two-plate capacitor as was done earlier in Sec. 9.1.2 for the derivation of the shot noise (see Fig. 9.17). Let the width be d.

The currents generated by the electron and the hole are i_e and i_h, given respectively by

$$i_e = \frac{eu_e}{d} \tag{9.65}$$

and

$$i_h = \frac{eu_h}{d} \tag{9.66}$$

In Eqs. (9.65) and (9.66), u_e and u_h are the electron and hole velocities. The total charge flow q due to one generated electron-hole pair will therefore be

$$q = \int dt(i_e + i_h) = \frac{e}{d}d_2 + \frac{e}{d}d_1 = \frac{d_1 + d_2}{d}e = e \tag{9.67}$$

It is therefore desirable to maximize the generation of electron-hole pairs in region II as opposed to regions I and III. This has several

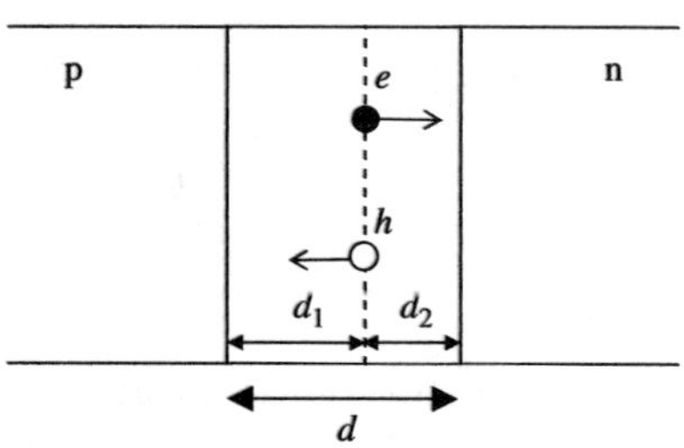

Figure 9.17 Photo generation of carriers inside the depletion region of a semiconductor p-n junction.

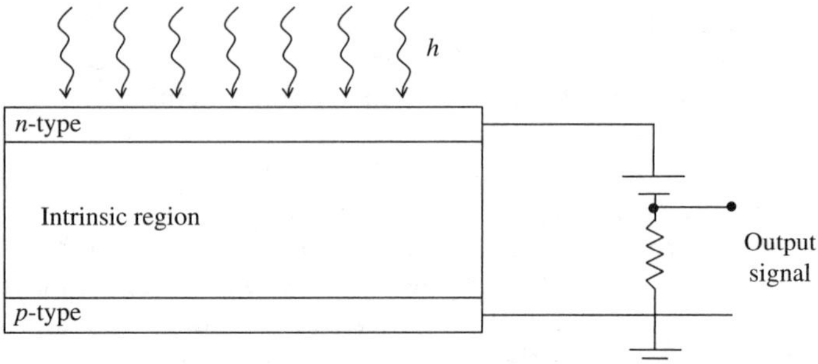

Figure 9.18 Sketch of a *p-i-n* photodiode.

advantages. First, each absorbed photon produces a charge of e since diffusion and recombination do not impede the charge flow. Second, the response time is much shorter since drift motion under the electric field is much faster than diffusion. To optimize the speed and responstivity of the detector, it is common to use what is termed a *p-i-n* diode structure which is sketched in Fig. 9.18.

In a *p-i-n* photodiode, the doped *n*- and *p*-type regions are thin with a wider intrinsic region in between, where most of the incoming light is absorbed. By making the intrinsic region large compared to *n*-type and *p*-type end segments, the generated photocurrent is maximized and the speed is increased. The overall speed of the *p-i-n* diode is limited by the RC constant of the combined detector and measurement system. The capacitance C varies linearly with detector area. Hence, high-speed *p-i-n* photodiodes typically have a small area. Response times can be in the picosecond to nanosecond range, depending on the detector area.

The spectral response of the photodiode is similar to that of other quantum detectors, and the generated photocurrent I_s can be calculated from

$$I_s = \eta \frac{e}{h\nu} P_s \tag{9.68}$$

where P_s is the signal power, η is the quantum efficiency, and ν is the light frequency. The responsivity grows linearly with the wavelength up to the bandgap wavelength λ_g and then abruptly drops for wavelengths greater than λ_g. Finally, reverse-biasing the *p-i-n* diode increases the electric field across the intrinsic layer and makes the detector faster.

A semiconductor p-n junction detector can be designed to operate under a large negative bias. Such a detector is commonly termed an avalanche photodiode. If the reverse bias applied to a p-n junction is very large, dielectric breakdown can cause permanent damage to the

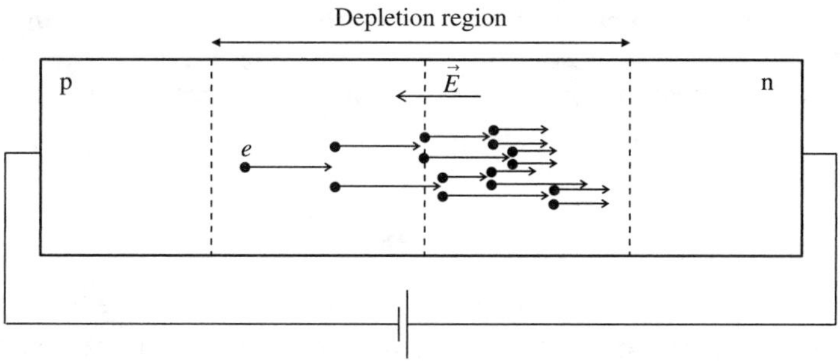

FIGURE 9.19 Charge carrier multiplication in an avalanche photodiode.

structure. However, if the reverse bias is increased in a controlled manner and kept below the breakdown voltage, the generated charge carriers can gain sufficient kinetic energy to cause impact ionization, as a result of which, additional electrons are produced. Subsequent ionization events produced by the charges can lead to an avalanche multiplication of the photocurrent as sketched in Fig. 9.19. Note that the depletion region in Fig. 9.19 was drawn with a large width for the sake of clarity. When operated in this mode, the avalanche photodiode can become very sensitive even to few photons arriving at a time, and is used in single-photon detection applications. See Refs. 3 through 5 and 7 for further discussion on different types of detectors.

9.2.4 Noise Characteristics of Semiconductor Detectors

Noise Characteristics of Photoconductive Detectors

As in the earlier cases, shot noise and thermal noise contributions should be taken into account for a photoconductive detector. The shot noise formula is further complicated by the fact that there are two types of randomness involved in the case of photoconductive detectors: randomness in the generation of the train of emitted charge carriers and randomness in the time when recombination occurs. The former was already taken into account when we derived the shot noise formula [Eq. (9.26) in Sec. 9.1.2]. The latter factor can be taken into account by developing a probability distribution $P(t)dt$ which gives the likelihood that the generated charge carrier does not undergo recombination up to a time t and then recombines between t and $t + dt$. This probability distribution is similar to what we obtained earlier for collisional broadening in Chap. 4 and can be expressed as

$$P(t)dt = \frac{1}{\tau_r}\exp\left(\frac{-t}{\tau_r}\right) \qquad (9.69)$$

where τ_r is the average recombination time. The resulting expression for the weighted shot noise formula is commonly referred to as the generation/recombination noise and is characterized by the mean-squared noise current[3] (see Prob. 9.6)

$$\left\langle i_{\text{recomb}}^2 \right\rangle = \frac{4e\bar{I}_s(\tau_r/\tau_T)\Delta\nu}{1+4\pi^2\nu^2\tau_r^2} \tag{9.70}$$

Above, $\bar{I}_s$ is the average signal current, τ_r is the average recombination time, τ_T is the average drift time, ν is the signal frequency, and $\Delta\nu$ is the detector bandwidth.

Noise Characteristics of p-n Junction Detectors

In the case of semiconductor p-n junctions, we can have several noise sources, including

1. Johnson noise
2. Noise due to background radiation
3. Dark current noise

Both the dark current as well as the background radiation may contribute to the shot noise, resulting in

$$\left\langle i_{\text{shot}}^2 \right\rangle = 2e\left[I_d + \frac{\eta e}{h\nu}(P_s + P_b)\right]\Delta\nu \tag{9.71}$$

where I_d is the dark current, P_s is the signal current, and P_b is the background radiation. The shot-noise-limited signal-to-noise ratio S/N becomes

$$\frac{S}{N} = \frac{\left\langle i_s^2 \right\rangle}{\left\langle i_{\text{shot}}^2 \right\rangle} = \frac{(\eta e P_s/h\nu)^2}{2e\left[I_d + \frac{\eta e}{h\nu}(P_s + P_b)\right]\Delta\nu} \tag{9.72}$$

In an avalanche detector, the usual photo-induced current $I_s = \eta e P_s/h\nu$ is increased by a factor, commonly known as the magnification factor M, whose value depends on the reverse-bias voltage. Since both the current as well as the number of charge carriers are magnified by the same amount, the mean-squared shot noise current is given by

$$\left\langle i_{\text{shot}}^2 \right\rangle = 2(Me)(MI_s)\Delta\nu = 2M^2 e I_s \Delta\nu \tag{9.73}$$

and the shot-noise-limited signal-to-noise ratio becomes

$$\frac{S}{N} = \frac{\left\langle i_s^2 \right\rangle}{\left\langle i_{\text{shot}}^2 \right\rangle} = \frac{M^2 I_s^2}{2M^2 e I_s \Delta\nu} = \frac{I_s}{2e\Delta\nu} = \eta\frac{P_s}{2h\nu\Delta\nu} \tag{9.74}$$

9.3 Background-Limited Infrared Detection

Let us now investigate the noise performance of detectors where thermal Johnson noise is negligible and the lowest detectable power is limited by the ambient background radiation. This analysis is applicable to cooled photoconductive detectors which operate in the infrared region for imaging applications. Suppose that the power of the incident signal is P_s. As was discussed earlier, the incoming signal will generate a photocurrent of

$$I_s = \eta_s e \frac{P_s}{h\nu_s}\left(\frac{\tau_r}{\tau_T}\right) \tag{9.75}$$

where η_s is the quantum efficiency at the signal wavelength, τ_r is the average recombination time, τ_T is the carrier transit time across the detector, and ν_s is the signal frequency. The background radiation will have a distribution of different frequency components as, for example, in the case of blackbody radiation. To simplify our analysis we will assume that the average frequency of the background radiation is ν_B. If the background radiation intensity falling on the detector is S_B (S_B here denotes the magnitude of the Poynting vector of the background radiation), then the steady-state population density N_B of radiation-induced photo-carriers will be given by

$$N_B = \eta_B \frac{S_B a}{h\nu_B V}\tau_r = \eta_B \frac{S_B \tau_r}{h\nu_B d} \tag{9.76}$$

where V is the effective volume in which the background signal is absorbed, a is the surface area of the detector, τ_r is the recombination time, and d is the thickness of the detector. This is a rather simplified picture, since in general η_B is a function of frequency. We also assume here that the radiation photons at the frequency ν_B are energetic enough to cause band-to-band or band-to-donor/acceptor transitions. The background-induced current I_B will then be given by

$$I_B = J_B a = N_B e u a \tag{9.77}$$

where J_B is the current density and u is the average drift velocity of the photo-induced carriers. The generation recombination noise current for the detector contains two current contributions and, by using Eq. (9.70), is given by

$$\langle i_n^2 \rangle = \frac{4e(I_B + I_s)(\tau_r/\tau_T)\Delta\nu}{1 + 4\pi^2\nu^2\tau_r^{\,2}} \tag{9.78}$$

Let us assume that $\nu\tau_r << 1$, and that the background-induced current I_B dominates—hence the name background-limited infrared detection. This gives

$$\langle i_n^2 \rangle = 4eI_B \frac{\tau_r}{\tau_T} \Delta\nu = \frac{4e\eta_B S_B \tau_r}{h\nu_B d} eua \frac{\tau_r}{\tau_T} \Delta\nu \tag{9.79}$$

Since $d/u = \tau_T$, we get

$$\langle i_n^2 \rangle = 4e^2 \eta_B \frac{S_B a}{h\nu_B} \left(\frac{\tau_r}{\tau_T} \right)^2 \Delta\nu \tag{9.80}$$

To determine the minimum detectable signal power in the background-limited detection regime, we set $I_s^2 = \langle i_n^2 \rangle$ to obtain

$$\eta_s^{\,2} e^2 \frac{P_s^2}{h^2 \nu_s^{\,2}} \left(\frac{\tau_r}{\tau_T} \right)^2 = 4e^2 \eta_B \frac{S_B a}{h\nu_B} \left(\frac{\tau_r}{\tau_T} \right)^2 \Delta\nu \tag{9.81}$$

By using Eq. (9.81), the minimum detectable signal power, also referred to as the noise equivalent power (NEP), is given by

$$(P_s)_{\min} = \text{NEP} = \sqrt{\frac{4\eta_B}{\eta_s^{\,2}} h \left(\frac{\nu_s^{\,2}}{\nu_B} \right) S_B} \sqrt{a\Delta\nu} \tag{9.82}$$

As can be seen from the above equation, the factors affecting NEP can be grouped into two parts. The first part contains frequency-dependent terms which are fixed for a given signal frequency, average background radiation frequency, and detector material (η_B and η_s). The second factor containing $\sqrt{a\Delta\nu}$, depends on the design of the detector and the detection circuit, both of which can be controlled during the design stage. To bring out this important dependence of NEP on $\sqrt{a\Delta\nu}$, the equation for NEP is commonly written in the form

$$\text{NEP} = \frac{\sqrt{a\Delta\nu}}{D^*} \tag{9.83}$$

where the constant of proportionality D^* is known as the detectivity. Note that D^* has the units of $\sqrt{\text{area} \cdot \text{frequency}}/\text{power}$ or cm-$\sqrt{\text{Hz}}$/W. The larger the value of D^*, the lower NEP is, and the higher the detection capability of the detector.

9.4 Noise Figure (F_N)

In a practical detection system consisting of a detector and an amplifier, additional noise is typically introduced during the amplification stage. The noise characteristics of the overall system can be described by introducing the parameter F_N, which is called the noise figure, and which can be defined as follows:

$$F_N = \frac{\langle i_n^2 \rangle_{\text{actual}}}{\langle i_n^2 \rangle_{\text{ampl}}} \tag{9.84}$$

where $\langle i_n^2 \rangle_{\text{actual}}$ and $\langle i_n^2 \rangle_{\text{ampl}}$ represent the actual mean-squared noise current (or power) at the output of the detection system and the amplified mean-squared noise current. This ratio is typically larger than 1 due to the added noise from the amplifier, and is expressed in decibels (dB) which is defined for two power levels P_1 and P_2 as

$$\text{dB} = 10\log\frac{P_2}{P_1} \tag{9.85}$$

The following example serves the purpose of combining most of the concepts introduced in this chapter for the noise analysis of optical detectors.

Example: Consider a silicon photodiode operated at the wavelength of 0.8 μm with a quantum efficiency of 30%. The incident signal power is 30 μW. The frequency bandwidth and the average gain of the detection system are 250 MHz and 10, respectively. The effective resistance of the receiver is 50 Ω. The noise figure of the circuit is 8 dB and the dark current is 12 nA. The detection system is at room temperature. Let us calculate the major sources of noise and estimate the signal-to-noise ratio for the detector. This can be done by using the following steps:

1. The responsivity can be calculated from

$$\frac{I_s}{P_s} = \eta\frac{e}{h\nu_s} = \frac{(0.3)(1.6\times10^{-19})(0.8\times10^{-6})}{(6.626\times10^{-34})(3\times10^{8})} = 0.19\frac{A}{W}$$

Hence the signal current becomes

$$I_s = 0.19(30) = 5.7\ \mu\text{A}$$

2. The corresponding signal power in the circuit is given by

$$P_{\text{cir}} = I_s^2 R = (5.7\times10^{-6})^2(50) = 1.62\times10^{-9}\ \text{W}$$

3. The shot noise power becomes

$$P_{\text{shot}} = \langle i_{\text{shot}}^2 \rangle R = 2e(I_s + I_d)\Delta\nu R$$
$$= 2(1.6\times10^{-19})(12\times10^{-9} + 5.7\times10^{-6})(250\times10^{6})(50) = 22.85\times10^{-15}\ \text{W}$$

Note that the noise contribution arising from the dark current is negligible in this case.

4. The Johnson noise power P_j is given by

$$P_j = \langle i_j^2 \rangle R = 4k_B T\Delta\nu = 4(1.38\times10^{-23})(300)(250\times10^{6}) = 4.14\times10^{-12}\text{ W}$$

5. The total noise power will be the sum of the shot noise and the Johnson noise, given by

$$P_n = P_{\text{shot}} + P_j = 22.85\times10^{-15} + 4.14\times10^{-12} = 4.16\times10^{-12}\text{ W}$$

6. After the detection system, the amplified signal power becomes $10\times1.62\times10^{-9} = 16.2\times10^{-9}$ W. To find the effective noise power after the amplifier, we note that 8 dB $= 10\log P_2/P_1$, giving $P_2/P_1 = 6.3$. In other words, the amplified noise power becomes $10\times6.3\times4.16\times10^{-12} = 262.5\times10^{-12}$ W.

7. Finally, the signal-to-noise ratio S/N for the detector becomes

$$\frac{S}{N} = \frac{16.2\times10^{-9}}{262.5\times10^{-12}} = 62$$

The signal-to-noise ratio in this case can be improved by reducing the bandwidth of the detection system or by cooling the detector. If we can cool the detector to go to the shot noise limit, the signal-to-noise ratio becomes

$$\frac{S}{N} = \frac{16.2\times10^{-9}}{(22.85\times10^{-15})(6.3)(10)} = 11.25\times10^{3}$$

9.5 Homodyne and Heterodyne Detection

So far, we have only discussed direct or so-called homodyne detection schemes with different detectors. The main attribute of homodyne detection is that it is sensitive to changes in amplitude. However, it cannot detect phase variation. To detect phase modulated (or frequency modulated) signals, we resort to what is referred to as heterodyne detection, where the incoming field is mixed with a local oscillator to generate the signal on the detector. The heterodyne detection scheme is schematically shown in Fig. 9.20.

Ignoring the space dependence of the fields, we take the signal and local electric fields after the mixing beamsplitter as

$$\vec{E}_s(t) = \vec{E}_s \exp[i\omega_s t] \tag{9.86}$$

and

$$\vec{E}_o(t) = \vec{E}_o \exp[i\omega_o t] \tag{9.87}$$

respectively. Above, ω_s is the signal frequency and ω_o is the local oscillator frequency. The electric field on the detector will be the superposition of these two fields, given by

$$\vec{E}_T = \vec{E}_s \exp[i\omega_s t] + \vec{E}_o \exp[i\omega_o t] \tag{9.88}$$

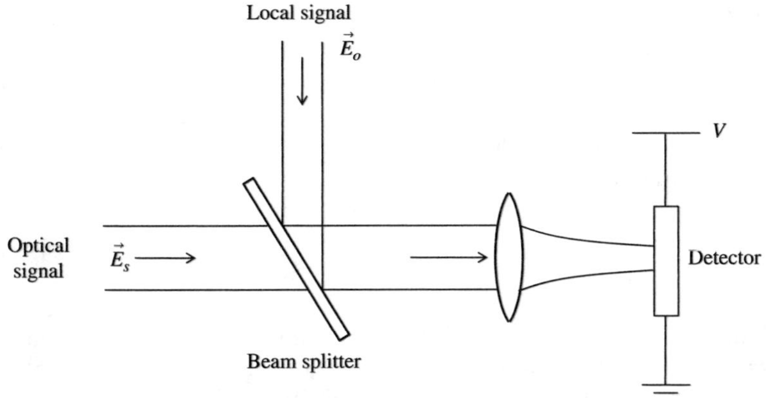

FIGURE 9.20 Heterodyne detection scheme.

The cycle-averaged power on the detector becomes

$$P_{av} = a\frac{\vec{E}_T \cdot \vec{E}_T^*}{2Z_0} = \frac{a}{2Z_0}(\vec{E}_s \exp[i\omega_s t] + \vec{E}_o \exp[i\omega_o t]) \cdot (\vec{E}_s^* \exp[-i\omega_s t]$$
$$+ \vec{E}_o^* \exp[-i\omega_o t])$$
$$= \frac{a}{2Z_0}\left[\vec{E}_s \cdot \vec{E}_s^* + \vec{E}_o \cdot \vec{E}_o^* + 2\mathrm{Re}[\vec{E}_s \cdot \vec{E}_o^* \exp[i(\omega_s - \omega_o)t]]\right] \tag{9.89}$$

where a is the detector area and $Z_0 = 377\ \Omega$ is the impedance of free space. Taking $\vec{E}_s$ and $\vec{E}_o$ real and parallel, the expression for P_{av} becomes

$$P_{av} = \frac{a}{2Z_0}\left[E_s^2 + E_o^2 + 2E_s E_o \cos(\omega_s - \omega_o)t\right] \tag{9.90}$$

The current detected on the detector can be calculated from (assuming ω_s and ω_o are close)

$$I(t) = \frac{e\eta P_{av}}{\hbar\omega_s} = \left(\frac{e\eta}{\hbar\omega_s}\right)\left(\frac{a}{2Z_0}\right)\left(E_s^2 + E_o^2 + 2E_s E_o \cos(\omega_s - \omega_o)t\right) \tag{9.91}$$

Denoting the signal and oscillator currents as I_s and I_o, we see that

$$I_s = \frac{e\eta}{\hbar\omega}\frac{a}{2Z_0}E_s^2 \tag{9.92}$$

and

$$I_o = \frac{e\eta}{\hbar\omega}\frac{a}{2Z_0}E_o^2 \tag{9.93}$$

Hence, $I(t)$ becomes

$$I(t) = I_s + I_o + 2\sqrt{I_s I_o}\cos(\omega_s - \omega_o)t \tag{9.94}$$

Note that if the incoming field is frequency modulated, the third term, which depends on the difference of the signal and oscillator frequencies, will vary in time. Hence, the desired signal information is contained in this third, mixed term. To determine the signal-to-noise ratio, we assume that the local field is much larger than the signal field. The shot-noise power is then given by

$$\left\langle i_{\text{shot}}^2 \right\rangle = 2eI_o\Delta\nu \tag{9.95}$$

The mean-squared signal current will be $(1/2)(4)(I_s I_o) = 2I_s I_o$, giving

$$\frac{S}{N} = \frac{2I_s I_o}{2eI_o\Delta\nu} = \frac{I_s}{e\Delta\nu} = \frac{\eta P_s}{h\nu\Delta\nu} \tag{9.96}$$

which turns out to be twice as large as the signal-to-noise ratio for direct detection. Also, note that the signal intensity which is related to $\sqrt{I_s I_o}$ can be increased by increasing the intensity of the local oscillator field. This can enable the detection of very weak signals.

Problems

9.1 Show that $I_{sT}^*(\omega)I_{sT}(\omega)$ is an even function of frequency.

9.2 Show that when a single electron is momentarily located at a distance x from the cathode as shown in Fig. 9.4, the expressions for the induced charges q_a and q_c are given by Eqs. (9.14) and (9.15).

9.3 Calculate the mean values of N and N^2 by using the Poisson distribution given by Eq. (9.36). Also calculate the variance defined by Eq. (9.38).

9.4 Show that the high-temperature or the low-frequency limit of the Johnson noise formula is given by Eq. (9.53).

9.5 Show that the mean-squared current fluctuation due to shot noise in a PMT is given by Eq. (9.57).

9.6 Show that the mean-squared current fluctuation due to generation/recombination noise in a photoconductive detector is given by Eq. (9.70).

9.7 A silicon *p-i-n* photodiode is operated in the shot-noise limit (no other noise sources are present) to detect light at 800 nm. The quantum efficiency of the detector is 40% and the detection bandwidth is 100 kHz. What is the minimum detectable signal power?

9.8 A photoconductive detector with a quantum efficiency of 50% is used to detect radiation at 2 μm. The incident power is 4 mW. The parameters of the detector are as follows: the electron-hole recombination time = 5 ns, electron drift time = 30 ns, detector bandwidth = 10 kHz, load resistance = 100 Ω, temperature = 30°C.

(*a*) Calculate the signal current.
(*b*) Calculate the powers due to major sources of noise.

References

1. H. A. Haus, *Electromagnetic Noise and Quantum Optical Measurements* (Springer, Berlin, 2000).
2. H. A. Haus, *Waves and Fields in Optoelectronics* (Prentice-Hall, Englewood Cliffs, N.J., 1984).
3. A. Yariv, *Optical Electronics in Modern Communications*, 5th ed. (Oxford University Press, New York, 1997).
4. C. R. Pollock, *Fundamentals of Optoelectronics* (Irwin, Chicago, 1995).
5. C. C. Davis, *Lasers and Electro-Optics* (Cambridge University Press, Cambridge, 1996).
6. K. F. Brennan, *The Physics of Semiconductors* (Cambridge University Press, Cambridge, 1999).
7. B. E. A. Saleh and M. C. Teich, *Fundamentals of Photonics*, 2d ed. (John Wiley & Sons, Inc., Hoboken, N.J., 2007).
8. H. A. Haus, M. Margalit, and C. X. Yu, "Quantum noise of a mode-locked laser," *Journal of the Optical Society of America B-Optical Physics*, **17**(7), 1240–1256 (2000).
9. H. A. Haus and A. Mecozzi, "Noise of mode-locked lasers," *IEEE Journal of Quantum Electronics*, **29**(3), 983–996 (1993).
10. L. A. Jiang, M. E. Grein, H. A. Haus, and E. P. Ippen, "Noise of mode-locked semiconductor lasers," *IEEE Journal of Selected Topics in Quantum Electronics*, 7(2), 159–167 (2001).
11. R. Paschotta, "Noise of mode-locked lasers (Part I): Numerical model," *Applied Physics B-Lasers and Optics*, **79**(2), 153–162 (2004).
12. R. Paschotta, "Noise of mode-locked lasers (Part II): Timing jitter and other fluctuations," *Applied Physics B-Lasers and Optics*, **79**(2), 163–173 (2004).
13. F. Quinlan, S. Gee, S. Ozharar, and P. J. Delfyett, "Ultralow-jitter and amplitude-noise semiconductor-based actively mode-locked laser," *Optics Letters*, **31**(19), 2870–2872 (2006).
14. T. R. Schibli, J. Kim, O. Kuzucu, J. T. Gopinath, S. N. Tandon, G. S. Petrich, L. A. Kolodziejski, J. G. Fujimoto, E. P. Ippen, and F. X. Kaertner, "Attosecond active synchronization of passively mode-locked lasers by balanced cross correlation," *Optics Letters*, **28**(11), 947–949 (2003).
15. W. Schottky, "Regarding spontaneous current fluctuation in different electricity conductors," *Annalen Der Physik*, **57**(23), 541–567 (1918).
16. J. B. Johnson, "Thermal agitation of electricity in conductors," *Physical Review*, **32**(1), 97–109 (1928).

17. H. Iams and B. Salzberg, "The secondary emission phototube," *Proceedings of the Institute of Radio Engineers*, **23**(1), 55–64 (1935).
18. K. Krane, *Modern Physics*, 2d ed. (John Wiley & Sons, Inc., New York, 1996).
19. A. Einstein, "Generation and conversion of light with regard to a heuristic point of view," *Annalen Der Physik*, **17**(6), 132–148 (1905).
20. Hamamatsu, *Handbook of Photomultiplier Tubes: Basics and Applications*, 3d ed. (© 2006 Hamamatsu Photonics K. K., Electron Tube Division).

CHAPTER 10

Dielectric Waveguides and Optical Fibers

In this chapter, we discuss the electromagnetics of translationally invariant, dielectric waveguide structures. Starting with Maxwell's equations, we derive the scalar wave equation satisfied by the longitudinal components of the electric and magnetic fields. Once the longitudinal components are determined for a given set of boundary conditions, the transverse fields can then be calculated as well by following a straightforward set of steps. We show how this can be done for cross sections with rectangular or cylindrical symmetry. As applications, we employ this method to analyze the rectangular dielectric slab waveguide and the circular step-index optical fiber. Spatial profiles of the guided modes are derived. Practical design aspects of step-index optical fibers are also discussed. We end the chapter by giving a concise overview of emerging novel fiber technologies, including double-clad fibers and photonic crystal fibers. References 1 to 9 provide further discussion on the theory and applications of optical dielectric waveguides and optical fibers. For early work on the development of optical fibers, see Refs. 7, 10, and 13.

10.1 General Formulation of Optical Waveguides

10.1.1 Calculation of the Transverse Field Components

We will consider the propagation of an electromagnetic wave in a dielectric medium whose optical properties are uniform along the direction of propagation, and which has a nonuniform refractive index distribution in the plane perpendicular to the direction of propagation as sketched in Fig. 10.1. This plane will be referred to as the transverse plane. We can use Faraday's law to derive the general relations between the transverse and axial components of the electric and magnetic fields for the propagating electromagnetic mode. Let us take the direction of propagation as the z-axis. Hence, the dielectric

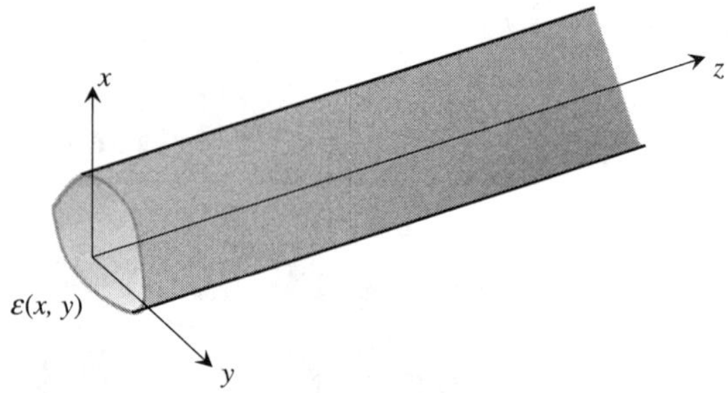

FIGURE 10.1 Sketch of an optical dielectric waveguide structure with translational invariance along the direction of propagation.

permittivity is only a function of the coordinates in the transverse plane. As an example, the dielectric permittivity function becomes $\varepsilon = \varepsilon(x,y)$ and $\varepsilon = \varepsilon(r,\theta)$ in Cartesian (x,y,z) and cylindrical (r,θ,ϕ) coordinates, respectively. It turns out that this geometry will be suitable in the discussion of the propagation modes of rectangular and circular waveguides discussed in this chapter.

We first derive the general wave equations for this guiding geometry. It turns out that knowledge of the axial field components E_z and H_z is sufficient to determine the remaining transverse field components. We will also show that the axial components satisfy the scalar wave equation. In this analysis, we will assume that the electric and magnetic fields, consisting of axial and transverse components, can be expressed in the form

$$\vec{E}(\vec{r},t) = [\vec{E}_t(\vec{r},t) + \hat{z}E_z(\vec{r},t)] \tag{10.1}$$

and

$$\vec{H}(\vec{r},t) = [\vec{H}_t(\vec{r},t) + \hat{z}H_z(\vec{r},t)] \tag{10.2}$$

For the sake of simplicity, we drop the explicit spatial and time dependence of the fields in the following analysis. Since the medium properties are invariant along the z-axis, we will assume that all of the field components have $\exp[i(\omega t - \beta z)]$ dependence for propagation along the positive z-direction. Here, ω and β are the angular frequency and the propagation parameter, respectively. From Maxwell's equations, we have

$$\nabla \times \vec{E} = -\mu \frac{\partial \vec{H}}{\partial t} \tag{10.3}$$

and

$$\nabla \times \vec{H} = \varepsilon \frac{\partial \vec{E}}{\partial t} \tag{10.4}$$

where the magnetic permeability μ is constant but the dielectric function ε depends on the transverse coordinates. We will further separate the del operator ∇ into its transverse and axial components according to

$$\nabla = \nabla_t + \hat{z}\frac{\partial}{\partial z} \tag{10.5}$$

By using Eqs. (10.1) to (10.5), Maxwell's curl equations become

$$\left(\nabla_t + \hat{z}\frac{\partial}{\partial z}\right)\times(\vec{E}_t + \hat{z}E_z) = -i\omega\mu(\vec{H}_t + \hat{z}H_z) \tag{10.6}$$

and

$$\left(\nabla_t + \hat{z}\frac{\partial}{\partial z}\right)\times(\vec{H}_t + \hat{z}H_z) = i\omega\varepsilon(\vec{E}_t + \hat{z}E_z) \tag{10.7}$$

We can separate Eqs. (10.6) and (10.7) into four equations, two for the transverse and two for the axial components. It is easy to see which terms become transverse or axial by using the Cartesian expression for ∇_t given by

$$\nabla_t = \hat{x}\frac{\partial}{\partial x} + \hat{y}\frac{\partial}{\partial y} \tag{10.8}$$

Note that in this case, the terms containing $\nabla_t \times (\hat{z}E_z)$ and $\hat{z}(\partial/\partial z)\times\vec{E}_t$ are transverse, whereas the term $\nabla_t \times \vec{E}_t$ is axial. Also noting that $\hat{z}(\partial/\partial z)\times\vec{E}_t = \hat{z}\times(\partial\vec{E}_t/\partial z)$, we can separate Eqs. (10.6) and (10.7) into the following four equations:

Transverse components:

$$\nabla_t \times (\hat{z}E_z) + \hat{z}\times\left(\frac{\partial\vec{E}_t}{\partial z}\right) = -i\omega\mu\vec{H}_t \tag{10.9}$$

$$\nabla_t \times (\hat{z}H_z) + \hat{z}\times\left(\frac{\partial\vec{H}_t}{\partial z}\right) = i\omega\varepsilon\vec{E}_t \tag{10.10}$$

Axial (longitudinal) components:

$$\nabla_t \times \vec{E}_t = -i\omega\mu(\hat{z}H_z) \tag{10.11}$$

$$\nabla_t \times \vec{H}_t = i\omega\varepsilon(\hat{z}E_z) \tag{10.12}$$

Now, we go back to Eq. (10.9) and operate on each term with $\hat{z}(\partial/\partial z)$ to get

$$\left(\hat{z}\frac{\partial}{\partial z}\right)\times[\nabla_t \times (\hat{z}E_z)] + \left(\hat{z}\frac{\partial}{\partial z}\right)\times\left(\hat{z}\times\frac{\partial\vec{E}_t}{\partial z}\right) = -i\omega\mu\left(\hat{z}\frac{\partial}{\partial z}\right)\times(\vec{H}_t) \tag{10.13}$$

By applying the well-known vector identity $\vec{A}\times(\vec{B}\times\vec{C}) = \vec{B}(\vec{A}\cdot\vec{C}) - \vec{C}(\vec{A}\cdot\vec{B})$ to the vector-valued operators, we see that

$$\left(\hat{z}\frac{\partial}{\partial z}\right)\times[\nabla_t \times (\hat{z}E_z)] = \nabla_t\frac{\partial E_z}{\partial z} \tag{10.14}$$

and

$$\left(\hat{z}\frac{\partial}{\partial z}\right)\times\left(\hat{z}\times\frac{\partial \vec{E}_t}{\partial z}\right)=-\frac{\partial^2 \vec{E}_t}{\partial z^2} \tag{10.15}$$

Hence, we have, by using Eq. (10.10)

$$\nabla_t\left(\frac{\partial E_z}{\partial z}\right)-\frac{\partial^2 \vec{E}_t}{\partial z^2}=-i\omega\mu\hat{z}\times\left(\frac{\partial \vec{H}_t}{\partial z}\right)=-i\omega\mu\left[-\nabla_t\times(\hat{z}H_z)+i\omega\varepsilon\vec{E}_t\right] \tag{10.16}$$

Since $\partial^2\vec{E}_t/\partial z^2=-\beta^2\vec{E}_t$, we can rearrange Eq. (10.16) to obtain

$$\vec{E}_t=\frac{1}{\omega^2\mu\varepsilon-\beta^2}\left[\nabla_t\left(\frac{\partial E_z}{\partial z}\right)-i\omega\mu\nabla_t\times(\hat{z}H_z)\right] \tag{10.17}$$

This is the first key result we obtain. We leave it as an exercise for the reader (see Prob. 10.1) to repeat the above analysis for $\vec{H}_t$, in which case, we obtain

$$\vec{H}_t=\frac{1}{\omega^2\mu\varepsilon-\beta^2}\left[\nabla_t\left(\frac{\partial H_z}{\partial z}\right)+i\omega\varepsilon\nabla_t\times(\hat{z}E_z)\right] \tag{10.18}$$

The most important result of the foregoing analysis is that we can determine the transverse components $\vec{E}_t$ and $\vec{H}_t$ of the electromagnetic mode from a knowledge of the axial fields H_z and E_z alone. Now, the question is, how can we determine the axial components H_z and E_z? To do this, we need to derive the wave equation satisfied by H_z and E_z. To simplify our analysis, we consider the rectangular and cylindrical geometries separately.

10.1.2 Derivation of the Wave Equation for the Field Components in Rectangular Coordinates

Equations (10.17) and (10.18) are valid for any coordinate system chosen to describe the transverse refractive index distribution. To obtain the specific forms applicable in rectangular coordinates, we note that in this case, the transverse component of the del operator becomes

$$\nabla_t=\hat{x}\frac{\partial}{\partial x}+\hat{y}\frac{\partial}{\partial y} \tag{10.19}$$

so that

$$\nabla_t\left(\frac{\partial E_z}{\partial z}\right)=\left(\hat{x}\frac{\partial}{\partial x}+\hat{y}\frac{\partial}{\partial y}\right)(-i\beta E_z)=\hat{x}(-i\beta)\frac{\partial E_z}{\partial x}+\hat{y}(-i\beta)\frac{\partial E_z}{\partial y} \tag{10.20}$$

Similarly, it can be shown that (see Prob. 10.2)

$$\nabla_t \times (\hat{z}H_z) = -\hat{y}\frac{\partial H_z}{\partial x} + \hat{x}\frac{\partial H_z}{\partial y} \tag{10.21}$$

Since

$$\vec{E}_t = \hat{x}E_x + \hat{y}E_y \tag{10.22}$$

in Cartesian coordinates, we have, from Eq. (10.17)

$$(\hat{x}E_x + \hat{y}E_y) = \frac{1}{\omega^2\mu\varepsilon - \beta^2}\left[\hat{x}(-i\beta)\frac{\partial E_z}{\partial x} + \hat{y}(-i\beta)\frac{\partial E_z}{\partial y} - i\omega\mu\left(-\hat{y}\frac{\partial H_z}{\partial x} + \hat{x}\frac{\partial H_z}{\partial y}\right)\right] \tag{10.23}$$

Hence, separating the x- and y-components of Eq. (10.23) and simplifying, we obtain

$$E_x = \frac{-i}{\omega^2\mu\varepsilon - \beta^2}\left[\beta\frac{\partial E_z}{\partial x} + \omega\mu\frac{\partial H_z}{\partial y}\right] \tag{10.24}$$

and

$$E_y = \frac{-i}{\omega^2\mu\varepsilon - \beta^2}\left[\beta\frac{\partial E_z}{\partial y} - \omega\mu\frac{\partial H_z}{\partial x}\right] \tag{10.25}$$

Again, we leave it as an exercise for the reader (see Prob. 10.3) to show that H_x and H_y are given by

$$H_x = \frac{-i}{\omega^2\mu\varepsilon - \beta^2}\left[\beta\frac{\partial H_z}{\partial x} - \omega\varepsilon\frac{\partial E_z}{\partial y}\right] \tag{10.26}$$

and

$$H_y = \frac{-i}{\omega^2\mu\varepsilon - \beta^2}\left[\beta\frac{\partial H_z}{\partial y} + \omega\varepsilon\frac{\partial E_z}{\partial x}\right] \tag{10.27}$$

Now, we go back to Eq. (10.12), express it in rectangular coordinates, and obtain

$$\left(\hat{x}\frac{\partial}{\partial x} + \hat{y}\frac{\partial}{\partial y}\right) \times (H_x\hat{x} + H_y\hat{y}) = \hat{z}i\omega\varepsilon E_z \tag{10.28}$$

or

$$\frac{\partial H_y}{\partial x} - \frac{\partial H_x}{\partial y} = i\omega\varepsilon E_z \tag{10.29}$$

Substituting Eqs. (10.26) and (10.27) into Eq. (10.29) and simplifying, we get

$$\frac{\partial^2 E_z}{\partial x^2} + \frac{\partial^2 E_z}{\partial y^2} + (\omega^2\mu\varepsilon - \beta^2)E_z = 0 \tag{10.30}$$

So, as we noted earlier, the longitudinal component of the electric field satisfies the simplified scalar wave equation given in Eq. (10.30). Carrying out a similar analysis, it can be shown that (see Prob. 10.4), the longitudinal component H_z of the magnetic field further satisfies a similar equation given by

$$\frac{\partial^2 H_z}{\partial x^2} + \frac{\partial^2 H_z}{\partial y^2} + (\omega^2 \mu\varepsilon - \beta^2) H_z = 0 \tag{10.31}$$

In practice, we first solve for E_z and H_z by using Eqs. (10.30) and (10.31) and the specific boundary conditions. Then, Eqs. (10.24) to (10.27) are used to compute the transverse field components. We will consider the rectangular dielectric slab waveguide as an example in the next section. Beforehand, we also outline the derivation of the equations satisfied by the field components in cylindrical coordinates.

10.1.3 Derivation of the Wave Equation for the Field Components in Cylindrical Coordinates

In cylindrical coordinates, the transverse component ∇_t of the del operator has r- and θ-components, given by

$$\nabla_t = \hat{r}\frac{\partial}{\partial r} + \hat{\theta}\frac{1}{r}\frac{\partial}{\partial \theta} \tag{10.32}$$

To derive the expressions for the transverse electric-field components, we note that

$$\nabla_t\left(\frac{\partial E_z}{\partial z}\right) = \left(\hat{r}\frac{\partial}{\partial r} + \hat{\theta}\frac{1}{r}\frac{\partial}{\partial \theta}\right)(-i\beta E_z) = \hat{r}(-i\beta)\frac{\partial E_z}{\partial r} + \hat{\theta}(-i\beta)\frac{1}{r}\frac{\partial E_z}{\partial \theta} \tag{10.33}$$

and

$$\nabla_t \times (\hat{z}H_z) = \left(\hat{r}\frac{\partial}{\partial r} + \hat{\theta}\frac{1}{r}\frac{\partial}{\partial \theta}\right) \times (\hat{z}H_z) = -\hat{\theta}\frac{\partial H_z}{\partial r} + \hat{r}\frac{1}{r}\frac{\partial H_z}{\partial \theta} \tag{10.34}$$

Since the transverse electric field can be written in the form

$$\vec{E}_t = \hat{r}E_r + \hat{\theta}E_\theta \tag{10.35}$$

we get, from Eq. (10.17),

$$(\hat{r}E_r + \hat{\theta}E_\theta) = \frac{1}{\omega^2\mu\varepsilon - \beta^2}\left[\hat{r}(-i\beta)\frac{\partial E_z}{\partial r} + \hat{\theta}(-i\beta)\frac{1}{r}\frac{\partial E_z}{\partial \theta} - i\omega\mu\left(-\hat{\theta}\frac{\partial H_z}{\partial r} + \hat{r}\frac{1}{r}\frac{\partial H_z}{\partial \theta}\right)\right] \tag{10.36}$$

Separating into the r- and θ-components finally yields

$$E_r = \frac{-i}{\omega^2\mu\varepsilon - \beta^2}\left[\beta\frac{\partial E_z}{\partial r} + \omega\mu\frac{1}{r}\frac{\partial H_z}{\partial \theta}\right] \tag{10.37}$$

and

$$E_\theta = \frac{-i}{\omega^2\mu\varepsilon - \beta^2}\left[\beta\frac{1}{r}\frac{\partial E_z}{\partial \theta} - \omega\mu\frac{\partial H_z}{\partial r}\right] \tag{10.38}$$

A similar analysis can be done for the transverse components of the magnetic field (see Prob. 10.5) giving

$$H_r = \frac{-i}{\omega^2\mu\varepsilon - \beta^2}\left[\beta\frac{\partial H_z}{\partial r} - \omega\varepsilon\frac{1}{r}\frac{\partial E_z}{\partial \theta}\right] \tag{10.39}$$

and

$$H_\theta = \frac{-i}{\omega^2\mu\varepsilon - \beta^2}\left[\beta\frac{1}{r}\frac{\partial H_z}{\partial \theta} + \omega\varepsilon\frac{\partial E_z}{\partial r}\right] \tag{10.40}$$

We can start with Eqs. (10.11) and (10.12) and use the curl operator in cylindrical coordinates to derive the wave equation satisfied by E_z and H_z. This is a very good exercise in applying one's knowledge of vector calculus in orthogonal curvilinear coordinates. We leave it as an exercise for the reader (see Probs. 10.6 and 10.7) to show that E_z and H_z satisfy the scalar wave equations

$$\frac{\partial^2 E_z}{\partial r^2} + \frac{1}{r}\frac{\partial E_z}{\partial r} + \frac{1}{r^2}\frac{\partial^2 E_z}{\partial \theta^2} + (\omega^2\mu\varepsilon - \beta^2)E_z = 0 \tag{10.41}$$

and

$$\frac{\partial^2 H_z}{\partial r^2} + \frac{1}{r}\frac{\partial H_z}{\partial r} + \frac{1}{r^2}\frac{\partial^2 H_z}{\partial \theta^2} + (\omega^2\mu\varepsilon - \beta^2)H_z = 0 \tag{10.42}$$

10.2 Guided Modes of a Rectangular Dielectric Slab Waveguide

As an application of the general theory discussed in Sec. 10.1, we consider the rectangular dielectric slab waveguide shown in Fig. 10.2 and discuss how the propagation constants and the field distributions of the propagating eigenmodes are calculated. TE and TM cases are treated separately.

We will only consider the guided modes of the slab waveguide. Two types exist, designated as TE (transverse electric) and TM (transverse magnetic) modes. For both cases, since the structure is uniform along the y-direction, we can take $\partial/\partial y = 0$.

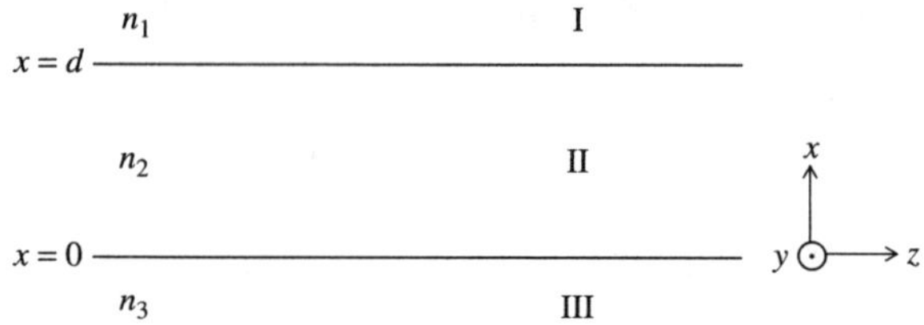

FIGURE 10.2 Rectangular dielectric slab waveguide.

10.2.1 TE Modes of the Slab Waveguide

For the TE modes, as the name suggests, there is no component of the electric field along the direction of propagation. Hence $E_z = 0$. By using Eqs. (10.24) to (10.27), we have

$$E_x = 0 \tag{10.43}$$

$$E_y = \frac{i\omega\mu}{\omega^2\mu\varepsilon - \beta^2}\frac{\partial H_z}{\partial x} \tag{10.44}$$

$$H_x = \frac{-i\beta}{\omega^2\mu\varepsilon - \beta^2}\frac{\partial H_z}{\partial x} \tag{10.45}$$

and

$$H_y = 0 \tag{10.46}$$

Furthermore, the axial magnetic field satisfies

$$\frac{\partial^2 H_z}{\partial x^2} + (\omega^2\mu\varepsilon - \beta^2)H_z = 0 \tag{10.47}$$

For a guided wave, we want an oscillatory wave solution in region II and decaying solutions in regions I and III (see Fig. 10.2). It is therefore suitable to define the following constants to be used in the wave equation [Eq. (10.47)]:

$$\gamma_1^2 \equiv \beta^2 - \omega^2\mu\varepsilon_1 = \beta^2 - \left(\frac{\omega}{c}\right)^2 n_1^2 \tag{10.48}$$

$$\kappa_2 \equiv \omega^2\mu\varepsilon_2 - \beta^2 = \left(\frac{\omega}{c}\right)^2 n_2^2 - \beta^2 \tag{10.49}$$

and

$$\gamma_3^2 \equiv \beta^2 - \left(\frac{\omega}{c}\right)^2 n_3^2 \tag{10.50}$$

In region I, we have, from Eq. (10.47),

$$\frac{\partial^2 H_z}{\partial x^2} - \gamma_1^2 H_z = 0 \tag{10.51}$$

whose solution has the form

$$H_z = A\exp(-\gamma_1 x) \tag{10.52}$$

for $x \geq d$. Here, A is a constant. In region II, the wave equation becomes

$$\frac{\partial^2 H_z}{\partial x^2} + \kappa_2^2 H_z = 0 \tag{10.53}$$

which has the oscillatory solution

$$H_z = B\sin(\kappa_2 x) + C\cos(\kappa_2 x) \tag{10.54}$$

for $0 \leq x \leq d$. B and C are constants. Finally, Eq. (10.47) becomes

$$\frac{\partial^2 H_z}{\partial x^2} - \gamma_3^2 H_z = 0 \tag{10.55}$$

in region III with the solution

$$H_z = D\exp(\gamma_3 x) \tag{10.56}$$

for $x \leq 0$. Above, we introduced four constants (A, B, C, and D) into the solutions. It turns out that by using the continuity equations discussed back in Chap. 1, we can eliminate three of these constants and find the solution for the field up to a multiplicative factor, which can, in principle, be set from a knowledge of how much power is carried by the mode. To find the relations between the constants, we first use the fact that H_z is tangential and it should be continuous at $x = 0$ and $x = d$. This gives

$$D - C \qquad (x - 0) \tag{10.57}$$

and

$$A\exp(-\gamma_1 d) = B\sin(\kappa_2 d) + C\cos(\kappa_2 d) \qquad (x = d) \tag{10.58}$$

Note that E_y is also tangential and must be continuous at the edges of the different slabs. From Eq. (10.44), E_y is given by

$$E_y = \begin{cases} \dfrac{i\omega\mu}{-\gamma_1^2} A(-\gamma_1)\exp(-\gamma_1 x) & x \geq d \\ \dfrac{i\omega\mu}{\kappa_2^2}[\kappa_2 B\cos(\kappa_2 x) - \kappa_2 C\sin(\kappa_2 x)] & 0 \leq x \leq d \\ \dfrac{i\omega\mu}{-\gamma_3^2}(\gamma_3) D\exp(\gamma_3 x) & x \leq 0 \end{cases} \tag{10.59}$$

Applying the continuity relations, we obtain

$$\frac{A}{\gamma_1}\exp(-\gamma_1 d)=\frac{1}{\kappa_2}[B\cos(\kappa_2 d)-C\sin(\kappa_2 d)] \qquad (x=d) \tag{10.60}$$

and

$$-\frac{D}{\gamma_3}=\frac{B}{\kappa_2} \qquad (x=0) \tag{10.61}$$

We leave it as an exercise for the reader (see Prob. 10.8) to show that, based on Eqs. (10.57), (10.58) and (10.60), (10.61), the characteristic equation for the guided TE modes of the slab waveguide becomes

$$\tan(\kappa_2 d)=\frac{\gamma_1+\gamma_3}{\kappa_2\left(1-\dfrac{\gamma_1\gamma_3}{\kappa_2^2}\right)} \tag{10.62}$$

Equation (10.62) can only be solved numerically. For typical waveguide dimensions, we get a finite number of guided wave modes. We will look at a particular example later.

10.2.2 TM Modes of the Dielectric Slab Waveguide

For the TM case, $H_z=0$. As before, we further have $\partial/\partial y=0$, since the structure is also invariant in the y-direction. From Eqs. (10.24) to (10.27), we get

$$\left.\begin{aligned} E_x &= \frac{-i\beta}{(\omega^2\mu\varepsilon-\beta^2)}\frac{\partial E_z}{\partial x} \\ E_y &= 0 \\ H_x &= 0 \\ H_y &= \frac{-i\omega\varepsilon}{\omega^2\mu\varepsilon-\beta^2}\frac{\partial E_z}{\partial x} \end{aligned}\right\} \tag{10.63}$$

By using the definitions given in Eqs. (10.48) to (10.50), we get

$$E_z=\begin{cases} A\exp(-\gamma_1 x) & x\ge d \\ (B\sin(\kappa_2 x)+C\cos(\kappa_2 x)) & 0\le x\le d \\ D\exp(\gamma_3 x) & x\le 0 \end{cases} \tag{10.64}$$

To obtain the eigenvalue equation for the propagation constant of the guided modes, we first find the expression for H_y in different regions of the waveguide and then apply the boundary conditions on the tangential field components at $x = 0$ and $x = d$. Leaving it as an exercise for the reader (see Probs. 10.9 and 10.10), we arrive at the result

$$\tan(\kappa_2 d) = \frac{\kappa_2\left(\gamma_1 + \gamma_3\left(\frac{n_1}{n_3}\right)^2\right)}{\kappa_2^2 - \gamma_1\gamma_3\left(\frac{n_2}{n_3}\right)^2} \tag{10.65}$$

10.2.3 Example on the Rectangular Dielectric Waveguide

Let us consider a rectangular dielectric waveguide with $n_1 = 1$ (air), $n_2 = 1.50$, and $n_3 = 1.40$. Furthermore, $\lambda = 1.3\ \mu\text{m}$ and $d = 3\ \mu\text{m}$. For the TE case, we need to solve the eigenvalue equation

$$\tan(\kappa_2 d) = \frac{\gamma_1 + \gamma_3}{\kappa_2\left(1 - \frac{\gamma_1\gamma_3}{\kappa_2^{\,2}}\right)} \tag{10.66}$$

To proceed with the numerical solution, we first introduce the dimensionless parameters $\bar{x}$ and x_0, defined according to

$$\bar{x} = \kappa_2 d \tag{10.67}$$

and

$$x_0 = k_0 d = 2\pi\frac{d}{\lambda} \tag{10.68}$$

Then

$$\bar{x}^2 = n_2^2 x_0^2 - (\beta d)^2 \tag{10.69}$$

Furthermore,

$$(\gamma_1 d)^2 = (\beta d)^2 - x_0^2 n_1^2 = x_0^2\left(n_2^2 - n_1^2\right) - \bar{x}^2 \tag{10.70}$$

and

$$(\gamma_3 d)^2 = (\beta d)^2 - x_0^2 n_3^2 = x_0^2\left(n_2^2 - n_3^2\right) - \bar{x}^2 \tag{10.71}$$

Let us assume that $n_2 > n_3 > n_1$ as in the specific example considered here. The propagation parameter β will be in the range $[k_0 n_3, k_0 n_2]$ for guided modes. Also, $\bar{x}$ will vary from 0 to $x_0\sqrt{n_2^2 - n_3^2}$.

Figure 10.3 shows the variation of the left-hand side (l.h.s.) and right-hand side (r.h.s.) of Eq. (10.66) as a function of the normalized parameter $\bar{x} = \kappa_2 d$ for the slab wave guide with $n_1 = 1$, $n_2 = 1.50$, $n_3 = 1.40$, $\lambda = 1.3\ \mu\text{m}$, and $d = 3\ \mu\text{m}$. There are three solutions for $\bar{x}$ which can be obtained numerically. We leave it as an exercise for the reader

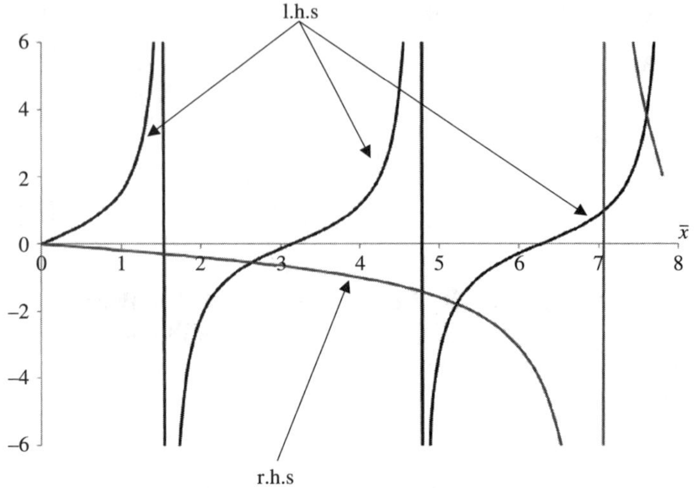

FIGURE 10.3 Variation of the left-hand side (l.h.s.) and right-hand side (r.h.s.) of Eq. (10.66) as a function of the normalized parameter $\bar{x} = \kappa_2 d$ for the slab waveguide with $n_1 = 1$, $n_2 = 1.50$, $n_3 = 1.40$, $\lambda = 1.3\ \mu\text{m}$, and $d = 3\ \mu\text{m}$. This is the TE case.

(see Prob. 10.11) to determine these solutions, which turn out to be $\bar{x} = 2.634, 5.223, 7.598$. The corresponding values of β/k_0 are 1.489, 1.456, 1.405. We can further determine the electric field distribution for E_y for each propagating mode. By using the boundary conditions, it can be shown that (see Prob. 10.12) up to a multiplicative constant (the constant is D in this case), the expression for E_y can be written as

$$\Rightarrow E_y = \begin{cases} \dfrac{A}{\gamma_1}\exp(-\gamma_1 x) & x \geq d \\ \dfrac{D}{\kappa_2}\left(-\dfrac{\kappa_2}{\gamma_3}\cos\kappa_2 x - \sin\kappa_2 x\right) & 0 \leq x \leq d \\ -\dfrac{D}{\gamma_3}\exp(\gamma_3 x) & x \leq 0 \end{cases} \tag{10.72}$$

where

$$A = \gamma_1 \exp(+\gamma_1 d)\frac{D}{\kappa_2}\left(-\frac{\kappa_2}{\gamma_3}\cos\kappa_2 d - \sin\kappa_2 d\right) \tag{10.73}$$

The field distributions for the three guided modes are shown in Fig. 10.4*a* to *c* as a function of the normalized distance (x/d). Note that the field distributions shown in Fig. 10.4 are the only guided modes

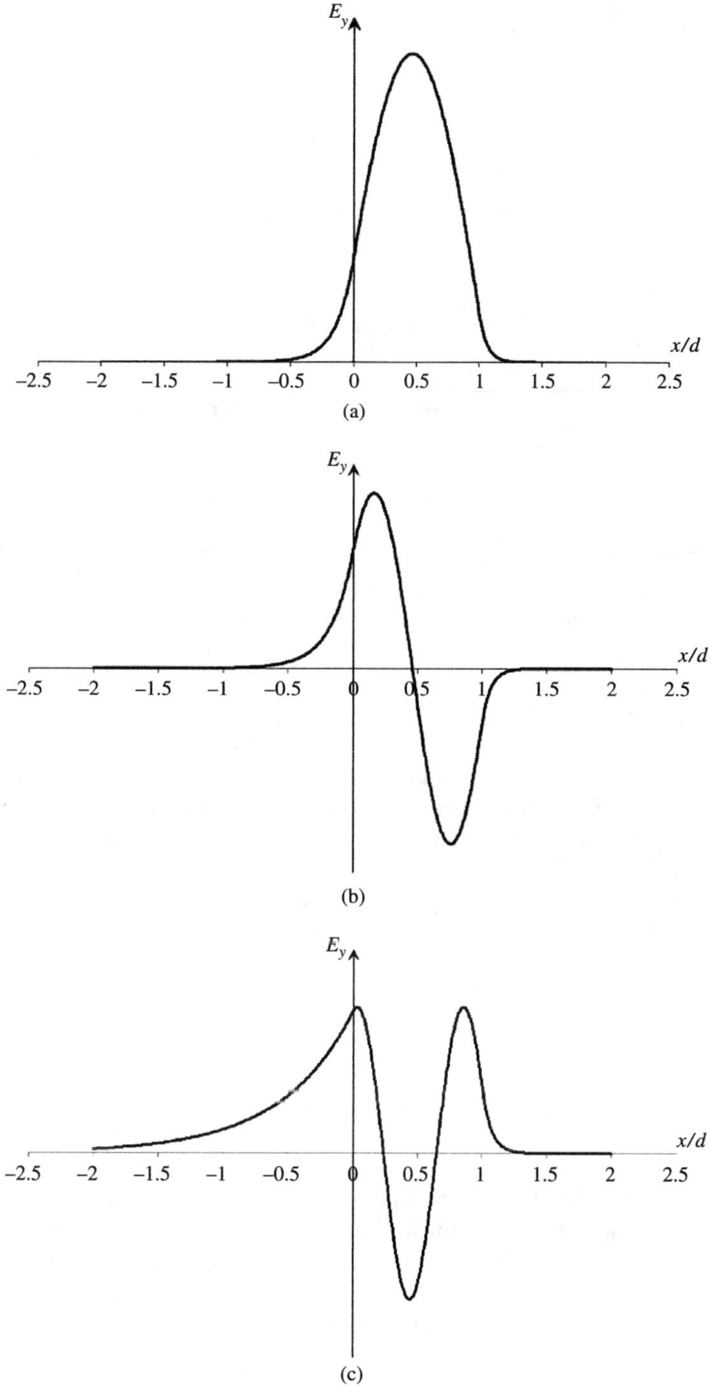

FIGURE 10.4 Variation of the electric field for the three guided TE modes of the rectangular dielectric slab waveguide as a function of the normalized distance x/d. The waveguide parameters are $n_1 = 1$, $n_2 = 1.50$, $n_3 = 1.40$, $\lambda = 1.3$ μm, and $d = 3$ μm.

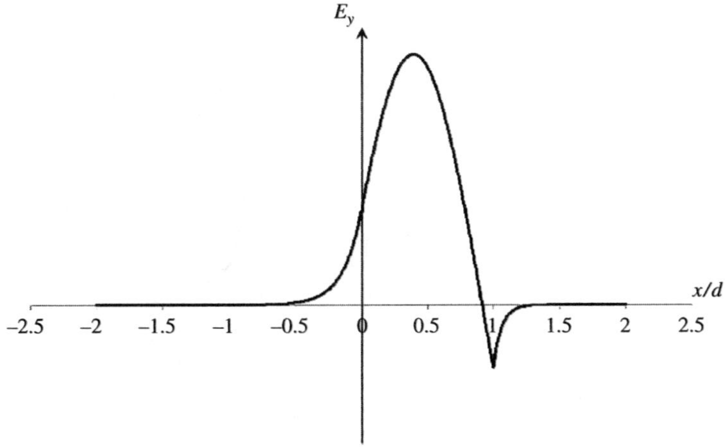

FIGURE 10.5 Plot of the TE electric field at an arbitrarily chosen $\bar{x} = \kappa_2 d$ value of 3. Note that the field has a discontinuity in its derivative at $x = d$. This does not represent a guided mode of the slab waveguide!

which satisfy the prescribed boundary conditions. To appreciate this better, let's plot the field distribution for the lowest-order TE mode (i.e., what is shown in Fig. 10.4*a*) at a value of $\kappa_2 d$ which is not a solution of the characteristic equation. As an example, let's take $\kappa_2 d = 3$ instead of the actual eigenvalue 2.634. This is shown in Fig 10.5. The discontinuity in the derivative of the field is clearly visible at $x = d$.

For the TM mode, we need to solve the equation

$$\tan(\kappa_2 d) = \frac{\left(\gamma_1 + \gamma_3\left(\frac{n_1}{n_3}\right)^2\right)}{\kappa_2 - \left(\frac{\gamma_1\gamma_3}{\kappa_2}\right)\left(\frac{n_2}{n_3}\right)^2} \tag{10.74}$$

numerically. For the same waveguide parameters as in the above example, Fig. 10.6 shows the left-hand side (l.h.s) and right-hand side (r.h.s) of Eq. (10.74). The allowed values of $\kappa_2 d$ and β/k_0 become:

$$\kappa_2 d = 2.739,\ 5.345,\ 7.608$$

$$\beta/k_0 = 1.488,\ 1.454,\ 1.405$$

Again, we leave it as an exercise (see Prob. 10.13) for the reader to show that up to a multiplicative constant (taken as C_0 in the following equation), the component H_y of the TM mode can be expressed as

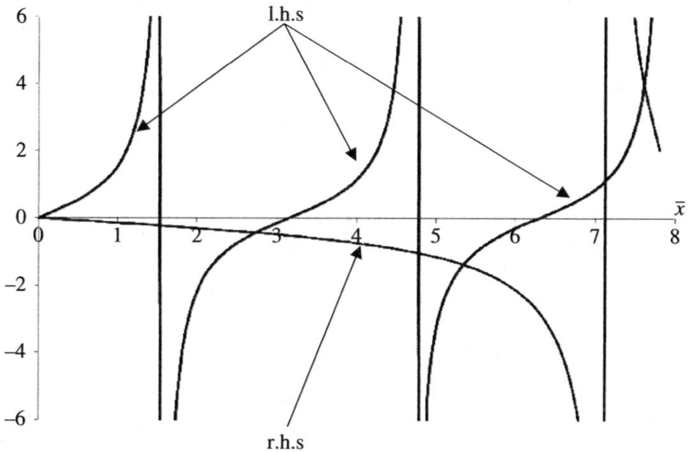

Figure 10.6 Variation of the left-hand side (l.h.s.) and right-hand side (r.h.s.) of Eq. (10.74) as a function of the normalized parameter $\bar{x} = \kappa_2 d$ for a slab wave guide with $n_1 = 1$, $n_2 = 1.50$, $n_3 = 1.40$, $\lambda = 1.3$ µm, and $d = 3$ µm. This is the TM case.

$$H_y = \begin{cases} -C_0 \dfrac{n_2^2}{\kappa_2}\left(-\dfrac{\kappa_2}{\gamma_3}\dfrac{n_3^2}{n_2^2}\cos\kappa_2 d - \sin(\kappa_2 d)\right)\exp(\gamma_1(d-x)) & x \geq d \\ -C_0 \dfrac{n_2{}^2}{\kappa_2}\left(-\dfrac{\kappa_2}{\gamma_3}\dfrac{n_3{}^2}{n_2{}^2}\cos\kappa_2 x - \sin(\kappa_2 x)\right) & 0 \leq x \leq d \\ C_0 \dfrac{n_3^2}{\gamma_3}\exp(\gamma_3 x) & x \leq 0 \end{cases} \tag{10.75}$$

The magnetic field distributions for the TM modes of the above example are plotted in Fig. 10.7*a* to *c*.

10.3 Step-Index Optical Fibers

In this section, we give a detailed electromagnetic analysis of the circular step-index optical fiber. Equations satisfied by the propagation constants of different classes of guided modes are derived. We introduce the normalized *V*-number and derive the cut-off conditions for different modes. We discuss the condition under which a single guided mode, referred to as the fundamental HE_{11} mode is allowed. On the practical side, we give the Gaussian approximation of the HE_{11} mode in terms of the mode field diameter which is a very useful parameter in the description of single-mode fibers. Coupling into single-mode and multi-mode fibers is then discussed. Further discussion on the electromagnetic properties of the propagating modes may be found in Refs. 1, 4 to 6, 12 and 14.

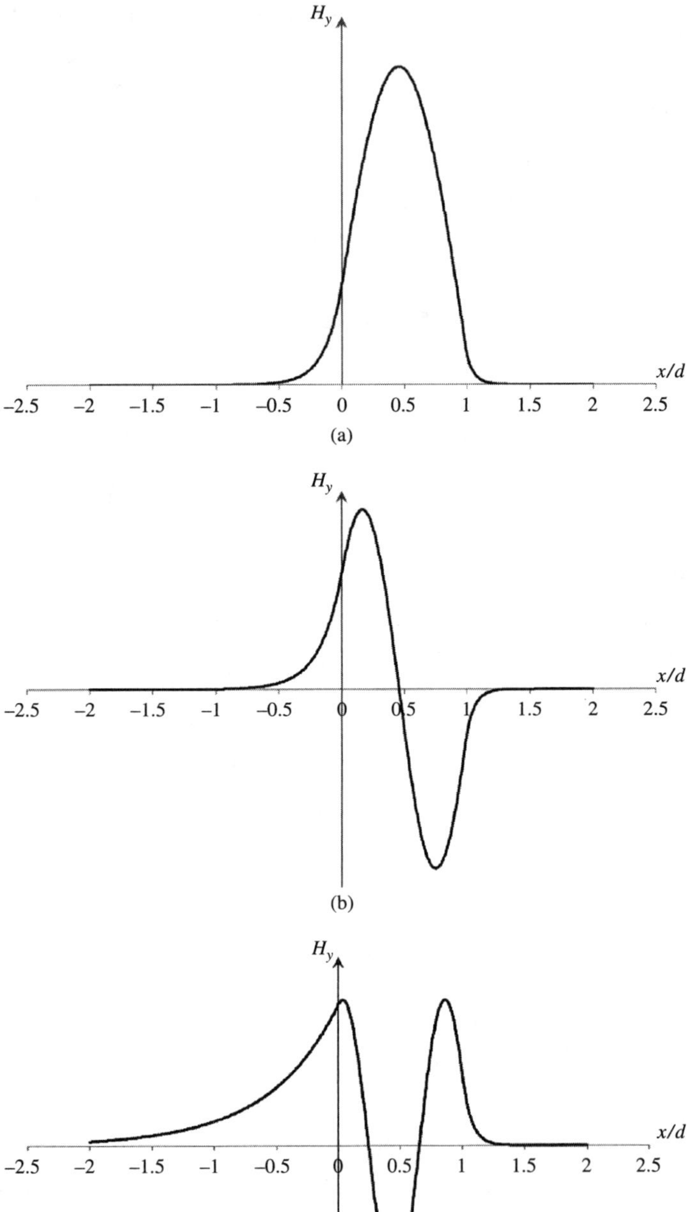

FIGURE 10.7 Variation of the magnetic field H_y for the three-guided TM modes of the rectangular dielectric slab waveguide as a function of the normalized distance x/d. The waveguide parameters are $n_1 = 1$, $n_2 = 1.50$, $n_3 = 1.40$, $\lambda = 1.3\ \mu\text{m}$, and $d = 3\ \mu\text{m}$.

10.3.1 General Aspects of Optical Fibers

In a crude sense, we can think of an optical fiber as a glass cable for transmitting light over long distances. From the simple ray picture of light, we know that total internal reflection confines the light beam within the glass cable and enables its propagation over long distances. The throughput of an optical fiber is typically limited by the fraction of light that can be coupled into and out of the fiber, as well as the various attenuation mechanisms including absorption, scattering, and bending losses. Earlier, in Chap. 4, we discussed the wavelength-dependent scattering and absorption losses, which are among the leading loss mechanisms in optical fibers as well.

There are many possible guiding geometries which are used in state-of-the-art optical fibers. By guiding geometry, we mean the refractive index variation over the cross section that is perpendicular to the transmission axis of the fiber. Under most cases, we can model the fiber as a translationally invariant structure and the solution techniques developed in Sec. 10.1 can be readily used in the electromagnetic analysis of the guided modes. An exception is the tapered fiber, where the cross-sectional properties vary with position. One of the simplest and most important fiber types is the so-called step-index optical fiber which is sketched in Fig. 10.8. In this section, we will mainly focus on the characteristics of the step-index optical fibers.

The step-index fiber has a central circular core region (shown with a radius of a in Fig. 10.8) with a refractive index (n_1) which is greater than that of the surrounding cladding region (shown with a refractive index of n_2 in Fig. 10.8). The corresponding refractive index variation as a function of the radial coordinate r is depicted in Fig. 10.9. The plastic jacket serves to protect the fiber and increases its flexibility. The refractive index of the core and the cladding can be adjusted by using different dopants during the synthesis stage. In silica (SiO_2) optical fibers, for example, adding germanium or phosphorous oxide into the core region increases the refractive index. On the other hand, fluorine may be added to lower the refractive index.[14] Having $n_1 > n_2$ ensures that the guided mode is predominantly in the core region. Of course, since both the core and cladding layers are dielectric, a portion of

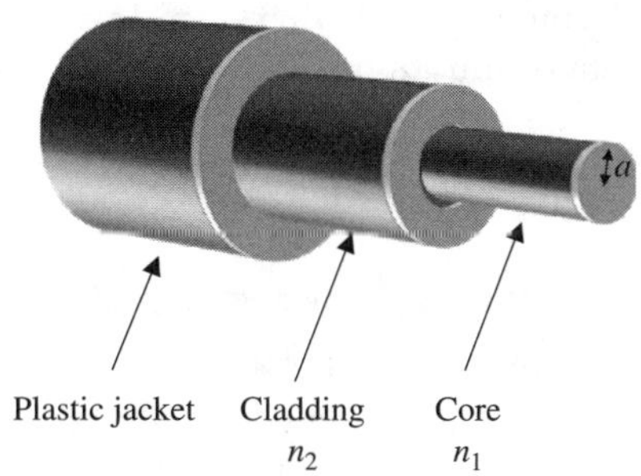

FIGURE 10.8 Step-index optical fiber.

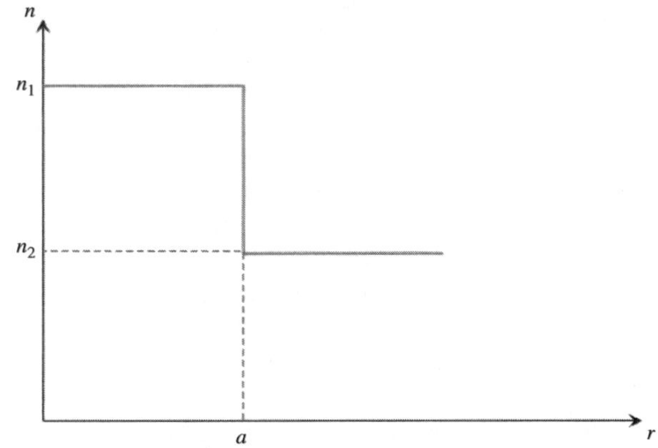

FIGURE 10.9 Variation of the refractive index as a function of the radial coordinate for a step-index optical fiber.

the guided field inevitably leaks into the cladding region. Soon, we will solve Maxwell's equations to determine the propagating electromagnetic modes of the optical fiber. Just to refresh our memory once again, each mode refers to a particular electromagnetic field distribution that satisfies both Maxwell's equations and the specific boundary conditions in our problem. To simplify our analysis, we will assume that the cladding radius is much larger than a so that fields decay to negligible levels before the cladding-jacket interface is reached.

Depending on the number of propagating modes, we categorize optical fibers into two groups:

1. **Single-mode optical fibers** In this class of optical fibers, only one electromagnetic field distribution can propagate. In this regime, trying to analyze the fiber with the simple ray picture of light is often insufficient. We present the full electromagnetic analysis of this case in the next section and derive a simple analytical condition that must be satisfied in order for the fiber to be single mode.
2. **Multi-mode optical fibers** In this case, several modes can propagate simultaneously. If the number of propagating modes is large, the fiber can be analyzed by using the simple ray model of light. We will discuss this later in this chapter.

10.3.2 Electromagnetic Analysis of the Step-Index Optical Fiber

As before, we take the direction of propagation to be the z-axis. The equations satisfied by the field components in a translationally invariant waveguide were derived in Sec. 10.1. In the particular case of cylindrical coordinates, E_z and H_z satisfy Eqs. (10.41) and (10.42). Note that both of these equations have the same mathematical form and will admit similar solutions. We start with the equation satisfied by E_z:

$$\frac{\partial^2 E_z}{\partial r^2} + \frac{1}{r}\frac{\partial E_z}{\partial r} + \frac{1}{r^2}\frac{\partial E_z}{\partial \theta^2} + (\omega^2\mu\varepsilon - \beta^2)E_z = 0 \tag{10.76}$$

If we assume a solution of the form

$$E_z(r,\theta) = R(r)\,\Theta(\theta) \tag{10.77}$$

and substitute into Eq. (10.76), we get, after multiplying by $r^2/R(r)\Theta(\theta)$,

$$\frac{1}{R(r)}\left(r^2\frac{d^2R}{dr^2} + r\frac{dR}{dr}\right) + \frac{1}{\Theta(\theta)}\frac{d^2\Theta}{d\theta^2} + r^2(\omega^2\mu\varepsilon - \beta^2) = 0 \tag{10.78}$$

or

$$\frac{1}{R(r)}\left(r^2\frac{d^2R}{dr^2} + r\frac{dR}{dr}\right) + r^2(\omega^2\mu\varepsilon - \beta^2) = -\frac{1}{\Theta(\theta)}\frac{d^2\Theta}{d\theta^2} = \ell^2 \tag{10.79}$$

In Eq. (10.79), the left-hand side is a function of r only, whereas the right-hand side is a function of θ only. This is possible only if each side is constant. We follow the standard convention and choose the separation constant to be ℓ^2. The solution of the equation for θ is readily given by

$$\Theta(\theta) = \exp[\pm i\ell\theta] \tag{10.80}$$

Here, ℓ takes only integer values (0, 1, 2, 3, and so on) to make the function $\Theta(\theta)$ single-valued. This is necessary since the cross section of the fiber includes the full range $[0,2\pi]$ for θ. After rearranging, Eq. (10.79) becomes

$$\frac{d^2R(r)}{dr^2} + \frac{1}{r}\frac{dR(r)}{dr} + \left(\omega^2\mu\varepsilon - \beta^2 - \frac{\ell^2}{r^2}\right)R(r) = 0 \tag{10.81}$$

Equation (10.81) is the well-known Bessel's equation. Since Eq. (10.81) is a second-order differential equation, it has two linearly independent solutions. Furthermore, the specific mathematical form of these solutions depends on whether $\omega^2\mu\varepsilon - \beta^2$ is positive or negative. Let us look at the solutions in the core and cladding separately.

Field Solutions in the Core

For a guided mode, we want the field solutions to be oscillatory inside the core and decaying in the cladding. This is guaranteed if $\omega^2\mu\varepsilon - \beta^2$ is greater than 0 in the core. By using the definition

$$\kappa^2 \equiv k_0^2 n_1^2 - \beta^2 \tag{10.82}$$

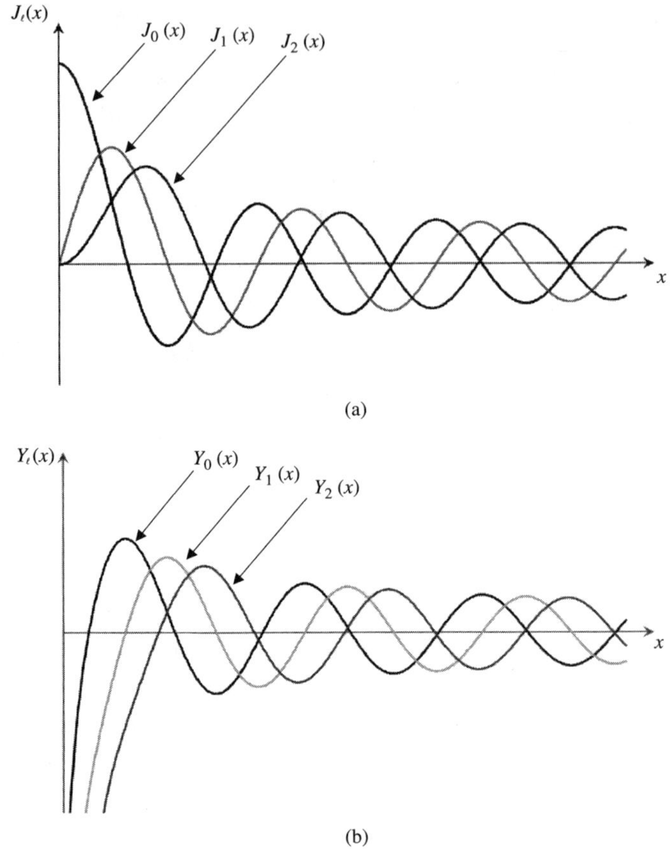

FIGURE 10.10 Plot of the lowest three orders ($\ell = 0,1,2$) of the Bessel functions (*a*) $J_\ell(x)$ and (*b*) $Y_\ell(x)$ as a function of x.

where $k_0 = \omega/c$ is the vacuum wave number, the solution for $R(r)$ can be written in the form

$$R(r) = A_1 J_\ell(\kappa r) + A_2 Y_\ell(\kappa r) \tag{10.83}$$

for $r \le a$. Above, $J_\ell(\kappa r)$ and $Y_\ell(\kappa r)$ are the Bessel functions of the first and second kind, respectively, of order ℓ. Figure 10.10*a* and *b* shows the variation of $J_\ell(x)$ and $Y_\ell(x)$, respectively, for $\ell = 0,1,2$. We further want the field solutions to remain finite everywhere inside the core. Since $Y_\ell(\kappa r)$ has a singularity at $r = 0$ (see Fig. 10.10*b*), we disregard these solutions. Recalling that H_z satisfies a similar differential equation and also must be finite, the expressions for the spatial part of the fields inside the core become

$$E_z(\vec{r}) = A J_\ell(\kappa r) \exp[i\ell\theta - i\beta z] \tag{10.84}$$

and

$$H_z(\vec{r}) = B J_\ell(\kappa r) \exp[i\ell\theta - i\beta z] \tag{10.85}$$

for $r < a$. Here, A and B are constants to be determined from the boundary conditions.

Field Solutions in the Cladding

Here, we want decaying solutions for the guided modes. This is guaranteed if $\omega^2\mu\varepsilon - \beta^2 < 0$. With the definition

$$\gamma^2 \equiv \beta^2 - k_0^2 n_2^2 \tag{10.86}$$

the solution for $R(r)$ in the cladding can be written as

$$R(r) = B_1 I_\ell(\gamma r) + B_2 K_\ell(\gamma r) \tag{10.87}$$

where $I_\ell(\gamma r)$ and $K_\ell(\gamma r)$ are the modified Bessel functions of the first and second kind of order ℓ. B_1 and B_2 in Eq. (10.87) are constants. Figure 10.11*a* and *b* shows the lowest three orders ($\ell = 0,1,2$) of $I_\ell(x)$

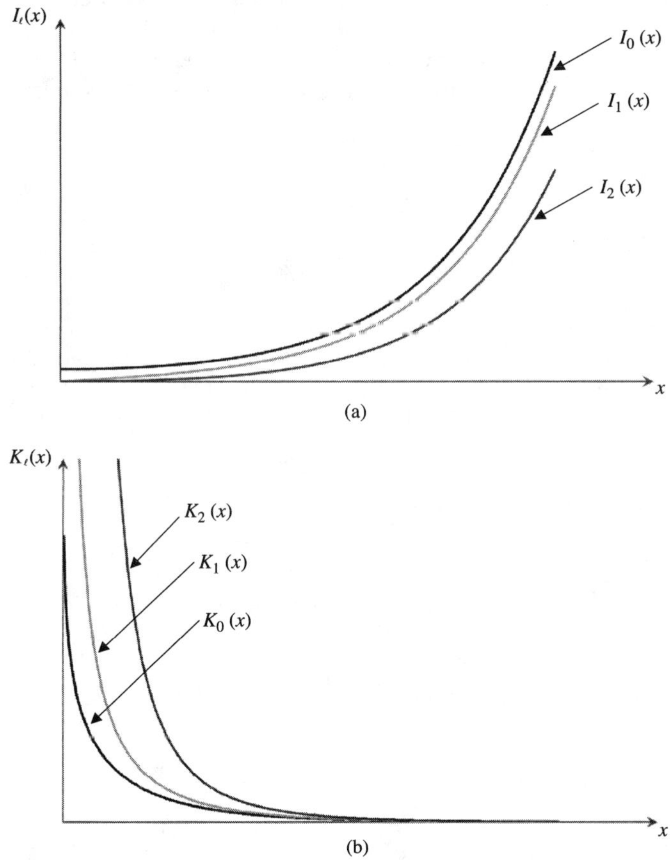

FIGURE 10.11 Plot of the lowest three orders ($\ell = 0,1,2$) of (*a*) $I_\ell(x)$ and (*b*) $K_\ell(x)$ as a function of x.

and $K_\ell(x)$, respectively. Among these solutions, I_ℓ grows exponentially at large values of r. Hence, we retain only the decaying solution $K_\ell(\gamma r)$ to guarantee that the fields remain bounded everywhere within the fiber.

Hence, the solution for E_z and H_z in the cladding region can be written in the form

$$E_z(\vec{r}) = CK_\ell(\gamma r)\exp\left[i\ell\theta - i\beta z\right] \tag{10.88}$$

and

$$H_z(\vec{r}) = DK_\ell(\gamma r)\exp[i\ell\theta - i\beta z] \tag{10.89}$$

for $r > a$. Again, the constants C and D are to be determined from the boundary conditions as we will discuss in the next section.

Boundary Conditions at the Core-Cladding Interface

From this point on, we apply the continuity equations to determine the constants A, B, C, and D in Eqs. (10.84), (10.85) and (10.88), (10.89). Since E_z and H_z are both tangential at the core-cladding interface ($r = a$), we require them to be continuous. This gives the following two relations:

$$AJ_\ell(\kappa a) = CK_\ell(\gamma a) \tag{10.90}$$

and

$$BJ_\ell(\kappa a) = DK_\ell(\gamma a) \tag{10.91}$$

We need two more equations since there are four unknowns. Note that among the transverse components, E_θ and H_θ are also tangential at the interface and hence must be continuous. E_θ and H_θ can be calculated in terms of E_z and H_z by using Eqs. (10.38) and (10.40). E_θ becomes

$$E_\theta = \frac{-i}{\omega^2\mu\varepsilon - \beta^2}\left[\beta\frac{1}{r}\frac{\partial E_z}{\partial\theta} - \omega\mu\frac{\partial H_z}{\partial r}\right]$$

$$= \begin{cases} \dfrac{-i}{\kappa^2}\left(\dfrac{i\beta\ell}{r}AJ_\ell(\kappa r) - \omega\mu B\dfrac{dJ_\ell(\kappa r)}{dr}\right)\exp[i\ell\theta - i\beta z] & r \le a \\[2ex] \dfrac{i}{\gamma^2}\left(\dfrac{i\beta\ell}{r}CK_\ell(\gamma r) - \omega\mu D\dfrac{dK_\ell(\gamma r)}{dr}\right)\exp[i\ell\theta - i\beta z] & r \ge a \end{cases} \tag{10.92}$$

Similarly, H_θ is given by

$$H_\theta = \frac{-i}{\omega^2\mu\varepsilon - \beta^2}\left[\frac{\beta}{r}\frac{\partial H_z}{\partial\theta} + \omega\varepsilon\frac{\partial E_z}{\partial r}\right]$$

$$= \begin{cases} \dfrac{-i}{\kappa^2}\left(\dfrac{i\beta\ell}{r}BJ_\ell(\kappa r)+\omega\varepsilon_0 n_1^2 A\dfrac{dJ_\ell(\kappa r)}{dr}\right)\exp[i\ell\theta-i\beta z] & r\le a \\ \dfrac{i}{\gamma^2}\left(\dfrac{i\beta\ell}{r}DK_\ell(\gamma r)+\omega\varepsilon_0 n_2^2 C\dfrac{dK_\ell(\gamma r)}{dr}\right)\exp[i\ell\theta-i\beta z] & r\ge a \end{cases} \tag{10.93}$$

In what follows, we will use the notation

$$\frac{dJ_\ell}{dr}=\kappa\frac{dJ_\ell}{d(\kappa r)}=\kappa J_\ell'(\kappa r) \tag{10.94}$$

and

$$\frac{dK_\ell}{dr}=\gamma\frac{dK_\ell}{d(\gamma r)}=\gamma K_\ell'(\gamma r) \tag{10.95}$$

in the derivation of the eigenvalue equation for the propagating modes. Based on the above equations, continuity of E_θ at $r = a$ gives the relation

$$\frac{-1}{\kappa^2}\left(\frac{i\beta\ell}{a}AJ_\ell(\kappa a)-\omega\mu B\kappa J_\ell'(\kappa a)\right)=\frac{1}{\gamma^2}\left(\frac{i\beta\ell}{a}CK_\ell(\gamma a)-\omega\mu D\gamma K_\ell'(\gamma a)\right) \tag{10.96}$$

Similarly, continuity of H_θ at $r = a$ can be expressed in the form

$$\frac{-1}{\kappa^2}\left(\frac{i\beta\ell}{a}BJ_\ell(\kappa a)+\omega\varepsilon_0 n_1^2 A\kappa J_\ell'(\kappa a)\right)=\frac{1}{\gamma^2}\left(\frac{i\beta\ell}{a}DK_\ell(\gamma a)+\omega\varepsilon_0 n_2^2 \gamma C K_\ell'(\gamma a)\right) \tag{10.97}$$

Derivation of the Eigenvalue Equation for the Guided Modes of a Step-Index Fiber

Equations (10.90), (10.91) and (10.96), (10.97) relate the four unknown constants A, B, C, and D in the expressions for the fields in the core and the cladding. By rearranging to make the right-hand side of each equation equal to 0, these four equations can be expressed in matrix form as

$$\begin{bmatrix} J_\ell(\kappa a) & 0 & -K_\ell(\gamma a) & 0 \\ \dfrac{i\ell\beta}{\kappa^2 a}J_\ell(\kappa a) & -\dfrac{\omega\mu}{\kappa}J_\ell'(\kappa a) & \dfrac{i\ell\beta}{\gamma^2 a}K_\ell(\gamma a) & -\dfrac{\omega\mu}{\gamma}K_\ell'(\gamma a) \\ 0 & J_\ell(\kappa a) & 0 & -K_\ell(\gamma a) \\ \dfrac{\omega\varepsilon_0 n_1^2 J_\ell'(\kappa a)}{\kappa} & \dfrac{i\ell\beta}{\kappa^2 a}J_\ell(\kappa a) & \dfrac{\omega\varepsilon_0 n_2^2 K_\ell'(\gamma a)}{\gamma} & \dfrac{i\ell\beta}{\gamma^2 a}K_\ell(\gamma a) \end{bmatrix}\begin{bmatrix} A \\ B \\ C \\ D \end{bmatrix}=\begin{bmatrix} 0 \\ 0 \\ 0 \\ 0 \end{bmatrix} \tag{10.98}$$

The only way we can get a nontrivial solution to this system of equations is if the determinant of the matrix appearing in Eq. (10.98) is zero. Now, computing the determinant of this 4×4 matrix is not trivial but it can be done. Following the approach given in Ref. 1, we define the following parameters:

$$\bar{a} \equiv J_\ell(\kappa a) \tag{10.99}$$

$$\bar{b} \equiv \frac{i\ell\beta}{\kappa^2 a} \tag{10.100}$$

$$\bar{c} \equiv -\frac{\omega}{\kappa} J'_\ell(\kappa a) \tag{10.101}$$

$$\bar{d} \equiv K_\ell(\gamma a) \tag{10.102}$$

$$\bar{e} \equiv \frac{i\ell\beta}{\gamma^2 a} \tag{10.103}$$

$$\bar{f} \equiv K'_\ell(\gamma a)\left(\frac{\omega}{\gamma}\right) \tag{10.104}$$

Hence, the matrix (let's call it **M**) appearing in Eq. (10.98) becomes

$$\mathbf{M} = \begin{bmatrix} \bar{a} & 0 & -\bar{d} & 0 \\ \bar{b}\bar{a} & \mu\bar{c} & \bar{e}\bar{d} & -\mu\bar{f} \\ 0 & \bar{a} & 0 & -\bar{d} \\ -\varepsilon_1\bar{c} & \bar{b}\bar{a} & \varepsilon_2\bar{f} & \bar{e}\bar{d} \end{bmatrix} \tag{10.105}$$

We leave it as an exercise for the reader (see Prob. 10.17) to show that the relation

$$\det \mathbf{M} = 0 \tag{10.106}$$

can be written in terms of the parameters defined above in Eqs. (10.99) to (10.104) as

$$\left(\frac{\bar{c}}{\bar{a}} - \frac{\bar{f}}{\bar{d}}\right)\left(\frac{\bar{f}}{\bar{d}}\mu\varepsilon_2 - \frac{\bar{c}}{\bar{a}}\mu\varepsilon_1\right) = \left(\bar{e} + \bar{b}\right)^2 \tag{10.107}$$

By using the definitions of $\bar{a} - \ldots - \bar{f}$ given above, we can further express Eq. (10.107) as

$$\left(-\frac{\omega}{\kappa}\frac{J'_\ell(\kappa a)}{J_\ell(\kappa a)} - \frac{\omega}{\gamma}\frac{K'_\ell(\gamma a)}{K_\ell(\gamma a)}\right)\left(\frac{K'_\ell(\gamma a)}{K_\ell(\gamma a)}\frac{\omega}{\gamma}\mu\varepsilon_2 + \frac{\omega J'_\ell(\kappa a)}{\kappa J_\ell(\kappa a)}\mu\varepsilon_1\right)$$
$$= \left(\frac{i\ell\beta}{\gamma^2 a} + \frac{i\ell\beta}{\kappa^2 a}\right)^2 = -\left(\frac{\ell\beta}{a}\right)^2\left(\frac{\kappa^2 + \gamma^2}{\gamma^2\kappa^2}\right)^2 \tag{10.108}$$

To proceed, we introduce the normalized parameters

$$x = \kappa a \tag{10.109}$$

$$x_0 = k_0 a = \frac{2\pi a}{\lambda} \tag{10.110}$$

$$x_1 = k_1 a = \frac{2\pi}{\lambda} n_1 a \tag{10.111}$$

and

$$x_2 = k_2 a = \frac{2\pi}{\lambda} n_2 a \tag{10.112}$$

Note that since

$$(\kappa a)^2 = (k_1 a)^2 - (\beta a)^2 \tag{10.113}$$

we have

$$\beta a = \sqrt{x_1^2 - x^2} \tag{10.114}$$

Furthermore, $y = \gamma a$ becomes

$$\gamma a = \sqrt{(\beta a)^2 - (k_2 a)^2} = \sqrt{x_1^2 - x_2^2 - x^2} \tag{10.115}$$

Since γa is real, x must be in the range $\left[0, \sqrt{x_1^2 - x_2^2}\right]$. Furthermore, we leave it as an exercise for the reader (Prob. 10.18) to show that Eq. (10.108) can now be expressed in terms of the normalized parameters as

$$\frac{n_1^2}{x^2}\left(\frac{J_\ell'(x)}{J_\ell(x)}\right)^2 + \frac{n_1^2 + n_2^2}{xy}\frac{K_\ell'(y)}{K_\ell(y)}\frac{J_\ell'(x)}{J_\ell(x)} + \frac{n_2^2}{y^2}\left(\frac{K_\ell'(y)}{K_\ell(y)}\right)^2$$
$$- \ell^2\left(\frac{1}{x^2} + \frac{1}{y^2}\right)^2 \frac{\left(x_1^2 - x^2\right)}{x_0^2} = 0 \tag{10.116}$$

Since Eq. (10.116) is quadratic in $\left(J_\ell'(x)/xJ_\ell(x)\right)$, it can be readily solved to yield

$$\frac{J_\ell'(x)}{xJ_\ell(x)} = -\frac{\left(n_1^2 + n_2^2\right)}{2n_1^2}\frac{K_\ell'(y)}{yK_\ell(y)}$$
$$\pm\sqrt{\left(\frac{n_1^2 - n_2^2}{2n_1^2}\right)^2\left(\frac{K_\ell'(y)}{yK_\ell(y)}\right)^2 + \frac{\ell^2}{n_1^2}\left(\frac{1}{x^2} + \frac{1}{y^2}\right)^2\frac{\left(x_1^2 - x^2\right)}{x_0^2}} \tag{10.117}$$

10.3.3 Classification of the Modes in a Step-Index Optical Fiber

The modes that can propagate in an optical fiber are categorized into four groups depending on the value of ℓ and the particular root of the eigenvalue equation [Eq. (10.117)]. In this section, we describe how the propagation constant of a mode can be numerically determined and discuss the mode cut-off conditions.

TM Modes

For TM modes, $\ell = 0$ and the eigenvalue equation contains the positive root in Eq. (10.117):

$$\frac{J_0'(x)}{xJ_0(x)} = \frac{-\left(n_1^2 + n_2^2\right)}{2n_1^2}\frac{K_\ell'(y)}{yK_\ell(y)} + \frac{\left(n_1^2 - n_2^2\right)}{2n_1^2}\frac{K_\ell'(y)}{yK_\ell(y)} = \frac{-n_2^2}{n_1^2}\frac{K_0'(y)}{yK_0(y)} \tag{10.118}$$

Since the argument y of $K_\ell(y)$ should be real, the right-hand side of this eigenvalue equation is defined only for x values in the interval $[0, V]$, where V is known as the V-number of the fiber and is given by

$$V = \sqrt{x_1^2 - x_2^2} = \frac{2\pi a}{\lambda}\sqrt{n_1^2 - n_2^2} \tag{10.119}$$

The V number is a very important parameter of a step-index optical fiber and we will use it in the discussion of the mode cut-off conditions in this section.

Figure 10.12 shows the numerical plot of the left-hand (l.h.s) and right-hand side (r.h.s) of Eq. (10.118). Note that the right-hand side has a singularity at $x = V$ and is not defined for $x > V$. We will see the

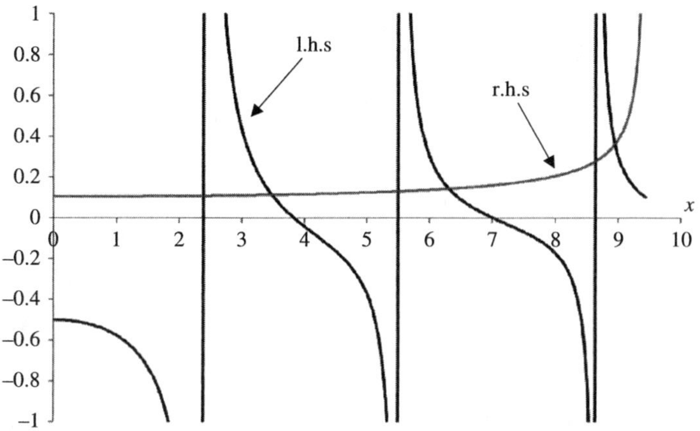

FIGURE 10.12 Numerical plot of the left-hand (l.h.s) and right-hand side (r.h.s) of the eigenvalue equation [Eq. (10.118)] for TM modes. The parameters are $a = 5\ \mu\text{m}$, $\ell = 0$, $\lambda = 1.3\ \mu\text{m}$, $n_1 = 1.55$, and $n_2 = 1.5$.

same behavior for all of the modes. So, if there are any possible mode solutions, they are to be found in the x interval $[0, V]$. The plot of Figure 10.12 was generated by using the parameters $a = 5\ \mu m$, $\ell = 0$, $\lambda = 1.3\ \mu m$, $n_1 = 1.55$, and $n_2 = 1.5$, giving a V value of 9.44. Note that there are three solutions for x at 3.48, 6.31, and 8.95. The propagation constant βa can then be determined from Eq. (10.114). Note that the left-hand side of Eq. (10.118) has repeating singularities which occur at the zeroes of the zeroth-order Bessel function $J_0(x)$. The zeroes of $J_0(x)$ occur at 2.405, 5.52, and so on. If V is less than 2.405, the first zero of $J_0(x)$, we see that no solution is possible for the eigenvalue Eq. (10.118). This is because, for $V < 2.405$ the TM mode is no longer evanescent in the cladding region and, for this combination of fiber parameters and wavelength, we do not get a confined propagating mode. We say in this case that the cut-off V number for TM_{01} mode is 2.405.

TE Modes

Similar to TM modes, $\ell = 0$ for the TE modes. The eigenvalue equation now includes the negative root and is given by

$$\frac{J_0'(x)}{xJ_0(x)} = \frac{-\left(n_1^2 + n_2^2\right)}{2n_1^2}\frac{K_0'(y)}{yK_0(y)} - \frac{\left(n_1^2 - n_2^2\right)}{2n_1^2}\frac{K_0'(y)}{yK_0(y)} = \frac{-K_0'(y)}{yK_0(y)} \tag{10.120}$$

The eigenvalue equation for TE modes looks very similar to that for TM modes, as can be seen from Eqs. (10.118) and (10.120) as well as Figs. 10.12 and 10.13. Figure 10.13 shows the numerically computed left-hand and right-hand sides of Eq. (10.120) for the parameters $a = 5\ \mu m$, $\ell = 0$, $\lambda = 1.3\ \mu m$, $n_1 = 1.55$, and $n_2 = 1.5$. Again the V number

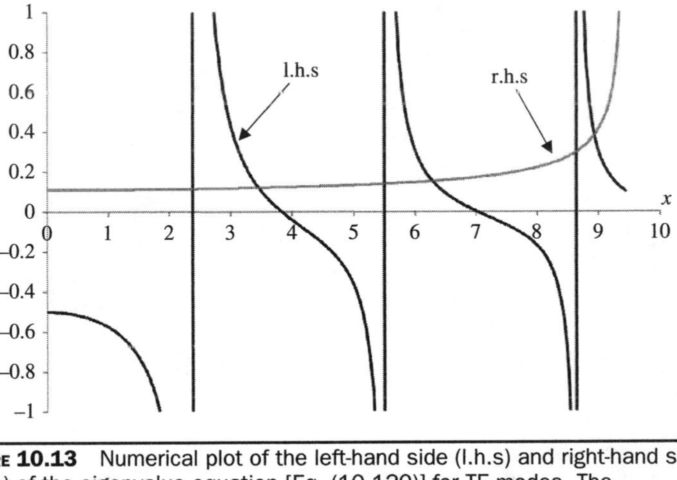

FIGURE 10.13 Numerical plot of the left-hand side (l.h.s) and right-hand side (r.h.s) of the eigenvalue equation [Eq. (10.120)] for TE modes. The parameters are $a = 5\ \mu m$, $\ell = 0$, $\lambda = 1.3\ \mu m$, $n_1 = 1.55$, and $n_2 = 1.5$.

is 9.44 as in Fig. 10.12. There are three solutions for x at 3.46, 6.29, and 8.95. Furthermore, the cut-off value of V for the TE_{01} mode is also 2.405.

EH Modes

Here, $\ell \neq 0$ and the eigenvalue equation is given by

$$\frac{J_\ell'(x)}{xJ_\ell(x)} = \frac{-\left(n_1^2+n_2^2\right)}{2n_1^2}\frac{K_\ell'(y)}{yK_\ell(y)} + \sqrt{\left(\frac{n_1^2-n_2^2}{2n_1^2}\right)^2\left(\frac{K_\ell'(y)}{yK_\ell(y)}\right)^2 + \frac{\ell^2}{n_1^2}\left(\frac{1}{x^2}+\frac{1}{y^2}\right)^2\frac{x_1^2-x^2}{x_0^2}} \tag{10.121}$$

For $\ell \neq 0$, the left-hand side blows up as x goes to zero. We can remove this singularity by subtracting the limiting value of $J_\ell'(x)/xJ_\ell(x)$ from both sides of Eq. (10.121). To see this, note that, (see, for example, Ref. 15)

$$\lim_{x\to 0} J_\ell(x) = \frac{x^\ell}{2^\ell \ell!} \tag{10.122}$$

Hence, as x goes to zero, we have

$$\lim_{x\to 0}\frac{J_\ell'(x)}{xJ_\ell(x)} = \frac{\ell}{x^2} \tag{10.123}$$

Hence, we rewrite the eigenvalue equation for the EH mode as

$$\frac{J_\ell'(x)}{xJ_\ell(x)} - \frac{\ell}{x^2} = \frac{-\left(n_1^2+n_2^2\right)}{2n_1^2}\frac{K_\ell'(y)}{yK_\ell(y)} - \frac{\ell}{x^2} + \sqrt{\left(\frac{n_1^2-n_2^2}{2n_1^2}\right)^2\left(\frac{K_\ell'(y)}{yK_\ell(y)}\right)^2 + \frac{\ell^2}{n_1^2}\left(\frac{1}{x^2}+\frac{1}{y^2}\right)^2\frac{x_1^2-x^2}{x_0^2}} \tag{10.124}$$

Figure 10.14 shows the left-hand side and right-hand side of Eq. (10.124) for $\ell = 1$, $a = 5\ \mu\text{m}$, $\lambda = 1.3\ \mu\text{m}$, $n_1 = 1.55$, and $n_2 = 1.5$ ($V = 9.44$). As can be seen, we have two solutions at $x = 4.64$ (EH_{11} mode) and 7.52 (EH_{12} mode). If $V < 3.83$, the first zero of $J_1(x)$, we will have no propagating EH modes. So 3.83 is the cut-off V-number for the EH_{11} mode.

HE Modes

Similar to the case of EH modes, $\ell \neq 0$ for the HE modes and the eigenvalue equation, with the singularity at $x = 0$ removed, reads

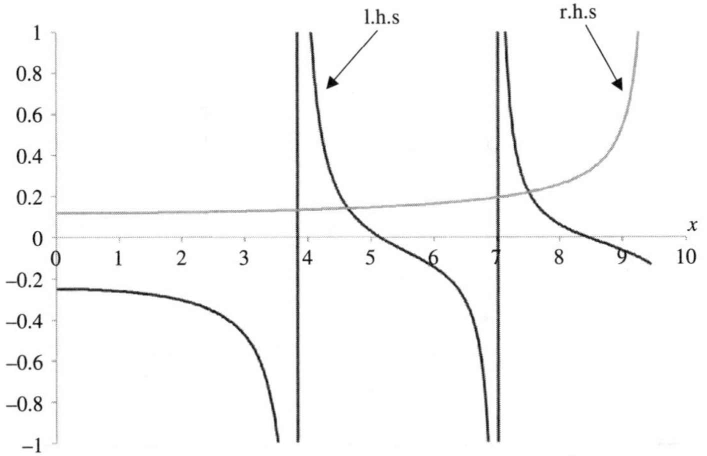

FIGURE 10.14 Numerical plot of the left-hand side (l.h.s) and right-hand side (r.h.s) of the eigenvalue equation [Eq. (10.124)] for the EH modes. The parameters are $\ell = 1$, $a = 5\ \mu\text{m}$, $\lambda = 1.3\ \mu\text{m}$, $n_1 = 1.55$, and $n_2 = 1.5$.

$$\frac{J'_\ell(x)}{xJ_\ell(x)} - \frac{\ell}{x^2} = \frac{-\left(n_1^2 + n_2^2\right)}{2n_1^2}\frac{K'_\ell(y)}{yK_\ell(y)} - \frac{\ell}{x^2}$$

$$- \sqrt{\left(\frac{n_1^2 - n_2^2}{2n_1^2}\right)^2 \left(\frac{K'_\ell(y)}{yK_\ell(y)}\right)^2 + \frac{\ell^2}{n_1^2}\left(\frac{1}{x^2} + \frac{1}{y^2}\right)^2 \frac{x_1^2 - x^2}{x_0^2}} \tag{10.125}$$

A graphical solution of Eq. (10.125) is shown in Fig. 10.15 for $\ell = 1$, $a = 5\ \mu\text{m}$, $\lambda = 1.3\ \mu\text{m}$, $n_1 = 1.55$, and $n_2 = 1.5$ ($V = 9.44$). Three solutions of x are at 2.18 (HE_{11} mode), 4.98 (HE_{12} mode), and 7.71 (HE_{13} mode). As can also be seen from Fig. 10.15, the HE modes have a unique feature which is distinct from the other three mode types discussed earlier. Here, a propagating mode can be obtained for V values all the way to 0. In this case, we say that the cut-off V number is 0 for the HE_{11} mode. The foregoing example clearly shows that if the V number of the step-index fiber is less than 2.405, only the HE_{11} mode will propagate. In this case, the fiber becomes single-mode since only one confined electromagnetic mode can propagate. This configuration is very important for the transmission of light waves with a definite and high-quality beam profile, and minimizes the fluctuations that can occur in the transmitted light due to unwanted coupling between multiple modes.

LP modes

In the previous section, we obtained the exact solution of the eigenvalue equation for the step-index fiber. If the refractive index difference between the core and the cladding is very small, (i.e., $n_1 - n_2 << 1$),

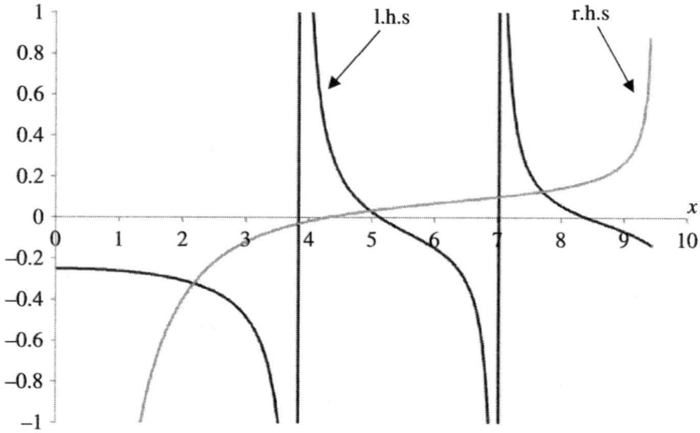

FIGURE 10.15 Numerical plot of the left-hand side (l.h.s) and right-hand side (r.h.s) of the eigenvalue equation [Eq. (10.125)] for HE modes. The parameters are $\ell = 1$, $a = 5\ \mu\text{m}$, $\lambda = 1.3\ \mu\text{m}$, $n_1 = 1.55$, and $n_2 = 1.5$.

then a much simpler solution can be obtained in Cartesian coordinates. These are referred to as the linearly polarized or LP modes of the fiber. Fibers which satisfy this approximation are also known as weakly guiding fibers. The detailed analysis of LP modes may be found in Refs. 1, 4, 5, and 13. As shown in Ref. 4, the eigenvalue equation of the LP modes, each of which is doubly degenerate, becomes

$$x\frac{J_{\ell+1}(x)}{J_\ell(x)} = y\frac{K_{\ell+1}(y)}{K_\ell(y)} \tag{10.126}$$

Each mode is designated as $LP_{\ell m}$. Here, ℓ takes the values 0, 1, 2, 3, whereas the integer m (m = 1, 2, 3,...) gives the mth solution of Eq. (10.126). $LP_{\ell m}$ mode is a superposition of the $HE_{\ell+1,m}$ and $EH_{\ell-1,m}$ modes. The lowest order LP mode corresponding to HE_{11} is the LP_{01} mode. The cut-off value of V for the LP_{01} mode is again 0.

10.3.4 Practical Aspects of Optical Fibers

Mode-Field Diameter and Coupling into Single-Mode Fibers

As we discussed in the electromagnetic analysis of fibers, HE_{11} mode is the fundamental mode of a step-index fiber and has no cut-off condition. In other words, if the core radius a is sufficiently small (for given values of n_1 and n_2) so that the condition

$$V = \frac{2\pi a}{\lambda}\sqrt{n_1^2 - n_2^2} < 2.405 \tag{10.127}$$

is satisfied, then only the HE_{11} mode can be excited and the fiber becomes single-mode. The HE_{11} mode is radially symmetric with decaying wings in the cladding region and can be well approximated by a Gaussian profile of the form

$$E(r) = E_o \exp\left[-\left(\frac{r}{w}\right)^2\right] \tag{10.128}$$

Above, w is chosen in such way as to obtain the best fit with the actual field distribution for the HE_{11} mode. The parameter $2w$ is known as the mode-field diameter and is tabulated for single-mode fibers. It can be shown that, if a Gaussian beam of spotsize w_i is incident on a single-mode fiber with a mode-field diameter of $2w$, the fraction η of the incident power coupled into the fiber, in the absence of reflection and offset/tilt losses, is given by[3]

$$\eta = \frac{4w^2 w_i^2}{\left(w^2 + w_i^2\right)^2} = \frac{4\left(\dfrac{w_i}{w}\right)^2}{\left(1+\left(\dfrac{w_i}{w}\right)^2\right)^2} \tag{10.129}$$

Figure 10.16 shows a plot of η as a function of w_i/w. Note that the maximum amount of coupling occurs when $w_i = w$. Hence, to obtain the maximum coupling into a single-mode fiber, the focusing optics should be designed in such a way as to match the beam waist of the focused beam to the mode field radius of the optical fiber.

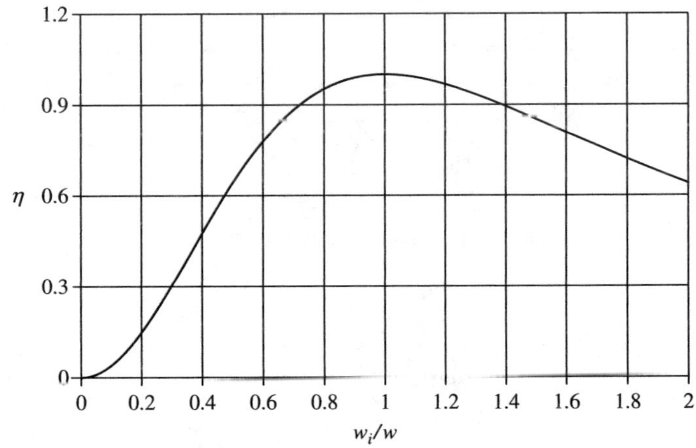

FIGURE 10.16 Variation of the coupling efficiency into a single-mode optical fiber as a function of w_i/w, where w_i is the spotsize of the incident Gaussian beam and $2w$ is the mode-field diameter.

Ray Picture of Multimode Fibers

Figure 10.17 shows a sketch of a step-index fiber with a core diameter of d. If $d >> \lambda$, where λ is the wavelength of light, then the fiber becomes multimode and the ray picture provides a fairly accurate description of how light is guided by the fiber. In this case, rays with an angle of incidence $\theta_i > \theta_c$ will undergo total internal reflection and get guided along the fiber. This is schematically shown in Fig. 10.18. Here, θ_c is the critical angle at the core-cladding interface given by

$$\theta_c = \sin^{-1}\left(\frac{n_2}{n_1}\right) \tag{10.130}$$

The concept of numerical aperture is very useful in describing the guiding properties of multimode fibers. The numerical aperture, abbreviated as N.A., is defined as the sine of the largest incidence angle that gives rise to total internal reflection at the core-cladding interface. Rays with larger angle of incidence escape into the cladding and are not guided. The limiting case is shown in Fig. 10.19.

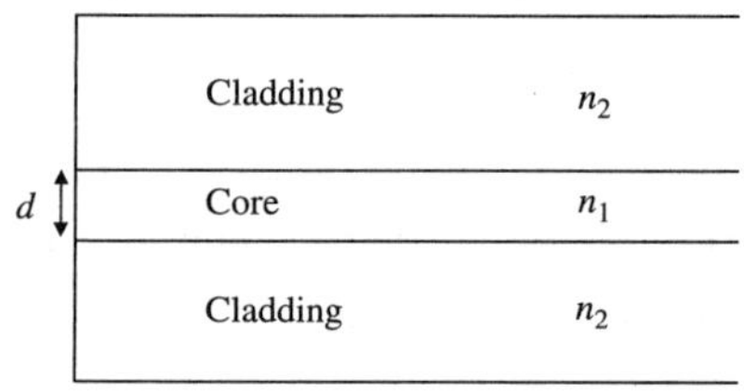

FIGURE 10.17 Sketch of a step-index fiber with a core diameter of d.

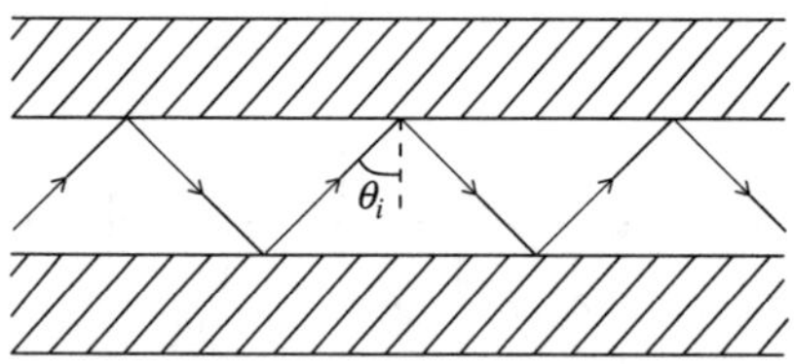

FIGURE 10.18 Internal rays with an incidence angle of $\theta_i > \theta_c$ will undergo total internal reflection and get guided along the fiber.

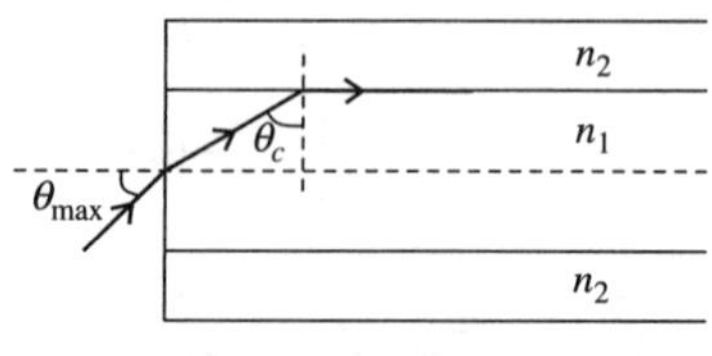

FIGURE 10.19 An incident ray with an incidence angle of θ_{max} hits the cladding-core interface at the critical angle θ_c. The numerical aperture N.A. is given by $\sin\theta_{max}$.

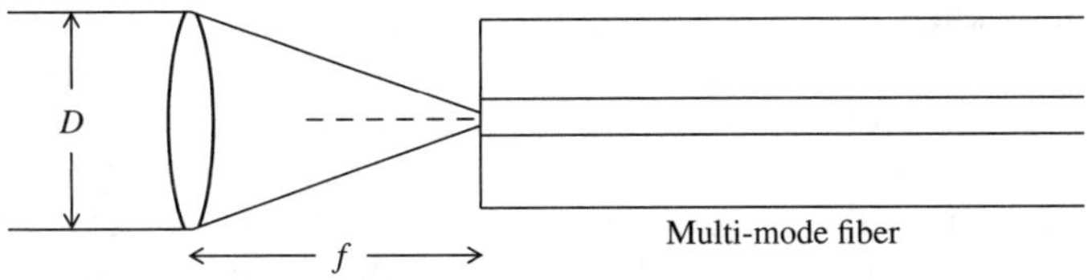

FIGURE 10.20 Optical setup for coupling into a multi-mode fiber.

It can be shown that (see Prob. 10.25) the numerical aperture is given by

$$\text{N.A.} = \sin\theta_{\text{max}} = \sqrt{n_{\text{core}}^2 - n_{\text{clad}}^2} \tag{10.131}$$

To couple light into a multi-mode fiber efficiently, it is important to match the numerical aperture of the focused beam to the numerical aperture of the fiber. Figure 10.20 shows the setup that can be used for coupling into a multi-mode fiber. For efficient coupling, we would require

$$\frac{D}{f} \approx 2\text{N.A.} \tag{10.132}$$

where D is the diameter of the beam incident on the lens, and f is the focal length of the lens. Equation (10.132) gives an approximate design rule. In practice, one optimizes the coupling efficiency by trying lenses with different focal lengths around the value given by Eq. (10.132).

10.4 Emerging Novel Fiber Technologies

There are numerous emerging technologies based on optical fibers, including fiber lasers, double-clad fibers, photonic crystal fibers, and so on. In this section, we provide a brief overview of some of these new and exciting areas in guided-wave photonics.

10.4.1 Fiber Lasers

Fibers doped with active rare-earth ions such as Nd^{3+}, Yb^{3+}, Er^{3+}, Tm^{3+}, and others have been used as lasers[16–20] and optical amplifiers.[21–24] Operation wavelengths vary over a wide spectral range (typically in the 1–3 μm range, although systems such as upconversion fiber lasers, operating outside this range, also exist[25]) depending on the type of glass host and the active ion. For example, Yb^{3+}-doped fiber lasers emit around 1 μm, whereas the output of Er^{3+}-doped fiber lasers is near 1.5 μm. Fibers have several advantages that enable efficient, high-power laser operation. These include high optical gain (since the guided modes can be confined to small volumes over much longer lengths than what is possible in bulk solid-state lasers where limitations arise due to diffractive spreading), rapid dissipation of the pump-induced unused heat (since the fiber diameters are very small in relation to the length of the active medium and the heat load is distributed over longer lengths, again, in comparison with solid-state gain media). In fact, 4% Fresnel reflection from the fiber ends typically provides sufficient feedback to obtain laser oscillation. Since the modes are confined to small volumes, nonlinear effects (which we will study in Chap. 11), such as stimulated Raman scattering, can also play an important role and can be used to obtain amplification and oscillation at wavelengths not overlapping with the emission bands of the laser-active ions.[26,27] Raman fiber amplifiers are also becoming indispensable in long-haul (300–800 km) and ultra-long-haul (>800 km) fiber-optic transmission systems since they enable high-bit-rate communication and improved noise performance.[28]

In the case of high-power fiber lasers, development of double-clad fiber geometries has played an important role in the scaling of the output power levels.[29–31] In this case, the fiber consists of a core region surrounded by two claddings as sketched in Fig. 10.21. What is shown in Fig. 10.21 is a circularly symmetric double-clad fiber. Off-center, rectangular, and other cross-sectional geometries also exist.[32] Only the core region contains the laser-active ions. A high-power pump diode with relatively poor mode quality is coupled into the larger cladding. Over repeated passes through the active core, the pump gets absorbed and generates optical gain for the laser mode which is confined to the core region. When the fiber parameters are properly adjusted, single-mode laser output with high beam quality can be obtained even at very high pumping levels. So, such double-clad fiber geometries can also be thought of as spatial mode converters, whereby the output of a laser with low beam quality can be converted to basically diffraction-limited coherent radiation at the lasing wavelength. Improvements in resonator geometries and pumping schemes yielded high-output powers from fiber lasers and amplifiers entering the hundreds of watts to multi-kW regime.[33–35] Other

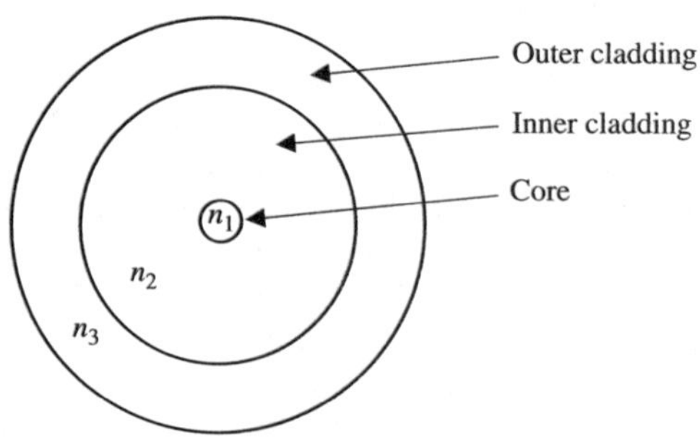

FIGURE 10.21 Sketch showing the cross section of a circularly symmetric, double-clad fiber. The central core region is surrounded by two cladding layers each with a different index.

competitors in the race toward high output powers include thin-disk,[36] slab,[37] and ceramic lasers.[38]

Fiber lasers are also suitable for the generation of ultrashort optical pulses, since the doped active ions provide gain over large spectral windows. Various mode-locking techniques have been employed to generate femtosecond optical pulses from fiber lasers.[39–43] Mode-locked fiber lasers have found applications in spectroscopy, materials processing, and high-precision optical metrology.[44, 45]

10.4.2 Novel Fiber Architectures

Since their demonstration in 1996, photonic crystal fibers (PCF) [46–48] have been widely used in many areas of photonics. Similar to photonic crystals,[49] PCFs have a periodic, microstructured cross section such as the one shown in Fig. 10.22. Here, black circular regions designate holes (filled with air) and the white background is pure glass. Microstructuring can be done in other ways as well. For example, the air-filled holes in Fig. 10.22 can be replaced by doped glass to provide index contrast. What is amazing about PCFs is that by adjusting the ratio of the hole diameter to the spacing between the holes, in other words, by engineering the geometry of the microstructured cross section, the optical characteristics of the fiber can be dramatically modified. One important case is what is termed as an endlessly single-mode (ESM) PCF,[46] where no matter how small the wavelength is, the fiber remains single-mode at all times. It is also possible to adjust the PCF parameters to obtain a desired amount of dispersion in a particular spectral region.[50] In addition, PCFs are ideal for laser applications. In this case, laser-active rare-earth ions are doped into the glass during the preparation of the preform. Furthermore, by properly designing

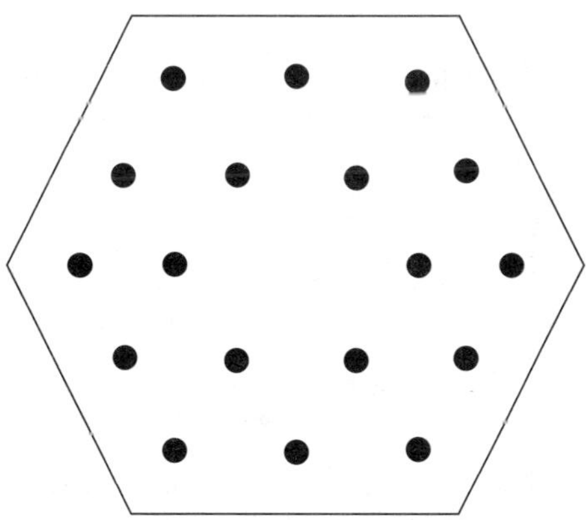

FIGURE 10.22 Sketch of a photonic crystal fiber with a microstructured cross section (adapted from Ref. 48). Here, black circular regions designate air-filled holes and the white background is pure glass.

the cross-section parameters, large-area cores can be obtained to enable the propagation of high-power modes, which facilitates power scaling.[35] We will not present a detailed electromagnetic analysis of PCFs here. The reader should consult Refs. 46 to 48 for a detailed treatment.

Another method of obtaining large mode areas for power scaling is through use of what has been termed as "gain-guided, index anti-guided" (GG+IAG) optical fibers.[51–53] Here, we have a conventional step-index fiber where the refractive index of the core is lower than the refractive index of the cladding (hence the term "index anti-guided"). It can be shown that when the core region of this anti-guiding fiber is doped with active ions and pumped to generate a sufficiently large optical gain, guiding of electromagnetic modes with a large effective area becomes possible. One advantage of (GG+IAG) fibers is that complex microstructuring of the cross section is not needed to obtain large mode areas. Ref. 52 provides a theoretical analysis of this class of fibers.

Recently, omni-directional dielectric mirror fibers[54–57] have been developed for applications in telecommunications, high-power beam delivery, and the development of fabrics as radiation barriers. Here, the core region (which can be hollow) is surrounded by alternating layers of materials with different refractive indices. The operation of this device is similar to the multilayer high-reflection mirrors discussed in Chap. 1. One added feature is that, when properly designed, the multilayer structure provides high reflection at all incidence angles and all polarizations in a particular wavelength range and confines the radiation in the core region.[55] The fibers can be designed to enable single-mode transmission of the TE_{01} mode with a large core area due to the larger losses experienced by the higher-order modes.[56]

Problems

10.1 Show that for a translationally invariant dielectric waveguide, the transverse magnetic field can be expressed in terms of the axial electric and magnetic field by using Eq. (10.18).

10.2 Derive Eq. (10.21) in Cartesian coordinates.

10.3 Show that in Cartesian coordinates, H_x and H_y are given in terms of the axial field components by Eqs. (10.26) and (10.27).

10.4 Show that in Cartesian coordinates, the longitudinal component H_z of the magnetic field satisfies the scalar wave equation given by Eq. (10.31).

10.5 Show that in cylindrical coordinates, the transverse components of the magnetic field are given by Eqs. (10.39) and (10.40).

10.6 Show that the wave equation satisfied by E_z in cylindrical coordinates is given by Eq. (10.41).

10.7 Show that the wave equation satisfied by H_z in cylindrical coordinates is given by Eq. (10.42).

10.8 Show that, based on Eqs. (10.57), (10.58) and (10.60), (10.61), the characteristic equation for the guided TE modes of the slab waveguide is given by Eq. (10.62).

10.9 Based on Eqs. (10.63) and (10.64), obtain the expression for H_y of the TM mode in different regions of the rectangular dielectric slab waveguide.

10.10 Apply the boundary conditions on the tangential field components of the TM mode at $x = 0$ and $x = d$ of the dielectric waveguide to derive the eigenvalue equation given in Eq. (10.65).

10.11 Write a numerical code to determine the solutions for $\bar{x}$ and β/k_0 for the TE modes of the rectangular dielectric waveguide with $n_1 = 1$, $n_2 = 1.50$, $n_3 = 1.40$, $\lambda = 1.3\ \mu\text{m}$, and $d = 3\ \mu\text{m}$. The solution points are depicted in Fig. 10.3.

10.12 Show that the electric field E_y for the TE mode of the rectangular dielectric waveguide is given by Eqs. (10.72) and (10.73) up to a multiplicative constant. Write a numerical code to plot E_y as a function of x/d for each of the guided modes determined in Prob. 10.11. This is further shown in Fig. 10.4*a* to *c*.

10.13 Find the allowed β/k_0 values for the TE modes of a rectangular dielectric slab waveguide with the parameters $n_1 = 1.3$, $n_2 = 1.35$, $n_3 = 1.32$, $\lambda = 1\ \mu\text{m}$, and $d = 8\ \mu\text{m}$. For each guided mode, plot the field distribution as a function of x/d.

10.14 Write a numerical code to determine the allowed values of $\kappa_2 d$ and β/k_0 for the TM modes of the rectangular dielectric waveguide with $n_1 = 1$, $n_2 = 1.50$, $n_3 = 1.40$, $\lambda = 1.3\ \mu\text{m}$, and $d = 3\ \mu\text{m}$. The graphical solution is depicted in Fig. 10.6.

10.15 Show that the magnetic field H_y for the TM mode of the rectangular dielectric waveguide is given by Eq. (10.75) up to a multiplicative constant. Write a numerical code to plot H_y as a function of x/d for each of the guided modes determined in Prob. 10.14. This is further shown in Fig. 10.7*a* to *c*.

10.16 Use a numerical computation program such as Mathematica to generate the plots of the three lowest-order Bessel functions given in Figs. 10.10 and 10.11.

10.17 By computing the determinant of the matrix given in Eq. (10.105), show that the relation $\det \mathbf{M} = 0$ can be expressed as Eq. (10.107).

10.18 Show that the eigenvalue equation for the step-index fiber can be expressed in terms of the normalized parameters as Eq. (10.116).

10.19 Plot the left-hand side and right-hand side of Eq. (10.118) for the TM modes of the step-index optical fiber with the parameters $a = 5\ \mu\text{m}$, $\ell = 0$, $\lambda = 1.3\ \mu\text{m}$, $n_1 = 1.55$, and $n_2 = 1.5$. This should generate a plot similar to

Fig. 10.12. Write a numerical code to determine the solutions of the normalized propagation parameter x.

10.20 Plot the left-hand side and right-hand side of Eq. (10.120) for the TE modes of the step-index optical fiber with the parameters $a = 5\ \mu\text{m}$, $\ell = 0$, $\lambda = 1.3\ \mu\text{m}$, $n_1 = 1.55$, and $n_2 = 1.5$. This should generate a plot similar to Fig. 10.13. Write a numerical code to determine the solutions of the normalized propagation parameter x.

10.21 Plot the left-hand and right-hand sides of Eq. (10.124) for the EH modes of the step-index optical fiber with the parameters $a = 5\ \mu\text{m}$, $\ell = 1$, $\lambda = 1.3\ \mu\text{m}$, $n_1 = 1.55$, and $n_2 = 1.5$ (see Fig. 10.14). Write a numerical code to determine the solutions of the normalized propagation parameter x.

10.22 Plot the left-hand and right-hand sides of Eq. (10.125) for the HE modes of the step-index optical fiber with the parameters $a = 5\ \mu\text{m}$, $\ell = 1$, $\lambda = 1.3\ \mu\text{m}$, $n_1 = 1.55$, and $n_2 = 1.5$ (see Fig. 10.15). Write a numerical code to determine the solutions of the normalized propagation parameter x.

10.23 A step-index fiber has a core index of 1.45 and cladding index of 1.43. What is the maximum radius above which the fiber becomes multimode? Take the wavelength as 1.26 μm.

10.24 A Gaussian beam with a spotsize of 7 μm is incident on a single-mode fiber whose mode-field diameter is 8 μm. What fraction of the incident power is coupled into the fiber neglecting Fresnel losses?

10.25 Show that the numerical aperture of a multi-mode optical fiber is given by Eq. (10.131).

References

1. A. H. Cherin, *An Introduction to Optical Fibers* (McGraw-Hill, New York, 1983).
2. J. A. Kong, *Electromagnetic Wave Theory* (EMW, Cambridge, Mass., 2000).
3. A. W. Snyder and J. D. Love, *Optical Waveguide Theory* (Chapman and Hall, London, 1983).
4. A. Yariv, *Optical Electronics in Modern Communications*, 5th ed. (Oxford University Press, New York, 1997).
5. C. R. Pollock, *Fundamentals of Optoelectronics* (Irwin, Chicago, 1995).
6. B. E. A. Saleh and M. C. Teich, *Fundamentals of Photonics*, 2d ed. (Wiley, Hoboken, N.J., 2007).
7. C. R. Doerr and H. Kogelnik, "Dielectric waveguide theory," *Journal of Lightwave Technology*, **26**(9–12), 1176–1187 (2008).
8. W. A. Gambling, "The rise and rise of optical fibers," *IEEE Journal of Selected Topics in Quantum Electronics*, **6**(6), 1084–1093 (2000).
9. J. D. Jackson, *Classical Electrodynamics*, 3d ed. (John Wiley & Sons, Inc., New York, 1998).
10. K. C. Kao and G. A. Hockham, "Dielectric-fibre surface waveguides for optical frequencies," *Proceedings of the Institution of Electrical Engineers-London*, **113**(7), 1151–1158 (1966).
11. F. P. Kapron, D. B. Keck, and R. D. Maurer, "Radiation losses in glass optical waveguides," *Applied Physics Letters*, **17**, 423–425 (1970).
12. E. Snitzer, "Cylindrical dielectric waveguide modes," *Journal of the Optical Society of America*, **51**, 491–498 (1961).

13. D. Gloge, "Weakly guiding fibers," *Applied Optics*, **10**, 2252–2258 (1971).
14. G. P. Agrawal, *Nonlinear Fiber Optics* (Academic Press, Boston, 1989).
15. G. B. Arfken and H. J. Weber, *Mathematical Methods for Physicists*, 5th ed. (Harcourt Academic Press, Amsterdam, 2001).
16. C. J. Koester, "Laser action by enhanced total internal reflection," *IEEE Journal of Quantum Electronics*, **QE 2**(9), 580–584 (1966).
17. R. J. Mears, L. Reekie, S. B. Poole, and D. N. Payne, "Neodymium-doped silica single-mode fiber lasers," *Electronics Letters*, **21**(17), 738–740 (1985).
18. H. M. Pask, R. J. Carman, D. C. Hanna, A. C. Tropper, C. J. Mackechnie, P. R. Barber, and J. M. Dawes, "Ytterbium-doped silica fiber lasers: Versatile sources for the 1-1.2 Mu-M region," *IEEE Journal of Selected Topics in Quantum Electronics*, **1**(1), 2–13 (1995).
19. A. Mori, Y. Ohishi, and S. Sudo, "Erbium-doped tellurite glass fibre laser and amplifier," *Electronics Letters*, **33**(10), 863–864 (1997).
20. P. F. Moulton, G. A. Rines, E. V. Slobodtchikov, K. F. Wall, G. Frith, B. Samson, and A. L. G. Carter, "Tm-doped fiber lasers: Fundamentals and power scaling," *IEEE Journal of Selected Topics in Quantum Electronics*, **15**(1), 85–92 (2009).
21. C. J. Koester and E. Snitzer, "Amplification in fiber laser," *Applied Optics*, **3**(10), 1182–1186 (1964).
22. R. J. Mears, L. Reekie, I. M. Jauncey, and D. N. Payne, "Low-noise erbium-doped fiber amplifier operating at 1.54 micron," *Electronics Letters*, **23**(19), 1026–1028 (1987).
23. C. R. Giles and E. Desurvire, "Modeling erbium-doped fiber amplifiers," *Journal of Lightwave Technology*, **9**(2), 271–283 (1991).
24. W. J. Miniscalco, "Erbium-doped glasses for fiber amplifiers at 1500 nm," *Journal of Lightwave Technology*, **9**(2), 234–250 (1991).
25. J. Y. Allain, M. Monerie, and H. Poignant, "Blue upconversion fluorozirconate fiber laser," *Electronics Letters* **26**(3), 166–168 (1990).
26. R. H. Stolen, "Fiber Raman lasers," *Fiber and Integrated Optics*, **3**(1), 21–52 (1980).
27. R. H. Stolen and E. P. Ippen, "Raman gain in glass optical waveguides," *Applied Physics Letters*, **22**(6), 276–278 (1973).
28. M. N. Islam, "Raman amplifiers for telecommunications," *IEEE Journal of Selected Topics in Quantum Electronics*, **8**(3), 548–559 (2002).
29. J. D. Kafka, "Laser diode pumped fiber lasers without pump cavity," United States Patent 4829529 (May 9, 1989).
30. E. Snitzer, H. Po, R. P. Tumminelli, and F. Hakimi, "Optical fiber lasers and amplifiers," United States Patent 4815079 (March 21, 1989).
31. L. Zenteno, "High-power double-clad fiber lasers," *Journal of Lightwave Technology*, **11**(9), 1435–1446 (1993).
32. A. P. Liu and K. Ueda, "The absorption characteristics of circular, offset, and rectangular double-clad fibers," *Optics Communications*, **132**(5–6), 511–518 (1996).
33. Y. Jeong, J. K. Sahu, D. N. Payne, and J. Nilsson, "Ytterbium-doped large-core fiber laser with 1.36 kW continuous-wave output power," *Optics Express*, **12**(25), 6088–6092 (2004).
34. G. D. Goodno, L. D. Book, and J. E. Rothenberg, "Low-phase-noise, single-frequency, single-mode 608 W thulium fiber amplifier," *Optics Letters*, **34**(8), 1204–1206 (2009).
35. J. Limpert, F. Roser, S. Klingebiel, T. Schreiber, C. Wirth, T. Peschel, R. Eberhardt, and A. Tunnermann, "The rising power of fiber lasers and amplifiers," *IEEE Journal of Selected Topics in Quantum Electronics*, **13**(3), 537–545 (2007).
36. A. Giesen, H. Hugel, A. Voss, K. Wittig, U. Brauch, and H. Opower, "Scalable concept for diode-pumped high-power solid-state lasers," *Applied Physics B-Lasers and Optics*, **58**(5), 365–372 (1994).
37. T. S. Rutherford, W. M. Tulloch, E. K. Gustafson, and R. L. Byer, "Edge-pumped quasi-three-level slab lasers: Design and power scaling," *IEEE Journal of Quantum Electronics*, **36**(2), 205–219 (2000).

38. A. Ikesue and Y. L. Aung, "Ceramic laser materials," *Nature Photonics*, **2**(12), 721–727 (2008).
39. M. E. Fermann and I. Hartl, "Ultrafast fiber laser technology," *IEEE Journal of Selected Topics in Quantum Electronics*, **15**(1), 191–206 (2009).
40. K. Tamura, E. P. Ippen, H. A. Haus, and L. E. Nelson, "77-Fs pulse generation from a stretched-pulse mode-locked all-fiber ring laser," *Optics Letters*, **18**(13), 1080–1082 (1993).
41. H. A. Haus, K. Tamura, L. E. Nelson, and E. P. Ippen, "Stretched-pulse additive-pulse mode-locking in fiber ring lasers: theory and experiment," *IEEE Journal of Quantum Electronics*, **31**(3), 591–598 (1995).
42. H. A. Haus, E. P. Ippen, and K. Tamura, "Additive-pulse modelocking in fiber lasers," *IEEE Journal of Quantum Electronics*, **30**(1), 200–208 (1994).
43. F. O. Ilday, J. R. Buckley, W. G. Clark, and F. W. Wise, "Self-similar evolution of parabolic pulses in a laser," *Physical Review Letters*, **92**(21), 2133902 (2004).
44. T. R. Schibli, I. Hartl, D. C. Yost, M. J. Martin, A. Marcinkevicius, M. E. Fermann, and J. Ye, "Optical frequency comb with submillihertz linewidth and more than 10 W average power," *Nature Photonics* **2**(6), 355–359 (2008).
45. M. E. Fermann, A. Galvanauskas, and G. Sucha, (eds), *Ultrafast Lasers: Technology and Applications* (Marcel Dekker, New York, 2005).
46. T. A. Birks, J. C. Knight, and P. S. Russell, "Endlessly single-mode photonic crystal fiber," *Optics Letters*, **22**(13), 961–963 (1997).
47. J. C. Knight, "Photonic crystal fibres," *Nature*, **424**(6950), 847–851 (2003).
48. P. S. J. Russell, "Photonic-crystal fibers," *Journal of Lightwave Technology* **24**(12), 4729–4749 (2006).
49. E. Yablonovitch, "Inhibited spontaneous emission in solid-state physics and electronics," *Physical Review Letters*, **58**(20), 2059–2062 (1987).
50. T. A. Birks, D. Mogilevtsev, J. C. Knight, and P. S. Russell, "Dispersion compensation using single-material fibers," *IEEE Photonics Technology Letters*, **11**(6), 674–676 (1999).
51. A. E. Siegman, Y. Chen, V. Sudesh, M. C. Richardson, M. Bass, P. Foy, W. Hawkins, and J. Ballato, "Confined propagation and near single-mode laser oscillation in a gain-guided, index antiguided optical fiber," *Applied Physics Letters*, **89**(25), 251101 (2006).
52. A. E. Siegman, "Gain-guided, index-antiguided fiber lasers," *Journal of the Optical Society of America B-Optical Physics*, **24**(8), 1677–1682 (2007).
53. V. Sudesh, T. McComb, Y. Chen, M. Bass, M. Richardson, J. Ballato, and A. E. Siegman, "Diode-pumped 200 mu m diameter core, gain-guided, index-antiguided single mode fiber laser," *Applied Physics B-Lasers and Optics*, **90**(3-4), 369–372 (2008).
54. Y. Fink, J. N. Winn, S. H. Fan, C. P. Chen, J. Michel, J. D. Joannopoulos, and E. L. Thomas, "A dielectric omnidirectional reflector," *Science*, **282**(5394), 1679–1682 (1998).
55. S. D. Hart, G. R. Maskaly, B. Temelkuran, P. H. Prideaux, J. D. Joannopoulos, and Y. Fink, "External reflection from omnidirectional dielectric mirror fibers," *Science*, **296**(5567), 510–513 (2002).
56. S. G. Johnson, M. Ibanescu, M. Skorobogatiy, O. Weisberg, T. D. Engeness, M. Soljacic, S. A. Jacobs, J. D. Joannopoulos, and Y. Fink, "Low-loss asymptotically single-mode propagation in large-core OmniGuide fibers," *Optics Express* **9**(13), 748–779 (2001).
57. J. N. Winn, Y. Fink, S. H. Fan, and J. D. Joannopoulos, "Omnidirectional reflection from a one-dimensional photonic crystal," *Optics Letters*, **23**(20), 1573–1575 (1998).

CHAPTER 11

Nonlinear Optics

Earlier in this book we have investigated two types of photonic materials:

1. Isotropic, homogeneous, linear materials
2. Anisotropic linear materials

The first class of materials is described by a scalar electric susceptibility function $\chi^{(1)}$. As such, the induced polarization of the medium is linearly proportional and parallel to the applied electric field. Furthermore, $\chi^{(1)}$ in general, has frequency dependence, as we discussed in Chap. 4. In the second group, each component of the dipole moment can be expressed as a linear combination of the electric field components, but the direction of the induced polarization is in general not the same as that of the applied electric field. As we saw in Chap. 8, such birefringent materials can be characterized by a susceptibility tensor $\chi_{ij}^{(1)}$ and used in the design of many useful polarization control devices. Going one step forward, we find that as the strength of the applied electric field is increased, the induced polarization now contains terms that depend on the square, cube, and/or higher orders of the electric field components. Consider, for example, the particular case of second-order contributions. Here, the induced polarization will depend on the square of the applied electric field. In an anisotropic medium, we will in general have the ith component of the polarization vector depend on various product combinations involving two components of the electric field vector. Motivated by this observation, we can express the ith component of the polarization vector P_i as

$$P_i = \varepsilon_o \left(\chi_{ij}^{(1)} E_j + \chi_{ijk}^{(2)} E_j E_k + \chi_{ijkl}^{(3)} E_j E_k E_l + \cdots \right) \tag{11.1}$$

where, $\chi_{ijk}^{(2)}$ and $\chi_{ijkl}^{(3)}$ are the second-order and third-order susceptibility tensors, and Einstein summation convention is assumed.

Nonlinear optics deals with light-matter interactions in media where the higher-order susceptibilities $\chi_{ijk}^{(2)}$ and $\chi_{ijkl}^{(3)}$ are no longer negligible. In other words, the electric field strength in this case is strong enough that linear dependence between $\vec{P}_i$ and $\vec{E}$ is no longer sufficient to account for all of the observed effects. These effects were much harder to observe before the advent of lasers. However, following the invention of lasers, it has become commonplace to produce ultrabright, coherent light and observe nonlinear optical phenomena.

Investigating nonlinear phenomena is not just an academic curiosity. Many useful, practical devices based on nonlinear interactions exist and are employed in photonic systems. To give an example, let us look at second-order effects. Here, by sending two intense light beams at two different frequencies of ω_1 and ω_2, it is possible to generate additional wave components at $2\omega_1$, $2\omega_2$, $\omega_1 + \omega_2$, $\omega_1 - \omega_2$. These processes are respectively known as second harmonic, sum frequency, and difference frequency generation. Typically, the propagation direction is adjusted to optimize one of these processes as we will discuss later. For example, with an infrared laser operating near 1 μm and with a suitable nonlinear crystal, it is possible to generate, by using a second harmonic crystal, light near 500 nm. Whereas the original beam is invisible to the eye, passage through the nonlinear crystal generates visible green radiation. Not all of the 1 μm light will be converted to green. In a practical second-harmonic generation setup, it is important to know how the conversion efficiency will vary with the intensity of the fundamental light and the properties of the second harmonic crystal.

Our concise treatment of nonlinear optics will consist of two major parts. In the first part, we discuss second-order nonlinearities and related phenomena including second harmonic generation, parametric amplification, and parametric oscillation. A simple model will be developed to determine the conversion efficiency for plane-wave and Gaussian beam input. Phase matching calculation in uniaxial crystals will also be described. The second part concentrates on third-order nonlinearities, where we provide a survey of different phenomena without going into a detailed investigation. We discuss self-focusing, Raman effect, self-phase modulation, and pulse propagation in nonlinear dispersive media. For further treatment of nonlinear optical effects, the reader is referred to Refs. 1 to 10.

11.1 Second-Order Nonlinear Optics

11.1.1 Classical Derivation of the Nonlinear Polarization and Susceptibility

Our treatment of second-order nonlinear effects will be classical. We will consider a crystal which lacks inversion symmetry and investigate the motion of a bound electron under the action of a driving oscillatory electric field. As we will show, the presence of the cubic term in the potential leads to second-order optical effects.

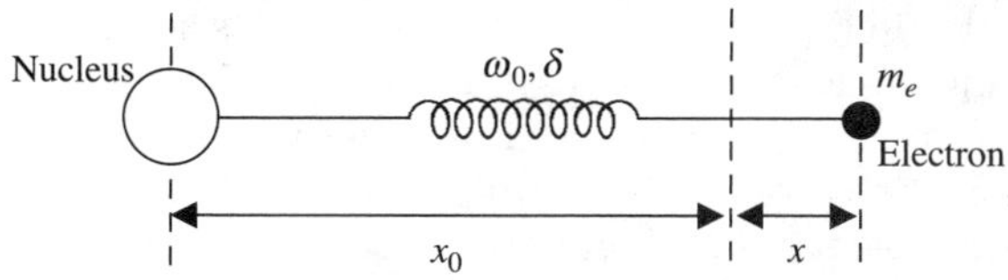

FIGURE 11.1 Electron bound with a nonlinear spring-like restoring force to an atom.

To derive the nonlinear polarization, we consider an atom with one bound electron of mass m_e. We further assume that the bound electron can be modeled as a classical damped harmonic oscillator as was done in Chap. 4. This is sketched in Fig. 11.1. We will further limit ourselves to one-dimensional motion.

Let x be the instantaneous deviation of the electron position from the equilibrium point. The potential energy $V(x)$ around the equilibrium point can be expressed as

$$V(x) = V_0 + \left.\frac{dV}{dx}\right|_{x=0} x + \frac{1}{2}\left.\frac{d^2V}{dx^2}\right|_{x=0} x^2 + \frac{1}{6}\left.\frac{d^3V}{dx^3}\right|_{x=0} x^3 + h.o.t \tag{11.2}$$

Above, *h.o.t* represents higher-order terms. Since only changes in the potential energy are important, we take $V_0 = 0$. Furthermore, since $x = 0$ is the equilibrium point, the restoring force $F = -dV/dx = 0$ at $x = 0$. In crystals with no inversion symmetry, the potential will not have even symmetry any more. In other words, $V(x) \neq V(-x)$. This means that in addition to the quadratic term, the first higher order contribution to $V(x)$ will be cubic in x, giving

$$V(x) = \frac{1}{2}\kappa x^2 + \frac{1}{3}\kappa_1 x^3 + \cdots \tag{11.3}$$

where

$$\kappa = \left.\frac{d^2V}{dx^2}\right|_{x=0} \tag{11.4}$$

and

$$\kappa_1 = \frac{1}{2}\left.\frac{d^3V}{dx^3}\right|_{x=0} \tag{11.5}$$

are the linear and nonlinear spring constants. The restoring force acting on the electron hence becomes

$$\vec{F}_r = -(\kappa x + \kappa_1 x^2)\hat{x} \tag{11.6}$$

The first term in Eq. (11.6) is the familiar Hooke's law contribution to the restoring force. We further assume that the electron is subject to a damping force $-\gamma m_e(dx/dt)\hat{x}$ and an external driving force due to electromagnetic waves at the frequencies ω_1 and ω_2. From Newton's second law, the equation of motion can be written as

$$\frac{d^2x}{dt^2}+\gamma\frac{dx}{dt}+\omega_0^2x+\delta x^2=-\frac{e}{2m_e}\big(E_{10}\exp(i\omega_1 t)+E_{10}^*\exp(-i\omega_1 t)$$
$$+E_{20}\exp(i\omega_2 t)+E_{20}^*\exp(-i\omega_2 t)\big) \tag{11.7}$$

where

$$\omega_0^2=\frac{\kappa}{m_e} \tag{11.8}$$

and

$$\delta=\frac{\kappa_1}{m_e} \tag{11.9}$$

In Eq. (11.7), ω_0 is the natural oscillation frequency at small amplitudes, m_e is the mass of the electron, δ gives the cubic contribution to the potential energy, and E_{10} and E_{20} are the electric field amplitudes at the frequencies ω_1 and ω_2, respectively. As we did in Chap. 4, we are neglecting here the magnetic contribution to the Lorentz force by assuming that electron velocities are small compared to the speed of light. To solve Eq. (11.7), we assume that δ is small and employ the perturbation method to expand the solution as a power series in δ:

$$x(t)=x_1(t)+\delta x_2(t)+\delta^2x_3(t)+\cdots \tag{11.10}$$

Substituting into Eq. (11.7), we get

$$\frac{d^2}{dt^2}(x_1+\delta x_2+\cdots)+\gamma\frac{d}{dt}(x_1+\delta x_2+\cdots)+\omega_0^2(x_1+\delta x_2+\cdots)$$
$$+\delta(x_1+\delta x_2+\cdots)^2=-\frac{e}{2m_e}\big[E_{01}\exp(i\omega_1 t)$$
$$+E_{10}^*\exp(-i\omega_1 t)+E_{20}\exp(i\omega_2 t)+E_{20}^*\exp(-i\omega_2 t)\big] \tag{11.11}$$

Collecting the terms which are independent of δ, we obtain

$$\frac{d^2x_1}{dt^2}+\gamma\frac{dx_1}{dt}+\omega_0^2x_1=-\frac{e}{2m_e}\big[E_{10}\exp(i\omega_1 t)+E_{10}^*\exp(-i\omega_1 t)$$
$$+E_{20}\exp(i\omega_2 t)+E_{20}^*\exp(-i\omega_2 t)\big] \tag{11.12}$$

Equating the coefficients of the terms containing δ further gives

$$\frac{d^2x_2}{dt^2}+\gamma\frac{dx_2}{dt}+\omega_o^2x_2+x_1^2=0 \tag{11.13}$$

To find the zeroth-order solution, we assume that x_1 can be expressed in the form

$$x_1(t)=\frac{1}{2}[x_{10}(\omega_1)\exp(i\omega_1 t)+x_{10}(\omega_2)\exp(i\omega_2 t)+c.c] \tag{11.14}$$

Above, *c.c.* denotes the complex conjuate of the preceding terms.

Substituting into Eq. (11.12), we note that to zeroth order in δ, the two solutions at the frequencies ω_1 and ω_2 remain decoupled. This is in fact very similar to the classical damped oscillator model which we saw back in Chap. 4. Collecting coefficients of $\exp(i\omega_1 t)$, we get

$$\left(-\omega_1^2+i\gamma\omega_1+\omega_0^2\right)\frac{x_{10}(\omega_1)}{2}=-\frac{e}{2m_e}E_{10} \tag{11.15}$$

or

$$x_{10}(\omega_1)=-\frac{e}{m_e}E_{10}\frac{1}{\omega_0^2-\omega_1^2+i\gamma\omega_1} \tag{11.16}$$

Similarly, $x_{10}(\omega_2)$ becomes

$$x_{10}(\omega_2)=\frac{-eE_{20}}{m_e}\frac{1}{\omega_0^2-\omega_2^2+i\gamma\omega_2} \tag{11.17}$$

From these relations, we can derive the linear dielectric susceptibility χ at the frequency ω. To zeroth order in δ, the polarization vector at ω_1, for example, will be given by

$$P(\omega_1,t)=-\frac{Ne}{2}[x_{10}(\omega_1)\exp(i\omega_1 t)+c.c.]=\frac{\varepsilon_o}{2}[\chi(\omega_1)E_{10}\exp(i\omega_1 t)+c.c.] \tag{11.18}$$

where N is the atomic density and ε_0 is the vacuum permittivity. Hence, by using Eqs. (11.16) and (11.18), the linear susceptibility at ω_1 becomes

$$\chi(\omega_1)=-\frac{Nex_{10}(\omega_1)}{\varepsilon_0 E_{10}}=\frac{Ne^2}{m_e\varepsilon_o}\frac{1}{\omega_0^2-\omega_1^2+i\gamma\omega_1} \tag{11.19}$$

This is exactly what we obtained earlier by using the Lorentz model of the atom. Now, it represents the zeroth-order solution of the perturbation expansion given in Eq. (11.10). A similar relation can be obtained at the frequency ω_2.

Now, we examine Eq. (11.13) to derive the solution which is first order in δ. Note that now, we have the term $x_1(t)^2$ in this equation. By using Eqs. (11.16) and (11.17), we can express $x_1(t)^2$ as

$$x_1(t)^2 = \frac{1}{4}\begin{bmatrix} (x_{10}(\omega_1))^2\exp(2i\omega_1 t) + (x_{10}(\omega_2))^2\exp(2i\omega_2 t) + 2x_{10}(\omega_1)x_{10}(\omega_2)\exp(i(\omega_1+\omega_2)t) \\ + (x_{10}(\omega_1)^*)^2\exp(-2i\omega_1 t) + (x_{10}(\omega_2)^*)^2\exp(-2i\omega_2 t) \\ + 2x_{10}(\omega_1)^* x_{10}(\omega_2)^*\exp(-i(\omega_1+\omega_2)t) + 2x_{10}(\omega_1)x_{10}(\omega_1)^* \\ + 2x_{10}(\omega_1)x_{10}(\omega_2)^*\exp(i(\omega_1-\omega_2)t) + 2x_{10}(\omega_1)^* x_{10}(\omega_2)\exp(i(\omega_2-\omega_1)t) \\ + 2x_{10}(\omega_2)x_{10}(\omega_2)^* \end{bmatrix} \tag{11.20}$$

The above equation is somewhat crowded but it goes to show the nature of the material response for second-order interactions. As can be seen from Eq. (11.20), we get electron displacements, and hence polarization terms, oscillating at the frequencies $2\omega_1$, $2\omega_2$, $(\omega_1+\omega_2)$, and $(\omega_1-\omega_2)$. We also get a nonlinear DC displacement. For example, to find the second-order response of the medium, we assume that the displacement at the frequency $2\omega_1$ can be written in the form

$$x_2(2\omega_1, t) = \frac{1}{2}[x_{20}(2\omega_1)\exp(2i\omega_1 t) + c.c.] \tag{11.21}$$

By using Eqs. (11.13) and (11.20) with $\omega = 2\omega_1$, we obtain

$$x_{20}(2\omega_1) = \frac{-\frac{1}{2}x_{10}(\omega_1)^2}{\omega_0^2 - 4\omega_1^2 + i2\omega_1\gamma} = \frac{-\frac{1}{2}\frac{e^2E_{10}^2}{m_e^2}}{\left[(\omega_0^2-\omega_1^2)+i\gamma\omega_1\right]^2\left(\omega_0^2-4\omega_1^2+i2\gamma\omega_1\right)} \tag{11.22}$$

The corresponding polarization at $2\omega_1$ can be written in the form

$$\begin{aligned} P(2\omega_1, t) &= -\frac{Ne}{2}[\delta x_{20}(2\omega_1)\exp(i2\omega_1 t) + c.c.] \\ &= \frac{1}{2}\left[d(2\omega_1)E_{10}^2\exp(i2\omega_1 t) + c.c.\right] \\ &= \frac{1}{2}[P_0(2\omega_1)\exp(i2\omega_1 t) + c.c.] \end{aligned} \tag{11.23}$$

where the d-coefficient which relates the second-order polarization amplitude at $2\omega_1$ to E_{10}^2 is now given by

$$d(2\omega_1) = \frac{P_0(2\omega_1)}{E_{10}^2} = \frac{\delta N e^3}{2m_e^2\left[(\omega_0^2 - \omega_1^2) + i\gamma\omega_1\right]^2(\omega_0^2 - 4\omega_1^2 + i2\omega_1\gamma)} \tag{11.24}$$

Note that it is also common to define an effective d-coefficient as $d_{\text{eff}} = d/2\varepsilon_0$, where ε_0 is the vacuum permittivity (see, for example, Ref. 3). For typical nonlinear materials, d_{eff} is of the order of 10^{-12} m/V, or, pm/V. In the remainder of this chapter, we will work with the d-coefficient as defined in Eq. (11.24). As we will see shortly, the polarization term oscillating at $2\omega_1$ acts as the source of the wave generated at the frequency $2\omega_1$ and this process is known as second-harmonic generation. In addition to second harmonic generation, we also have other nonlinear wave mixing processes as listed below:

1. Second harmonic generation at $2\omega_2$
2. Sum frequency generation at $(\omega_1 + \omega_2)$
3. Difference frequency generation at $(\omega_1 - \omega_2)$
4. DC induced polarization known as optical rectification

What we discussed above is not the whole story. We need to match the phase velocities as well, which requires propagation in a particular direction of the anisotropic crystal in order to enhance a particular process. We will discuss this in more detail later for second harmonic generation. For example, if a wave is moving in a particular direction that produces second harmonic, it is very unlikely that other nonlinear processes will also simultaneously satisfy the phase-matching condition, although special cases may occur where more than one nonlinear process is simultaneously phase matched.[11]

We leave it as an exercise for the reader (see Prob. 11.1) to show that the expression for the d-coefficient for sum frequency generation is given by

$$d(\omega_1 + \omega_2) =$$

$$\frac{\delta N e^3}{m_e^2\left[\omega_0^2 - (\omega_1 + \omega_2)^2 + i\gamma(\omega_1 + \omega_2)\right](\omega_0^2 - \omega_1^2 + i\gamma\omega_1)(\omega_0^2 - \omega_2^2 + i\gamma\omega_2)} \tag{11.25}$$

So, let's summarize some of the key findings of this section once again. First, note that the second-order d-coefficients that we learned how to derive have complicated frequency dependence. Hence, the d-coefficient for sum frequency generation will in general be different than that for second harmonic generation and so on. Most importantly, note one more time that if the potential energy of the bound electron did not have the cubic nonlinearity (in other words, when $\delta = 0$), we would get none of the frequency mixing processes discussed here.

The analysis we presented above was simplified by the fact that the induced polarization of the medium was assumed to be a scalar. In the case of second harmonic generation, for example, we could express the induced second-order polarization at $\omega = 2\omega_1$ in the form

$$P_0(2\omega_1) = d(2\omega_1)E_{10}E_{10} \tag{11.26}$$

Above, the product $E_{10}E_{10}$ was deliberately not written as E_{10}^2 in order to facilitate the generalization to anisotropic nonlinear crystals in which things are far more complicated. As such, each component of the nonlinear polarization will depend on all possible product combinations involving two components of the electric field. Furthermore, the strength of the nonlinearity will also be different for different combinations of components. Hence, for example, the x-component $P_{0x}(2\omega_1)$ of the second-order polarization vector can be expressed in terms of the electric field components at the fundamental frequency in the form

$$\begin{aligned} P_{0x}(2\omega_1) = {} & d_{xxx}(2\omega_1)E_{10x}E_{10x} + d_{xyy}(2\omega_1)E_{10y}E_{10y} + d_{xzz}(2\omega_1)E_{10z}E_{10z} \\ & + 2d_{xzy}(2\omega_1)E_{10z}E_{10y} + 2d_{xzx}(2\omega_1)E_{10z}E_{10x} + 2d_{xxy}(2\omega_1)E_{10x}E_{10y} \end{aligned} \tag{11.27}$$

The above relation can be generalized to the ith component of the nonlinear polarization vector by using

$$P_{0i}(2\omega_1) = d_{ijk}(2\omega_1)E_{10j}E_{10k} \tag{11.28}$$

Again, the Einstein summation convention is used in Eq. (11.28). Note that, in general, the second-order susceptibility tensor $d_{ijk}(2\omega_1)$ exhibits permutation symmetry because exchanging E_{10j} with E_{10k} leaves the polarization invariant. In other words, we have

$$d_{ijk}(2\omega_1) = d_{ikj}(2\omega_1) \tag{11.29}$$

Equation (11.29) summarizes a very important result in second-order nonlinear optics. On the practical side, it reduces the number of independent elements of the $d_{ijk}(2\omega_1)$ tensor from 27 ($3 \times 3 \times 3$) to 18 (3×6). The 18 independent elements of the $d_{ijk}(2\omega_1)$ tensor can be written by using the standard piezoelectric contraction notation where the subscript jk is replaced according to the convention: $xx = 1$, $yy = 2$, $zz = 3$, $yz = zy = 4$, $xz = zx = 5$, and $xy = yx = 6$. Hence, the relation between the second order polarization and fundamental field components can be expressed in matrix form as

$$\begin{bmatrix} P_{0x}(2\omega_1) \\ P_{0y}(2\omega_1) \\ P_{0z}(2\omega_1) \end{bmatrix} = \begin{bmatrix} d_{11} & d_{12} & d_{13} & d_{14} & d_{15} & d_{16} \\ d_{21} & d_{22} & d_{23} & d_{24} & d_{25} & d_{26} \\ d_{31} & d_{32} & d_{33} & d_{34} & d_{35} & d_{36} \end{bmatrix} \begin{bmatrix} E_{10x}{}^2 \\ E_{10y}{}^2 \\ E_{10z}{}^2 \\ 2E_{10z}E_{10y} \\ 2E_{10z}E_{10x} \\ 2E_{10x}E_{10y} \end{bmatrix} \tag{11.30}$$

The specific form of the second-order nonlinear susceptibility tensor $d_{il}(2\omega_1)$ depends on the symmetry of the crystal and group theoretic techniques can be employed to find the relation among the different elements. We will not discuss this approach further in this text. Refs. 3, 5 and 6 provide a detailed tabulation of different crystal classes and the form of the associated second-order susceptibility tensor. To give an example, for the commonly used nonlinear crystal KH_2PO_4 which has the abbreviation KDP, the nonlinear susceptibility tensor takes the form

$$d_{il} = \begin{bmatrix} 0 & 0 & 0 & d_{14} & 0 & 0 \\ 0 & 0 & 0 & 0 & d_{14} & 0 \\ 0 & 0 & 0 & 0 & 0 & d_{36} \end{bmatrix} \tag{11.31}$$

Note that in this case, only three of the 18 elements are non-zero due to the symmetry of the KDP crystal ($\bar{4}2m$ in this particular case for KDP), and elements d_{14} and d_{25} are equal. As a second example, in a centrosymmetric crystal which has inversion symmetry, all of the tensor coefficients vanish. Such is the case for crystals with cubic symmetry. Hence, second harmonic generation is not possible in this class of crystals.

Based on the foregoing discussion, we see that the analysis of second-order interactions in a real crystal is far more complicated than the simple scalar model we used to derive the second-order d-coefficients. The strength of the nonlinearity will in general vary with the orientation of the fields as well as the direction of propagation. Since we merely want to give the reader a flavor of what type of electromagnetic effects can arise in a nonlinear medium, we will continue with this simplified scalar model in the remainder of the chapter, except when we discuss phase matching in uniaxial crystals. For a detailed treatment of anisotropy effects in nonlinear interactions, the reader is referred to Refs. 1, 3, and 4.

11.1.2 Electromagnetics of Three-Wave Mixing

Consider the case where two waves at the frequencies of ω_1 and ω_2 are launched into a nonlinear medium. As discussed in Sec. 11.1.1, nonlinear mixing will generate other frequency components. As a specific example, assuming that conditions are satisfied to enable sum frequency generation, we will have an induced polarization that will radiate a wave at the frequency $\omega_1 + \omega_2$. In turn, difference frequency generation between the waves at frequencies $\omega_1 + \omega_2$ and ω_2 will produce a wave at ω_1. Hence, the nonlinear medium will produce a coupling between the electromagnetic fields at the frequencies ω_1, ω_2, and $\omega_3 = \omega_1 + \omega_2$. We could come up with a similar argument for the case of difference frequency generation. Here, we focus specifically on sum frequency generation. To derive the coupled wave equations satisfied by the propagating fields, we assume that we have a lossless, nonmagnetic insulator with no free charges or currents. Maxwell's curl equations become

$$\nabla \times \vec{E}(\vec{r},t) = -\frac{\partial \vec{B}(\vec{r},t)}{\partial t} = -\mu \frac{\partial \vec{H}(\vec{r},t)}{\partial t} \tag{11.32}$$

and

$$\nabla \times \vec{H}(\vec{r},t) = \frac{\partial \vec{D}(\vec{r},t)}{\partial t} \tag{11.33}$$

The constitutive relation for the medium accounting for linear and second-order effects can be written as

$$\vec{D}(\vec{r},t) = \varepsilon_0 \vec{E}(\vec{r},t) + \vec{P}(\vec{r},t) \tag{11.34}$$

where the polarization vector which contains both the linear and nonlinear contributions is given by

$$\vec{P}(\vec{r},t) = \vec{P}^{(1)}(\vec{r},t) + \vec{P}^{(2)}(\vec{r},t) \tag{11.35}$$

Above, the polarization vectors $\vec{P}^{(1)}(\vec{r},t)$ and $\vec{P}^{(2)}(\vec{r},t)$ represent the linear and nonlinear response of the medium, respectively. As discussed earlier, $\vec{P}^{(1)}(\vec{r},t)$ can be expressed in the form

$$\vec{P}^{(1)}(\vec{r},t) = \varepsilon_0 \chi \vec{E}(\vec{r},t) \tag{11.36}$$

where the linear electric susceptibility is given by Eq. (11.19). Taking the curl of Eq. (11.32) and using Eqs. (11.33) to (11.36), we obtain

$$\begin{aligned}\nabla \times \nabla \times \vec{E}(\vec{r},t) &= \nabla[\nabla \cdot \vec{E}(\vec{r},t)] - \nabla^2 \vec{E}(\vec{r},t) \\ &= -\mu_0 \nabla \times \left(\frac{\partial \vec{H}(\vec{r},t)}{\partial t} \right) = -\mu_0 \frac{\partial}{\partial t}[\nabla \times \vec{H}(\vec{r},t)] \\ &= -\mu_0 \frac{\partial^2}{\partial t^2}[\varepsilon_0 (1+\chi)\vec{E}(\vec{r},t) + \vec{P}^{(2)}(\vec{r},t)]\end{aligned} \tag{11.37}$$

Assuming that the dielectric constant is not position-dependent, we further take $\nabla \cdot \vec{E}(\vec{r},t) = 0$. For the sake of simplicity, we assume that we have a linearly polarized electric field and we consider only one of the Cartesian components, hereafter denoted as $E(\vec{r},t)$. Then, the vector Laplacian in Eq. (11.37) can be replaced by the scalar form, giving the scalar wave equation

$$\nabla^2 E(\vec{r},t) - \mu_0 \varepsilon \frac{\partial^2 E(\vec{r},t)}{\partial t^2} - \mu_0 \frac{\partial^2 P^{(2)}(\vec{r},t)}{\partial t^2} = 0 \tag{11.38}$$

with

$$\varepsilon = \varepsilon_0 (1+\chi) \tag{11.39}$$

We assume that there are three plane waves moving in the medium due to the coupling described above. Taking the direction of propagation as the z-axis, the total electric field $E(\vec{r},t)$ becomes $E(z,t)$ and can be expressed as

$$E(z,t) = E_1(z,t) + E_2(z,t) + E_3(z,t) \tag{11.40}$$

In Eq. (11.40), $E_1(z,t)$, $E_2(z,t)$, and $E_3(z,t)$ represent the respective electric fields at the frequencies ω_1, ω_2, and ω_3. In order to analyze the effect of the coupling, we further allow the amplitude of each electric field component to have position dependence. Hence, we write, for example, $E_1(z,t)$ in the form

$$E_1(z,t) = \frac{1}{2}\{E_{10}(z)\exp[i(\omega_1 t - k_1 z)] + c.c\} \tag{11.41}$$

Similar expressions can be written for $E_2(z,t)$ and $E_3(z,t)$. Note that if $P^{(2)}(z,t)$ were zero, we would get no coupling between the three electric fields and each would move independently through the medium.

To derive the coupled wave equations, let us first concentrate on the field $E_1(z,t)$ at the frequency ω_1. To obtain the differential equation satisfied by $E_{10}(z)$, we substitute the total field $E(z,t)$ given by Eq. (11.40) into Eq. (11.38) and equate the coefficients of the terms containing $\exp[i(\omega_1 t - k_1 z)]$. The third term in Eq. (11.38) containing $P^{(2)}(z,t)$ will be proportional to $[E_1(z,t) + E_2(z,t) + E_3(z,t)]^2$. When we consider the differential equation satisfied by $E_{10}(z)$, we only need to include the nonlinear polarization term which will be oscillating at the frequency ω_1. As we discussed earlier, this will be the term due to the difference frequency mixing of the waves at $\omega_3 = \omega_1 + \omega_2$ and ω_2. The corresponding nonlinear polarization component, $P^{(2)}_{\omega_1}(z,t)$ oscillating at ω_1 will be given by

$$P^{(2)}_{\omega_1}(z,t) = dE_{30}(z)E^*_{20}(z)\exp\{i[(\omega_3 - \omega_2)t - (k_3 - k_2)z]\} \tag{11.42}$$

where d is the nonlinear susceptibility coefficient for difference frequency generation. Above, $E_{30}(z)$ and $E_{20}(z)$ are the amplitudes of the waves at ω_3 and ω_2, respectively. To complete the derivation of the equation for $E_{10}(z)$, we next simplify the Laplacian term $\nabla^2 E_1(z,t)$ by using the slowly varying envelope approximation. Since, we are considering plane waves with no variation along x and y, we get

$$\nabla^2 E_1(z,t) = \frac{\partial^2}{\partial z^2}\frac{1}{2}\left[E_{10}(z)\exp(i(\omega_1 t - k_1 z)) + c.c\right]$$
$$= \frac{1}{2}\frac{\partial}{\partial z}\left[\left(\frac{dE_{10}(z)}{dz} - ik_1 E_{10}(z)\right)\exp(i(\omega_1 t - k_1 z)) + c.c\right]$$
$$= \frac{1}{2}\left[\left(\frac{d^2E_{10}(z)}{dz^2} - 2ik_1\frac{dE_{10}(z)}{dz} - k_1^2 E_{10}(z)\right)\exp(i(\omega_1 t - k_1 z)) + c.c\right] \tag{11.43}$$

Neglecting the second derivative in relation to $k_1 dE_{10}(z)/dz$, we obtain

$$\nabla^2 E_1(z,t) \approx -\frac{1}{2}\left(2ik_1\frac{dE_{10}(z)}{dz} + k_1^2 E_{10}(z)\right)\exp(i(\omega_1 t - k_1 z)) + c.c \tag{11.44}$$

Putting all this into Eq. (11.38) and gathering those terms oscillating at ω_1, we get

$$-\frac{1}{2}\left(2ik_1\frac{dE_{10}(z)}{dz} + k_1^2 E_{10}(z)\right) + \omega_1^2\mu_0\varepsilon_1\frac{E_{10}(z)}{2}$$
$$+\frac{\mu_0\omega_1^2 d}{2}E_{30}(z)E_{20}^*(z)\exp(-i\Delta kz) = 0 \tag{11.45}$$

where

$$\Delta k = k_3 - k_2 - k_1 \tag{11.46}$$

and

$$\varepsilon_1 = \varepsilon_0(1 + \chi(\omega_1)) \tag{11.47}$$

Furthermore, we know from before that $k_1^2 = \mu_0\omega_1^2\varepsilon_1$. Hence, the equation for $E_{10}(z)$ simplifies to

$$\frac{dE_{10}(z)}{dz} = d\frac{-i\mu_0\omega_1^2 E_{30}(z)E_{20}^*(z)\exp(-i\Delta kz)}{2ik_1} \tag{11.48}$$

The above equation can be further simplified by noting that

$$\frac{\omega_1^2}{k_1} = \frac{\omega_1}{\sqrt{\mu_0 \varepsilon_1}} \tag{11.49}$$

and

$$\sqrt{\frac{\mu_0}{\varepsilon_1}} = \sqrt{\frac{\mu_0}{\varepsilon_0}} \frac{1}{\sqrt{\frac{\varepsilon_1}{\varepsilon_0}}} = \frac{\eta_0}{n_1} \tag{11.50}$$

where η_0 is the vacuum impedance (377 Ω), and n_1 is the refractive index at the frequency ω_1. The final form of the differential equation satisfied by $E_{10}(z)$ now reads

$$\frac{dE_{10}(z)}{dz} = -id\frac{\omega_1}{2}\frac{\eta_0}{n_1} E_{30}(z)E_{20}^*(z)\exp(-i\Delta kz) \tag{11.51}$$

The equation for $E_{20}(z)$ will be very similar, except the subscripts 1 and 2 will be interchanged:

$$\frac{dE_{20}(z)}{dz} = -id\frac{\omega_2}{2}\frac{\eta_0}{n_2} E_{30}(z)E_{10}^*(z)\exp(-i\Delta kz) \tag{11.52}$$

In the case of $E_{30}(z)$, the polarization term driving the wave at ω_3 will be due to sum frequency mixing of the fields at ω_1 and ω_2. We leave it as an exercise for the reader (see Prob. 11.2) to show that $E_{30}(z)$ will satisfy the equation

$$\frac{dE_{30}(z)}{dz} = -id\frac{\omega_3}{2}\frac{\eta_0}{n_3} E_{10}(z)E_{20}(z)\exp(+i\Delta kz) \tag{11.53}$$

By using Eqs. (11.51) to (11.53), we can investigate different types of second-order interactions in a nonlinear optical medium. We note, once again, that in the derivation of Eqs. (11.51) to (11.53), losses were neglected. This can be remedied by introducing a phenomenological attenuation constant in each of the three equations, if necessary. From a practical point of view, however, this approximation is often valid since we choose to work with nonlinear materials in their transparency range and the surfaces are typically coated to eliminate reflection losses.

Before ending this section, we use Eqs. (11.51) to (11.53) to derive Manley-Rowe relations,[12] which give the conservation of energy principle in second-order nonlinear interactions. We start out by calculating the cycle-averaged Poynting vector of the wave at ω_1 given by

$$\langle S_1 \rangle = \frac{n_1}{2\eta_0} E_{10}(z)E_{10}^*(z) \tag{11.54}$$

By differentiating Eq. (11.54) with respect to z and using Eq. (11.51), we obtain

$$\begin{aligned}\frac{d\langle S_1\rangle}{dz} &= \frac{n_1}{2\eta_0}\left(E_{10}(z)\frac{dE_{10}^*}{dz} + \frac{dE_{10}}{dz}E_{10}^*(z)\right)\\ &= \frac{n_1}{2\eta_0}\left[E_{10}(z)d\left(\frac{i\omega_1}{2}\right)\frac{\eta_0}{n_1}E_{20}(z)E_{30}^*(z)\exp(i\Delta kz)\right.\\ &\quad \left. + E_{10}^*(z)d\left(-\frac{i\omega_1}{2}\right)\frac{\eta_0}{n_1}E_{30}(z)E_{20}^*(z)\exp(-i\Delta kz)\right]\\ &= \frac{i}{2}d\,\mathrm{Im}\left[E_{10}(z)E_{20}(z)E_{30}^*(z)\exp(i\Delta kz)\right]\omega_1 \end{aligned} \tag{11.55}$$

Above, Im[ ] represents the imaginary part. We leave it as an exercise for the reader (see Prob. 11.3) to show that for the Poynting vectors at ω_2 and ω_3, we have

$$\frac{d\langle S_2\rangle}{dz} = \frac{i}{2}d\,\mathrm{Im}[E_{10}(z)E_{20}(z)E_{30}^*(z)\exp(i\Delta kz)]\omega_2 \tag{11.56}$$

and

$$\frac{d\langle S_3\rangle}{dz} = -\frac{i}{2}d\,\mathrm{Im}[E_{10}(z)E_{20}(z)E_{30}^*(z)\exp(i\Delta kz)]\omega_3 \tag{11.57}$$

Using the above relations, we can obtain two useful relations. First, note that

$$\frac{d\langle S_1\rangle}{dz} + \frac{d\langle S_2\rangle}{dz} + \frac{d\langle S_3\rangle}{dz} = 0 \tag{11.58}$$

Next, examining Eqs. (11.55) to (11.57), we see that

$$\frac{1}{\omega_1}\frac{d\langle S_1\rangle}{dz} = \frac{1}{\omega_2}\frac{d\langle S_2\rangle}{dz} = -\frac{1}{\omega_3}\frac{d\langle S_3\rangle}{dz} \tag{11.59}$$

Recalling that the photon energy is given by $\hbar\omega$, we can interpret $\langle S_i\rangle/\omega_i$ as being proportional to the photon flux at ω_i, in other words, the number of photons passing per unit cross-sectional area per unit time. Hence, the rate of change of the photon flux at ω_1 equals the rate of change of the photon flux at ω_2, both being the negative of the rate of change of photon flux at ω_3. Stated in an equivalent way, every photon generated at ω_3, results from the combination (and hence the decrease) of one photon at ω_1 and another at ω_2.

11.1.3 Second Harmonic Generation

One of the most widely used practical applications of second-order nonlinear interactions is optical second harmonic generation (SHG). A possible experimental setup for observing SHG is shown in Fig. 11.2. The output of a laser operating at the wavelength λ is focused with a lens (L_1 in Fig. 11.2) into a second harmonic crystal which is positioned at the optimum orientation to maximize the conversion of light at λ to its second harmonic at $\lambda/2$. To minimize reflection losses, the crystal surfaces are coated to be antireflecting at λ and $\lambda/2$. Both beams emerging from the crystal are then collimated by a second lens (L_2). A dichroic beam splitter (M_1 in Fig. 11.2) filters out the unused fundamental light by selectively transmitting most of the light at λ and reflecting predominantly at $\lambda/2$. The conversion is maximized by adjusting the orientation of the crystal and by adjusting its location with respect to the focusing lens L_1. Historically, second harmonic generation was first demonstrated in 1961 by passing the output of a ruby laser through a quartz crystal.[13]

We can use the formalism developed in the previous sections to obtain an analytical expression for the generated second harmonic power. Note that SHG is a special case of three-wave mixing with $\omega_1 = \omega_2 = \omega$ and $\omega_3 = \omega_1 + \omega_2 = 2\omega$. Furthermore, $E_{10}(z) = E_{20}(z)$. From Eq. (11.53), we have

$$\frac{dE_{30}(z)}{dz} = -id\frac{2\omega}{2}\frac{\eta_0}{n_3}[E_{10}(z)]^2 \exp(+i\Delta kz) \tag{11.60}$$

where

$$\Delta k \equiv k_3 - k_1 - k_2 = k^{(2\omega)} - 2k^{(\omega)} \tag{11.61}$$

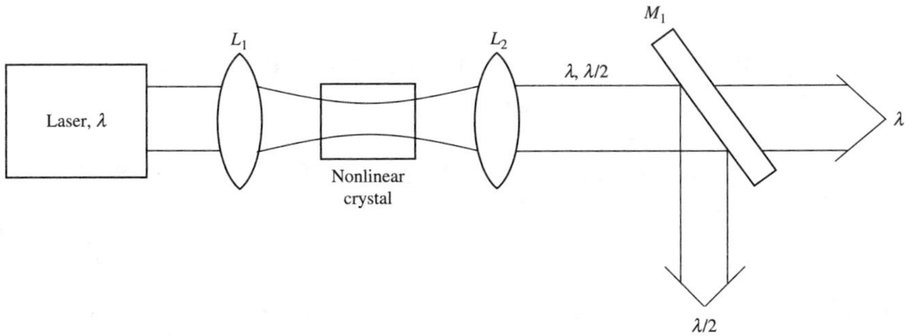

FIGURE 11.2 Experimental setup for observing second harmonic generation.

For the case where the conversion efficiency is small, we can neglect the depletion of the input light intensity at λ and assume that $E_{10}(z)$ is constant throughout the crystal, equal to its initial value at $z = 0$. Note that $E_{30}(0) = 0$. Taking the crystal length as ℓ, we can readily integrate Eq. (11.60) to get

$$\begin{aligned} E_{30}(\ell) &= -id\omega\frac{\eta_0}{n_3}(E_{10}(0))^2\frac{1}{i\Delta k}\exp(i\Delta kz)\Big]_0^\ell \\ &= -id\omega\frac{\eta_0}{n_3}(E_{10}(0))^2\frac{\exp(i\Delta k\ell)-1}{i\Delta k} \\ &= -id\omega\frac{\eta_0}{n_3}(E_{10}(0))^2\exp\left(i\frac{\Delta k\ell}{2}\right)\frac{\sin(\Delta k\ell/2)}{\Delta k/2} \end{aligned} \tag{11.62}$$

If the effective area of the beam is a, the generated second harmonic power $P^{2\omega}$ will be given by

$$P^{2\omega} = a\frac{1}{2}\frac{n_3}{\eta_0}E_{30}(\ell)E_{30}(\ell)^* = a\frac{1}{2}d^2\omega^2\frac{\eta_0}{n_3}\left|E_{10}(0)\right|^4\left(\frac{\sin(\Delta k\ell/2)}{\Delta k/2}\right)^2 \tag{11.63}$$

The input power P^ω at ω is further given by

$$P^\omega = a\frac{1}{2}\frac{n_1}{\eta_0}\left|E_{10}(0)\right|^2 \tag{11.64}$$

Combining Eqs. (11.63) and (11.64), we see that the SHG conversion efficiency η_{SHG} becomes

$$\eta_{SHG} = \frac{P^{2\omega}}{P^\omega} = 2\eta_0^3\frac{\omega^2 d^2\ell^2}{n_1^2 n_3}\frac{P^\omega}{a}\left(\frac{\sin(\Delta k\ell/2)}{\Delta k\ell/2}\right)^2 \tag{11.65}$$

11.1.4 Phase Matching in Second Harmonic Generation

Figure 11.3 shows a calculated plot of the normalized second harmonic power as a function of the wave vector mismatch $\Delta k = k^{(2\omega)} - 2k^{(\omega)}$. Clearly, Δk should be zero in order to maximize the conversion efficiency. In other words, we need

$$k^{(2\omega)} = 2k^{(\omega)} \tag{11.66}$$

Initially, this may appear to be in contradiction with what we learned about dispersion in Chap. 4. As we saw earlier, the fundamental and second harmonic waves move with different phase velocities due to

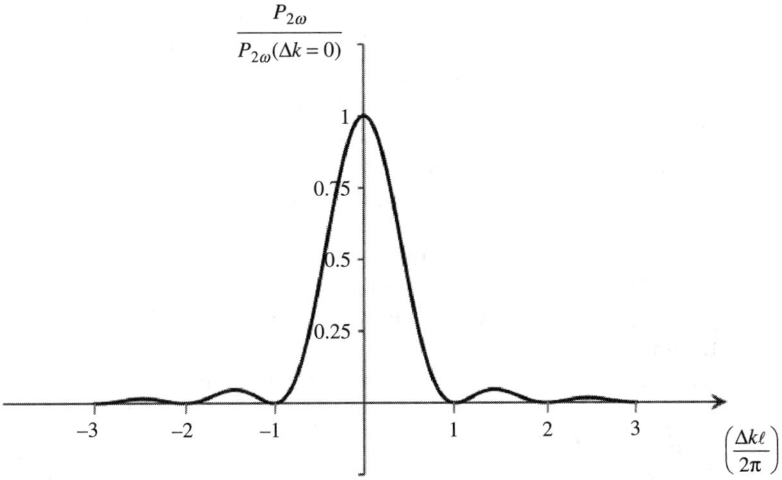

FIGURE 11.3 Variation of the normalized second harmonic power as a function of the normalized wave vector mismatch $\Delta k\ell/2\pi$.

the dispersion of the crystal and, in general, do not remain in phase over the length of the crystal. Hence, although some second harmonic signal may be produced, it will vary strongly with crystal length as shown in Fig. 11.3. It turns out that the natural birefringence of the crystal can be exploited to our advantage to remedy this problem by using a technique known as phase matching. We will discuss this method shortly in this section.

First, let us further explore the case where we have a non-zero wave vector mismatch. Note that the relative phase difference between the fundamental and the second harmonic waves will become π after propagating a distance $\ell = \ell_c$, where ℓ_c is known as the coherence length and satisfies

$$\Delta k \ell_c = \pi \tag{11.67}$$

Based on this relation, we leave it as an exercise for the reader (see Prob. 11.4) to show that ℓ_c is given by

$$\ell_c = \frac{\lambda}{4(n_3 - n_1)} \tag{11.68}$$

where $n_1 = n(\omega)$ and $n_3 = n(2\omega)$ are the refractive indices at the fundamental and second harmonic frequencies, respectively. From Fig. 11.3, we see that when $\ell = 2\ell_c$, the second harmonic power completely vanishes. Typically, ℓ_c is of the order of microns. Hence, if the condition in Eq. (11.66) is not met, the practical crystal lengths that can be used in second harmonic generation are very short and high conversion efficiencies cannot be achieved. One approach that can overcome this limitation is known as quasi-phase matching (QPM).[14] In QPM, the sign of the nonlinearity is periodically flipped every coherence length. This way, the accumulated phase difference between the fundamental and second harmonic waves is periodically reset to enable the growth of the second harmonic signal over relatively long crystal lengths. In practice, this can be achieved in ferroelectric nonlinear crystals (such as $LiNbO_3$) by using the method of poling to form regions of periodically reversed spontaneous polarization. Numerous nonlinear frequency converters based on this scheme have been demonstrated.[14-16]

Now, let's discuss the method of phase matching. It turns out that we can use the natural birefringence of the anisotropic crystals to satisfy the condition $\Delta k = 0$. Equivalently, we want to make the refractive index $n(\omega)$ at the fundamental frequency equal to the refractive index $n(2\omega)$ at the second harmonic. To describe how this is done, we consider a uniaxial crystal. If we are in the normal dispersion regime of the crystal, the refractive index increases with increasing frequency. This is typically the case for both the ordinary and the extraordinary indices. Hence, if both the fundamental and the second harmonic waves are of the same type, in other words, both ordinary or both extraordinary, we cannot satisfy the desired condition. However, if the waves are of different types, it may become possible to have the same effective index for both the fundamental and the second harmonic wave. As a specific example, consider a negative uniaxial crystal, where, as we saw in Chap. 8, $n_e < n_o$. Here, n_e and n_o are the extraordinary and ordinary indices of refraction. If $n_e(2\omega,\theta) < n_o(\omega)$, then there is an angle θ_0, where

$$n_e(2\omega,\theta_0) = n_o(\omega) \tag{11.69}$$

This case is schematically shown in Fig. 11.4, which depicts the normal surfaces for the ordinary and extraordinary waves at the frequencies ω and 2ω. As a reminder, the normal surface for the ordinary wave is a sphere, whereas for the extraordinary wave, we have an ellipsoid of revolution. Since the crystal is assumed to have normal dispersion, for the ordinary wave, the radius for the sphere at 2ω is larger than that for the sphere at ω. At each frequency, the corresponding extraordinary index surface has a smaller intercept on the y-axis since the crystal is negative uniaxial. The fact that we have an intersection point of the extraordinary ellipsoid at 2ω with the ordinary sphere at ω means that when we launch the fundamental light as an ordinary wave along the direction making an angle of θ_0 with the optic axis, the second harmonic wave will be generated in the same direction but as an extraordinary wave.

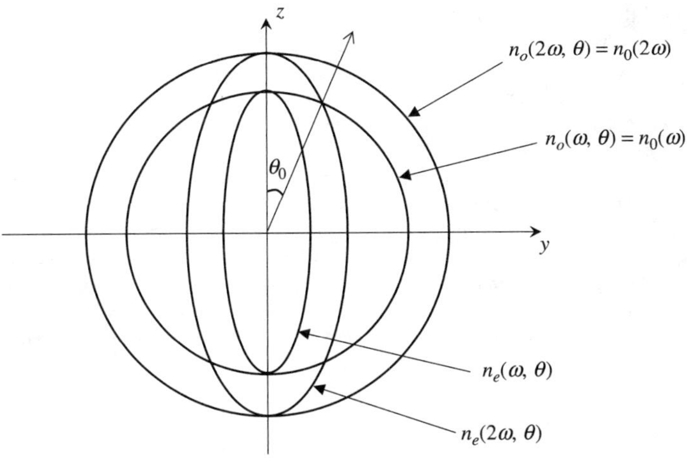

FIGURE 11.4 Normal surfaces for the ordinary and extraordinary waves at the frequencies ω and 2ω for a negative uniaxial crystal. Phase matching occurs if the fundamental wave is launched as an ordinary wave along the direction making an angle of θ_0 with the optic axis.

To obtain the expression for θ_0, we set $n_e(2\omega,\theta_0) = n_o(\omega)$. By using Eq. (8.56), we get

$$\frac{1}{(n_e(2\omega,\theta))^2} = \frac{\cos^2\theta}{(n_o(2\omega))^2} + \frac{\sin^2\theta}{(n_e(2\omega))^2} = \left(\frac{1}{n_o(\omega)}\right)^2 \tag{11.70}$$

Solving Eq. (11.70) yields

$$\sin^2\theta_0 = \frac{\left(\frac{1}{n_o(\omega)}\right)^2 - \left(\frac{1}{n_o(2\omega)}\right)^2}{\left(\frac{1}{n_e(2\omega)}\right)^2 - \left(\frac{1}{n_o(2\omega)}\right)^2} \tag{11.71}$$

Example: Phase matching in beta-barium borate
Beta-barium borate (BBO, β-BaB_2O_4) is a widely used nonlinear crystal for second harmonic generation and three-wave mixing. The wavelength dependence of the ordinary and extraordinary indices of refraction can be calculated by using the empirical Sellmeier equations[17]

$$n_o^2(\lambda) = 2.7359 + \frac{0.01878}{\lambda^2 - 0.01822} - 0.01354\lambda^2 \tag{11.72}$$

and

$$n_e^2(\lambda) = 2.3753 + \frac{0.01224}{\lambda^2 - 0.01667} - 0.01516\lambda^2 \tag{11.73}$$

where the wavelength λ is in microns. Figure 11.5*a* and *b* shows the calculated variation of the ordinary and extraordinary indices for BBO [based on Eqs. (11.72) and (11.73)] as a function of wavelength between 300 and 2000 nm.

As an example, let us calculate the phase-matching angle for second harmonic generation at 1064 nm. From Eqs. (11.72) and (11.73), we have $n_o(1064\text{ nm}) = 1.6545$, $n_e(1064\text{ nm}) = 1.5392$, $n_o(532\text{ nm}) = 1.6742$, and $n_e(532\text{ nm}) = 1.5547$. So, BBO is indeed negative uniaxial at these wavelengths and furthermore, $n_e(532\text{ nm}) < n_o(1064\text{ nm})$. Straightforward application of Eq. (11.71) gives $\theta_0 = 22.8°$, What we have described in this section is known as Type-I phase matching. Viewed as a three-wave mixing process, the waves at the frequencies ω_1 and ω_2 are ordinary, whereas the sum frequency wave at ω_3 is extraordinary. We could also have the case where the wave at ω_1 is ordinary while the waves at ω_2 and ω_3 are extraordinary, in which case we have Type-II phase matching.

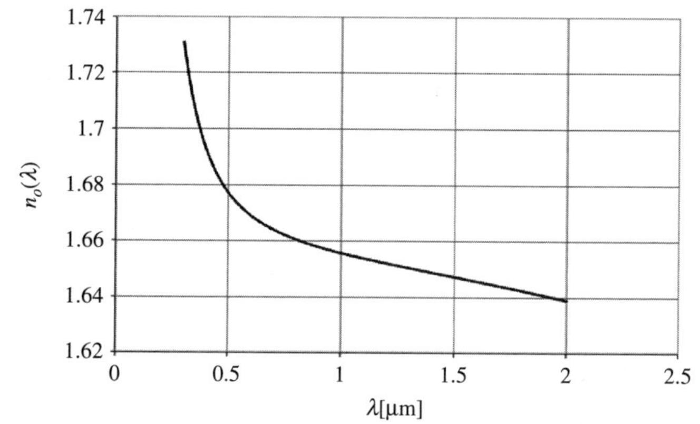

(a)

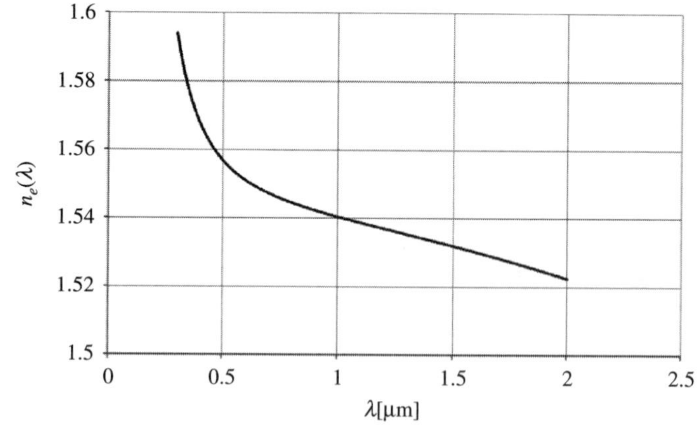

(b)

FIGURE 11.5 Calculated variation of the (*a*) ordinary and (*b*) extraordinary refractive index for beta-barium borate (BBO, β-BaB_2O_4) between 300 and 2000 nm. The empirical Sellmeier equations were taken from Ref. 17.

For the latter case, the phase-matching angle can be calculated numerically. Problem 11.5 provides additional practice in doing Type I and Type II phase-matching calculations for BBO.

11.1.5 Second Harmonic Generation with Gaussian Beams

In most practical applications, the laser beams used in nonlinear interactions are not plane waves; rather, they have a Gaussian intensity distribution in the plane perpendicular to the direction of propagation. It is therefore important to see how our results for second harmonic generation will be modified in this case. This gives useful guidelines for experimentalists. Our approximate analysis is similar to that in Ref. 5. A more accurate treatment that investigates the parametric interaction of focused Gaussian beams may be found in Refs. 18 and 19. Consider the case where the electric field of the fundamental wave is given by

$$E_{10}(r,z) = E_0 \exp(-r^2/w(z)^2) \tag{11.74}$$

As we saw in Chap. 3, the spot-size function $w(z)$ varies with position according to

$$w^2(z) = w_0^2\left(1+\left(\frac{z-z_f}{z_0}\right)^2\right) \tag{11.75}$$

where w_0 is the beam waist, z_f is the beam focus location, and z_0 is the Rayleigh range given by

$$z_0 = \frac{n_1 \pi w_0^2}{\lambda} \tag{11.76}$$

Above, n_1 is the refractive index at the fundamental wavelength λ. Typically, to obtain the highest conversion efficiency, one focuses the Gaussian beam inside the nonlinear crystal as shown in Fig. 11.6*a*.

Reference 18 describes in detail how the general case can be analyzed. Here, we will make the approximation that the beam is weakly focused so that $z_0 >> \ell$, where ℓ is the crystal length. Hence, the intensity will vary across the beam cross section, but the transverse intensity distribution will be the same at different locations along the direction of propagation (z in our case). Now, we can write Eq. (11.62) as

$$E_{30}(r,\ell) = -id\omega\frac{\eta_0}{n_3}E_0^2 \exp(-2r^2/w(z)^2)\exp\left(i\,\frac{\Delta k\ell}{2}\right)\frac{\sin(\Delta k\ell/2)}{\Delta k/2} \tag{11.77}$$

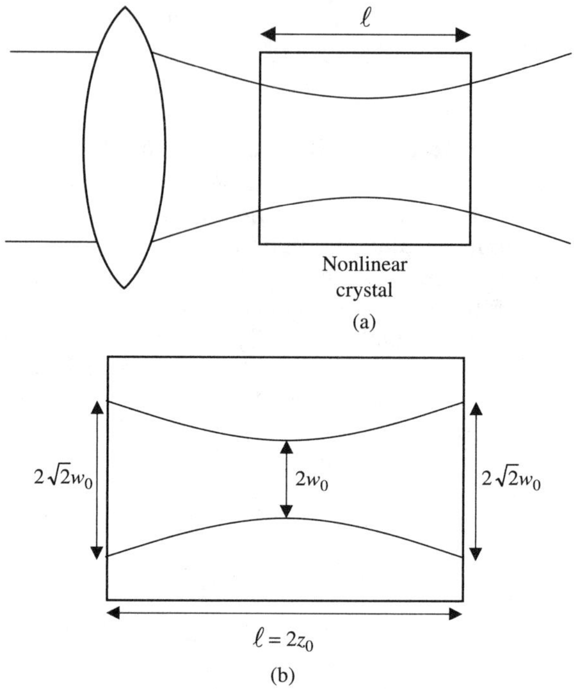

FIGURE 11.6 (*a*) A Gaussian beam focused inside a nonlinear crystal of length ℓ. (*b*) Confocal focusing case.

We leave it as an exercise for the reader (see Prob. 11.8) to show that the SHG conversion efficiency in this case becomes

$$\eta_{\text{SHG}} = \frac{P^{\omega}}{P^{2\omega}} = 2\eta_0^3 \frac{\omega^2 d^2 \ell^2}{n_1^2 n_3} \frac{P^{\omega}}{\pi w_0^2} \left(\frac{\sin(\Delta k \ell/2)}{\Delta k \ell/2} \right)^2 \tag{11.78}$$

Note that for the same input fundamental power, we can increase the efficiency by decreasing the beam waist w_0. However, this cannot go on indefinitely, since the diffractive spreading of the beam will also play a role once the Rayleigh range z_0 becomes comparable to or less than the crystal length ℓ. In practice, it is commonplace to design the focusing optics to give $\ell = 2z_0$. This configuration is known as confocal focusing and is sketched in Fig. 11.6*b*. Again, we leave it as an exercise for the reader (see Prob. 11.8) to show that the conversion efficiency for confocal focusing becomes

$$\eta_{\text{SHG}} = \frac{P^{\omega}}{P^{2\omega}} = \frac{2}{\pi c} \eta_0^3 \frac{\omega^3 d^2 \ell}{n_1 n_3} P^{\omega} \left(\frac{\sin(\Delta k \ell/2)}{\Delta k \ell/2} \right)^2 \tag{11.79}$$

The most important conclusion to draw from this analysis is related to the dependence of the conversion efficiency on crystal length. Different from the plane-wave case, we find that the conversion efficiency is linearly proportional to the crystal length in the case of confocal focusing.

11.1.6 Second Harmonic Generation in the Presence of Pump Depletion

If the incident fundamental beam is intense enough or if the crystal length is long, we can no longer assume that the fundamental beam intensity remains nearly the same inside the nonlinear crystal. As the second harmonic intensity grows, the fundamental beam intensity gets depleted due to the conservation of energy. The case of pump depletion can be treated by using the coupled equations we derived earlier for three-wave mixing. To recall, we now have $\omega_3 = 2\omega_1 = 2\omega$ and $E_{10}(z) = E_{20}(z)$. Assuming perfect phase matching, we get, by using Eqs. (11.51) and (11.53),

$$\frac{dE_{10}}{dz} = -id\frac{\omega}{2}\frac{\eta_0}{n_1}E_{30}(z)E_{10}(z)^* \tag{11.80}$$

and

$$\frac{dE_{30}}{dz} = -id\omega\frac{\eta_0}{n_3}E_{10}^2(z) \tag{11.81}$$

We define

$$\gamma = \omega\frac{\eta_0}{n_1}d \tag{11.82}$$

and let $E'_{30}(z) = iE_{30}(z)$. This gives (assuming phase matching with $n_1 = n_3$)

$$\frac{dE_{10}}{dz} = -\frac{\gamma}{2}E'_{30}(z)E_{10}(z) \tag{11.83}$$

and

$$\frac{dE'_{30}}{dz} = \gamma E_{10}(z)^2 \tag{11.84}$$

From Manley-Rowe relations, we have

$$\frac{d}{dz}(< S_1(z) > + < S_3(z) >) = 0 \tag{11.85}$$

or

$$E_{10}^2(z) + E_{30}'(z)^2 = \text{constant} \tag{11.86}$$

Since we only have the ω_1 field at $z = 0$, this gives

$$E_{10}^2(z) + E_{30}'(z)^2 = E_{10}^2(0) \tag{11.87}$$

Hence, the differential equation satisfied by $E_{30}(z)$ becomes

$$\frac{dE_{30}'}{dz} = \gamma\left(E_{10}(0)^2 - E_{30}'(z)^2\right) \tag{11.88}$$

We leave it as an exercise for the reader (see Prob. 11.10) to show that the solution for $E_{30}'(z)$ takes the form

$$E_{30}(z) = -iE_{10}(0)\tanh(\gamma E_{10}(0)z) \tag{11.89}$$

The power conversion efficiency was earlier calculated in the absence of pump depletion [see Eq. (11.65)]. Due to pump depletion, the conversion efficiency is modified to

$$\eta_{\text{SGH}} = \frac{P^{2\omega}(z)}{P^{\omega}(0)} = \tanh^2(\gamma E_{10}(0)z) \tag{11.90}$$

Figure 11.7 shows the variation of the fundamental and second harmonic powers as a function of distance within the nonlinear crystal.

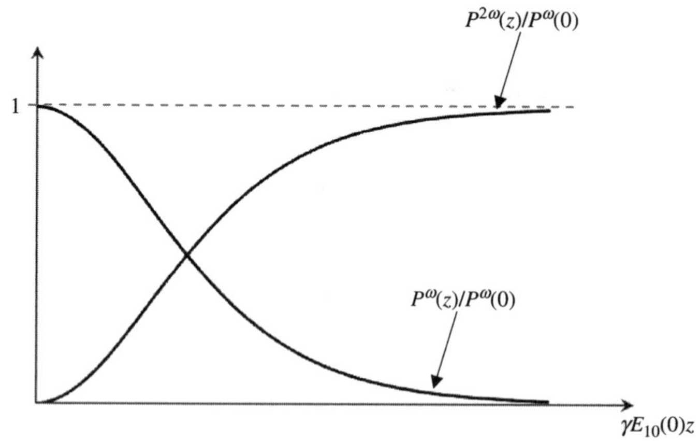

FIGURE 11.7 Variation of the fundamental and second harmonic powers as a function of distance within the nonlinear crystal in the presence of pump depletion.

Notice, first of all, that the conversion efficiency does not exceed unity in the presence of pump depletion, which makes physical sense. The corresponding fundamental power monotonically decreases with increasing distance and is given by

$$\frac{P^{\omega}(z)}{P^{\omega}(0)} = 1 - \tanh^2(\gamma E_{10}(0)z) = \mathrm{sech}^2(\gamma E_{10}(0)z) \tag{11.91}$$

Second, whether pump depletion is important or not is determined by the initial value of the pump electric field $E_{10}(0)$. If $E_{10}(0)$ is very large, conversion efficiency approaches unity over a short crystal length. On the other hand, for very small initial pump intensity, one remains in the quadratic regime, where conversion efficiency is proportional to the incident power. We leave it as an exercise (see Prob. 11.11) for the reader to show that in the small-signal limit, one can actually recover Eq. (11.65) from Eq. (11.90).

11.1.7 Optical Parametric Amplification

Let us recall the coupled differential equations we derived earlier in Sec. 11.1.2:

$$\frac{dE_{10}(z)}{dz} = -id\frac{\omega_1}{2}\frac{\eta_0}{n_1}E_{30}(z)E_{20}^*(z)\exp(-i\Delta kz) \tag{11.92}$$

$$\frac{dE_{20}(z)}{dz} = -id\frac{\omega_2}{2}\frac{\eta_0}{n_2}E_{30}(z)E_{10}^*(z)\exp(-i\Delta kz) \tag{11.93}$$

$$\frac{dE_{30}(z)}{dz} = -id\frac{\omega_3}{2}\frac{\eta_0}{n_3}E_{10}(z)E_{20}(z)\exp(+i\Delta kz) \tag{11.94}$$

Now, we assume that the wave at the frequency ω_3 is sent into the medium to generate waves at ω_1 and ω_2 such that

$$\omega_3 = \omega_1 + \omega_2 \tag{11.95}$$

This process is called optical parametric generation/amplification. In the jargon of nonlinear optics, the ω_3-wave is referred to as the pump, the ω_1-wave as the signal, and the ω_2-wave as the idler. In addition to the conservation of energy requirement given by Eq. (11.95), the phase-matching condition $\Delta k = 0$ must be simultaneously satisfied to maximize the efficiency of this nonlinear process. The phase-matching condition now becomes

$$\frac{n_3}{\lambda_3} = \frac{n_1}{\lambda_1} + \frac{n_2}{\lambda_2} \tag{11.96}$$

where n_1, n_2, and n_3 are the effective indices at the wavelengths λ_1, λ_2, and λ_3, respectively. In the case of a negative uniaxial crystal, the phase-matching condition can be satisfied in two different polarization configurations termed Type-I and Type-II phase matching. In Type-I phase matching, both signal and the idler waves are ordinary, whereas the pump is extraordinary and the phase-matching condition can be written as

$$n_e(\lambda_3,\theta_0) = \frac{\lambda_3}{\lambda_1} n_o(\lambda_1) + \frac{\lambda_3}{\lambda_2} n_o(\lambda_2) \tag{11.97}$$

In Type-II phase matching, one of the signal and idler waves is also extraordinary, resulting in

$$n_e(\lambda_3,\theta_0) = \frac{\lambda_3}{\lambda_1} n_e(\lambda_1,\theta_0) + \frac{\lambda_3}{\lambda_2} n_o(\lambda_2) \tag{11.98}$$

Above, θ_0 is the phase, matching angle. Recall that the angular dependence of the extraordinary index in a uniaxial crystal was given by Eq. (8.56) in Chap. 8.

To solve the coupled differential equations, we assume that the pump beam at the frequency ω_3 remains nearly constant inside the crystal. Again, this corresponds to the case of no pump depletion. Hence, $E_{30}(z) \cong E_{30}(0)$. Assuming perfect phase matching and real E_{30}, and differentiating Eq. (11.92), we get

$$\frac{d^2E_{10}}{dz^2} = -d\frac{i\omega_1}{2}\frac{\eta_0}{n_1}E_{30}(0)\frac{dE_{20}^*}{dz} = d^2\frac{\omega_1}{2}\frac{\eta_0}{n_1}E_{30}(0)\left(\frac{\omega_2}{2}\right)\frac{\eta_0}{n_2}E_{30}^*(0)E_{10}(z) \tag{11.99}$$

or

$$\frac{d^2E_{10}}{dz^2} = \left(\frac{g}{2}\right)^2 E_{10}(z) \tag{11.100}$$

where

$$g^2 = \eta_0^2\frac{\omega_1\omega_2 d^2}{n_1 n_2}|E_{30}(0)|^2 \tag{11.101}$$

Since the intensity I_3 at the frequency ω_3 is given by

$$I_3 = \frac{|E_{30}(0)|^2}{2\eta_0}n_3 \tag{11.102}$$

we obtain

$$g^2 = 2\frac{\eta_0^3\omega_1\omega_2 d^2}{n_1 n_2 n_3}I_3 \tag{11.103}$$

A similar differential equation can be obtained for $E_{20}(z)^*$. We leave it as an exercise for the reader (see Prob. 11.12) to show that the solution for $E_{10}(z)$ and $E_{20}(z)^*$ can be written in the form

$$E_{10}(z) = E_{10}(0)\cosh\left(\frac{g}{2}z\right) - i\sqrt{\frac{n_2}{n_1}\frac{\omega_1}{\omega_2}}E_{20}^*(0)\sinh\left(\frac{g}{2}z\right) \tag{11.104}$$

and

$$E_{20}^*(z) = E_{20}^*(0)\cosh\left(\frac{g}{2}z\right) + i\sqrt{\frac{n_1}{n_2}\frac{\omega_2}{\omega_1}}E_{10}(0)\sinh\left(\frac{g}{2}z\right) \tag{11.105}$$

It can be further shown that (see Prob. 11.13), for the case where the initial idler field $[E_{20}(0)]$ is negligible, both of the fields grow exponentially with position given by

$$E_{10}(z) = \frac{E_{10}(0)}{2}\exp\left(\frac{gz}{2}\right) \tag{11.106}$$

and

$$E_{20}^*(z) = \frac{iE_{10}(0)}{2}\sqrt{\frac{n_1}{n_2}\frac{\omega_2}{\omega_1}}\exp\left(\frac{gz}{2}\right) \tag{11.107}$$

The parameter g which we introduced in Eqs. (11.101) and (11.103) is known as the differential parametric gain coefficient. Theoretical analysis of parametric amplification dates back to 1962.[20, 21] Experimental demonstration of optical parametric amplification was first reported in 1965.[22]

11.1.8 Optical Parametric Oscillation

Now, we can envision placing this parametric gain medium within an optical cavity, similar to what we did in the case of a laser, and obtain oscillation at the frequencies ω_1 and/or ω_2. The parametric gain medium is oriented in such a way that when we use the pump at ω_3, the phase-matching condition allows the generation of the signal at ω_1 and the idler at ω_2. The device which produces coherent light as a result of optical parametric amplification and positive feedback is called an optical parametric oscillator (OPO). First demonstrated in 1965,[23] OPOs have become very important and versatile devices in photonics since they enable the generation of coherent light in wavelength ranges where direct generation with a laser is difficult. Furthermore, broad tunability can be achieved with these devices by rotating the nonlinear crystal so that different sets of wavelengths satisfy the phase-matching condition. The tuning range depends on the transparency window of the nonlinear crystal and the wavelength range over which the phase-matching condition can be satisfied. Two types of parametric oscillators exist:

1. Singly resonant OPO (SRO): Here, optical oscillation is obtained either at ω_1 or ω_2.
2. Doubly resonant OPO (DRO): In this case, oscillation occurs both at ω_1 and ω_2.

Analysis of Singly Resonant Optical Parametric Oscillators

A schematic of an SRO is shown in Fig. 11.8. The characteristics of the end mirrors are as follows:

1. M_1 is highly transmitting at the frequency ω_3 and highly reflecting at ω_1. We will assume that the amplitude reflectivity at ω_1 is 100%.
2. M_2 is partially reflecting at ω_1 with an amplitude reflectivity of

$$r_1 = \sqrt{R_1} \tag{11.108}$$

where we neglected the reflection phase factor. R_1 is the intensity reflectivity of mirror M_2.

Note that, different from the analysis of the laser resonators, we will work here with amplitude reflection coefficients, since all of the spatial evolution equations we derived [for example, Eqs. (11.92) to (11.94)] are for the electric fields and not intensities. To obtain the oscillation threshold at ω_1, we follow the wave at ω_1 once around the cavity and find the condition where the net round-trip amplitude gain is equal to unity. In the analysis of the SRO, we neglect $E_{20}^*(0)$. Let $E_{10}(m)$ be the field incident at $z = 0$ after m roundtrips. After one more roundtrip, $E_{10}(m+1)$ will be given by

$$E_{10}(m+1) = E_{10}(m) r_1 \cosh\left(\frac{g}{2}L\right) \tag{11.109}$$

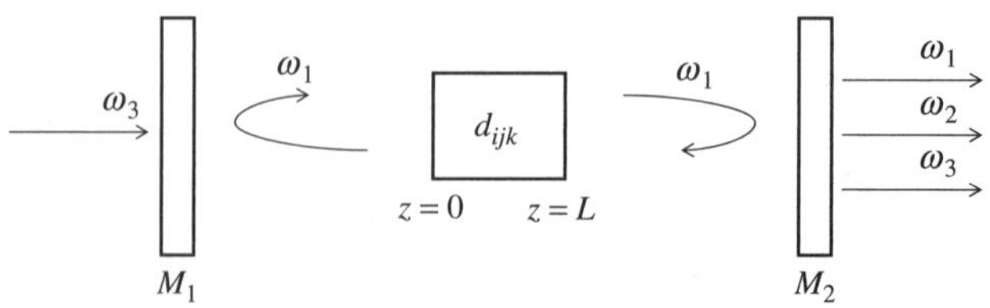

Figure 11.8 Schematic of a singly resonant optical parametric oscillator.

Here, L is the length of the nonlinear crystal and we make the assumption that the phase matching condition is satisfied only in the forward direction. At oscillation threshold, we need to have

$$E_{10}(m+1) = E_{10}(m) \tag{11.110}$$

giving

$$r_1 \cosh\left(\frac{g}{2}L\right) = 1 \tag{11.111}$$

If we now assume that the parametric gain is small with $gL << 1$, we get from Eq. (11.111)

$$\frac{1}{r_1} \approx 1 + \frac{1}{2}\frac{g^2L^2}{4} \tag{11.112}$$

or

$$g^2L^2 = \frac{8(1-r_1)}{r_1} \tag{11.113}$$

By using Eq. (11.103) for g, we see that the threshold pumping intensity $(I_3)_{th}$ needed to attain parametric oscillation in an SRO becomes

$$(I_3)_{th,\mathrm{SRO}} = \frac{4n_1n_2n_3(1-r_1)}{\eta_0^3\omega_1\omega_2 d^2L^2 r_1} \tag{11.114}$$

Clearly, the higher the nonlinear d-coefficient of the material, the lower will be the threshold pumping intensity. Furthermore, increasing the amplitude reflectivity of the output coupler also reduces the threshold pumping intensity, as expected. In this derivation, we have neglected the reflection or absorption losses occurring within the parametric gain medium as well as the finite beam size effects.

Analysis of the Doubly Resonant Optical Parametric Oscillator

In this case, the mirrors are coated to enable the oscillation of both the signal and the idler waves, as shown in Fig. 11.9. Now, we assume

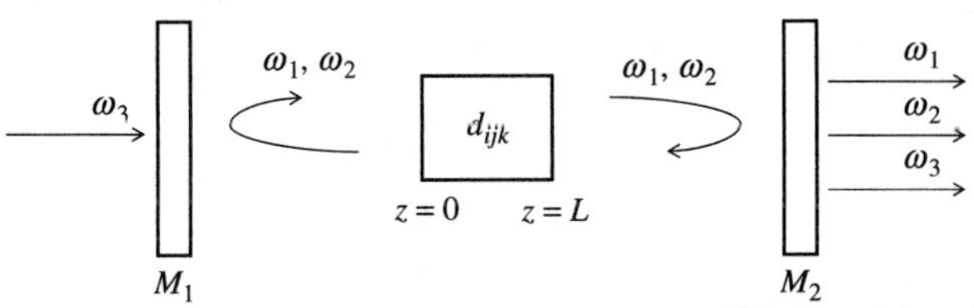

FIGURE 11.9 Schematic of a doubly resonant optical parametric oscillator.

that the mirror M_1 is perfectly reflecting at the frequencies ω_1 and ω_2. The amplitude reflectivities of the mirror M_2 at ω_1 and ω_2 are r_1 and r_2, respectively. To obtain oscillation, we need to have unity round-trip amplitude gain for both of the fields $E_{10}(z)$ and $E_{20}(z)$. With the notation used in Eq. (11.110), the threshold conditions for the signal and idler fields become

$$E_{10}(m+1, z=0) = r_1 E_{10}(m, z=L) = E_{10}(m) \tag{11.115}$$

and

$$E_{20}^*(m+1, z=0) = r_2 E_{20}^*(m, z=L) = E_{20}^*(m) \tag{11.116}$$

Again, m stands for the number of roundtrips that the circulating beam completes. By using Eqs. (11.104) and (11.105), we get

$$E_{10}(m) = \left(E_{10}(m)\cosh\left(\frac{g}{2}L\right) - i\sqrt{\frac{n_2}{n_1}\frac{\omega_1}{\omega_2}} E_{20}^*(m)\sinh\left(\frac{g}{2}L\right) \right) r_1 \tag{11.117}$$

and

$$E_{20}^*(m) = \left(E_{20}^*(m)\cosh\left(\frac{g}{2}L\right) + i\sqrt{\frac{n_1}{n_2}\frac{\omega_2}{\omega_1}} E_{10}(m)\sinh\left(\frac{g}{2}L\right) \right) r_2 \tag{11.118}$$

In matrix form, Eqs. (11.117) and (11.118) can be written as

$$\begin{pmatrix} 1 - r_1\cosh\left(\frac{g}{2}L\right) & i\sqrt{\frac{n_2}{n_1}\frac{\omega_1}{\omega_2}} r_1 \sinh\left(\frac{g}{2}L\right) \\ -i\sqrt{\frac{n_1}{n_2}\frac{\omega_2}{\omega_1}} r_2 \sinh\left(\frac{g}{2}L\right) & 1 - r_2\cosh\left(\frac{g}{2}L\right) \end{pmatrix} \begin{pmatrix} E_{10}(m) \\ E_{20}^*(m) \end{pmatrix} = \begin{pmatrix} 0 \\ 0 \end{pmatrix} \tag{11.119}$$

For nontrivial solutions, we require the determinant to vanish. We leave it as an exercise (see Prob. 11.14) for the reader to show that if the parametric gain is small as in the previous analysis of the SRO, the threshold pumping intensity for the DRO becomes

$$(I_3)_{th,\text{DRO}} = 2\frac{n_1 n_2 n_3 (1-r_1)(1-r_2)}{\eta_0^3 \omega_1 \omega_2 d^2 L^2 r_1 r_2} \tag{11.120}$$

Note that since the amplitude reflectivities r_1 and r_2 are typically close to unity, the threshold for the DRO is significantly reduced in relation to that of an SRO. Comparing Eqs. (11.114) and (11.120), the ratio of the threshold pumping intensities becomes

$$\frac{(I_3)_{th,\mathrm{DRO}}}{(I_3)_{th,\mathrm{SRO}}} = \frac{1}{2}\frac{(1-r_2)}{r_2} \tag{11.121}$$

11.2 Third-Order Nonlinear Phenomena

In this section, we discuss phenomena arising from the third-order nonlinear interaction of light with matter. Without going into the detailed derivation of the nonlinear susceptibilities, we provide a survey of the important effects including third harmonic generation, nonlinear refractive index, self-focusing, self-phase modulation, and beam propagation in a nonlinear dispersive medium.

11.2.1 Third Harmonic Generation

In third-order nonlinear effects, each component of the induced polarization will be related to the product of three electric field components. This then leads to four-wave mixing where, as an extension of three-wave mixing, three waves of frequency ω_1, ω_2, and ω_3 are coupled to produce a fourth wave at ω_4. A rich variety of possibilities exist. One important special case is sum frequency generation where

$$\omega_4 = \omega_1 + \omega_2 + \omega_3 \tag{11.122}$$

In the degenerate case, where $\omega_1 = \omega_2 = \omega_3 = \omega$, we have $\omega_4 = 3\omega$ and the nonlinear process leads to third harmonic generation. Classically, we can think of the third-order nonlinearity arising from the presence of a cubic restoring force proportional to x^3, and generally no symmetry requirements exist. Hence, four-wave mixing occurs in all media including liquids and gases, provided that the light intensity is sufficiently high. A further generalization of this where a beam with a very high intensity produces harmonics of even higher order as a result of different wave mixing processes is termed high harmonic generation (HHG) and currently stands as a very important technique for generating coherent radiation in the deep ultraviolet or the soft x-ray region.[24-26]

11.2.2 Intensity-Dependent Refractive Index

One special case of four-wave mixing leads to a third-order polarization at the frequency of the propagating wave and introduces an intensity-dependent contribution to the refractive index of the medium. Hence, the refractive index becomes

$$n = n_0 + n_2 I \tag{11.123}$$

where I is the intensity of the beam, n_0 is the low-intensity refractive index, and n_2 is known as the nonlinear refractive index coefficient or simply the nonlinear refractive index. This is the well-known Kerr effect first reported in 1875.[27] The light intensity can have spatial or temporal variation which leads to very interesting beam shaping effects in the spatial or temporal domain. In the next section, we discuss a special case of this known as self-focusing.

11.2.3 Self-Focusing

Self-focusing occurs when the nonlinear refractive index is positive ($n_2 > 0$). To see qualitatively what happens, consider a light beam with a Gaussian intensity distribution as sketched in Fig. 11.10*a*.

The induced change in the refractive index will be maximum at the center of the beam and will follow the transverse intensity distribution. The Kerr medium will then act as a positive lens to focus the beam. As the beam is focused, the lens will get stronger, leading to further focusing. How severe the effect will be depends on the starting power in the beam. If the beam power is very high above a critical limit, self-focusing can even lead to a permanent refractive index change or catastrophic damage in the medium. On the positive side, self-focusing has been employed in several important useful applications, including:

1. Kerr-lens mode locking (KLM): Here, the self-focusing effect within the gain medium of a laser is employed in generating ultrashort optical pulses. Suppose the medium is end pumped by another laser. Both the pump and the laser beams will have position dependent spotsizes within the gain medium. The resonator optics can be adjusted in such a way as to give a smaller mode volume for the high-intensity laser beam as

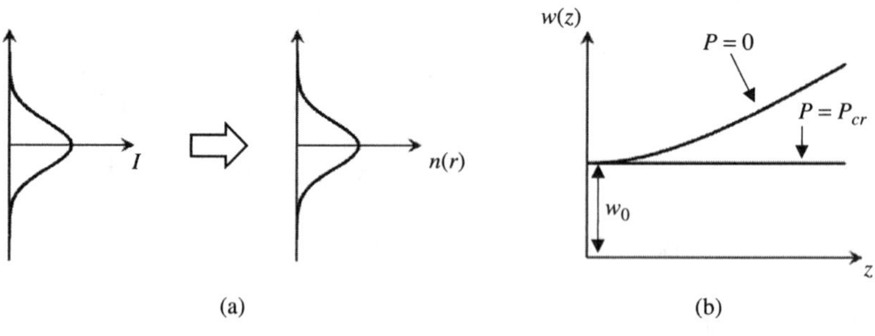

FIGURE 11.10 (*a*) Transverse intensity distribution of a Gaussian beam and the corresponding induced index change in a nonlinear Kerr medium. (*b*) Approximate variation of the spotsize near the waist location for a low-intensity beam ($P = 0$) and a high-intensity beam whose power is equal to the critical focusing power P_{cr}.

opposed to the low-intensity version. If a hard aperture of the correct size is placed inside the cavity, it can favor the oscillation of the high-intensity beam by providing larger losses at low intensities. This will favor the oscillation of high-intensity mode-locked pulses with durations in the picosecond and femtosecond time scales, depending on the available spectral bandwidth and the level of dispersion. Even without a hard aperture, KLM action can still take place due to the soft aperturing effect provided by the pump-induced spatial gain profile. Ever since its demonstration in 1991,[28] KLM has been applied to numerous solid-state laser media[28-33] to generate ultrashort optical pulses.

2. Laser micromachining: In this case, material processing can be done by employing self-focusing above the damage threshold of the material. This method has been used in cutting and drilling, as well as the fabrication of photonic devices.[34-36]

We next provide a crude, simplified analysis of self-focusing by using the ABCD formalism introduced in Chaps. 2 and 3. Consider a collimated Gaussian beam incident on a Kerr medium whose refractive index depends on the intensity as given by Eq. (11.123). The beam intensity can be expressed as

$$I(r) = \frac{2P}{\pi w_0^2}\exp\left(-\frac{2r^2}{w_0^2}\right) \tag{11.124}$$

where w_0 is the beam waist at the input plane ($z = 0$) of the medium, and P is the beam power. This leads to a radial variation in the refractive index $n(r)$ which can be expressed up to second order in r in the form

$$n(r) \cong n_0 + n_2\frac{2P}{\pi w_0^2}\left(1-\frac{2r^2}{w_0^2}\right) \approx n_0\left(1-\frac{1}{2}\beta^2 r^2\right) \tag{11.125}$$

Above, the quadratic coefficient β is given by

$$\beta = \sqrt{\frac{8n_2 P}{\pi w_0^4 n_0}} \tag{11.126}$$

In the actual case, β will vary with position due to the change in the spotsize. Neglecting this position dependence of the spotsize, we recognize that the radial refractive index variation given by Eq. (11.125) can be represented by the 2×2 ray transformation matrix $\mathbf{M}_\mathrm{T}$ given by [see Eq. (3.53)]

$$\mathbf{M}_\mathrm{T} = \begin{bmatrix} \cos\beta z & \frac{1}{\beta}\sin\beta z \\ -\beta\sin\beta z & \cos\beta z \end{bmatrix} \tag{11.127}$$

Now, consider a collimated Gaussian beam incident on the medium at $z = 0$ with a Rayleigh range of z_0. The transformed q-parameter at position z will be given by [see Eq. (3.79)]

$$q(z) \cong \frac{iz_0 \cos\beta z + \frac{1}{\beta}\sin\beta z}{iz_0(-\beta\sin\beta z) + \cos\beta z} \tag{11.128}$$

We leave it as an exercise for the reader (see Prob. 11.16) to show that in the limit where β tends to zero, the resulting position-dependent spotsize can be expressed in the form

$$w(z)^2 = w_0^2\left[1 + \left(1 - \beta^2 z_0^2\right)\left(\frac{z}{z_0}\right)^2\right] \tag{11.129}$$

Above, the product $\beta^2 z_0^2$ can be written as

$$\beta^2 {z_0}^2 = \frac{8n_2 P}{\pi\omega_0^4 n_0}\cdot\frac{n_0^2\pi^2{\omega_0}^4}{\lambda^2} = \frac{P}{P_{cr}} \tag{11.130}$$

where the critical power P_{cr} is given by

$$P_{cr} = \frac{\lambda^2}{8\pi n_0 n_2} \tag{11.131}$$

In practice P_{cr} is taken as[37]

$$P_{cr} = \frac{\alpha\lambda^2}{8\pi n_0 n_2} \tag{11.132}$$

where the dimensionless correction factor α, which has a value in the range of 3.77–6.4, accounts for beam distortions. The formula for the spotsize variation now has the form

$$w^2(z) = w_0^2\left(1 + \left(1 - \frac{P}{P_{cr}}\right)\left(\frac{z}{z_0}\right)^2\right) \tag{11.133}$$

Note that when $P = P_{cr}$, diffractive spreading is just balanced by self-focusing and the beam propagates with a nearly constant spotsize $w(z) \simeq w_0$, as sketched in Fig. 11.10*b*.

11.2.4 Electromagnetic Pulse Propagation in Nonlinear Dispersive Media

We employ a method similar to that used in Refs. 5, 38, and 39 to derive the partial differential equation satisfied by the pulse electric field in a nonlinear dispersive medium. Consider the propagation of an electromagnetic pulse in a dispersive, nonlinear medium where the wave vector k is a function of frequency [i.e., $k = k(\omega)$] and the refractive index n has a nonlinear intensity-dependent contribution given by Eq. (11.123). For simplicity, we consider one-dimensional motion where the electric field vector is linearly polarized and can be represented by one scalar component, say $E(z,t)$. We further assume that the spectral amplitude distribution is locally peaked about some central carrier frequency ω_0 so that k can be expanded about $\omega = \omega_0$ as a Taylor series in the form

$$k(\omega) = k_0 + \left.\frac{dk}{d\omega}\right|_0 (\omega - \omega_0) + \frac{1}{2}\left.\frac{d^2k}{d\omega^2}\right|_0 (\omega - \omega_0)^2 + \cdots \tag{11.134}$$

where $k_0 = k(\omega_0)$ and the derivatives with the subscript 0 are evaluated at $\omega = \omega_0$. We introduce a pulse envelope function $a(z,t)$ and express the electric field $E(z,t)$ as

$$\begin{aligned} E(z,t) &= \sqrt{\frac{2\eta}{A_{\text{eff}}}}\, a(z,t)\exp[i(\omega_0 t - k_0 z)] \\ &= \sqrt{\frac{2\eta}{A_{\text{eff}}}} \int_{-\infty}^{+\infty} d\omega A(\omega)\exp[i(\omega t - k(\omega)z)] \end{aligned} \tag{11.135}$$

Above, η is the wave impedance of the medium, A_{eff} is the effective beam area, and $A(\omega)$ is the spectral amplitude distribution function. Note that $|a(z,t)|^2$ directly gives the instantaneous power of the pulse. By using Eq. (11.134), we get

$$\begin{aligned} a(z,t)\exp[i(\omega_0 t - k_0 z)] &= \exp[i(\omega_0 t - k_0 z)] \int_{-\infty}^{+\infty} d\omega A(\omega) \\ &\times \exp\left[i(\omega - \omega_0)t - k_1 z(\omega - \omega_0) - \frac{1}{2}k_2 z(\omega - \omega_0)^2 + \cdots \right] \end{aligned} \tag{11.136}$$

where

$$k_1 = \left.\frac{dk}{d\omega}\right|_{\omega=\omega_0} \tag{11.137}$$

and

$$k_2 = \left.\frac{d^2k}{d\omega^2}\right|_{\omega=\omega_0} \tag{11.138}$$

Setting

$$\Omega = \omega - \omega_0 \tag{11.139}$$

we see that $a(z,t)$ can be expressed as

$$a(z,t) = \int_{-\infty}^{+\infty} d\Omega A(\Omega) \exp\left[i\Omega t - i \sum_{m=1}^{\infty} \frac{1}{m!} k_m \Omega^m z \right] \tag{11.140}$$

where

$$k_m = \left. \frac{d^m k}{d\Omega^m} \right|_{\Omega=0} \tag{11.141}$$

Clearly, $\Omega = 0$ corresponds to the center frequency $\omega = \omega_0$. Going from z to $z + dz$ in the medium, $a(z + dz, t)$ becomes

$$\begin{aligned} a(z+dz,t) &= \int_{-\infty}^{+\infty} d\Omega A(\Omega) \exp\left[i\Omega t - i(z+dz) \sum_{m=1}^{\infty} \frac{k_m}{m!} \Omega^m \right] \\ &\cong \int_{-\infty}^{+\infty} d\Omega A(\Omega) \exp\left[i\left(\Omega t - z \sum_{m=1}^{\infty} \frac{k_m}{m!} \Omega^m \right) \right] \left(1 - idz \sum_{m=1}^{\infty} \frac{k_m}{m!} \Omega^m \right) \end{aligned} \tag{11.142}$$

Noting that

$$\frac{\partial^m a(z,t)}{\partial t^m} = \int_{-\infty}^{+\infty} d\Omega (i\Omega)^m A(\Omega) \exp\left[i\left(\Omega t - z \sum_{m=1}^{\infty} \frac{k_m}{m!} \Omega^m \right) \right] \tag{11.143}$$

we get

$$a(z+dz,t) = a(z,t) - idz \sum_{m=1}^{\infty} \frac{(-i)^m}{m!} k_m \frac{\partial^m a}{\partial t^m} \tag{11.144}$$

Hence,

$$\frac{\partial a}{\partial z} = \frac{a(z+dz,t) - a(z,t)}{dz} = \sum_{m=1}^{\infty} \frac{(-i)^{m+1}}{m!} k_m \frac{\partial^m a}{\partial t^m} \tag{11.145}$$

Equation (11.145) summarizes how the pulse envelope function $a(z,t)$ evolves in the presence of different orders of dispersion.

As a specific example, let us consider what happens when dispersion effects up to second order are dominant. We then have, using Eq. (11.145),

$$\frac{\partial a}{\partial z} = -\frac{1}{u_g} \frac{\partial a}{\partial t} + \frac{i}{2} k_2 \frac{\partial^2 a}{\partial t^2} \tag{11.146}$$

where

$$\frac{1}{u_g} = k_1 = \left. \frac{dk}{d\Omega} \right|_{\Omega=0} \tag{11.147}$$

and

$$k_2 = \left.\frac{d^2k}{d\Omega^2}\right|_{\Omega=0} \tag{11.148}$$

Recall from our earlier discussions that u_g corresponds to the group velocity. Defining

$$T = t - \frac{z}{u_g} \tag{11.149}$$

where T represents the time measured in the reference frame that moves at the group velocity of the pulse, we obtain

$$\frac{\partial a}{\partial z} = \frac{i}{2}k_2 \frac{\partial^2 a}{\partial T^2} \tag{11.150}$$

Use of Eq. (11.150) shows us what happens to the shape of the pulse envelope during propagation, since we are watching in the rest frame of the pulse. Now, to include the effect of self-phase modulation, we note that over a small distance dz, and ignoring dispersive effects for a moment, we have

$$a(z+dz,t) = a(z,t)\exp[-i\Phi_{NL}] \tag{11.151}$$

where the nonlinear phase shift Φ_{NL} is given by

$$\Phi_{NL} = n_2 I k_0 dz \tag{11.152}$$

Here, I is the beam intensity and n_2 is the nonlinear index introduced in Sec. 11.2.2. For a small nonlinear phase shift, we can expand the exponential term up to first order to get

$$a(z+dz,t) \cong a(z,t)[1 - in_2 I k_o dz] \tag{11.153}$$

or

$$\frac{a(z+dz,t) - a(z,t)}{dz} = \frac{\partial a}{\partial z} = -in_2 I k_o a(z,t) \tag{11.154}$$

Since $|a(z,t)|^2$ represents the power, the instantaneous intensity becomes

$$I(z,t) = \frac{|a(z,t)|^2}{A_{\text{eff}}} \tag{11.155}$$

where A_{eff} is the effective beam area. Hence, Eq. (11.154) can be written as

$$\frac{\partial a}{\partial z} = -i\gamma a(z,t)|a(z,t)|^2 \tag{11.156}$$

Above, the Kerr nonlinearity coefficient γ has the form

$$\gamma = \frac{2\pi}{\lambda}\frac{n_2}{A_{\text{eff}}} \tag{11.157}$$

Now, putting together the effects of second-order dispersion and self-phase modulation, we obtain for $a(z,T)$

$$\frac{\partial a}{\partial z} - i\frac{k_2}{2}\frac{\partial^2 a}{\partial T^2} + i\gamma a(z,T)\left|a(z,T)\right|^2 = 0 \tag{11.158}$$

11.2.5 Effect of Dispersion on Pulse Propagation

If we only have dispersion up to second order and no significant Kerr nonlinearity, Eq. (11.158) simplifies to

$$\frac{\partial a}{\partial z} = i\frac{k_2}{2}\frac{\partial^2 a}{\partial T^2} \tag{11.159}$$

We can solve Eq. (11.159) by introducing the Fourier transform pair:

$$a(z,T) = \frac{1}{\sqrt{2\pi}}\int_{-\infty}^{\infty} d\omega A(z,\omega)\exp(i\omega T) \tag{11.160}$$

and

$$A(z,\omega) = \frac{1}{\sqrt{2\pi}}\int_{-\infty}^{\infty} dT a(z,t)\exp(-i\omega T) \tag{11.161}$$

By using Eqs. (11.159) and (11.160), we get

$$\frac{\partial a}{\partial z} = \frac{1}{\sqrt{2\pi}}\int_{-\infty}^{\infty} d\omega \frac{\partial A}{\partial z}\exp(i\omega T) = i\frac{k_2}{2}\int_{-\infty}^{\infty} d\omega(-\omega^2)A(\omega)\exp(i\omega T) \tag{11.162}$$

or

$$\frac{\partial A}{\partial z} = -i\frac{k_2}{2}\omega^2 A(\omega) \tag{11.163}$$

We leave it as an exercise for the reader (see Prob. 11.17) to show that the solution of Eq. (11.163) is given by

$$A(z,\omega) = A(0,\omega)\exp\left(-i\frac{k_2}{2}\omega^2 z\right) \tag{11.164}$$

where $A(0,\omega)$ is the initial spectral amplitude distribution of the pulse at $z=0$. As an example, let us consider as input, a Gaussian pulse with a temporal profile of

$$a(0,T) = a_0 \exp(-\alpha T^2) \tag{11.165}$$

at $z=0$. It can be easily shown that (see Prob. 11.18) the full-width half-maximum duration $T_{1/2}$ of the pulse satisfies

$$T_{1/2}^2 = \frac{2\ln(2)}{\alpha} \tag{11.166}$$

The procedure for finding the pulse profile at a finite value of z is straightforward but requires a fair amount of algebra. It involves determining the initial spectral amplitude distribution $A(0,\omega)$, then using Eq. (11.164) to find the amplitude distribution at z, and finally, taking the inverse Fourier transform [Eq. (11.160)] to calculate $a(z,T)$. Leaving this as an exercise (see Prob. 11.18), the pulse envelope $a(z,T)$ becomes

$$a(z,T) = \frac{a_0}{\sqrt{(1+i2\alpha k_2 z)}} \exp\left(-\frac{\alpha T^2}{(1+i2\alpha k_2 z)}\right) \tag{11.167}$$

Note that we can write $a(z,T)$ in the form

$$a(z,T) = |a(z,T)| \exp[i\Phi(z,T)] \tag{11.168}$$

It turns out that the pulse maintains its Gaussian profile and becomes broader with increasing distance. The corresponding peak intensity at the pulse center also decreases monotonically. It can be shown that the full-width half-maximum duration at z becomes (see Prob. 11.19)

$$\frac{T_{1/2}(z)}{T_{1/2}} = \sqrt{1+4\alpha^2 k_2^2 z^2} = \sqrt{1+\left(\frac{4k_2 z\ln(2)}{T_{1/2}^2}\right)^2} \tag{11.169}$$

Equation (11.169) can be written in a somewhat more compact form as

$$\frac{T_{1/2}(z)}{T_{1/2}} = \sqrt{1+\left(\frac{z}{z_D}\right)^2} \tag{11.170}$$

where z_D is the characteristic dispersion length given by

$$z_D = \frac{T_{1/2}^2}{4|k_2|\ln 2} \tag{11.171}$$

There is one more commonly used version of Eq. (11.169) expressed in terms of the group delay dispersion parameter D (also referred to as the second-order dispersion parameter) which, following the standard convention in ultrafast optics, is defined according to

$$D = \frac{d^2\Phi}{d\omega^2} = \ell \frac{d^2k}{d\omega^2} = \ell k_2 = \ell \frac{d}{d\omega}\frac{1}{u_g} = \frac{d}{d\omega}\left(\frac{\ell}{u_g}\right) \tag{11.172}$$

Here, ℓ is the length of the medium traversed by the pulse, and ℓ/u_g gives the group delay. D is related to the second derivative of the acquired phase with respect to the angular frequency ω and leads to the temporal broadening that we just described. Typical units of D are fs^2. Also note that the definition of D given here is different from that for the group velocity dispersion parameter D_{GV} introduced in Chap. 4, Eq. (4.117). In terms of D, Eq. (11.169) becomes

$$\frac{T_{1/2}(z=\ell)}{T_{1/2}} = \sqrt{1+\left(\frac{4D\ln(2)}{T_{1/2}^2}\right)^2} \tag{11.173}$$

It can be shown that (see Prob. 11.20) if the wavelength dependence of the refractive index is known, D can be calculated by using

$$D = \ell \frac{\lambda^3}{2\pi c^2}\frac{d^2n}{d\lambda^2} \tag{11.174}$$

Now, what about the phase of the field in Eq. (11.168)? This is also very important because, as it turns out, the presence of second-order dispersion adds a frequency chirp to the propagating pulse. To see this in detail, we examine the phase $\Phi(z,T)$, which at position z is given by (see Prob. 11.21)

$$\Phi(z,T) = \frac{2\alpha^2T^2k_2z}{1+4\alpha^2k_2^2z^2} - \frac{1}{2}\tan^{-1}(2\alpha k_2 z) \tag{11.175}$$

The dynamic carrier frequency ω, which now varies across the pulse, can be calculated from

$$\omega = \omega_0 + \frac{\partial\Phi(z,T)}{\partial T} = \omega_0 + \Delta\omega \tag{11.176}$$

where ω_0 is the center frequency and $\Delta\omega$ is the dispersion-induced frequency shift. By using Eq. (11.175), $\Delta\omega$ becomes

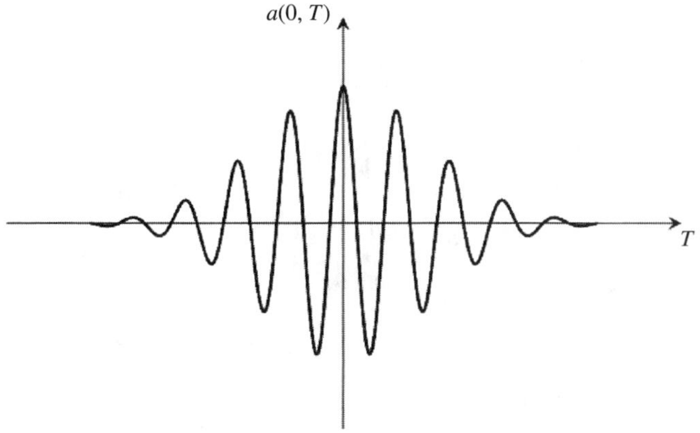

Figure 11.11 Electric field of an incident Gaussian pulse.

$$\Delta\omega = \frac{4\alpha^2 T k_2 z}{1+4\alpha^2 k_2^2 z^2} = \frac{4\left(\dfrac{2\ln(2)}{T_{1/2}^2}\right)^2 D}{1+\left(\dfrac{4D\ln(2)}{T_{1/2}^2}\right)^2} T = \frac{16D(\ln(2))^2}{T_{1/2}^4 + (4D\ln(2))^2} T \tag{11.177}$$

Alternatively, we can express Eq. (11.177) in terms of the characteristic dispersion length z_D as

$$\Delta\omega = \frac{(4\ln 2)\operatorname{sgn}(D)(z/z_D)}{1+(z/z_D)^2}\frac{T}{T_{1/2}^2} \tag{11.178}$$

Above, sgn(D) is −1 when $D < 0$, and +1 when $D > 0$. Figure 11.11 shows the electric field of an incident Gaussian pulse at $z = 0$. If the second-order dispersion is positive (i.e., $D > 0$), the dynamic frequency is downshifted in the leading edge ($T < 0$) and upshifted in the trailing edge ($T > 0$) of the pulse as shown in Fig. 11.12*a*. The corresponding variation of the frequency shift across the pulse is displayed in Fig. 11.12*b*. The frequency shift has the opposite trend when $D < 0$, as shown in Fig. 11.13*a* and *b*.

11.2.6 Self-Phase Modulation

As the name implies, self-phase modulation (SPM) refers to the phase variation induced across a pulse when its own intensity is high enough to create a time-dependent index change in the medium. To

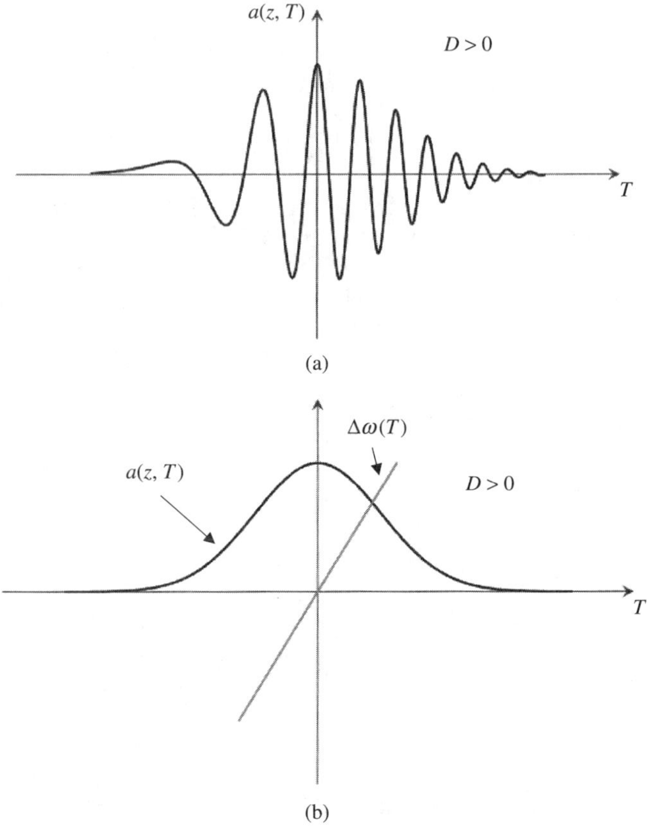

FIGURE 11.12 (*a*) Electric field of a pulse propagating in a medium with positive group delay dispersion ($D > 0$). (*b*) The corresponding variation of the frequency shift across the pulse for $D > 0$.

analyze SPM alone, we neglect second-order dispersion, in which case Eq. (11.158) becomes

$$\frac{\partial a}{\partial z} = -i\gamma a(z,T)\left|a(z,T)\right|^2 \tag{11.179}$$

Let us assume that $\left|a(z,T)\right|^2$ appearing in Eq. (11.179) remains nearly constant and equal to its initial value $\left|a(0,T)\right|^2 = P(0,T)$ over short distances, giving

$$a(z,T) \cong a(0,T)\exp(-i\gamma P(0,T)z) \tag{11.180}$$

Here, the nonlinear phase shift is

$$\Phi_{NL}(z,T) = -\gamma P(0,T)z \tag{11.181}$$

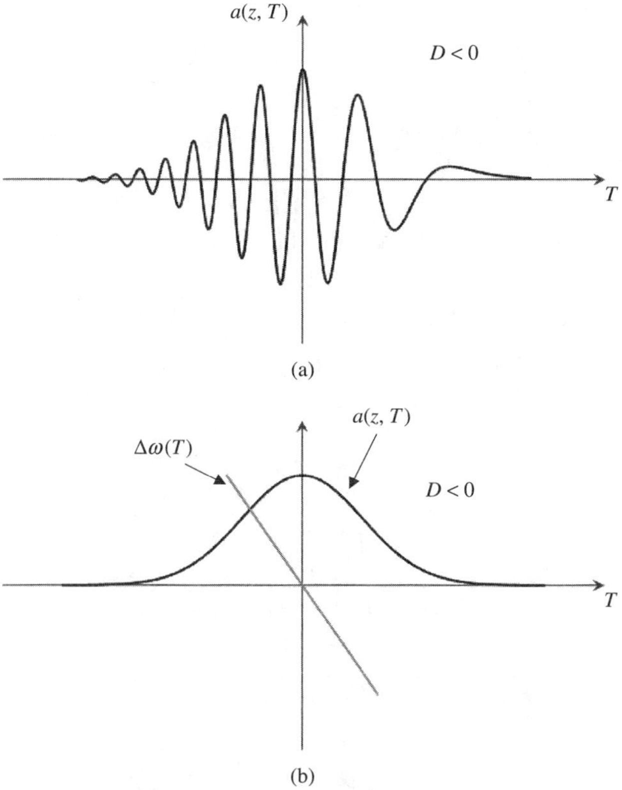

FIGURE 11.13 (*a*) Electric field of a pulse propagating in a medium with negative group delay dispersion ($D < 0$). (*b*) The corresponding variation of the frequency shift across the pulse for $D < 0$.

As an example, let us consider an input Gaussian pulse with the envelope

$$a(0,T) = a_0 \exp[-\alpha T^2] \tag{11.182}$$

The peak power P_0 of the pulse is equal to $|a_0|^2$. In this case, the nonlinear phase shift can be written in the form

$$\Phi_{NL}(z,T) = -\Phi_{\max} \exp[-2\alpha T^2] \tag{11.183}$$

where the maximum phase shift is given by

$$\Phi_{\max} = \gamma z |a_0|^2 \tag{11.184}$$

Figure 11.14*a* to *d* shows the pulse profiles for the cases where the maximum nonlinear phase shift is 0, π/2, π, and 2π. It becomes more

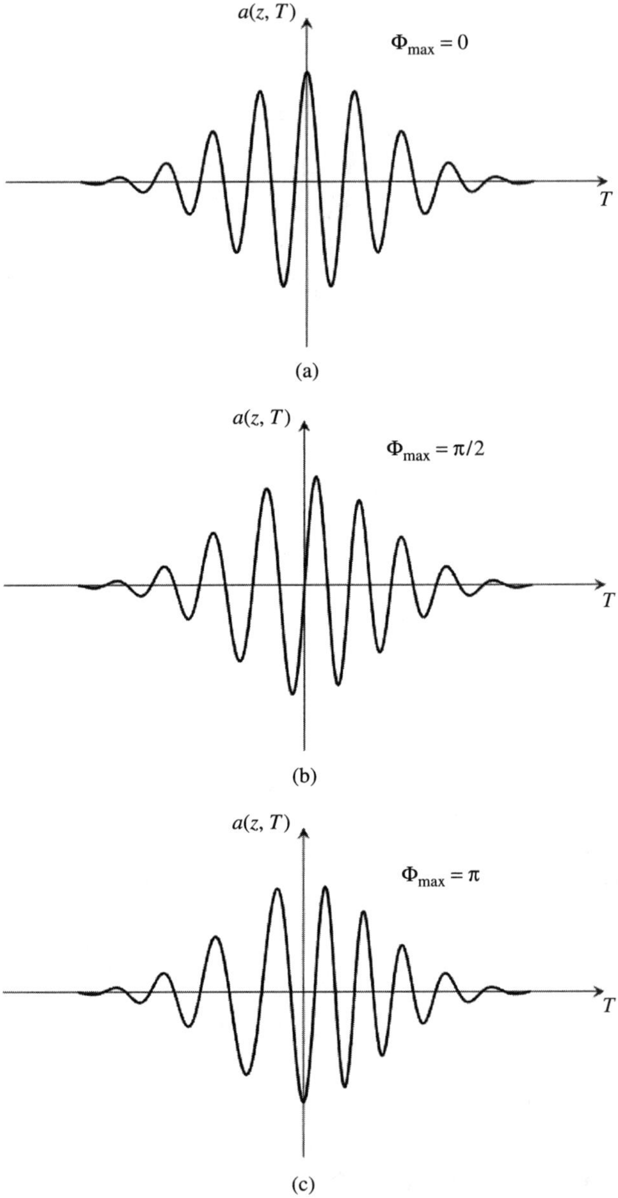

FIGURE 11.14 Pulse profiles in the presence of self-phase modulation with a maximum nonlinear phase shift of (*a*) 0, (*b*) $\pi/2$, (*c*) π, and (*d*) 2π.

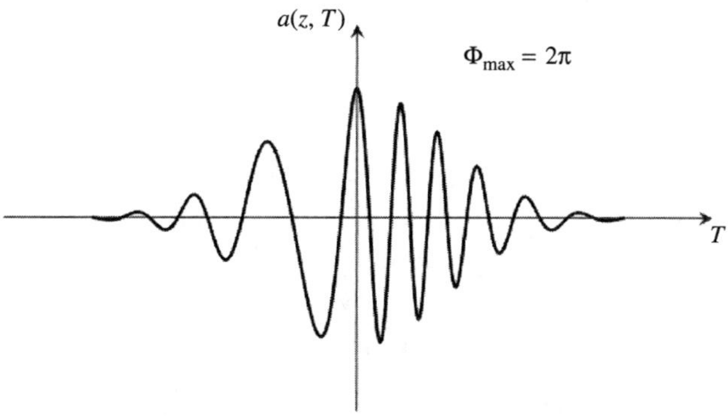

(d)

Figure 11.14 *(Continued)*

apparent with an increasing nonlinear phase shift that the leading edge of the pulse is downshifted, while the trailing edge is upshifted in frequency, as in the case of positive dispersion. The corresponding frequency shift $\Delta\omega$ can be calculated from

$$\Delta\omega = \frac{\partial \Phi}{\partial T} = -\gamma z \frac{\partial P(0,T)}{\partial T} \tag{11.185}$$

It can be shown that (see Prob. 11.23) for a Gaussian pulse, SPM-induced frequency shift becomes

$$\Delta\omega = 4 \ln 2 \frac{\Phi_{max}}{T_{1/2}} \exp[-2\alpha T^2] \frac{T}{T_{1/2}} \tag{11.186}$$

Above, $T_{1/2}$ is the initial pulse duration (FWHM).

11.2.7 Solitary Optical Pulses

It is beyond the scope of this text to provide a rigorous and comprehensive account of nonlinear wave equations and solitary optical pulses. However, just to provide some physical insight about why these pulses are special, we look at the combined effect of self-phase modulation and second-order dispersion. Again, we consider a Gaussian pulse, although, as we will comment shortly, actual solitary pulses are not Gaussian. From the analysis of the previous sections, we can see that if $D < 0$, one can obtain cancellation of the frequency chirp near the central part of the pulse provided that the $\Delta\omega - T$ curves have equal but opposite slopes near $T = 0$. This is schematically shown in Fig. 11.15*a* and *b*. If this condition is met, the frequency chirp vanishes near the pulse center (see Fig. 11.15*b*) and the pulse distortion due to SPM is balanced by negative dispersion. Hence, the pulse

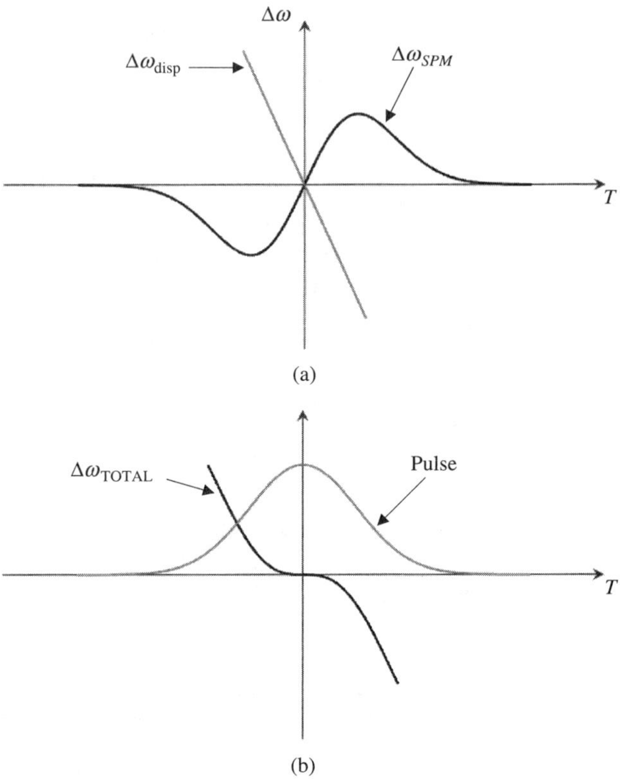

FIGURE 11.15 (*a*) $\Delta\omega - T$ curves for self-phase modulation (SPM) and negative second-order dispersion for a Gaussian input pulse. (*b*) Combination of self-phase modulation and negative second-order dispersion can lead to cancellation of the frequency chirp if the input peak power is $P_0 = 4\ln 2|k_2| / \gamma T_{1/2}^2$.

propagates with minimum distortion and maintains its shape. Of course, for a given level of negative dispersion, this requires a particular peak power to begin with. Setting the magnitudes of the slopes equal to each other, we get

$$\frac{4\left(\dfrac{2\ln 2}{T_{1/2}^2}\right)^2 D}{1+\left(\dfrac{4D\ln 2}{T_{1/2}^2}\right)^2} T = \frac{4\gamma z P_0 \ln 2}{T_{1/2}^2} \exp[-2\alpha T^2] T \tag{11.187}$$

Near the pulse center, we can take $\exp[-2\alpha T^2] \cong 1$ and also assume that $T_{1/2}^2$ is very long compared to $4D\ln 2$. Also, recalling that $D = k_2 z$, Eq. (11.187) then simplifies to

$$\frac{16(\ln 2)^2}{T_{1/2}^2} k_2 z = 4\gamma z P_0 \ln 2 \tag{11.188}$$

or

$$P_0 = \frac{4|k_2|\ln 2}{\gamma T_{1/2}^2} \approx \frac{2.77|k_2|}{\gamma T_{1/2}^2} \tag{11.189}$$

So, if we launch our pulse with the initial peak power given by Eq. (11.189), then, provided that attenuation is negligible, the pulse propagates with minimal distortion.

In fact, the actual pulse shape and the required input power can be determined analytically by solving Eq. (11.158). To outline the method, we follow Ref. 38 and express Eq. (11.158) in dimensionless form by using the transformations

$$\tau = \frac{T}{\tau_0} \tag{11.190}$$

$$x = \frac{|k_2|}{\tau_0^2} z \tag{11.191}$$

and

$$\psi = \tau_0 \sqrt{\frac{\gamma}{|k_2|}} a \tag{11.192}$$

We leave it as an exercise for the reader (see Prob. 11.24) to show that Eq. (11.158) can then be put into the dimensionless form

$$-i\frac{\partial \psi}{\partial x} - \frac{\mathrm{sgn}(k_2)}{2}\frac{\partial^2 \psi}{\partial \tau^2} + \psi|\psi|^2 = 0 \tag{11.193}$$

where

$$\mathrm{sgn}(k_2) = \begin{cases} 1 & k_2 > 0 \\ -1 & k_2 < 0 \end{cases} \tag{11.194}$$

We are particularly interested in the case where $k_2 < 0$, giving

$$-i\frac{\partial \psi}{\partial x} + \frac{1}{2}\frac{\partial^2 \psi}{\partial \tau^2} + \psi|\psi|^2 = 0 \tag{11.195}$$

Equation (11.195) is the well-known nonlinear Schrödinger equation which has been extensively studied (see Refs. 5, 38, and 40 to 42 and references therein). It can be shown that the fundamental soliton solution has the analytical form

$$\psi(x,\tau) = \text{sech}(\tau)\exp(ix/2) \tag{11.196}$$

For a sech-shaped pulse, the pulse duration $T_{1/2}$ (FWHM) is equal to 1.76 T_0. Furthermore, it can be shown that the peak power of the fundamental soliton which moves without any distortion is given by[38]

$$P_1 = \frac{3.11|k_2|}{\gamma T_{1/2}^2} \tag{11.197}$$

Note that our approximate expression for the peak power in Eq. (11.189) comes close to the exact expression in Eq. (11.197).

Infinitely many higher-order soliton solutions also exist. In particular, if the input pulse has the form

$$\psi(0,\tau) = N\,\text{sech}(\tau) \tag{11.198}$$

where N is an integer, one can obtain higher-order solutions of Eq. (11.195). Note that typically, the pulse shape of these higher-order solutions varies with position in a complicated way. However, all higher-order solitons, irrespective of their functional form, are periodic with a period of $x_0 = \pi/2$. In other words, although their shape is not preserved during propagation, they return to their initial shape periodically at integer multiples of x_0. To launch the Nth-order soliton, one requires an initial peak power of N^2P_1, where P_1 is given by Eq. (11.197). By using the methods of Kerr-lens mode locking, additive-pulse mode locking, or passive mode locking with semiconductor saturable absorbers, it is possible to generate solitary optical pulses with sech profiles and with durations in the picosecond and femtosecond time scales from different laser systems.[28-31, 43-46] Propagation of solitons in optical fibers has also been widely studied.[47-49]

Example Let us assume that we have a silica fiber with an effective core radius of 4.5 μm operating at 1550 nm. Based on the Sellmeier equation for silica given in Chap. 4, k_2 comes to -2.79×10^{-26} s^2/m. Taking the nonlinear index as 3.2×10^{-20} m^2/W, we need a peak power of 4.7 W for the fundamental soliton, if the input pulsewidth is assumed to be 3 ps. The soliton period for higher-order solitons becomes 163 m. To launch a second-order soliton, one would then need a sech pulse with a peak power of about $4 \times 4.7 = 19$ W.

11.3 Raman Effect

When light is scattered by a gas, liquid, or solid, there occurs a shift in the frequency in comparison with the frequency of the incident light. The frequency shift is characteristic of the material and provides a spectral signature with which constituents of a compound can be identified. This is the famous Raman effect, first discovered by Sir Chandrasekhara Venkata Raman in the 1920s.[50, 51] We will not go into a detailed analysis of the Raman effect here but just comment that the process is actually an example of four-wave mixing.

To grasp the essence of the Raman effect, consider an electromagnetic wave at frequency ω_p incident on a material whose low-lying energy levels are shown in Fig. 11.16*a*. If some of the molecules are excited from the ground state ($|1\rangle$ in Fig. 11.16*a*) but end up at a higher-lying vibrational level (such as level $|2\rangle$ in Fig. 11.16*a*), a portion of the scattered radiation will be at a lower frequency, given by

$$\omega_S = \omega_p - \omega_{21} \tag{11.199}$$

where ω_S is commonly referred to as the Stokes component. The frequency ω_{21} corresponds to one of the characteristic resonance frequencies due to the vibrational or rotational degrees of freedom.

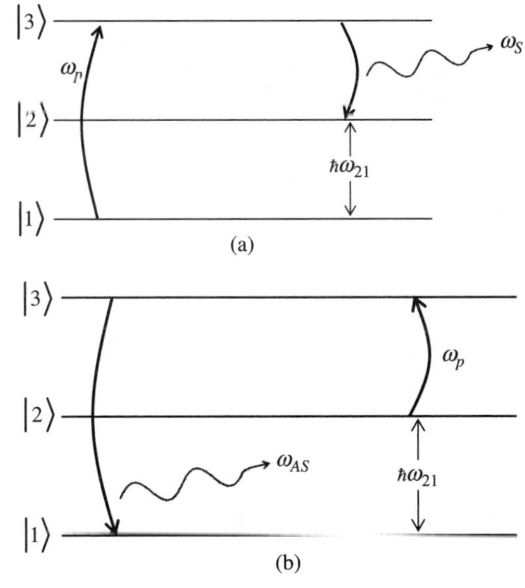

FIGURE 11.16 Low-lying energy levels of a molecule, showing the transitions involved in the generation of (*a*) the Stokes and (*b*) the anti-Stokes waves during Raman scattering.

Alternatively, some of the molecules may already be at the higher-lying level $|2\rangle$ to begin with due to thermal excitations. In this case, absorption of the pump photon and subsequent transition down to the ground state $|1\rangle$ (see Fig. 11.16*b*) generates a scattered wave at a higher frequency given by

$$\omega_{AS} = \omega_p + \omega_{21} \tag{11.200}$$

Here, ω_{AS} is known as the anti-Stokes signal.

Needless to say, the actual energy-level structure of a material is much more complicated than what is shown in Fig. 11.16. Supposing that we only have this vibrational resonance at the frequency ω_{21} and look at the light emerging from the material with a spectrometer, we typically would see a central peak at the frequency ω_p due to elastic scattering, and two neighboring peaks due to the Stokes and anti-Stokes components. This is schematically shown in Fig. 11.17. The beauty of the Raman effect is that the scattered signals will be seen ideally for any pump frequency, provided that the pump as well as the scattered frequencies fall within the transparency regime of the material. Usually, the anti-Stokes component is weaker, because in thermal equilibrium we have a larger number of molecules in the ground state than in the upper levels.

The growth of the Raman signal is exponential and is given by

$$I_S(z) = I_S(0)\exp(g_R z) \tag{11.201}$$

where $I_S(z)$ is the Stokes signal intensity at position z within the material, $I_S(0)$ is the initial Stokes intensity, and g_R is the differential Raman gain coefficient which depends on the intensity I_p of the pump signal according to

$$g_R = g_0 I_p \tag{11.202}$$

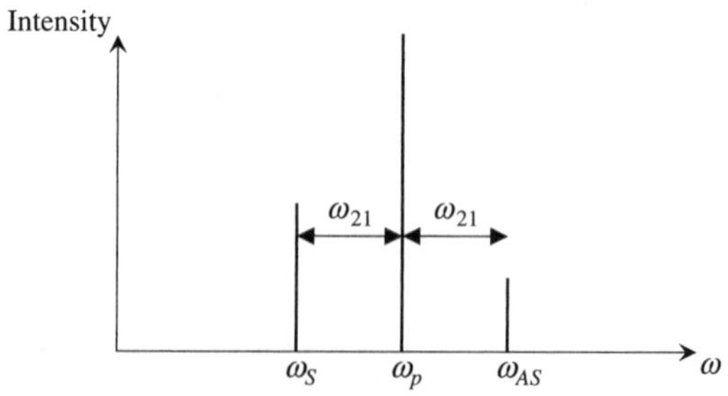

FIGURE 11.17 Spectrum of the light scattered from a material showing the pump, Stokes, and anti-Stokes components.

The Raman gain coefficient g_0 which appears in Eq. (11.202) is a material parameter and is typically given in units of cm/GW. Stimulated Raman scattering was first reported in 1962.[52, 53] Raman amplification in glass has been utilized in the development of fiber Raman amplifiers [54, 55] as we mentioned in Chap. 10. A medium with a sufficiently high Raman gain can be placed inside an optical resonator to obtain coherent oscillation of the Stokes component. Such oscillators, based on amplification due to the Raman effect, are known as Raman lasers. This has become a widely used technique in generating coherent light at new wavelengths.[56-61]

Example: Raman gain in $BaNO_3$ crystal
$BaNO_3$ has a high Raman gain coefficient and is widely used as a Raman amplifier or a laser. For $BaNO_3$, the Stokes shift is 1046 cm^{-1} and $g_0 = 11$ cm/GW.[59] Suppose that we send a pump wave at the wavelength of 1064 nm. The corresponding pump wave number is given by

$$\bar{\lambda}_p = \frac{1}{1064\times 10^{-7}} = 9398.5 \text{ cm}^{-1}$$

The Stokes wave number and wavelength will then be given by

$$\bar{\lambda}_s = 9398.5 - 1046 = 8352.5 \text{ cm}^{-1}$$

and

$$\bar{\lambda}_s = 1.197 \ \mu\text{m}$$

If I_p at 1064 nm is 0.5 GW/cm^2, then over a 2 cm-long sample, we will get a Raman gain of

$$\frac{I_s(2\text{ cm})}{I_s(0)} = \exp[2(11)(0.5)] \cong 60\times 10^3$$

Problems

11.1 Show that the second-order *d*-coefficient for sum frequency generation is given by Eq. (11.25).

11.2 Show by using the scalar wave equation that the differential equation satisfied by the amplitude $E_{30}(z)$ at the frequency ω_3 is given by Eq. (11.53).

11.3 Show that the derivatives of the Poynting vectors at the frequencies ω_2 and ω_3 are given by Eqs. (11.56) and (11.57).

11.4 Show that the coherence length for second harmonic generation is given by Eq. (11.68).

11.5 BBO (β-BaB_2O_4, or in words beta-barium borate) is a negative uniaxial crystal with the following wavelength-dependent indices of refraction:[17]

$$n_o^2(\lambda) = 2.7359 + \frac{0.01878}{\lambda^2 - 0.01822} - 0.01354\lambda^2$$

$$n_e^2(\lambda) = 2.3753 + \frac{0.01224}{\lambda^2 - 0.01667} - 0.01516\lambda^2$$

Above, the wavelength λ is in microns. For sum-frequency generation ($\omega_1 + \omega_2 -> \omega_3$), two possible polarization cases exist. These are termed Type-I and Type-II phase matching as defined below:

Type I: $\omega_1 = o-\text{wave},\ \omega_2 = o-\text{wave},\ \omega_3 = e-\text{wave}$
Type II: $\omega_1 = e-\text{wave},\ \omega_2 = o-\text{wave},\ \omega_3 = e-\text{wave}$

(*a*) Write down the index relations that must be satisfied for Type-I and Type-II phase matching. Do not bother to solve for the angle explicitly. In fact, it may not be possible to do this in general.

(*b*) Write a simple computer code to plot the variation of the ordinary and extraordinary indices between 300 and 2000 nm for BBO. These are sketched in Fig. 11.5*a* and *b*.

(*c*) Write a simple computer code to calculate the phase-matching angle for sum-frequency generation ($\omega_1 + \omega_2 -> \omega_3$) for the two possible polarization cases. Calculate numerically the phase-matching angles for the BBO crystal for the following cases and compare your answers with those in the above reference.

λ_1	λ_2	λ_3	Type
1.064	1.064	0.532	ooe
1.064	1.064	0.532	eoe
1.064	0.532	0.355	ooe
1.064	0.532	0.355	eoe
1.064	0.532	0.355	oee
0.532	0.532	0.266	ooe
0.532	0.532	0.266	eoe

11.6 A fundamental beam ($\lambda = 0.694\ \mu m$) with an intensity of $10^8\ W/cm^2$ is sent through a quartz crystal to generate second harmonic light at 0.347 μm. Take $n = 1.5$ and $d = 4 \times 10^{-24}$ in mks units. The crystal length is 1 cm. Estimate the second harmonic conversion efficiency by assuming perfect phase matching.

11.7 In a positive uniaxial crystal with normal dispersion, phase matching can be achieved with an extraordinary wave at the fundamental frequency and an ordinary wave at the second harmonic frequency. Derive the expression for the phase-matching angle.

11.8 Show that the SHG efficiency for a weakly focused Gaussian beam is given by Eq. (11.78). Also show that the expression for the efficiency reduces to Eq. (11.79) in the case of confocal focusing.

11.9 Find the beam waist w_0 needed to get confocal focusing in a nonlinear crystal with $n_1 = 1.8$, $\lambda = 1064$ nm, and $\ell = 10$ mm.

11.10 Show that the solution of Eq. (11.88) is given by Eq. (11.89).

11.11 Show that in the small-signal regime, Eq. (11.90) reduces to Eq. (11.65).

11.12 Show that in a parametric interaction where the depletion of the pump at the frequency ω_3 is negligible, the solution for the fields $E_{10}(z)$ and $E_{20}(z)^*$ is given by Eqs. (11.104) and (11.105).

11.13 Show that when the initial idler field is negligible, the solution for the fields $E_{10}(z)$ and $E_{20}(z)^*$ reduces to Eqs. (11.106) and (11.107).

11.14 Show that the threshold pumping intensity for a doubly resonant optical parametric oscillator is given by Eq. (11.120) when the differential parametric gain coefficient g is small.

11.15 Analysis of nonlinear frequency upconversion. Another interesting three-wave mixing process is nonlinear frequency upconversion where a strong pump beam is used at ω_2 to convert the signal at ω_1 to a wave with higher frequency ω_3. Again, the conditions $\omega_1 + \omega_2 = \omega_3$ and $k_3 = k_1 + k_2$ need to be satisfied. Analyze this situation in the presence of neglible pump depletion and show that in a crystal of length ℓ, the conversion efficiency can be written in the form

$$\frac{P^{\omega_3}(\ell)}{P^{\omega_1}(0)} = \frac{\omega_3}{\omega_1}\sin^2\left(\frac{g}{2}\ell\right)$$

and find the expression for g.

11.16 Show that in a Kerr medium with a weak nonlinear quadratic index coefficient β, the spotsize variation is given by Eq. (11.129).

11.17 Show that when nonlinearity is negligible, the solution for the spectral amplitude distribution $A(z,\omega)$ is given by Eq. (11.164).

11.18 Show that the full-width half-maximum duration of the input Gaussian pulse is given by Eq. (11.166). Furthermore, use the procedure described in the text to show that the pulse envelope in the presence of second-order dispersion (and no nonlinearity) has the form given by Eq. (11.167).

11.19 Show that for the solution obtained in Prob. 11.18, the full-width half-maximum duration at z is given by Eq. (11.169).

11.20 Show that the group delay dispersion parameter D can be calculated by using Eq. (11.174).

11.21 Go to the limit where D is large in Eq. (11.173) and compare the result with the approximate formula we obtained in Chap. 4, when we introduced the group velocity dispersion parameter D_{GV}.

11.22 Derive the expression given in Eq. (11.175) for the phase of the Gaussian pulse in the presence of second-order dispersion.

11.23 Show that the frequency shift of a Gaussian pulse due to self-phase modulation is given by Eq. (11.186).

11.24 Show that by using the transformations in Eqs. (11.190) to (11.192), Eq. (11.158) can be put into the dimensionless form given by Eq. (11.193).

11.25 Write a numerical code to calculate k_2 as a function of wavelength for silica by using the Sellmeier equations given in Chap. 4. Use your results to verfy the results for the soliton peak power and soliton period calculated in the example of Sec. 11.2.7.

11.26 A certain crystal has a Raman shift of 1200 cm^{-1}. It is excited by a pump laser operating at a wavelength of 532 nm.

(*a*) Find the wavelength (in nm) of the Stokes signal.

(*b*) The Raman gain coefficient at the Stokes wavelength is $g_R = 15$ cm/GW. If the pump electric field is 10^5 V/cm, calculate the total Raman gain in dB for a crystal of length 50 mm. Take the index of refraction of the crystal to be 1 and treat the pump as a plane wave.

References

1. Y. R. Shen, *The Principles of Nonlinear Optics* (Wiley, Hoboken, N.J., 2003).
2. N. Bloembergen, *Nonlinear Optics* (World Scientific, Singapore, 1996).
3. R. W. Boyd, *Nonlinear Optics*, 3d ed. (Academic Press, San Diego, 2008).
4. A. Yariv and P. Yeh, *Optical Waves in Crystals* (John Wiley & Sons, Inc., New York, 1984).
5. A. Yariv, *Optical Electronics in Modern Communications*, 5th ed. (Oxford University Press, New York, 1997).
6. A. Yariv, *Quantum Electronics* (Wiley, New York, 1975).
7. B. E. A. Saleh and M. C. Teich, *Fundamentals of Photonics*, 2d ed. (Wiley, Hoboken, N.J., 2007).
8. C. C. Davis, *Lasers and Electro-Optics* (Cambridge University Press, Cambridge, 1996).
9. N. Bloembergen, "Nonlinear optics: Past, present, and future," *IEEE Journal of Selected Topics in Quantum Electronics*, **6**(6), 876–880 (2000).
10. R. L. Byer, "Nonlinear optics and solid-state lasers: 2000," *IEEE Journal of Selected Topics in Quantum Electronics*, **6**(6), 911–930 (2000).
11. T. Kartaloglu, K. G. Koprulu, and O. Aytur, "Phase-matched self-doubling optical parametric oscillator," *Optics Letters*, **22**(5), 280–282 (1997).
12. J. M. Manley and H. E. Rowe, "General energy relations in nonlinear reactances," *Proceedings of the Institute of Radio Engineers*, **47**(12), 2115–2116 (1959).
13. P. A. Franken, G. Weinreich, C. W. Peters, and A. E. Hill, "Generation of optical harmonics," *Physical Review Letters*, **7**(4), 118–119 (1961).
14. M. M. Fejer, G. A. Magel, D. H. Jundt, and R. L. Byer, "Quasi-phase-matched second harmonic-generation: Tuning and tolerances," *IEEE Journal of Quantum Electronics*, **28**(11), 2631–2654 (1992).
15. G. D. Miller, R. G. Batchko, W. M. Tulloch, D. R. Weise, M. M. Fejer, and R. L. Byer, "42%-efficient single-pass cw second-harmonic generation in periodically poled lithium niobate," *Optics Letters*, **22**(24), 1834–1836 (1997).
16. L. E. Myers, R. C. Eckardt, M. M. Fejer, R. L. Byer, W. R. Bosenberg, and J. W. Pierce, "Quasi-phase-matched optical parametric oscillators in bulk periodically poled $LiNbO_3$," *Journal of the Optical Society of America B-Optical Physics*, **12**(11), 2102–2116 (1995).

17. K. Kato, "2nd-harmonic generation to 2048 A in Beta-BaB_2O_4," *IEEE Journal of Quantum Electronics*, **22**(7), 1013–1014 (1986).
18. G. D. Boyd and D. A. Kleinman, "Parametric interaction of focused gaussian light beams," *Journal of Applied Physics*, **39**(8), 3597–3639 (1968).
19. R. Sarrouf, T. Badr, and J. J. Zondy, "Intracavity second-harmonic generation of diode-pumped continuous-wave, single-frequency 1.3 micron Nd:YLiF4 lasers," *Journal of Optics A-Pure and Applied Optics*, **10**(10), 104011 (2008).
20. J. A. Armstrong, N. Bloembergen, J. Ducuing, and P. S. Pershan, "Interactions between light waves in a nonlinear dielectric," *Physical Review*, **127**(6), 1918–1939 (1962).
21. R. H. Kingston, "Parametric amplification and oscillation at optical frequencies," *Proceedings of the Institute of Radio Engineers*, **50**(4), 472–474 (1962).
22. C. C. Wang and G. W. Racette, "Measurement of parametric gain accompanying optical difference frequency generation," *Applied Physics Letters*, **6**(8), 169–171 (1965).
23. J. A. Giordmaine, and R. C. Miller, "Tunable coherent parametric oscillation in $LiNbO_3$ at optical frequencies," *Physical Review Letters*, **14**(24), 973–976 (1965).
24. T. Brabec and F. Krausz, "Intense few-cycle laser fields: Frontiers of nonlinear optics," *Reviews of Modern Physics*, **72**(2), 545–591 (2000).
25. A. Rundquist, C. G. Durfee, Z. H. Chang, C. Herne, S. Backus, M. M. Murnane, and H. C. Kapteyn, "Phase-matched generation of coherent soft X-rays," *Science*, **280**(5368), 1412–1415 (1998).
26. P. B. Corkum, "Plasma perspective on strong-field multiphoton ionization," *Physical Review Letters*, **71**(13), 1994–1997 (1993).
27. J. Kerr, "A new relation between electricity and light: Dielectrified media birefringent," *Philosophical Magazine*, **50**, 337–348 (1875).
28. D. E. Spence, P. N. Kean, W. Sibbett, "60-fsec pulse generation from a self-mode-locked Ti:sapphire laser," *Optics Letters*, **16**(1), 42–44 (1991).
29. A. Seas, V. Petricevic, R. R. Alfano, "Self-mode-locked chromium-doped forsterite laser generates 50-fs pulses," *Optics Letters*, **18**(11), 891–893 (1993).
30. A. Sennaroglu, Pollock, C. R., Nathel, H., "Continuous-wave self-mode-locked operation of a femtosecond Cr4+:YAG laser," *Optics Letters*, **19**(6), 390–392 (1994).
31. H. A. Haus, J. G. Fujimoto, and E. P. Ippen, "Analytic theory of additive pulse and Kerr lens mode-locking," *IEEE Journal of Quantum Electronics*, **28**(10), 2086–2096 (1992).
32. R. Ell, U. Morgner, F. X. Kartner, J. G. Fujimoto, E. P. Ippen, V. Scheuer, G. Angelow, T. Tschudi, M. J. Lederer, A. Boiko, and B. Luther-Davies, "Generation of 5-fs pulses and octave-spanning spectra directly from a Ti:sapphire laser," *Optics Letters*, **26**(6), 373–375 (2001).
33. M. N. Cizmeciyan, H. Cankaya, A. Kurt, and A. Sennaroglu, "Kerr-lens mode-locked femtosecond Cr2+:ZnSe laser at 2420 nm," *Optics Letters*, **34**(20), 3056–3058 (2009).
34. C. B. Schaffer, A. Brodeur, J. F. Garcia, and E. Mazur, "Micromachining bulk glass by use of femtosecond laser pulses with nanojoule energy," *Optics Letters*, **26**(2), 93–95 (2001).
35. X. Liu, D. Du, and G. Mourou, "Laser ablation and micromachining with ultrashort laser pulses," *IEEE Journal of Quantum Electronics*, **33**(10), 1706–1716 (1997).
36. C. B. Schaffer, A. Brodeur, and E. Mazur, "Laser-induced breakdown and damage in bulk transparent materials induced by tightly focused femtosecond laser pulses," *Measurement Science & Technology*, **12**(11), 1784–1794 (2001).
37. D. Huang, M. Ulman, L. H. Acioli, H. A. Haus, and J. G. Fujimoto, "Self-focusing-induced saturable loss for laser mode-locking," *Optics Letters*, **17**(7), 511–513 (1992).
38. G. P. Agrawal, *Nonlinear Fiber Optics* (Academic Press, Boston, 1989).
39. H. A. Haus, "Mode-locking of lasers," *IEEE Journal of Selected Topics in Quantum Electronics*, **6**(6), 1173–1185 (2000).

40. V. E. Zakharov and A. B. Shabat, "Exact theory of 2-dimensional self-focusing and one-dimensional self-modulation of waves in nonlinear media," *Soviet Physics JETP-USSR*, **34**(1), 62–69 (1972).
41. P. G. Drazin and R. S. Johnson, *Solitons: An Introduction* (Cambridge Texts in Applied Mathematics, Cambridge, 1989).
42. A. Hasegawa, *Optical Solitons in Fibers*, 2d ed. (Springer, Berlin, 1989).
43. H. A. Haus, J. G. Fujimoto, and E. P. Ippen, "Structures for additive pulse mode-locking," *Journal of the Optical Society of America B-Optical Physics*, **8**(10), 2068–2076 (1991).
44. U. Keller, K. J. Weingarten, F. X. Kartner, D. Kopf, B. Braun, I. D. Jung, R. Fluck, C. Honninger, N. Matuschek, and J. A. der Au, "Semiconductor saturable absorber mirrors (SESAMs) for femtosecond to nanosecond pulse generation in solid-state lasers," *IEEE Journal of Selected Topics in Quantum Electronics*, **2**(3), 435-453 (1996).
45. U. Morgner, F. X. Kartner, S. H. Cho, Y. Chen, H. A. Haus, J. G. Fujimoto, E. P. Ippen, V. Scheuer, G. Angelow, and T. Tschudi, "Sub-two-cycle pulses from a Kerr-lens mode-locked Ti:sapphire laser," *Optics Letters*, **24**(6), 411–413 (1999).
46. F. X. Kartner, I. D. Jung, and U. Keller, "Soliton mode-locking with saturable absorbers," *IEEE Journal of Selected Topics in Quantum Electronics*, **2**(3), 540–556 (1996).
47. L. F. Mollenauer and R. H. Stolen, "The soliton laser," *Optics Letters*, **9**(1), 13–15 (1984).
48. L. F. Mollenauer, R. H. Stolen, and J. P. Gordon, "Experimental-observation of picosecond pulse narrowing and solitons in optical fibers," *Physical Review Letters*, **45**(13), 1095–1098 (1980).
49. L. F. Mollenauer, J. P. Gordon, and M. N. Islam, "Soliton propagation in long fibers with periodically compensated loss," *IEEE Journal of Quantum Electronics*, **22**(1), 157–173 (1986).
50. C. V. Raman, "A change of wave-length in light scattering," *Nature*, **121**, 619–619 (1928).
51. C. V. Raman and K. S. Krishnan, "A new type of secondary radiation," *Nature*, **121**, 501–502 (1928).
52. G. Eckhardt, S. E. Schwarz, F. J. McClung, R. W. Hellwarth, E. J. Woodbury, and D. Weiner, "Stimulated Raman scattering from organic liquids," *Physical Review Letters*, **9**(11), 455–457 (1962).
53. E. J. Woodbury and W. K. Ng, "Ruby laser operation in near IR," *Proceedings of the Institute of Radio Engineers*, **50**(11), 2367 (1962).
54. R. H. Stolen and E. P. Ippen, "Raman gain in glass optical waveguides," *Applied Physics Letters*, **22**(6), 276–278 (1973).
55. M. N. Islam, "Raman amplifiers for telecommunications," *IEEE Journal of Selected Topics in Quantum Electronics*, **8**(3), 548–559 (2002).
56. H. S. Rong, A. S. Liu, R. Jones, O. Cohen, D. Hak, R. Nicolaescu, A. Fang, and M. Paniccia, "An all-silicon Raman laser," *Nature*, **433**(7023), 292–294 (2005).
57. K. J. Vahala, "Optical microcavities," *Nature*, **424**(6950), 839–846 (2003).
58. P. Cerny, H. Jelinkova, P. G. Zverev, and T. T. Basiev, "Solid state lasers with Raman frequency conversion," *Progress in Quantum Electronics*, **28**(2), 113–143 (2004).
59. H. M. Pask, "The design and operation of solid-state Raman lasers," *Progress in Quantum Electronics*, **27**(1), 3–56 (2003).
60. A. Sennaroglu, A. Kiraz, M. A. Dundar, A. Kurt, and A. L. Demirel, "Raman lasing near 630 nm from stationary glycerol-water microdroplets on a super-hydrophobic surface," *Optics Letters*, **32**(15), 2197–2199 (2007).
61. R. H. Stolen, "Fiber raman lasers," *Fiber and Integrated Optics*, **3**(1), 21–52 (1980).

APPENDIX A

Background on Hermite Polynomials

Hermite polynomials were used in the analysis of higher-order laser beams (Chap. 3), the quantum mechanical harmonic oscillator (Chap. 5), and AM mode locking of lasers (Chap. 6). Their general characteristics can be found in many sources on quantum mechanics and mathematical methods (see, for example, Refs. 1 through 4). In this appendix, we provide a concise review of Hermite polynomials. After a discussion of double infinite series, we derive some of the important mathematical properties of Hermite polynomials including Rodrigue's formula, the differential equation, recurrence formulas, and normalization relations. Further information about these polynomials can be obtained from the references listed at the end of the appendix.

Double Infinite Series

Let us investigate a double infinite series s_1 which can be obtained from the product of two single infinite series in the form

$$s_1 = \sum_{m=0}^{\infty} V_m \sum_{k=0}^{\infty} W_k = \sum_{m=0}^{\infty}\sum_{k=0}^{\infty} c_{m,k} \tag{A.1}$$

Above, V_m and W_k are the mth and kth terms of the single series and $c_{m,k} = V_m W_k$. If the double infinite sum is absolutely convergent, we can often express it as a single infinite series consisting of finite sums. To see how this can be done, we introduce a change of indices according to

$$m = n - k \tag{A.2}$$

Since $m \geq 0$, this means that $n-k \geq 0$ or $k \leq n$. By using this change of indices, s_1 can be written as

$$s_1 = \sum_{m=0}^{\infty}\sum_{k=0}^{\infty} c_{m,k} = \sum_{n=0}^{\infty}\sum_{k=0}^{n} c_{n-k,k} \tag{A.3}$$

We can also employ a different change of variables according to

$$m = n - 2k \tag{A.4}$$

in which case, k is limited to values between 0 and $n/2$. To avoid fractions, we introduce the following notation:

$$[n/2] = \begin{cases} n/2 & \text{if} \quad n = \text{even} \\ (n-1)/2 & \text{if} \quad n = \text{odd} \end{cases} \tag{A.5}$$

Hence, we have

$$\sum_{m=0}^{\infty}\sum_{k=0}^{\infty} c_{m,k} = \sum_{n=0}^{\infty}\sum_{k=0}^{[n/2]} c_{n-2k,k} \tag{A.6}$$

Combining the above two equivalent forms for the series sum s_1, we obtain the relation

$$\sum_{n=0}^{\infty}\sum_{k=0}^{n} A_{n-k,k} = \sum_{n=0}^{\infty}\sum_{k=0}^{[n/2]} A_{n-2k,k} \tag{A.7}$$

Generating Function for Hermite Polynomials

To derive the properties of Hermite polynomials, we define them in terms of a generating function and show that these polynomials satisfy the differential equation derived earlier in Chaps. 3, 5, and 6. Hermite polynomials can be defined as the expansion coefficients of a generating function $S(y,z)$ defined according to

$$S(y,z) = \exp[2zy - z^2] \equiv \sum_{n=0}^{\infty} H_n(y)\frac{z^n}{n!} \tag{A.8}$$

To obtain the series expression for $H_n(y)$, the nth-order Hermite polynomial, we first expand $S(y,z)$ as a double infinite series and use the results of the section "Double Infinite Series." We note that

$$\exp[2zy - z^2] = \exp(2yz)\exp(-z^2) = \sum_{m=0}^{\infty} \frac{1}{m!}(2yz)^m \sum_{k=0}^{\infty} \frac{(-z^2)^k}{k!}$$

$$= \sum_{m=0}^{\infty}\sum_{k=0}^{\infty} \frac{(-1)^k 2^m}{m!k!} y^m z^m z^{2k} \tag{A.9}$$

By making a change indices to $m = n - 2k$ where $k = 0, \ldots, [n/2]$, we obtain

$$S(y,z) = \sum_{n=0}^{\infty}\sum_{k=0}^{[n/2]} \frac{(-1)^k (2y)^{n-2k} z^{n-2k} z^{2k}}{k!(n-2k)!} = \sum_{n=0}^{\infty}\left(\sum_{k=0}^{[n/2]} \frac{(-1)^k (2y)^{n-2k}}{k!(n-2k)!}\right) z^n \tag{A.10}$$

Comparing with the definition given in Eq. (A.1), we see that $H_n(y)$ can be expressed as a finite series in the form

$$H_n(y) = \sum_{k=0}^{[n/2]} \frac{(-1)^k n!}{k!(n-2k)!}(2y)^{n-2k} \tag{A.11}$$

We can already see some general properties of $H_n(y)$ from Eq. (A.11):

1. $H_n(y)$ is a polynomial of degree n.
2. $H_n(y)$ contains either odd or even powers of y.
3. If n is even, $H_n(y)$ is even. If n is odd, $H_n(y)$ is odd. This can be combined into the relation

$$II_n(-y) - (-1)^n H_n(y) \tag{A.12}$$

Derivation of Rodrigue's Formula

Note that

$$\lim_{z\to 0} \frac{\partial^n S(y,z)}{\partial z^n} = \lim_{z\to 0} \sum_{n=0}^{\infty} H_n(y) \frac{\partial^n}{\partial z^n}\left(\frac{z^n}{n!}\right) = H_n(y) \tag{A.13}$$

Differentiating the generating function, we get

$$\lim_{z\to 0} \frac{\partial^n S(y,z)}{\partial z^n} = \lim_{z\to 0} \frac{\partial^n}{\partial z^n}(\exp(2zy - z^2))$$

$$= \lim_{z\to 0} \frac{\partial^n}{\partial z^n}(\exp(-(-2zy + z^2 + y^2) + y^2))$$

$$= \lim_{z\to 0} \frac{\partial^n}{\partial z^n}(\exp(y^2 - (z-y)^2))$$

$$= \exp(y^2) \lim_{z\to 0} \frac{\partial^n}{\partial z^n}(e^{-(z-y)^2}) \tag{A.14}$$

We apply the change of variables $z' = z - y$ with $dz' = dz$ to obtain

$$\lim_{z\to 0}\frac{\partial^n S(y,z)}{\partial z^n} = \exp(y^2)\lim_{z'\to(-y)}\frac{\partial^n}{\partial z'^n}(\exp(-z'^2))$$

$$= \exp(y^2)\frac{\partial^n}{\partial(-y)^n}(\exp(-y^2))$$

$$= (-1)^n \exp(y^2)\frac{d^n}{dy^n}(\exp(-y^2)) \qquad \text{(A.15)}$$

Hence, we get

$$H_n(y) = (-1)^n \exp(y^2)\frac{d^n}{dy^n}(\exp(-y^2)) \qquad \text{(A.16)}$$

This is the well-known Rodrigue's formula for Hermite polynomials. Note that since the differentiation operator leaves a factor of $\exp(-y^2)$ with each term, multiplication with $\exp(y^2)$ leaves just a polynomial of y of order n.

Recurrence Relations

The important recurrence relations can be derived by using the generating function given in Eq. (A.8). By using Eq. (A.8), we get

$$\frac{\partial S}{\partial z} = -2(z-y)\exp(2zy - z^2) = -2(z-y)S \qquad \text{(A.17)}$$

Equivalently,

$$\frac{\partial S}{\partial z} = \frac{\partial}{\partial z}\sum_n \frac{H_n(y)}{n!}z^n = \sum_n \frac{H_n(y)z^{n-1}}{(n-1)!} = -2(z-y)\sum_n \frac{H_n(y)}{n!}z^n$$

$$= \sum_n \frac{-2H_n(y)z^{n+1}}{n!} + \sum_n \frac{2H_n(y)}{n!}yz^n \qquad \text{(A.18)}$$

This gives

$$\sum_n \frac{H_n(y)z^{n-1}}{(n-1)!} + \sum_n \frac{2H_n(y)z^{n+1}}{n!} - \sum_n \frac{2H_n(y)yz^n}{n!} = 0 \qquad \text{(A.19)}$$

Now, we can look at the coefficients of the terms containing z^n to obtain

$$\sum_n \frac{H_{n+1}(y)z^n}{n!} + \sum_n \frac{2H_{n-1}(y)z^n}{(n-1)!} - \sum_n \frac{2H_n(y)yz^n}{n!} = 0 \qquad \text{(A.20)}$$

Multiplying through by $n!$, we get

$$H_{n+1}(y) + 2nH_{n-1}(y) - 2yH_n(y) = 0 \tag{A.21}$$

Next, we differentiate $S(y,z)$ with respect to y to obtain the differential equations satisfied by the Hermite polynomials. Differentiating $S(y,z)$ with respect to y, we get

$$\frac{\partial S}{\partial y} = 2z\exp(2zy - z^2) = 2zS(y,z) \tag{A.22}$$

Equivalently,

$$\frac{\partial S}{\partial y} = \frac{\partial}{\partial y}\sum_n \frac{H_n(y)}{n!}z^n = \sum_n \frac{1}{n!}\frac{dH_n}{dy}z^n$$

$$= 2z\sum_n \frac{H_n(y)}{n!}z^n = \sum_n \frac{2H_n(y)z^{n+1}}{n!} = \sum_n \frac{2H_{n-1}(y)z^n}{(n-1)!} \tag{A.23}$$

Hence, we get

$$\frac{dH_n}{dy} = 2nH_{n-1}(y) \tag{A.24}$$

From Eq. (A.24),

$$\frac{d^2H_n}{dy^2} = 2n\frac{dH_{n-1}}{dy} = 2n(2)(n-1)H_{n-2}(y) \tag{A.25}$$

From Eq. (A.21), by changing the index from n to $(n-1)$, we get

$$H_n(y) + 2(n-1)\frac{1}{4n(n-1)}\frac{d^2H_n}{dy^2} - 2y\frac{1}{2n}\frac{dH_n}{dy} = 0 \tag{A.26}$$

or, multiplying by $2n$, we get

$$\frac{d^2H_n}{dy^2} - 2y\frac{dH_n}{dy} + 2nH_n(y) = 0 \tag{A.27}$$

Normalization of Hermite Polynomials

Normalization properties of Hermite polynomials are important, especially in the derivation of the wave functions for the quantum mechanical harmonic oscillator. We now define two generating functions $S_1(y,s_1)$ and $S_2(y,s_2)$ according to

$$S_1(y,s_1) = \sum_n \frac{H_n(y)}{n!}s_1^n = \exp\left[y^2 - (s_1 - y)^2\right] \tag{A.28}$$

and

$$S_2(y,s_2) = \sum_m \frac{H_m(y)}{m!} s_2^m = \exp\left[y^2 - (s_2 - y)^2\right] \tag{A.29}$$

Next, we investigate the integral I given by

$$I = \int_{-\infty}^{+\infty} dy\, S_1(y,s_1)\, S_2(y,s_2) \exp(-y^2) \tag{A.30}$$

This can be expressed in two equivalent forms as

$$\begin{aligned} I &= \sum_n \sum_m s_1^n s_2^m \int_{-\infty}^{+\infty} dy \frac{H_n(y) H_m(y)}{n!\, m!} \exp(-y^2) \\ &= \int_{-\infty}^{+\infty} dy \exp\left[y^2 - (s_1 - y)^2 + y^2 - (s_2 - y)^2 - y^2\right] \\ &= \int_{-\infty}^{+\infty} dy \exp\left[-s_1^2 + 2s_1 y + 2s_2 y - s_2^2 - y^2\right] \end{aligned} \tag{A.31}$$

We note that

$$\begin{aligned} -s_1^2 + 2s_1 y + 2s_2 y - s_2^2 - y^2 &= -\left[s_1^2 + y^2 + s_2^2 - 2s_1 y - 2s_2 y\right] \\ &= -\left[(y - s_1 - s_2)^2 - 2s_1 s_2\right] \end{aligned} \tag{A.32}$$

Hence, I becomes

$$\begin{aligned} I &= \int_{-\infty}^{+\infty} dy\, S_1(y,s_1)\, S_2(y,s_2) \exp(-y^2) \\ &= \int_{-\infty}^{+\infty} dy \exp(2s_1 s_2) \exp\left(-(y - s_1 - s_2)^2\right) = \sqrt{\pi} \exp(2s_1 s_2) \\ &= \sqrt{\pi} \sum_n \frac{(2s_1 s_2)^n}{n!} = \sum_m \sum_n \frac{\delta_{mn} \sqrt{\pi}\, 2^n s_1^n s_2^m}{n!} \\ &= \sum_m \sum_n s_1^n s_2^m \int_{-\infty}^{+\infty} dy \frac{H_n(y) H_m(y)}{n!\, m!} \exp(-y^2) \end{aligned} \tag{A.33}$$

Above, δ_{mn} is the Kroenecker delta. Hence,

$$\int_{-\infty}^{+\infty} dy H_n(y) H_m(y) \exp(-y^2) = \delta_{mn} 2^n n! \sqrt{\pi} \tag{A.34}$$

Normalized Wave Functions of the Harmonic Oscillator

In Chap. 5, we found that the wave equation for the nth eigenstate of the harmonic oscillator can be written in the form

$$u_n(x) = A_n H_n(y) \exp\left({-y^2}/{2} \right) \tag{A.35}$$

where $y = \beta x$ and

$$\beta = \sqrt{\frac{m\omega}{\hbar}} \tag{A.36}$$

Then, assuming that A_n is real, we get

$$\int_{-\infty}^{+\infty} dx \left| u_n(x) \right|^2 = A_n^2 \int_{-\infty}^{+\infty} \frac{dy}{\beta} \exp(-y^2) H_n^2(y) = \frac{A_n^2}{\beta} 2^n n! \sqrt{\pi} = 1 \tag{A.37}$$

Hence, A_n is given by

$$A_n = \sqrt{\frac{\beta}{\sqrt{\pi}} \frac{1}{2^n n!}} \tag{A.38}$$

The full time-dependent wave function of the nth state becomes

$$\psi_n(x,t) = \sqrt{\frac{1}{2^n n!} \sqrt{\frac{m\omega}{\pi\hbar}}} H_n\left(\sqrt{\frac{m\omega}{\hbar}} x \right) \exp\left(-\frac{m\omega x^2}{2\hbar} \right) \exp\left(-i\omega \left(n + \frac{1}{2} \right) t \right) \tag{A.39}$$

Some Hermite Polynomials

Hermite polynomials can be generated by using Rodrigue's formula or the series expression. Looking at the series expression in Eq. (A.11), we see that the $k = 0$ term gives the highest polynomial power y^n. Furthermore, the coefficient of y^n becomes $[(-1)^0 n!/0!\, n!] \times 2^n = 2^n$. By using Rodrigue's formula

$$H_0(y) = (-1)^0 e^{y^2} \frac{d^0}{dy^0} \exp(-y^2) = 1 \tag{A.40}$$

To get $H_1(y)$, we use

$$H_1(y) = (-1)^n \exp(y^2) \frac{d}{dy} (\exp(-y^2)) = 2y \tag{A.41}$$

n	$H_n(y)$
0	1
1	$2y$
2	$4y^2 - 2$
3	$8y^3 - 12y$
4	$16y^4 - 48y^2 + 12$
5	$32y^5 - 160y^3 + 120y$

TABLE A.1 The First Six Hermite Polynomials

This procedure can be repeated to generate higher-order Hermite polynomials. Table A.1 lists the first six Hermite polynomials.

References

1. L. Pauling and E. B. Wilson, *Introduction to Quantum Mechanics with Applications to Chemistry* (Dover, New York, 1963).
2. R. H. Dicke and J. P. Wittke, *Introduction to Quantum Mechanics* (Addison Wesley, Reading, Mass., 1960).
3. G. B. Arfken and H. J. Weber, *Mathematical Methods for Physicists*, 5th ed. (Harcourt Academic Press, Amsterdam, 2001).
4. M. Abramowitz and I. A. Stegun, *Handbook of Mathematical Functions* (Dover, New York, 1965).

APPENDIX B

Some Fundamental Constants

Speed of light in vacuum	$c = 2.99792 \times 10^{8}$ m/s
Vacuum permittivity	$\varepsilon_0 = 8.85419 \times 10^{-12}$ F/m
Vacuum permeability	$\mu_0 = 4\pi \times 10^{-7}$ H/m
Vacuum impedance	$\eta_0 = 376.730\ \Omega$
Planck's constant	$h = 6.62607 \times 10^{-34}$ J · s
h-bar	$\hbar = h/2\pi = 1.054 \times 10^{-34}$ J · s
Planck's constant × speed of light	$hc = 1240$ eV · nm
Electron charge	$e = 1.60218 \times 10^{-19}$ C
Electron rest mass	$m_e = 9.10938 \times 10^{-31}$ kg
Avogadro's number	$N_A - 6.02214 \times 10^{23}$/mole
Boltzmann constant	$k_B = 1.38065 \times 10^{-23}$ J/K
Stefan-Boltzmann constant	$\sigma_{SB} = 5.67040 \times 10^{-8}$ W/m^2-K^4
Wien's displacement law constant	$b = 2.89777 \times 10^{-3}$m · K

Reference

1. http://physics.nist.gov/cuu/Constants/index.html.

Index

Note: Page numbers followed by "*f*" indicate figures.

B

E

F

G

H

I

J

K

L

S

T